Pattern Recognition and String Matching

COMBINATORIAL OPTIMIZATION

VOLUME 13

Through monographs and contributed works the objective of the series is to publish state of the art expository research covering all topics in the field of combinatorial optimization. In addition, the series will include books which are suitable for graduate level courses in computer science, engineering, business, applied mathematics, and operations research.

Combinatorial (or discrete) optimization problems arise in various applications, including communications network design, VLSI design, machine vision, airline crew scheduling, corporate planning, computer-aided design and manufacturing, database query design, cellular telephone frequency assignment, constraint directed reasoning, and computational biology. The topics of the books will cover complexity analysis and algorithm design (parallel and serial), computational experiments and applications in science and engineering.

The titles published in this series are listed at the end of this volume.

Pattern Recognition and String Matching

edited by

Dechang Chen

University of Wisconsin – Green Bay,
Green Bay, WI, U.S.A.

and

Xiuzhen Cheng

The George Washington University,
Washington DC, U.S.A.

KLUWER ACADEMIC PUBLISHERS

DORDRECHT / BOSTON / LONDON

A C.I.P. Catalogue record for this book is available from the Library of Congress.

ISBN 1-4020-0953-4

Published by Kluwer Academic Publishers,
P.O. Box 17, 3300 AA Dordrecht, The Netherlands.

Sold and distributed in North, Central and South America
by Kluwer Academic Publishers,
101 Philip Drive, Norwell, MA 02061, U.S.A.

In all other countries, sold and distributed
by Kluwer Academic Publishers,
P.O. Box 322, 3300 AH Dordrecht, The Netherlands.

Printed on acid-free paper

Printed in the Netherlands.

Contents

Foreword

The research and development of pattern recognition have proven to be of importance in science, technology, and human activity. Many useful concepts and tools from different disciplines have been employed in pattern recognition. Among them is string matching, which receives much theoretical and practical attention. String matching is also an important topic in combinatorial optimization.

This book is devoted to recent advances in pattern recognition and string matching. It consists of twenty eight chapters written by different authors, addressing a broad range of topics such as those from classification, matching, mining, feature selection, and applications. Each chapter is self-contained, and presents either novel methodological approaches or applications of existing theories and techniques.

The aim, intent, and motivation for publishing this book is to provide a reference tool for the increasing number of readers who depend upon pattern recognition or string matching in some way. This includes students and professionals in computer science, mathematics, statistics, and electrical engineering.

We wish to thank all the authors for their valuable efforts, which made this book a reality. Thanks also go to all reviewers who gave generously of their time and expertise.

Dechang Chen
in Green Bay, WI

Xiuzhen Cheng
in Washington, DC

Correcting the Training Data

Ricardo Barandela
Instituto Tecnológico de Toluca, 52140 Metepec, México
Instituto de Geografía Tropical, La Habana, Cuba
E-mail: `rbarandela@hotmail.com`

Eduardo Gasca
Instituto Tecnológico de Toluca, 52140 Metepec, México

Roberto Alejo
Instituto Tecnológico de Toluca, 52140 Metepec, México

Contents

D. Chen and X. Cheng (eds.),
Pattern Recognition and String Matching, 1–41.

1 Introduction

Traditionally, learning algorithms and pattern recognition methods have been sorted into two broad groups: supervised and unsupervised (predictive and informative in Data Mining terminology) whether training data is available or not. Supervised classifier design is based on the information supplied by a training sample (TS): a set of training patterns, instances or prototypes that are assumed to represent all the relevant classes and to bear correct class labels.

In several practical applications, however, class identification of the prototypes is a difficult and very costly task. As a consequence, some imperfectly or incorrectly labeled instances may be present in the TS, leading to situations lying in between supervised and unsupervised methods, or as they have been called: imperfectly supervised environments [13]. Examples have been reported in drawing of mineral deposit maps and, particularly, in the interpretation of remotely sensed data. There is another source of distortion in the training data: instances with errors in some attribute values and patterns that are atypical or exceptional cases.

Generalization accuracy of the learning algorithm may be degraded by the presence of incorrectness or imperfections in the training data. Particularly sensitive to these deficiencies are nonparametric classifiers whose training is not based upon any assumption about probability density functions. That explains the emphasis given in the machine learning and pattern recognition communities to the evaluation of procedures used to collect and to clean the training sample, critical aspects for the effective automation of discrimination tasks.

The present paper presents a methodology for correcting and decontaminating training samples while employing nonparametric classifiers. This methodology can be regarded as a cleaning process removing some elements of the training sample and correcting the labels of several others while retaining them. It has been conceived specifically for the Nearest Neighbor (NN) rule although exploratory results with a Multi-Layer Perceptron point to a broader applicability. Experimental results with both simulated and real datasets are presented to support the proposed methodology, to clarify the ideas behind it, to illustrate how the methodology has evolved and its outcomes on the training data. These experiments show that our methodology allows coping with all types of imperfections (mislabeled, noisy, atypical or exceptional) in the set of training instances, improving the classifier's performance and lowering its computational cost.

2 The NN Rule and Some Related Techniques

The Nearest Neighbor (NN) rule is one of the oldest and better-known algorithms for performing nonparametric classification. The entire training sample is stored in the computer memory. To classify a new instance, its distance is computed to each one of the stored training cases. The new instance is then assigned to the class represented by its nearest neighboring training pattern. More formally:

Let $TS = \{(x_1, \varphi_1), (x_2, \varphi_2), ..., (x_n, \varphi_n)\}$ be the training sample. That is, TS is the set of n pairs of random samples (x_i, φ_i) $(i = 1, 2, ..., n)$, where the label φ may take values in $\{1, 2, ..., c\}$ and φ_i designates the class of x_i among the c possible classes. For classifying an unknown pattern Y with the NN rule, it is necessary to determine first the nearest neighbor x' of Y in the TS. That is, it is necessary to find x' in TS such that:

$d(Y, x') = min\ d(Y, x_i) \quad x_i \in TS$

where $d(,)$ means any suitable metric defined in the feature space. Then, the pattern Y is assigned to the class identified by the label associated to x'.

The NN rule is very popular because of:

1. Conceptual simplicity
2. Easy implementation
3. Known error rate bounds
4. Potentiality to compete favorably with other classification methods in real data applications.

Devijver and Kitler [17] have expressed that "the basic idea behind the NN rule is that samples which fall close together in feature space are likely to belong to the same class".

From the above definition, it is obvious that this classifier suffers of a main drawback: large memory requirement to store the whole training sample (that one wishes to be as big as possible) and the large response time needed. The computational burden of this classifier, very high with brute-force searching methods, has been considerably cut down by developing suitable data structures and associated non-exhaustive search algorithms (e.g., [29]) or by reducing the TS size (e.g., [2][31]). The idea of Hart [30] about a consistent subset (of the training set) has been the pioneer in the latter research line, stimulating a sequel of new algorithms aimed at eliminating as many training cases as possible without seriously affecting the discriminatory power of the classifier. A consistent subset is defined as a set of prototypes such that when it is employed to classify the entire training sample (with the NN rule) it guarantees 100% of correct assignment. The algorithms that have been proposed try to attain a consistent subset as small as possible since the minimal consistent subset is not, in general, obtained. The algorithm proposed by Hart (the Condensed Subset) does not provide a unique solution because of the strong dependence on the order the training patterns are examined in.

Another drawback of the NN rule is its sensitivity to the presence of erroneous or noisy prototypes in the training set. The NN rule shares this characteristic with other non-parametric methods such as the Neural Networks models. Imperfections in the training sets are the usual situation in many real-world applications. It is widely accepted that the presence of these imperfections may produce a serious decrease in classification accuracy. Editing techniques are aimed at improving the performance of the resulting systems by discarding outliers and cleaning the overlap among classes. There have been also a few proposals whose main goal is to modify the structure of the training sample by correcting the label of some prototypes supposedly with a wrong initial identification.

Tomek [45] very early mentioned the convenience of combining Editing with a method of the consistent subset approach for obtaining a much more significant reduction of the training sample size. Combined employment of the editing-condensing techniques has been often promoted (e.g., [43]) and proved in several practical applications. For instance, Barandela [5] employed this combination with the NN rule and Gasca and Barandela [22] with the learning set for a Multi-Layer Perceptron. In both cases, improvement of the classification accuracy was also observed. In those experiments to be reported in Section 4, when reduction of the training sample size was advisable (number of training instances was high enough), the combination of Wilson's editing and the Modified Selective [2] was used.

In the following, several algorithms that have been proposed to cope with these two drawbacks are outlined. Many of them are employed in the experiments reported in Section 4.

2.1 Editing Techniques

As already explained, editing techniques are mainly aimed at improving the performance of the NN rule by filtering the training prototypes. As a byproduct, they also obtain a decrease in the training set size and, consequently, a reduction of the computational burden of the classification method. The first work of editing corresponds to Wilson [50] and several others have followed.

2.1.1 Wilson's Editing Procedure

This technique consists of applying the k-NN ($k > 1$) classifier to estimate the class label of every prototype in the training set and discard those instances whose class label does not agree with the class associated to the majority of the k neighbors. The procedure is:

1. *Let $S = X$ (X is the original training set and S will be the edited TS)*
2. *For each x_i in X do:*
   (a) *Find the k nearest neighbors of x_i in $X \backslash \{x_i\}$*
   (b) *Discard x_i from S if its label disagrees with the class associated with the largest number of the k neighbors.*

The benefits –improvements of the generalization accuracy- of Wilson's algorithm have been supported by theoretical and empirical evaluations (e.g., [3]).

2.1.2 All k-NN Editing Algorithm [45]

This algorithm can be seen as a different realization of the original Wilson's editing. It employs all the h-NN classification rules, with h ranging from 1 to k, for a predetermined value of k.

1. *Let $S = X$.*
2. *For each x_i in X do:*
   (a) *Set $h = 1$*
   (b) *While $h < k + 1$ do:*
      i. *Find the h nearest neighbors of x_i in $X \backslash \{x_i\}$*

ii. *Discard* x_i *from* S *if its label disagrees with the class associated with the largest number of the* h *neighbors; go to Step 2.*

iii. *Set* $h = h + 1$

2.1.3 Generalized Editing (GE: [35])

This is another modification of the Wilson's algorithm. Koplowitz and Brown were concerned with the possibility of too many prototypes being removed from the training set because of Wilson's editing procedure. This approach consists of removing some suspicious prototypes and changing the class labels of some other instances. Accordingly, it can be regarded as a technique for modifying the structure of the training sample (through re-labeling of some training instances) and not only for eliminating atypical instances. In Generalized editing two parameters have to be defined: k and k' in such a way that

$(k+1)/2 \leq k' \leq k$

This editing algorithm can be written as follows:

1. *Let* $S = X$.

2. *For each* x_i *in* X *do:*

   (a) *Find the* k *nearest neighbors of* x_i *in* $X \backslash \{x_i\}$

   (b) *If a class has at least* k' *representatives among those* k *neighbors, then identify* x_i *according to that class (independently of its original class label). Otherwise, discard* x_i *from* S.

2.2 Condensing (Pruning) Algorithms

The main goal of these algorithms is the reduction of the size of the stored set of training instances while trying to maintain (or even improve) generalization accuracy. Hart [30] who presented the Condensing algorithm for obtaining a Consistent subset of the training sample first introduced these techniques. The shortcomings of Hart's algorithm were promptly pointed out: results depending on the order the instances are assessed, no preservation of the prototypes nears the decision boundaries and no elimination of all the superfluous inner training cases. Several methods attempting to remedy the disadvantages of Hart's algorithm have been proposed. Many of these proposals maintain as a basic goal to attain a consistent subset.

2.2.1 Selective Subset [41]

This is a remarkable approach because of a clear and precise formulation of the desired goals and the way to reach them. It is an extension of the Condensing algorithm. In the Selective Subset, every member of X (the original training set) must be closer to a member of S (the pruned set) of the same class than to any member of X (instead of S, as in the Condensing algorithm) of a different class. The purpose is to eliminate the order-dependence of the building algorithm. Although the importance of selecting instances near the decision boundaries was stated, this requisite is not included in the criteria that are used as a basis for the Selective Subset. R_i is described as the set of all related neighbors to the training case x_i (x_j is related neighbor of x_i if both are of the same class and x_j is closer to x_i than the nearest neighbor of x_i of a different class in X). But then, the Selective Subset is merely defined as the smallest subset containing at least one member of R_i for each x_i in X. As a natural consequence, the algorithm does not guarantee the best approximation to the decision boundary.

As Ritter et al. [41] have recognized, their method does not necessarily conduct to a unique solution. Besides, their algorithm is much more complex (in terms of memory and execution time needed) than most other reduction techniques. It requires the manipulation of an nxn matrix (where n is the number of instances in X) and occasional recursion. For these reasons, this method is mentioned here only as a reference for a better understanding of the Modified Selective.

2.2.2 Modified Selective [2]

This method rests upon a modification of the Selective Subset definition. The Modified Selective is defined as that subset of X containing, for every x_i in X, that element of its R_i that is the nearest to a class different than that of x_i. Although the main purpose of this modification is to strengthen the condition to be fulfilled by the reduced subset in order to attain a best approximation to the decision boundaries as a byproduct the resulting algorithm is much simpler than that proposed by Ritter et al. The Modified Selective method has proved to produce better classification accuracy than the obtained by several other pruning algorithms, including the original (Hart's) condensing algorithm and the Selective Subset [7].

The algorithm, for the two-class case, consists of the following steps. Only the class 1 is considered, class 2 being afterwards processed in a similar way:

1. *Let* $S = \varnothing$.

2. *Let* $x_1, x_2, ..., x_{n1}$ *be the training cases of class 1. These instances are ordered such that* $D_1 < D_2 < ... < D_{n1}$ *where* D_i *stands for the minimum distance from* x_i *to the class 2.*

3. *Place* x_i *in* S. *Put* $KN = n_1 - 1$.

   For $i = 2$ *to* n_1 *do:*

   If $d(x_1, x_i) < D_i$ *then* $\{K_i = 0;\ KN = KN - 1\}$

   Else $K_i = 1$;

4. *For* $i = 2$ *to* n_i *begin I*

   (a) *Put* $IND = 0$; *If* $KN = 0$ *then exit;*

   Else $\{$*If* $K_i = 1$ *then* $\{K_i = 0;\ KN = KN - 1;$

   Put x_i *in* S;

   $IND = 1\}\}$

   (b) *For* $j = i + 1$ *to* n_1 *begin II*

   If $(K_j = 1$ *and* $d(x_i, x_j) < D_j)$ *then* $\{K_j = 0;\ KN = KN - 1;$

   If $IND = 0$ *then* $\{IND = 1;$ *Put* x_i

   in $S\}\}$

   End II;

   End I;

The first training instance, after they have been ordered in step 2, is always selected for the pruned subset (step 3) since it is the nearest to the other class and, at least, its own related neighbor. Also in step 3, all the other sets R_i where x_1 is present are detected and marked by letting $K_i = 0$. KN is used for the current number of training cases still not represented (that is, none of its related neighbors is currently included) in S. In step 4 it is searched through the rest of the training prototypes (in increasing order of their distances to class 2). The purpose is to look after those that must be incorporated to S either because they have not been represented by any previous instance or because they are related neighbors of a posterior (more far to the other class) not yet represented training case. IND is a flag that prevents duplication of an instance in S.

2.3 Methods for Changing Some Class Labels

Up to our knowledge, only three published proposals have been involved with the change of the class label of some training prototypes. One of them (Generalized Editing) was already outlined in subsection 2.1.3. The other two are hereafter explained.

2.3.1 Mutual Neighborhood Value [25]

This method is based on the concept of Mutual Nearest Neighborhood and on a similarity measure called the Mutual Neighborhood Value (MNV). The MNV is defined as the sum of the conventional nearest neighbors' ranks of two training prototypes, with respect to each other. Let x_j be the pth nearest neighbor of x_i and let x_i be the qth nearest neighbor of x_j. Then the MNV between x_i and x_j is given by

$MNV(x_i, x_j) = p + q$

where p and q take values ranging from 0 to $n-1$ (n is the number of instances in the training set) and 0 is used when $i = j$.

If only the k nearest neighbors of each training case are considered then, if either x_i or x_j or both are not found in each other's k nearest neighborhood, it is said that x_i and x_j are not mutual neighbors. Their MNV is taken as a large number greater than $2k$.

The MNV method can be regarded as a kind of distance-weighted k-NN rule. The mutual neighborhood values determine the weights attributed to the k nearest neighbors of a training case. The weight V_{ij} assigned to x_j with respect to x_i is defined as:

$V_{ij} = 1/MNV(x_i, x_j)$ provided that $x_i \neq x_j$

The MNV method consists of two steps. Only the two-class case is considered here.

First step.

1. *Let* $S = \varnothing$.
2. *For each* x_i *in* X *do:*
   (a) *Find the* k *nearest neighbors of* x_i *in* $X \backslash \{x_i\}$.
   (b) *Compute*
      $V_{i1} = \sum_j 1/MNV(x_i, x_j)$ *where* x_j *is one of the* k *nearest neighbors of* x_i *belonging to class 1.*
      $V_{i2} = \sum_t 1/MNV(x_i, x_t)$ *where* x_t *is one of the* k *nearest neighbors of* x_i *belonging to class 2.*

(c) *Generate a label LMN_i to x_i, such that $LMN_i = 1$ if $V_{i1} > V_{i2}$. Otherwise, $LMN_i = 2$.*

(d) *If LMN_i coincides with the original label of x_i, then x_i is transferred from X to S. Otherwise, x_i remains in X.*

Second step.

The set S as conformed in the First step is assumed to contain training instances that all bear correct class identifications. Now, all the remaining prototypes that still are located in X are moved also to S. This incorporation to S can be either with the original label each instance has or with a new one, according to the following procedure:

1. *For each x_i still in X do:*

   (a) *Find the k nearest neighbors of x_i among the prototypes in S.*

   (b) *Generate a new label, $NLMN_i$, for x_i according to the majority of these k nearest neighbors.*

   (c) *If $LMN_i = NLMN_i$ then x_i is given this label. Otherwise, x_i retains its original label.*

2.3.2 Re-identification [11]

After a slight modification to make it order-independent, this algorithm results in a scheme very similar to Generalized Editing (subsection 2.1.3.) but without removal of training cases. As in Generalized Edition, the parameters k and k' are defined under the same conditions. Then:

1. *Let $S = X$.*

2. *For each x_i in X do:*

   (a) *Find the k nearest neighbors of x_i in $X \backslash \{x_i\}$.*

   (b) *If a class has at least k' representatives among those k neighbors, then identify x_i according to that class (independently of its original class label). Otherwise, x_i keeps its original label.*

3 Incorrectness in the Training Data and Related Works

Approaches to supervised learning methods require the fulfillment of two basic assumptions concerning training samples in order to guarantee accurate identification of new cases:

1. The set of the c classes considered (with representatives) in the training sample covers the whole space of the relevant classes.

2. The training instances used to teach the classifier how to identify each class are actually members of that class (that is, they are correctly labeled).

Practical experience has shown that in many real applications one or both of these hypothesis do not entirely hold. Violations of these assumptions may strongly degrade classification accuracy, particularly when employing nonparametric methods. In conformity with this view, the number of papers and proposals for handling this subject has significantly increased in the last years.

Outlier data is a concept that has been considered in Statistics for some time. It can be defined as a case that differs significantly from the rest of the instances in its class or group. Difficulties for isolating these cases and to cope with them have long been discussed. Now the term has come to the fore also in the Machine Learning, Pattern Recognition and Data Mining areas. Reports about the effect of these "noisy" or atypical patterns when included in the training sample and how to counteract it have been published (e.g., [40][46]). Gopalakrishnan et al. [24] are concerned by the long training time required usually by some neural networks models. They propose the use of a clustering technique for identifying noisy training cases that slow down the learning phase of these models. John [33] addresses the identification of outliers in categorical data and promotes an iterative procedure to clean the training data. He works with the C4.5 tree induction algorithm and removes those training instances linked to the nodes eliminated during the pruning phase (that is, those cases whose removal from the training set will most increase the model's estimate of its own performance). Afterwards, the model is built again up considering only the cleaned training data. His purpose is to obtain a smaller and more accurate decision tree. Even for unsupervised methods outlier data has been the concern of several researchers, e.g., [26][34].

In the interpretation of remotely sensed data, training field selection and the yielding of suitable training statistics have been the concern of several researchers and practitioners, e.g., [9][23][34][48]. Foody [20] and Foody et al. [21] discuss to some extent the difficulties introduced into the classification process by those prototypes that represent more than one class (e.g., those allocated on the border between classes) or by unknown patterns that belong to a class not considered when the training sample has been collected. Mather [36] refers to atypical elements in the TS that may belong to another class or may be hybrid or mixed elements.

Brodley and Friedl [10] employ a combination of classifiers to filter the training patterns looking for detection and elimination of wrong-labeled training cases prior to applying the chosen learning algorithm. They are very concerned with the difficulty of correctly separating instances that are merely exceptions to the general rule from noisy prototypes that are incorrectly labeled. They considered that the elimination of exceptional training cases must be avoided and mention some previous proposals aimed at discerning among true exceptions and noisy data. Some of these proposals require the participation of a human expert.

Presence of noise in the training set may cause problems for many pruning algorithms in Instance-based learning methods. When reducing the size of a training sample that contains noisy instances two problems can occur:

1. Size reduction will be very small since many instances would be needed to keep the too complex (because of the noisy prototypes) decision boundaries.

2. Generalization accuracy can be importantly degraded, particularly if noisy instances are retained while good training cases are removed.

That is the reason explaining the benefits of the combination editing-condensing for processing the training sample. After the application of the editing technique, the pruning algorithm can work in a cleaner training set. Therefore, both size reduction and generalization accuracy are increased. Another interesting approach is due to Aha [1] who proposed a pruning algorithm (IB3) with a reduced sensitivity to noisy instances.

The terms outlier, noisy and atypical are generally employed to cover a broad range of circumstances That reflects some confusion among dissimilar situations and a lack of a rigorous and unified concept of outlier data. In general, there are three of these potential situations:

1. Noisy data that are usually produced by errors (measuring, recording, etc), an unfortunate property of many real-world databases.

2. New unidentified patterns appearing in the classification phase and that do not belong to any of the classes represented in the TS (partially exposed environments: [14][38]). These cases are usually handled by a reject option [16][27][44].

3. Some authors employ the term outlier for denoting mislabeled instances in the TS, what constitutes one of the focuses of the present work.

Key differences of the just mentioned works with the procedure presented in this paper are:

1. The previous proposals do not consider correcting labels of some imperfectly identified training data (exceptions are the above mentioned Generalized editing technique and those algorithms explained in the subsection 2.3). Most of them try to solve the problem only through elimination of those instances that appear to be "doubtful". Another line (e.g., [15]) rests upon the measurement of a confidence level or " typicality" for each training instance. Then, this information is employed when classifying an unknown pattern with a weighted k-NN rule.

2. In some of these publications, real data is used for demonstration purposes. But then, the labels of some training patterns are intentionally modified to simulate an imperfectly supervised situation (e.g., [10][51]). Since there is not control or knowledge of the original contamination level in those real datasets it is possible, by working in that way, to produce a dataset cleaner than the original one.

4 Decontamination of Training Samples

The NN rule, as other nonparametric pattern recognition methods, suffers from an extreme sensitivity to the presence of contaminated training sets. Hence, the importance of building procedures to contend with imperfectly supervised environments and with incorrectness in the training data. Hereafter, we present a methodology for correcting or decontaminating the training data. The procedure is elucidated through application with multiple datasets, both simulated and from real domains. As the applications are discussed in the chronological order they were realized in, it is possible to observe how the methodology (Decontamination) has evolved until its present form.

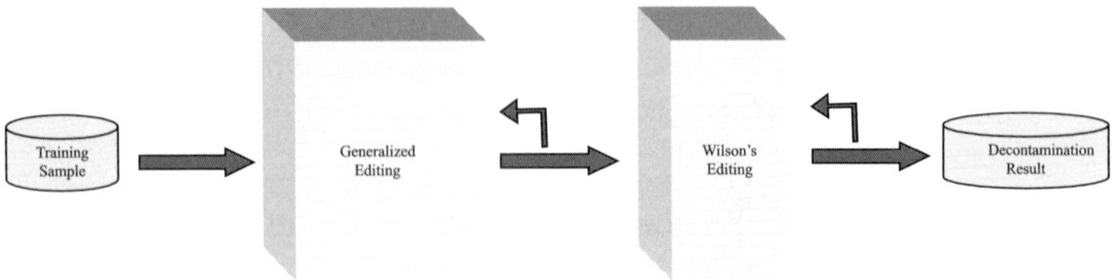

Figure 1: The Decontamination methodology.

The Decontamination methodology involves several applications of the Generalized Editing technique followed by the employment, also repeatedly, of the Wilson's Editing algorithm (see Fig. 1). Repetition in the application of each one of these techniques stops if one of the following criteria is fulfilled:

1. Stability in the structure of the training sample has been reached (no more removals and no more re-labeling).

2. Estimate of the misclassification rate (leave-one-out method; see [27]) has begun to increase.

3. One class has resulted emptied (all its representatives in the training sample have been removed or transferred to another class) or has resulted with too few prototypes (less than five training instances).

4.1 Experiments with Simulated Data.

Barandela [2] reports a considerable amount of Monte Carlo experiments to assess several procedures usable for decontaminating training samples that contain some erroneous instance identifications. The purpose was to gain some insight about the benefits of these procedures and to elucidate which of them is the most adequate for the imperfectly supervised situations. The methods included in the evaluation were those explained in subsection 2.1 (that is, the three editing techniques: Wilson's, All k-NN and Generalized Editing) and in subsection 2.3 (Mutual Neighborhood Value and Re-identification).

For comparing these procedures, experiments with simulated patterns from two bivariate Gaussian populations with different mean vectors ($m_1 = (0.0, 0.0)^t$ and $m_2 = (2.58, 0.0)^t$, respectively) and an equal covariance (identity) matrix were carried out. This is a model with a not negligible amount

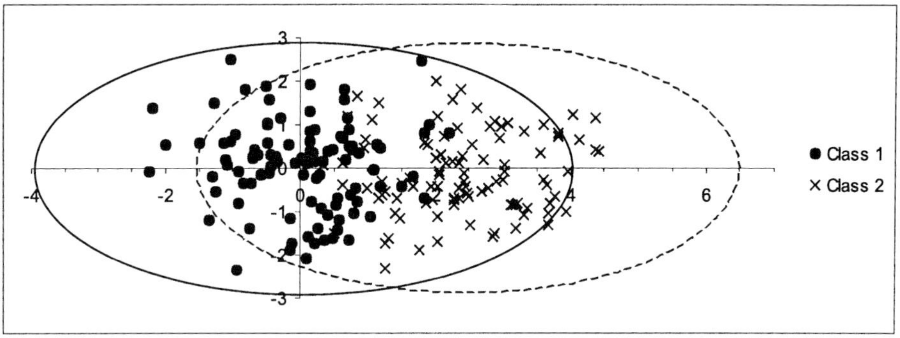

Figure 2: Probabilistic model employed for random generation of the artificial data

of overlap between the two classes (see Fig. 2). To work with simulated data is the only possible way to control the level of mislabeled instances, since with real data there is always uncertainty about the actual presence of erroneous labels or in which degree they exist. In every experiment, the TS consisted of 200 prototypes and the independent test sample (used for validation purposes) contained 500 elements, always a half from each class. Five different levels (percentages) of members of the second class were intentionally wrong labeled as belonging to the first class in each one of the training sets. The test set remained uncontaminated. For each of these combinations (the training set before and after being processed by the five techniques and the five contamination levels) 30 replications were done. According to Wilson's recommendation and to our own previous experiences, $k = 3$ was used for Wilson's and All k-NN editing methods as well as for Mutual Neighborhood technique. For a two-class case, Generalized Editing with $k = 3$ and $k' = 2$ would produce only transferences of prototypes from one class to another and no removals. Since it was intended to observe the combination of both effects, $k = 5$ and $k' = 4$ were employed in these experiments. The parameters of Re-identification received these same values, because of the similarity with Generalized Editing. The averaged results, misclassification percents on the test set, are shown in Table 1.

The effect of Generalized Editing on the contaminated training samples was remarkable and its superiority over the rest of the evaluated methods was recorded in more than 98% of the individual replications. Mutual Neighborhood and Re-identification are procedures that yield only re-labeling of some instances and no eliminations from the training sample. On the other hand, Wilson's Editing and All k-NN Editing produce only removal of some prototypes and not identification changes. Although these two techniques compare with Generalized Editing when the level of wrong labels is low (5%), they cannot compete in more extreme degrees of contamination. From the results in Table 1, it is evident that Generalized Editing excels all the other methods because of its combination of effects on the training data (both some removals and some re-labeling). Generalized Editing without removals would be equivalent to the Re-identification algorithm. Without re-labeling, the Generalized Editing is exactly the Wilson's technique.

Results in Table 1 also show that the greater the percent of initially mislabeled prototypes (in the original training set) the greater the percent of test patterns erroneously classified after the Generalized Editing (and after all the other methods) application. It should be noted that, for the decision about the labeling of a prototype (and perhaps the transference to another class or re-labeling) or about its elimination from the TS, Generalized Editing takes into account the labels of the k nearest neighbors of this prototype. That is, for evaluating the correctness of this training pattern label, the algorithm rests on the information supplied by the labels of other prototypes. Paradoxically, this supporting information might be wrong since these other prototypes can be incorrectly identified because of the contamination in the training set. Therefore, Generalized Editing can produce some improper identification changes, with a probability that increases as the level of noise in the whole TS grows. However, Generalized Editing obviously improves the classifier's performance, therefore Generalized Editing must have reduced the number of wrong labels in the training set. Consequently, performing Generalized Editing repeatedly might be beneficial: an ever increasing number of correct labels in the cleaned training sample improves the chances for repairing even more labels during further GE steps.

This idea of repeating the GE application was explored with the same artificial data sets and the same experimental design. Obtained results are showed in Table 2. It was found unnecessary to carry out more than three successive applications. In the third application, very few removals and no transferences at all were produced. The details are in Table 3. Already after the second application -and, more clearly after the third one- misclassification rates tend to reach stability. It is noteworthy also that these misclassification rates attain figures almost coincident after the third application. Since these two facts –stability and coincidence of the misclassification rates- are independent of the original amount of wrong labels in the training set, adequacy of the reiteration of Generalized Editing is confirmed. Besides,

Table 1: Comparison of methods for decontamination (averaged misclassification rates).

	Wrong labels in class 1				
	5%	15%	25%	35%	45%
Original TS (contaminated)	15.0	17.8	20.4	22.9	25.4
Wilson's Editing (k=3)	10.4	11.0	12.7	14.4	16.7
Generalized Editing (k=5, k'=4)	10.2	10.4	11.5	12.4	14.3
Mutual Neighborhood (k=3)	12.8	13.7	15.0	16.4	18.3
Re-identification (k=5, k'=4)	10.9	11.6	13.5	15.6	18.3
All k-NN (k=3)	10.4	10.9	12.6	14.4	17.8

Table 2: Reiteration of Generalized Editing (k=5, k'=4) – averaged error rates.

Wrong labels in Class 1	Original Training Set contaminated	Generalized Editing		
		1st app	2nd app	3rd app
5%	15.0	10.2	10.1	10.1
15%	17.8	10.4	10.1	10.1
25%	20.4	11.5	10.2	10.2
35%	22.9	12.4	10.5	10.3
45%	25.4	14.3	10.8	10.6

these results and those of the practical applications below indicate that the procedure will always stop after a finite number of iterations.

Table 3: Modifications in contaminated training sets structure due to reiteration of Generalized Editing (k=5, k'=4). Averaged number of prototypes.

Wrong Labels in class 1	Label changes			Removals		
	1^{st}app	2^{nd}app	3^{rd}app	1^{st}app	2^{nd}app	3^{rd}app
5%	15.6	0.4	0.0	26.3	2.3	0.4
15%	21.3	0.7	0.0	37.4	2.4	0.3
25%	24.8	1.2	0.0	52.0	2.6	0.4
35%	28.0	1.9	0.0	66.0	3.4	0.6
45%	30.1	3.1	0.0	80.7	5.0	0.6

Experimental results in Tables 1-3 contain enough evidence of the superiority of Generalized Editing (repeated) for handling imperfectly supervised situations (see also Fig. 3). Afterwards, the suitability of this technique was

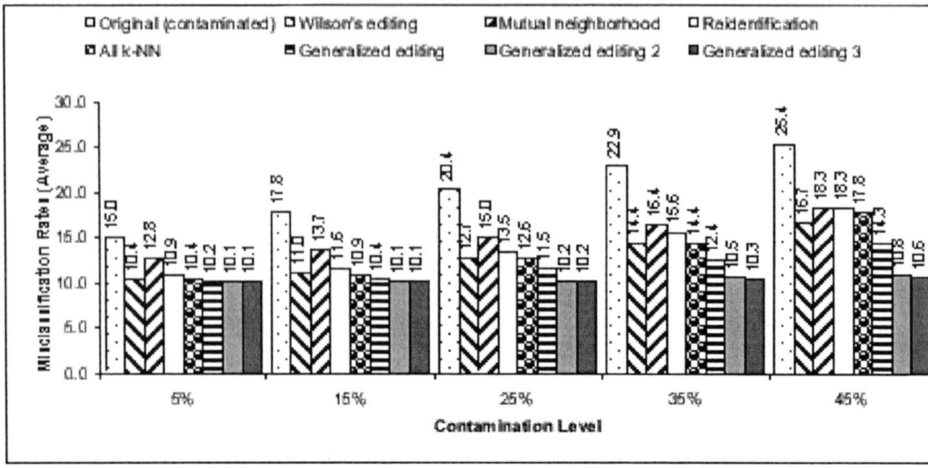

Figure 3: Performance of the Repeated Generalized Editing in contaminated training sets (Comparison with the other four explored techniques).

confirmed by the experience accumulated in several real-world applications with geophysical data. The interaction with several domain experts was also very helpful and instructive in these practical tasks.

The possibility of very few or no prototypes remaining in one or more classes is always present when processing training sets with editing algorithms. With Generalized Editing, it is true, in particular, when high numbers are assigned to the parameter values. Therefore, we have used $k = 3$ and $k' = 2$ as the parameter values in the practical applications. One of the acquired lessons is that the same results can be obtained if Generalized Editing is applied once with $k = 5$ and $k' = 4$ or if the technique is repeated two or three times with smaller parameter values. The latter process is slower but allows a more careful control of the effects on the training set and helps to avoid drastic situations.

Another learned lesson has consisted in the observation that, after the first few applications of Generalized Editing, the elimination of training instances stops and the number of transferences (re-labeling) decreases gradually. From the second application onwards, these transferences affect only to prototypes whose class labels had been already changed in previous iterations. That produces an oscillatory movement for several training cases: some prototypes are passed from one class to another in an application of Generalized Editing and then moved backwards in the following application.

As the Generalized Editing technique is repeated, this movement involves less and less training instances. That is, in each application of GE, some of these prototypes are anchored at the class in which they currently are located while other instances –in a gradually decreasing number- keep jumping from one class to another. Then, when repetition of Generalized Editing is stopped, Wilson's Editing removes from the training sample all or an important part of these prototypes that have remained oscillating until the last iterations of GE.

4.2 Geophysical Applications

The just mentioned and other characteristics of the Decontamination methodology will become more evident after a brief outline of some of the already mentioned practical applications. In it important to remark that the three practical applications that are explained in this subsection were not conceived as designed experiments. Consequently, no proper control sets were available and the assessment of the results was of a qualitative nature.

Application No. 1 [4]. Six features were measured on a set of 268 unlabeled patterns (strata belonging to a well log data). A clustering method was used to sort the data set into 4 groups that, since the ultimate purpose was to explore possibilities of gas-oil manifestations, have been regarded as:

- Classes 1 and 3, conformed by no prospective strata,
- Classes 2 and 4, with those prototypes that present strong association with the prospective strata of the area.

This configuration was taken as the initial training set to be used afterwards for classifying additional strata. This situation corresponds to a semi-supervised classification system [28][32]. Obviously, this training set should be assumed as imperfectly supervised since, by the very nature of clustering algorithms, some prototypes have received unreliable labels. Hence, the necessity of employing the Decontamination procedure was considered for improving the structure of the training sample. Generalized Editing was applied twice and the second application produced no removals and only a few transferences (about 4% of the prototypes). On the other side, with these two applications the error estimate decreased more than 75%. Wilson's Editing got also a remarkable improvement in the error estimate, yet with a small size reduction. As a final task, the processed training sample was used to classify, with the NN rule, another 628 strata coming from three different well logs. The class identifications assigned to these 628 patterns were verified using ancillary information and evaluated (by the domain experts) as more accurate than those got when the classification was done with the NN rule and the whole original TS.

Application No.2 [42]. This work was aimed to characterize the stratigraphic sequences present in an oil deposit, on the basis of an evaluation of several petrophysical and geophysical parameters, adding up to 12 features. A learning set with 139 prototypes belonging to six classes was available. Peculiarities of the area under study gave place to vagueness in the conceptual definition of the classes and ambiguity in the identification of the prototypes. Skill and intuition of the human interpreter played a decisive role in those identifications. Consequently, the training sample at hand was adopted as imperfectly supervised and the Decontamination methodology was employed. Most of the resulting transferences (re-labeling) affected prototypes located on the physical borders between classes. Consequently, these changes of label have shifted these physical borders up or down in the studied well log. As already explained, accuracy of these borders definitions had been mistrusted. Geophysicists in charge of the study of the area accepted the depurated TS as more consistent and more convenient for modeling purposes than the original one.

Application No. 3 [5]. Data employed in this case came from a previous work [47], aimed to study the processes conditioning the shape of an ophiolitic sequence. The collected sample consisted of 187 prototypes that were sorted into four classes (according to the lithology). Seven features were recorded for each pattern. Geophysicists well acquainted with the data and with the procedures that were employed to collect and prepare these data regarded this training set as perfectly supervised. Consequently, they pronounced themselves strongly against any modification. When they accepted to collaborate in the evaluation of the Decontamination process, merely with exploratory purposes, they demanded severe restrictions about amount and type of transferences (re-labeling) that could be allowed. Nevertheless, after the methodology was applied, these same domain experts accepted the decontaminated TS without any hesitation as more accurate and better structured than the original one. Here again, an important part of the transferences involved prototypes located near the geographical borders between the different lithologies (classes). Besides, undesirable transferences were not recorded or they occurred in a very low (accepted by the domain specialists) level.

Concerning these practical applications, some issues may be remarked. Firstly, Decontamination has evidenced its benefits in the three possible environments: unsupervised (as a complement to the cluster algorithm in

Application No. 1), imperfectly supervised, and supervised. In the latter situation, success of Decontamination could be explained by the lack of a precise and well-defined distinction (physical and conceptual) among classes. That is a rather common situation in the praxis, at least in the geosciences fields, and is the cause of frequent imprecision and incorrectness in the identification of prototypes. As a byproduct, the Decontamination process yielded a significant reduction of the TS size and, accordingly, of the computational time required for subsequent works involving these data. With the exception of Application No. 2 when the low Dimensionality TS-size ratio compelled the elimination of superfluous features at the very beginning of the process, feature selection was easier when implemented as an intermediate step. Employment of estimate L (leave-one-out) for error probability as a guide for conducting the process showed itself as suitable.

Some details about the gradual impact of Decontamination in the structure of the training set employed in Application No. 3 can help to illuminate and to support the statements at the end of the preceding subsection, concerning the oscillatory movement. In Application No. 3, it was required to apply Generalized Editing five consecutive times. Only in the first two of these applications, removal of prototypes was produced. The amount of label changes decreased slowly along these five reiterations (76, 34, 13, 8 and 6). Next, Wilson's Editing was employed twice, yielding elimination of 29 instances, 23 of them corresponding to prototypes that had changed their labels more than once. On the other hand, an attempt to employ Wilson's Editing directly (that is, without the previous use of Generalized Editing) on this dataset could not be fulfilled since, from the very first application of Wilson's Editing, the number of instances eliminated was too high and class 3 was emptied (all its representatives were removed).

4.3 Remote Sensing Application

The Decontamination methodology was applied on a training sample comprising 3015 pixels (instances) selected from a Landsat-TM image [6]. These prototypes represented to 11 classes of geographical objects. The work was aimed at corroborating that the NN rule can compete, with an adequate preprocessing of the training sample, with the widely accepted Maximum Likelihood classifier [19] for image interpretation tasks. The two classification methods were evaluated through classification of an independent test set (2933 cases). Three choices for the NN rule were assessed: with the original training set, after the training set has been processed with the Decontamination methodology and after the employment of the Modified Selective

Table 4: Results with remote sensing data.

Method	*TS* size	Misclassifications (%)	Kappa coefficient
Maximum Likelihood	3015	12.0	0.86
NN rule	3015	8.5	0.90
NN rule-Decontamination	2973	6.9	0.92
NN rule-Decont+ Mod.Selec	234	6.9	0.92

pruning algorithm (see subsection 2.2.2). Domain specialists with a remarkable knowledge of the studied region have collected both samples. Results are shown in Table 4.

The Kappa coefficient (fourth column) is the usual measure to evaluate classification performance in remote sensing applications [12]. It is computed on the basis of a so-called confusion or error matrix and takes into consideration the classification results on the instances of each individual class. Statistical comparison of the pairs of estimates of the Kappa coefficient: Decontamination procedure -with or without the application of the Modified Selective technique- versus the NN rule with the original training set and versus the Maximum Likelihood classifier, showed significant improvements in the performance of the classifier ($p < 0.01$ and $p < 0.005$, respectively). The difference in performance (as measured by the Kappa coefficient) between NN rule with the original training sample and with the Maximum Likelihood method was also statistically significant ($p < 0.05$).

It is noteworthy to emphasize that the domain specialists have collected this dataset with a careful selection. That has produced instances identification and a structure of the *TS* (all the classes are represented in the training set in the same proportion as they occur in the whole image) with an unusual high quality. This statement is corroborated by the not frequent —in the remote sensing domain– large values for classification accuracies (third and fourth columns in Table 4). Even so, the processing methodology was effective both in accuracy improvement and in computational cost reduction. Analysis of the Decontamination effects on the training data was supported by ancillary information concerning the spatial distribution of the training instances. This analysis reflected that all the training instances that received label modification corresponded to pixels located at the edges of the selected training fields, particularly in those classes that correspond to very narrow spatial objects (e.g., roads). These pixels that are located on the border between different classes can be considered as mislabeled or noisy training instances. Their inclusion in practical tasks is due mainly to the difficulties to single small training fields out when marking them with the computer mouse and to the fact that some of these pixels are mixed elements that combine characteristics of the two adjacent classes. Because of the not too small size of the training set, the Modified Selective technique for eliminating superfluous instances (reduction of the training sample size) was employed. Consequently, less than 8% of the original prototypes have remained, with a substantial decrease in the computational cost of the classifier while preserving the obtained accuracy improvement.

4.4 A New Artificial Experiment

In the above real data applications, explanations of the domain specialists motivated the belief in the possibility of some mislabeled training cases, because of the characteristics of the given tasks, the way in which the classes are defined and the procedures employed to collect the training data. The idea of the Decontamination methodology (Generalized editing followed by Wilson's algorithm, also repeated) was completed as the result of the exchanges with several experts in the application areas and of the careful joint analysis of the outcomes. The results obtained by the Decontamination methodology, in generalization accuracy and in the resulting TS structures, indicate that the premise (some contamination was present in these training sets) was correct and that the methodology was appropriate. But, in general, it is not possible in actual applications with real data to know the precise level of existent contamination in training sets.

Since we wanted to evaluate, as exactly as possible, the Decontamination methodology with total control of the contamination present in every step, experiments with artificial datasets were carried out again. These new experiments were done to address the assessment of the methodology as a whole and not to evaluate individual techniques as in subsection 4.1. Ten datasets were pseudo randomly generated following the same model as in Subsection 4.1. Each dataset (200 instances, 100 from each class) was employed as training sample and received 20 percent of intentional contamination (20 cases from class 1 were identified as belonging to class 2) to simulate imperfectly supervised situations. Then, each training set was exposed to the Decontamination procedure. Averaged results (and standard deviation) of the 10 replications appear in Table 5.

Deterioration of classifier's performance was striking after contamination of the TS. The average value of the required applications of Generalized Editing was 5.6 times and the level of misidentified instances was decreas-

Table 5: Results of the methodology on artificial datasets. Results of the Decontamination procedure are shown separating the first phase (Generalized Editing) from the second one (Wilson's Editing). Use of Modified Selective is optional.

	Class 1 size	Class 2 size	Misclassification Mean (st. dev.)
Original TS (not contaminated)	100	100	12.3 (2.02)
Contaminated TS	80	120	20.9 (2.06)
After Generalized Editing	94	106	12.0 (2.04)
After Wilson's Editing	87.6	100.2	10.7 (2.13)
After Modified Selective	8.5	7.8	10.7 (1.79)

ing gradually, dropping from 20% to only 6%. The resulting classification accuracy excelled that of the original (not contaminated) TS. Then, in each training set, Wilson's Editing was employed once or at most twice, vanishing the small remaining contamination in Class 2 and yielding an additional accuracy improvement. Modified Selective got a very satisfactory size reduction, again without affecting classification accuracy. All these assessments were done with an independent test set consisting of 500 instances (250 from each class) generated in a similar manner as the training samples but without contamination. Mean misclassification percentage of the Decontamination procedure (with and without the application of the Modified Selective technique) was significantly better than that obtained with the Original Training Set Contaminated ($p < 0.001$). Also, the comparison with the non-contaminated training set produced a statistically significant difference ($p < 0.01$).

It is important to emphasize that Decontamination obtains better performance than the original no contaminated TS -not to speak of the contaminated one– because it is not limited to the removal or to the re-identification of some training instances. The methodology can take advantage of an adequate combination of both actions, correcting a considerable amount of wrong labels while eliminating most of the atypical cases. Because of the way in which they were generated for these experiments, original (no-contaminated) training sets are assumed to be free of mislabeled instances. But the possibility of existent atypical or exceptional cases is always present. Besides, as already mentioned, there is a considerable amount of overlap between the two classes.

Table 6: Description of the real datasets employed in the experiments.

Dataset	No. classes	No. features	Training set size	Test set size
Breast-cancer	2	9	546	137
Glass	6	9	174	40
Heart	2	13	216	54
Iris	3	4	120	30
Liver	2	6	276	69
Pima	2	6	615	153
Vehicle	4	18	678	168
Vowel	11	10	429	99
Wine	3	13	144	34

4.5 Some Additional Experiments with Real Data.

As an additional evaluation of the Decontamination methodology and to follow a rather common practice, experiments covering some public datasets taken from the UCI Repository [37] where also carried out. Nine datasets were selected among those that received more attention in the machine learning literature. Only datasets with a not too large size were considered. Descriptions of these datasets are in Table 6.

Five-fold cross-validation was employed (80% of the data as training set and 20% as test set). In each case, we have used Decontamination Methodology (Generalized Editing + Wilson's Editing). For comparison purposes, Wilson's Editing (that is, not correcting any label, only eliminations) was also employed. Experimental results, in mean misclassification percentage and standard deviation, are shown in Table 7. Since nothing is known or can be presumed about existence of mislabeled instances or atypical elements, the obtained results are merely exploratory

Table 7 also shows the results of the statistical tests for the comparisons of the Decontamination methodology with the NN rule when working with the original training sample. Decontamination is also statistically compared with the sample processed by the Wilson's editing technique. Decontamination produced significantly better or equal classification accuracy than the Original Training Sample in six of the nine datasets and has a inferior performance in three of them. On the other hand, Decontamination excels or equals to Wilson's editing in all datasets with only one exception.

Table 8 presents a more detailed analysis of the characteristics of those four datasets where Decontamination got the worst results. It is widely accepted that the appropriate number of training instances must depend on

Table 7: Experimental results with public datasets. Decontamination and Wilson's editing.

Dataset	Original training set	Decontamination procedure	Decontamination vs. Original TS	Wilson's editing	Decontamination vs. Wilson's editing
Cancer	4.4	3.3	p<0.005	4.0	p<0.005
Glass	30.0	33.0	(p<0.005)	37.0	p<0.005
Heart	41.8	30.7	p<0.001	35.6	p<0.001
Iris	3.3	3.3	no significant	3.3	no significant
Liver	34.8	29.7	p<0.001	30.7	no significant
Pima	36.1	24.1	p<0.001	28.0	p<0.001
Vehicle	35.7	38.7	(p<0.005)	40.4	p<0.005
Vowel	2.4	16.8	(p<0.001)	13.4	(p<0.005)
Wine	27.7	29.4	no significant	28.2	no significant

Table 8: Size/dimensionality rate when considering the smallest class, after Decontamination.

Dataset	No. features	Smallest class size	Size/dimensionality rate
Glass	9	7	0.78
Vehicle	18	92	5.11
Vowel	10	38	3.80
Wine	13	29	2.23

the model complexity (the number of the considered features). That is, the training sample must hold a size /dimensionality rate high enough (5-10 prototypes for each attribute). In fact, adequacy of the training set size must be measured by considering the number of training cases from the smallest class (the less represented) and not that of the whole training sample. Fourth column of Table 8 shows a rather low rate for these datasets (in the smallest class) after they were processed by the Decontamination methodology. Reduction of the number of attributes by a convenient selection might be an adequate previous step.

4.6 Experiments with Real Datasets and with the Multi-Layer Perceptron

Table 9 shows the results obtained when the experiments with the same real datasets (see Table 6 above) as in the previous subsection were carried out

Table 9: Experimental results with the Multi-Layer Perceptron model and public datasets.

Dataset	Original Training set	Decontamination procedure	Decontamination vs. Original *TS*	Wilson's editing	Decontamination vs. Wilson's editing
Cancer	6.0	3.4	p <0.001	3.4	no significant
Glass	40.0	34.5	p <0.001	35.5	no significant
Heart	33.0	30.4	p <0.05	33.7	p <0.001
Iris	5.3	2.7	p <0.01	2.7	no significant
Liver	37.5	29.0	p <0.001	28.7	no significant
Pima	25.9	25.9	no significant	26.9	p <0.005
Vehicle	25.0	40.9	(p <0.001)	40.2	no significant
Vowel	21.2	16.8	p <0.001	13.1	(p <0.001)
Wine	4.1	29.4	(p <0.001)	25.7	no significant

for the learning of a Multi-Layer Perceptron neural model. In each case, the number of hidden nodes (only one hidden layer) was set as the number of features plus one. Stopping criteria for the learning phase were: 30000 as the maximum number of iterations in the training phase or a system error less than 0.0001.

The results are very similar to those obtained with the Nearest Neighbor rule. Decontamination procedure yields a significant improvement of the classification accuracy in six of the datasets, when compared with the Original training set. Again, Decontamination excels or equals to Wilson's editing in all datasets with only one exception.

4.7 An Empirical Comparison with the Ensemble Filter.

Brodley and Friedl [10] proposed an innovative approach (Ensemble filter) for identifying training instances that are mislabeled. After identification, these prototypes are removed from the training set. Ensemble classifiers combine the outputs of several individual learning algorithms [39]. Employment of ensemble methods is considered as one of the most promising research lines [18] in machine learning and pattern recognition. The idea of Brodley and Friedl is to create an Ensemble filter to process the training sample. Their filter works by identifying an instance as mislabeled if h of the m individual classifiers cannot classify it correctly.

Since this approach offers a great number of possibilities, we intend to explore it thoroughly. In this subsection, we present an initial empirical comparison with the Decontamination procedure. Our ensemble is built with four individual classifiers from machine learning and statistical pattern recognition: 1-NN (one nearest neighbor), Maximum Likelihood, Multi-Layer Perceptron and a Decision Tree (C4.5). Experiments were done with the same real datasets and the same design as in subsections 4.5 and 4.6. Results in Table 10 show the misclassification rates (and the standard deviations) of the Decontamination methodology and of the two variants of the Ensemble Filter that have been proposed by Brodley and Friedl. These two variants are:

- Ensemble filter 1 (majority filter): an instance is considered as mislabeled if more than half of the m individual classifiers classify it incorrectly.
- Ensemble filter 2 (consensus filter): an instance is considered mislabeled if all of the m individual classifiers classify it incorrectly.

The Decontamination procedure has less accuracy than the ensemble filters in three of the four datasets that were discussed at the end of subsection 4.5. That is, those where the size/dimensionality rate is too low. In two of these datasets (Vehicle and Wine), Ensemble filter 2 has not produced any modification of the training sets. The same situation was recorded with the majority filter (Ensemble 1) and the Wine dataset. Accordingly, misclassification rate in these datasets is the same after employing the Ensemble filters than when working with the original training sets.

The size/dimensionality rate is only slighter better after the employment of the Ensemble filter approaches when compared with the obtained after Decontamination (see Table 11). Ensemble filters seem to be more indifferent to the " curse of the dimensionality" . Our ensemble includes two learning algorithms that contribute to this reduced sensitivity: C4.5 and Maximum Likelihood (we have used the linear discriminant function, that is, a pooled covariance matrix for all the classes). In the other five datasets, however, Decontamination produced better or equal classification accuracy than the Ensemble filters.

4.8 The Decontamination Methodology in Brief

After the above illustrative examples, it is convenient (for a better understanding) to assemble in this subsection, the main elements of our methodology. As already stated at the beginning of the present Section, the Decontamination methodology involves several applications of the Generalized

Table 10: Empirical comparison of the Decontamination procedure with the two variants of the Ensemble filter technique. Misclassification rates and statistical comparisons.

Datasets	Decontamination procedure	Ensemble filter 1	Ensemble filter 2	Decontamination vs. E. filter 1	Decontamination vs. E. filter 2
Cancer	3.3	3.5	4.0	no significant	$p < 0.01$
Glass	33.0	29.5	31.0	($p < 0.005$)	($p < 0.05$)
Heart	30.7	40.7	41.9	$p < 0.001$	$p < 0.001$
Iris	3.3	3.4	4.0	no significant	$p < 0.05$
Liver	29.7	38.0	35.9	$p < 0.001$	$p < 0.005$
Pima	24.1	28.0	31.4	$p < 0.001$	$p < 0.001$
Vehicle	38.7	35.9	35.7	($p < 0.005$)	($p < 0.005$)
Vowel	16.8	3.1	2.8	($p < 0.001$)	($p < 0.001$)
Wine	29.4	27.7	27.7	no significant	no significant

Table 11: Size/dimensionality rate (when considering the smallest class) of the four real datasets after application of the Decontamination methodology and the Ensemble filters.

	Size/dimensionality rate (smallest class)		
Dataset	Decontamination	Ensemble Filter 1	Ensemble Filter 2
Glass	0.78	0.89	0.89
Vehicle	5.11	8.84	8.84
Vowel	3.80	3.80	3.90
Wine	2.23	3.00	3.00

Editing technique followed by the employment, also repeatedly, of the Wilson's Editing algorithm.

It was there also explained that repetition in the application of each one of these techniques stops if one of the following criteria is fulfilled:

a) Stability in the structure of the training sample has been reached (no more removals and no more re-labeling).

b) Estimate of the misclassification rate (leave-one-out method) has begun to increase.

c) One class has resulted emptied (all its representatives in the training sample have been removed or transferred to another class) or has resulted with too few prototypes (less than five training instances).

The experience acquired in some practical applications, has motivated a modification of criterion c) to take into account the possibility of a not satisfactory size/dimensionality rate. Then, this criterion has been reworded in the following way:

c) One class has resulted emptied (all its representatives in the training sample have been removed or transferred to another class) or has resulted with a very small size/dimensionality rate (less than five training instances for each considered feature).

When this size/dimensionality rate is not adequate, feature selection must be done to allow further decontamination of the training sample.

As for the parameters k and k', it was already explained that their most suitable values are 3 and 2, respectively. These values might force to a slower procedure, requiring more iterations than other values (5 and 4, for instance). But at the end, the same results are reached and they facilitate a more careful control of the number of prototypes remaining in the training sample.

5 The Role of Decontamination in a Classification System with On-Going Learning Capacity

The Decontamination procedure has shown its benefits not only for cleaning and correcting training samples. It can be used also to help in other situations such as the one outlined in this section.

Supervised classification methods operate usually in two chronologically non-overlapping stages: the learning or training phase and the classification or operational phase. The second phase is not started before completion of the first one and thereafter no new knowledge acquisition is attempted.

In this section, an implementation of a classification system with an on-going learning capability [8] is discussed. That is, a system that not only can learn with the manipulation of a training set but could also benefit from the experience acquired when working in the classification of new patterns. The goal is to create an environment to facilitate a steady increase in the knowledge of the computer system and, consequently, to enhance the classification accuracy. The system is built around the Nearest Neighbor rule.

Since the idea is to develop a procedure that works as automatically as possible (that is, without human participation), it is necessary to cope with two potential dangers. The system is to enrich its knowledge by incorporating to the training sample those patterns identified by the own system during the classification phase. However, it is evident that this method can be self-defeating. These new training elements would have the class label assigned by the classifier. Therefore, there is the risk to incorporate several wrong labeled patterns to the TS and, consequently, to degrade the system accuracy. The implemented procedure attempts to overcome this difficulty by employing two complementary automatic tools: a reject option (to be explained below) and the Decontamination methodology. Reject options have been proposed for reducing the number of classification errors and, recently, to detect new patterns that belong to classes not represented in the TS [44]. However, in our procedure the reject option has the goal to restrict incorporation of new patterns to the training sample and does not take part in the classification decisions. That is, although every new pattern is a candidate to become member of the TS after being identified by the system, only those that are not refused by the reject option are accepted for updating the knowledge of the classifier. This is the first resource against the possible contamination of the training sample. After incorporation of some of the new patterns to the TS, the Decontamination is applied as a second filtering step, to amend (by removal or re-identification) labels that could have been incorrectly assigned by the system to those new patterns that were not detected by the reject option.

The other problem is related with the possibility for the training sample size to become so big as to make it impractical to be handled. As for this concern, it is to remark that the Decontamination methodology has the property, as a byproduct, of reducing the training sample size (although in a not considerable amount). The Modified Selective algorithm (subsection 2.2.2) can be employed when, even after the Decontamination application, the TS size is still a too high number.

The employed reject option [2] is:

a) For every new pattern X to be classified, its two nearest neighbors are searched into the training sample. If these two neighbors are both from the same class, assign X to that class.

b) If the two NN's labels do not coincide, then compare the rate of these two neighbors' distances to X (distance of X to its first nearest neighbor / distance of X to its second nearest neighbor) with a predefined threshold value. If smaller, then classify X as member of the class of its first nearest neighbor. Otherwise, reject X.

In summary, the on-going learning procedure consists of the following steps (see Fig. 4):

Table 12: Information about the real datasets employed with the on-going learning procedure.

Dataset	#Classes	#Features	*TS* size	CS size
Liver	2	6	138	207
Ionosphere	2	34	140	210
Sonar	2	60	83	125
Vowel	11	10	209	325

1. Initial *TS* is stored in memory.

2. Classification phase starts (1-NN rule, not reject option). After identification of a certain number (for example: 100) of new patterns, this classification phase is temporarily stopped.

3. The just identified new patterns are assessed for being incorporated to the *TS*. To minimize the risk of introducing contamination (by wrong labels) into the *TS*, the reject option filters the candidates to decide which of them are worthy to be joined. Then, Decontamination is employed as a second filter to re-label or remove some of the incorporated patterns. When the *TS* size is too high, reduction algorithm (Modified Selective) is applied.

4. If no new pattern remains unidentified, end. Otherwise, the procedure goes to step 2.

The on-going learning procedure was assessed through experiments with four real datasets that were taken from the UCI Repository. Datasets were divided in, approximately, 40% for the training sample and 60% for an independent control set used for validation purposes. Characteristics of these datasets are in Table 12. For the experiments, three different partitions were randomly produced for each dataset, all with the same proportion between training and control set, as described in Table 12. To simulate the sequence required for developing the potentiality of the ongoing learning capacity, the control set of each partition was divided into 10 lots or layers. That was also the reason for selecting medium-sized datasets. Each layer contained, approximately, the same proportion of patterns of every class. Since the classification phase stopped temporarily after identification of the members in each of these lots, the system had 10 opportunities to increase its knowledge.

The experimental comparison included two variants:

Table 13: Experimental results of the on-going learning procedure with real datasets.

	Averaged misclassification (%)		
Dataset	Variant 1	Variant 2	Stat. significance
Liver	40.2	38.2	$p<0.005$
Ionosphere	25.4	26.0	no significant
Sonar	61.0	44.8	$p<0.001$
Vowel	60.7	47.2	$p<0.001$

Variant 1: the usual procedure. All the test patterns are classified in a single batch, without knowledge increase in the classification phase.

Variant 2: the on-going learning procedure. The control set was partitioned in several groups. After identification of the patterns of each group, operational phase stopped temporarily to incorporate to the TS those new patterns not filtered by the reject option, to Decontaminate the resulting training sample and to reduce its size when needed.

Results in Table 13 present the misclassification rates –both alternatives-averaged over the three replications of each dataset. Confidence levels for the statistical significance (one-tailed t tests) of the difference in accuracy of the on-going learning procedure with the usual variant are also reported. The on-going learning system was significantly better than the traditional alternative in three of the four cases and there was not significant difference in the Ionosphere application.

The Decontamination procedure was an important help for the good performance obtained by the on-going learning method. In Table 14, it can be observed that the role of Decontamination as a second filter contributed to avoid contamination in the training set. Second column of Table 14 indicates the number of those prototypes that were accepted (not refused) by the reject option although they had been labeled by the system with an incorrect identification (different from the one they initially had). Third column shows that more than half of these incorrectly accepted instances were eliminated by the Decontamination methodology.

6 Some Concluding Comments

Wilkinson et al. [49] mention adequacy of the training data as one of the factors with a great influence on the performance of any learning algorithm.

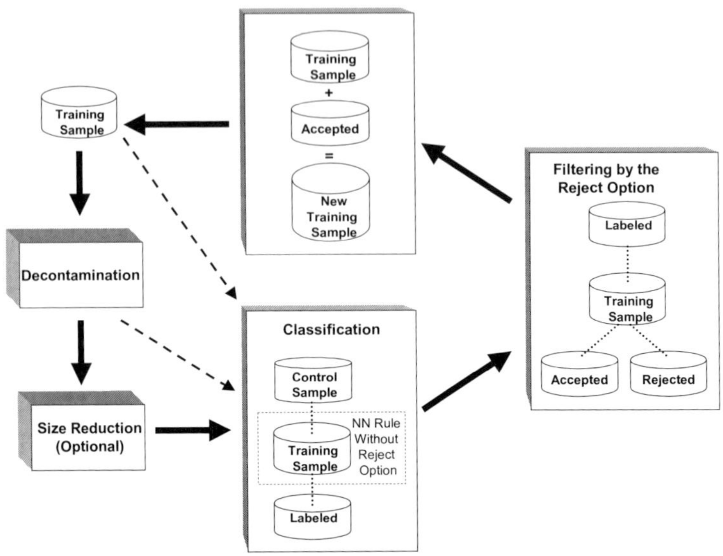

Figure 4: The on-going learning procedure.

They manifest that this issue is "very often outside the control of the data analyst". This paper presents a methodology that offers an important contribution to amend deficiencies of the available training set and to increase its usefulness. Experimental results with artificial data have revealed that the methodology can cope with contamination levels up to 45% and that it permits generalization accuracies even greater than those obtained with the original (without mislabeled instances) training sets. Practical experiences with geophysical and remote sensing data corroborate the ability of the methodology to correct an important part of wrong class labels and to eliminate most of the noisy and atypical data. In those cases, the methodology has not only yielded more accurate classification results but also has provided a better model for the phenomena under study. The latter has been corroborated when correction of some labels has allowed a deeper understanding of the studied classes and of the physical (geological and geographical) borders among them.

The Decontamination methodology has been developed with the experiences acquired in these experiments and by the analysis of their outcomes. This has involved: first, the realization that Generalized Editing (originally proposed with other purpose) could be useful in imperfectly supervised situ-

Table 14: Analysis of the contamination produced in the training sets after the reject option filtering and after the Decontamination.

Dataset	No. of instances accepted by the reject option with incorrect label	No. of instances with incorrect label that remained after Decontamination
Liver	16	5
Ionosphere	7	1
Sonar	15	2
Vowel	62	41

ations. Secondly, the execution of empirical comparisons where GE showed supremacy over other four techniques. Then, the idea of applying GE repeatedly to handle those situations with high levels of mislabeled instances. Finally, the methodology that combines Generalized Editing repeated and Wilson's algorithm also repeated. The procedure includes the computation of an error estimate (we have used leave-one-out) to be used as one of the elements that helps to conduct the process.

Because of this combination, Decontamination produces elimination of several training instances and modification of some class labels. Accordingly, it is applicable for all the possible causes of contamination in the training sample: mislabeled, noisy (with errors in some attribute values) and atypical or exceptional cases. In that way, the procedure manages quite well those situations highlighted in the above mentioned practical applications: misidentification of some prototypes due to the difficulty and the high cost of collecting the data or because of some characteristics of the application domain that induce imprecision and a lack of clear separation among classes. In many geophysical tasks, distinction among classes is only by qualitative factors. That makes almost unavoidable the introduction of subjectivity when identifying training instances. The methodology also produces the elimination of suspicious instances, what has been a focus of research in pattern recognition and machine learning, to improve the performance of instance-based and other learning algorithms.

It is true that this automatic removal includes, without distinction, noisy cases (meaning those with errors in some feature measurements) as well as prototypes that are merely exceptions to the general rule. But, although it is arguable whether to keep exceptional instances is important for modeling purposes, it is evident that it is not a straightforward task without a significant human intervention. Besides, when concern is about generalization accuracy and simplification of the classification process, all the experimental evidence points to the convenience of removing these exceptions from the training data. It is important to bear in mind that if these instances are exceptions it is because they appear with a relatively low frequency. For this reason, classification errors that exceptional cases can provoke if they remain in the training set are, in general, much more than those that their elimination can produce.

In real-world tasks, it is not easy to conclude a priori when the Decontamination methodology can be helpful. Domain knowledge has an important role to play in this determination. In our experience, discussion with domain experts may give hints to assume the presence of contamination in the training sample. But the amount of wrong-labeled cases cannot be asserted. The Decontamination methodology, being an iterated process, does not require information about this amount. Situations as those described in Section 5 are also clear candidates for the use of the methodology. It is evident that, by trying to increase the knowledge incorporated originally to the training set, some degree of contamination is introduced in those situations.

In those experiments with public data (UCI Repository) there was not the opportunity of examine (together with domain experts) elements of the respective domains that might throw light on the possibility for these datasets to contain noisy or mislabeled instances. Therefore, no definitive conclusion can be obtained from these experiments. However, comparison with Wilson's editing (a widely accepted technique for pre-processing training sets) has been positive for the Decontamination methodology when classifying with the NN rule as well as with the Multi-Layer Perceptron. Decontamination produces, in general, less instances elimination than Wilson's algorithm.

The Ensemble filter approach is very interesting, in particular in those cases where the size/dimensionality rate is not satisfactory. We intend to go deeper into this approach and to experiment with modifications to add to this approach the combination of removals and re-labeling.

Acknowledgement 1 *We acknowledge support from Grant 32016-A of Consejo Nacional de Ciencia y Tecnología and from Grant 744.99-P of Consejo Nacional de Enseñanza Tecnológica, México.*

References

[1] Aha, D.: 1992, 'Tolerating Noisy, Irrelevant and Novel Attributes in Instance-Based Learning Algorithms'. *International Journal of Man-*

Machine Studies **36**, 267–287.

[2] Barandela, R.: 1987, 'The NN Rule: An Empirical Study of its Methodological Aspects'. Ph.D. thesis, Unpublished.

[3] Barandela, R.: 1990a, 'La Regla NN con Muestras de Entrenamiento No Balanceadas'. *Investigación Operacional* **X**, 45–56.

[4] Barandela, R.: 1990b, 'Métodos de Reconocimiento de Patrones en la Solución de Tareas Geólogo- Geofísicas'. *Ciencias de la Tierra y el Espacio* **19**, 1–7.

[5] Barandela, R.: 1995, 'Una Metodología para el Reconocimiento de Patrones en Tareas Geólogo-Geofísicas'. *Geofísica Internacional* **34**(4), 399–405.

[6] Barandela, R. and E. Castellanos: 1996, 'La Regla NN para la Interpretación de Imágenes de Percepción Remota'. In: *Tercer Taller Iberoamericano Geociencias e Informática*. La Habana.

[7] Barandela, R., N. Cortes, and A. Palacios: 2001, 'The Nearest Neighbor Rule and the Reduction of the Training Sample Size'. In: *Proceedings of the 9th Spanish Symposium on Pattern Recognition and Image Analysis*. pp. 103–108.

[8] Barandela, R. and M. Juárez: 2001, 'Ongoing Learning for Supervised Pattern Recognition'. In: *Proceedings of the 14th Brazilian Symposium on Computer Graphics and Image Processing*. pp. 41–50.

[9] Bolstad, P. and T. Lillesand: 1991, 'Semi-Automated Training Approaches for Spectral Class Definition'. *International Journal of Remote Sensing* **13**(16), 3157–3168.

[10] Brodley, C. and M. Friedl: 1999, 'Identifying Mislabeled Training Data'. *Journal of Artificial Intelligence Research* **11**, 131–167.

[11] Chitinenni, C.: 1979, 'Learning with Imperfectly Labeled Patterns'. In: *Proceedings Conference on Pattern Recognition and Image Processing*. Chicago.

[12] Congalton, R. and K. Green: 1999, *Assessing the Accuracy of Remotely Sensed Data: Principles and Practices*. Boca Raton: CRC Press.

[13] Dasarathy, B.: 1979, 'All You Need to Know About the Neighbors'. In: *Proceedings International Conference on Cybernetics and Society*. Denver.

[14] Dasarathy, B.: 1993, 'Is Your Near Enough Neighbor Friendly Enough? Recognition in Partially Exposed Fuzzy Learning Environments'. In: *Proceedings of the North American Fuzzy Information Processing Society*.

[15] Dasarathy, B. and B. Sheela: 1979, 'Design of Composite Classifier Systems in Imperfectly Supervised Environments'. In: *Proceedings of the IEEE Computer Society on Pattern Recognition and Image Processing*. Chicago.

[16] Denouex, T.: 1995, 'A K-Nearest Neighbor Classification Rule Based on Dempster-Shafer Theory'. *IEEE Transactions on Systems, Man and Cybernetics* **25**(5), 804–813.

[17] Devijver, P. and J. Kittler: 1982, *Pattern Recognition - a Statistical Approach*. London: Prentice Hall.

[18] Dietterich, T.: 1997, 'Machine Learning Research: Four Current Directions'. *AI Magazine* **68**(4), 97–136.

[19] Duda, R. and P. Hart: 1973, *Pattern Classification and Scene Analysis*. New York: Wiley.

[20] Foody, G.: 1990, 'Directed Ground Survey for Improved Maximum Likelihood Classification of Remotely Sensed Data'. *International Journal of Remote Sensing* **11**(10), 1935–1940.

[21] Foody, G., N. Campbell, N. Trodd, and T. Wood: 1992, 'Derivation and Application of Probabilistic Measures of Class Membership from the Maximum Likelihood Classification'. . *Phot. Eng. and Remote Sensing* **58**(9), 1335–1341.

[22] Gasca, E. and R. Barandela: 2000, 'Influencia del Preprocesamiento de la Muestra de Entrenamiento en el Poder de Generalización del Perceptron Multicapa'. In: *Proceedings of the 6th Brazilian Symposium on Neural Networks*. Río de Janeiro.

[23] Gopal, S. and C. Woodcock: 1994, 'Theory and Methods for Accuracy Assessment of Thematic Maps Using Fuzzy Sets'. *Phot. Eng. and Remote Sensing* **60**(2), 181–188.

[24] Gopalakrishnan, M., V. Sridhar, and H. Krishnamurthy: 1995, 'Some Applications of Clustering in the Design of Neural Networks'. *Pattern Recognition Letters* **16**, 59–65.

[25] Gowda, K. and G. Krishna: 1979, 'Learning with a Mutualistic Teacher'. *Pattern Recognition* **11**, 387–390.

[26] Guha, S., R. Rastogi, and K. Shim: 1998, 'CURE: An Efficient Clustering Algorithm for Large Databases'. In: *Proceedings of the ACM-SIGMOD International Conference On Management of Data*. Seattle, Washington.

[27] Hand, D.: 1997, *Construction and Assessment of Classification Rules*. Chichester: John Wiley and Sons.

[28] Hardin, P.: 1994, 'Parametric and Nearest Neighbor Methods for Hybrid Classification: A Comparison of Pixel Assignment Accuracy'. *Phot. Eng. and Remote Sensing* **60**(12), 1439–1448.

[29] Hardin, P. and C. Thomson: 1992, 'Fast Nearest Neighbor Classification Methods for Multi-Spectral Imagery'. *The Professional Geographer* **44**(2), 191–201.

[30] Hart, P.: 1968, 'The Condensed Nearest Neighbor Rule'. *IEEE Transactions on Information Theory* **IT-14**, 505–516.

[31] Huang, Y., K. Liu, and C. Suan: 1995, 'A New Method of Optimizing Prototypes for Nearest Neighbor Classifiers Using a Multi-Layer Network'. *Pattern Recognition Letters* **16**, 77–82.

[32] Hung, C.: 1993, 'Competitive Learning Networks for Unsupervised Training'. *International Journal of Remote Sensing* **14**(12), 2411–2415.

[33] John, G.: 1997, 'Enhancements to the Data Mining Process'. Ph.D. thesis, Stanford University.

[34] Kershaw, C. and R. Fuller: 1992, 'Statistical Problems in the Discrimination of Land Cover from Satellite Images: A Case Study in Lowland Britain'. *International Journal of Remote Sensing* **13**(16), 3085–3104.

[35] Koplowitz, J. and T. Brown: 1978, 'On the Relation of Performance to Editing in Nearest Neighbor Rules'. In: *Proc. 4th International Joint Conference on Pattern Recognition*. Japan.

[36] Mather, P.: 1999, *Computer Processing of Remotely Sensed Images - an Introduction.* Chichester: Wiley and Sons, second edition.

[37] Merz, C. and P. Murphy: 1998, 'UCI Repository of Machine Learning Databases'. University of California at Irvine. http://www.csi.uci.edu/ mlearn.

[38] Muzzolini, R., Y. Yang, and R. Pierson: 1998, 'Classifier Design with Incomplete Knowledge'. *Pattern Recognition* **31**(4), 345–369.

[39] Optiz, D. and R. Maclin: 1999, 'Popular Ensemble Methods: An Empirical Study'. *Journal of Artificial Intelligence Research* **11**, 169–198.

[40] Ritter, G. and M. Gallegos: 1997, 'Outliers in Statistical Pattern Recognition and an Application to Automatic Chromosome Classification'. *Pattern Recognition Letters* **18**, 525–539.

[41] Ritter, G., H. Woodritz, S. Lowry, and T. Isenhour: 1975, 'An Algorithm for Selective Nearest Neighbor Rule'. *IEEE Transactions on Information Theory* **IT-21**, 665–669.

[42] Rodriguez, M. and R. Barandela: 1989, 'Aplicación de Algunas Técnicas de Reconocimiento de Patrones en la Caracterización Estratigráfica Del Yacimiento Varadero'. *Serie Geológica* **2**, 29–38.

[43] Sánchez, J., F. Pla, and F. Ferri: 1997, 'Prototype Selection for the Nearest Neighbor Rule Through Proximity Graphs'. *Pattern Recognition Letters* **18**(6), 507–513.

[44] Tax, D. and R. Duin: 1998, 'Outlier Detection Using Classifier Instability'. In: A. Amin, D. Dori, P. Pudil, and H. Freeman (eds.): *Advances in Pattern Recognition, Lecture Notes in Computer Science, Vol.1451.* Berlin: Springer.

[45] Tomek, I.: 1976, 'An Experiment with the Edited Nearest Neighbor Rule'. *IEEE Transactions on Systems, Man and Cybernetics* **SMC-6**, 448–452.

[46] Urahama, K. and Y. Furukawa: 1995, 'Gradient Descent Learning of Nearest Neighbor Classifiers with Outlier Rejection'. *Pattern Recognition* **28**(5), 761–768.

[47] Valladares, S.: 1986, 'Metodología para la Evaluación de los Colectores y sus Propiedades en las Rocas Pertenecientes al Complejo Aloctono Eugeosinclinal'. Ph.D. thesis, La Habana.

[48] Warren, S., M. Johnson, W. Goran, and V. Diersing: 1990, 'An Automated Objective Procedure for Selecting Representative Field Sample Sites'. *Phot. Eng. and Remote Sensing* **56**(3), 333–335.

[49] Wilkinson, G., F. Feriens, and I. Kenellopoulos: 1995, 'Integration of Neural and Statistical Approaches in Spatial Data Classification'. *Geographical Systems* **2**, 1–20.

[50] Wilson, D.: 1972, 'Asymptotic Properties of Nearest Neighbor Rules Using Edited Data Sets'. *IEEE Transactions on Systems, Man and Cybernetics* **SMC-2**, 408–421.

[51] Wilson, D. and T. Martinez: 2000, 'Reduction Techniques for Instance-Based Learning Algorithms'. *Machine Learning* **38**(3), 257–286.

Context Free Grammars and Semantic Networks for Flexible Assembly Recognition

Christian Bauckhage
Faculty of Technology
Bielefeld University, 33501 Bielefeld, Germany
E-mail: `cbauckha@techfak.uni-bielefeld.de`

Gerhard Sagerer
Faculty of Technology
Bielefeld University, 33501 Bielefeld, Germany
E-mail: `sagerer@techfak.uni-bielefeld.de`

Contents

D. Chen and X. Cheng (eds.),
Pattern Recognition and String Matching, 43–72.

1 Motivation and Context

This contribution presents a syntactic approach to assembly recognition from vision. It emerged from a research project on advanced man-machine communication in an assembly scenario where computer scientists, linguists, and psychologist develop a machine that simulates human sensing and behavior and learns about its surroundings [19]. Instructed by a naive user the machine assembles toy models from parts of the baufix® construction-kit (see Fig. 1). As interaction takes place in a non standardized environment with arbitrarily arranged objects, flexible but robust techniques for object recognition are required. This is a problem of knowledge representation.

Knowledge representation is a key issue in high-level computer vision and many approaches exist. For instance, dealing with a fixed set of objects, an expert might model characteristic object features using a semantic network. Or, applying techniques like neural networks, features might be learned from examples. For the recognition of assembled objects from our scenario, however, methodologies like these seem rather inapplicable. Since the instructor is free to decide what to construct, the machine must be equipped with comprehensive knowledge of assemblies. But baufix® parts are multifunctional objects which can be assembled into infinitely many configurations. It is thus impossible to model or learn all feasible assemblies in advance. Instead, a flexible and generic representation is needed.

Before we address this problem, we shall familiarize us with the notion and properties of mechanical assemblies as well as with established modeling techniques. Then, we will point out that simple methods from formal language theory can solve our problem. We will introduce string generating grammars to represent assembly structures and we will present a parsing algorithm to detect assemblies in image data. This framework is not restricted to baufix® assemblies but we will use them to set forth our ideas. In order to stress the generality and suitability of our approach we will discuss another application and present an evaluation.

(a) (b) (c)

Figure 1: Examples of baufix® objects and assemblies and an illustration of how to construct bolted assemblies.

1.1 Mechanical Assemblies

Mechanical assembly is a process of constructing complex objects from a set of separate parts. It results in mechanical assemblies which Rabemanantsoa and Pierre [17] define as follows:

> A *mechanical assembly* is a set of interconnected parts representing a stable unit in which each part is a solid object. Surface contacts between parts reduce the degree of freedom for relative motion. Attachments by means of screw, glue or pressure act on surface contacts and eliminate all degrees of freedom for relative motion.

Assembly usually happens in a sequence of *assembly tasks* in each of which *subassemblies* are joined to form larger units. Subassemblies either consist of a single element or a number of rigidly connected ones. They can thus be treated as if they were a part.

Mechanical assembly imposes structure on a set of previously unrelated parts. In fact, it yields several structures for it establishes different relations among the parts. Common criteria to characterize assemblies are, for instance, the relations *in-contact-with* and *consists-of*.

The surface features of parts which reduce the degree of freedom for mutual motion are known as *mating features*. They can be considered as a *functional property* of mechanical parts: from an abstract point of view a stable connection of parts, i.e. an assembly, consists of several functional units each of which contributes a mating feature. Simultaneously, an assembly is composed of subassemblies which may be elementary parts or assembled objects. Thus, from a functional viewpoint elementary objects and complex subassemblies are equivalent; both must provide mating features to be assembled.

Assemblies can be classified according to the predominant attachment; baufix® assemblies, for example, are bolted assemblies where a bolt and a nut hold together a series of parts. This is a common structure and draftsmen know various techniques to establish rigid connections between bolts and nuts. Screw connections between threaded cylindrical shafts and holes, however, are probably the most common method. baufix® assemblies are bolted assemblies with screw connections. Bolts have a threaded shaft while cubes and rhomb-nuts provide threaded holes. Parts like bars or rings, which we will call *miscellaneous objects* in the following, have simple cylindrical holes and can be put onto bolts. However, unless there is a fixating nut, connections of bolts and miscellaneous objects will not be stable.

1.2 Methods and Applications of Assembly Modeling

Throughout the past decade there has been extensive research in automatic reasoning for mechanical assembly. Mostly, it was dedicated to assembly sequence planning for efficient manufacturing but literature also knows contributions to assembly recognition from sensor data. Although our interest lies in recognition, we will briefly review relevant approaches to recognition *and* planning for ideas developed for planning can be adopted to pattern recognition.

Known approaches to assembly modeling consider either geometric or structural properties. Geometric models describe spatial extensions, positions, and orientations of the objects comprised in an assembly. Structural representations, in contrast, are more abstract. They usually reflect topological or semantical relations among assembly parts and reveal all sorts of discrete structures within composite objects.

1.2.1 Structural Modeling

Among the most popular methods of structural assembly modeling for sequence planning are *liaison graphs* as introduced by Bourjault [2]. They are used to model *in-contact-with* relations: their vertices represent elementary assembly parts and edges are drawn between physically connected parts. This neglects geometric details but describes the topological structure of assemblies. Stadzisz and Henrioud [22] point out that families of products have similar liaison graphs and apply *graph grammars* to generate them automatically.

Another widespread method in assembly modeling was proposed by Homem de Mello and Sanderson [9] who suggested the use of *AND/OR graphs* to summarize all feasible assembly sequences of a composite object.

AND/OR graphs are well known from heuristical problem solving (cf. [16]). They are characterized by *hyperlinks* which connect a parent node with a set of successors. The elements of a hyperlink are called *AND links.* In problem solving, they correspond to subproblems of a problem represented in the parent node. If there are several hyperlinks emanating from a node, they are called *OR links* and represent alternatives to subdivide the corresponding problem. The goal of heuristical problem solving based on AND/OR graphs is to find *solutions*, i.e. trees in a graph with the following properties:

1. The tree contains the root node.
2. All its leaves are nodes in the original AND/OR graph with no edges leaving them.
3. If it contains an AND link a of a hyperlink h it must contain all other AND links of h.

Homem de Mello and Sanderson proposed to identify the nodes of an AND/OR graph with an assembly and its subassemblies and to draw hyperlinks such that every hyperlink corresponds to a mechanically feasible decomposition of an assembly. If this is done correctly, every solution of the AND/OR graph represents a feasible sequence to disassemble (or vice versa to assemble) the composite object in the root node.

As AND/OR graphs are essential for the next section, we shall have a look at an example. Figure 2 depicts a baufix® assembly and its AND/OR graph. Each node shows a sketch of the (sub)assembly it represents. Hyperlinks appear as bundles of edges connected by an arch. If there are several hyperlinks emanating from a node, there are several possibilities to loose a rigid connection. Hence, for the assembly in the upper most node, there are two possibilities for disassembly. Since the graph contains two trees, there are two possible sequences to assemble the root node object.

1.2.2 Geometric Modeling

Geometric models represent the shape of physical objects. Common techniques are *Boundary Representations* and *Constructive Solid Geometry* [18].

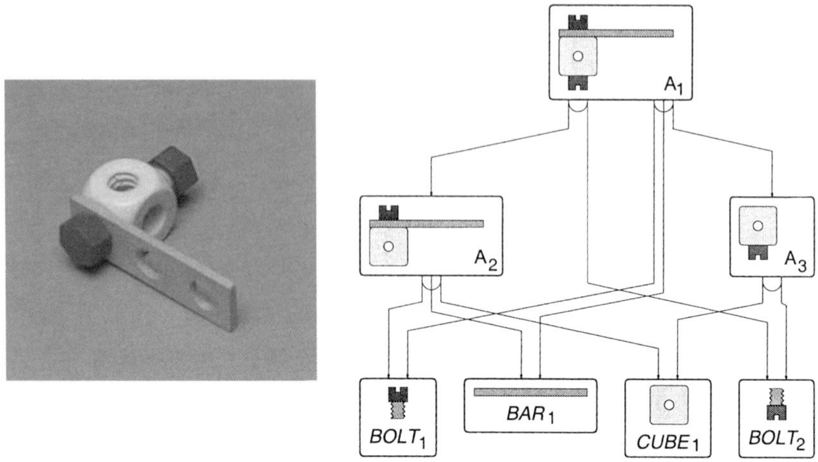

Figure 2: An exemplary assembly and its feasible decompositions represented by means of an AND/OR graph.

Within both frameworks composite objects are modeled by transforming and rotating elementary models such that they form feasible assemblies.

Geometric modeling is often used for sequence planning (cf. e.g. [17, 18, 23]) but it is also widely applied in assembly monitoring. Ikeuchi and his coworkers [11, 15] use cameras to observe assembly tasks and a laser range finder to determine how objects are positioned afterwards. By fitting geometric object models into the accurate range data newly established surface contacts can be classified and manipulator sequences to reproduce the observed actions are derived. Tung and Kak [24] describe a system that searches its surrounding for mechanical parts and assembles them. As the parts are simply shaped and carry markers, 2D vision based on geometric models is sufficient for recognition and pose estimation.

1.3 Discussion

So far, we introduced a scenario of intelligent man-machine interaction in construction and presented basic facts concerning mechanical assemblies. As we also learned about diverse methods to represent assembly knowledge, we must next verify to what extent these methods can be applied to visual assembly recognition in a non standardized environment.

Since our scenario requires flexibility in recognition, structural modeling as discussed above seems inappropriate. Except of the graph grammar approach by Stadzisz and Henrioud [22], structural models only represent individual assemblies. Flexible recognition, however, would require comprehensive sets of models which are cumbersome to provide. Geometric models, in contrast, are generic. Numerous complex models can be composed from elementary ones. But geometry based recognition would require exact information about locations and orientations of assembly parts which is difficult to extract from image data. Note that the cited contributions to assembly recognition actually circumvent this problem by employing accurate sensors like laser range finders or by concentrating on marked objects.

Graph grammars are a generic approach to assembly modeling as well. Given rules for composition, assembly models can be generated from a set of graph based object models. In fact, graph grammars are a very general concept and their suitability for pattern recognition has been known for long (cf. e.g. [5]). But they are rather difficult to implement and thus did not quite prevail. Hence, all cited methods either lack the flexibility or ease required for our application. Nevertheless, grammatical techniques seem to provide an avenue to flexible assembly modeling for recognition and in fact, we shall see that simple concepts from formal language theory are sufficient for our purpose.

2 Formal Languages and Assembly Modeling

The notion of formal languages was first introduced by Chomsky [3]. A formal language L is a set of words over an alphabet Σ, i.e. $L \subseteq \Sigma^*$. The elements of L are selected by a finite set of rules called a formal grammar. A tuple $G = (N, T, P, S)$ defines a Chomsky-grammar, if

N is an alphabet of non terminal symbols (also called variables)

T is an alphabet of terminal symbols where $N \cap T = \emptyset$

$P \subseteq ((N \cup T)^* \cdot N \cdot (N \cup T)^*) \times (N \cup T)^*$ is a non empty set of productions

$S \in N$ is a variable called the start symbol.

To denote that (u, v) is a production in P we write $u \rightarrow v$. If $u \rightarrow v$ is a production in P and a and b are strings in $(N \cup T)^*$ we write $aub \Rightarrow avb$ and say that aub directly derives avb. If $a_1, a_2, \ldots a_n$, $n \geq 1$, are strings in $(N \cup T)^*$ and $a_1 \Rightarrow a_2$, $a_2 \Rightarrow a_3$, $\ldots$, $a_{n-1} \Rightarrow a_n$ we write $a_1 \overset{*}{\Rightarrow} a_n$ and say that a_1 derives a_n. Given this terminology the formal language $L(G)$ generated by a grammar G is defined as follows:

$$L(G) = \{x \in T^* \mid S \overset{*}{\Rightarrow} x\}$$

In his work on linguistic structures Chomsky identified different classes of formal languages which are characterized by the productions of their generating grammars. The above definition presented the most general type of grammar that generates the so called *unrestricted* languages. Formal languages of other classes are derived from grammars with less flexible productions. Dealing with assemblies, the class of *context free* languages and the corresponding *context free grammars* (CFGs) are of special interest. A CFG is a Chomsky-grammar with productions $u \rightarrow v$ where $u \in N$, $v \in (N \cup T)^* \backslash \{\epsilon\}$, and ϵ denotes the empty word.

Next, we will see that mechanical assemblies can be represented as words of context free languages. Even complex assemblies can be structured according to simple CFGs and thus be represented by means of strings.

2.1 Context Free Grammatical Assembly Models

Above, we discussed work by Homem de Mello and Sanderson [9] who proposed the use of AND/OR graphs to represent *all* feasible decompositions of an assembly. However, already almost 30 years ago Hall [7] noted that there is a one to one correspondence between AND/OR graphs and context free grammars. The example in Fig. 2 should illustrate this. The depicted AND/OR graph of the exemplary assembly corresponds to a CFG G_{A} with

$$G_{\mathsf{A}} = (N_{\mathsf{A}} = \{\mathsf{A}_1, \mathsf{A}_2, \mathsf{A}_3\}, T_{\mathsf{A}} = \{BOLT_1, BOLT_2, CUBE_1, BAR_1\}, P_{\mathsf{A}}, S_{\mathsf{A}} = \mathsf{A}_1)$$

where P_{A} contains the following productions:

$$\begin{array}{lcl} \mathsf{A}_1 & \rightarrow & BOLT_2\ \mathsf{A}_2 \mid BOLT_1\ BAR_1\ \mathsf{A}_3 \\ \mathsf{A}_2 & \rightarrow & BOLT_1\ BAR_1\ CUBE_1 \\ \mathsf{A}_3 & \rightarrow & BOLT_2\ CUBE_1. \end{array}$$

Hence, nodes with descending hyperlinks are mapped onto grammatical variables, leaf nodes are equal to terminal symbols, and hyperlinks correspond to the productions of a grammar. The assembly and its complex

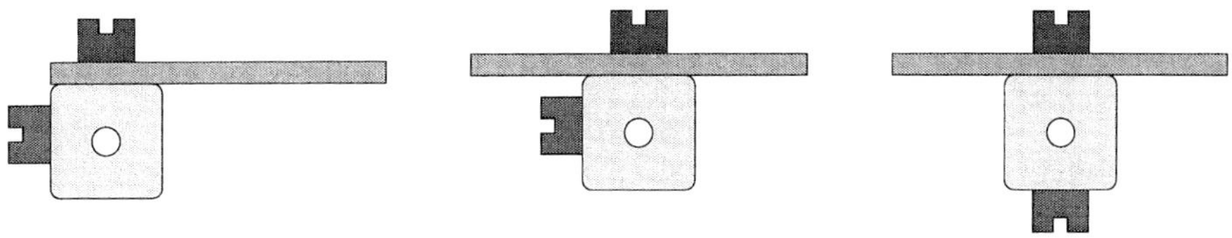

Figure 3: Examples of different assemblies that can be characterized by the context free language $L(G_{\mathsf{A}})$.

subassemblies thus appear as the variables of G_{A} while its elementary components are the terminal symbols; the language derivable from G_{A} is

$$\begin{aligned} L(G_{\mathsf{A}}) &= \{x \mid \mathsf{A}_1 \stackrel{*}{\Rightarrow} x\} \\ &= \big\{BOLT_2\ BOLT_1\ BAR_1\ CUBE_1,\ BOLT_1\ BAR_1\ BOLT_2\ CUBE_1\big\}. \end{aligned}$$

Thus, each tree of the AND/OR graph corresponds to a word of the grammar and the component structures of the assembly can be represented as a string of symbols[1]. This observation is not restricted to assemblies from our domain. As *any* mechanical assembly can be structured by means of an AND/OR graph, *any* mechanical assembly of *any* domain can be represented by a context free language.

2.2 Generalizing Context Free Assembly Models

Since string representations of assemblies strongly abstract from geometric details and since a given set of objects might be assembled into geometrically different configurations with yet the same component structure, context free languages generally characterize a whole set of assemblies. Figure 3 shows that even for our simple example it is easy to find further configurations represented by $L(G_{\mathsf{A}})$. However, although our exemplary grammar describes a set of assemblies structurally equivalent to the assembly A_1, it does not derive words characterizing assemblies with more parts. Nor does it derive words for assemblies with less parts not even for the complex subassemblies A_2 and A_3. Furthermore, G_{A} does not reveal functional aspects we discussed in section 1.1. However, three simple steps of substitution which do not change its Chomsky type will integrate functional significance into the

[1]Hall pointed out that mapping an AND/OR graph to a grammar imposes an ordering on the edges of the hyperlinks but argued that this is no loss of generality and that both techniques are equivalent, anyway. We adopt his point of view and will henceforth treat AND/OR graphs and CFGs as equivalent mathematical formalisms

grammar and simultaneously generalize it so that it will represent a whole class of structurally different assemblies.

First, we introduce new variables such that for each symbol $a \in N_{\mathsf{A}} \cup T_{\mathsf{A}}$ on the right hand side of a production there is a variable $\mathsf{f}(a)$. Thus, there will be six new variables $\mathsf{f}(\mathsf{A}_2), \ldots, \mathsf{f}(CUBE_1)$. For each new variable we introduce productions $\mathsf{f}(a) \rightarrow a$ and replace each old production of the form $l \rightarrow r_1 \ldots r_n$ by $l \rightarrow \mathsf{f}(r_1) \ldots \mathsf{f}(r_n)$. Therefore, the grammar G_{A}

$$
\begin{array}{lcl}
\mathsf{A}_1 & \rightarrow & BOLT_2\ \mathsf{A}_2 \mid BOLT_1\ BAR_1\ \mathsf{A}_3 \\
\mathsf{A}_2 & \rightarrow & BOLT_1\ BAR_1\ CUBE_1 \\
\mathsf{A}_3 & \rightarrow & BOLT_2\ CUBE_1
\end{array}
$$

will become a grammar G'_{A}:

$$
\begin{array}{rcl}
\mathsf{A}_1 & \rightarrow & \mathsf{f}(BOLT_2)\,\mathsf{f}(\mathsf{A}_2) \mid \mathsf{f}(BOLT_1)\,\mathsf{f}(BAR_1)\,\mathsf{f}(\mathsf{A}_3) \\
\mathsf{A}_2 & \rightarrow & \mathsf{f}(BOLT_1)\,\mathsf{f}(BAR_1)\,\mathsf{f}(CUBE_1) \\
\mathsf{A}_3 & \rightarrow & \mathsf{f}(BOLT_2)\,\mathsf{f}(CUBE_1) \\
\mathsf{f}(\mathsf{A}_2) & \rightarrow & \mathsf{A}_2 \\
\mathsf{f}(\mathsf{A}_3) & \rightarrow & \mathsf{A}_3 \\
\mathsf{f}(BOLT_1) & \rightarrow & BOLT_1 \\
\mathsf{f}(BOLT_2) & \rightarrow & BOLT_2 \\
\mathsf{f}(BAR_1) & \rightarrow & BAR_1 \\
\mathsf{f}(CUBE_1) & \rightarrow & CUBE_1.
\end{array}
$$

We thus introduced *functional variables* into the grammar. Each $\mathsf{f}(a)$ represents a functional unit and the (sub)assembly variables now derive functional units instead of elementary or complex subassemblies. Next, we shall substantiate the meaning of the functional variables. This is done by replacing the symbols $\mathsf{f}(a)$ with identifiers indicating the mechanical function of a component. The objects symbolized by $BOLT_1$ and $BOLT_2$, for instance, certainly act as a bolt; both functional variables $\mathsf{f}(BOLT_1)$ and $\mathsf{f}(BOLT_2)$ could hence be renamed to BOLT-PART. As all other components either act as a MISC-PART or a NUT-PART, this step yields a grammar G''_{A}:

$$
\begin{array}{rcl}
\mathsf{A}_1 & \rightarrow & \text{BOLT-PART NUT-PART} \mid \text{BOLT-PART MISC-PART NUT-PART} \\
\mathsf{A}_2 & \rightarrow & \text{BOLT-PART MISC-PART NUT-PART} \\
\mathsf{A}_3 & \rightarrow & \text{BOLT-PART NUT-PART} \\
\text{BOLT-PART} & \rightarrow & BOLT_1 \mid BOLT_2
\end{array}
$$

$$\begin{aligned}\mathsf{MISC\text{-}PART} &\rightarrow BAR_1\\ \mathsf{NUT\text{-}PART} &\rightarrow \mathsf{A}_2 \mid \mathsf{A}_3 \mid CUBE_1.\end{aligned}$$

Even though G''_{A} reflects the functional role of all components and allows to derive words for the subassemblies A_2 and A_3, it still is of marginal generality. Its terminals identify objects instead of object types. But the name of an object should be irrelevant for a structural description; rather, it would suffice to know what type of objects are assembled. Thus, we substitute terminal symbols denoting individual objects with terminals denoting object types. Furthermore, we introduce a new variable A and map all variables A_i onto A to annihilate the distinction between assemblies and subassemblies. Finally, we chose $S'_{\mathsf{A}} = \mathsf{A}$ to be the new start symbol and obtain G'''_{A}:

$$\begin{aligned}\mathsf{A} &\rightarrow \mathsf{BOLT\text{-}PART\ NUT\text{-}PART} \mid \mathsf{BOLT\text{-}PART\ MISC\text{-}PART\ NUT\text{-}PART}\\ \mathsf{BOLT\text{-}PART} &\rightarrow BOLT\\ \mathsf{MISC\text{-}PART} &\rightarrow BAR\\ \mathsf{NUT\text{-}PART} &\rightarrow \mathsf{A} \mid CUBE.\end{aligned}$$

One might argue that this step could have been carried out earlier and that is unnecessary to introduce recursion over functional units for there could have been a production like $\mathsf{A} \rightarrow \mathsf{A} \mid CUBE$. However, recursion by means of functional parts is more general and more efficient. Productions like the mentioned one would simply mean to enumerate all decompositions of a certain example.

G'''_{A} generates the infinite set of words

$$L(G'''_{\mathsf{A}}) = \{BOLT, BOLT\ BAR\}^+ \cdot \{CUBE\}$$

which obviously includes elements without physically feasible counterparts like, for instance, words composed of infinitely many bolts and a single cube. However, the language certainly represents every feasible assembly composable from a single cube and an arbitrary number of bolts and bars. Therefore, we say that the exemplary assembly is a *generator* of a context free grammar that derives the class of all bolted assemblies containing bolts and bars and a single cube. It is the class of assemblies that contain a composite nut. There will be other assemblies that generate the class of assemblies containing composite bolts or misc parts (see Fig. 1(c)). The resulting grammars can be unified so that the general syntactic structure of bolted assemblies can be derived from examples. Of course, this generalization scheme is neither restricted to the baufix® scenario nor to bolted assemblies. It applies to other domains as well, i.e. in any assembly domain comprehensive assembly grammars can be acquired from adequate examples.

A CFG that allows to describe the component structure of *all* baufix® assemblies is $G = (N, T, P, S)$ with the following set of variables

$$N = \{\text{ASSEMBLY, BOLT-PART, MISC-PART, NUT-PART}\},$$

the set of terminals

$$T = \{BOLT, BAR, RING, FELLY, SOCKET, CUBE, RHOMBNUT\},$$

the start symbol $S =$ ASSEMBLY and the productions P chosen to be:

$$
\begin{aligned}
\text{ASSEMBLY} &\rightarrow \text{BOLT-PART NUT-PART} \mid && (1)\\
&\quad\ \text{BOLT-PART MISC-PART NUT-PART} && (2)\\
\text{BOLT-PART} &\rightarrow \text{ASSEMBLY} \mid BOLT && (3)\\
\text{NUT-PART} &\rightarrow \text{ASSEMBLY} \mid CUBE \mid RHOMBNUT && (4)\\
\text{MISC-PART} &\rightarrow \text{ASSEMBLY} \mid BAR \mid FELLY \mid RING \mid SOCKET \mid && (5)\\
&\quad\ \text{ASSEMBLY MISC-PART} \mid && (6)\\
&\quad\ BAR\ \text{MISC-PART} \mid RING\ \text{MISC-PART} \mid && (7)\\
&\quad\ FELLY\ \text{MISC-PART} \mid SOCKET\ \text{MISC-PART}. && (8)
\end{aligned}
$$

Due to its recursive properties this grammar provides a very compact representation of abstract assembly knowledge. Productions 1 and 2 describe the composition of a bolted assembly. They point out that bolted assemblies at least consist of a bolt and a nut and that the miscellaneous-part is optional (see Fig. 1). The next productions state that the bolt- and the nut-part consist either of a (sub)assembly or an elementary object. Rules 5 – 8 describe how the miscellaneous-part of an assembly is constructed: it is either given by a single part or a subassembly or a sequence of those. Note that this grammar, too, derives words which do not correspond to feasible assemblies. We will return to this problem in the next section where we shall also see that this simple grammar is sufficient for recognition from vision.

3 Syntactic Approaches to Assembly Detection

Analyzing and classifying patterns of elementary components with regard to a grammar is called *parsing*. Dealing with strings parsing is usually

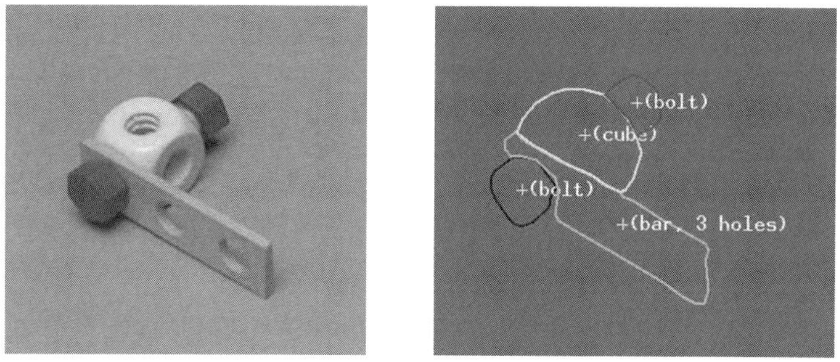

(a) (b)

Figure 4: The assembly known from Fig. 2 and corresponding results from elementary object recognition.

done from *left* to *right* and performed by *LR(k) parsers* [10]. But in image data assemblies appear as two-dimensional entities. Therefore, even though assemblies correspond to 1D words of context free languages, simple left to right parsing cannot work and we need more sophisticated mechanisms for syntactic image analysis.

Following a hybrid approach that combines neural and semantic nets Kummert et al. [13] realized a reliable method for elementary object recognition from image signals. Their algorithm yields labeled image regions as shown in Fig. 4(b). From this example we see that assemblies appear as clusters of regions which are either adjacent to or nested in each other. Mathematically, however, the presence of a region cluster does not provide a sufficient criterion for the presence of an assembly because elementary objects may form clusters without being assembled. Camera perspective may even cause clusters to occur even if there are only separated objects in the scene. But the existence of a region cluster is a necessary precondition for finding assembly structures. Thus we will assume 2D objects to be syntactic primitives and clusters of them to be the data that must be parsed.

3.1 Semantic Networks as Parsers

This section introduces the idea of implementing context free grammatical assembly models as semantic networks using the semantic network language ERNEST [20]. Similar to other formalisms like KL-ONE or PSN, semantic networks implemented in ERNEST consist of *concepts* and *links*. A concept represents an object, an event, or an idea. Dealing with image understanding interest lies in interpreting the sensor signal by means of given concepts. *Part links* decompose a concept into its natural components. If a concept has several possible component structures, they can be modeled as different *modalities*. Moreover, concepts are associated with *attributes* that represent numerical or symbolic features. Relations defining constraints for attributes can be specified and must be satisfied before valid *instances* can be generated. Instances are copies of concepts where common properties of the concept are substituted by values derived from the signal. Thus, the knowledge modeled in an ERNEST network is utilized in processes of creating and assessing instances. Assuming that complex object recognition means to detect all object parts, an object is considered to be recognized as soon as instances for all its parts were generated.

Actually, these are only some of the features ERNEST provides. For our purpose, however, it is interesting that ERNEST networks with no other than part links are formally equivalent to context free grammars[2]. Thus, implementing context free assembly grammars as semantic networks is perfectly reasonable. Moreover, in syntactic image analysis semantic networks can cope with problems that cannot be dealt with by a simple CFG. Attributes assigned to concepts may introduce additional knowledge or restrictions into a syntactic model[3] and thus prevent the acceptance of structures that might coincide with the CFG but do not correspond to feasible assemblies.

3.2 Assembly Parsing and Discourse Theory

As the research project we introduced at the beginning concerns human perception and communication, efforts are not only spent on computer vision but also on speech recognition and discourse parsing. Intense cooperation among researchers from both fields revealed a similarity between discourse and assembly structures and led to the idea to adopt strategies from discourse processing to visual assembly detection. Before we will discuss this, we shall outline some basic facts about discourse.

It is generally agreed that discourse has recursive structure, i.e. in discourse smaller segments with common reference points are recursively re-

[2]This is easy to see if we consider the AND/OR graph representation of a CFG: nodes correspond to concepts, AND links correspond to part links, and each hyperlink corresponds to a modality.

[3]This actually provides opportunities known from attributed grammars.

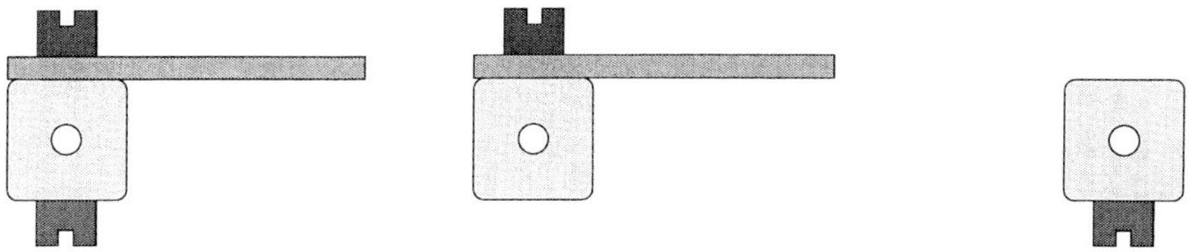

(a) Assembly A_1 (b) Subassembly A_2 (c) Subassembly A_3

Figure 5: The assembly A_1 and its two complex subassemblies. Since both subassemblies contain the cube, they are said to be *conjoined* in A_1.

lated. Therefore, a discourse grammar consists of a sentence grammar, usually chosen to be unification based, and a set of discourse productions which describe discourse relations [21]. This defines a framework in which both intra- and intersentential constraints can be expressed for it integrates discourse constraints as well as syntactic and semantic restrictions. Moreover, often the processed discourse structure is only partially available for further processing. This means that only subparts are admissible for subsequent interpretation and that partial interpretations must be brought together at later stages of a parsing process [25].

In order to transform discourse theory into a framework for assembly recognition, the structural properties of assemblies must match with those of discourse. That is, assemblies must have recursive structure where complex units result from recursively composing less complex ones. Furthermore, there must be a relation between subassemblies that allows to unify their structures into larger units. As we saw above, mechanical assemblies have recursive component structures. A relation between subassemblies has not been introduced yet, but there actually is one. If two subassemblies share elementary objects, we say they are *conjoined.* Figure 5 exemplifies this by means of the familiar assembly A_1. Since its complex subassemblies A_2 and A_3 both contain the cube they are *conjoined* in A_1.

3.3 Implementation Issues

Our assembly grammar was mapped onto an ERNEST knowledge base and domain specific attributes were defined to deal with context sensitive conditions. For example, each instance of an elementary object has an attribute to register the parts it was associated with during an assembly recognition attempt. This actually solves a problem mentioned earlier: since the CFG of baufix® assemblies also derives infeasible assembly structures, a corresponding parser might accept infeasible structures, too. But gathering information like for instance the number of objects put onto a bolt ensures that image parsing yields only feasible structures.

The structural similarities between discourse and assemblies inspired an assembly detection strategy that adapts an approach to discourse parsing due to Kronenberg [12] who combined a LR(1)-parser with a unification grammar. Her unification grammar puts most of the syntactic information into a lexicon that represents words by feature structures. This resembles the definition of concepts and attributes of elementary objects in an ERNEST network. Thus, the part of our knowledge base that describes elementary objects can be identified with a lexicon as used for discourse parsing.

In Kronenberg's LR(1)-parser every *reduce* action which reduces right hand symbols of a grammatical production to the left hand symbol includes a unification of feature structures and yields structural descriptions of larger elements. This corresponds to the instantiation of ERNEST concepts. Instantiating a concept requires to (re)calculate its attributes and is only possible after all its parts have been instantiated. Since we defined attributes that register the states of all mating features comprised in an assembly, the instantiation of an assembly therefore includes a unification of feature structures as well.

The second action carried out by a LR(1)-parser is called *shift*. It pushes symbols from an input string onto a processing stack. In the semantic network implementation, this action is simulated by a set of attributes which control the instantiation process. These attributes register, for instance, which objects were already examined or select the objects that should be analyzed next. If, like in our case, one is dealing with higher dimensional data the control attributes must realize a mechanism to sequentialize the input. This mechanism will certainly be domain dependent and might require expert knowledge. Fortunately, for the object clusters of our domain a suitable heuristic is easy to find; it will be discussed in the next section.

Given these emulations of shift and reduce actions, the basic idea for assembly detection is to try to find simple assembly structures in an object cluster (similar to discourse segments in speech parsing) and to conjoin them with earlier found ones, if possible. Thus from the analysis of subparts of a cluster the instantiation process yields syntactical descriptions of subassemblies which are conjoined with already derived descriptions. Subcluster parsing terminates if the cluster is completely analyzed or if the

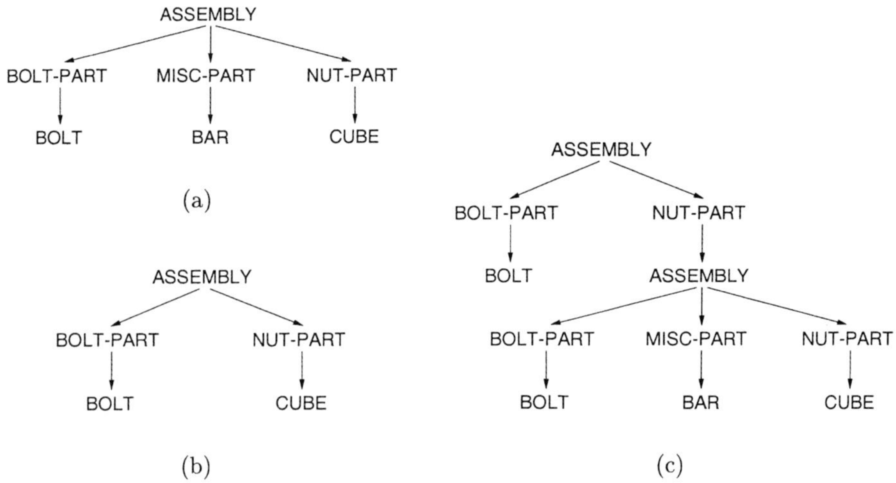

Figure 6: 6(a) Current assembly tree describing subassembly A_2 of assembly A_1. 6(b) Incoming assembly tree describing subassembly A_3 of assembly A_1. 6(c) Assembly tree representing A_1 as $\mathsf{A}_3 \triangle \mathsf{A}_2$.

unification fails. The latter occurs if two objects should be instantiated as parts of an assembly but the feature structure of one of these objects does not allow further unification. If, for example, the thread of a bolt is already completely occupied it must not be related with another object.

Merging of subassemblies is done by a *conjoin* operator that manipulates derivation trees (also denoted *assembly trees*). It integrates already derived trees (which are called *current assembly trees* according to discourse theory terminology) into *incoming assembly trees* (which is another name for the recently derived assembly tree).

Definition: The *conjoin* operator $\triangle$ combines two assembly trees X and Y if both share an elementary object. In $\mathsf{X} \triangle \mathsf{Y}$ node n of X is substituted by Y, if and only if n represents a common object (see Fig. 6 for an example). If X and Y do not share an object, $\mathsf{X} \triangle \mathsf{Y}$ is undefined.

Thus, each time an assembly is instantiated, the ERNEST algorithm for cluster analysis tests if other assemblies have been instantiated so far which share an elementary object with the recent one. If so, a conjoin operation is performed. An exemplary parsing process and its intermediate results are visualized in Fig. 7. Given the cluster from Fig. 4(b), the analysis started at the lower bolt and yielded the result shown Fig. 7(b). The depicted

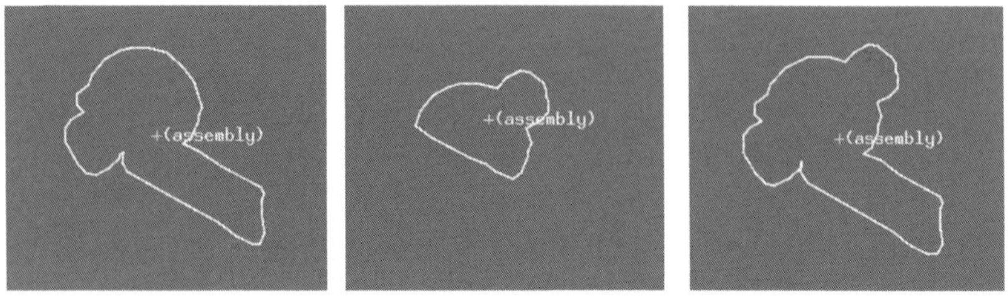

(a) (b) (c)

Figure 7: A sequence of recognition results obtained from the input depicted in Fig. 4(b). The syntactic structures represented by these sreenshots correspond to the ones shown in Fig. 6.

polygon comprises a subset of parts that, according to the grammar, are arranged such that they form an assembly. Since the there were unexamined cluster object left after the subassembly was found, parsing restarted at a yet unexamined part, i.e. the upper bolt. This led to the result in Fig. 7(a). And as this subassembly shares an object with the earlier found one, they were conjoined as we see in Fig. 7(c). Afterwards, no unexamined parts were left and parsing terminated.

3.4 Mating Relations from Vision

In bolted assemblies bolts impose partial orders on the set of compound objects. A set A is said to be partially ordered by a relation $\sqsubseteq$ (pronounced *weaker than*) if the following properties hold for every $x, y, z \in A$:

1. $x \sqsubseteq x$.

2. $x \sqsubseteq y$ and $y \sqsubseteq x$ implies $x = y$.

3. $x \sqsubseteq y$ and $y \sqsubseteq z$ implies $x \sqsubseteq z$.

A bolt and a set of objects attached to it are subject to an ordering relation if earlier attached objects are interpreted to be weaker than later attached ones. And as a bolt's thread enforces rigid linear orders of objects, all properties of partially ordered sets are met. The familiar assembly A_1 consisting of the set of objects $\{BOLT_1, BOLT_2, BAR_1, CUBE_1\}$, for instance, comes along with the following two partial orders:

$$BOLT_1 \sqsubseteq_1 BAR_1 \sqsubseteq_1 CUBE_1 \text{ and } BOLT_2 \sqsubseteq_2 CUBE_1$$

where $\sqsubseteq_1$ and $\sqsubseteq_2$ are ordering relations with respect to $BOLT_1$ and $BOLT_2$, respectively.

Since bolted assemblies necessarily consist of at least two objects, each partial order within an assembly has a weakest and a second weakest element which define a line. This leads to a linearization heuristic for image parsing. Since photography preserves topology such lines must also be found in images of baufix® assemblies. And indeed, in the image plane the coordinates of mating features of objects attached to a bolt are more or less aligned. Along these lines syntactic cluster analysis can be performed: as a bolted assembly necessarily contains a bolt, each parsing attempt starts with a region representing a bolt and the bolt part of an assembly will be instantiated. Then, an adjacent part is analyzed and if it complies with the assembly grammar a corresponding concept will be instantiated. If there are several parts adjacent to the bolt, one of them is chosen arbitrarily to continue the analysis. If this does not lead to a valid syntactic structure, the other candidates are tested in a backtracking manner. After two parts of an assembly have been instantiated, a direction to search for further parts is given. Again, if this does not lead to a feasible structure, backtracking will be performed.

For objects like rhomb-nuts or rings mating feature coordinates are assumed to be the center points of the corresponding image regions. For bars they are assumed to be equally spaced along the principle axis of the corresponding region and dealing with cubes the positions of mating features are approximated by the points of intersection between a search direction and the border of a cube's region (see Fig. 8(a)).

Given their image coordinates, it is possible to estimate which mating features of an object participate in a connection. Attributes defined in object concepts verify whether a feature is close to a line due to a bolt and if so, it is assumed to be attached to that bolt. The parser can therefore generate syntactic descriptions of assemblies which reflect hierarchical component structures and simultaneously describe relations among mating features.

Figure 8 displays such an augmented hierarchical description. Its overall structure corresponds to the syntactic structure in Fig. 6(c), mating relations are characterized as follows: objects with holes occur with (lists of) slots representing their mating features. Although the cube has six surfaces, in

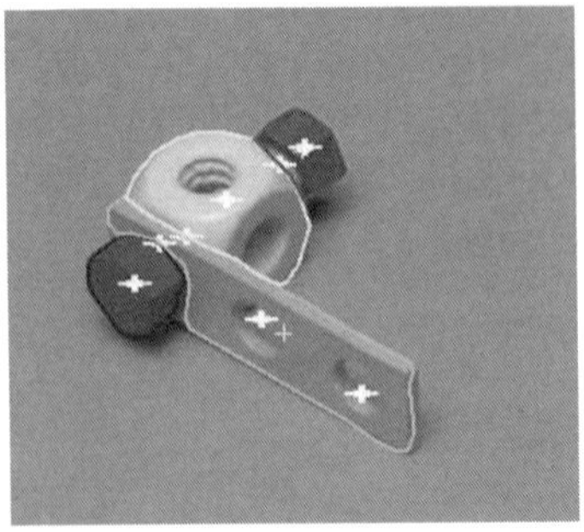

(a)

```
ASSEMBLY(Highest Level)
  BOLT2
  ASSEMBLY(Complex Nut)
    BOLT1
    BAR1 (BOLT1,a) () ()
    CUBE1 (BOLT1) () (BOLT2) ()
```

(b)

Figure 8: Calculated positions of mating features and a high-level assembly sequence plan for the exemplary assembly A_1.

the plan it only has four slots since baufix® cubes feature only four threaded holes. If a feature is occupied, it is labeled with the corresponding bolt. Holes of a bar are additionally marked with a flag indicating whether the corresponding bolt is inserted from above (a) or below (b). The order of the slots of bars and cubes is defined intrinsically. For example, the hole of a cube that was considered first during analysis always appears as the leftmost slot; the other holes are counted in a clockwise manner. Further details on this labeling scheme are explained in [1].

Syntactic structures augmented with mating feature relations are *topologically* unique descriptions of an assembly. Therefore, syntactic ambiguities as mentioned in section 2.2 do not disturb assembly recognition. Since for each assembly component details about its connections to other parts are known, syntactically equivalent assemblies can be distinguished by their feature structures.

4 Another Domain

Next, we will exemplify that our method developed for modeling mechanical assemblies is not restricted to physically connected objects: compact context free grammatical models with recursion on functional components can be found for any set of interrelated objects. This is rather obvious because any non trivial set contains subsets, i.e. it *consists-of* of smaller units and can thus be structured according to AND/OR graphs or CFGs, respectively.

Furthermore, we shall corroborate the suitability of semantic networks as grammatical parsers. To this end we will briefly introduce a domain of more abstractly related objects and discuss a suitable and comprehensive context free model and its corresponding ERNEST implementation.

4.1 Syntactic Models of Furniture Scenes

Within a research project on service robotics we preliminarily tested methods of syntactical environment modeling. The project does not primarily aim at traditional mobile robotics problems like autonomous exploration or path planning but is again focused on man-machine interaction. And similar to the baufix® scenario, for communication and navigation the intended service robot should rely on visual and acoustic input.

Global and local localization are well studied problems and literature on these issues is vast (cf. [6]). Computer vision based approaches usually try to identify landmarks which may be objects characteristic for a location. Thus, questions of self-localization from vision motivated preparatory work on 3D furniture recognition [8]. However, some pieces of furniture might be present in several rooms and thus are of no use for localization. Others might be characteristic but unrecognizable due to perspective occlusion. This led to the idea of taking contextual information into account and to model typical groups of furniture syntactically.

A group of chairs arranged around several tables, for instance, is a typical arrangement that can be recursively decomposed into smaller units. Assuming binary partitions resulted in the following CFG:

$$\begin{aligned} \text{TABLE-SCENE} &\rightarrow \text{TABLE-PART CHAIR-PART} \mid \text{TABLE-PART TABLE-PART} \\ \text{TABLE-PART} &\rightarrow \text{TABLE-SCENE} \mid TABLE_1 \mid TABLE_2 \mid \ldots \\ \text{CHAIR-PART} &\rightarrow CHAIR_1 \mid CHAIR_2 \mid \ldots \end{aligned}$$

A TABLE-SCENE may thus consist of a TABLE-PART and a CHAIR-PART or of two TABLE-PARTs. And while the CHAIR-PART cannot be of complex nature, a TABLE-PART may have recursive structure.

In contrast to mechanical assemblies, the syntactic model of a TABLE-SCENE is not bound to functions of mating features. The functional relations between the variables are rather abstract and the model was deduced from experience. While mechanical mating features thus seem superfluous in deriving context free models of classes of complex entities, we will see that they still are advantageous when implementing a parser.

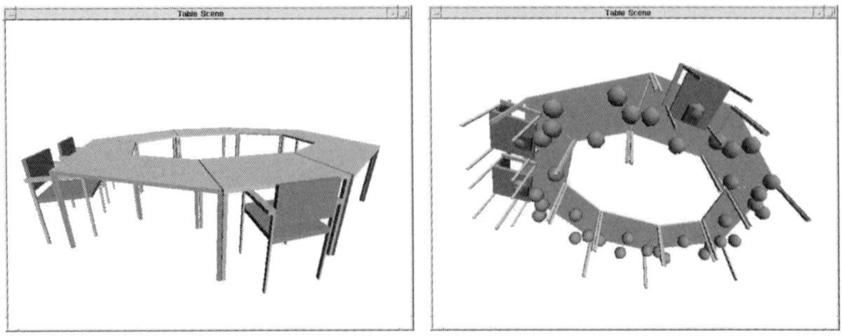

(a) (b)

Figure 9: A table scene consisting of several chairs and tables and corresponding positions of *virtual mating features.*

4.2 Implementation Issues

Implementing the grammar as an ERNEST knowledge base is straightforward. However, the data that should be searched for syntactic structures results from 3D object recognition and there are no obvious partial orders in a cluster of furniture that might guide parsing. But since relations among pieces of furniture can be stated geometrically, a practicable parse strategy for sets of 3D geometric object models can be provided. A chair, for instance, has an intrinsic front usually oriented towards a table. Thus, although there are no rigid connections there are spatial constraints that apply to well formed table scenes. And these can easily be modeled as attributes or restrictions of an ERNEST network. In analogy to mechanical assemblies we assigned *virtual mating features* to the 3D furniture models as shown in Fig. 9(b). Each table is associated with a set of 3D points which are indicated by balls. They represent prototypical positions near whom one would expect to find a chair or another table. Chairs come along with points indicated by cubes which represent the part of a chair most likely to be found close to a table.

Constraints for well-formedness could be stated using any measure of distance between virtual features and like in the case of baufix® assemblies positions of mating features can guide the parsing process performed by the semantic network. Figure 10(b) exemplifies how detected structures are visualized. It depicts a result obtained from the scene in Fig. 10(a). In

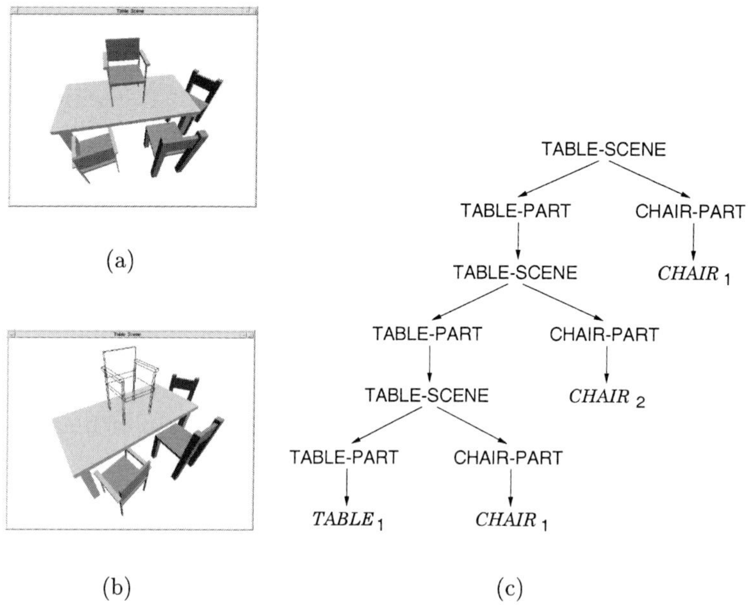

Figure 10: 10(a) A scene of several pieces of furniture. 10(b) The table scene detected within this configuration. 10(c) Syntactic structure resulting from a parsing process.

this example, the chair upon the table is not close enough to any of the table's virtual features. Consequently, it was not instantiated as a part of a TABLE-SCENE and thus appears as a wireframe in the output. The syntactic structure yielded for this scene is shown in Fig. 10(c). Two of the three chairs are of type $CHAIR_1$. The one at the right front of the table is represented on the lowest level of the parsing tree. The $CHAIR_2$ type chair at the left front can be found on the next level and the chair in the back is integrated on the highest hierarchical level.

5 Evaluation

Now that we know how classes of composite objects can be modeled by means of simple context free grammars and how simple parsing strategies can detect assembly structures in non-linear input data, it remains to examine what accuracy this approach does reach in practice.

Figure 12 displays prototypical recognition results for the baufix® domain. Assembly structures found in the images are enclosed by white polygons. These examples illustrate that the semantic network implementation really copes with context sensitive conditions. On the left, for instance, only two of the four cubes were associated with the bolt which is correct because its thread is not long enough to carry more than two.

The examples also indicate that we have to deal with images of varying complexity. Hence, it is no surprise to observe that the performance in assembly recognition depends on the *difficulty of the task*. This figure is influenced by several factors. For instance, assembly recognition will depend on the visibility of the objects that form an assembly. The more parts are (partially) occluded the more likely it is to gain a fairly wrong description. This dependency on the input complicates the provision of meaningful test sets to asses our algorithm. If a test set contained too many simple cases, the parser might be overrated. If it contained too many difficult images, the observable performance might be below average. We thus decided for an evaluation with respect to the difficulty of the input.

The basic idea to assess the task difficulty was to consider *influencing factors* measurable from an image. And as only clusters of labeled image regions are parsed for assembly structures it must be properties of object clusters which influence assembly recognition. Thus, assuming a cluster C of labeled image regions to be given, we consider the following four influencing factors:

- the number $\mathcal{B}(C)$ of visible bolts in C
- the number $\mathcal{P}(C)$ of visible parts in C
- the mean number $\mathcal{A}(C)$ of adjacent objects in C
- the mean degree $\mathcal{O}(C)$ of perspective occlusion in C

Each of these indices was chosen with regard to facts known from literature: the number of parts of an assembly influences the problem of finding assembly structures [9]. The numbers of visible bolts and visible parts will take this into account. Likewise, it is generally accepted that assembly topology determines how many syntactic structures there may be [9]. The mean number of adjacent objects in a cluster is supposed to be an according measure of connectedness. And finally, we just discussed the influence of part visibility on recognition.

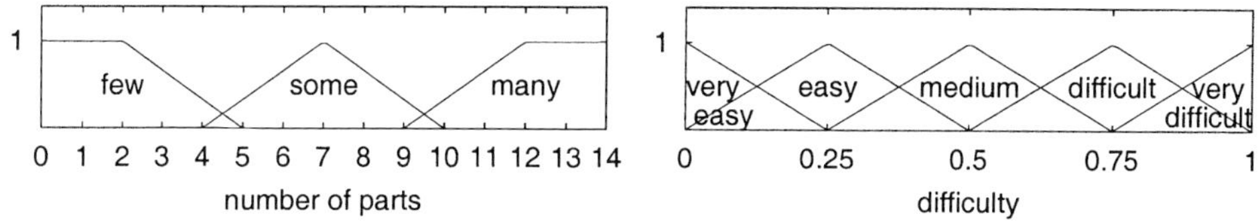

Figure 11: Examples of linguistic variables used to asses the task difficulty.

$\mathcal{P}(C)$	$\mathcal{B}(C)$	$\mathcal{A}(C)$	$\mathcal{O}(C)$	difficulty $\mathcal{D}$
few	few	low	low	very easy
few	few	low	medium	easy
⋮	⋮	⋮	⋮	⋮

Table 1: Examples of fuzzy rules to asses the task difficulty.

Obtaining values of the first three factors from an input cluster is trivial. The mean degree of perspective occlusion is computed as follows:

$$\mathcal{O}(C) = 1 - \frac{1}{\mathcal{P}(C)} \sum_{p \in C} v(p)$$

where

$$v(p) = \begin{cases} \frac{s(p)}{\overline{s}(p)} & \text{if } \frac{s(p)}{\overline{s}(p)} < 1 \\ 1 & \text{otherwise} \end{cases}$$

denotes the degree of visibility of a cluster part p and $s(p)$ denotes its size in pixels. The denominator $\overline{s}(p)$ represents the average size a part p would have if it was not occluded by any other part; suitable values for $\overline{s}$ were determined empirically.

In order to quantify the task difficulty given the values of influencing factors, we apply a fuzzy classification scheme originally introduced by Mamdani and Assilian [14]. Figure 11 shows the fuzzy partitions of one of the factors and the difficulty. Examples of fuzzy rules that characterize interdependencies between task attributes and difficulty are listed in Table 1. Both, the partitions as well as the rules were adjusted according to a training set of 70 images.

Figure 13 summarizes the results obtained from this evaluation scheme. To produce this figure 164 images from an independent test set were attributed to difficulties between 0.1 and 0.8 (for difficulties 0.0 and 0.9 no

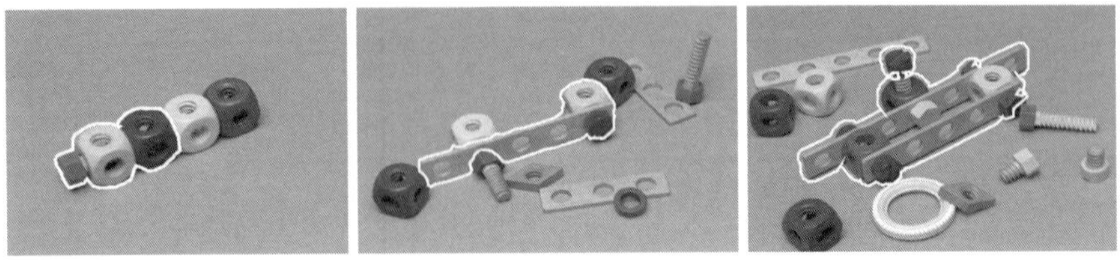

Figure 12: Detection results obtained from input images of difficulties 0.32, 0.53, and 0.72, respectively.

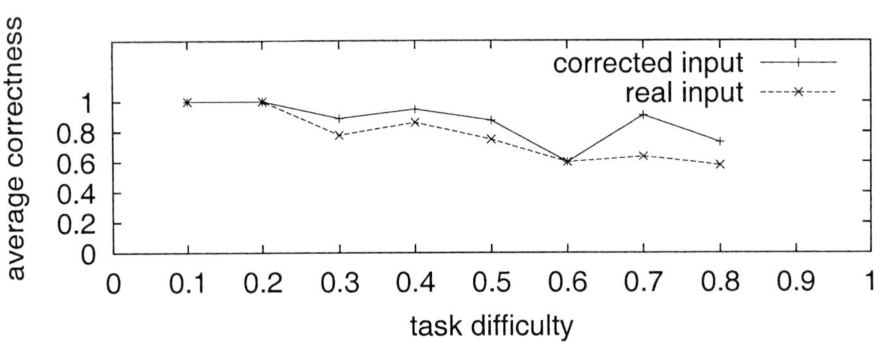

Figure 13: Average correctness with respect to the difficulty of the input.

examples were in the set). In the first series of experiments, labels yielded by elementary object recognition were corrected manually so that the input was always correct. This was done to estimate the ideal performance in assembly detection for if there were errors in the input, assembly recognition simply cannot produce correct results. A second series was done without manual corrections. An experiment was counted a success if it resulted in a correct topological description. Evidently, easy data was always processed correctly whereas correctness diminishes with increasing difficulty. But even in difficult cases it is quite accurate and the difference between results obtained from manually corrected data and real input is small. The latter confirms the robustness and practical appropriateness of the introduced approach: using semantic networks for syntactic parsing is not just of academic interest. Rather, even applied to cluttered and complex data resulting from several steps of image processing, it yields useful and reliable results.

6 Summary

This contribution dealt with mechanical assemblies from a pattern recognition point of view. Faced with the requirement of visual detection of baufix® assemblies in a dynamic and unpredictable environment we introduced a simple but flexible method of syntactic assembly modeling. We demonstrated that context free grammars deriving strings are sufficient to model the component structure not only of individual mechanical assemblies but of whole classes of composite objects. The fundamental idea behind this approach was to bring together concepts first introduced by Homem de Mello and Sanderson and a discovery by Hall.

Dealing with assembly sequence planning, Homem de Mello and Sanderson proposed AND/OR graphs to represent all feasible decompositions of mechanical assemblies. And Hall pointed out that AND/OR graphs and context free grammars as introduced by Chomsky are equivalent formalisms.

We also demonstrated that consideration of functional relations among subassemblies lead to especially compact representations of assembly knowledge. Understanding mechanical assemblies to be composed of functional units results in very general grammatical models which actually can be derived from prototypical examples. Grammars with recursion over functional variables allow to derive descriptions of all feasible assemblies of a certain class.

However, context free grammatical models of composite objects are rather coarse and ignore geometric details. They cannot model even simple mechanical restrictions and thus may derive structures which are mechanically infeasible. Therefore, more sophisticated grammatical approaches like graph grammars [22] or plex grammars [4] seem more appropriate since they cope with context sensitive and geometrical constraints. However, we saw that CFGs are structurally equivalent to a certain type of semantic networks which can be implemented using ERNEST . ERNEST allows to define attributes which can model context sensitive constraints and thus help to avoid the derivation of infeasible assembly structures from image analysis.

We considered ideas developed for discourse parsing and derived a suitable algorithm for assembly detection in higher dimensional input. We discussed that parsing requires sequential input and saw that bolted assemblies can easily be sequentialized for they can be understood as partially ordered sets. Due to this property, the coordinates of mating features involved in a bolted attachment are aligned and parsing can be performed along these lines. However, using ERNEST , data that represents assemblies of non partially ordered nature can be parsed as well. Attributes of concepts in an ERNEST network may encode domain dependent peculiarities and procedural knowledge and thus provide a means to steer parsing processes in any domain. I.e. after instantiating an assembly component, an attribute of this instance can select the part of the input that should be examined next. The applicability of this approach was corroborated by means of two exemplary implementations.

The practical performance of the proposed methods was evaluated in a flexible assembly scenario where a machine must assemble parts from the baufix® construction kit. As syntactic methods crucially depend on the number, the arrangement, and the correctness of the underlying terminals, we applied an evaluation scheme that takes into account how complex the respective input is. A couple of factors that influence the difficulty of a recognition task were identified and a fuzzy function was used to estimate the difficulty therefrom. In cases of simple input our method always reached 100 percent correctness but also for input of high complexity it yielded satisfactory high rates. From these results we can conclude that simple syntactical techniques and their implementation as semantic networks provide an easy and suitable approach to flexible and robust assembly recognition.

References

[1] C. Bauckhage, F. Kummert, and G. Sagerer. Learning Assembly Sequence Plans Using Functional Models. In *Proc. IEEE Int. Symposium on Assembly and Task Planning*, pages 1–7, 1999.

[2] A. Bourjault. *Contribution à une approche méthodologique de l'Assemblage Automatisé: Elaboration Automatique des Séquences Opératoires.* PhD thesis, Université de Franche-Comté, 1984.

[3] N. Chomsky. *Syntactic Structures.* Mouton and Company, The Hague, 1957.

[4] J. Feder. Plex Languages. *Information Sciences*, 3:225–241, 1971.

[5] K.S. Fu, editor. *Syntactic Pattern Recognition, Applications.* Springer-Verlag, Berlin, 1977.

[6] G. Dudek and M. Jenkin. *Computational Principles of Mobile Robotics.* Cambridge University Press, New York, 2000.

[7] P.A.V. Hall. Equivalence Between AND/OR Graphs and Context-Free Grammars. *Communications of the ACM*, 16(7):444–445, 1973.

[8] M. Hanheide. Objektbezogene 3D-Erkennung automatisch generierter Konturmodelle in Intensitätsbildern. Master Thesis (Diplomarbeit), Bielefeld University, 2001.

[9] L.S. Homem de Mello and A.C. Sanderson. AND/OR Graph Representation of Assembly Plans. *IEEE Transactions on Robotics and Automation*, 6(2):188–199, 1990.

[10] J.E. Hopcroft and J.D Ullman. *Introduction to Automata Theory, Languages and Computation.* Addison-Wesley, Reading, MA, 1979.

[11] K. Ikeuchi and T. Suehiro. Towards an Assembly Plan from Observation Part I: Task Recognition with Polyhedral Objects . *IEEE Transactions on Robotics and Automation*, 10(3):368–385, 1994.

[12] S. Kronenberg and F. Kummert. Soft unification: Towards robust parsing of spontanous speech. In *Int. Conf. on Artificial Intelligence and Soft Computing*, pages 381–385, 1999.

[13] F. Kummert, G.A. Fink, G. Sagerer, and E. Braun. Hybrid Object Recognition in Image Sequences. In *Proc. Int. Conf. on Pattern Recognition*, volume II, pages 1165–1170, 1998.

[14] E.H. Mamdani and S. Assilian. An Experiment in Linguistic Synthesis with a Fuzzy Controller. *Int. J. of Man Machine Studies*, 7:1–13, 1975.

[15] J. Miura and K. Ikeuchi. Task Oriented Generation of Visual Sensing Strategies in Assembly Tasks. *IEEE Transactions on Pattern Analysis and Machine Intelligence*, 20(2):126–138, 1998.

[16] J. Pearl. *Heuristics: Intelligent Search Strategies for Computer Problem Solving.* Addison-Wesley, Reading, MA, 1984.

[17] M. Rabemanantsoa and S. Pierre. An artificial intelligence approach for generating assembly sequences in CAD/CAM. *Artificial Intelligence in Engineering*, 10:97–107, 1996.

[18] A.A.G. Requicha and T.W. Whalen. Representations for Assemblies. In L.S. Homem de Mello and S. Lee, editors, *Computer-aided mechanical assembly planning*, pages 15–39. Kluwer Academic Press, 1991.

[19] G. Rickheit and I. Wachsmuth. Collaborative Research Centre "Situated Artificial Communicators" at the University of Bielefeld. *Artificial Intelligence Review*, 10(3–4):165–170, 1996.

[20] G. Sagerer and H. Niemann. *Semantic Networks for Understanding Scenes.* Plenum Publishing Corporation, New York, 1997.

[21] R. Scha and L. Polanyi. An augmented context free grammar for discourse. In *Proc. Int. Conf. of the Association for Computational Linguistics*, pages 573–577, 1988.

[22] P.C. Stadzisz and J.M. Henrioud. An integrated approach for the design of multiple-product assembly systems. *Computers in Industry*, 36:21–29, 1998.

[23] J.P. Thomas, N. Nissanke, and K.D Baker. Boundary Models for Assembly Knowledge Representation. *IEEE Transactions on Robotics and Automation*, 12(2):302–312, 1996.

[24] C.-P. Tung and A.C. Kak. Integrating Sensing, Task Planning and Execution for Robotic Assembly. *IEEE Transactions on Robotics and Automation*, 12(2):187–201, 1996.

[25] B.L. Webber. Structure and ostension in the interpretation of discourse deixis. *Language and Cognition Processes*, 6(2):107–135, 1991.

Stochastic Recognition of Occluded Objects

Bir Bhanu, Yingqiang Lin and Bing Tian
Center for Research in Intelligent Systems
University of California, Riverside, CA 92521
Email: bhanu@cris.ucr.edu

Contents

D. Chen and X. Cheng (eds.),
Pattern Recognition and String Matching, 73–103.

Abstract

Recognition of occluded objects in synthetic aperture radar (SAR) images is a significant problem for automatic target recognition. Stochastic models provide some attractive features for pattern matching and recognition under partial occlusion and noise. In this paper, we present a discrete hidden Markov modeling (HMM) based approach for recognizing objects in SAR images. We identify the peculiar characteristics of SAR sensors and using these characteristics we develop feature based multiple models for a given SAR image of an object. The models exploiting the relative geometry of feature locations *or* the amplitude of SAR radar return are based on sequentialization of scattering centers extracted from SAR images. In order to improve performance we integrate these models synergistically using their probabilistic estimates for recognition of a particular target at a specific azimuth. Experimental results are presented using both synthetic and real SAR images.

1 Introduction

One of the critical problems for object recognition is that the recognition approach should be able to handle partial occlusion of an object and spurious or noisy data [1, 2, 3, 4]. In most of the object recognition approaches, the spatial arrangement of structural information of an object is the crucial part with the most important information. Under partial occlusion situations the recognition process must be able to work with only portions of the *correct* spatial information. Rigid template matching and shape-based recognition approaches depend on good prior segmentation results. The structural primitive (e.g., line segments, point-like features, etc.) extracted from occluded and noisy images, however, may not have sufficient reliability, which will directly undermine the performance of those recognition approaches.

We want to suggest an object recognition mechanism that effectively makes use of all available structural information. Based on the nature of the problems caused by occlusion and noise, we view the spatial arrangement of structural information as a whole rather than view the spatial primitives individually. Because of its stochastic nature, a hidden Markov model (HMM) is quite suitable for characterizing patterns. Its nondeterministic model structure makes it capable of collecting useful information from distorted or partially unreliable patterns. Many successful applications of

HMM in speech recognition [5, 6, 7] and character recognition [8, 9] attest to its usefulness. Thus, it is potentially an effective tool to recognize objects with partial occlusion and noise.

However, the limit of traditional HMMs is that they are basically one dimensional models. So how to appropriately apply this approach to two dimensional image problems becomes the key. It has been largely an unsolved problem. In this paper we use the features based on the image formation process to encode the 2-D image into 1-D sequences. These 1-D sequences or strings can be formed in many different ways. HMM models built off-line are used for matching during testing. In this paper, we use information from *both* the relative positions of the scattering centers and their relative magnitude in synthetic aperture radar (SAR) images to address the fundamental issues of building object models and using them for robust recognition of objects in SAR images.

1.1 Overview of the Approach

Figure 1 provides an overview of the HMM based approach for recognition of occluded objects in SAR imagery. During an off-line phase, scattering centers are extracted from training SAR images by finding local maxima of intensity. Both locations and magnitudes of these peak features, viewed as *emitting patterns* of some hidden stochastic process, are used to generate multiple observation sequences. These multiple observation sequences, based on both the *relative* geometry and amplitude of SAR return signal (obtained as a result of the physics of the SAR image formation process), are used to build the bank of stochastic models. From each training SAR image containing a particular object under a particular azimuth angle, five kinds of observation sequences are extracted to build five HMM models with one model taking one kind of observation sequence as input. These models provide robust recognition in the presence of occlusion and unstable features caused by scintillation phenomena where some of the features may appear/disappear at random in an image. At the end of the off-line phase, hidden Markov recognition models for various objects and azimuths are obtained. Similar to the off-line phase, during the on-line phase, features are extracted from SAR images and observation sequences based on these features are matched by the HMM forward process with the stored models obtained previously. Then a maximum likelihood decision is made on the classification results. Finally the results obtained from multiple models are combined in a voting kind of approach that uses both the object, azimuth

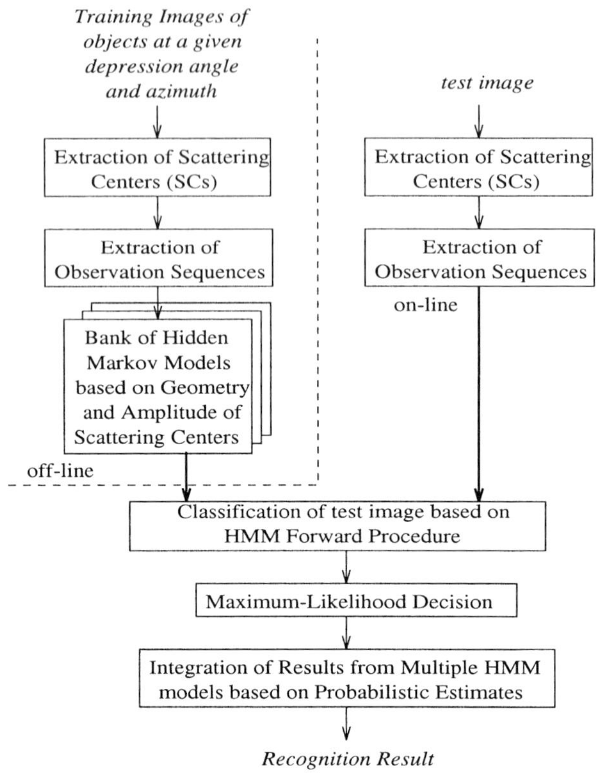

Figure 1: The HMM-Based approach for recognition of occluded objects.

label and its probability of classification. This produces a rank ordered list of classifications of the test image and associated confidences.

1.2 Related Research and Our Contribution

Fielding and Ruck [10] have used HMM models for spatio-temporal pattern recognition to classify moving objects in image sequences. Rao and Mersereau [11] have attempted to merge HMM and deformable template approaches for image segmentation. Kottle et al.[12] propose a HMM-Based SAR automatic target recognition system. They first segment a SAR image, and then extract features followed by Radon transforms. The feature sequences so obtained are input to HMMs. Template matching [13] and major axis based approaches [14] have been used to recognize and index objects in SAR images, however, they are not suitable to recognize occluded objects. This approach to recognizing occluded objects in SAR images uses significantly fewer models (typically 5) of an object at some aspect than our geometric hashing approach [15, 16] (which uses 20 - 40 models).

The contributions of this paper are:

- Hidden Markov modeling approach, commonly used for recognizing 1-D speech signals, is applied in a novel manner to 2-D SAR images to solve the occluded object recognition problem.

- Unlike most of the work for model building in pattern recognition and computer vision, our recognition models using hidden Markov modeling concepts are based on the peculiar characteristics of SAR images where the number of models used for recognition is justified by the quantification of the azimuthal variance in SAR images.

- Multiple models derived from various observation sequences, based on both the relative geometry and signal amplitude (four based on geometry and one based on amplitude) are used to capture the unique characteristics of patterns to recognize objects.

- Extensive amounts of data are used to test the approach for recognition of objects with various amounts of occlusion for both synthetic and real data.

2 Hidden Markov Modeling Approach

It is well known that HMM can model speech signals well [5, 6, 7]. It is a model used to describe a doubly stochastic process which has a set of states, a set of output symbols and a set of transitions. Each transition is from state to state and associated with it are a probability and an output symbol. The word 'hidden' means that although we observe an output symbol, we cannot determine which transition has actually taken place. At each time step t, the state of the HMM will change according to a transition probability distribution which depends on the previous state and an observation y_t is produced according to a probability distribution which depends on the current state.

Formally, a HMM is defined as a triple $\lambda = (A, B, \pi)$, where a_{ij} is the probability that state i transits to state j, $b_{ij}(k)$ is the probability that we observe symbol k in a transition from state i to state j, and π_i is the

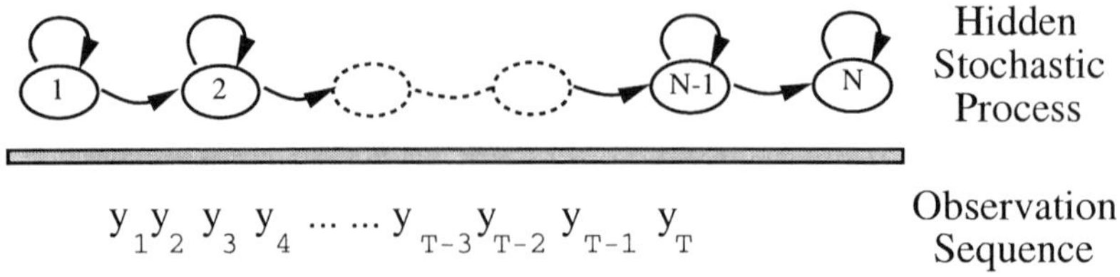

N: the number of states.
M: the number of distinct observable symbols.
A: a_{ij} is the probability that state i will transit to state j.
B: $b_{ij}(k)$ is the probability that symbol k will be observed when there is a transition from state i to state j.
Π : π_i is the probability that state i is the initial state.

Figure 2: A N states forward-type HMM.

probability of i being the initial state. Figure 2 shows an example of a N states HMM.

Recognition Problem – Forward Procedure: The HMM provides us a useful mechanism to solve the problems we face for robust object recognition. Given a model and a sequence of observations, the probability that the observed sequence was produced by the model can be computed by the forward procedure [17]. Suppose we have a HMM $\lambda = \{A, B, \pi\}$ and an observation sequence $y_1^T = \{y_1, y_2, ..., y_t, ..., y_T\}$. We define $\alpha_i(t)$ as the probability that the Markov process is in state i, having generated $y_1^t = \{y_1, y_2, ..., y_t\}$.

$$\begin{aligned}
\alpha_i(t) &= 0, \text{ when t=0 and i is not an initial state.} \\
\alpha_i(t) &= 1, \text{ when t=0 and i is an initial state.} \qquad (1) \\
\alpha_i(t) &= \Sigma_j[\alpha_j(t-1)a_{ji}b_{ji}(y_t)], \text{ when } t > 0.
\end{aligned}$$

The probability that the HMM stopped at the final state and generated y_1^T is $\alpha_{S_F}(T)$. The forward procedure is given below:

Let T be the length of an observation sequence and N is the number of states in the HMM.

1. Initialize $\alpha_i(0)$, where i = 1, 2, ..., N.

2. Compute $\alpha_i(t)$ inductively (equation 1), where t = 1, 2, ..., T. At each step, the previously computed $\alpha_i(t-1)$ is used. Repeat this process until t reaches T.

3. Output $\alpha_{S_F}(T)$, where $\alpha_{S_F}(T)$ is the probability that the HMM stopped at the final state and generated the observation sequence.

Usually, α becomes too small to be represented in a computer after several iterations. We take the logarithm of the α value in the computation.

Training Problem – Baum-Welch Algorithm: To build a HMM is actually an optimization of the model parameters so that it can describe the observation better. This is a problem of training. The Baum-Welch re-estimation algorithm is used to calculate the maximum likelihood model. But before we use the Baum-Welch algorithm, we must introduce the counterpart of $\alpha_i(t)$; $\beta_i(t)$, which is the probability that the Markov process is in state i and will generate $y_{t+1}^T = \{y_{t+1}, y_{t+2}, ..., y_T\}$.

$$\begin{aligned} \beta_i(t) &= 0, \text{ when t=T and i is not a final state.} \\ \beta_i(t) &= 1, \text{ when t=T and i is a final state.} \\ \beta_i(t) &= \Sigma_j[a_{ij} b_{ij}(y_{t+1})\beta_j(t+1)], \text{ when } 0 \leq t < T. \end{aligned} \tag{2}$$

The probability of being in state i at time t and state j at time $t+1$ given observation sequence y_1^T and the model λ is defined as follows:

$$\begin{aligned} \gamma_{ij}(t) &= P(X_t = i, X_{i+1} = j \mid y_1^T) \\ &= \frac{\alpha_i(t-1)a_{ij}b_{ij}(y_t)\beta_j(t)}{\alpha_{S_F}(T)} \end{aligned} \tag{3}$$

Now the expected number of transitions from state i to state j given y_1^T at any time is simply $\Sigma_{t=1}^T \gamma_{ij}(t)$ and the expected number of transitions from state i to any state at any time is $\Sigma_{t=1}^T \Sigma_k \gamma_{ik}(t)$. Then, given some initial parameters, we could recompute $\overline{a_{ij}}$, the probability of taking the transition from state i to state j as:

$$\overline{a_{ij}} = \frac{\Sigma_{t=1}^T \gamma_{ij}(t)}{\Sigma_{t=1}^T \Sigma_k \gamma_{ik}(t)} \tag{4}$$

Similarly, $\overline{b_{ij}(k)}$ can be re-estimated as the ratio between the frequency that symbol k is emitted and the frequency that any symbol is emitted:

$$\overline{b_{ij}(k)} = \frac{\Sigma_{t:y_t=k} \gamma_{ij}(t)}{\Sigma_{t=1}^T \gamma_{ij}(t)} \tag{5}$$

It can be proved that the above equations are guaranteed to increase $\alpha_{S_F}(T)$ until a critical point is reached, after which the re-estimate will remain the same. In practice, we set a threshold as the ending condition for re-estimation.

So the whole process of training a HMM is as follows:

1. Initially, we have only an observation sequence y_1^T and blindly set (A, B, π).
2. Use y_1^T and (A, B, π) to compute α and β (equations (1), (2)).
3. Use α and β to compute γ (equation (3)).
4. Use y_1^T, (A, B, π), α, β and γ to compute A and B (equations 4, 5). Go to *step* 2.

A HMM is able to handle pattern distortions and the uncertainty of the locally observed signals, because of its nondeterministic nature. However, a HMM is primarily suited for sequential, one-dimensional patterns and it is not obvious that how a HMM can be used on 2-D patterns in object recognition. The basic ideas to apply a HMM for our purpose are (a) training the HMM λ by samples of SAR images of a certain object, and (b) recognizing an unknown object in a given SAR image. These two problems are addressed in the following. The key questions are what we shall use as observation data and how we get the observation sequences.

3 Hidden Markov Models for SAR Object Recognition

3.1 SAR Imagery and Target Recognition

The differences between SAR and other imagery include the source of illumination, image resolution, imaging geometry, and noise. SAR imagery provides its own source of illumination. However, microwaves reflect off of objects in a scene differently (specular vs. diffuse reflections) than sunlight. The signal returned from an object depends upon its microwave reflectivity, as well as relative position, size, and texture. The reflected microwaves are measured to produce the image. One of the useful properties of SAR imagery is that the spatial resolution is independent of the distance. In the range direction, the time resolution determines slant range resolution, and both are dependent only on bandwidth. The azimuth resolution is dependent only on the length of the antenna in the azimuth direction. The ability

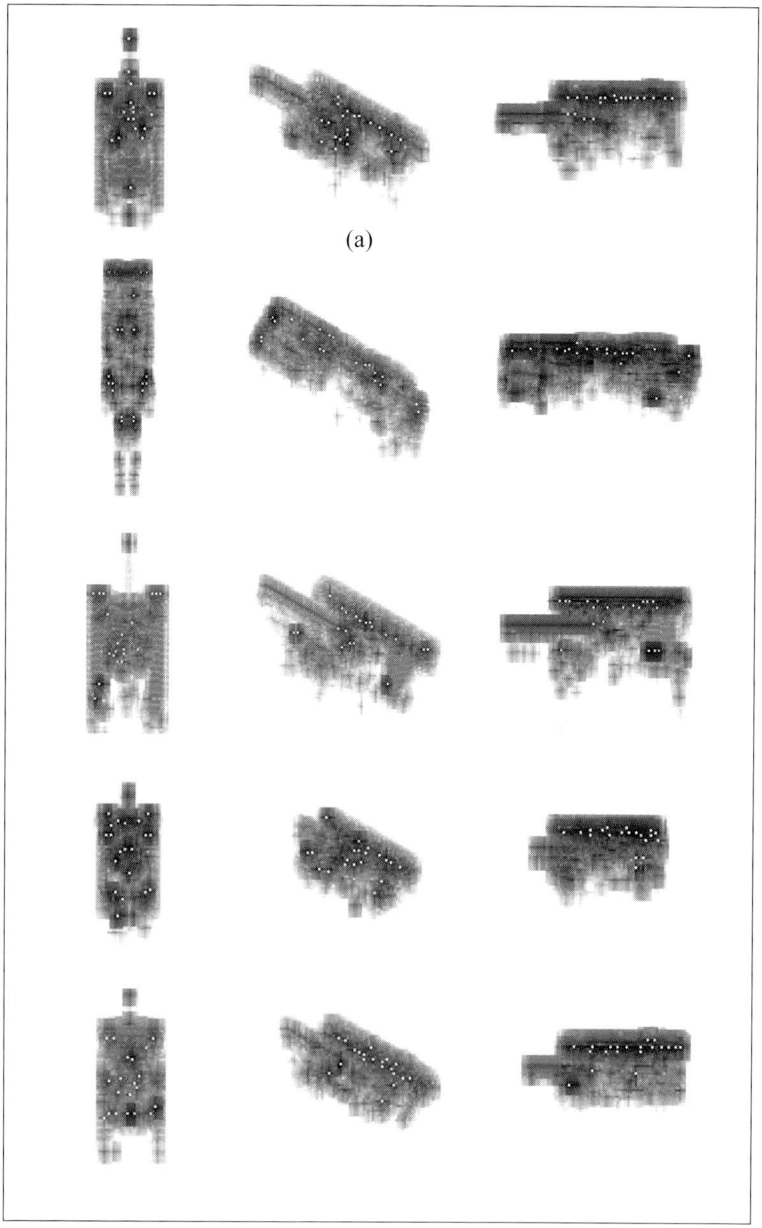

Figure 3: Examples of scattering centers (white dots) extracted from SAR images at azimuths $0°, 60°, 90°$. (a) Fred tank, (b) SCUD launcher with missile down, (c) T72 tank, (d) T80 tank, (e) M1a1 tank.

to collect high-resolution images from extreme distances is a particularly attractive characteristic of SAR. Although there are many positive aspects to the properties of SAR, there are negative aspects as well. For example, the imaging geometry makes understanding SAR imagery and its formation unintuitive and unnatural. In addition, the interaction of microwaves on surfaces is not entirely understood. Another obstacle for SAR imagery applications is speckle which results from the interference of many scatterers within the same resolution cell. Although speckle can be considered as multiplicative noise, when the image is converted to a dB image speckle statistics become additive and techniques for additive noise removal, such as spatial averaging, can be applied to remove the noise. However, spatial averaging degrades resolution which is important for many applications which use SAR imagery. Because of the fundamental differences between SAR and other types of imagery, algorithms developed for application to other imagery are not directly applicable to SAR.

The goal of automatic target recognition in SAR imagery is to recognize the identity (ID) of a target, its location and azimuth or pose. The challenge is to recognize occluded targets, articulated targets, and targets with different configurations[2]. In this paper, we focus on the occlusion problem and assume that the potential targets have already been extracted. Thus, we are given the region-of-interest (or chip) of a target and the task is to determine the ID and pose or only the ID of the target.

3.2 Extraction of Scattering Centers

Scattering centers (location and magnitude) extracted from SAR images are used to train and test models for recognition. At six to twelve inch resolution (See Figure 3), there exist a large number of peaks corresponding to scattering centers. We have selected peaks as features in this work since we wanted to evaluate the limits of our approach using six inch resolution synthetic SAR images (generated using the XPATCH code [18]) and real SAR images at one foot resolution from the MSTAR data [19] before more complicated features are used. We consider a pixel as a scattering center if the magnitude of SAR return at this pixel is larger than all its eight neighbors. Figure 3 shows some examples of top 20 scattering centers (based on their magnitude) extracted from XPATCH generated SAR images of various objects at $15°$ depression angle and azimuths at $0°$, $60°$, and $90°$. The squint angle, the angle between the flight path and radar beam, is known ($90°$ here) and kept fixed for all the images used in this paper. Figure 4 shows

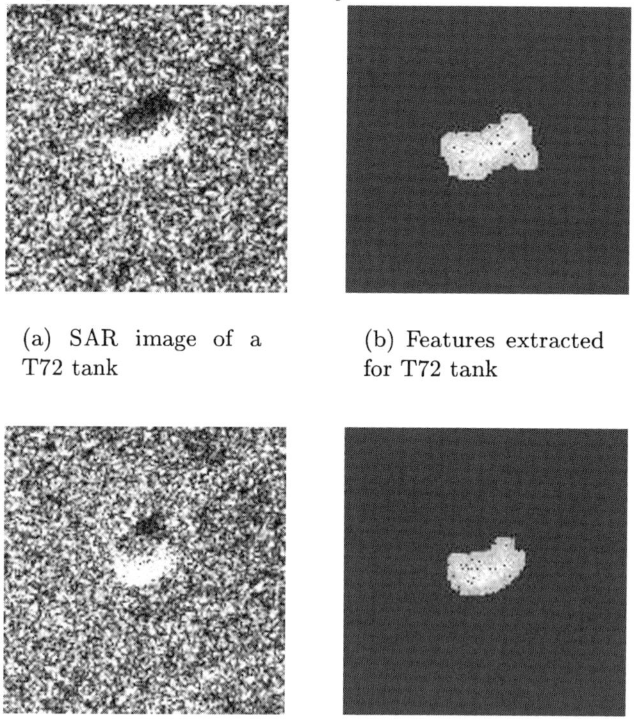

(a) SAR image of a T72 tank

(b) Features extracted for T72 tank

(c) SAR image of a ZSU gun

(d) Features extracted for ZSU gun

Figure 4: Real SAR images and region-of-interests (ROIs) (with peaks shown as black dots superimposed on the ROI) for T72 tank #a64 and ZSU anti-aircraftgun #d08.

examples of scattering centers extracted for MSTAR SAR data. It is to be noted that in this paper we have used different number of scattering centers for different experiments for variety. It is possible to determine an optimal number of scattering centers for a given task.

3.3 Rotation Variance of Scatterers and Representation of 3-D Objects

In order to determine the number of azimuths needed to represent a 3-D object in a systematic manner, we evaluate the characteristics of SAR images. Unlike the visible images, SAR images are extremely sensitive to slight changes in viewpoint (azimuth and depression angle) and are not affected by scale [20]. We evaluate [15] the characteristics of scattering centers to

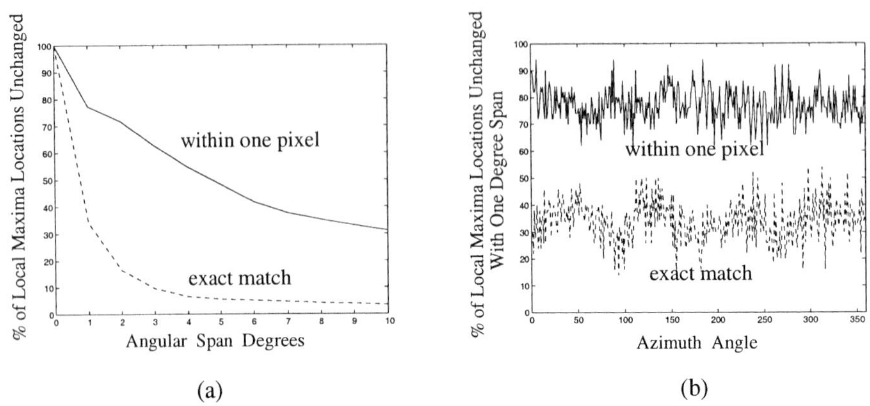

Figure 5: XPATCH Data (a)T72 tank rotational invariance.(b)T72 tank rotational invariance with 1° Angular Span.

find out what kind of location invariance exists among scattering centers. Figure 5 shows the rotation invariance of the T72 tank with XPATCH data. The data is obtained by rotating the image at azimuth i° (for a fixed depression angle) by x° (angular span), and comparing the rotated image with the image of $(i+x)^\circ$ to see how many scattering centers do not change their location. Since the object chip is 256×256 pixels, we rotate the image with respect to the center point $(127.5, 127.5)$. The distance measurement criteria "exact match" and "within one pixel" are defined in the following:

$$\begin{cases} \mathbf{x_r} \textbf{ exactly matches } \mathbf{x} \text{ if } MAX(|x-x_r|, |y-y_r|) < \frac{1}{2} \text{ pixel} \\ \mathbf{x_r} \textbf{ and } \mathbf{x} \textbf{ are within one pixel} \text{ if } MAX(|x-x_r|, |y-y_r|) < 1\frac{1}{2} \text{ pixel} \end{cases}$$

Figure 5(a) shows the average (over all the 360 azimuth angles) results for the number of scattering centers that remain unchanged within some error tolerance (as defined in the above) for T72 tank images for angular span of 1° to 10°. The 50 strongest scattering centers are used for each image. Note that for a 1° change in azimuth (angular span), approximately 1/3 and 3/4 scattering centers remain unchanged for the exact match case and within one pixel case, respectively. Figure 5(b) gives the percentage of scattering center locations unchanged vs. azimuth angle, for each of the 360 azimuths, with 1° angular span for the exact match and within one pixel match. For example, after rotating the image at azimuth 100° by 1° and comparing the rotated image with the image at azimuth 101° using "exact match" criterion, we get the point whose horizontal coordinate is 100 on the "exact match" curve in

Figure 5(b). The results given in Figure 5 show that scattering centers for XPATCH SAR images for all azimuths vary greatly with relatively small changes of angles. Similar results are obtained with real MSTAR SAR data at one foot resolution[21]. Therefore, to recognize occluded objects, we need to get SAR images of an object at a given depression angle under various finely sampled azimuth angles. It is possible to collect SAR images which are $1°$ apart in azimuth. Thus, under ideal conditions, we can have 360 SAR images of an object with one image corresponding to one and only one azimuth angle between $0°$ and $359°$. Consequently, we treat an object under different azimuth angles separately and build a model for each azimuth angle whenever possible. Note that when building models for an object azimuth, a single HMM model is inadequate because of noise, articulation, occlusion, etc. Therefore, to increase robustness, we build multiple HMM models for a given object at a specific azimuth with one model taking a particular kind of observation sequence as input, as discussed in the next section.

3.4 Extraction of Observation Sequences

After the scattering centers are extracted, we need to encode the data into a 1-D sequence as the input to a model based HMM process for recognition. It is one of the *key* factors which affects the performance of HMM approach for object recognition. There are many ways to choose sequences, but we want to use information from both the magnitude and the relative spatial location of the scattering centers extracted from a SAR image. Also the sequentialization method should not be affected by distortion, noise, or partial occlusion and should be able to represent the image efficiently. Based on the above considerations, we build five kinds of sequences.

- Sequences based on amplitudes:
  $O_1 = \{Magnitude_1, Magnitude_2, ..., Magnitude_n\}$,
  where $Magnitude_i$ is the amplitude of ith scattering center and $Magnitude_i \geq Magnitude_{i+1}$, $\forall\ i = 1, ..., n-1$.

- Sequences based on *relative* geometrical relationships:
  $O_2 = \{d(1,2), d(2,3), ..., d(n,1)\}$ $(length\ n)$
  $O_3 = \{d(1,2), d(1,3), ..., d(1,n)\}$ $(length\ n-1)$
  $O_4 = \{d(2,1), d(2,3), ..., d(2,n)\}$ $(length\ n-1)$
  $O_5 = \{d(3,1), d(3,2), ..., d(3,n)\}$ $(length\ n-1)$,
  where $d(i,j)$ is the Euclidean distance between scattering centers i and

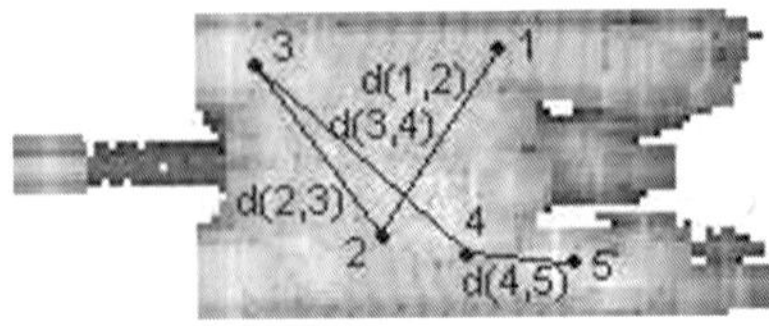

Figure 6: Example of an observation sequence superimposed on an image of T72 tank.

> *j*. The scattering centers are sorted in descending order (1, ..., n) of their magnitude.

Figure 6 gives an example to illustrate how we get the sequences. Sequence O_1 is obtained by sorting the scattering centers by their magnitude. We label the scattering centers 1 through n in descending order. So in this approach, we do not use the location information and thus can avoid the instability caused by the error in localization of scattering centers. Sequences O_2 through O_5 are obtained based on the relative locations of the scattering centers and the magnitude of the scattering centers only affects the ordering. Note that they are not scale dependent, since the SAR images are scale independent [20]. In experiments described in section 4, we only consider the top 20 scattering centers for XPATCH data and the top 30 scattering centers for MSTAR data (sorted in descending order of their magnitude). This method of ordering is done for convenience, for repeatability and because we expect that the scattering centers with larger magnitude are generally more stable than the weaker ones.

Since we use discrete HMMs, each element in the sequence should be converted to an observation symbol. Labels from 1 to M represent the symbols which can be observed for a HMM. We use the K-means algorithm [22] to classify the magnitude values (or distance values) of all the scattering centers in the database into K classes (here K = M, it is experimentally determined, see Section 4.1.1). Once we know to which class each of the elements of a sequence belongs, we label the element with the label of its class. Thus, for a given sequence, we obtain a sequence of observation symbols and call the resulting sequence as observation sequence.

3.5 Off-line Training and On-line Recognition Phases

The procedure for building a model based on a particular kind of observation sequence is described as follows:

1. Loop (for a given depression angle) lines 2-4 for each object and each azimuth angle.

2. Generate images which simulate occlusion and clutter with scattering centers occluded from different directions (see Section 4).

3. Extract one observation sequence from each image generated in (2), including the original image.

4. Run Baum-Welch algorithm on all the observation sequences generated in (3) to estimate the HMM parameters. First, randomly set the parameter values and run Baum-Welch algorithm on the first observation sequence (Use equations (1) to (5)). Then the parameter values resulting from the first observation sequence are used as the initial parameter values when the second observation sequence is input and so on. Exit the loop when there is no further change in parameter values.

The recognition procedure based on a particular kind of observation sequence is described as follows:

1. Extract the observation sequence from the testing image.

2. Loop line 3 for all the models in the model base.

3. Feed the observation sequence into the model, $(A, B, \Pi)_{(M_i^*, a_j^*)}$, where (M_i^*, a_j^*) represents the model of object i with azimuth angle j. Use Forward algorithm to compute the probability that this sequence is produced by this model.

4. The models with the H highest probabilities of generating the observation sequence are selected as the best possible matches to be used by the next integration step. The model with maximum probability is considered the best match.

3.6 Integration of Results from Multiple Sequences

Since a model based on a particular kind of observation sequence for a particular object and azimuth cannot provide optimal recognition performance under occlusion, noise, etc., we improve the recognition performance by combining the results obtained from multiple models corresponding to multiple

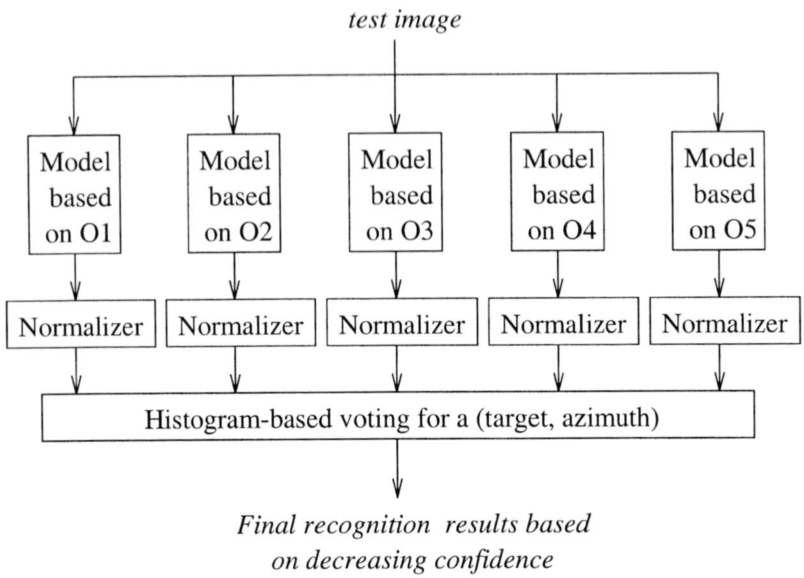

Figure 7: Integration of results by histogram-based method.

kinds of observation sequences. Thus, when a testing image cannot be recognized correctly by a model based on a particular observation sequence, say O_1, it may be recognized correctly by models based on other kinds of observation sequences.

We have developed a voting-like method shown in Figure 7 to integrate the results from models based on five kinds of observation sequences. The algorithmic steps are:

1. For each kind of observation sequence, we collect the H highest possibilities in the test results. Each possibility is the probability that the test image is the image of that object at that azimuth.

2. A normalization is done to the H probabilistic estimates for each kind of observation sequence so that estimates from five kinds of observation sequences can be compared.

3. We draw a histogram that sums the normalized probabilities of the various object/azimuth instances for the results obtained in step (2).

4. If the object associated with the highest frequency in the histogram is the same as the ground truth, we count it as one correct recognition.

4 Experiments

Both simulated and real-world data is used to test the performance of our approach.

4.1 XPATCH SAR Data

We use the XPATCH SAR simulator [18] to generate the data to perform controlled experiments. XPATCH is considered the best radar signature prediction code available; it has been extensively validated against primitive objects and complex full scale vehicles.

In our research, we selected six inch resolution data to get a significant number of "pixels-on-target" to facilitate solving the occluded target recognition problem. It is possible to collect real data at this resolution and also super resolution techniques [23] exist to achieve such data. We generate one set of SAR images of 5 objects (Fred tank, SCUD missile launcher, T72 tank, T80 tank and M1a1 tank, shown in Figure 8.) at 15° depression angle, and 90° squint angle (fixed), at each of the azimuth angles from 0° to 359°. We extract the 20 scattering centers (local maxima) with largest magnitudes. In the experiments, since we want to test the performance of our approach under partial occlusion and spurious data, we simulate realistic occlusion situations and generate images for training and testing.

Simulating occlusion and clutter: There are no accepted empirical models of object occlusion in SAR imagery. So, we consider the occlusion to occur possibly from 9 different directions as shown in Figure 9. Scattering centers being occluded are not available. Moreover, we add some spurious data into the image to simulate noise or clutter. For instance, 20 scattering centers are shown in each image of Figure 9. They are obtained by removing 4 scattering centers (20% occlusion) from the center of one object or from one particular direction (simulated occlusion) and adding 4 spurious scattering centers (clutter or noise) into the image. The spurious scattering centers are added based on the following rules:

- The location of the scattering center is generated as a pair of random numbers.

- The magnitude of the scattering center depends on a random number r between 1 and 50. We use the magnitude of the rth brightest scattering center as the magnitude of the spurious scattering center. Top 50 vs top 20 biases clutter to low values.

(a) Fred tank

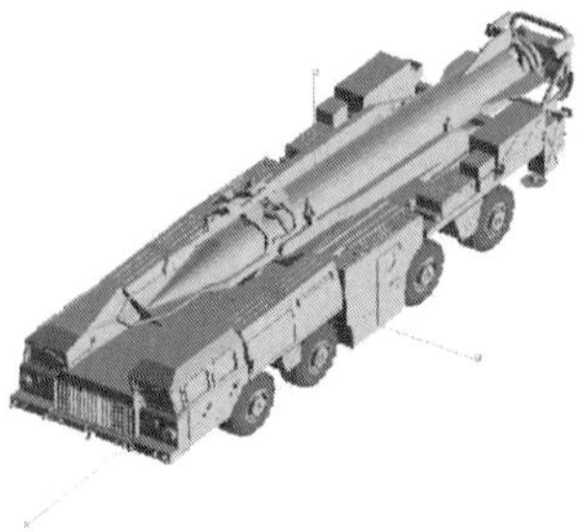

(b) SCUD missile launcher

(c) T72 tank

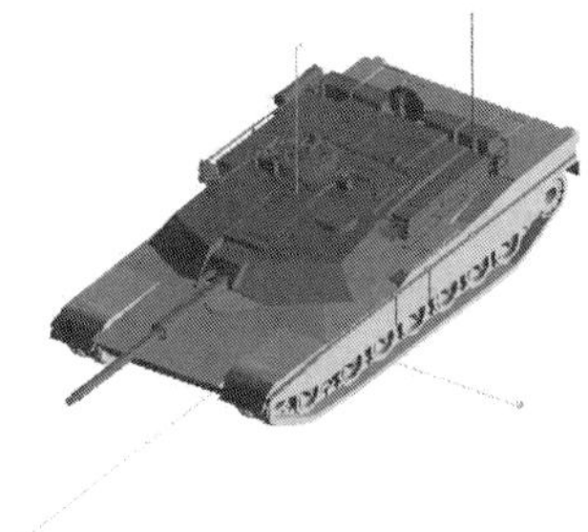

(d) T80 tank

(e) M1a1 tank

Figure 8: Targets.

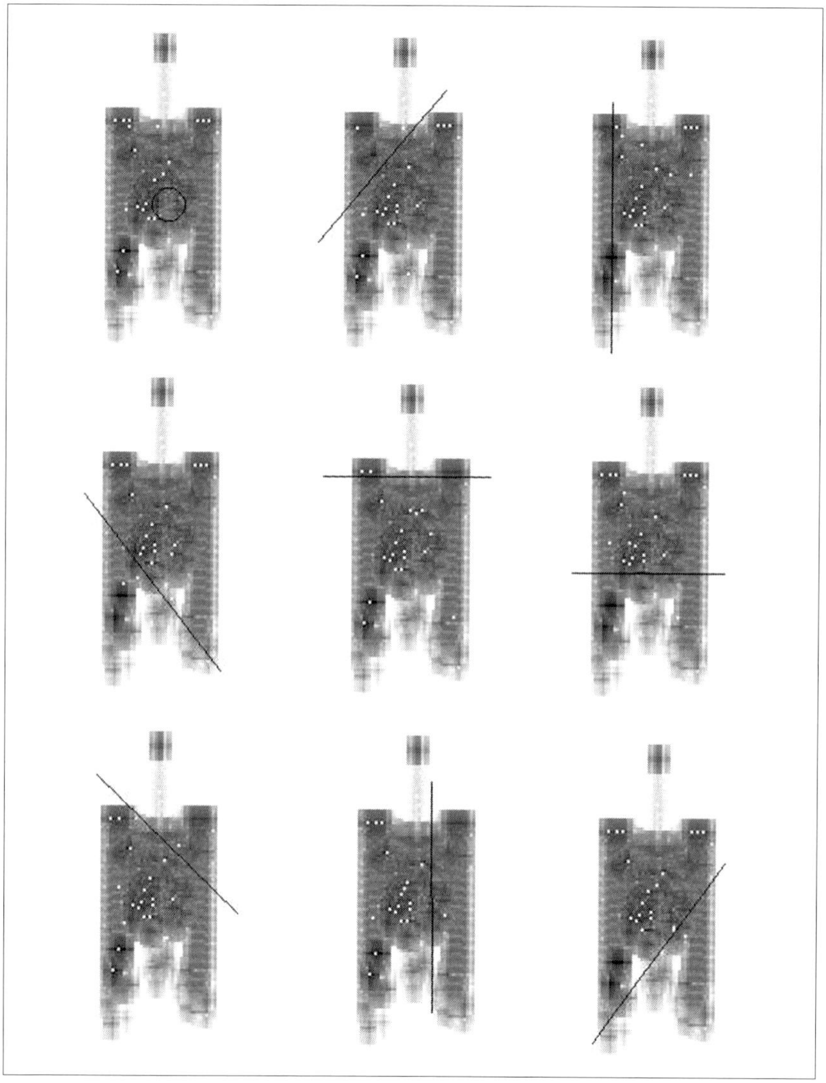

Figure 9: Scattering centers of T72 tank at azimuth 0°, part of scattering centers are occluded from a particular direction (0-8, left to right, top to bottom).

Training Data: Based on the method of simulating occlusion and clutter described above, we generate 90 images from the original image (10 samples

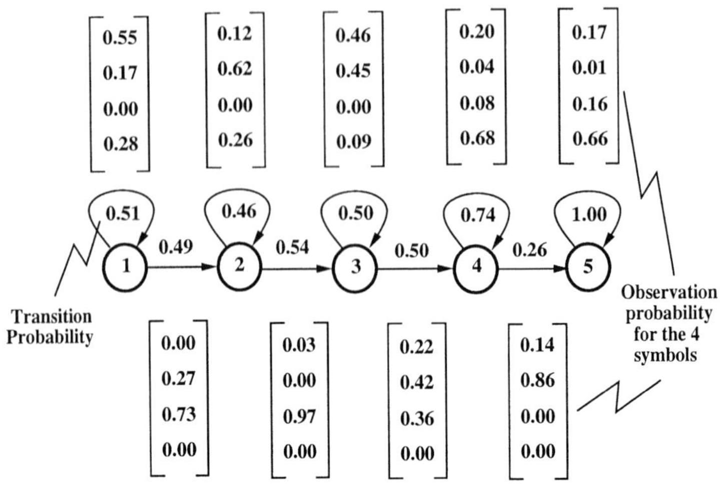

Figure 10: An example: parameters of a 5 states, 4 symbols HMM. The number on edges represents the transition probability, and the vector associated with each transition represents $b_{ij}(k)$. In our case, we use HMM with 8 states and 32 symbols.

for each of 9 directions) at 5% occlusion and another 90 images at 10% occlusion. Including the original image, we have 181 images per object per azimuth angle to train multiple HMM models. Thus, we have a total of 325, 800 (5 objects, 360 azimuths, 181 occluded images) samples for training.

Testing Data: We generate one image with o scattering centers occluded ($o = 2, 4, 6, 8$ or 10) from direction d ($d = 0, 1, ..., 8$) per azimuth angle per object. So there are 1800 images (5 objects $\times$ 360 degrees) generated for testing of occlusion with o scattering centers occluded from direction d. Thus, we have a total of 81,000 (5 objects, 360 azimuths, 5 different occlusions 10% – 50%, and 9 directions) samples for testing.

4.1.1 Training – Building Bank of HMM Models for Recognition

The results of experiments to choose the appropriate number of states and symbols of the HMM are shown in Table 1. We use data from five azimuth angles of five objects (Fred tank, SCUD missile launcher, T80 tank, T72 tank, and M1a1 tank). Table 1 shows that with an increase in the number of states and symbols, recognition performance increases. Considering both the recognition performance and the computation cost, we choose 8 states

Table 1: Recognition rate of HMM with different number of states and symbols. N = # of states. M = # of symbols. R = Recognition rate % (top answer is correct). I = Indexing rate % (correct answer is in the top 5).

		id only		*id with pose*	
N	M	R	I	R	I
4	8	76.1	96.5	62.6	79.9
4	16	89.6	98.4	85.4	93.1
4	24	95.1	99.3	91.8	97.3
4	32	96.6	99.9	94.8	99.0
4	64	99.7	100.0	99.6	100.0
5	8	80.1	97.4	67.3	84.0
5	16	91.9	98.6	86.7	93.7
5	24	96.6	99.7	94.6	98.6
5	32	97.8	99.8	96.7	99.3
5	64	99.9	100.0	99.9	100.0
6	8	82.5	96.9	71.7	84.8
6	16	93.8	99.5	90.1	96.7
6	24	98.5	99.8	97	99.7
6	32	98.9	100.0	98.5	99.9
6	64	100.0	100.0	100.0	100.0
8	8	84.3	97.6	77.4	87.6
8	16	96.4	99.8	94.6	98.3
8	24	99.4	100.0	99	99.9
8	32	99.8	100.0	99.8	100.0
8	64	100.0	100.0	100.0	100.0
10	8	100.0	100.0	100.0	100.0
10	16	98.3	99.9	97.3	99.6
10	24	99.9	100.0	99.9	99.9
10	32	100.0	100.0	99.9	100.0
10	64	100.0	100.0	100.0	100.0

and 32 symbols as the "desired" number of states and symbols for our HMM models. Figure 10 illustrates an example of a 5 state, 4 symbol HMM.

Using the algorithm in Section 3.5 we built recognition models. For a selected sequence type we have 1800 (= 360 azimuths × 5 object classes) HMM models. Since we have defined five kinds of observation sequences

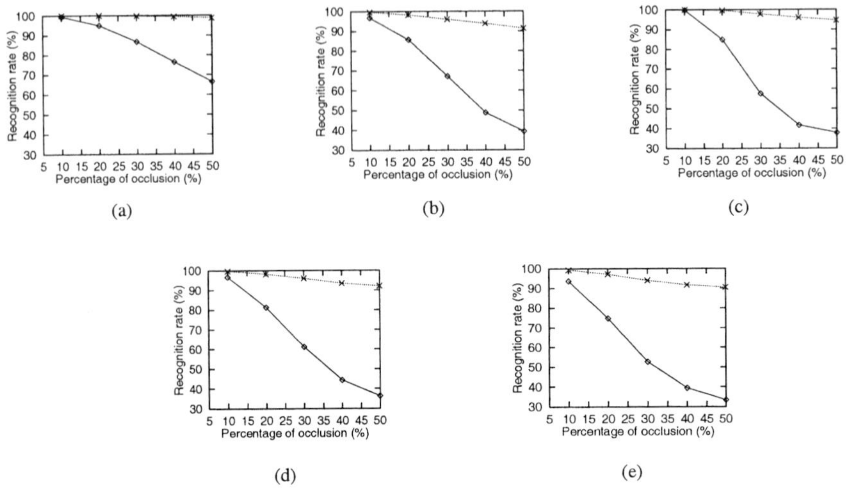

Figure 11: Recognition rate vs. percentage of occlusion for HMM models based on (a) O_1, (b) O_2, (c) O_3, (d) O_4, and (e) O_5.

$(O_1, O_2, O_3, O_4, O_5)$ for each image, we get models based on each kind of observation sequence.

4.1.2 Testing Results

During the testing phase, for a given observation sequence type each of the 81,000 testing images is tested against all models (1800 models: 5 objects, each has 360 models for each azimuth angle). If the model with the maximum probability is associated with the object and azimuth angle from which the observation sequence is extracted, we count it as one correct recognition (object type and its pose). Otherwise, we count it as one incorrect recognition. After we get the results on all the images with a given percentage of occlusion (occluded from all 9 directions), we average these results and associate this recognition performance with the selected model for this particular percentage of occlusion.

Figure 11 shows the testing results taken directly from the output of each of the five kinds of observation sequences: $O_1, O_2, ..., O_5$ (section 3.4). The top curve, a dotted line, is the percentage that the test case object and pose is among the top ten recognition results, and the lower curve, in solid line, indicates the percentage that the recognition result with the highest probability is the same as the test case object and pose.

4.1.3 Integration of Results from Multiple Observation Sequences

In order to understand the integration process better, we have done a set of testing experiments in which both geometric (O_2 to O_5) and magnitude (O_1) observation sequences are used. The results are shown in Table 2 where each row, for a given percentage of occlusion, shows the average results for occlusion that takes place from 9 different directions. The table shows how many incorrect recognitions, made by using models based on observation sequence O_2 can be correctly recognized ("captured") by models based on other observation sequences. The table demonstrates that each of these observation sequences add complimentary information. This implies the independent nature of these observation sequences and their results are integrated by means of a histogram based voting mechanism. We *define* the "upper bound" as the highest possible recognition performance that can be achieved using the five observation sequences, considering *only* the top candidate. The percentage of correct recognition corresponding to "upper bound" is shown in the 8th column. The results shown in column 9 are obtained by integrating the results from observation sequences O_1 to O_5. The integrated recognition performance is 86.8%. Note that these results are for both object and azimuth (object/azimuth), that is, both the object identity and pose should be correctly recognized.

In the rest of this section, we demonstrate the results using 5 kinds of observation sequences (O_1 to O_5). We draw two curves (Figure 12(a)) to show the "upper bound" and "lower bound" of recognition rate. The curve on the top is obtained by considering all 5 kinds of observation sequences, if a model based on one of them correctly recognizes the test data, we count it as a correct recognition. The "lower bound" or the bottom curve is the worst recognition result out of the five models.

The second curve from the bottom in Figure 12(b) is the result for integrated recognition. The corresponding confusion matrix for various amounts of occlusion is shown in Table 3. On the average, we find 86.8% correct recognition performance when the objects are occluded from 10–50%. The second curve from the top in Figure 12(b) is obtained by counting a correct indexing result when the ground truth object/azimuth is in the objects/azimuth associated with the highest 5 probabilities in the histogram. For the purpose of comparison, we have also superimposed the curves of Figure 12(a) into Figure 12(b) with "lower/upper" bounds. Considering the correct indexing answer in the top 5 responses, the average performance is 94.4% for 5 objects occluded from 10% – 50%. The difference between top 5 and top 1

Table 2: Testing results for occluded object recognition using 81,000 testing cases. Average results of 9 directions based on integration of O_1 to O_5.

Percent. occlusion	*Errors with model* O_2	*Errors Captured by models*				*Errors using model* O_1 *to* O_5	*% Correct Recognition ("upper " "bound")*	*% Based on Integrated Recognition*	*% Based on Integrated Indexing*
		O_1	O_3	O_4	O_5				
10%	4	0	1	0	1	2	100.0	99.9	99.9
20%	271	19	53	74	101	7	99.6	98.9	99.6
30%	763	111	294	339	418	25	98.6	93.4	97.6
40%	1050	265	580	629	675	79	95.6	79.4	91.8
50%	1119	397	726	755	784	148	91.8	62.2	83.3
Average Recognition Rate							97.1	86.8	94.4

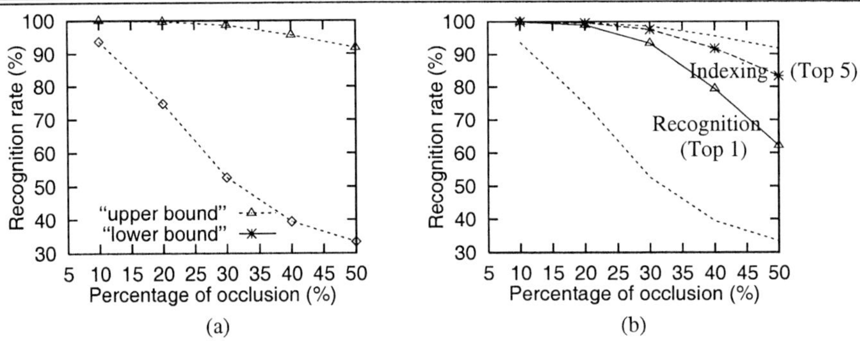

Figure 12: (a) "Upper" and "lower" bound of recognition rate vs. percentage of occlusion. (b)Performance of integrated models: using integrated models O_1 to O_5. The results for recognition (Top 1) and indexing (Top 5 candidates) are superimposed on the figure shown in (a).

recognition performance is 7.6%.

4.2 Real SAR Data

The methods used here are the same as those used in the previous subsection. The only difference is that here we experiment on real data instead of XPATCH data. We use MSTAR public real SAR images (at one foot resolution and depression angle 15°) of 2 objects (T72 tank with serial number #a64 and ZSU anti-aircraft gun with serial number #d08). Ideally, we should have 360 object models for each azimuth for each object. However, we don't have 360 SAR images for each object in the MSTAR data set. For

Table 3: Confusion Matrix for 5 object classes at varying amounts of occlusion (10% – 50%).

	% Occlusion	*Fred*	*SCUD*	*T72*	*T80*	*M1a1*
Fred	10	100.0	0.0	0.0	0.0	0.0
	20	99.2	0.0	0.1	0.4	0.3
	30	95.9	0.2	0.6	1.9	1.4
	40	87.1	0.7	2.8	5.5	3.9
	50	73.2	1.6	7.1	12.1	6.0
SCUD	10	0.0	100.0	0.0	0.0	0.0
	20	0.0	99.7	0.2	0.1	0.0
	30	0.9	97.3	1.2	0.4	0.3
	40	3.1	88.8	4.9	1.9	1.3
	50	5.6	77.9	11.9	2.7	1.9
T72	10	0.0	0.0	100.0	0.0	0.0
	20	0.4	0.2	99.2	0.1	0.2
	30	2.4	0.5	95.3	1.1	0.6
	40	9.1	2.1	82.5	3.8	2.4
	50	16.8	5.2	65.9	6.8	5.4
T80	10	0.0	0.0	0.0	100.0	0.0
	20	1.2	0.0	0.1	98.6	0.1
	30	6.9	0.0	0.6	91.1	1.4
	40	21.5	0.1	1.6	72.6	4.2
	50	37.4	0.8	3.1	50.9	7.8
M1a1	10	0.0	0.0	0.0	0.0	100.0
	20	1.6	0.0	0.1	0.3	98.0
	30	8.5	0.2	0.7	2.9	87.8
	40	22.5	0.8	2.0	8.5	66.1
	50	36.9	1.1	5.2	13.8	42.9

the T72 tank, there are 288 images available for different azimuths. Also for the ZSU gun, 288 images are available. Thus, each object consists of 288 azimuths (or aspects) which we call object models. Each object model consists of HMM models based on observation sequences (O_1 to O_5). We extract 30 scattering centers with the largest magnitudes from each SAR image. Figure 4 shows some examples of SAR imagery, regions-of-interest and scattering centers superimposed on the SAR image. ROIs are obtained using an automatic thresholding and dilation/erosion process [21].

We consider the occlusion to occur possibly from 9 different directions (center, 4 sides and 4 corners of the image). Scattering centers being occluded are not available. Moreover, we add back into the image at random locations a number of spurious scattering centers, equal to the number of occluded scatterers, of random magnitude. The random magnitude could be equal to the magnitude of any of the top 30 scatterers. This simulates noise or clutter from other objects. For example, for 30% occlusion, we remove 9 scattering centers from the center of one object or from one particular direction and add randomly 9 spurious scattering centers back into the image with magnitude randomly determined as stated in the above. We compute the observation sequences based on the scattering centers available after the occlusion process has taken place. Note that we assume there is only one object present in one image. However, the technique we use to synthetically generate occluded data adds an equivalent number of clutter features which could be viewed as a surrogate for multiple objects. No real SAR data with multiple objects exist in the public domain.

4.2.1 Training and Testing Data

Training Data: We generate 91 training images from each SAR image. The first one is the original SAR image without occlusion. Then we occlude the SAR image from 9 directions. For each direction, the occlusion level is 5% and 10%. For each occlusion level and direction, we extract 5 sample images. So 91 images are obtained from each original testing SAR image. We have two objects and 288 SAR images of each object, thus, the number of training images is 52,416. Since there are 5 kinds of observation sequences that are extracted from each image, the total number of observation sequences is 262,080.

Testing Data: From each SAR image, we generate 36 testing images. We occlude the SAR image from 9 directions. For each direction, the occlusion level is from 20% to 50% with 10% increments. That is, the numbers of occluded scattering centers are 6, 9, 12, and 15 respectively. Thus, we have total 20736 ($2 \times 288 \times 36$) testing images and corresponding to each image, there are 5 types of observation sequences.

4.2.2 Experimental Results

• **Training:** For consistency, we use the same 8 states and 32 symbols used for XPATCH data for our HMM models for the real SAR data. Using the algorithm presented in Section 3.5, we build recognition models using the real SAR training data described in Section 4.2.1. For one observation

Table 4: Confusion matrix for various occlusions using MSTAR data.

Target Type	Occlusion Level			
	20%		30%	
	T72	ZSU	T72	ZSU
T72	280(97.2%)	8(2.8%)	233(80.9%)	55(19.1%)
ZSU	5(1.7%)	283(98.3%)	64(22.2%)	224(77.8%)
Target Type	Occlusion Level			
	40%		50%	
	T72	ZSU	T72	ZSU
T72	200(69.4%)	88(30.6%)	164(56.9%)	124(43.1%)
ZSU	98(34.0%)	190(66.0%)	117(40.6%)	171(59.4%)

sequence type we have 576 (288 azimuths $\times$ 2 object classes) HMM models. Since we have defined five kinds of observation sequences $(O_1, O_2, O_3, O_4, O_5)$ for each image, we get models based on each kind of observation sequence.

• **Testing Results:** During the testing phase, we apply the recognition algorithm (described in Section 3.6) to the real SAR test data (described in Section 4.2.1). Here, we consider only the identity of an object, which we call ID. We count a recognition result as a correct recognition if the HMM model with the maximum probability is associated with the object from which the testing observation sequence was extracted, we do not consider the corresponding azimuth angles of the HMM models.

Tables 4 and 5 show the experimental results on MSTAR real data. These results are obtained by integrating the results from 5 different types of observation sequences O_1, O_2, O_3, O_4, and O_5. Table 4 shows the results for recognizing 20% to 50% occluded T72 tank and ZSU gun. The confusion matrix shows how many of them are correctly identified and how many are incorrectly recognized. Table 5 shows the results similar to Table 4. The difference is that here we use "probability ratio threshold" instead of considering only the maximum probability. Only when the ratio between the maximum and second maximum (other object type) probabilities is greater than the probability ratio threshold (1.01 used here) we accept the recognition result. Otherwise, it is rejected and the test data is labeled as unknown object.

Table 5: Recognition results for various occlusions using MSTAR data.

Target Type	Occlusion Level 20%		
	correct	error	rejection
T72	279(96.9%)	7(2.4%)	2(0.7%)
ZSU	281(97.6%)	4(1.4%)	3(1.0%)
Target Type	Occlusion Level 30%		
	correct	error	rejection
T72	230(79.9%)	47(16.3%)	11(3.8%)
ZSU	216(75.0%)	59(20.5%)	13(4.5%)
Target Type	Occlusion Level 40%		
	correct	error	rejection
T72	188(65.3%)	75(26.0%)	25(8.7%)
ZSU	168(58.3%)	91(31.6%)	29(10.1%)
Target Type	Occlusion Level 50%		
	correct	error	rejection
T72	156(54.2%)	110(38.2%)	22(7.6%)
ZSU	164(56.9%)	102(35.5%)	22(7.6%)

We also performed an additional experiment to determine the optimal number of observation sequences. Table 6 shows the results when only observation sequences O_1 and O_2 are used in integration. The results of Table 6 based on O_1 and O_2 are slightly better than the results of Table 4 based on all the five observation sequences (O_1 to O_5). The reason is that the results from observation sequences O_3, O_4, and O_5 are less reliable than others. Each of these observation sequences measures distances from a specific scattering center, for example, O_3 measures distances of other scattering centers from (reference) scattering center 1, O_4 measures distances of other scattering centers from scattering center 2, and so on. There is a significant probability that these prominent scattering centers (used as a reference) get occluded at higher levels of occlusion. If the first scattering center is occluded, then the entire observation sequence (O_3, O_4 or O_5) is subject to error, which is reflected in the integration results. An alternative is to use all the scattering centers as reference points to build models to achieve increased performance at the expense of increased computation, not just the top three scattering centers that we used in observation sequences O_3, O_4, and O_5.

From Tables 4, 5, 6, we can see that the performance degrades as the occlusion level increases. These results show that reasonable recognition performance can be achieved when the occlusion is as much as 30% - 40%. The time needed to recognize a test case is 6ms on a SUN Ultra 2.

Table 6. Results of Integration (Observation Sequences O_1 and O_2).

Occlusion Level	# of Correct Recognition				# of test sequences
	20%	30%	40%	50%	
T72 tank	284(98.6%)	249(86.5%)	208(72.2%)	186(64.6%)	288
ZSU gun	285(99.0%)	238(82.6%)	194(67.4%)	175(60.8%)	288

5 Conclusions

Recognition of occluded objects has been a significant problem for automatic target recognition. In this paper, we have presented a novel conceptual approach based on hidden Markov modeling for the recognition of occluded objects in SAR images. The approach uses multiple models for various observation sequences that are chosen based on the SAR image formation and account for both the geometry and magnitude of SAR image features. We have shown the results on both XPATCH and real SAR data. The number of observation sequences and the number of features are design parameters which can be optimized by following the approach presented in the paper. We have demonstrated that the HMM approach makes use of available structural information to solve the problem caused by occlusion and noise. It takes the spatial arrangement of structural information as a whole and is able to collect useful information from distorted or partially unreliable patterns. Reasonable recognition can be achieved for up to 30% - 40% occlusion with both the simulated and real SAR data.

Acknowledgments: This research is supported in part by grants F49620-97-1-0184 and DAAD 19-01-1-0357. The contents and information do not necessarily reflect the positions and policies of the U. S. Government.

References

[1] B. Bhanu, Automatic target recognition: state-of-the-art survey, *IEEE Trans. Aerospace and Electronic System* Vol. 22, No. 4 (July 1986) pp. 364-379.

[2] B. Bhanu, D. E. Dndgeon, E. G. Zelnio, A. Rosenfeld, D. Casasent and I. S. Reed, Introduction to the special issue on automatic target detection and recognition, *IEEE Trans. Image Processing* Vol. 6, No. 1 (January 1997) pp. 1-6.

[3] B. Bhanu and T. L. Jones, Image understanding research for automatic target recognition, In *Proc. ARPA Image Understanding Workshop* (January 26-29, 1992) pp. 249-254.

[4] D. E. Dndgeon and R. T. Lacoss, An overview of automatic target recognition, *The Lincoln Laboratory Journal* Vol. 6, No. 1 (1993) pp. 3-9.

[5] L. R. Rabiner, A tutorial on hidden Markov models and selected applications in speech recognition, *Proc. of the IEEE* Vol. 77, No. 2 (Febrary 1989) pp. 257-285.

[6] L. R. Rabiner and B. H. Juang, An introduction to hidden Markov models, *IEEE ASSP Magazine* Vol. 3, No. 1 (January 1986) pp. 4-16.

[7] L. R. Rabiner, B. H. Juang, S. E. Levinson and M. M. Sondhi, Recognition of isolated digits using hidden Markov models with continuous mixture densities, *AT&T Technical Journal* Vol. 64, No. 6 (July-August 1985) pp. 1211-1233.

[8] O. E. Agazzi and S. S. Kuo, Hidden Markov model based optical character recognition in the presence of deterministic transformations, *Pattern Recognition* Vol. 26, No. 12 (November 1993) pp. 1813-1826.

[9] O. E. Agazzi and S. S. Kuo, Pseudo two-dimensional hidden Markov models for document recognition, *AT&T Technical Journal* Vol. 72, No. 5 (1993) pp. 60-72.

[10] K. H. Fielding and D. W. Ruck, Spatio-temporal pattern recognition using hidden Markov models, *Journal of Combinatorial Optimization* Vol. 31, No. 4 (1995) pp. 1292-1300.

[11] R. R. Rao and R. M. Mersereau, On merging hidden Markov models with deformable templates, *Proc. International Conf. on Image Processing* (1995) pp. 556-559.

[12] D. P. Kottle, P. D. Fiore, K. L. Brown and J. K. Fwu, A Design for HMM-Based SAR ATR In *SPIE Conf. on Algorithms for Synthetic Aperture Radar Imagery V* (April 1998) pp. 541-551.

[13] L. M. Novak, G. J. Owirka and C. M. Netishen, Performance of a high-resolution polarimetric SAR automatic target recognition system, *The Lincoln Laboratory Journal* Vol. 6, No. 1 (1993) pp. 11-24.

[14] J. H. Yi, B. Bhanu and M. Li, Target indexing in SAR images using scattering centers and Hausdorff distance, *Pattern Recgnition Letters* Vol. 17 (September 1996) pp. 1191-1198.

[15] G. Jones III and B. Bhanu, Recognition of articulated and occluded objects, *IEEE Transactions on Pattern Analysis and Machine Intelligence* Vol. 21, No. 7 (July 1999) pp. 603-613.

[16] G. Jones III and B. Bhanu, Recognizing occluded objects in SAR images, *IEEE Transactions on Aerospace and Electronic Systems* Vol. 37, No. 1 (January 2002) pp. 316-328.

[17] K. F. Lee, *Automatic Speech Recognition - The development of the SPHINX system*, (Kluwer Academoc Publishers, Boston, 1989).

[18] D. J. Andersh, S. W. Lee, H. Ling and C. L. Yu, XPATCH: A high frequency electromagnetic scattering prediction code using shooting and bouncing ray, In *Proc. Ground Target Modeling and Validation Conference* (August 1994) pp. 498-507.

[19] T. Ross, S. Worrell, V. Velten, J. Mossing and M. Bryant, Standard SAR ATR evaluation experiments using the MSTAR public release data set, *SPIE Proceedings: Algorithms for Synthetic Aperture Radar Imagery V* Vol. 3379 (April 1998) pp. 566-573.

[20] J. P. Fitch, *Synthetic Aperture Radar*, (New York: Springer-Verlag, 1988).

[21] B. Bhanu and G. Jones III, Recognizing target variants and articulations in synthetic aperture radar images, *Optical Engineering* Vol. 39, No. 3 (March 2000) pp. 712-723.

[22] J. T. Tou and R. C. Gonzalez, *Pattern Recognition Principles*, (Addison-Wesley, London, 1974).

[23] L. Novak, G. Benitz, G. Owirka and L. Bessette, ATR performance using enhanced resolution SAR, In *SPIE Conf. on Algorithms for Synthetic Aperture Radar Imagery* (April 10-12, 1996).

Approximate String Matching for Angular String Elements with Applications to On-line and Off-line Handwriting Recognition

Sung-Hyuk Cha
Department of Computer Science
Pace University, Pleasantville, New York, 10570
E-mail: scha@pace.edu

Sargur N. Srihari
Center of Excellence for Document Analysis and Recognition
State University of New York at Buffalo
E-mail: srihari@cedar.buffalo.edu

Contents

D. Chen and X. Cheng (eds.),
Pattern Recognition and String Matching, 105–130.

Abstract

String distance measures are useful in both on-line and off-line character recognition for comparing on-line stroke and off-line contour sequence strings. Since stroke and contour string elements are angular in that they represent a circular measurement ($0° \sim 360°$), usual *edit distances* with cost matrix are inadequate for this type of strings. For this reason, we extend *edit distances*, previously defined for nominal type, to handle angular and magnitude types. The newly defined measure utilizes the *"turn"* concept in place of *substitution* for angular string elements and takes local context into account in *indels* (insertions and deletions). The approximate string matching, besides being of interest in itself, provides solutions to *writer verification, on-line* and *off-line character recognition* problems. We also discuss the string concatenation and reverse operations to recognize on-line characters when written unnaturally.

Key words: *Approximate String Matching, Chain-code, Character Recognition, Edit distance, Pattern Matching, Similarity Measure, Writer Verification*

1 Introduction

A string is a sequence of symbols drawn from the alphabet Σ and it is one of the most popular representations of pattern [1]. The approximate string matching algorithm is often used to discriminate between two patterns and it is one of widely studied areas in computer science due to a variety of applications such as genetics and DNA sequence analysis, a spelling corrector, etc [2, 3, 4]. Earlier definition and solution for the traditional *approximate string matching* problem are found in the literature [5, 3] and extensive surveys on various techniques are shown in [6]. It computes the *edit distance*, also known as *Levenshtein distance* that is the minimum number of *indels* (insertions and deletions) and substitutions needed to transform one string to another.

In the earlier definition, the symbol types are assumed to be nominal while the measurement types made on the symbols are various. We consider four versions of the string distance measure, corresponding to whether the type of measurement is nominal, angular, magnitude and cost-matrix. Since the original *Levenshtein* or the *Levenshtein with cost-matrix* edit distances are inadequate for either angular or magnitude, we propose a modified edit distance. The algorithm for the newly defined edit distance uses the *dynamic programming* method of computing the edit distance between two strings [7].

"Find all letters that look like this letter." Such a query has received a great deal of attention in *character recognition* and *handwriting analysis.* This problem has been formalized by defining a distance metric between characters and finding the nearest characters from the reference set. One promising distance measure is using the edit distance between strings after extracting the on-line stroke and off-line contour sequence strings. This approach has been studied by many researchers [8, 9, 10, 11] dating as far back as 1975 [12]. *Fujimoto et al* developed the *OCR* using the idea of *"Non-linear Elastic Matching"* to read hand-printed alpha-numerics and Fortran programs [12]. The edit distance with cost matrix technique [5] was used to solve the on-line [9] and off-line [8] character recognition problems. Since symbols in both *Freeman style* on-line stroke and off-line contour sequence strings are not nominal type but angular type, we extend *Levenshtein edit distance* to handle this measurement type utilizing the *"turn"* concept in place of *substitution* for angular strings and takes local context into account in *indels.* We compare the newly defined edit distance with the edit distance with cost matrix which uses predetermined costs for *indels.*

We claim that the smaller the edit distance is, the more similar two characters look like to each other. To appreciate the usage of the approximate string matching for angular string elements in *pattern classification,* we consider three applications: *writer verification, on-line* and *off-line character recognitions.* First, in *writer verification,* one would like to find out whether the given two letters are similar or whether they were written by the same author [13, 14, 15]. The *form* or *shape* is an important feature for characterizing individual handwriting as it is quite consistent with most writers in normal undisguised handwriting [14]. The form can be described by a sequence of directional strokes. For the *on-line character recognition* problem(see [16, 17, 18] for comprehensive and detailed survey on successful techniques and applications of on-line handwriting recognition), the stroke sequence string is obtained from the movements of a mouse or a pen-based device. As a stroke sequence signifies the shape of the individual letters, a letter *"a"* is distinguished from a letter *"b"* by its different stroke sequences. Finally, in *off-line character recognition,* contour sequence strings derived from a *Freeman style chain-code* are used.

The rest of this chapter is organized as follows. In section 2, four types of strings are defined according to their element type: nominal, angular, linear and cost-matrix type strings. Section 3 presents the newly defined approximate string matching algorithm for stroke direction sequence strings. Also, the difference between the newly proposed edit distance and Levenshtein edit distance with cost matrix is stated. In section 4, we discuss three applications: *writer verification*, *on-line* and *off-line character recognitions*. Finally, section 5 concludes this work.

2 Type of Strings

A string is a sequence of symbols drawn from the alphabet Σ. We will use the following notations and symbols throughout the rest of this chapter.

$$\begin{aligned} S_1 &= (s_{1,1}, s_{1,2}, \cdots, s_{1,n_1}) \\ S_2 &= (s_{2,1}, s_{2,2}, \cdots, s_{2,n_2}) \end{aligned}$$

S_1 and S_2 are strings and n_1 and n_2 denote the length of each string. Each symbol in the string, $s_{x,y}$ has two index labels where the first index, x indicates the string to which it belongs and the second index, y indicates the location of the symbol in the string $(1 \leq y \leq n_x)$. We consider four types of strings based on the measurement type of the alphabet: *nominal, angular, linear* and *cost-matrix.*

Different types of strings have been considered in the literature of string matching according to three models of alphabet: constant size alphabet, unbounded alphabet and integer alphabet. In this taxonomy, algorithmists divided strings to find the efficient computational time matching algorithms. Briefly to summarize the history, Landau and Vishkin suggested an efficient algorithm; after building suffix trees in $O(f(n))$, the constant time *Least Common Ancestor* algorithm [19] is applied k times for every location of the text [20]. The time complexity is $O(f(n) + kn)$; the time complexity of building a suffix tree depends on three models of alphabet: a constant size alphabet, integer alphabet and unbounded alphabet. Weiner, who introduced the suffix tree, gave $O(n)$ suffix tree construction algorithm for the constant size alphabet [21]. Farach showed an algorithm to build a suffix tree in linear time when symbols are drawn from an integer alphabet [22]. Finally, $\Omega(n \log n)$ algorithm is known for the unbounded alphabet. In our version of categorization, however, we consider four types of strings according to their measurement to design suitable edit distances for each type not to design the efficient algorithm.

2.1 Nominal type string

In a nominal type string, each symbol value is named, e.g. string of English alphabets or DNA sequences $[A-A-C-T-G\cdots-A-A]$ where the letters A, G, C, and T stand for the *nucleic acids adenine, guanine, cytosine*, and *thymine*. The distance between two nominal type symbols are either a match or a mismatch:

$$d(s_{1,i}, s_{2,j}) = \begin{cases} 0, & \text{if } s_{1,i} = s_{2,j} \\ 1, & \text{otherwise} \end{cases} \tag{1}$$

Traditional approximate string matching problem takes nominal type strings as inputs [3, 6]; given a text string of length n, a pattern string of length m, the number k of differences (*indels* and substitutions) allowed in a match, find every location in the text where a match occurs. The *edit distance*, also known as *Levenshtein distance*, is the minimum number of *indels* and substitutions needed to transform one string to another. Nominal type strings are useful in *spelling correction* that is a post-processing stage of word recognition [2].

2.2 Angular type string

In the angular type string, symbol values are angular degrees. Stroke directions are quantized into r directional values as shown in Figure 1, e.g. *Freeman style chain-code scheme* has $r = 8$ directional values. One element is different from another by a number of *turn* operations of changing direction by $\frac{360°}{r}$. The turning distance between two individual directional elements is defined as follows:

$$d(s_{1,i}, s_{2,j}) = \begin{cases} |s_{1,i} - s_{2,j}|, & \text{if } |s_{1,i} - s_{2,j}| \leq \frac{r}{2} \\ r - |s_{1,i} - s_{2,j}|, & \text{otherwise} \end{cases} \tag{2}$$

One can edit a stroke direction to make the other stroke direction by turning it clockwise or counter-clockwise whichever is shorter. For example of $r = 8$, $d(\uparrow, \rightarrow) = 2$ as one can turn $\uparrow$ two clockwise steps to make $\rightarrow$. Therefore, the term $d(s_{1,i}, s_{2,j})$ denotes the minimum number of necessary steps in turning.

The equally weighted substitution operation in nominal type strings is not suitable in angular type strings as a distance between two elements, $d(s_{1,i}, s_{2,j})$, is not simply a match or mismatch. Note that element values are categorical in nominal type strings whereas they represent a scale from small to large in angular type strings. For example, the difference between symbols "A" and "B" is the same as that between "A" and "C" whereas the difference

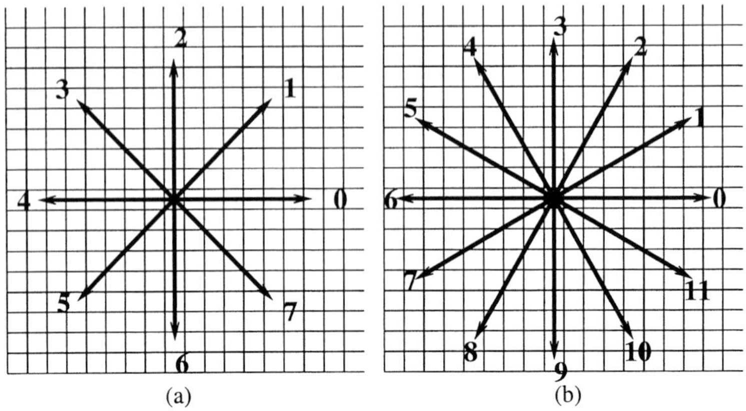

Figure 1: Number of stroke directions and length along horizontal and vertical axes: (a)Strokes with 8-directions and 7 pixel length, (b)Strokes with 12-directions and 8 pixel length.

between $\rightarrow$ and $\uparrow$ is smaller than that between $\rightarrow$ and $\leftarrow$. Motivated from the aforementioned argument, we utilize the turning distance between two angular string elements in place of substitution to compute the edit distance between two angular type strings. This idea was originally implemented by Fugimoto and et al in 1975 and they called it *"Nonlinear Elastic Matching"*.

As shown in Figure 2, a character image can be represented in a piecewise linear stroke sequence string and we will use A_x and A_y examples the data structures and algorithms throughout the rest of this chapter. The string of arrows is called a *Stroke Direction Sequence String*, or an *SDSS* in short and an *SDSS* is an angular type string. As a character may have more than one contiguous stroke, contiguous strokes are represented in a pair of parentheses, e.g. a capital letter *"A"*, $A_x = (A'_x) + (A''_x)$; a certain character letter is consisted of a sequence of one or more *SDSS*'s.

2.2.1 Choice of Number of Quantization Levels for Stroke Direction and Length

A digital off-line handwriting is represented by rectangular arrays of picture elements called *pixels*. A digital on-line handwriting is represented by a sequence of mouse positions where the positions are also pixels. Each pixel is a rectangle. If one moves from one pixel to another by k pixels in diagonal direction, the exact length of the path is not k but $k\sqrt{2}$. Unlike *Freeman code*, we would like to have all directional arrows to be the same length.

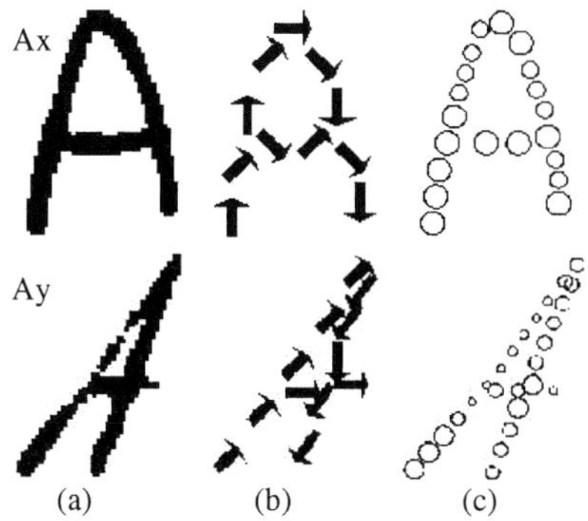

$$
\begin{array}{lcl}
\mathrm{SDSS}(A_x) & = & \mathrm{SDSS}(A'_x) + \mathrm{SDSS}(A''_x) \\
\mathrm{SDSS}(A'_x) & = & [\uparrow \;\; \nearrow\!\uparrow \;\; \nearrow \;\; \rightarrow \;\; \searrow \;\; \downarrow \;\; \searrow \;\; \downarrow] \\
\mathrm{SDSS}(A''_x) & = & [\searrow \;\; \nearrow] \\
\mathrm{SPSS}(A_x) & = & \mathrm{SPSS}(A'_x) + \mathrm{SPSS}(A''_x) \\
\mathrm{SPSS}(A'_x) & = & [\bigcirc \;\; \bigcirc \;\; \bigcirc \;\; \circ \;\; \bigcirc \;\; \bigcirc \;\; \mathrm{O} \;\; \bigcirc \;\; \mathrm{O}] \\
\mathrm{SPSS}(A''_x) & = & [\bigcirc \;\; \mathrm{O}] \\
\mathrm{SDSS}(A_y) & = & \mathrm{SDSS}(A'_y) + \mathrm{SDSS}(A''_y) \\
\mathrm{SDSS}(A'_y) & = & [\nearrow \;\; \nearrow \;\; \nearrow \;\; \nearrow \;\; \nearrow \;\; \swarrow \;\; \swarrow \;\; \downarrow \;\; \swarrow \;\; \swarrow] \\
\mathrm{SDSS}(A''_y) & = & [\rightarrow \;\; \rightarrow] \\
\mathrm{SPSS}(A_y) & = & \mathrm{SPSS}(A'_y) + \mathrm{SPSS}(A''_y) \\
\mathrm{SPSS}(A'_y) & = & [\bigcirc \;\; \bigcirc \;\; \circ \;\; \circ \;\; \mathrm{O} \;\; \mathrm{O} \;\; \mathrm{O} \;\; \bigcirc \;\; \mathrm{O} \;\; \mathrm{O}] \\
\mathrm{SPSS}(A''_y) & = & [\mathrm{O} \;\; \circ]
\end{array}
$$

Figure 2: Sample Stroke Direction and Pressure Sequences: (a)original character images (b) Angular Stroke Direction (c) Stroke Width(Pressure).

Let $s(l, r, i)$ be a stroke whose length is l and the number of directions is r. The last index, i, indicates the arrow number and thus $i \in \{0, \cdots, r-1\}$. For example, $s(7, 8, 1) = \nearrow$. To illustrate, suppose the arrow begins from the origin. Then the coordinate of $s(l, r, 1)$ must be $(l \cos \frac{2\pi}{r}, l \sin \frac{2\pi}{r})$. Due to the rectangular grid representation of pixels, the actual coordinate in a digital image is $(\lceil l \cos \frac{2\pi}{r} \rceil, \lceil l \sin \frac{2\pi}{r} \rceil)$. These differences introduce two errors in diagonal stroke arrows: length errors and angle errors.

$$
\text{Length error} = \left| l - \sqrt{\left\lceil l \cos \frac{2\pi}{r} \right\rceil^2 + \left\lceil l \sin \frac{2\pi}{r} \right\rceil^2} \right| \tag{3}
$$

$$\text{Angle error} = \left| \frac{2\pi}{r} - \tan^{-1} \frac{\left\lceil l \sin \frac{2\pi}{r} \right\rceil}{\left\lceil l \cos \frac{2\pi}{r} \right\rceil} \right| \qquad (4)$$

Figure 3 shows various combinations of the number of stroke directions and lengths with their length errors and angle errors. For the $s(x, 16, i)$'s, we

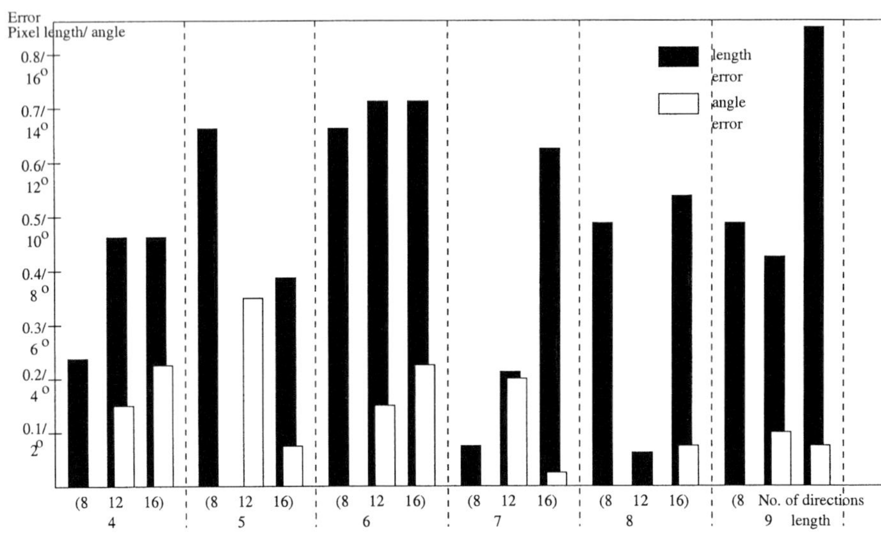

Figure 3: Error in representing stroke direction and length for various levels of direction quantization (8,12,16) and length quantization (4-9).

show only $\frac{\pi}{8}$ stroke length and angle errors as the those of $\frac{\pi}{4}$ stroke are given in $s(x, 8, i)$'s. The figure suggests that there are only two choices whose length and angle errors are lowest at $s(7, 8, i)$ and $s(8, 12, i)$.

One used here is $s(7, 8, i)$. We chose it because the approximate 7 length of diagonal arrows can be achieved by moving 5 pixel to the right and 5 to the north; the exact distance is 7.071 as shown in Figure 1. An alternative choice is $s(8, 12, i)$. This gives the approximate length for the 1 o'clock direction can be obtained by moving 4 pixels to the right and 7 pixels to the north; the exact distance is 8.062. Although the length error is smaller than the previous case, it has the angle error. The exact angle between the 0 stroke and 1 stroke is 29.74° whereas the desired and intended one is 30°.

2.3 Linear type string

Next, in a linear type string, the type of an element value is an integer(discrete) or a real(continuous). The distance between element values $s_{1,i}$ and $s_{2,j}$ is:

$$d(s_{1,i}, s_{2,j}) = |s_{1,i} - s_{2,j}| \quad (5)$$

It is called a linear type because a measurement value, $s_{x,y}$, represents a scale strictly from small to large. For instance, a *Stroke Pressure Sequence String* or *SPSS* in short, shown in Figure 2 falls into this category.

Pressure information is regarded one of the most important features in *writer verification*. The pressure information is readily available in *on-line signature verification* whereas it is hard to extract in off-line writer verification. In ink pen based handwriting, however, pressure usually appears as a thickness of the stroke in character binary images and an *SPSS* is a sequence of stroke thickness although the pressure can also appear as different grey scale levels in grey scale images.

An *SPSS* is obtained after its *SDSS* is obtained. This is because there is no reason to measure the distance between *SPSS*'s if their *SDSS*'s do not match. Thus, the thickness is measured for each character directional stroke. As shown in Figure 11, vertical and horizontal strokes have 7 candidates as

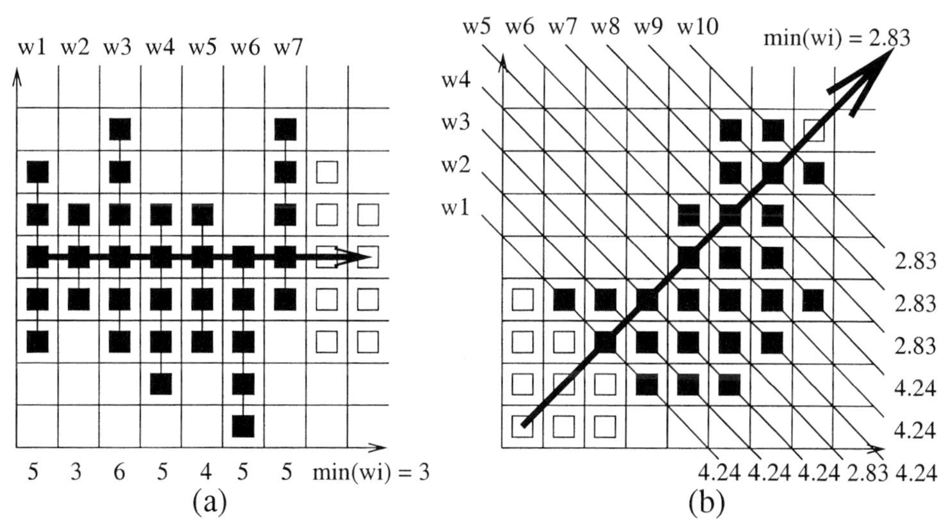

Figure 4: Stroke Width: (a) vertical and horizontal stroke width (b) diagonal stroke width.

their stroke lengths are 7 pixel long: $w_1 \sim w_7$. The width of these strokes is the minimum value of w_i's. For example of Figure 4 (a), the stroke width is $\min(w_i) = w_2 = 3$. Diagonal strokes have 10 candidates, $w_1 \sim w_{10}$, as illustrated in Figure 4 (b) and $\min(w_i) = w_4, w_8, w_9, w_{10} = 2.83$.

As with the *SDSS*, the *SPSS* is also a pseudo pressure information in off-line images. Suppose two strokes cross each other and the entire part of one stroke is completely covered by the other stroke or strokes. Then the detected width of the hidden stroke, $\min(w_i)$ is the length of the other stroke or more. This extreme case is, however, detectable and can be avoided by replacing it with the interpolated value of the previous and subsequent stroke widths. Now consider a looped letter *"l"* without a visible hole or a letter with a retrace as illustrated in Figure 5. The width is almost doubled

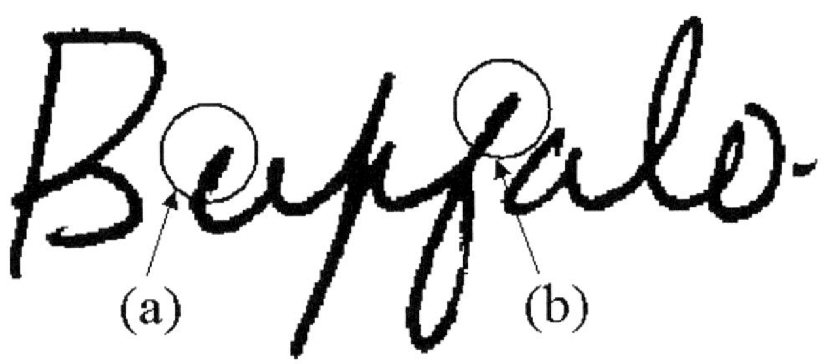

Figure 5: (a) a letter with a retrace, and (b) a looped letter without a visible hole.

because of the retrace. In a grey scale image, the latter stroke width can be detected by a segmentation technique but there is no way to find the exact width of the bottom stroke. However, we use the width information even for the letter with a retrace as shown in the Figure 4 as this erroneous width information may be a unique peculiarity of one's handwriting style. In all, *SPSS* is only a pseudo pressure information.

In computing the distance between linear type strings, *Euclidean* distance is used after aligning them. Aligning process is necessary because *SPSS*'s are of disparate lengths and we would like to utilize the standard vector distance measures. By aligning two strings into the equal length l, the problem becomes the distance between two vectors of the same dimensionality with components of numerical values. Now standard vector norms such as *city block, Euclidean* or *Minkowski* distances can be used as distances between two aligned *SPSS's*. The detailed algorithm is given in the following section 4.1.

2.4 Cost-Matrix type string

Finally, the Levenshtein distance can be extended to cost-matrix type strings. A proper assignment of cost values for substitution between alphabets is used instead of a uniform penalty value 1 or 2. Levenshtein used 2 since the substitution between $d(s_{1,i}, s_{2,j})$ can be regarded as deleting $s_{1,i}$ from the i'th position of S_1 string and inserting $s_{2,j}$ value in that position. Different substitution cost values can be assigned to different alphabets. As long as the size of the alphabet is finite, the cost-matrix can be built.

Wagner and Fischer [5] used the following notation: $\gamma(a \longrightarrow b)$ meaning the cost of changing the alphabet a to b. If $a = \Lambda$ where Λ is a null string, it is the cost of inserting b to the string. If $b = \Lambda$, it is the cost of deleting a from the string. If none of them is Λ, it is the substitution cost. Thus the size of the cost-matrix is $(|\Sigma|+1) \times (|\Sigma|+1)$ where Σ is a finite alphabet; the null value Λ must be included. The section 3.2 discusses how our algorithm differs from the edit distance with cost matrix.

3 Stroke Direction Sequence String Matching

The approximate string matching for *SDSS*'s is that given two *SDSS*'s, find the minimum edit distance between them by allowing *turn* and *indels*. In this section, we describe the algorithm and compare it with the edit distance with cost matrix. The algorithm for the newly defined edit distance uses the dynamic programming method of computing the edit distance between two strings [7].

3.1 Algorithm

Consider two "A" letters in Figure 2. As discussed earlier, the type of stroke direction element, $s_{x,y}$ is angular and the distance between two stroke direction elements, $d(s_{1,i}, s_{2,j})$ is given in equation (2) as a turning distance. This operation, *turn* is a very important difference between the conventional nominal type strings and angular type strings. While the former allows the substitution with the uniform cost of 1 or 2, the later allows the turn with various cost values.

The costs for insertion and deletion are defined in the following equation (6).

$$T[i,j] = \min \begin{cases} T[i-1,j-1] + d(s_{1,i-1}, s_{2,j-1}) & \Leftrightarrow \text{ turn} \\ T[i-1,j] + 1 + d(s_{1,i-1}, s_{2,j-1}) & \Leftrightarrow \ s_{1,i-1} \text{ is missing} \\ T[i,j-1] + 1 + d(s_{1,i-1}, s_{2,j-1}) & \Leftrightarrow \ s_{2,j-1} \text{ is missing} \end{cases} \quad (6)$$

Figure 6 illustrates the distance computing table for the first parts of sample letters and shows how each entry is computed. Let S_1 lie on the top of the

	s1	2	1	2	1	0	7	6	7	6
s2	00	02	03	05	06	08	11	15	18	22
1	02	01	02	04	05	07	10	14	17	21
1	04	03	01	03	04	06	09	13	16	20
1	06	05	02	02	03	05	08	12	15	19
1	08	07	03	03	02	04	07	11	14	18
1	10	09	04	04	03	03	06	10	13	17
5	14	13	09	07	08	06	05	07	10	12
5	18	17	14	11	11	10	08	06	09	11
6	23	22	18	16	14	13	10	07	07	08
5	27	26	23	20	19	17	13	09	09	08
5	31	30	28	24	24	21	16	11	11	10

(a)

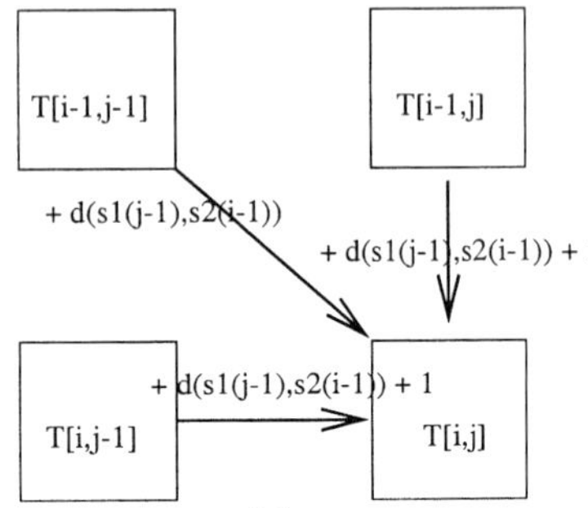

(b)

Figure 6: (a) Computing edit distance table (b) cell computation.

table and S_2 in the left side of the table. The first row is initialized by inserting a stroke, $s_{1,0}$ to the head of the string S_2 as shown in Algorithm 1 line 4 and 5. The left-most column is initialized by inserting a stroke, $s_{2,0}$ to the head of the string S_1 as shown in Algorithm 1 line 6 and 7. Now the rest of the table entries $T[i,j]$'s where $i = 2 \cdots n_1$ and $j = 2 \cdots n_2$ are computed by taking the minimum value of three values computed in equation 6 in line 8 to 10. The distance between two *SDSS*'s is achieved from the table in Figure 6 and it is 10.

Algorithm 1 Edit Distance (S_1, S_2)

1 begin
2 $\quad n_1 \leftarrow length\ (S_1),\ n_2 \leftarrow length\ (S_2)$
3 $\quad T[0,0] \leftarrow 0$
4 $\quad$ *for* $i = 1$ *to* n_1
5 $\qquad T[0,i] = T[0,i-1] + d(s_{1,i-1}, s_{2,0})$
6 $\quad$ *for* $i = 1$ *to* n_2
7 $\qquad T[i,0] = T[i-1,0] + d(s_{1,0}, s_{2,i-1})$
8 $\quad$ *for* $i = 1$ *to* n_2
9 $\qquad$ *for* $i = 1$ *to* n_1
10 $\qquad\quad T[i,j] = \min(T[i-1,j-1] + d(s_{1,i-1}, s_{2,j-1}), T[i-1,j] + 1 + d(s_{1,i-1}, s_{2,j-1}), T[i,j-1] + 1 + d(s_{1,i-1}, s_{2,j-1}))$
11 $\quad$ *return*$(T[n_2, n_1])$
12 end

The computational time complexity for the algorithm 1 is $O(n_1 n_2)$ and the space requires only $O(n_1)$ because only the entries in the previous column need be stored.

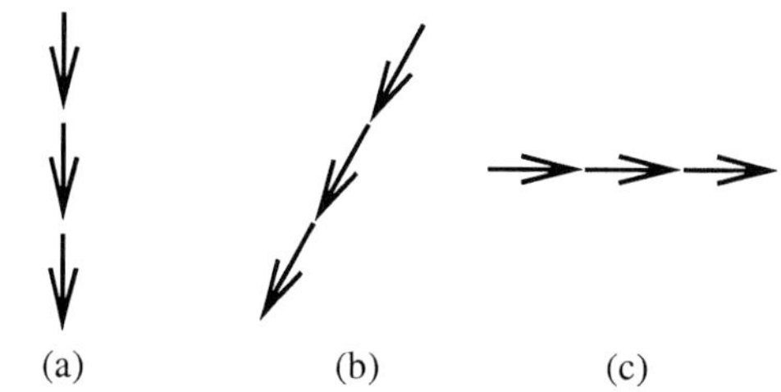

(a) (b) (c)

Figure 7: Sample Characters (a) "1", (b) skewed "*1*" (c) "-".

To illustrate, consider Figure 7. We would like $D($"1", "*1*"$)$ to be smaller than $D($"1", "-"$)$. The equal weighted Levenshtein edit distance is incapable of differentiating these. The D_T and D_L are the distance matrices of the newly defined edit distance with turning concept and the equal weighted Levenshtein distance, respectively. The row and column indices are characters in Figure 7 in the same order: (a),(b),(c).

$$D_T = \begin{pmatrix} 0 & 3 & 6 \\ 3 & 0 & 9 \\ 6 & 9 & 0 \end{pmatrix}, D_L = \begin{pmatrix} 0 & 3 & 3 \\ 3 & 0 & 3 \\ 3 & 3 & 0 \end{pmatrix}$$

This inadequacy of the equal weighted Levenshtein edit distance can be improved by simply defining different costs for each element of strings so called a cost-matrix. The next section discusses the difference between the newly proposed edit distance and Levenshtein edit distance with cost matrix.

3.2 Comparison with cost-matrix string version

The conventional edit distance table used in spelling correction is computed by the following equation:

$$T[i,j] = \min \begin{cases} T[i-1,j-1] + \gamma(s_{1,i-1} \longrightarrow s_{2,j-1}) & \Leftrightarrow \text{ substitution} \\ T[i-1,j] + \gamma(\Lambda \longrightarrow s_{2,j-1}) & \Leftrightarrow s_{1,i-1} \text{ is missing} \\ T[i,j-1] + \gamma(s_{1,i-1} \longrightarrow \Lambda) & \Leftrightarrow s_{2,j-1} \text{ is missing} \end{cases} \quad (7)$$

The substitution part is the same in both equations (6) and (7) and we use the turn concept as a proper assignment of cost values to the substitution part of the Levenshtein matrix. The difference between the previous edit distance with a cost matrix and the newly defined edit distance is the *indel* part.

In equation (7), the insertion cost is added by pre-defined costs between the null element Λ and the corresponding alphabets in the other string. For example, consider our example from Figure 2:

$$\begin{array}{rcccccccccc} \text{SDSS}(S_1) = & \uparrow & \nearrow & \uparrow & \nearrow & \rightarrow & \searrow & \downarrow & \Lambda & \searrow & \downarrow \\ \text{SDSS}(S_2) = & \nearrow & \nearrow & \nearrow & \nearrow & \nearrow & \swarrow & \swarrow & \downarrow & \swarrow & \swarrow \end{array} \quad (8)$$

The penalty for the insertion in S_1 is given by the predefined $\gamma(\Lambda \longrightarrow \downarrow)$. Λ is an arbitrary element that can be any one of eight directional arrows. Despite this fact, using the predefined penalty value is quite absurd.

We have a different approach that uses the local context. Instead of using the predefined penalty value, we first change $\Lambda \longrightarrow s_{1,i-1}$ whose penalty value is 1 and then add the turn distance between them, $d(s_{1,i-1}, s_{2,j-1})$. In the example, we first assign $\Lambda \longrightarrow \downarrow = 1$ that is the left stroke of the Λ, $s_{1,i-1}$ in *SDSS*(S_1) not $s_{2,j-1}$ and then add $d(\downarrow, \downarrow) = 0$. The penalty value is not uniform but varies depending on the stroke in the left side of Λ. We use only one side of local context instead of the average value between $s_{1,i-1}$ and $s_{1,i}$. This is because the average cannot be defined for directions, e.g. ave$(\uparrow, \downarrow) = \leftarrow$ or $\rightarrow$?

Marzal and Vidal showed the normalized edit distance with a cost-matrix and applied it to recognize the off-line digit images using the *Freeman style* contour chain-code strings [8]. *M. Parizeau et al* used genetic algorithms to optimize the cost matrix to recognize the on-line digits [9]. Table 1 shows the comparison of various cost-matrices used on angular type strings. Note that

Table 1: Comparison of various cost-matrices

	Marzal's cost-matrix [8]	Parizeau's cost-matrix [9]	New Turning distance
$\gamma(\swarrow, \uparrow)$	9.26	9.7	3
$\gamma(\swarrow, \nwarrow)$	∞	11.0	2
$\gamma(\swarrow, \leftarrow)$	6.43	0.1	1
$\gamma(\swarrow, \swarrow)$	0	0	0
$\gamma(\swarrow, \downarrow)$	6.27	0.1	1
$\gamma(\downarrow, \swarrow)$	5.76	0.1	1
$\gamma(\downarrow, \Lambda)$	3.34	1.9	Varies

Marzal's matrix is not symmetric. In Parizeau's matrix, $\gamma(\swarrow, \nwarrow) > \gamma(\swarrow, \uparrow)$.

In all, the idea of *turn* and utilizing the local context is more suitable for the stroke direction sequence strings of character images.

4 Applications

The approximate string matching technique has found its applications in pattern classification [4], and it utilizes the nearest-neighbor algorithm for the classification as shown in Figure 8. We consider three important appli-

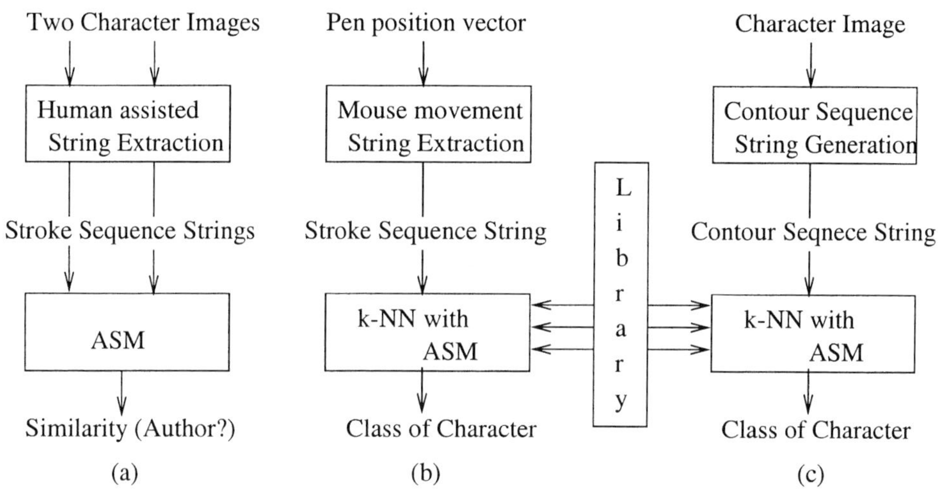

Figure 8: Applications of the string distance measure: (a) Writer Verification (b) On-line Recognition (c) Off-line Recognition.

cations: *writer verification* [13, 14, 15, 23, 24], *on-line* [16, 25, 17, 18] and *off-line character recognitions* [16, 26].

4.1 Writer Verification

In Figure 8 (a) *writer verification*, one would like to find out whether two characters were written by the same author based on their similarity. Among many features being currently used by the handwriting analysis practitioners or the document examiners, the *form* or *shape* is an important one for characterizing individual handwriting as it is quite consistent with most writers in normal undisguised handwriting [14]. The form can be described by a sequence of directional strokes and the similarity measure can be a key feature in handwriting analysis. Among many features, we consider two types of stroke sequence strings of a character as character level image signatures for writer verification: stroke direction sequence string(*SDSS*) and stroke pressure sequence strings(*SPSS*).

To extract stroke directional/pressure sequence strings for both letters to be compared, a trained human, well known as a document examiner, is involved during the examination. Albeit many methods have been proposed in the literature for accomplishing the task automatically [27, 13], it is impossible to extract the exact stroke sequence of the author of the letter. Hence, we develop a semi-automatic writer verification system as it is fairly easy for a human, especially for document examiners, to extract the pseudo on-line information. A stroke direction sequence string extractor allows an examiner to load a digitally scanned questioned letter or word image and extract the author's pseudo on-line information while he or she is tracing the strokes on screen. Hence, an *SDSS* is a pseudo on-line stroke directional sequence string. A pen based input device is highly recommended rather than a mouse device.

Next, computing the distance between *SPSS*'s first requires the aligning process because two *SPSS*'s may be of disparate lengths. To align them, we utilize Algorithm 1 as a stepping stone. The algorithm constructs the computing edit distance table as shown in Figure 6 and then the editing sequence is obtained. New *SDSS*'s with the exactly same lengths are derived according to the editing sequence as expressed in equation (8). They are of same lengths by inserting Λ in the respected places. One only needs to consider insertion since deletion in one string means insertion in the other string. Once the *SDSS*'s with Λ symbols are obtained, one can replace the directional arrow elements with respected magnitude values. Two *SPSS*'s in Figure 2 are aligned accordingly and they are denoted as SPSS_l's.

$$\begin{array}{rcccccccccc} \text{SPSS}_l(p_1) = & 3.0 & 3.0 & 3.0 & 1.0 & 3.0 & 3.0 & 2.0 & 2.5 & 3.0 & 2.0 \\ \text{SPSS}_l(p_2) = & 3.0 & 3.0 & 1.0 & 1.0 & 2.0 & 2.0 & 2.0 & 3.0 & 2.0 & 2.0 \end{array} \quad (9)$$

Note that we use the notation $p_{x,y}$ instead of $s_{x,y}$ to distinguish the pressure of a stroke from its direction. The value in the inserted position, Λ, is the average value between the pressures of the adjacent strokes. By aligning two strings into the equal length l, the problem becomes the distance between two vectors of the same dimensionality with components of numerical values. Now standard vector norms such as *city block, Euclidean* or *Minkowski* distances can be used as distances between two aligned *SPSS's*. Thus, the distance between *SPSS*'s is:

$$D[\text{SPSS}_l(p_1), \text{SPSS}_l(p_2)] = \frac{\sqrt{\sum_{i=1}^{l}(p_{1,i} - p_{2,i})^2}}{l} \text{ where } l = \max(n_1, n_2) \quad (10)$$

4.1.1 Demonstration

45 people provided their handwriting exemplars. All writers used the same writing materials: pen, papers and table. Each writer provided three exemplars of a word list that contains all capital letter in the beginning of the word and all small letters in all three positions: beginning, middle and terminal positions of a word. They are {April, Bob, California, December, English, February, Greg, Halloween, Iraq, June, Kentucky, Los Angeles, Markov, November, October, Pennsylvania, Queen, Raj, States, Texas, United, Virginia, What, Xray, York, Zorro, alumni, boy, come, date, enjoy, false, great, have, interest, jazz, keep, leave, millennium, now, of, picnic, question, run, six, time, unique, video, where, xenophobia, you, zero}.

The test was designed with 10 exemplars. 10 writers are selected randomly and we chose one word from each writer except for one writer with 2 exemplars since we need one query word. 10 such sample tests [1] are prepared and the system corrected 9 out of 10 whereas expert and non-expert human beings corrected 9 and 7 on average, respectively. One such sample test is given in Figure 9. The word image shown in the program window is the query handwriting and the right-side 9 words are known handwritings. Note that the query image is slightly scaled up. Since writers 2, 3 and 9 have two contiguous strokes for the letter *"B"* while the query writer has only 1, they are eliminated immediately from further investigation. Now one can load each image on the *SDSS* extractor. After extracting the *SDSS*'s, the edit distances are computed.

Let $D(W_q, W_x) = \sum_{i=\text{“B”}}^{\text{“b”}} D(\text{SDSS}(S_q(i)), \text{SDSS}(S_{W_x}(i)))$. Note that there are three contiguous strokes: $S_q(\text{“B”})$, $S_q(\text{“o”})$ and $S_q(\text{“b”})$. The authorship of W_q is determined by that of $W_x = \arg\min_{i=1\cdots n} D(W_q, W_i)$. The results of the edit distances are $D(W_q, W_1) = 175$, $D(W_q, W_2) = \times$, $D(W_q, W_2) = \times$, $D(W_q, W_4) = 197$, $D(W_q, W_5) = 101$, $D(W_q, W_6) = 53$, $D(W_q, W_7) = 98$, $D(W_q, W_8) = 131$, $D(W_q, W_9) = \times$. We have $W_x = W_6$. Therefore, the author of the query handwriting is more likely to be the writer 6 than other given samples.

4.1.2 Discussion

The purpose of the previous experiment is to demonstrate the procedure of comparing two handwriting sequences. The proposed method is not the solution to the *writer verification* dealing with real cases, but only a part of the entire process. Since the detailed *writer verification* is in [28], we

[1] "http://www.cedar.buffalo.edu/NIJ"

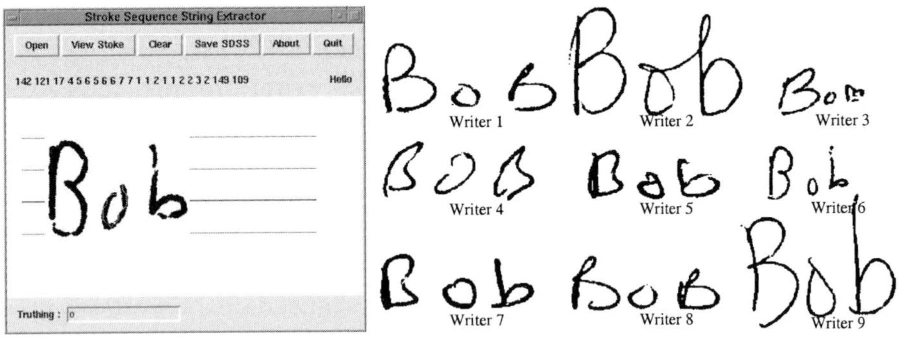

$q(B) =$ (↓↓↓↓↓↓↓↓↓↓↓↓↓↑↖↗↑↑↑↑↑↗↑↑↗↑↗→↗↘↘↘↘↓↙↓↙↙↙↖↘↓↓↘↘↘↓↓↙↓↙↙←←←↑↑)
$q(o) =$ (↙↙↓↓↓↓→→↗↑↗↑↑↑↑)
$q(b) =$ (↓↓↓↘↓↓↓↓↓↓↙↓↓↙↗↑↗→↗→→↘↘↙↓↙←←←)

$w_1(B) =$ (↓↓↓↘↓↓↘↓↓↘↓↑↑↑↖↑↑↑↑↖↖↑↑↑↑↗↑↑↑↑↗↗↑↑→→→↘→↘↘↘↓↓↓↓↓↓↙←↙←↙↙←↗→
↗→→→→→↘↘→↘→↘↘↓↘↓↙↙↙↙←↙←↙←←↙←←←)
$w_1(o) =$ (←↓↙↙↓↓→↘→↘→↗→→↑↑↑↗↑↖←↖←↖←←)
$w_1(b) =$ (↑↑↗↑↗↑↑↗↗↗↗↗↗↗↘↙↙↓↙←↙↙↙↘→→→→→→→→→↘↘↘↘↓↓↙↙↙←←←↙
←←←←←←←)
$w_2(B) =$ (↓↓↙↓↓↓↓↓↓↓↓↓↓↓↓↓↓↓↙↓↓↓↓↓↘↓↓↙↓↘↓↓)(↑↑↗↑↑↑↗↑↗↑↑↑↗↗↑↑↗→→↗↗→→→→→↘
↘↘↓↓↓↓↙↓↙↓↙↓↙↙↙↙←↙↙←←↙↙↗↗↗→↗↗→→→↘→↘↘↘↘↘↘↘↓↓↓↓↙↓↙
↙↓↓↙↙↙↙↙←←↙←←←←↖↖)
$w_2(o) =$ (↓↓↙↓↓↓↓↙↓↙↓↓↘↓↘→↘→↗→↗→↗↗↑↗↗↑↑↖↑↖↖↑↖↖↖←↖↖↖←←←←↘↘↘→
↗→→→→→↗→→)
$w_2(b) =$ (↗↗↗↗↗↗↗↗↗↑↑↗↑↗↑↖↑↖←↙↓↓↓↓↓↓↓↓↓↓↓↓↓↓↓↓↓↓↓↓↓↘↓↓↓↙↘↓↓↓↓↓↘→→→↗→↗
→↗↗↗↑↗↗↗↗↗↑↑↖↖↖↑↑↖↖↖↖←↖↙←↙↙↙↙↙↓↙↙↓)
$w_3(B) =$ (↙↙↓↓↙↓↓↙↓↙↓)(↗↗↗→↗→↗↗→↗→→↘↙↓↙↙↙↙↙↙←↙↙↗→→→→→→↘→↓↓↙↓
↙←↙←←←↙↙←←←)
$w_3(o) =$ (↙↙↙↓↘→→↗↑↑↖)
$w_3(b) =$ (↑↑↑↑↑→↘↙↘↑→→→↘↙↓↙↙←)
$w_4(B) =$ (↑↖↑↗↑↑↑↑↗↑↑↗↗↑↗↗↗↗→↗↗→↗→→→↓↙↓↙↙↙←↙↓↙↙↓↘↘↓↘↘↓↙↙←←←
←←←)
$w_4(o) =$ (↙↙↙↙↙↙↓↘→→→→↗→↗↗→↗↑→↗↗↗↗↑↑↑↖↑↖←↖←↙←↙↙↙↓↑)
$w_4(b) =$ (↙↙←←↑↑↑↗↑↑↗↗↑↗↑↗↗↗↗↗↗↗↗→↓↙↙↙↓↓↓↘↓↘↘→↓↓↓↙←↙↙↙←↙↙)
$w_5(B) =$ (↓↓↘↓↓↓↙↓↓↓↓↓↓↓←↑↖↑↑↗↑↗↑↗↑↑↑↗↗↗↗→→→→↓↓↓↙↓↙↓↓↘↓↘↘↓↙↙←←↙←←
←↙)
$w_5(o) =$ (←←↙←←↙↙↘↘↘↘→→→→↗↑↑↖↖↑↑↖↖←←↓)
$w_5(b) =$ (←↖←↙↓↓↙↓↓↘↓↓→→↗↗↗→↗→→→→→↘↘↓↓↙↙↙←←←←←←←←↖↖)
$w_6(B) =$ (↓↘↓↓↓↙↓↓↓↓↓↓↓↑↗↖↑↑↗↑↑↑↑↑↑↑↑↗↑↖↑→→↗↘→↘↓↘↘↓↙↓↙↓↙←↖←→↘↘↓↘↓↓↙↓
↓↙←↙←←←)
$w_6(o) =$ (↙↙↓↙↘↓→→↗↑↑↑↖)
$w_6(b) =$ (↓↙↓↓↘↓↓↓↙↓↓↓↓↓↗→↗↗→→↘↙←←)
$w_7(B) =$ (↓↘↓↓↓↓↘↓↓↓↓↓↓↙↑↖↑↑↑↑↑↗↑↑↑↑↗↑↗↑↗↗↗→→→→↘↓↓↓↓↙↓↙↓↙↓↓↘↘↘↘↘↙←↙
←←←←←)
$w_7(o) =$ (↙←↙↓↓↘↘↘↘↘→→→→→↗↗↗↑↖↑↖↖↑↖←↖←↙←↙)
$w_7(b) =$ (↓↘↓↓↓↘↓↓↘↓↓↘↓↓↓↗↑↗↗↗↗↗→→→↘↘↓↙↙↙↙←↙←↙←←←←←←)
$w_8(B) =$ (↓↓↙↓↓↓↓↓↓↘↖↑↑↗↑↑↗↑↑↑↑↑↑↗↑↗↗↑↑↗→→↘↙↓↙↙↙↙↙←↙↙→↘→→↘→→→→↘
→→↘↘↘↘↓↓↙↓↙←←↙←
←←←↖↑)
$w_8(o) =$ (↙←↙↙↙↓↙↓↘→↘→→↗→↗↑↗↗↑↖↑↖)
$w_8(b) =$ (↓↓↙↓↓↘↓↓↗↑↑↗↑↑↗↑↗↑↗↗↗↓↙↓↙↙←→→→→→↘→↘↘↘↘↓↓↙↓←←←←←←←↙
←←↖←)
$w_9(B) =$ (↘↘↘↓↘↓↓↓↘↘↓↓↓↓↘↓↓↓↘↓↓↙↓↓↓↓↓↙↓↙)(↑↗↑↗↗↗↗↗→→→→↘→↓↘↓↓↓↓↙↙↓↙
↙↙↙↙↙↙←↙←↙←→↗→→→→→→→→↘↘↘→↘↘↓↘↘↘↓↙↓↙↓↓↙↓↙↓↙↙←↙
↙←↙←↙←←)
$w_9(o) =$ (↙↓↓↙↓↓↓↓↓↓↓↓↓↘↓↘→→→↗↗↗→↑↗↑↖↑↑↖↑↑↖↖↖↖↖↖↖↖↑↖↖↑)
$w_9(b) =$ (↙↓↙↓↓↓↓↓↓↓↓↘↓↓↓↓↓↓↓↓↓↓↓↓↓↓↓↓↓↓↓↘↓↓↓↓↓↓↓↑↑↑↑↗↑↑↗↑↑↗↑↗↗↑↗↗↗↗→→↘↘↓↘
↘↓↘↓↓↓↓↙↓↓↓↙↙↙↙←←↙
←↙←←←←↖↖↖)

Figure 9: GUI for *SDSS* extractor, Sample writings and their *SDSS*'s.

give a sketchy outline of the experiment. The full and detailed experimental report and its integration with multiple features can be found in [24, 23] and the complete procedure to assess the authorship confidence of arbitrary handwritten items can be found in [29]. *SDSS* is the most effective one among features that we consider in [24] and gives very low 4.9% *type I error* and 19.5% *type II error* where *type I error* occurs when the same author's handwritings are identified as different writers and *type II error* occurs when the two different writers' handwritings are identified as the same writer. When integrated with many other features, we achieve 97% overall correctness performance.

A problem that arises with the proposed method is that of human involvement during the feature extraction phase. It is an open problem to make the writing sequence extraction completely automatic. Notwithstanding, the proposed semi-automatic system is of great interest because it is less subjective. Interaction with questioned document is generally personal and subjective. The proposed measure can reduce the subjectivity significantly.

4.2 On-line Character Recognition

The problem of *on-line character recognition* can be formalized by defining a distance between characters and finding the nearest neighbor in the reference set. We obtain stroke directional sequence strings from the movements of a mouse or a pen-based device. To do so requires that we convert the mouse position vector that is a non-uniform length direction string into the *SDSS* that is a string of uniform length (7-pixel long) directions.

$$\begin{aligned}
\text{mouse position} &= \{(0,0),(0,14),(3,15),(5,19),(12,19)\} \\
\overline{(0,0),(0,14)} &= \rightarrow\ \rightarrow \qquad (11)\\
\overline{(0,14),(3,15)} + \overline{(3,15),(5,19)} &= \searrow \qquad (12)\\
\overline{(5,19),(12,19)} &= \downarrow \\
\text{SDSS} &= \rightarrow\ \rightarrow\ \searrow\ \downarrow
\end{aligned}$$

When the mouse moves fast, it creates a few of mouse positions and more strokes are filled in between these positions. Such case is depicted in equation (11). When the mouse moves slow, on the other hands, it creates many tiny strokes. These tiny strokes are merged into fewer 7-pixel long strokes as in Equation (12). In all, geometrically best fitting strokes are selected to create an *SDSS* from the mouse movement vector.

To recognize an unknown on-line character, one measures the edit distances between the input string and reference strings [25, 9, 10, 11]. Next, the class of an input string is determined by votes on k-nearest neighbors.

As a stroke sequence signifies the shape of the individual letters, a letter *"a"* is distinguished from a letter *"b"* by its different stroke sequences.

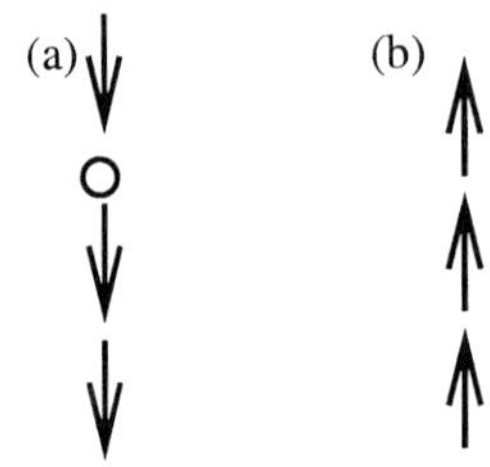

Figure 10: Sample Characters "1" (a) a break in the middle (b) written backward.

Many on-line character recognizers would not recognize a character if it were written in different order of drawing or there exists a break in the middle as shown in Figure 10. This problem can be diminished by string operations: *concat* and *reverse*. *Concat* operation will concatenate two strings whose ends points are near. *Reverse* operation will reverse the order of the elements as well as complement each string element. One or several breaks may occur in drawing a single line, curve or circle. This must be considered as a single string rather than multiple strings. To obviate this inadequacy, the *concat* operation is necessary. This break correction technique and various desirable invariance properties appear in the literature [30, 16, 17]. The *reverse* operation is also needed because people may write a letter in the reverse order.

In this experiment, 20 subjects provided each numeral five times in their own natural handwriting. When performing the nearest neighbor classification, each numeral is correctly classified without an error. When the unnatural handwriting is given, the classifier did not recognize many unnatural handwritings. Although the two simple string manipulation techniques diminish the effect of unnaturally written handwritings, much more complex string manipulation operations are required to handle unnaturally written handwriting.

4.3 Off-line Character/Digit Matching

The off-line character/digit recognizer takes a character or digit image as an input and classifies its class as shown in Figure 8 (c). Since it is difficult to extract the stroke directional sequence string in the off-line images automatically, we use a *contour directional sequence string* or *CDSS* in short, instead. A *CDSS* is defined using the idea of the chain-code. Like a chain-code, a contour sequence is a representation of boundaries of objects in the image. Note that a stroke is 7 pixel long as defined in Def. 2. Geometrically best fitting strokes are selected to replace pixel-based strokes in the chain-code. Figure 11 shows the contour sequence representation of a character "*A*". There are one outer contour sequence and one inner contour sequence

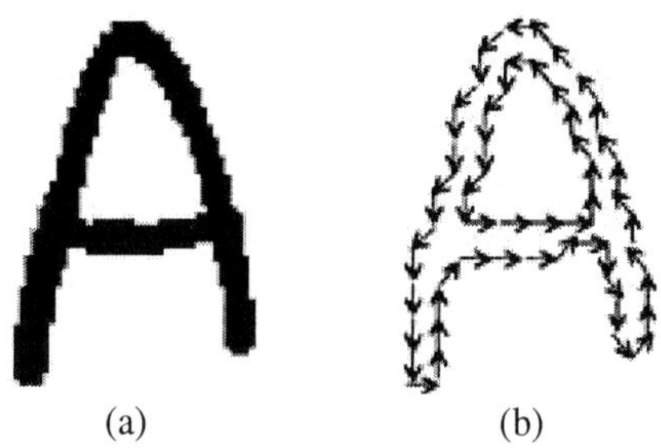

Figure 11: (a) original character image "A" (b) contour sequence representation.

stings.

We define the abstract data type of a contour sequence string as follows:

Definition 1 Contour sequence string
struct Contour_Seq_String {

cctype	:	*boolean* {*inner or outer*}
cclength	:	*integer*
cclist	:	*a list of strokes*
centroid	:	*a coordinate,* $(\bar{x}, \bar{y})$

}

A centroid is the center of a contour sequence:

$$(\bar{x}, \bar{y}) = \left(\frac{\sum_x \sum_y xB(x,y)}{\sum_x \sum_y B(x,y)}, \frac{\sum_x \sum_y yB(x,y)}{\sum_x \sum_y B(x,y)} \right).$$

$B(x, y) = 1$ if the pixel is labeled as an element of the segment and $B(x, y) = 0$ otherwise. Note that the centroid is computed after the character image is resized by adjusting the height to a fixed size. Both *cctype* and *centroid* are important because a character image has multiple contour sequence strings and the decision for selecting a corresponding contour sequence string must be made.

To extract contour sequence strings, several procedures are preceded such as noise removal, connected component analysis and chain-code generation. Although we expect the input character or digit image to be a single clean image, there are many undesired noises. Some broken or smeared character images occur due to the degradation of document images. Moreover, poor segmentation and improper background removal create certain noises. While the image restoration is beyond the scope of this chapter, we use simple unary and binary filters that handle salt and pepper noises. The *i-dots* in i and j are preserved.

Second, perform a connected component analysis and find the centroid of every connected component. Next, generate chain-codes by following component outer most boundaries. Finally, starting from the top of each chain-code, generate the contour sequence by fitting strokes to the chain-code. A stroke is 7 pixel long as defined in Def. 2. Geometrically best fitting strokes are selected to replace pixel-based strokes in the chain-code.

Next, we find the closest pair from two contour directional sequence strings by considering the inner or outer type, centroid and length of a string as criteria. In order to speed up the search time, only those reference characters with the same number of inner and outer type *CDSS*'s as the query character. We only need to compare the strings of the same type. When there are multiple *CDSS*'s of the same type, the distance between their centroids are computed to select the candidate pair.

After finding all corresponding contour sequence strings, compute each edit distance and accumulate all distances. If there is a string without a corresponding one, add the length of the string times the penalty value 2 to the total edit distance. Finally, select top 5 similar templates whose edit distances are smallest. The class of a query image is determined from a vote of these five templates.

We consider one of *CEDAR digit databases*, called *BR training set* which consists of 18,465 digit images. The database is divided into 18 different sets resulting in 1,000 instances each set. The first set is used as the reference or prototype images and the rest sets are used as test sets. Using the *k-NN* classifier, 17 experiments were performed and the average accuracy is about 96.08%. We achieve significant improvement over the original Levenshtein edit distance where the accuracy is about 94.02%. From the error analysis, we find an interesting consistency that the majority of errors are due to the broken characters resulting unexpected discontinuity in the *CDSS*. It is expected that provision on the broken characters will improve the performance significantly.

5 Conclusion

In this chapter, we categorized strings into four types: nominal, angular, magnitude and cost-matrix. We extended the *Levenshtein edit distance* to handle the angular and linear type strings. It is good for matching stroke and contour directional sequence strings. It takes *turn* and local context into account to compute the edit distance. This technique performs better than Levenshtein edit distance with cost matrix.

We also presented string distance measures to solve writer verification, on-line and off-line character recognition. To perform this, we converted a two-dimensional image to one-dimensional strings and then we measured the edit distance between strings. The smaller the edit distance is, the more similar they look like to each other. The fundamental idea underlying pattern recognition using the approximate string matching is based on the nearest neighbor classification. During training, we stored a full training set of strings and their associated category labels. During classification, a test string was compared to each stored string and the edit distance was computed. The test string was then assigned the category label of the nearest string in the training set.

We used two very important features, *SDSS* and *SPSS* to solve the *writer verification* problem. It is an unique approach to represent the *pressure* of handwriting as a sequence of pressure extracted from an off-line character image. In all, the proposed semi-automatic method provides a distinction or similarity between two handwritings figuratively and numerically. It is expected to help greatly document examiners and signature verifiers to compare handwritings and signatures.

Another major contribution in on-line character recognition is to diminish the effect of unnaturally written characters. Two string manipulation operators, concatenation and reverse, are used as a pre-processing to reduce the effect.

References

[1] R. C. Gonzalez and M. G. Thomason, *Syntactic Pattern Recognition: an Introduction.* Addison-Wesley, 1978.

[2] S. N. Srihari, *Computer Text Recognition and Error Correction.* IEEE Computer Society, 1984.

[3] G. A. Stephen, *String Searching Algorithms.* Singapore:World Scientific, 1994.

[4] R. O. Duda, P. E. Hart, and D. G. Stork, *Pattern Classification.* John Wiley & Sons, Inc., 2nd ed., 2000.

[5] R. A. Wagner and M. J. Fischer, "The string-to-string correction problem," *Journal of the ACM*, vol. 6, pp. 168–178, January 1974.

[6] P. A. V. Hall and G. R. Dowling, "Approximate string matching," *ACM Computing Surveys*, vol. 12, pp. 381–402, December 1980.

[7] P. Sellers, "The theory and computation of evolutionary distances: pattern recognition," *Journal of Algorithms*, vol. 1, pp. 359–373, 1980.

[8] A. Marzal and E. Vidal, "Computation of normalized edit distance and applications," *IEEE Transactions on Pattern Analysis and Machine Intelligence*, vol. 15, pp. 926–932, September 1993.

[9] M. Parizeau, N. Ghazzali, and J.-F. Hebert, "Optimizing the cost matrix for approximating string matching using genetic algorithms," *Pattern Recognition*, vol. 31, no. 4, pp. 431–440, 1998.

[10] S.-H. Cha, Y.-C. Shin, and S. N. Srihari, "Approximate character string matching algorithm," in *Proceedings of Fifth International Conference on Document Analysis and Recognition*, pp. 53–56, IEEE Computer Society, September 1999.

[11] S.-H. Cha, Y.-C. Shin, and S. N. Srihari, "Approximate string matching for stroke direction and pressure sequences," in *Proceedings of SPIE, Document Recognition and Retrieval VII*, vol. 3967, pp. 2–10, January 2000.

[12] Y. Fugimoto, S. Kadota, S. Hayashi, M. Yamamoto, S. Yajima, and M. Yasuda, "Recognition of handprinted characters by nonlinear elastic matching," in *The Third International Joint Conference on Pattern Recognition*, pp. 113–118, November 1976.

[13] M. L. Simmer, C. G. Leedham, and A.J.W.M-Thomassen, *Handwriting and Drawing Research: Basic and Applied Issues.* Amsterdam: IOS Press, 1996.

[14] R. R. Bradford and R. B. Bradford, *Introduction to Handwriting Examination and Identification.* Nelson-Hall Publishers: Chicago, 1992.

[15] R. Plamondon and G. Lorette, "Automatic signature verification and writer identification - the state of the art," *Pattern Recognition*, vol. 22, no. 2, pp. 107–131, 1989.

[16] R. Plamondon and S. N. Srihari, "On-line and off-line handwriting recognition: A comprehensive survey," *IEEE Transactions on Pattern Analysis and Machine Intelligence*, vol. 22, no. 1, pp. 63–84, 2000.

[17] F. Nouboud and R. Plamondon, "On-line recognition of handprinted characters. survey and beta tests," *Pattern Recognition*, vol. 23, no. 9, pp. 1031–1044, 1990.

[18] C. C. Tappert, C. Y. Suen, and T. Wakahara, "The state of the art in on-line handwriting recognition," *IEEE Transactions on Pattern Analysis and Machine Intelligence*, vol. 12, no. 8, pp. 787–808, 1990.

[19] D. Harel and R. Tarjan, "Fast algorithms for finding nearest common ancestors," *SIAM Journal on Computing*, vol. 13, pp. 338–355, 1984.

[20] G. M. Landau and U. Vishkin, "Introducing efficient parallelism into approximate string matching and a new serial algorithm," in *Proceedings of the 18th Annual ACM Symposium on Theory of Computing*, pp. 220–230, 1986.

[21] P. Weiner, "Linear pattern matching algorithm," in *Proceedings 14th IEEE Symposium on Switching and Automata Theory*, 1973.

[22] M. Farach, "Optimal suffix tree construction with large alphabets," in *38th Annual Symposium on Foundations of Computer Science*, (Miami Beach, Florida), pp. 137–143, IEEE, Oct. 1997.

[23] S.-H. Cha and S. N. Srihari, "Writer identification: Statistical analysis and dichotomizer," in *Proceedings of SS&SPR 2000 LNCS - Advances in Pattern Recognition*, vol. 1876, pp. 123–132, Springer-Verlag, September 2000.

[24] S.-H. Cha and S. N. Srihari, "Multiple feature integration for writer verification," in *Proceedings of 7th IWFHR 2000*, pp. 333–342, September 2000.

[25] R. Plamondon and F. Nouboud, "On-line character recognition system using string comparison processor," in *Proceedings of International Conference on Pattern Recognition*, pp. 460–463, IEEE, June 1990.

[26] C. Y. Suen, M. Berthod, and S. Mori, "Automatic recognition of hand-printed characters - the state of the art," in *Proceedings of the IEEE*, vol. 68, pp. 469–487, April 1980.

[27] C. Y. Suen and P. S. P. Wang, *Thinning Methodologies for Pattern Recognition.* World Scientific, 1994.

[28] S. N. Srihari, S.-H. Cha, H. Arora, and S. Lee, "Individuality of handwriting," *Journal of Forensic Sciences*, vol. 47, June 2002.

[29] S.-H. Cha and S. N. Srihari, "Assessing the authorship confidence of handwritten items," in *Proceedings of WACV 2000*, pp. 42–47, IEEE Computer Society, December 2000.

[30] P. J. Ye, H. Hugli, and F. Pellandini, "Techniques for on-line chinese character recognition with reduced writing constraints," in *Proceedings of 7th ICPR*, pp. 1043–1045, IEEE CS Press, 1984.

Uniform, Fast Convergence of Arbitrarily Tight Upper and Lower Bounds on the Bayes Error

Dechang Chen
Department of Natural and Applied Sciences
University of Wisconsin, Green Bay, WI 54311
E-mail: `chend@uwgb.edu`

Michael A Fries
School of Computer Science, Telecommunications and Information Systems
DePaul University, Chicago, IL 60604
E-mail: `mfries@cti.depaul.edu`

YingChang Xiang
Rizhao Vocational & Technical College
Yantai Road
Rizhao, Shandong 276826
P. R. China

Contents

D. Chen and X. Cheng (eds.),
Pattern Recognition and String Matching, 131–134.

1 Introduction

This chapter studies the convergence behavior of the arbitrarily tight upper and lower bounds on the Bayes error proposed by Avi-Itzhak and Diep [1]. We show that these bounds converge to the Bayes error uniformly with a very fast convergence rate.

2 Bayes Error

Consider the pattern recognition problem of two classes C_1 and C_2. Suppose for class C_i the prior probability is π_i and the probability density function is $f_i(x)$. It is known that in classifying C_1 and C_2, the Bayes error, denoted by $P(e)$, is

$$P(e) = \int_x \min(p, 1-p) f(x) dx, \tag{1}$$

where

$$p = \text{Prob}(C_1|x), \quad f(x) = \pi_1 f_1(x) + \pi_2 f_2(x).$$

Due to the term $\min(p, 1-p)$, theoretical evaluation of $P(e)$ may become impossible. Thus it is interesting to find elementary functions bounding $\min(p, 1-p)$, from which the corresponding bounds on $P(e)$ will be derived. Avi-Itzhak and Diep [1] introduced the following upper and lower bounds for $P(e)$:

$$\int_x L_\alpha(p) f(x) dx \leq P(e) \leq \int_x U_\alpha(p) f(x) dx, \tag{2}$$

where for any $\alpha > 0$ and $p \in [0, 1]$,

$$L_\alpha(p) = \frac{1}{\alpha} \ln \left[\frac{1 + e^{-\alpha}}{e^{-\alpha p} + e^{-\alpha(1-p)}} \right] \leq \min(p, 1-p) \tag{3}$$

and

$$U_\alpha(p) = L_\alpha(p) + \left[1 - 2L_\alpha\left(\frac{1}{2}\right) \right] C(p) \geq \min(p, 1-p). \tag{4}$$

In (4), $C(p)$ is any bounded function satisfying

$$\begin{cases} C(p) \geq \min(p, 1-p) \text{ for all } p \in [0, 1], \\ C(p) = C(1-p) \text{ for all } p \in [0, 1], \\ C(0) = C(1) = 0, \ C\left(\frac{1}{2}\right) = \frac{1}{2}\,. \end{cases} \tag{5}$$

Examples of function $C(p)$ include the Bhattacharyya bound $\sqrt{p(1-p)}$ ([5]), the equivocation bound $\frac{1}{2}[-p\log_2(p) - (1-p)\log_2(1-p)]$ ([4]), the Bayesian distance bound $2p(1-p)$ ([2]), and the Gaussian-Sinusoidal bound $\frac{1}{2}\sin(\pi p)\exp\left[\alpha\left(p-\frac{1}{2}\right)^2\right]$ ([3]). These bounds were introduced as the upper bounds of $\min(p, 1-p)$.

Avi-Itzhak and Diep [1] claimed that $\int_x L_\alpha(p)f(x)dx$ and $\int_x U_\alpha(p)f(x)dx$ can be arbitrarily tight by showing $\lim_{\alpha\to\infty} L_\alpha(p) = \min(p, 1-p)$ and $\lim_{\alpha\to\infty} U_\alpha(p) = \min(p, 1-p)$. In this paper, we provide a much stronger result in support of their claim. We show the following inequalities hold for any $\alpha > 0$.

$$0 \le P(e) - \int_x L_\alpha(p)f(x)dx \le \frac{2M\ln 2}{\alpha}, \tag{6}$$

$$0 \le \int_x L_\alpha(p)f(x)dx - P(e) \le \frac{2M\ln 2}{\alpha}, \tag{7}$$

where $M = \sup_{p\in[0,1]} C(p)$. Therefore as $\alpha \to \infty$ both $\int_x L_\alpha(p)f(x)dx$ and $\int_x U_\alpha(p)f(x)dx$ converge to $P(e)$ uniformly with a convergence rate $O(\frac{1}{\alpha})$. This fact implies that these bounds will become arbitrarily close to the Bayes error very quickly, as α increases.

3 Proof of (6) and (7)

By the definition of $L_\alpha(p)$ and $U_\alpha(p)$, we have

$$\begin{aligned}
U_\alpha(p) - L_\alpha(p) &= \left[1 - 2L_\alpha\left(\frac{1}{2}\right)\right]C(p) \\
&\le \left[1 - \frac{2}{\alpha}\ln\left(\frac{1+e^{-\alpha}}{2e^{-\alpha/2}}\right)\right]M \\
&= \left[1 - \frac{2}{\alpha}\left(\frac{\alpha}{2} + \ln\left(\frac{1+e^{-\alpha}}{2}\right)\right)\right]M \\
&= \frac{2M}{\alpha}\ln\left(\frac{2}{1+e^{-\alpha}}\right) \\
&\le \frac{2M\ln 2}{\alpha}.
\end{aligned}$$

Using the above estimate and the fact that $L_\alpha(p) \le \min(p, 1-p) \le U_\alpha(p)$, one has

$$0 \le \min(p, 1-p)f(x) - L_\alpha(p)f(x) \le \frac{2M\ln 2}{\alpha}f(x),$$

and

$$0 \leq U_\alpha(p)f(x) - \min(p, 1-p)f(x) \leq \frac{2M \ln 2}{\alpha} f(x).$$

Integrating with respect to x yields

$$0 \leq \int_x \min(p, 1-p)f(x)dx - \int_x L_\alpha(p)f(x)dx \leq \int_x \frac{2M \ln 2}{\alpha} f(x)dx. \quad (8)$$

and

$$0 \leq \int_x U_\alpha(p)f(x)dx - \int_x \min(p, 1-p)f(x)dx \leq \int_x \frac{2M \ln 2}{\alpha} f(x)dx. \quad (9)$$

Now (6) and (7) follow from the equations (1), (8) and (9) and the fact that $\int_x f(x)dx = 1$.

References

[1] H. Avi-Itzhak and T. Diep, Arbitrarily tight upper and lower bounds on the Bayesian probability of error, *IEEE Trans. Pattern Analysis and Machine Intelligence*, Vol. 18 No. 1 (Jan. 1996) pp. 89-91.

[2] P. A. Devijver, On a new class of bounds on Bayes risk in multihypothesis pattern recognition, *IEEE Trans. Computers* Vol. 23 (Jan. 1974) pp. 70-80.

[3] W. A. Hashlamoun, P. K. Varshney, and V. N. S. Samarasooriya, A tight upper bound on the Bayesian probability of error, *IEEE Trans. Pattern Analysis and Machine Intelligence*, Vol. 16, (Feb. 1994) pp. 220-224.

[4] M. E. Hellman and J. Raviv, Probability of error, equivocation, and Chernoff bound, *IEEE Trans. Information Theory* Vol. 16 (July 1970) pp. 368-372.

[5] T. Kailath, The divergence and Bhattacharyya distance measures in signal selection, *IEEE Trans. Communication Technology* Vol. 15 (Feb. 1967) pp. 52-60.

Building RBF Networks for Time Series Classification by Boosting[1]

Juan José Rodríguez Diez
Lenguajes y Sistemas Informáticos
Universidad de Burgos, Spain
E-mail: `jjrodriguez@ubu.es`

Carlos J. Alonso González
Departamento de Informática, Grupo de Sistemas Inteligentes
Universidad de Valladolid, Spain
E-mail: `calonso@uva.es`

Contents

[1]This work has been supported by the Spanish CYCIT project TAP 99–0344 and the "Junta de Castilla y León" project VA101/01

D. Chen and X. Cheng (eds.),
Pattern Recognition and String Matching, 135–153.

1 Introduction

This work presents a learning system for the classification of multivariate time series. This classification is useful in domains such as biomedical signals [9], continuous systems diagnosis [2] or data mining in temporal databases [3].

The learning system constructs a Radial Basis Function network, using the technique described in [15]. One of the features of this method for constructing RBF networks is that it can work with any dissimilarity function, therefore it can be used with dynamic time warping, DTW.

The RBF networks are constructed by boosting (a method for the generation of ensembles of classifiers) [19]. The task assigned to the base learner is to select a RBF, while the boosting algorithm combines linearly the different RBFs. For each iteration of boosting a new neuron is incorporated into the network.

The method for selecting each RBF is based on randomly selecting several examples as the centers, considering the distances to these centers as attributes of the examples and selecting the best split on one of these attributes. This selection of the best split is done in the same way than in the construction of decision trees. The center of the RBF is given by the attribute selected and its width depends on the threshold.

This work is a follow-up of our research in boosting similarity literals for time series classification [14]. In that work, following the good results of boosting very simple classifiers (i.e., *stumps*) for several data sets [5], we proposed to use similarity literals as base classifiers. The format of these literals were:

[not] <*distance*>_le(Example, Reference, Variable, Value)

which is true if the <*distance*> between the considered Example and the Reference example, for the chosen Variable, is less or equal (_le) than Value. The argument Variable is necessary because multivariate data sets are considered.

Each base classifier had only one literal, and their result was simply true or false. One of the improvements to the original ADABOOST algorithm is the use of confidence based predictions [20], where each base classifier also returns, for each example, a confidence (a real number) on its prediction.

A natural question is how to combine confidence based predictions with similarity literals. The first option was to consider a given literal as a boolean attribute and, using the methods of [20] for domain-partitioning weak classifiers, to assign a confidence value corresponding to the values true and false of the literal. Nevertheless, when using distance literals it is natural to use, somehow, the value of the distance from the current example to the reference example, and the threshold value to obtain a confidence. In fact, an obvious option is a radial basis function. On the other hand, the result of AdaBoost is a linear combination of the predictions of the base classifiers, and if the base classifiers are RBFs, then the result of boosting is a RBF Network [11, 12].

This work is also related with the methods for constructing RBFNs from decision trees [10]. These works share the idea of constructing the network from a symbolic machine learning method. For instance, the selection of each RBF is based on the one for the split of a node in a decision tree. A difference with that methods is that in the case of the decision trees there are two steps: first, constructing the symbolic classifier and second, "upgrading" it to a network. In our case, although there are also two parts, the base learner and the proper boosting method, they work in cooperation.

The rest of the paper is organized as follows. The specific details of the boosting variant used are described in section 2. Section 3 explains the operation of the base learner. A classifier is shown, as an example, in section 4. Section 5 presents experimental results when using this method. Finally, we give some concluding remarks in section 6.

2 Boosting

At present, an active research topic is the use of *ensembles* of classifiers. They are obtained by generating and combining base classifiers, constructed using other machine learning methods. The target of these ensembles is to increase the accuracy with respect to the base classifiers.

One of the most popular methods for creating ensembles is boosting [19], a family of methods, of which AdaBoost is the most prominent member. Figure 1 shows a generalized version of AdaBoost [20]. Each *instance* x_i belongs to a domain $\mathcal{X}$ and has an associated label. For binary problems, $+1$ or -1.

Every example has an associated weight. Initially, all the examples have the same weight. In each iteration a *base* (also named *weak*) classifier is constructed, according to the distribution of weights. Afterwards, the weight of each example is readjusted, based on the correctness of the class assigned to the example by the base classifier. The final result is obtained by weighted votes of the base classifiers.

Given $(x_1, y_1), \ldots, (x_m, y_m)$ where $x_i \in \mathcal{X}, y_i \in \{-1, +1\}$
Initialize $D_1(i) = 1/m$
For $t = 1, \ldots, T$:

- Train weak learner using distribution D_t
- Get weak hypothesis $h_t : \mathcal{X} \to \mathbf{R}$
- Choose $\alpha_t \in R$
- Update
$$D_{t+1}(i) = \frac{D_t(i)\exp(-\alpha_t y_i h_t(x_i))}{Z_t}$$
where Z_t is a normalization factor (chosen so that D_{t+1} will be a distribution).

Output the final hypothesis:

$$H(x) = \operatorname{sign}\left(\sum_{t=1}^{T} \alpha_t h_t(x)\right)$$

Figure 1: AdaBoost [20]

2.1 Selecting α

In [20] several methods are proposed for selecting the weight (α) associated to each base classifier. For two class problems, the better value for α is obtained by minimizing

$$Z = \sum_i D(i)e^{-\alpha u_i}$$

Where $D(i)$ is the weight of the example x_i, $u_i = y_i h(x_i)$, $y_i \in \{-1, +1\}$ is the class of the example and $h(x_i)$ is the confidence given by the base classifier to the example. For the original AdaBoost this expression is approximated, given $u_i \in [-1, +1]$, by

$$Z \leq \sum_i D(i) \left(\frac{1+u_i}{2} e^{-\alpha} + \frac{1-u_i}{2} e^{\alpha} \right)$$

And the minimum α for this expression can be calculated analytically:

$$\alpha = \frac{1}{2} \ln \left(\frac{1+r}{1-r} \right)$$

Where $r = \sum_i D(i) u_i$.

Nevertheless, they suggest that it is possible to select other upper bounds, and we use

$$Z \leq \sum_{i:u_i \geq 0} D(i)(u_i e^{-\alpha} - u_i + 1) + \sum_{i:u_i < 0} D(i)(-u_i e^{+\alpha} + u_i + 1)$$

which gives a tighter approximation. For this bound, the calculated value of α is

$$\alpha = \frac{1}{2} \ln \left(-\frac{W_+}{W_-} \right)$$

where $W_+ = \sum_{i:u_i \geq 0} D(i) u_i$, $W_- = \sum_{i:u_i < 0} D(i) u_i$, and $W_+ + W_- = r$.

Note that the two ways of calculating α are equivalent if $h(x)$ is restricted to $\{-1, +1\}$ because for this case $W_+ - W_- = 1$.

2.2 Multiclass Problems

AdaBoost is only for binary problems, but there are several methods of extending AdaBoost to the multiclass case. In previous works [15, 13] we had used AdaBoost.OC [18], because it can be used with binary base classifiers, e.g., one literal, even for multiclass problems. One of the possible limitations of AdaBoost.OC is that each base learner, assigns one weight for each class, but this weight has only two possible values. Intuitively, it seems better to have a different value for each class. Then, for this work, we used AdaBoost.MH.

Figure 1 shows AdaBoost.MH[2] [20]. AdaBoost.MH associates a weight to each combination of examples and labels. The base learner generates a base hypothesis h_t, according to the weights. A real value, α_t, the weight of the base classifier, is selected. Then, the weights are readjusted. For $y, l \in \mathcal{Y}$, $y[l]$ is defined as

[2] Although AdaBoost.MH also considers multilabel problems (an example can be simultaneously of several classes), the version presented here does not include this case

$$y[l] = \begin{cases} +1 \text{ if } l = y \\ -1 \text{ if } l \neq y \end{cases}$$

Given $(x_1, y_1), \ldots, (x_m, y_m)$ where $x_i \in \mathcal{X}, y_i \in \mathcal{Y}$
Initialize $D_1(i, l) = 1/(mk)$
For $t = 1, \ldots, T$:

- Train weak learner using distribution D_t
- Get weak hypothesis $h_t : \mathcal{X} \times \mathcal{Y} \rightarrow \mathbf{R}$
- Choose $\alpha_t \in R$
- Update

$$D_{t+1}(i, l) = \frac{D_t(i, l) \exp(-\alpha_t y_i[l] h_t(x_i, l))}{Z_t}$$

where Z_t is a normalization factor (chosen so that D_{t+1} will be a distribution).

Output the final hypothesis:

$$H(x) = \arg\max_y \sum_{t=1}^{T} \alpha_t h_t(x, y)$$

Figure 2: AdaBoost.MH [20]

AdaBoost.MH uses a multiclass base learner, and we want to use binary learners (only one literal). Then, for each iteration, we train the weak learner using a binary problem: one class (selected randomly) against the others. The output of this weak learner, $h_t^B(x)$ is binary. On the other hand, we do not generate a unique α_t, but for each class, l, an α_{tl} is selected. They are selected considering how good is the weak learner for discriminating between the class l and the rest of classes. This is a binary problem so the selection of the values can be done as indicated in [20]. Now, we can define $h_t(x, l) = \alpha_{tl} h_t^B(x)$ and $\alpha_t = 1$ and we use AdaBoost.MH.

3 Base Classifiers

3.1 Dynamic Time Warping

Dynamic Time Warping (DTW) aligns a time series to another reference series in a way that a distance function is minimized, using a dynamic programming algorithm [3]. If the two series have n points, the execution time is $O(n^2)$. The predicate dtw_le uses the minimized value obtained from the DTW as a similarity function between the two series.

3.2 Literals Selection

The base learner works as follows. First, several examples are selected, randomly, as possible references. The number of reference examples considered (r) is a parameter, and even could be possible to use different number of positive and negative examples.

For each reference example, the distance to all the other training examples (e) is computed. The time necessary for this process is $e\,t(n)$, where $t(n)$ is the time necessary for calculating the distances between two examples with n attributes. In the case of the euclidean distance, $t(n) \in O(n)$.

Then the best threshold for the distances is computed in a similar way as done with decision trees. First, all the distances are sorted (time $e \lg e$). All the values are considered, from left to right, keeping into account the number and weight of positive and negative examples at the left from the current value. For each distance value it is computed the error of selecting this threshold. This can be done, for each value, in $O(1)$ because it only involve to calculate a function of the weight of positive and negative examples at the left and at the right of the threshold. For e distances the time necessary is $O(e)$, which is smaller than $O(e \lg e\ )$. Hence, the base learner requires a time of $re(t(n) + \lg e)$.

3.3 Assigning Confidences

Given a literal <*distance*>_le(Example, Reference, Variable, Value), the RBF selected is

$$h(x) = 2 \exp\left(- \left(\frac{d_V(x,c)}{t} \right)^2 \ln 2 \right) - 1$$

where x is the Example, c the Reference example, t the threshold Value and d_V the <*distance*> for the Variable V.

This function has the following properties:

- $h(c) = 1$
- $h(x) = 0$ if $d_V(x, c) = t$
- $-1 \leq h(x) \leq 1$, given $d_V(x, c) \geq 0$
- This function monotonically decreases as $d_V(x, c)$ increases

If the literal is negated, then the function is multiplied by -1. These functions are radial basis functions [11], and a linear combination of them is a RBF Network.

4 Classifier Example

Table 1 shows a classifier. It was obtained from the data set CBF (section 5.1). The shown classifier was obtained without transforming the literals in RBF. This classifier is composed by 10 base classifiers. The first column shows the class that was selected randomly. The task of the base learner is to obtain a good literal for discriminating between this class and the others. The second column shows the literal. Normally, for positive literals, the selected reference example is an example of the class. For instance, the class for the first base classifier is "Funnel", and the reference example is the number 167 of this class. For each class in the data set there is another column, with the weight associated to the literal for that class. Normally, the greatest weight is for the class that the weak learner must discriminate.

In order to classify a new example, a weight is calculated for each class, and then the example is assigned to the class with greater weight. In the classifier, there is a weight for each base classifier and literal. For each class, its weight is the sum of the weights of the true literals for the example minus the sum of the weights of the false literals.

5 Experimental Validation

The characteristics of the data sets are summarized in table 2. If a training test partition it specified, the results are obtained by averaging 5 experiments. In other case, 10 fold stratified cross validation was used.

The experimental settings were the following:

- For all the data sets, boosting DTW literals.
- For fixed length data sets (*CBF* and *Control*), boosting Euclidean literals.

Class	Literal	Cylinder	Bell	Funnel
Funnel	dtw_le(E, f_167, x, 1.480433)	-0.474296	-0.580342	2.007847
Funnel	not dtw_le(E, b_38, x, 2.350229)	0.118574	-0.482637	1.133678
Bell	dtw_le(E, b_193, x, 3.130481)	-0.420418	1.118650	0.203333
Funnel	dtw_le(E, f_258, x, 0.962927)	-0.099700	-0.040386	1.685912
Bell	not dtw_le(E, f_70, x, 3.987856)	-0.339226	1.485537	-0.569564
Cylinder	dtw_le(E, c_61, x, 1.637973)	1.136820	-0.050223	-0.443976
Funnel	dtw_le(E, f_248, x, 1.137615)	-0.639028	0.289889	1.080814
Funnel	not dtw_le(E, c_67, x, 1.519279)	-0.210730	-0.270318	0.599596
Funnel	dtw_le(E, f_204, x, 1.198512)	-0.111483	-0.089113	0.770602
Cylinder	dtw_le(E, c_171, x, 2.753013)	0.734230	-0.616611	-0.232045

Table 1: Classifier example, for the CBF data set (section 5.1). The first column shows the class that the weak learner must discriminate from the other classes. The second column shows the literal selected by the weak learner. The last three columns show, for each literal, the weight associated to every class.

	Classes	Variables	Examples	Training / Test	Length Min	Length Average	Length Max
CBF	3	1	798	10 fold CV	128	128.00	128
CBF-Var	3	1	798	10 fold CV	64	95.75	128
Control	6	1	600	10 fold CV	60	60.00	60
Control-Var	6	1	600	10 fold CV	30	44.86	60
Trace	16	4	1600	800 / 800	67	82.49	99
Auslan	10	8	200	10 fold CV	33	49.34	102

Table 2: Characteristics of the data sets.

- For each of the above settings, two variants were considered: *discrete* and *real*. The first one corresponds to using directly the binary output of the literals, and the second one corresponds to using the RBF obtained from the literals.

The obtained results are shown in figure 3 and in table 3. The default maximum number of iterations (the number of literals) is 100, but the figure and the table also show the results with less iterations. The data sets *Control* and *Trace* are more complex (they have more classes, are multivariate) so 300 iterations were used. This results show that for these data set there arre not overfitting problems. For the data sets where the Euclidean distance could be used, it is clear that the results for DTW are much better. The real variant is not always the best, but

- It is the more adequate for the two most difficult data sets, *Control* and *Trace*.

- For the data sets *CBF* and *Control*, the real variant worsens the results for the Euclidean distance and improve the results for DTW. Since Euclidean gives worst results than DTW, the effect of using RBF is more important in the second case.

The rest of this section contains a detailed discussion for each data set, including its description, the obtained results and another known results for the data set.

5.1 CBF (Cylinder, Bell and Funnel)

This is an artificial problem, introduced in [17]. The learning task is to distinguish between three classes: cylinder (c), bell (b) or funnel (f). Examples are generated using the following functions:

$$\begin{aligned}
c(t) &= (6+\eta)\cdot\chi_{[a,b]}(t)+\epsilon(t)\\
b(t) &= (6+\eta)\cdot\chi_{[a,b]}(t)\cdot(t-a)/(b-a)+\epsilon(t)\\
f(t) &= (6+\eta)\cdot\chi_{[a,b]}(t)\cdot(b-t)/(b-a)+\epsilon(t)
\end{aligned}$$

where

$$\chi_{[a,b]}(t) = \begin{cases} 0 & \text{if} \quad t < a \lor t > b \\ 1 & \text{if} \quad a \le t \le b \end{cases}$$

and η and $\epsilon(t)$ are obtained from a standard normal distribution $N(0,1)$, a is an integer obtained from a uniform distribution in $[16,32]$ and $b-a$ is another integer obtained from another uniform distribution in $[32,96]$. The examples are generated evaluating those functions for $t = 1, 2 \ldots 128$. Figure 4 shows some examples of this data set.

The error reported in [8] is 1.9, using event extraction, event clustering and decision trees. Our results using DTW and RBF using 20 or more iterations are smaller than 1.9, and even the error is 0.0 when using 100 literals. On the other hand, the results using the Euclidean distance are clearly worst than 1.9.

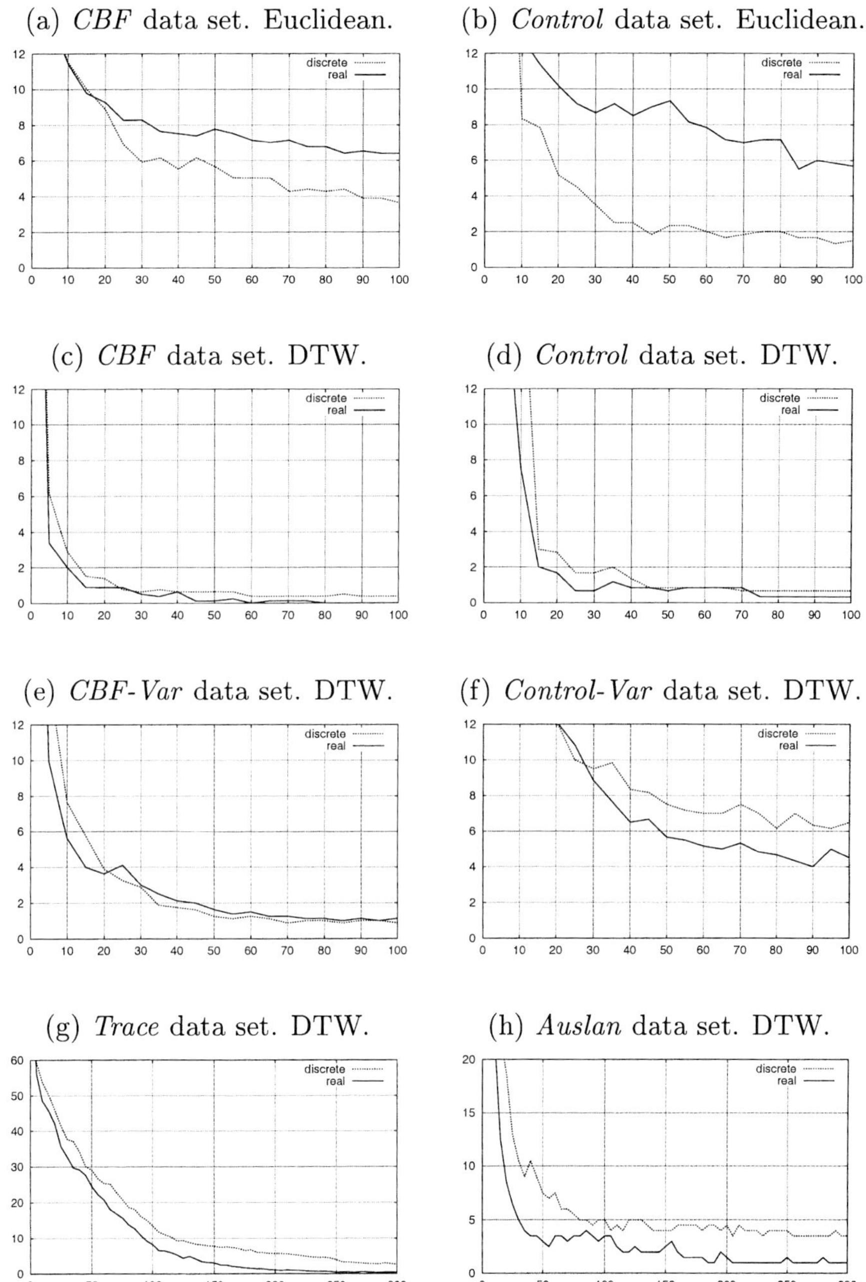

Figure 3: Error graphs for the different data sets.

(a) *CBF* data set. Euclidean.

Literals	10	20	30	40	50	60	70	80	90	100
Discrete	11.53	8.90	5.91	5.53	5.66	5.03	4.27	4.27	3.88	3.64
Real	11.42	9.28	8.29	7.53	7.78	7.15	7.16	6.78	6.53	6.40

(b) *CBF* data set. DTW.

Literals	10	20	30	40	50	60	70	80	90	100
Discrete	2.88	1.39	0.63	0.64	0.63	0.38	0.38	0.38	0.38	0.38
Real	2.00	0.88	0.50	0.63	0.12	0.00	0.13	0.00	0.00	0.00

(c) *CBF-Var* data set. DTW.

Literals	10	20	30	40	50	60	70	80	90	100
Discrete	7.65	3.90	2.88	1.75	1.25	1.25	0.87	1.00	1.00	0.87
Real	5.62	3.63	3.01	2.13	1.63	1.50	1.25	1.13	1.13	1.13

(d) *Control* data set. Euclidean.

Literals	10	20	30	40	50	60	70	80	90	100
Discrete	8.33	5.17	3.50	2.50	2.33	2.00	1.83	2.00	1.67	1.50
Real	12.83	10.17	8.67	8.50	9.33	7.83	7.00	7.17	6.00	5.67

(e) *Control* data set. DTW.

Literals	10	20	30	40	50	60	70	80	90	100
Discrete	19.67	2.83	1.67	1.33	0.83	0.83	0.67	0.67	0.67	0.67
Real	7.50	1.67	0.67	0.83	0.67	0.83	0.83	0.33	0.33	0.33

(f) *Control-Var* data set. DTW.

Literals	10	20	30	40	50	60	70	80	90	100
Discrete	24.83	12.17	9.50	8.33	7.50	7.00	7.50	6.17	6.33	6.50
Real	17.50	12.17	8.83	6.50	5.67	5.17	5.33	4.67	4.00	4.50

(g) *Trace* data set. DTW.

Literals	30	60	90	120	150	180	210	240	270	300
Discrete	37.63	25.35	16.18	9.30	7.85	6.68	5.55	4.60	2.95	2.58
Real	32.68	20.70	10.85	5.25	3.05	1.53	1.03	0.68	0.50	0.33

(h) *Auslan* data set. DTW.

Literals	30	60	90	120	150	180	210	240	270	300
Discrete	10.50	7.50	4.50	5.00	4.00	4.00	4.50	4.00	3.50	3.50
Real	5.00	3.50	3.50	2.00	2.50	1.50	1.00	1.00	1.00	1.00

Table 3: Results for the different data sets.

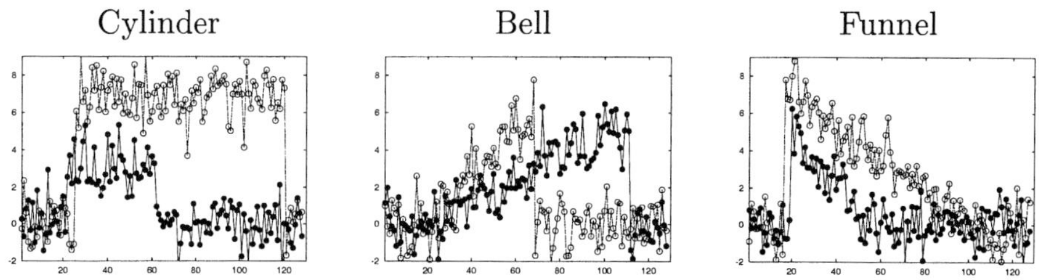

Figure 4: Examples of the *CBF* data set. Two examples of the same class are shown in each graph.

Variable Length Version All the examples of the *CBF* data set have the same length. In order to check the method for variable length series, the data set was altered. For each example, a random number of values were deleted from its end. The maximum number of values eliminated is 64, the half of the original length.

An interesting issue is that the obtained results with this altered data set are still better than the results reported in [8] using the original data set.

5.2 Control Charts

In this data set there are six different classes of control charts, synthetically generated by the process in [1]. Each time series is of length $n = 60$, and it is defined by $y(t)$, with $1 \leq t \leq n$:

1. Normal: $y(t) = m + rs$. Where $m = 30$, $s = 2$ and r is a random number in $[-3, 3]$.

2. Cyclic: $y(t) = m + rs + a\sin(2\pi t/T)$. a and T are in $[10, 15]$.

3. Increasing: $y(t) = m + rs + gt$. g is in $[0.2, 0.5]$.

4. Decreasing: $y(t) = m + rs - gt$.

5. Upward: $y(t) = m + rs + kx$. x is in $[7.5, 20]$ and $k = 0$ before time t_3 and 1 after this time. t_3 is in $[n/3, 2n/3]$.

6. Downward: $y(t) = m + rs - kx$.

Figure 5 shows two examples of each class. The used data was obtained from the UCI KDD Archive [7].

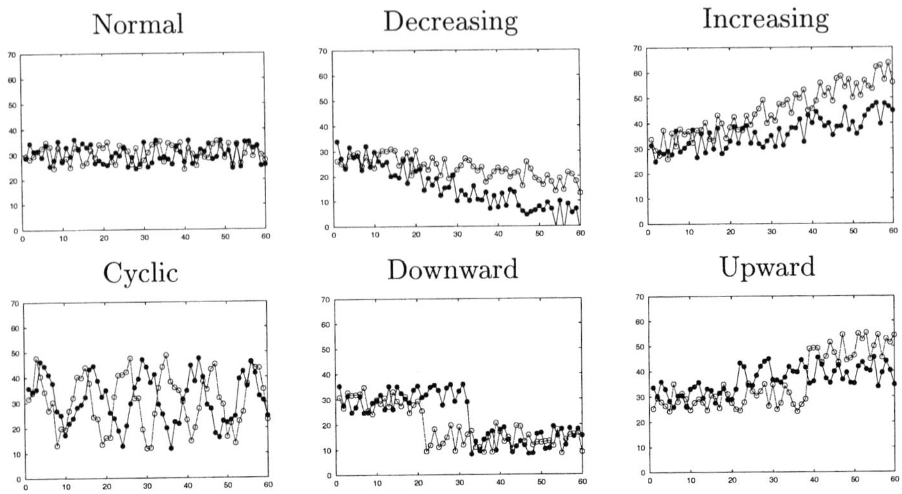

Figure 5: Examples of the *Control* data set. Two examples of the same class are shown in each graph.

Variable Length Version The *Control* data set was also altered to a variable length variant, *Control-Var*. In this case the resulting series have lengths between 30 and 60.

5.3 Trace

This dataset is introduced in [16]. It is proposed as a benchmark for classification systems of temporal patterns in the process industry. This data set was generated artificially. There are four variables, and each variable has two behaviors, as shown in figure 6. The combination of the behaviors of the variables produces 16 different classes. For this data set it is specified a training / test partition.

The length of the series is between 268 and 394. The running times of the algorithms depend on the length of the series, so the data set was preprocessed, averaging the values in groups of four. The lengths of the examples in the resulting data set have lengths between 67 and 99.

The result reported in [16], using recurrent neural networks and wavelets is an error of 1.4%, but 4.5% of the examples were not assigned to any class. The result for the discrete variant, 2.58, is worst than 1.4, but it classifies all the examples. The result for the real variant, 0.33, is better than 1.4. Moreover, this result can be improved using more iterations: for

500 iterations, the error is only 0.05.

5.4 Auslan

Auslan is the Australian sign language, the language of the Australian deaf community. Instances of the signs were collected using an instrumented glove [8]. Each example is composed by 8 series: x, y and z position, wrist roll, thumb, fore, middle and ring finger bend. There are 10 classes and 20 examples of each class. The minimum length is 33 and the maximum is 102.

The best result reported in [8] is an error of 2.50, using event extraction, event clustering and Naïve Bayes Classifiers. Our results for the discrete variant are worst than 2.5, but on the other hand, the results are better when using the real variant.

6 Conclusions

A time series classification system has been presented. It is based on boosting very simple classifiers. Boosting is a general framework for improving the results of any classification method, by the generation of an ensemble of classifiers. The base classifiers are not useful by themselves, but their use in boosting produces good classifiers. These base classifiers compute a distance between the series to classify with another reference series, and this value is compared with a threshold. The good results obtained are due to the use of an adequate function for measuring the similarity of two series, DTW.

The results are improved by the use of confidence rated predictions. Instead of using directly the binary output of the literals, a RBF is obtained from each literal, and a variant of boosting that uses confidence rated predictions it is used, in this case RBF. The obtained ensemble of classifiers is a linear combination of RBF, so it is a RBF network. Nevertheless, it is a rather unconventional network, because the use of DTW.

The method is nearly parameter-free. The only clear parameter is the number of iterations. Nevertheless, an interesting fact is that the classifiers (networks) obtained with a number of iterations are included (are subnetworks) in the ones obtained with more iterations. Hence, it is possible i) to select only an initial fragment of the obtained network and ii) to continue adding base classifiers (neurons) to a previously obtained classifier.

The obtained results for the data sets are clearly very competitive with the known previous results for these data sets.

The research on boosting has focused on classification, and this work follows this trend. On the other hand, RBFNs are frequently used for re-

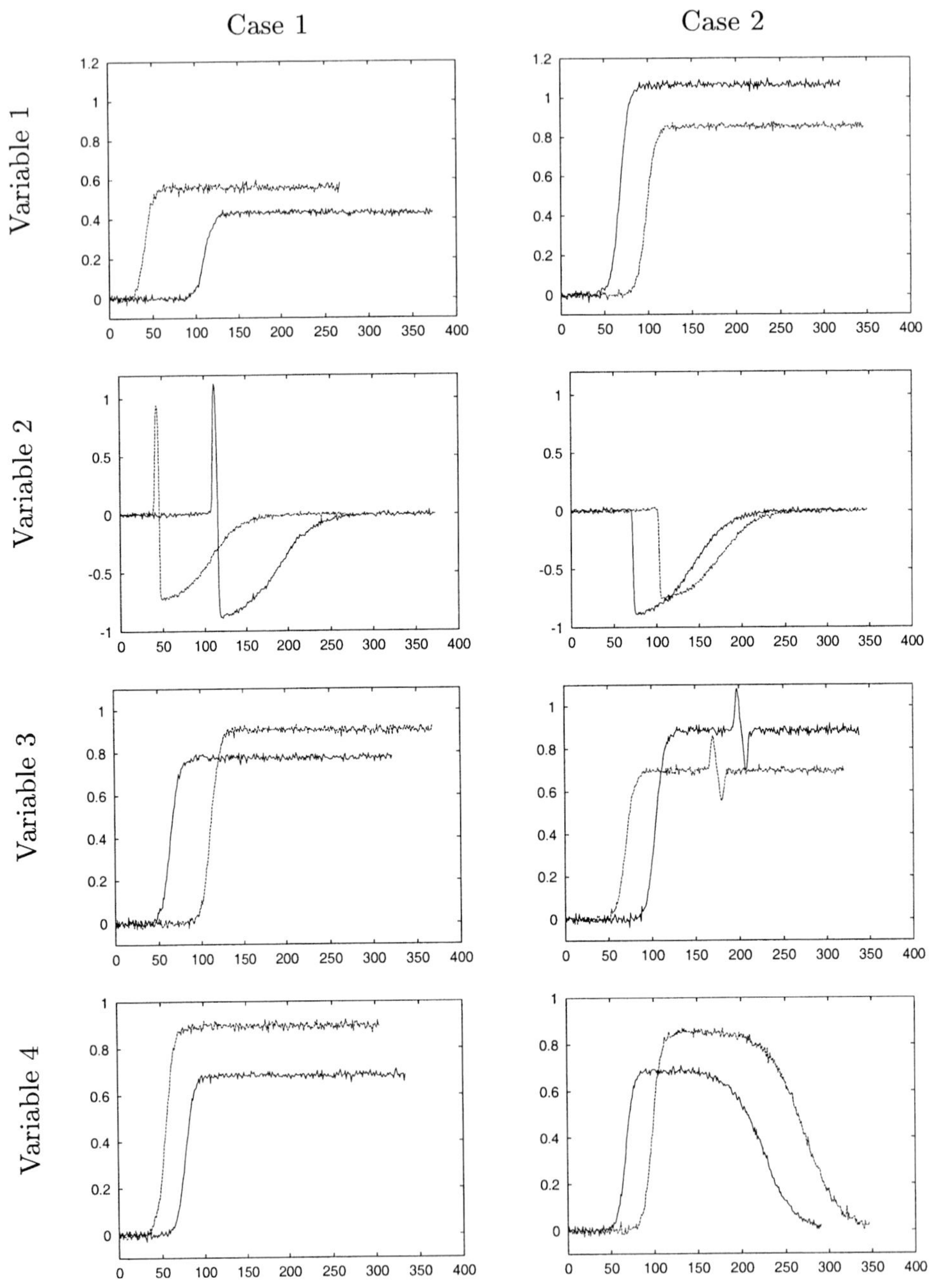

Figure 6: *Trace* data set. Each example is composed by 4 variables, and each variable has two possible behaviors. In the graphs, two examples of each behavior are shown.

gression. Hence, it is interesting to consider the construction of regression RBFN using one of the variants of AdaBoost for regression, such as AdaBoost.R [6] or ExpLev [4].

References

[1] Robert J. Alcock and Yannis Manolopoulos. Time-series similarity queries employing a feature-based approach. In *Proceedings of the 7th Hellenic Conference on Informatics*, Ioannina, Greece, 1999.

[2] Carlos J. Alonso González and Juan J. Rodríguez Diez. A graphical rule language for continuous dynamic systems. In *International Conference on Computational Intelligence for Modelling, Control and Automation – CIMCA'99*, Vienna, Austria, February 1999.

[3] D.J. Berndt and J. Clifford. Finding patterns in time series: a dynamic programming approach. In U.M. Fayyad, G. Piatetsky-Shapiro, P. Smyth, and R. Uthurusamy, editors, *Advances in Knowledge Discovery and Data Mining*, pages 229–248. AAAI Press /MIT Press, 1996.

[4] Nigel Duffy and David Helmbold. Leveraging for regression. In *Computational Learning Theory: 13th Conference (COLT 2000)*, Stanford University, 2000.

[5] Y. Freund and R. Schapire. Experiments with a new boosting algorithm. In *13th International Conference om Machine Learning (ICML-96)*, pages 148–156, Bari, Italy, 1996.

[6] Yoav Freund and Robert E. Scapire. A decision-theoretic generalization of on-line learning and an application to boosting. *Journal of Computer and System Sciences*, 55(1):119–139, August 1997.

[7] S. Hettich and S. D. Bay. The UCI KDD archive, 1999. `http://kdd.ics.uci.edu/`.

[8] Mohammed W. Kadous. Learning comprehensible descriptions of multivariate time series. In Ivan Bratko and Saso Dzeroski, editors, *Proceedings of the 16th International Conference of Machine Learning (ICML-99)*. Morgan Kaufmann, 1999.

[9] M. Kubat, I. Koprinska, and G. Pfurtscheller. Learning to classify biomedical signals. In R.S. Michalski, I. Bratko, and M. Kubat, edi-

tors, *Machine Learning and Data Mining*, pages 409–428. John Wiley & Sons, 1998.

[10] Miroslav Kubat. Decision trees can initialize radial-basis function networks. *IEEE Transactions on Neural Networks*, 9:813–821, 1998.

[11] Mark J. Orr. Introduction to radial basis function networks. Technical report, 1996. `http://www.anc.ed.ac.uk/~mjo/papers/intro.ps.gz`.

[12] Mark J. Orr. Recent advances in radial basis function networks. Technical report, 1999. `http://www.anc.ed.ac.uk/~mjo/papers/recad.ps.gz`.

[13] Juan J. Rodríguez, Carlos J. Alonso, and Henrik Boström. Boosting interval based literals. *Intelligent Data Analysis*, 5(3):245–262, 2001.

[14] Juan J. Rodríguez Diez and Carlos J. Alonso González. Applying boosting to similarity literals for time series classification. In Josef Kittler and Fabio Roli, editors, *Multiple Classifier Systems: 1st International Workshop; MCS 2000*, volume 1857 of *Lecture Notes in Computer Science*, pages 210–219. Springer, 2000.

[15] Juan J. Rodríguez Diez and Carlos J. Alonso González. Learning classification RBF networks by boosting. In Josef Kittler and Fabio Roli, editors, *Multiple Classifier Systems: second international workshop, MCS 2001*, Lecture Notes in Computer Science, pages 43–52, Cambridge, UK, July 2001. Springer.

[16] Davide Roverso. Multivariate temporal classification by windowed wavelet decomposition and recurrent neural networks. In *3rd ANS International Topical Meeting on Nuclear Plant Instrumentation, Control and Human-Machine Interface*, 2000.

[17] Naoki Saito. *Local Feature Extraction and Its Applications Using a Library of Bases*. PhD thesis, Department of Mathematics, Yale University, 1994.

[18] Robert E. Schapire. Using output codes to boost multiclass learning problems. In *14th International Conference on Machine Learning (ICML-97)*, pages 313–321, 1997.

[19] Robert E. Schapire. A brief introduction to boosting. In *16th International Joint Conference on Artificial Intelligence (IJCAI-99)*, pages 1401–1406. Morgan Kaufmann, 1999.

[20] Robert E. Schapire and Yoram Singer. Improved boosting algorithms using confidence-rated predictions. In *11th Annual Conference on Computational Learning Theory (COLT 1998)*, pages 80–91. ACM, 1998.

Similarity Measures and Clustering of String Patterns

Ana Fred
Telecommunications Institute
Instituto Superior Técnico, Technical University of Lisbon,
E-mail: afred@lx.it.pt

Contents

D. Chen and X. Cheng (eds.),
Pattern Recognition and String Matching, 155–193.

Abstract

Clustering is a powerful tool in revealing the intrinsic organization of data. A clustering of structural patterns consists of an unsupervised association of data based on the similarity of their structures and primitives. This chapter addresses the problem of structural clustering, and presents an overview of similarity measures used in this context. The distinction between string matching and structural resemblance is stressed. The hierarchical agglomerative clustering concept and a partitional approach are explored in a comparative study of several dissimilarity measures: minimum code length based measures; dissimilarity based on the concept of reduction in grammatical complexity; and error-correcting parsing.

1 Introduction

Consider a data set D consisting of N sequences of symbols (or strings), $D = \{s_1, \ldots, s_N\}$, defined over the alphabet Σ, where $s_i = (x_1^i x_2^i \ldots x_{L_i}^i)$ is a sequence of length L_i. The problem of sequence clustering is defined as follows: given the sample patterns in D, discover from the data a "natural"grouping of the sequences into K clusters, K being in general unknown.

The problem of clustering in multivariate feature spaces has been extensively studied [32, 33, 11, 66]. Many clustering algorithms have been proposed, each with its own approach for handling cluster validity [1, 27, 53, 21], number of clusters [40, 65], and structure imposed in the data [63, 44]. Directly using dissimilarity values or exploring point densities for the patterns, either emphasizing compactness or connectedness in feature space, leads to two main strategies for clustering: hierarchical methods and partitional methods. Partitional structure organizes patterns into a small number of clusters; a data partition is obtained as the result of an optimization process or by exploring local structure. Examples of techniques in this class include mixture decomposition [47, 57, 14], non-parametric density estimation based methods [54], central clustering [2], square-error clustering [49], shape fitting approaches [63, 44], and geometrical approaches [26]. The K-means is a very popular algorithm in this category. Assuming *a priori* knowledge about the number of classes and based on the square-error criterion, it is a computationally efficient clustering technique that identifies hyper-spherical clusters [32, 11].

Hierarchical methods produce a nesting of data clusterings, that can be represented graphically as a dendrogram. Mostly inspired by graph theory [67], both agglomerative [32, 12, 21] and divisive approaches [5] have been attempted, the first starting with many clusters that are successively merged in accordance with inter cluster similarity, and the later working in the opposite direction. Variations of the algorithms can be obtained by the definition of a similarity measure between patterns and clusters [20]. The single link algorithm is one of the most popular methods in this class [32]. Data partitioning is usually obtained by setting a threshold on the dendrogram; cluster validity studies have also been proposed [10, 1] for the *a posteriori* analysis of structures, in order to evaluate the clustering results and define meaningful clusters.

When patterns are described in symbolic form as strings of characters, clustering algorithms found in the literature are typically extensions of conventional clustering methods [32, 23] by introducing dissimilarity measures between strings [20]. This is non trivial since the sequences can be of different lengths; also it is not always clear what is a meaningful similarity/dissimilarity measure for sequence comparison. Viewing similarity computation as a matching process [23, 3, 4, 51, 46, 8, 56], references [23, 43] present sentence-to-sentence clustering procedures based on the comparison of a candidate string with sentences in previously formed clusters (clustering based on a nearest-neighbor rule) or with cluster center strings (cluster center technique). String editing operations are there used in the transformation of strings to perform the matching. Following the string matching paradigm while modelling clusters' structure using grammars, error-correcting parsing and grammatical inference are combined in a clustering algorithm described in [23, 24]. Basically, it implements a nearest-neighbor rule, where sentences are compared, not directly with the patterns included in previously formed clusters, but with the best matching elements in the languages generated by the grammars inferred from the clusters' data. Also, using grammars for modelling clusters' structure, a distinct approach, based on the concept of minimum description, is proposed in [16]. Structural resemblance between patterns is assumed to reflect common rules of composition; a normalized reduction in grammar complexity obtained by associating patterns gives the measure of similarity underlying the hierarchical clustering algorithm proposed there. In [18] the search of common sub-patterns by means of Solomonoff's coding [62, 19] forms the basis of a clustering algorithm that defines similarity between patterns as a ratio of decrease in code length. Other model-based clustering techniques have been proposed in the context of continuous valued discrete time series, making use of Markov chain models [60] or hidden Markov models (HMM) [50, 50, 61].

In this chapter the problem of clustering of string patterns is addressed, the distinction between string matching and structural resemblance being stressed. The first part of this paper presents an overview and discussion of dissimilarity measures for string patterns. These are further analyzed and compared in the context of clustering, based on a set of test examples, in the second part of the chapter, followed by conclusions and current trends in clustering.

2 Basic Definitions

2.1 Similarity, Dissimilarity and Proximity

Let $s_1 = (x_1^1 x_2^1 \dots x_{L_1}^1)$, $s_2 = (x_1^2 x_2^2 \dots x_{L_2}^2)$ and $s_3 = (x_1^3 x_2^3 \dots x_{L_3}^3)$ be strings defined over the alphabet Σ, and let $|s_i| = L_i$ denote the string's length. A *similarity* index, $S(s_1, s_2)$, measures the degree to which a pair of objects are alike; the larger its value, the closer or more alike we think the patterns are. Conversely, *dissimilarity* coefficients, $d(s_1, s_2)$, assess the degree to which patterns differ, smaller values meaning closer or higher resemblance. Distances, differences, reciprocal of similarities, all constitute examples of dissimilarity measures. Globally, similarity and dissimilarity are referred to as *proximity* measures – $Prox(s_1, s_2)$.

Proximity values are positive numbers, its range being either bounded, such as the interval $[0; 1]$, or right unbounded: $Prox(s_1, s_2) \in [0; +\infty]$. We shall refer to proximity coefficients that have a $[0; 1]$ range as *normalized.* Given the inverse relationship between similarity and dissimilarity, a simple way to transform a similarity into a dissimilarity, in the situation of bounded ranges, is:

$$d(s_1, s_2) = maximum\_similarity - S(s_1, s_2) \tag{1}$$

2.2 Metrics and Dissimilarity Properties

The concept of proximity presented previously is very broad, imposing no restrictions to its definition other than positivity and monotonicity associated with its semantic interpretation. However, in order to either comply with perceptual intuition or to observe more elaborated mathematical formalisms, restrictions have been imposed on its definition, in the form of a set of desirable properties.

Most of the existing theories share a common important concept, namely that of geometric distance. Objects are seen as points in a metric space, $d(s_1, s_2)$ being the distance function of this space.

A *distance* or *metric*, d, is a real-valued function of two points that obeys the following properties:

- *Positivity*: $d(s_1, s_2) > 0$ and $d(s_1, s_2) = 0 \Leftrightarrow s_1 = s_2$.
- *Symmetry property*: $d(s_1, s_2) = d(s_2, s_1)$.
- *Triangle inequality*: $d(s_1, s_3) \leq d(s_1, s_2) + d(s_2, s_3)$.

The positivity property subsumes the following two distance axioms:

- *Consistency of self similarity*: $d(s_1, s_1) = d(s_2, s_2)$.
- *Minimality of self-similarity*: $d(s_1, s_1) \leq d(s_1, s_2)$.

Of the above properties, the first is generally accepted. Symmetry is also usually considered a desirable property. Concerning the triangle inequality, it is an understandable restriction under the geometrical interpretation of dissimilarity, having been explored, for instance, in the design of search techniques in the context of image retrieval [13, 6]. It basically states that, given a set of three points, the distance between any two points is at the most given by the sum of the distances between each of them and the remaining one, which occurs in the situation of colinearity.

Psychologists tend to distinguish between *perceived similarity*, expressed by d, and *judged similarity*, δ, the latter being expressed as a suitable monotonically non-decreasing function of the former: $\delta(s_1, s_2) = g[d(s_1, s_2)]$. Experimental investigation over the judged similarity, however, led to the refutation of all three properties (and underlying axioms) (see [59] and the references therein). In spite of this, metric models have been widely used in psychology and engineering application areas.

3 String Patterns and Proximity Measures

String descriptions may arise as the natural form for pattern representation, such as in the context of text documents, coding theory, and molecular biology applications, or as a result of some pattern pre-processing. Examples of the second include speech processing applications, processing of handwriting material, and image processing.

A key feature when performing pattern analysis under this representation paradigm, is the definition of a proximity measure between patterns. This is useful for analysis (comparison of sequences), classification, and clustering purposes. In the remainder of this section we present and discuss some of the proximity measures found in the literature.

3.1 Measuring Similarity between Sequences

In a wide sense, a similarity index measures the degree to which a pair of objects are alike. Concerning structural patterns represented as strings or sequences of symbols, the concept of pattern resemblance has typically been viewed from three main perspectives:

- similarity as matching, according to which patterns are seen as different viewpoints, possible instantiations or noisy versions of the same object;
- structural resemblance, based on the similarity of their composition rules and primitives;
- content-based similarity.

The string matching approach [3, 4] falls into the first category and forms the basis of most of the similarity and dissimilarity indices between sequences of symbols reported in the literature [58]. Assuming the classical string edit operations – insertion, substitution and deletion of symbols [56] – or, additionally, transposition errors [51], to which costs are associated, dissimilarity indices are typically variations of the Levensthein distance, of which the probabilistic modelling is a particular instance. Examples of dissimilarity measures related to the Levensthein distance are the Hamming distance, counting the number of mismatches over strings of the same length, and dissimilarity based on the search of longest common substrings [7]. These measures have been applied, for instance, in the comparison of sequences [41] in error correction of noisy sentences [36, 52, 51] and in recognition tasks [46, 8, 9].

Structural dissimilarity (similarity) measures the extent to which two sequences differ (are alike) in terms of rules of composition of the string patterns. One such measure is presented in [18], where the underlying structure of strings is assessed as inter-symbol dependencies, identified by means of Solomonoff's coding [62, 19]; similarity, viewed as shared subsequences, is computed as a ratio of decrease in code length. In the work reported in [16] syntactic rules model patterns' structure, and grammar complexity measures the compactness of this representation. Similarity is then defined as the ratio of decrease in grammar complexity. Fu [23] defined a distance between strings based on the modelling of a string structure by means of a grammar and on the concept of error-correcting parsing (ECP) [42, 64]. According to this model, the distance between a string and a reference string is given by the error-correcting parser as the weighted Levensthein distance between the string and the nearest (in terms of edit operations) string generated by the grammar inferred from the reference string (thus exhibiting a similar structure). This method provides a link between the structural resemblance and the string matching approaches.

In some applications, textual information comparison underly a semantic interpretation, the concept of matching being replaced by content similarity evaluation. In this case, features are extracted from the text data (such as token words or sub-strings), and conventional set-theoretical or vector-based similarity measures are applied to the resulting descriptions. The n-gram method [15] and frequency-based similarity [30, 28], constitute examples of this approach, particularly explored in the fields of text documents analysis and information retrieval. For a discussion of similarity measures and weighting functions in the context of text documents ranking see, for instance, [28] and the references therein. In this chapter we are focusing on the distinction between string matching and structural resemblance. Content-based similarity measures will not be further addressed.

In order to produce a data partition, clustering algorithms often require the definition of a similarity measure between sets of patterns (clusters). Either pairwise comparisons are performed or cluster representatives or models are adopted. According to the first approach, resemblance between pairs of patterns in distinct clusters are computed, the final similarity/dissimilarity measure being a function of these. An example in this category is the nearest-neighbor rule used in the single-link algorithm, dissimilarity between two clusters being the distance between nearest neighbor patterns. Sets of similar sequences can also be represented by a single prototype, chosen within the set (taken as the cluster center [43, 24]) or may represent the average string, such as the generalized median, defined as

No.	String
s1	ABILIO DOS SANTOS
s2	ABILIO DA CONCAICAO ALVES DOS SANTOS
s3	ABEL SANTOS
s4	ABEL LOPES SANTOS
s5	ADRIANO EDSON A. FIUSA
s6	ADRIANO EDSON FIUZA
s7	ADELAIDE CONCEICAO RODRIGUES
s8	ADELAIDE DA CONCEICAO HENRIQUES
s9	ADELAIDE CONCEICAO NATARIO FERREIRA
s10	ADELAIDE FERREIRA
s11	MARIA ADELAIDE CONCEICAO NATARIO FERREIRA
s12	ANA MARIA FARINHO MOURAO
s13	ANA MARIA FARINO MOURAO
s14	ANA PATRICIA NUNES GOMES
s15	CLARA GOMES
s16	PAULO ANTONIO GOMES

Table 1: Dataset $D1$: sample names from a customer database.

the string that has the smallest sum of distances to all strings in the set [39, 45, 34]. In this situation, the proximity between clusters is given by the proximity between their prototypes. A set of strings can also be represented by a model, such as a grammar [25, 48], a Markov chain [60] or a Hidden Markov Model (HMM) [55, 61]. Similarity between clusters are hence computed based on their model representations, being tightly related with the underlying clustering strategy.

3.2 String Alignment and Structural Similarity: Examples

The concepts of string matching and of structural similarity are illustrated in the following examples, which will be used throughout the chapter. While the first example is clearly a string matching problem, examples in sections 3.2.2 to 3.2.4 require the use of structural information, expressed either as rules of composition or as frequently occurring sub-sequences. In the example in section 3.2.5 it is not clear which type of similarity is best suited for discriminating between contour classes. Post evaluation of the different measures in the context of clustering is needed in order to make a decision, which will be addressed in section 4.

3.2.1 Names Data Set

Consider the set of strings given in table 1, representing names in a customers database. Suppose we are given the task of identifying multiple entries for the same individual in the database. Repeated entries might have occurred for instance due to incorrect typing on a keyboard, misspelling of words, suppression of middle names, or recognition failure from a scanning device. Given the type of data involved, it is obvious that the order of words in the sequences is important, and alignment of sequences is needed in order to detect misspelling. Furthermore, suppression of names may be modelled by deletion of characters. The string matching approach therefore seems to be the right approach to deal with this type of patterns. Additional information about the data entry process may provide useful hints to quantify the type of errors more likely to occur, and characterize costs on string editing operations.

3.2.2 Fractal Images

Fractals can be used as a mathematical model of many biological objects which are self-similar within a certain interval of scales. Many fractal objects result from iteration processes, which are a natural way to describe growth processes in biology and physics [35]. In a growth process the last growth stage serves as input for the next growth step. A well known model for biological pattern formation, from botany, is the Lindenmayer grammar or L-system. It is similar to grammars in conventional formal language theory, as in Chomsky hierarchy [29], the main difference being that rewriting rules are applied simultaneously in the Lindenmayer grammar, strings being generated in an iteration process.

The L-system

$$\begin{aligned} L_{dragon\_sweep} = & (V, R, start\_symbol) \\ & V = \{a, b, +, -\}, \quad start\_symbol = a \\ & R: \quad a \rightarrow a + b, \quad b \rightarrow a - b, \quad + \rightarrow +, \quad - \rightarrow - \end{aligned} \tag{2}$$

condenses the morphological structure of the Dragon-Sweep fractal curve [35] shown in figure 1(b). The strings in figure 1(c) correspond to the first 5 iterations using this L-system; fractal curves are obtained from strings according to the drawing rules in fig. 1(a).

The data set $D2$ consists of the first 10 iterations with this L-system, the corresponding string lengths being given in table 2. Although the string descriptions have very dissimilar lengths, all these patterns share the same rules of composition, i.e. that are structurally similar. It would be desirable to have a similarity measure that recognizes this structural resemblance, which is not addressable by string alignment techniques. Ideally, one would like to identify all these structures as being the same.

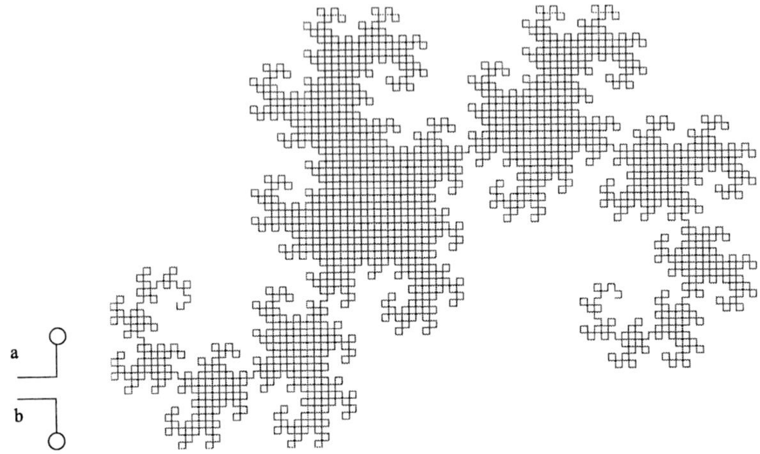

(a) (b) Dragon sweep curve corresponding to a string length of 4095 characters.

a+b
a+b+a-b
a+b+a-b+a+b-a-b
a+b+a-b+a+b-a-b+a+b+a-b-a+b-a-b
a+b+a-b+a+b-a-b+a+b+a-b-a+b-a-b+a+b+a-b+a+b-a-b-a+b+a-b-a+b-a-b

(c) First 5 strings in the iteration process.

Figure 1: Dragon sweep fractal curves. The construction of the curves, as in (b), from the string descriptions is as follows. The symbols a and b correspond to the graphical elements in figure (a); at the end point of each element (indicated as "o"), turns are made as indicated in the string. The symbol + indicates a turn of 90° to the right, and $-$ is a turn of 90° to the left.

s1	s2	s3	s4	s5	s6	s7	s8	s9	s10
3	7	15	31	63	127	255	511	1023	2047

Table 2: Data set $D2$: lengths of strings corresponding to the first 10 iterations of the L-system (2), that describes the Dragon sweep fractal.

3.2.3 Recursive Structure Line Patterns

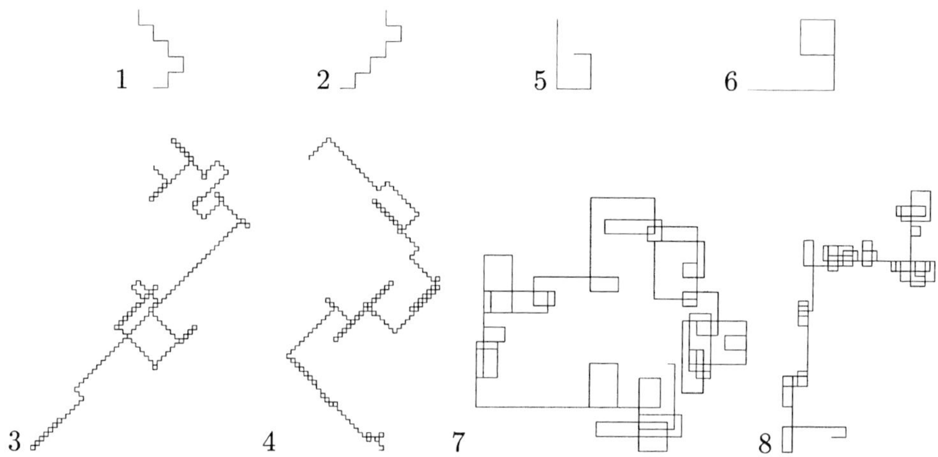

Figure 2: Data set *D*3. Graphical representation of strings of the form $(2^*6^*)^*$ (patterns 1 to 4) and $(0^*6)^*$ (patterns 5 to 8). The graphical interpretations of the symbols are as follows: 0 – maintain the line direction; 2 – turn right; 6 – turn left. The string's lengths for the several patterns are: (1)-9; (2)-9; (3)-389; (4)-409 (5)-18; (6)-9; (7)-487; (8)-411.

Another example of string patterns with underlying grammatical rules of composition is shown in figure 2, which represents a graphical description of instances, with various lengths, of strings of the form $(2^*6^*)^*$ or $(0^*6)^*$, with the symbol $*$ indicating an arbitrary repetition of the element on the left (parenthesis are used for delimiting elements with more than one character).

3.2.4 Email Messages

The next example consists of the first lines of text extracted from email messages, as shown in table 3, by removing space characters. Variable length samples using the English and the Portuguese languages were included. The purpose is the classification of content, in the presence of variable sized sequences. While associations of words may have a fundamental role in defining topics, the order of appearance of these sequences in the text is not relevant to characterize content. The string alignment concept is meaningless in this situation. On the other hand, no general rule of string composition can be identified that could be related to content. Key-word

ID	Mail Type	Cl.	Lang.	Size	Sample Text
1	call for papers	1	Eng.	142	ISSS'98 FINAL CALL FOR PAPERS (SUBMISSION ...
2	call for papers	1	Eng.	106	CALL FOR PAPERS WORLD MULTICONFERENCE ...
3	call for papers	1	Eng.	160	CALL FOR PAPERS RENSSELAER'S INTERNAT ...
4	call for papers	1	Eng.	87	CALL FOR PAPERS THE PRAGUE STRINGOLOGY ...
5	personal letter	2	Eng.	161	DEAR ANA, THANKS FOR THE EMAIL. PLEASE ...
6	personal letter	2	Eng.	129	DEAR ANA, IN THE BACKUP DIRECTORY ...
7	journal advert.	4	Eng.	137	CONTENTS DIRECT FROM ELSEVIER SCIENCE ...
8	MSc advert.	3	Port.	153	PROVAS DE MESTRADO EM ENGENHARIA ...
9	MSc advert.	3	Port.	280	PROVAS DE MESTRADO EM ENGENHARIA ...
10	MSc advert.	3	Port.	168	PROVAS DE MESTRADO EM ENGENHARIA ...

Table 3: Data set $D4$. Text from email messages.

based or sub-sequence based string comparisons seem more valuable tools for this problem.

3.2.5 Contour Images

The last data set, $D5$, used consists of string descriptions of contour images from three types of hardware tools, as follows (see table 4): 10 samples from tool type 1; tool type 2 has two poses (closed and half opened), 10 samples per pose being used; tool type 3 also has two poses (wide opened and half opened), 10 samples per pose. Each image was segmented to separate the object from the background, and the object boundary was sampled at 50 equally spaced points [22]. The strings correspond to encoding objects contours using an 8-directional differential chain code [31].

Class	Tool	Samples
1		61072620026270163702620026207261072601627026200262
		71172620026200162701637026207261073610726116370262
		71172620026270262002620737016370262002620026200162
2		00000000660000000171000000600000100000000750000000003
		00000000500000000000000076000001610000000660000000003
		000000005000000001710000005000001610000000760000000003
3		100000000500000000007207000701700000000005000000000012
		100000000500000000000107000702700000000005000000000012
		100000000570000000007107007702700000000005000000000012

Table 4: Data set $D5$. Sample string contour descriptions of hardware tools.

3.3 Dissimilarity based on String Editing Operations

A very well known measure of dissimilarity between strings is the *Levensthein distance.* It is based on the definition of string editing operations or error transformations, namely substitution, deletion and insertion of symbols:

- $T_S(b|a) \rightarrow$ substitution of a by b, $a \neq b$
- $T_S(a|a) \rightarrow$ maintaining symbol a
- $T_I(b) \rightarrow$ insertion of b
- $T_D(a) \rightarrow$ deletion of a

The Levensthein distance between two strings $s_1, s_2 \in \Sigma^*$, $D_L(s_1, s_2)$, is defined as the minimum number of editing operations needed in order to transform s_1 into s_2. An extension of the Levensthein distance by associating differentiated costs to the several editing operations is known as the *weighted Levensthein distance*, defined as

$$d_W(s_1, s_2) = \min\{\gamma(T)|T \text{ is an edit transformation of } s_1 \text{ into } s_2\} \tag{3}$$

where $\gamma(T) = \sum_{i=1}^{m} \gamma(T_i)$, T_i is the ith editing operation performed in the transformation T, and $\gamma(T_i)$ is the cost associated with that operation. Figure 3 illustrates the matching of two strings according to editing operations. A particular instance of this weighting function is the probabilistic modelling, according to which the following normalization condition must hold:

$$\sum_{b\in\Sigma} \gamma(T_S(b|a)) + \sum_{b\in\Sigma} \gamma(T_I(b)) + \gamma(T_D(a)) = 1. \tag{4}$$

In order to preserve the symmetry property, weights can not be assigned arbitrarily. The costs of inserting or deleting a given symbol must be equal, as for the substitution of symbols: $\gamma(T_S(b|a)) = \gamma(T_S(a|b))$. It can be shown that the triangle inequality is verified and thus it constitutes a metric (see for instance [8] and the references therein).

Normalization of the previous distances with respect to string lengths are obtained by post-normalization (normalized weighted Levensthein distance)

$$d_{NW}(s_1, s_2) = d_W(s_1, s_2)/\max\{|s_1|, |s_2|\}, \tag{5}$$

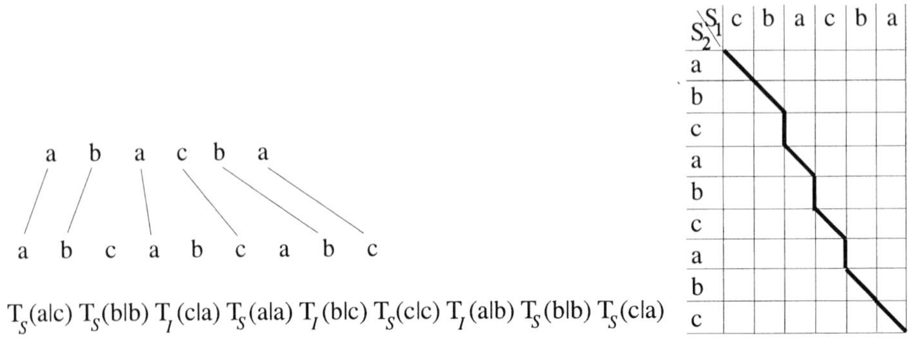

(a) String matching using editing operations: insertion ($T_I(.|.)$), deletion ($T_D(.)$)and substitution ($T_S(.|.)$).

(b) Editing path.

Figure 3: String matching. In (b) diagonal path segments represent substitutions, vertical segments correspond to insertions, and horizontal segments correspond to deletions.

or by optimal minimization of the distance normalized by the length of the editing path – normalized edit distance [46]

$$d_{NED}(s_1, s_2) = \min\left\{ \frac{\gamma(T)}{|T|} \mid T \text{ is an editing path between } s_1 \text{ and } s_2 \right\}, \tag{6}$$

where $|T|$ is the length of the editing path (see figure 3(b)). Vidal [46] has shown that the latter performs better in many situations. It should be emphasized that normalization leads to dissimilarity measures that are not metrics due to failure of the triangle inequality [8].

	$s2$	s3	s4	s5	s6	s7	s8	s9	s10
d_W	1.6	1.5	4	8.5	9.4	8.2	8.3	7.8	10.6
d_{NED} ($\times 1e-1$)	.52	1	2.3	3.4	3.9	2.7	2.6	2.2	4.3

Table 5: Distances between string $s1$ and strings $s2$ to $s10$ in data set $D1$.

Table 5 presents weighted Levensthein distances and normalized edit distances between string $s1$ and strings $s2$ to $s10$ in data set $D1$. In order to model omission of middle names, the deletion operation was given a low cost (0.1) in comparison to the other error transformations (cost=1).

As shown, without normalization, string $s1$ (ABILIO DOS SANTOS) is considered more similar to the third string (ABEL SANTOS) – $d_W = 1.5$, than to the second string (ABILIO DA CONCAICAO ALVES DOS SANTOS) – $d_W = 1.6$. By using normalization, however, the reverse situation occurs, with a more clear distinction between the two dissimilarities ($d_{NED}(s1, s2) = 0.052$ and $d_{NED}(s1, s3) = 0.1$), which is in better agreement with perceptual expectations.

For simplicity, and under the assumption of no *a priori* knowledge about the penalizing mechanisms of the error transformations, in the remaining of the chapter all edit operations will be assigned unitary costs, except for the operation of maintenance of a symbol, which will be assigned a null weight.

	$s2$	s3	s4	s5	s6	s7	s8	s9	s10
d_W	4	12	28	60	124	252	508	1020	2044
d_{NED} $(\times 1e-1)$	5.71	8.00	9.03	9.52	9.76	9.88	9.94	9.97	9.98

Table 6: Distances between string $s1$ and strings $s2$ to $s10$ in data set $D2$.

The dependency of these dissimilarity measures on the string lengths (even the normalized distances) is illustrated in table 6, with the fractal curve data set ($D2$).

3.4 Dissimilarity based on Error-Correcting Parsing

Fu [24, 23] defined a distance between strings based on the modelling of string structure by means of grammars and on the concept of error-correcting parsing [42, 64]. According to this model, the distance between a string and a reference string is given by the error-correcting parser as the weighted Levensthein distance between the string and the nearest (in terms of edit operations) string generated by the grammar inferred from the reference string (thus exhibiting a similar structure):

$$d_{ECP_f}(s_1, s_2) = \min_y \{ d_W(s_1, y) | y \in L(G_{s_2}) \qquad G_{s_2} := \text{ Grammar inferred from } s_2 \} \quad (7)$$

The computation of the dissimilarity between strings requires two steps: (1)- modelling of strings with grammars; (2)- computation of the dissimilarity by error-correcting parsing. The process is schematically described in figure 4. The grammatical inference procedure is responsible for the identification of regular patterns of composition that lead to the definition of

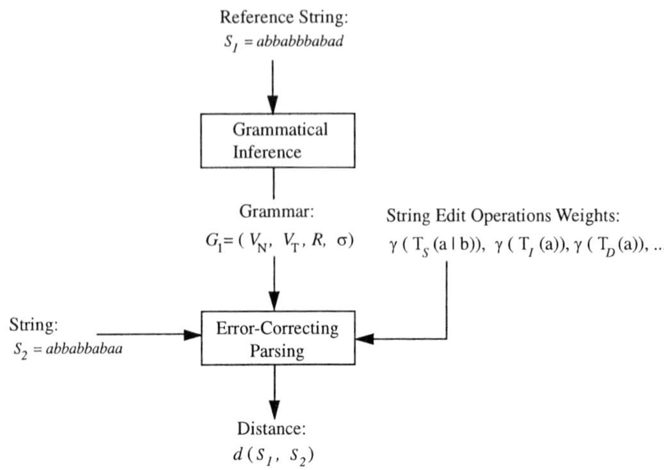

Figure 4: Computation of dissimilarity between strings using error-correcting parsing.

rules. Different inference algorithms will produce distinct results both quantitatively and qualitatively, depending on how far apart are the underlying heuristics or criteria supporting the methods. It is not possible to define the "optimal" grammatical inference algorithm. Several methods can be tested or the election of a specific algorithm has to be based on some assumption or prior knowledge about the underlying structure. Crespi-Reghizzi's method [25, 48] will be used for grammatical inference in this chapter.

It is important to notice that, even with adequate definition of the editing weights, the symmetry property cannot be ensured *a priori*. In fact, $d_{ECP_f}(s_1, s_2) = \min\{d_L(s_1, L(G_{s_2}))\}$, which, in general, will be different from $d_{ECP_f}(s_2, s_1) = \min\{d_L(s_2, L(G_{s_1}))\}$. For instance, given the strings $s1 = a + b$ and $s2 = a + b + a - b$, from the data set $D2$, $d_{ECP_f}(s1, s2) = 1$ (a recursive structure is identified by the inference algorithm on the second string) while $d_{ECP_f}(s2, s1) = 4$. In order to preserve the symmetry property, we will use a new definition:

$$d_{ECP}(s_1, s_2) = \min\{d_{ECP_f}(s_1, s_2), d_{ECP_f}(s_2, s_1)\} \tag{8}$$

We shall designate by d_{NECP} the previous dissimilarity measure normalized by the edit path. Neither d_{ECP} or d_{NECP} constitute a metric because they do not obey the triangle inequality (for instance in the data set $D1$, $d_{ECP}(s2, s1) = 5$, $d_{ECP}(s1, s6) = 11$ and $d_{ECP}(s2, s6) = 23$).

	$(s1, sj), j > 1$	$(s2, sj), j > 2$	$(si, sj), i, j > 2, i \neq j$
d_{ECP}	1	1	0
d_{NECP}	.33	.14	0

Table 7: Distances between pairs of strings in data set $D2$ using error-correcting parsing.

The capacity of these dissimilarity measures to capture structural resemblance, by means of inferred grammars, is shown in table 7, concerning the comparison of strings in the data set $D2$. Zero distance is obtained between pairs of strings produced after 3 or more steps of the L-system iteration process. The non-zero values for strings $s1$ and $s2$ result from the fact that these strings do not have the $-a$ or $-b$ structures, which are present in all remaining sequences, and therefore are included in the corresponding inferred grammatical models.

3.5 Grammar Complexity-based Dissimilarity

Grammars are compact representations of structural rules of composition in string patterns. Reference [16] assumes this formalism for modelling string patterns, exploring the concept of grammar complexity in the definition of a new measure of similarity between strings. The basic idea is that, if two sentences are structurally similar, then their joint description will be more compact than their isolated description due to sharing of rules of symbol composition; the compactness of the representation is quantified by the grammar complexity, and the similarity is measured by the *ratio of decrease in grammar complexity*, as follows:

$$RDGC(s_1, s_2) = \frac{C(G_{s_1}) + C(G_{s_2}) - C(G_{s_1,s_2})}{\min\{C(G_{s_1}), C(G_{s_2})\}}, \tag{9}$$

where $C(G_{s_i})$ denotes grammar complexity.

Let $G = (V_N, \Sigma, R, \sigma)$ be a context-free grammar, where V_N, Σ are the sets of nonterminal and terminal symbols, respectively, σ is the grammar start symbol and R is the set of productions written in the form:

$$\begin{array}{lll} A_1 & \rightarrow & \alpha_{11} \mid \ldots \mid \alpha_{1l_1} \\ \vdots & & \\ A_r & \rightarrow & \alpha_{r1} \mid \ldots \mid \alpha_{rl_r} \end{array}$$

Let $\alpha \in (V_N \bigcup \sigma)^*$, be a grammatical sentence of length n, in which the symbols $a_1, a_2, \ldots, a_m$ appear $k_1, k_2, \ldots, k_m$ times, respectively. The complexity of the sentence, $C(\alpha)$, is given by [25, 48]

$$C(\alpha) = (n+1)log(n+1) - \sum_{i=1}^{m} k_i log k_i. \tag{10}$$

The complexity of the grammar G, $C(G)$, is defined as [16]

$$C(G) = \sum_{i=1}^{r} \sum_{j=1}^{l} C(\alpha_{ij}) \tag{11}$$

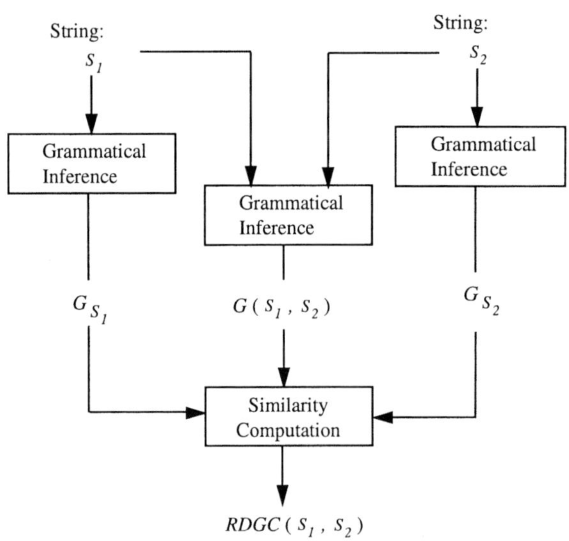

Figure 5: Computation of similarity between strings based on grammar complexity.

The computation of this similarity, as schematized in figure 5, is basically supported on grammatical inference and simple evaluation of the resulting rewriting rules. Computational complexity is therefore conditioned by the inference algorithm. When using Crespi-Reghizzi's method, strings are processed individually; the inference of a grammar for a pair of strings takes advantage of the previous computations, consisting of a simple merging of individually inferred grammars.

Similarity values obtained with this method, as well as for the preceding method, depend on the grammatical inference strategies adopted. However, the following characteristics hold, independently of the grammatical inference algorithm:

$$\begin{aligned} RDGC(s_1, s_2) &\geq 0 \\ &\leq 1 \\ &= 1 \text{ if } s_1 \equiv s_2 \\ &= 0 \text{ if } s_1 \text{ and } s_2 \text{ have non overlaping alphabets} \\ RDGC(s_1, s_2) &= RDGC(s_2, s_1) \end{aligned}$$

As indicated above, the symmetry property is verified; the triangle inequality, however, is not always preserved.

	$(s1, sj), j > 1$	$(s2, sj), j > 2$	$(si, sj), i, j > 2, i \neq j$
S_{RDGC}	0.15	0.85	1

Table 8: Minimum grammar complexity-based similarity between pairs of strings in data set $D2$.

The similarity values between pairs of strings in the data set $D2$ is summarized in table 8. As seen before, strings $s1$ and $s2$ do not contain enough variants of the L-system rules of composition, the inferred grammars not being able to generalize to the recognition of the remaining structures; on the other hand, grammars inferred from the longer strings also do not recognize $s1$ and $s2$. As a result, the similarity between strings involving $s1$ or $s2$ is low, identification of a unique structure being achieved for longer strings (similarity=1).

3.6 Minimum Code Length-based Similarity

A measure of structural similarity exploring the notion of compressibility of sequences and algorithmic complexity, is proposed in [18]. A string is said to be compressible if there exists a description shorter than the original sequence. Low compressibility is related with high complexity and randomness; on the other hand, sequences with inter-symbol dependencies, showing subpattern regularities, are likely to have compact descriptions. Solomonoff's code [62] is then used for the search of pattern regularities and sequence compression. According to this coding scheme, a sequence s_i is represented by the triplet:

$$alphabet\_description,\ special\_symbols\_definition,\ coded\_string,$$

where a coded string is obtained in an iterative procedure where, in each step, intermediate codes are produced by defining sequences of two symbols, which are represented by special symbols, and rewriting the sequences using them. Compact codes are produced when sequences exhibit local or distant inter-symbol interactions.

This coding scheme has been extended to sets of strings

$$alphabet,\ symbol\_definition\ ,\ s_1\_coded\ ,\ s_2\_coded\ ,\ \ldots\ ,\ s_m\_coded$$

and global compact codes are produced by considering the inter-symbol dependencies on the ensemble of the strings. Strings sharing subpattern regularities will therefore produce more compact codes than the gathering of the codes for the individual sequences. The quantification of this reduction in code length forms the basis of the similarity measure which will be designated by NRDCL. The length of the codes produced using Solomonoff's method (given by the sum of the lengths of the descriptions of the three part code above) is evaluated for the isolated strings and for the ensemble; NRDCL is defined as the normalized ratio of decrease in code length obtained by the association:

$$NRDCL(s_1, s_2) = \frac{code\_len(s_1) + code\_len(s_2) - code\_len(s_1, s_2)}{(as)/2 + abs(|s_1\_coded| - |s_2\_coded|)} \tag{12}$$

with

$$as = |alphabet\_s_1| + |symbol\_def\_s_1| + |alphabet\_s_2| + |symbol\_def\_s_2|$$

The order of appearance of sub-sequences is irrelevant for the computation of code lengths. This similarity measure therefore totally departs from string matching paradigms, structure being related to inter-symbol dependencies and not rules of string composition. Accounting for highly frequent sub-sequences, Solomonoff coding assumes long strings. Results with short strings may therefore be meaningless. This is illustrated in table 9, where similarities involving strings shorter than or equal to 31 characters (string $s4$) are very low.

	s1	s2	s3	s4	s5	s6	s7	s8	s9	s10
$s1$	1	0.4	0.19	0.10	0.05	0.19	0.14	0.12	0.09	0.08
$s10$	0.08	0.19	0.39	0.61	0.74	0.85	0.92	0.97	1	1

Table 9: Minimum code length based similarity between strings in data set $D2$.

4 Clustering of String Patterns

The previous proximity measures are evaluated in this section in the context of clustering. Clustering results are strongly conditioned by the underlying similarity measures, but also by the intrinsic cluster validity criteria and structure imposed on the data. Several clustering algorithms integrating the previous proximity measures will therefore be investigated. Table 10 characterizes the clustering algorithms tested. They are further detailed in the next section.

4.1 Clustering Algorithms Tested

a) Hierarchical Agglomerative Schemes			
Philosophy of the method	Proximity between clusters	String sim.	Acronym
String edit operations	Nearest-neighbor rule: $d(C_i, C_j) = \min\{d(s_k, s_l) \mid s_k \in C_i, s_l \in C_j\}$	d_{NED}	SL-NED
Grammar complexity	Ratio of decrease in grammar complexity: $RDGC(C_i, C_j) = \frac{C(G_{C_i}) + C(G_{C_j}) - C(G_{C_i,C_j})}{\min\{C(G_{C_i}), C(G_{C_j})\}}$	$RDGC$	MGC
Minimum code length	Normalized decrease in code length: $NRDCL(C_i, C_j) =$ $= \frac{code\_len(C_i) + code\_len(C_j) - code\_len(C_i, C_j)}{(as)/2 + abs(\|C_i\_coded\| - \|C_j\_coded\|)}$ $as = \sum_{k=i,j} \|alph\_C_k\| + \|symb\_def\_C_k\|$	$NRDCL$	MCL
b) Sentence to Sentence Clustering			
Error correcting parsing	Nearest-neighbor rule: $d_{NECP}(s_i, C_j) = \min\{d(s_i, y) \mid y \in L(G_{C_j},\}$	d_{NECP}	ECP-NED

Table 10: Characterization of clustering algorithms under study.

4.1.1 Hierarchical Agglomerative Clustering

The hierarchical agglomerative clustering method can be schematically described as follows:

Input: A set of n strings $D = \{s_1, s_2, \ldots, s_n\}$; a threshold th or the desired number of clusters N_c.

Output: A partition of D into m clusters ($m = N_c$ if the latter is specified), $C_1, C_2, \ldots, C_m$.

Steps:

1. Set each pattern into a single cluster. Set $m = n$.
2. Compute the proximity matrix between all pairs of strings.
3. If $m = N_c$ go to step 6.
4. Select the most similar pair of clusters, C_i, C_j, from the proximity matrix and let $simil$ be the value of this similarity. If $simil < th$ ($simil > th$, if it is a dissimilarity measure), go to step 6.
5. Merge the two clusters and update the proximity matrix accordingly. Set $m = m - 1$ and continue in step 3.
6. End the computation by returning the data partition encountered.

Different clustering algorithms are obtained from the previous method by defining the proximity measure between patterns and between clusters. The first algorithm in table 5 is based on string editing operations, and corresponds to using the normalized edit distance (see equation 6 in section 3.3) as dissimilarity measure between strings (step 2), and adopting the nearest-neighbor rule (single-link method) for defining distances between clusters (step 5):

$$d(C_i, C_j) = \min \{ d_{NED}(s_k, s_l) \mid s_k \in C_i, s_l \in C_j \} . \tag{13}$$

The second algorithm explores the minimum grammar complexity concept (section 3.5). Similarity between strings is computed using equation 9; this is extended to sets of strings, as given by the equation

$$RDGC(C_i, C_j) = \frac{C(G_{C_i}) + C(G_{C_j}) - C(G_{C_i,C_j})}{\min \left\{ C(G_{C_i}), C(G_{C_j}) \right\}}, \tag{14}$$

where G_{C_i} is the grammar inferred from strings in cluster C_i. Additionally to a data partition and the dendrogram expressing the relationships between string patterns, this method provides a model for each cluster: the corresponding grammar. Results reported in this chapter concern the application of Crespi-Reghizzi's method for grammatical inference. When dealing with long strings, the grammatical inference process is computationally more efficient (it is linear on strings length) than the computation of the normalized edit distance, and therefore this clustering method becomes faster than the first algorithm.

The last hierarchical clustering algorithm computes distances between strings by the normalized ratio of decrease in code length (see equation 12 in section 3.6). A code for the set of strings in a cluster is computed using the extension of Solomonoff's code described in section 3.6; similarity between clusters is obtained by replacing single strings by sets of strings (representing clusters) in equation 12, leading to

$$NRDCL(C_i, C_j) = \frac{code\_len(C_i) + code\_len(C_j) - code\_len(C_i, C_j)}{(as)/2 + abs(|C_i\_coded| - |C_j\_coded|)} \quad (15)$$

4.1.2 Sentence-to-Sentence Clustering

The sentence-to-sentence clustering algorithm is a partitional type method proposed by Fu [24, 23] based on the concept of error correcting parsing. Grammars are used to model cluster structure. The algorithm starts with a single cluster containing the first sample; the remaining data is classified by computing the (weighted) Levensthein distance between the candidate string and the most similar string (determined by error-correcting parsing) modelled by the grammars describing the clusters formed so far; if the minimal distance found is less than a given threshold, th, the string is included in the corresponding cluster and its grammar is updated; otherwise, a new cluster is formed with this sample.

This clustering strategy implements a nearest-neighbor rule, with sequences being compared, not directly with patterns previously included in clusters, but with the best matching elements in the languages generated by the grammars inferred from the clustered data. The implementation used in this chapter will apply Crespi-reghizzi's algorithm for grammatical inference, without assuming *a priori* information. The dissimilarity measure based on error-correcting parsing defined in equation 7 (section 3.4) is extended for the computation of dissimilarity between a string, s_i, and a cluster, C_j, as

$$d_{ECP}(s_i, C_j) = \min\{d_W(s_i, y) \mid y \in L(G_{C_j})\}, \quad (16)$$

where G_{C_j} is the grammar inferred from the strings in cluster C_j. Without *a priori* information, 0-1 costs will be assumed for the the string editing operations (corresponding to the Levensthein distance); in order to reduce the dependency on string lengths, distances will be normalized by the length of the editing path:

$$d_{NECP}(s_i, C_j) = \min\{d_{NED}(s_i, y) \mid y \in L(G_{C_j})\}. \quad (17)$$

Due to the nature of this similarity measure (asymmetric) and the clustering strategy, results obtained with the sentence-to-sentence clustering algorithm may depend on initialization and on the order of pattern presentation.

Concerning computational complexity, this clustering scheme requires fewer comparisons between strings than the hierarchical one. While hierarchical methods require the computation of $n(n-1)/2$ entries in the initial proximity matrix (the iterative process requires the update of this matrix, one line per iteration, as the size of the proximity matrix decreases), according to the sentence-to-sentence technique, each string is processed only once and compared with a, usually, reduced number of clusters. The processing of each string, however, involves grammatical inference (which can be made iterative and linear on the strings length, L) and error-correcting parsing, which in general has $O(L^2)$ time complexity. Thus, depending on the dimensions of the data sets and of the strings, one method or the other may be more advantageous from the computational complexity perspective.

4.2 Experimental Results

4.2.1 Names Data Set

Figure 6 presents the results of clustering of the names data set using several clustering algorithms. Methods based on string edit operations are strongly dependent on the costs assigned. This is illustrated in figures 6(a) and 6(b), showing the dendrograms produced by the SL-NED method for two different assignments for the deletion costs: (a)- $T_D(.) = 1$; (b)- $T_D(.) = 0.1$. The costs for all remaining error operations are set to 1 (no-error cost=0). Situation (b), reducing the penalty for deletion operations, gives the best pattern association results. For instance, the relations between patterns 1 to 4 in (a) are not "natural"given the context, string $s1$ being found more similar to string $s4$ (different first names) than to $s2$ (corresponding to the inclusion of middle names). This is better modelled in the dendrogram in (b). Since the lengths of the strings are not too different and normalized edit distances are being used, the results with error correcting parsing (ECP-NED) are similar to the ones produced by the string matching approach (SL-NED), the first directly proposing a data partition by cutting the minimum spanning tree at a given level, and the second producing a dendrogram. For example, the partition obtained with the ECP-NED with $th = 0.3$ (see figure 6(d)) corresponds to cutting the dendrogram in figure 6(a) at the same level.

The method based on minimum grammar complexity (RDGC, see figure 6(c)) gives inadequate pattern associations, significant similarity being found only between strings $s9$-$s10$-$s11$. The MCL technique, accounting for similar names, performs better (see figure 6(d)). The string matching paradigm,

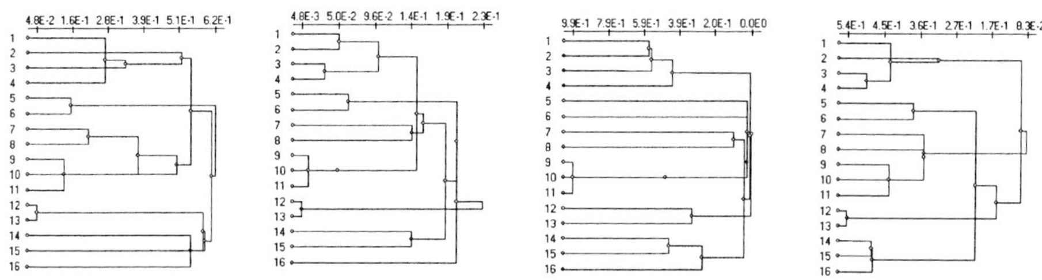

(a) SL-NED, $0-1$ costs. (b) SL-NED, $T_D(.) = 0.1$. (c) RDGC. (d) MCL.

(1 4), (2), (3), (5 6), (7 8), (9 11), (10), (12 13), (14), (15), (16)

(e) ECP-NED, $th = .3$..

Figure 6: Clustering of the names data set. Figures (a) to (d) represent dendrograms produced by the hierarchical clustering algorithms: the indices of the patterns are indicated on the left; the merging of clusters is graphically displayed by joining lines, and the corresponding proximity values are read from the scale on the top.

however, proves to be more appropriate for this problem, results being more in agreement with perceptual evaluation of the patterns.

4.2.2 Fractal Images

The structure of the Dragon-Sweep data set, based on grammatical rules of composition, is easily handled by the grammar-based clustering techniques, RDGC (see figure 7(b)) and ECP-NED. As seen before, both recognize maximum similarity between string patterns of length greater than 7 characters; shorter strings are not recognized as having the same structure because they do not make use of certain structures consistently identified in longer strings. Without *a priori* information, inference algorithms are not able to identify a common set of rules for all strings. As a result, both clustering algorithms will partition the data into two clusters $((s1), (s2 \ \ldots \ s10))$ or three clusters $((s1), (s2), (s3 \ \ldots \ s10))$, by an appropriate setting of the design parameter *th*. Neither SL-NED nor MCL methods can adequately handle these patterns, as shown in figures 7(a) and 7(c), depending on strings length in a linear (SL-NED) or non-linear way.

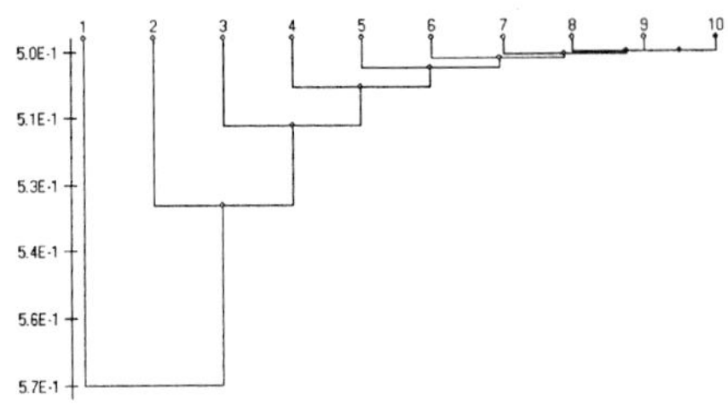

(a) SL-NED

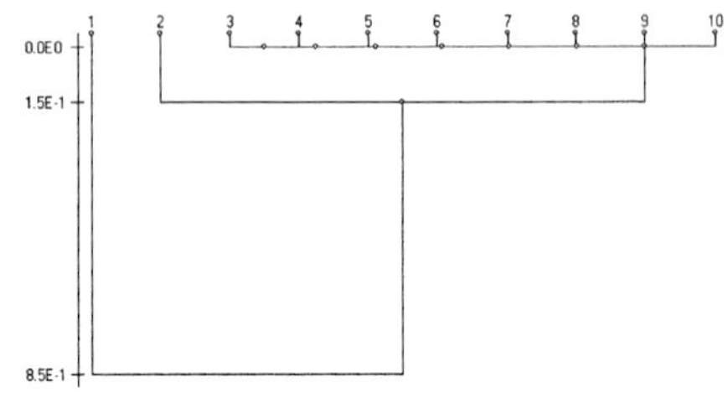

(b) RDGC.

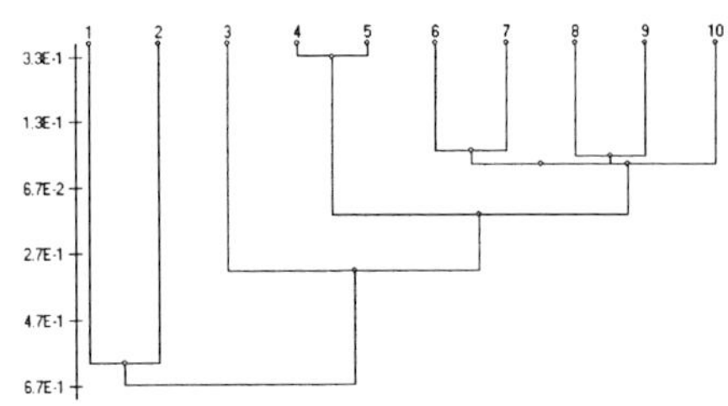

(c) MCL.

Figure 7: Clustering of the Dragon-Sweep data set.

4.2.3 Recursive Structure Line Patterns

The structural resemblance of string patterns in this example is correctly captured by the model-based techniques. The error-correcting parsing based clustering method correctly identifies two clusters in the data, for a wide range of values of th ($0.1 < th \leq 0.6$). Similar results are obtained with the minimum grammar complexity clustering algorithm, a clear separation between the two clusters being evident in the dendrogram in figure 8(b). The results obtained with the SL-NED show the dependency of the dissimilarity measure on the strings length, higher similarity being found between matching sized strings: the dendrogram in figure 8(a) shows that strings $s1$ and $s2$ are first joined with strings $s5$ and $s6$ (with different structure but similar length); strings $s3$ and $s4$ will join these when a single cluster with all strings is formed. The MCL method is unable to identify the correct structure of the data (see figure 8(c)).

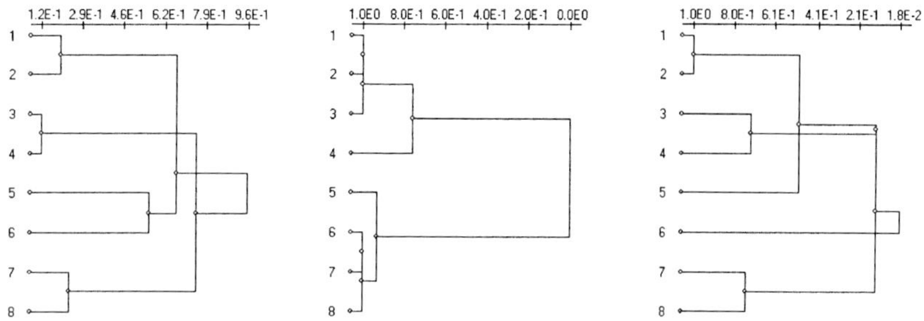

(a) SL-NED (b) RDGC. (c) MCL.

Figure 8: Clustering of the patterns in figure 2.

4.2.4 Email Messages

The categorization of email text is essentially dependent on content, the order of the narrative being arbitrary. In this case, only the MCL method was able to provide meaningful associations (figure 9(c)), correctly forming the classes 1 to 3; sample 7 (class 4) was assigned to class 1 due to the existence of a common theme in the conference and journal advertisements. Low similarity values (< 0.01) were found by the RDGC method at the character level, no significant associations being made. Methods based on error transformations failed to discover common patterns in the data (figures 9(a) and 9(b)). The normalized dissimilarity values of the SL-NED method were high and in narrow ranges, resulting in no associations for threshold values above 0.73; figure 9(a) corresponds to imposing 4 classes in the data.

4.2.5 Contour Images

Figure 10 shows the results of clustering of the contour images data set. The correspondence between string indices and tool classes is given by: $s1$ to $s10$ – tool 1; $s11$ to $s20$ – tool 2, pose closed; $s21$ to $s30$ – tool 2, pose half opened; $s31$ to $s40$ – tool 3, pose half opened; $s41$ to $s50$ – tool 3, pose wide opened. While tool 1 is quite different from the remaining, tools 2 and 3 are more similar to each other.

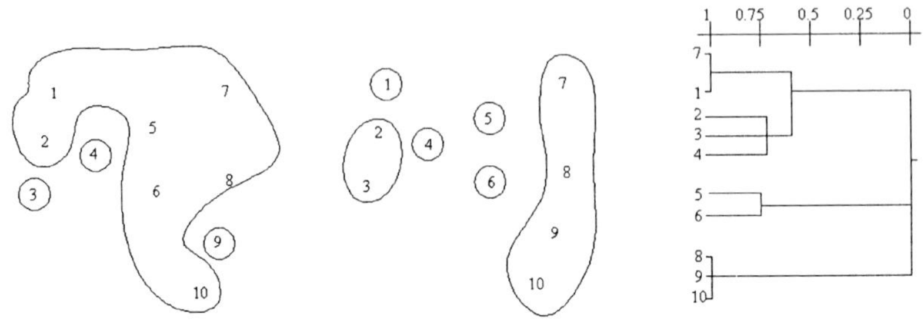

(a) SL-NED: partition obtained when forcing 4 clusters.

(b) ECP-NED ($th = 0.5$).

(c) MCL.

Figure 9: Clustering of the email data set.

Analysis of the dendrograms in figure 10 show that, based on a nearest-neighbor rule, all methods tend to organize string patterns based on pose rather than on tool shape: while tool 1 is well isolated from the remaining classes in all dendrograms, opened poses (wide open and half opened) are first gathered in the dendrograms (tool 3 first joined and then connected with tool 2 - half opened, with the SL-NED method, half-opened tools 2 and 3 being merged first with the RDGC technique, and no clear separation being noticed with the MCL algorithm) and only after that the strings from tool 2, pose closed, are joined. By thresholding on the dendrograms, two classes are obtained with the SL-NED method: tool 1 vs (tool 2 + tool 3). The RDGC method clearly separates tool 1 and tool 2, pose closed; the identification of tool 3, pose wide opened, would imply a splitting of the half-opened tools into several clusters. In order to avoid single element clusters, the threshold is selected to detect two clusters with the MCL tecnhique: tool 1 vs remaining tools.

Partitions produced with the error-correcting parsing based method (ECP-NED) are similar to the ones obtained with RDGC, except that no separation between classes is achieved without obtaining single-element clusters. Setting the threshold to 0.1, four clusters are obtained: an element from tool 1 is detected as an outlier (single-element cluster), the remaining strings forming another cluster; tool 2, pose closed also constitutes a cluster; the fourth cluster gathers the remaining strings. Although the strings have a fixed length (50 characters) and differentiation between classes are present

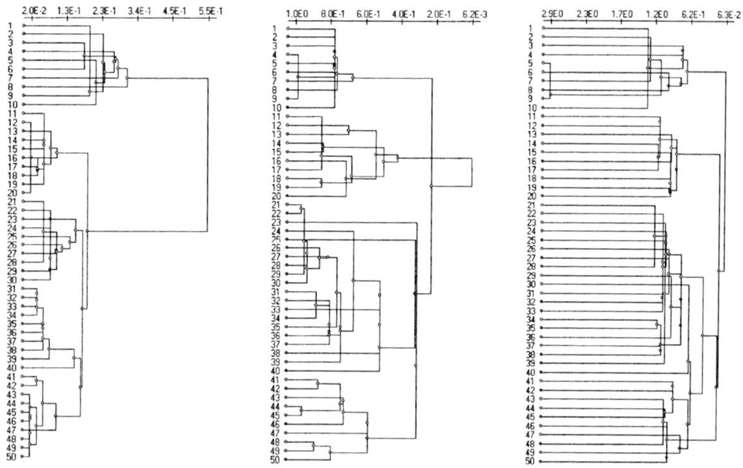

(a) SL-NED: dendrogram. (b) RDGC. (c) MCL.

($s1$), ($s2$ ... $s10$), ($s11$ ... $s20$), ($s21$ ... $s50$)

(d) ECP-NED. Partition obtained with $th = 0.1$

Figure 10: Clustering of the contour images data set. (a) to (c) represent dendrograms produced by the indicated hierarchical methods.

in the dendrograms using the string matching paradigm, the grammar complexity similarity measure leads to a better normalization of dissimilarities, making it simpler to detect classes by cutting the dendrograms.

4.3 Conclusions

Both model-based techniques, supported on grammatical models (ECP-NED and RDGC), are able to handle structural information where the order of sub-patterns in strings is important. The error-correcting parsing method combines the advantages of model based dissimilarity (being independent on strings length) with the string matching mechanism. Its major limitations are the computational complexity, and the lack of criteria to set the design parameter, th. Being a hierarchical method, cluster isolation criteria [10, 1, 21] can be adopted to extract a data partition from the dendrogram produced by the RDGC method. Additionally to a data partition, both methods provide a compact model describing the identity of patterns within a cluster.

The string matching paradigm, on the contrary, cannot handle structural similarity, being sensitive to strings length. The MCL method is the only method, among the ones under study, that recognizes similarity based on inter-pattern dependencies, where the order of sub-sequences is irrelevant.

4.4 Current Trends in Clustering

In the previous sections, several similarity measures have been described and discussed, being further analyzed in the context of clustering, either exploring nearest-neighbor based hierarchical agglomerative methods, or a partitional type method using the error-correcting parsing based dissimilarity. As seen, the similarity measures between string patterns strongly condition the final partitions produced. The similarity measure between clusters is another variant that, in particular in hierarchical methods, is related to the cluster shape concept. While the RDGC and the MCL method define a similarity between clusters by extrapolating the concept of similarity between strings, the SL-NED algorithm, which is a simple single-link method by using as proximity matrix the (weighted normalized) Levensthein distance, is easily adapted to other hypothesized clusters shapes. Figure 11(a) presents the dendrogram produced on the contour images data set, when adopting the farthest neighbor rule for updating the dissimilarity matrix, under the normalized string edit distance paradigm: complete link method over the same initial dissimilarity matrix (CL-NED). As shown, patterns tend now to be organized by tool type rather than pose. Thesholding on this dendrogram will, nevertheless, produce the same two-cluster solution. Still, using the hierarchical agglomerative technique, a cluster isolation criterion that is able to handle this distinct data sparseness situation has been recently proposed [21]. The cluster isolation criterion is based on the computation of a statistic of dissimilarity increments within a cluster, isolation of a cluster being produced when the increment towards further pattern associations is high in comparison with the cluster statistic (see [21] for a more detailed description of the method). Figure 11(b) shows the dendrogram obtained when adopting this cluster isolation criterion with the complete-link method over the NED matrix, with three clusters being formed, each corresponding to a different tool.

Other clustering algorithms based on the previous proximity measures could be developed. For instance, single link or complete link methods

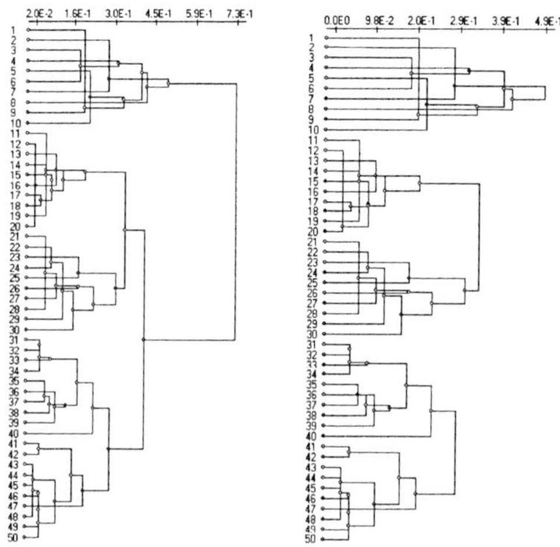

(a) CL-NED.

(b) CL-NED with cluster isolation criterion.

Figure 11: Clustering of the contour images data set. (a) dendrogram produced by the farthest neighbor rule applied to the normalized edit distance between pattern pairs; (b) corresponding dendrogram when adopting a cluster isolation criterion based on the analysis of dissimilarity increments statistics [21].

applied to an initial proximity matrix based on the ECP, RDGC or NRDCL paradigms; partitional type clustering, such as the K-means algorithm, using either of the proximity definitions, etc.

Different combinations of clustering strategies with dissimilarity paradigms will in general produce distinct results. The question that arises is: what is the best strategy to use? As shown in previous examples, the quality of a similarity measure/clustering method depends on the particular application problem. Knowledge about the situation modelled may dictate a particular measure as the most natural, and eventually determine underlying parameters and clustering strategies. Alternatively, several algorithms can be tested and checked for their performance.

Inspired by the work in sensor fusion and classifier combination techniques in pattern recognition [37, 38], the work reported in [17] proposes a combination of clusterings in order to devise a consistent data partition. The

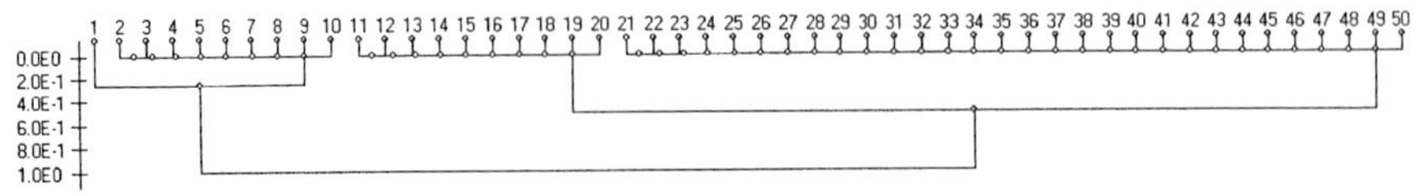

(a) Combination of the partitions obtained in section 4.2.5 (four clusterings) using a voting mechanism.

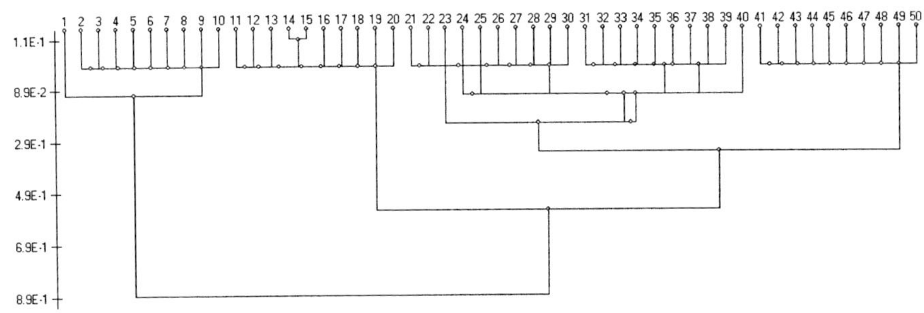

(b) Including more clustering results, exploring, for instance, the farthest neighbor rule in hierarchical clustering.

Figure 12: Combining clustering results.

idea of evidence accumulation clustering is to combine the results of multiple clusterings into a single data partition, by viewing each clustering result as an independent evidence of data organization. In [17] the K-means algorithm is used as the basic algorithm for decomposing the data into a large number, k, of compact clusters; evidence on pattern association is accumulated, by a voting mechanism, over multiple clusterings obtained by random initialization of the K-means algorithm. This produces a mapping of the clusterings into a new similarity measure between patterns, summarized in a matrix, co_assoc, where $co\_assoc(i,j)$ indicates the fraction of times the pattern pair (i,j) is assigned to the same cluster among N clusterings. The final data partition is obtained by applying the single-link method over this similarity matrix, using a fixed threshold, t.

Figure 12 presents results of the combination of data partitions, according to the evidence accumulation framework, on the contour images data set (dendrograms produced by the single-link method over the co_assoc matrix are shown). In figure 12(a) we combined results produced by the SL-NED,

ECP-NED, RDGC and MCL clustering algorithms using the voting mechanism proposed in [17]; by thresholding on this dendrogram we can identify the clusters formed by: (1)- tool 1; (2)- tool 2, pose closed; (3)- tool 2, pose half-opened, and tool 3. In a second experiment, more clusterings were added: the NED, ECP, RDGC and NRDCL similarity matrices between string patterns were used in a complete-link clustering strategy. The dendrogram resulting from this combination is shown in figure 12(b). We can now identify a more clear separation between classes. By adequate thresholding on the dendrogram, one obtains four clusters: (1)- tool 1; (2)- tool 2, pose closed; (3)- tool 2 and tool 3, pose half-opened; (4)- tool 3, pose wide-opened.

The combination of clusterings is therefore a new direction in cluster analysis, that is worth further investigation.

Acknowledgments

This work was partially supported by the Portuguese Foundation for Science and Technology (FCT), Portuguese Ministry of Science and Technology, and FEDER, under grant POSI/33143/SRI/2000.

References

[1] T. A. Bailey and R. Dubes. Cluster validity profiles. *Pattern Recognition*, 15(2):61–83, 1982.

[2] J. Buhmann and M. Held. Unsupervised learning without overfitting: Empirical risk approximation as an induction principle for reliable clustering. In Sameer Singh, editor, *International Conference on Advances in Pattern Recognition*, pages 167–176. Springer Verlag, 1999.

[3] H. Bunke. String matching for structural pattern recognition. In H. Bunke and A. Sanfeliu, editors, *Syntactic and Structural Pattern Recognition, Theory and Applications*, pages 119–144. World Scientific, 1990.

[4] H. Bunke. Recent advances in string matching. In H. Bunke, editor, *Advances in Structural and Syntactic Pattern Recognition*, pages 107–116. World Scientific, 1992.

[5] M. Chavent. A monothetic clustering method. *Pattern Recognition Letters*, 19:989–996, 1998.

[6] J-Y. Chen, C. A. Bouman, and J. P. Allebach. Fast image database search suing tree-structured VQ. In *Proc. of IEEE Int'l Conf. on Image Processing*, volume 2, pages 827–830, 1997.

[7] J. M. Coggins. Dissimilarity measures for clustering strings. In D. Sankoff and J. Kruskal, editors, *Time Warps, String Edits, and Macromolecules: The Theory and Practice of Sequence Comparison*, chapter 1, pages 311–321. Reprint, with a forward by J. Nerbonne, Stanford, CA: CLSI Publications, [1983] 1999.

[8] G. Cortelazzo, D. Deretta, G. A. Mian, and P. Zamperoni. Normalized weighted levensthein distance and triangle inequality in the context of similarity discrimination of bilevel images. *Pattern Recognition Letters*, 17:431–436, 1996.

[9] G. Cortelazzo, G. A. Mian, G. Vezzi, and P. Zamperoni. Trademark shapes description by string-matching techniques. *Pattern Recognition*, 27(8):1005–1018, 1994.

[10] R. Dubes and A. K. Jain. Validity studies in clustering methodologies. *Pattern Recognition*, 11:235–254, 1979.

[11] R. O. Duda, P. E. Hart, and D. G. Stork. *Pattern Classification*. Wiley, second edition, 2001.

[12] Y. El-Sonbaty and M. A. Ismail. On-line hierarchical clustering. *Pattern Recognition Letters*, pages 1285–1291, 1998.

[13] J. Barros et al. Using the triangle inequality to reduce the number of computations required for similarity-based retrival. In *Proc. of SPIE/IS&T, Conference on Storage and Retrieval for Still Image and Video Databases IV*, volume 2670, pages 392–403, 1996.

[14] M. Figueiredo and A. K. Jain. Unsupervised learning of finite mixture models. *IEEE Trans. Pattern Analysis and Machine Intelligence*, 24(3):381–396, 2002.

[15] W. B. Frakes. Stemming algorithms. In William B. Frakes and Ricardo Baeza-Yates, editors, *Information Retrieval: Data Structures and Algorithms*, chapter 8, pages 131–160. Prentice Hall, 1992.

[16] A. L. Fred. Clustering of sequences using a minimum grammar complexity criterion. In *Grammatical Inference: Learning Syntax from Sentence*, pages 107–116. Springer-Verlag, 1996.

[17] A. L. Fred. Finding consistent clusters in data partitions. In Josef Kittler and Fabio Roli, editors, *Multiple Classifier Systems*, volume LNCS 2096, pages 309–318. Springer, 2001.

[18] A. L. Fred and J. Leitão. A minimum code length technique for clustering of syntactic patterns. In *Proc. Of the 13th IAPR Int'l Conference on Pattern Recognition*, pages 680–684, Vienna, 1996.

[19] A. L. Fred and J. Leitão. Solomonoff coding as a means of introducing prior information in syntactic pattern recognition. In *Proc. of the 12th IAPR Int'l Conference on Pattern Recognition*, pages 14–18, 1994.

[20] A. L. Fred and J. Leitão. A comparative study of string dissimilarity measures in structural clustering. In Sameer Singh, editor, *Int'l Conference on Advances in Pattern Recognition*, pages 385–384. Springer, 1998.

[21] A. L. Fred and J. Leitão. Clustering under a hypothesis of smooth dissimilarity increments. In *Proc. of the 15th Int'l Conference on Pattern Recognition*, volume 2, pages 190–194, Barcelona, 2000.

[22] A. L. Fred, J. S. Marques, and P. M. Jorge. Hidden markov models vs syntactic modeling in object recognition. In *Int'l Conference on Image Processing, ICIP'97*, pages 893–896, Santa Barbara, October 1997.

[23] K. S. Fu. Syntactic pattern recognition. In *Handbook of Pattern Recognition and Image Processing*, pages 85–117. Academic Press, 1986.

[24] K. S. Fu and S. Y. Lu. A clustering procedure for syntactic patterns. *IEEE Trans. Systems Man Cybernetics*, 7(7):537–541, 1977.

[25] K. S. Fu and S. Y. Lu. Grammatical inference: Introduction and survey -part I and II. *IEEE Trans. Pattern Analysis and Machine Intelligence*, 8(5):343–359, 1986.

[26] J. A. Garcia, J. Valdivia, F. J. Cortijo, and R. Molina. A dynamic approach for clustering data. *Signal Processing*, 2:181–196, 1995.

[27] M. Har-Even and V. L. Brailovsky. Probabilistic validation approach for clustering. *Pattern Recognition*, 16:1189–1196, 1995.

[28] D. Harman. Ranking algorithms. In William B. Frakes and Ricardo Baeza-Yates, editors, *Information Retrieval: Data Structures and Algorithms*, chapter 14, pages 363–392. Prentice Hall, 1992.

[29] J. E. Hopcroft and J. D. Ullman. *Introduction to Automata Theory, Languages and Computation.* Addison-Wesley, London, 1979.

[30] Q. Huang, Z. Liv, and A. Rosenberg. Automated semantic structure reconstruction and representation generation for broadcast news. In *Proc. of IS&T/SPIE Conference on Storage and Retrieval for Image and Video Databases VII*, pages 50–62, 1999.

[31] A. Jain. *Fundamentals of Digital Image Processing.* Prentice-Hall, 1989.

[32] A. K. Jain and R. C. Dubes. *Algorithms for Clustering Data.* Prentice Hall, 1988.

[33] A.K. Jain, M. N. Murty, and P.J. Flynn. Data clustering: A review. *ACM Computing Surveys*, 31(3):264–323, September 1999.

[34] A. Juan and E. Vidal. Fast median search in metric spaces. In A. Amin and D. Dori, editors, *Advances in Pattern Recognition*, pages 905–912. Springer-Verlag, 1998.

[35] J. A. Kaandorp. *Fractal Modelling: Grouth and Form in Biology.* Springer-Verlag, 1994.

[36] R. L. Kashyap and B. J. Oommen. String correction using probabilistic models. *Pattern Recogition Letters*, pages 147–154, 1984.

[37] J. Kittler. Pattern classification: Fusion of information. In Sameer Singh, editor, *Int. Conf. on Advances in Pattern Recognition*, pages 13–22. Springer, 1998.

[38] J. Kittler, M. Hatef, R. P Duin, and J. Matas. On combining classifiers. *IEEE Trans. Pattern Analysis and Machine Intelligence*, 20(3):226–239, 1998.

[39] T. Kohonen. Median strings. *Pattern Recognition Letters*, 3:309–313, 1985.

[40] R. Kothari and D. Pitts. On finding the number of clusters. *Pattern Recognition Letters*, 20:405–416, 1999.

[41] J. Kruskal. An overview of sequence comparison. In D. Sankoff and J. Kruskal, editors, *Time Warps, String Edits, and Macromolecules: The Theory and Practice of Sequence Comparison*, chapter 1, pages 1–44. Reprint, with a forward by J. Nerbonne, Stanford, CA: CLSI Publications, [1983] 1999.

[42] S. Y. Lu and K. S. Fu. Stochastic error-correcting syntax analysis for the recognition of noisy patterns. *IEEE Trans. Computers*, C-26(12):1268–1276, December 1977.

[43] S. Y. Lu and K. S. Fu. A sentence-to-sentence clustering procedure for pattern analysis. *IEEE Trans. Systems Man Cybernetics*, 8(5):381–389, 1978.

[44] Y. Man and I. Gath. Detection and separation of ring-shaped clusters using fuzzy clusters. *IEEE Trans. Pattern Analysis and Machine Intelligence*, 16(8):855–861, 1994.

[45] C. D. Martinez-Hinarejos, A. Juan, and F. Casacuberta. Use of median string for classification. In *Proc. of the 15th Int'l Conf. on Pattern Recognition*, pages 907–910, 2000.

[46] A. Marzal and E. Vidal. Computation of normalized edit distance and applications. *IEEE Trans. Pattern Analysis and Machine Intelligence*, 2(15):926–932, 1993.

[47] G. McLachlan and K. Basford. *Mixture Models: Inference and Application to Clustering*. Marcel Dekker, New York, 1988.

[48] L. Miclet. Grammatical inference. In H. Bunke and A. Sanfeliu, editors, *Syntactic and Structural Pattern Recognition - Theory and Applications*, pages 237–290. Scientific Publishing, 1990.

[49] B. Mirkin. Concept learning and feature selection based on square-error clustering. *Machine Learning*, 35:25–39, 1999.

[50] T. Oates, L. Firoiu, and P. R. Cohen. Using time warping to bootstrap HMM-based clustering of time series. In R. Sun and C. L. Giles, editors, *Sequence Learning: Paradigms, Algorithms and Applications*, volume LNAI of *Lecture Notes in Computer Science*, pages 35–52. Springer-Verlag, 2000.

[51] B. J. Oomen and R. S. K. Loke. Pattern recognition of strings containing traditional and generalized transposition errors. In *Int'l Conf. on Systems, Man and Cybernetics*, pages 1154–1159, 1995.

[52] B. J. Oommen. Recognition of noisy subsequences using constrained edit distances. *IEEE Trans. Pattern Analysis and Machine Intelligence*, 9(5):676–685, 1987.

[53] N. R. Pal and J. C. Bezdek. On cluster validity for the fuzzy c-means model. *IEEE Trans. Fuzzy Systems*, 3:370–379, 1995.

[54] E. J. Pauwels and G. Frederix. Fiding regions of interest for content-extraction. In *Proc. of IS&T/SPIE Conference on Storage and Retrieval for Image and Video Databases VII*, volume SPIE Vol. 3656, pages 501–510, San Jose, January 1999.

[55] L. Rabiner. A tutorial on hidden Markov models and selected applications in speech recognition. *Proc. of IEEE*, 77(2):257–285, 1989.

[56] E. S. Ristad and P. N. Yianilos. Learning string-edit distance. *IEEE Trans. Pattern Analysis and Machine Intelligence*, 20(5):522–531, 1998.

[57] S. Roberts, D. Husmeier, I. Rezek, and W. Penny. Bayesian approaches to gaussian mixture modelling. *IEEE Trans. Pattern Analysis and Machine Intelligence*, 20(11), 1998.

[58] D. Sankoff and J. Kruskal. *Time Warps, String Edits, and Macromolecules: The Theory and Practice of Sequence Comparison.* Reprint, with a forward by J. Nerbonne, Stanford, CA: CLSI Publications, [1983] 1999.

[59] S. Santini and R. Jain. Similarity is a geometer. *Multimedia Tools and Applications*, 5(3):277–306, 1997.

[60] P. Sebastiani, M. Ramoni, and P. Cohen. Sequence clustering via bayesian clustering by dynamics. In R. Sun and C. L. Giles, editors, *Sequence Learning: Paradigms, Algorithms and Applications*, number 1828 in Lecture Notes in Computer Science, pages 11–34. Springer-Verlag, 2000.

[61] P. Smyth. Clustering sequences with hidden Markov models. In M. C. Mozer, M. I. Jordan, and T. Petsche, editors, *Advances in Neural Information Processing 9*, pages 72–93. MIT Press, Cambridge, 1997.

[62] R. J. Solomonoff. A formal theory of inductive inference (part I and II). *Information and Control*, 7:1–22,224–254, 1964.

[63] D. Stanford and A. E. Raftery. Principal curve clustering with noise. Technical report, University of Washington, http://www.stat.washington.edu/raftery, 1997.

[64] E. Tanaka. Parsing and error correcting parsing for string grammars. In *Syntactic and Structural Pattern Recognition – Theory and Applications*, pages 55–84. Scientific Publishing, 1990.

[65] H. Tenmoto, M. Kudo, and M. Shimbo. MDL-based selection of the number of components in mixture models for pattern recognition. In Adnan Amin, Dov Dori, Pavel Pudil, and Herbert Freeman, editors, *Advances in Pattern Recognition*, volume 1451 of *Lecture Notes in Computer Science*, pages 831–836. Springer Verlag, 1998.

[66] S. Theodoridis and K. Koutroumbas. *Pattern Recognition*. Academic Press, 1999.

[67] C. Zahn. Graph-theoretical methods for detecting and describing gestalt structures. *IEEE Trans. Computers*, C-20(1):68–86, 1971.

Pattern Recognition for Intrusion Detection in Computer Networks

Giorgio Giacinto and Fabio Roli
Department of Electrical and Electronic Engineering
University of Cagliari
09123 Cagliari, Italy
E-mail: {giacinto,roli}@diee.unica.it

Contents

1 Introduction

Nowadays an increasing number of commercial and public services are offered through the Internet, so that security is becoming a key issue. The so-called "attacks" on Internet service providers are carried out by exploiting both unknown weaknesses or bugs that are always contained in system and

D. Chen and X. Cheng (eds.),
Pattern Recognition and String Matching, 195–218.

application software, and complex unforeseen interactions between software components and/or network protocols [1], [2]. The objective of computer attacks is to obtain unauthorized access to the information stored in computer systems and/or to cause a temporary unavailability of its services. The so-called "first line" of defence against attacks is made up of a number of access restriction policies that act as a coarse grain filter. Intrusion detection systems (IDSs) are the fine grain filter placed inside the protected network, that look for known or potential threats in network traffic and/or in audit data recorded by hosts [2].

The first model proposed in the literature for a general purpose IDS relies on an expert system that monitors a system's audit records for abnormal patterns of system usage [3]. This approach is based on the knowledge of the behavior of system users, the knowledge being acquired from audit records according to a number of rules and being represented in terms of metrics and statistical models. An intrusion is detected when the actual usage profile deviates from the profile stored in the IDS. This approach originated a number of research and commercial IDSs usually called "anomaly detectors". This IDS model is capable of detecting intrusions regardless of the system type, the environment, the system vulnerabilities, and the type of intrusions, provided that the set of usage profiles stored in the system actually represents its "normal" working condition, and that any attack against the system involves its abnormal use. Unfortunately, the acquisition of profiles of "normal" activity is not an easy task. First of all, the audit records used to produce the profiles of normal activity may contain traces of intrusions not detected by the IDS. In this case intrusions are considered legitimate activities. In addition, imperfect normal profiles may cause a high rate of false alarms when the activity of legitimate users deviates from what is stored as being "normal" [4].

The difficulties in creating normal usage profiles for anomaly detectors forced researchers to follow a complementary approach based on the knowledge of known threats. Computer analysts have noted that each known intrusion leave a sort of "signature" in the stream of network packets and/or in audit records [5]. Known intrusions can thus be detected by analyzing network traffic and/or audit records in search for known "signatures". This process can be automated by coding the signatures into a set of rules and using an expert system to analyze the data. This approach, which is called "misuse detection", is the most common approach implemented in many commercial IDSs [1]. The effectiveness of signature-based IDSs is strictly related to the extent to which they are updated with the signatures of the latest attacks developed. This is currently a challenge since new attacks and new attack variants are constantly being developed. In particular attack variants exploit the vulnerability of current signature-based IDSs, i.e. the high specificity of attack signatures. At the time an attack signature is made publicly available, a number of attack variants are designed to produce the same effect as the original attack, but with a slightly different signature that is not detected by signature-based IDSs. The development of signatures with a limited scope is motivated by the following reasons: i) the difficulty in capturing the "root-cause" of an attack and ii) the requirement of a very small false alarm rate. If the signatures are too general, then high attack detection rates may be associated with unacceptable false alarm rates, as a relevant number of normal traffic events may match the signature. It is easy to see that the availability of signatures coding the "root-cause" of the attacks should protect from all attacks exploiting the same vulnerability. Unfortunately this is beyond the current state of the art of IDSs [2].

The above discussion points out that the two intrusion detection approaches are usually formulated in terms of "pattern matching" paradigms. "Anomaly detection" approaches classify as being intrusive all the activities not matching the normal profiles. On the other hand "misuse detection" approaches classify as being intrusive the activities matching one of the attack profiles. While pattern matching is effective when the objects to be classified exhibit quite a regular structure, this does not seem to be the most common case for current IDSs, because the development of pattern matching rules for both anomaly and misuse detection relies on the experience and intuition of human experts [6]. As a consequence, as pointed out in the above, such rules can hardly adapt to the high variability of "normal" activities or to the number of novel attacks constantly being developed. These difficulties in current IDSs lead researchers to apply statistical pattern recognition approaches based on the learning by examples paradigm [7]. The main motivation for using pattern recognition approaches for the development of advanced IDSs is their generalization capability, which may support the recognition of intrusions that have not been seen previously and have no previously described patterns. This formulation combines the advantages of signature-based and anomaly-based IDSs. It allows to design decision "boundaries" between normal and malicious network traffic. In addition learning by example paradigms avoid subjective human interpretation of intrusive behavior. At present research on advanced IDSs based on learning by example paradigms is at an early stage, so that a number of issues should be solved in order to be used in operational environments [6]. However the few works in the literature showed promising results, thus calling for further research [8],[9],[10],[11],[12],[13],[14].

This chapter aims at providing a review of the current state of the art in the development of IDSs based on pattern recognition techniques. In addition, a framework to developing pattern recognition techniques based on the Multiple Classifier System paradigm (MCS) is presented [15],[16]. MCSs are introduced to improve performances of pattern recognition techniques in terms of higher attack detection rates and lower false alarm rates. This result is achieved by training a number of classifiers on distinct feature representations, and then combining the results. While the distinct feature representations proposed in this chapter are extracted by a single data source, this paradigm may be used to perform the fusion of information from different sources, e.g. network traffic data, audit records, etc.

This chapter is organized as follows. A review of the current state of the art is given in Section 2. Section 3 presents a formulation of the intrusion detection problems as a pattern recognition task. In particular, three distinct feature representations are extracted from network traffic data. Section 4 illustrates an MCS approach to intrusion detection, while the results obtained on a public available data set are reported in section 5. Finally conclusions and future research directions are drawn in section 6.

2 Related Work on Pattern Recognition Approaches to Intrusion Detection

A recent technical report on current Intrusion Detection (ID) technology, where both commercial and research products are briefly reviewed, provides a discussion on the challenges to developing effective IDSs [6]. In particular it has been pointed out that advanced research issues on IDSs should involve the use of pattern recognition and learning by example approaches for the following three main reasons:

- the capability of learning by example approaches to generalize from a representative set of examples, allows to detect new types of intrusion;
- with learning by example approaches attack "signatures" can be extracted automatically from labeled traffic data, thus allowing to overcome the subjectivity of human interpretation of intrusive behavior, the latter being implemented in many current IDSs;
- learning by example approaches are able to adapt to new threats.

These issues have been addressed ever since the early years of IDS development. In particular the application of neural networks for IDSs has been investigated by a number of researchers. Neural networks provide a solution to the problem of modelling the users' behavior in anomaly detection because they do not require any explicit user model. In addition they can automatically learn attack signatures from attack samples. These signatures can be assumed to be more general than those coded by human experts thanks to the capability of neural networks to extract complex decision boundaries in the feature space.

Neural networks for intrusion detection have first been introduced as an alternative to statistical techniques in the IDES intrusion detection expert system to model users' behavior [8]. This system learns to predict the next command based on a sequence of previous commands by a specific user. An anomaly is detected if the command executed is not among the top k most likely commands. Similar approaches have been presented later [9]. Instead of learning to predict the sequence of commands, different solutions have been proposed where the distribution of commands executed by each user is learned. An anomaly is detected when based on the actual command distribution the user identified by the neural network is different from the actual user. A different approach to anomaly detection based on neural networks is proposed in [14]. While previous works have addressed the anomaly detection problem by analyzing the audit records produced by the operating system, in [14] anomalies are detected by looking at the usage of network protocols. The authors claim that the normal usage of network protocols is more predictable than user behavior. However, while a number of attacks can be detected by analyzing the protocol usage, in other cases it is difficult to separate legitimate from intrusive traffic [6].

A neural network model designed to perform anomaly and misuse detection has been proposed in [11]. The training set is made up of strings of events captured by the Base Security Module (BSM) that is part of many operating systems. If the training set is made up of strings related to normal behavior, neural networks act as an anomaly detector. On the other hand, if strings captured during an attack session are included in the training set, the network model can be modified to act as a misuse detection system.

Training sets made up of traffic data instead of audit records have also been used for misuse detection [10],[12]. Traffic features at different levels of abstraction have been used, from packet data [12] to very high level features, such as the security level of the source and destination machines, occurrences of suspicious strings, etc. [10]. For each neural network model, different numbers of output nodes have been selected according to the attack classification used. In some cases one network for each attack class has been used.

Recently neural networks have also been used for the improvement of network intrusion detection systems based on searching for attack-specific keywords in network traffic [13].

The above discussion points out that pattern recognition techniques based on neural networks are apt to provide a solution to some open issues in IDS development, e.g. the automatic extraction of normal and attack signatures from data and the ability to detect attacks not known at training time. In addition, the extensive evaluation of pattern classification techniques carried out on a sample data set of network traffic performed during the KDD'99 conference pointed out the feasibility of a pattern recognition approach to ID [17].

A common characteristic shared by all the pattern recognition approaches reviewed is the use of a feature space made up of features related to information at different abstraction levels. As an example, the detection of attacks typically requires the recognition of a particular sequence of events. Therefore an event (e.g., an audit record or a network connection) cannot be classified according to its intrinsic characteristics alone, but requires also the knowledge of past events. Except for the neural network hierarchy proposed in [14] where features at different abstraction levels are processed by distinct networks, the classification is usually performed in the feature space made up of all the features needed to detect the considered attack classes. Recent advances in pattern recognition theory suggest that classification performances may be improved by combining an ensemble of classifiers trained on different feature representations [15],[16]. First a decision is made by individual classifiers on the basis of partial information, e.g. intrinsic information, knowledge of past events, etc. Then the decisions are combined by means of suitable decision combination functions. Classifier combination performances are usually higher than those of individual classifiers. In addition, a combined system exhibits higher robustness with respect to uncertainties in training data. This is an important feature in IDSs because of the difficulties collecting a representative set of examples [4]. In the following section a formulation of the intrusion detection task as a pattern recognition problem is proposed. In section 4 an approach to intrusion detection based on the paradigm of multiple classifier systems is presented.

3 Intrusion Detection as a Pattern Recognition Task

Intrusions in computer networks basically exploit two different kinds of weaknesses: weaknesses of the network transmission protocol (network-based intrusions) and weaknesses and bugs exhibited by system and application software (host-based intrusions). As a consequence network-based intrusions can be traced by collecting statistical information about the traffic into and out of the network, while host-based intrusions can be traced by collecting information about the programs executed by each computer in the network. It is worth noting that the vast majority of host based information can also be extracted from the network traffic flow. Therefore intrusion detection systems based on network traffic processing are also capable of detecting a number of host-based intrusions. However, as the information is transmitted over the networks in small units of data called "packets", it is easy to see that the same information contained in few records of host audit data may need to process hundreds or thousands of packets [2]. As a consequence, the analysis of traffic data alone provides enough information to detect a large number of attacks though it requires a high computational cost. On the other hand the "fusion" of information collected by different network components is currently a challenge to computer network management and intrusion detection [19]. Accordingly, the proposed pattern recognition approach will be focused on network intrusion detection.

From the pattern recognition point of view, the network intrusion detection problem can be formulated as follows (see Figure 1). Given the information about network connections between pairs of hosts, assign each connection to one out of N data classes representing normal traffic or different categories of intrusions (e.g., Denial of Service, access to root privileges, etc.). The use of different attack categories places the proposed formulation among the misuse detection techniques. It is worth noting that various definitions of attack classes are possible [4].

Traffic data is first processed in order to identify network connections between hosts. The term "connection" refers to a sequence of data packets related to a particular service between a pair of hosts, e.g., the transfer of a web page via the http protocol. Multiple connections may be active for a given pair of hosts. For example, the visualization of a web page may require dozens of connections depending on the richness of the page content, e.g. images, forms, etc. As the aim of a network intrusion detector is to detect those connections that are related to malicious activities, each network connection can be defined as a "pattern" to be classified. This formulation is in agreement with the related work presented in Section 2.

The extraction of suitable features representing network connections is based on expert knowledge about the characteristics that better discriminate attacks from normal connections. These characteristics are at different

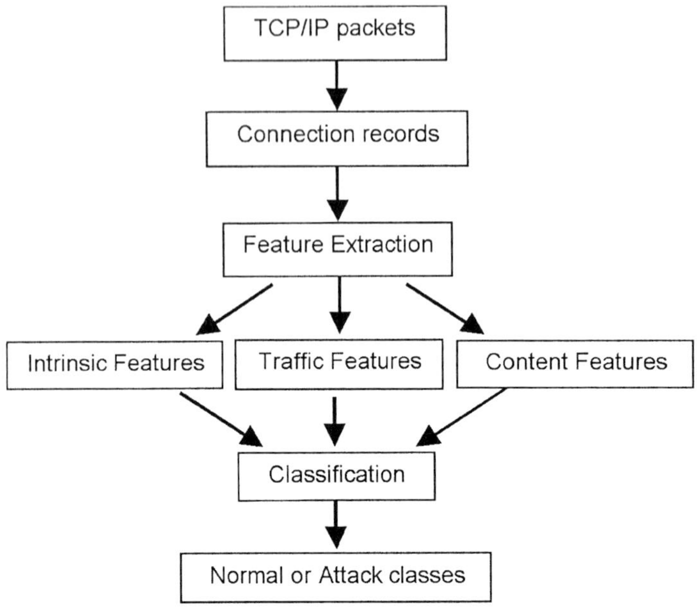

Figure 1: Intrusion Detection formulated as a Pattern Recognition problem

abstraction levels, so that their computation has different costs. Previous works on feature extraction for IDSs clearly identified three distinct feature representations [18]:

intrinsic features, i.e., general information related to the connection. These include the duration, type, protocol, flag, etc. of the connection;

traffic features, i.e., statistics related to past connections similar to the current one e.g., number of connections with the same destination host or connections related to the same service in a given time window or within a predefined number of past connections;

content features, i.e., features containing information about the data content of packets ("payload") that could be relevant to discover an intrusion, e.g., errors reported by the operating system, root access attempts, etc.

The first feature set is made up of measures related to each connection. This kind of information alone is able to provide a rough discrimination between normal traffic and some simple attack types. For example, normal connections typically involve the exchange of a significant amount of data, while connections related to some kinds of intrusions often transfer a small amount of data.

Other attack types cannot be detected on the basis of information related to individual connections because the effect of the attack depends on a sequence of similar connections, so that each connection alone is not an attack per se but only in relation to other connections. The measures belonging to the second set of features are aimed to detect these kinds of attacks and are currently used by human experts to code attack signatures.

The third feature set is related to the data portion of the packets and looks for particular strings that are "signatures" of attacks. This set of features is strictly related to the particular system and application software protected. It is worth noting that these kinds of intrusions can be better detected by host-based systems that analyze audit records and are designed on the particular software that must be protected. For some frequent intrusions, such as repeated trials for guessing the password of a system, they can be easily detected from packet data, while more specific intrusions, such as the exploitation of vulnerabilities related to a particular release of software installed on a host, can be detected more efficiently by host-based systems. It is worth remarking that effective intrusion detection requires the "fusion" of multiple sources and this is currently an open problem [19].

In our opinion, the proposed feature categorization is general enough to take into account the variability of network traffic and the different abstraction levels required to detect attacks. In addition it is worth noting that all the features used by current IDSs fall into one of the above feature categories. Usually a hand-coded signature involves the evaluation of a set of features belonging to different feature categories. It is worth noting that, except for very simple attacks, the choice of the more suitable features for signature-based attack detection is not an easy task, as "traces" of an attack can usually be detected by looking at different features. The choice depends essentially on human expert knowledge and intuition about the effectiveness of the signature in terms of high attack detection accuracy and low false alarms rates [6]. High false alarms rates may be generated if the selected features match attacks as well as some istances of normal traffic [5].

On the other hand the use of a pattern recognition approach may exploit the richness of information provided by all the available features extracted from network traffic data. More complex signatures may be coded, thus allowing for more effective separation of attacks from normal traffic. As a consequence, a pattern recognition approach may provide higher detection accuracy than by hand-coded signatures. In addition the generalization capability of pattern recognition algorithms allows for the detection of novel attacks.

4 A multiple classifier approach to network intrusion detection

In the previous section we have pointed out that three types of features can be extracted from network traffic data. Each feature category provides information that can be used to discriminate between attacks and normal traffic. In particular, when an attack is performed against a computer network, a "signature" related to that attack may be found in each feature category. As these "signatures" exhibit different discriminative power with respect to attack detection, network analysts produce an effective attack "signature" by selecting the more effective subsets of features according to their experience and intuition. On the other hand, pattern recognition tools have been designed to process the entire available feature set to extract more effective signatures than the ones hand-coded by network analysts (see Section 2). A pattern recognition approach based on the multiple classifier paradigm may exploit the above observation that attack evidence may be collected separately in different feature subspaces. First each feature subspace is used independently to perform attack detection. Then the evidences are combined in order to produce the final decision (see Figure 2). This process reflects the human analyst perspective that usually looks at different traffic statistics in order to produce reliable attack signatures, i.e. signatures providing effective attack detection and a very low false positive rate. It is worth noting that the reliability of an attack signature depends not only on its ability to detect all attacks for which the signature has been designed, but also on its specificity, so that very few normal connections match that signature. To make a signature more specific, multiple attack evidences must be checked in order to prevent normal connections from matching the signature and originating false alarms.

Recently, it has been pointed out that a very low false alarm rate is a key feature that advanced IDSs should exhibit in order to be effective [20]. This requirement can easily be explained by observing that the vast majority of network traffic is usually made up of "normal" connections, while attacks are rare events. As a consequence, small values of false alarm rates may originate a large number of false alarms per day thus making an IDS useless. On the other hand, as attacks are rare events, high detection rates are required. It is easy to see that there is a trade-off between the ability of an IDS to generalize

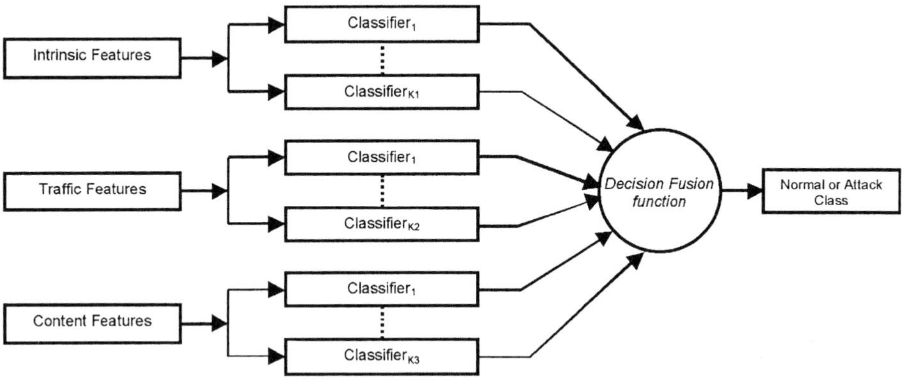

Figure 2: A Multiple Classifier System for Intrusion Detection

to novel attacks and the requirement of low false alarm rates, as the latter require specific signatures. The multiple classifier approach may satisfy this requirement as the combination of multiple evidences usually exhibits higher accuracies, i.e. lower false positives, than individual decisions. In addition, the generalization capabilities of pattern recognition algorithms allow for the detection of novel attacks.

In order to illustrate this approach, three simple fusion techniques may be used: the majority voting rule, the average rule, and the "belief" function. These fusion techniques compute the final decision from the set of decisions of an ensemble made up of K classifiers. The "majority voting rule" assigns a given input pattern to the majority class among the K outputs of the classifiers combined. If K=3 a network connection is classified as an attack if at least two classifiers agree on the same decision. The "average rule" assigns a given input pattern to the class with the maximum average posterior probability, the average being computed among the K classifiers (this rule can be applied if classifiers provide estimates of posterior probabilities, like multi-layer perceptron neural networks). The third fusion rule is based on the computation of a "belief" value for each data class, given the set of outputs of the K classifiers. Belief values are based on estimates of the probabilities that a pattern assigned to a given data class actually belongs to that class or to other classes. These probabilities can easily be computed from the confusion matrix on the training set. The classification is then performed by assigning the input pattern to the data class with the maximum "belief" value. The reader is referred to [21] for more details on the above combination methods.

These methods for combining classifier outputs are based on the assumption that the outputs of different classifiers are "independent". This assumption cannot be verified in practice, though in many real cases it may reasonably hold. In particular it may reasonably hold for the problem at hand, since the three feature sets are related to independent connection characteristics. For example, for a given set of values of the intrinsic features (such as the number of bytes transmitted), there is no relationship with the values assumed by the content features (e.g., an attempt to log in the system as user "root"). For this reason it can be argued that the combination of classifiers trained on different feature sets could provide better performances than each single classifier. In addition, performances should also be higher than those of classifiers based on a single feature vector containing all the available features. In particular it is expected that a low false alarm rate can be obtained by an MCS approach, as attack detection depends on multiple decisions.

5 Experimental Results

Experiments have been carried out on a subset of simulated data created by DARPA in the framework of the 1998 Intrusion Detection Evaluation Program [22]. This subset of data was pre-processed by the Columbia University and distributed as part of the UCI KDD Archive [23]. The available data set is made up of a large number of network connections related to normal and malicious traffic. Each connection is described by a 41 dimensional feature vector according to the feature extraction process illustrated in Section 3 [18]. Connections are also labelled as belonging to one of five classes, i.e. normal traffic, Denial of Service (DoS) attack, Remote to Local (R2L) attack, User to Root (U2R) attack, and Probing attack. The DoS attack type refers to attacks that cause the unavailability of a particular service (e.g., the http or ftp service). R2L attacks are those aimed at gaining unauthorized access to a local host from a remote machine, while U2R attacks are designed to gain unauthorized access to superuser privileges. Probing activities are not attacks in themselves since they simply aim at gathering information about the available services at the destination host. They are considered a potential hazard, since probing activities are carried out by attackers in order to check for the vulnerabilities of the target computer system.

In order to test the pattern recognition approach we restricted our investigation to connections related to the ftp service. Attacks against other services may be detected either by devoting distinct modules for different services or by using a unified pattern recognition module for all services, provided that at least one feature is related to the service to which the connection refers. Discussions on the development of the most suitable pattern recognition model to detect attacks against different services is beyond the scope of this chapter. Connections related to the ftp service are represented by a feature set containing 30 out of the 41 available features of the data set. It is worth recalling that the full feature set has been designed to detect attacks against a number of services. It was therefore possible to discard 11 features out of 41 because 9 of them were related to other services (e.g., the telnet or http service) and exhibit a constant value over all ftp connections, while the other 2 represent the type of protocol and service whose values are always TCP and ftp respectively for ftp connections. The description of the 30 features used to represent the connection is the following:

- 4 intrinsic features: duration of the connection; number of data bytes from source to destination; number of data bytes from destination to source; flag for normal or error status of the connection;

- 19 traffic features, subdivided into 9 "time-based" traffic features and 10 "host-based" traffic features. The 9 time-based traffic features are further subdivided into two sets: features related to the "same host", computed on the connections in the past 2 seconds that have the same destination host as the current connection; and features related to the "same service", computed on the connections in the past 2 seconds that have the same service as the current connection. The feature set related to the "same host" is made up of the number of connections, % of connections that have SYN errors, % of connections that have REJ errors, and % of connections to the same service % of connections to different services. The feature set related to the "same service" is made up of the number of connections, % of connections that have SYN errors, % of connections that have REJ errors, and % of connections to different hosts.
  The 10 host-based traffic features are a "mirror" set of the time-based traffic feature set based on a window of 100 connections with the same destination host as the current connection. This feature set is made up of the above 9 features and a new additional feature that takes into account the fraction of connections with the same source port;

- 7 content features: number of "hot indicators"; number of failed login attempts; binary value for successful or failed login attempt; number of file creation operations; number of write, delete and create operations on access control files; binary value for the login or not to the "hot" list (e.g., root, adm, etc.); binary value for the login or not to the "guest" login.

For more details about the feature extraction process the reader is referred to [18]. Feature values have been linearly normalized so that their value is always within the range 0,..,1.

A training set made up of 122 normal connections, 6 connections of U2R attacks, 539 connections of R2L attacks, 1 connection of probing, and 57 connections of DoS attacks, corresponding to a total of 725 ftp connections were extracted from the available data set. A test set made up of 7,436 connections was also extracted. In particular, it is made up of 5,128 connections related to normal traffic, 2,183 connections of attack types included in the training set, and 125 connections related to attack types not included in the training set. The classes in the training set are unbalanced due to the fact that different attacks are carried out by opening a different number of connections. This is a well-known problem for pattern recognition techniques. At present, we have solved the problem of unbalanced classes in the training set by augmenting the number of patterns related to poorly represented traffic types. In particular we carried out a number of preliminary classification experiments in order to identify which traffic types need to be augmented to produce reliable performances on the test set. Consequently, we populated the training set by a number of copies of these connections, so that the training set used in the experiments was made up of 157 normal samples, 46 U2R attacks, 598 R2L attacks, 25 probing, and 57 DoS attacks, corresponding to a total of 833 ftp connections. Details on the attack types included in the training and test sets are reported in Table 1. This heuristic approach is in agreement with the practical observation that if the same attack type is carried out a number of times, its connections exhibit the same feature values (i.e., these feature values correspond to the signature of that attack). Further work on the problem of unbalanced classes is obviously necessary.

The reported experiments were aimed at:

Assessing the performances of the multiple classifier paradigm applied to network intrusion detection;

Comparing the performances of pattern recognition techniques based on the evaluation of the entire feature vector with those of the combination

Data Classes	Attack Type	# training patterns	# test patterns
Normal		157	5128
U2R	Buffer Overflow	21	2
U2R	Rootkit	25	6
U2R	Ps	-	6
U2R	Xterm	-	2
R2L	Multihop	33	5
R2L	Warezmaster	535	537
R2L	Httptunnel	-	1
R2L	Guess Password	-	100
R2L	Named	-	2
R2L	Ftp write	30	3
Probing	Portsweep	25	16
Probing	Nmap	-	1
Probing	Satan	-	5
Probing	Saint	-	8
DoS	Neptune	57	1614

Table 1: Composition of the training and test sets used in the experiments on the FTP service. The test set was made up of attack types included and not in the training set

of classification results obtained by processing the three subsets of features separately;

Assessing the ability to detect correctly attack types not included at training time.

In order to compare performances, classification results will be evaluated according to the following measures:

Overall classification error;

Average classification cost, where the classification cost is obtained by summing each entry of the confusion matrix multiplied by the corresponding entry in the "cost" matrix reported in Table 2. Other researchers have used the cost matrix shown in Table 2 to weight errors according to their severity [17]. The classification cost is then averaged over the number of tested patterns. It is worth noting that "false positive" errors, i.e. classifying an attack as a normal traffic pattern, have a larger weight than misclassification among attacks. The only exception is the "Probing" class, as it is not an attack in itself but rather represents an attempt to discover the vulnerabilities of the probed system.

	Assigned Class				
True Class	**Normal**	**U2R**	**R2L**	**Probing**	**DoS**
Normal	0	2	2	1	2
U2R	3	0	2	2	2
R2L	4	2	0	2	2
Probing	1	2	2	0	2
DoS	2	2	2	1	0

Table 2: Cost matrix used to evaluate classification results. Each entry is used to weight the errors of the corresponding entry in the confusion matrix. Row labels refer to the true classes while column labels refer to classes assigned by the classifier

False alarm rate, i.e. the fraction of patterns assigned to one of the attack classes whose true class is the "normal" class. This is an important characteristic for IDSs, since they should exhibit a low rate of "false alarms", besides a high rate of detection of attacks;

Classification error related to attack types included in the training set and related classification cost, averaged over all patterns related to these attack types. These performances are aimed at evaluating the ability of the proposed technique to correctly detect attack types included in the training set;

Classification error related to attack types not included in the training set and related classification cost, averaged over all patterns related to these attack types. These performances are aimed at evaluating the ability to generalize to attack types not included in the training set.

Experiments have been carried out using neural networks as pattern classifiers. Neural networks have been trained using the backpropagation algorithm, with different learning rates and random starting weights. The reported results represent the best performances attained on the test set.

The first experiment was aimed at assessing the performances of neural networks designed to classify patterns represented by a 30-dimensional feature vector, i.e. by considering a single feature vector made up of all the available features. The results of this experiment will be compared with the proposed multiple classifier approach. Two neural network architectures have been considered. Both architectures are made up of three neuron layers: an input layer with 30 neurons, a hidden layer with 15 neurons, and an output layer with 5 neurons, one neuron for each of the data classes. One

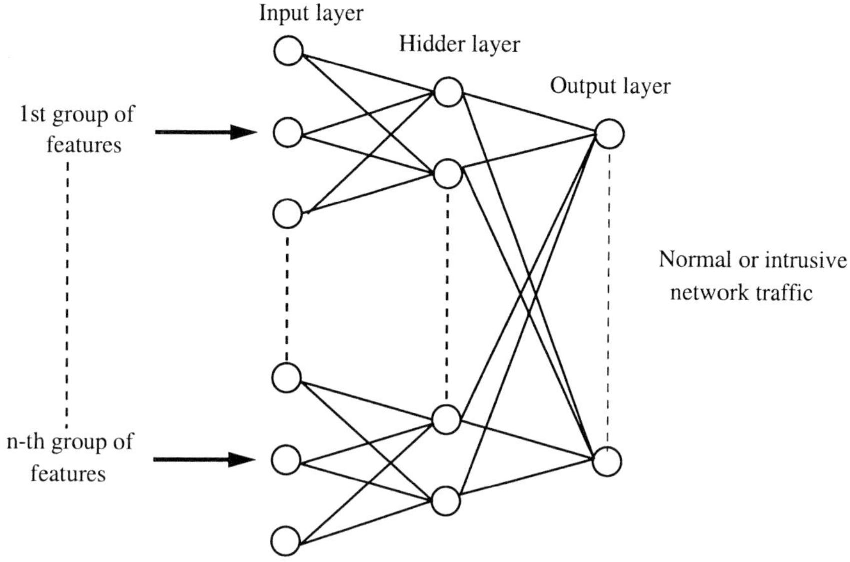

Figure 3: Structured neural network for intrusion detection

architecture is a fully connected multi-layer perceptron (MLP), where each node of the input layer is connected with all the nodes of the hidden layer, and each node of the hidden layer is connected to all the output nodes.

The other architecture is "structured" according to the feature categorization presented in Section 3. As shown in Figure 3, the input nodes related to each feature set are mapped on separate sets of nodes of the hidden layer whose number is equal to the number of output classes. The outputs of the hidden layer are then "combined" in the nodes of the output layer. In particular, the first 4 input neurons corresponding to the first group of features are fully connected to the first five neurons of the hidden layer; a second group made up of the successive 7 input neurons are fully connected to other 5 hidden neurons; the remaining 19 input neurons are fully connected to the remaining 5 neurons of the hidden layer; finally all the 15 hidden neurons are fully connected to the output neurons. This architecture may be interpreted as follows. Connections between input nodes and hidden nodes act as separate single layer perceptrons (SLPs), each providing outputs for all the data classes. Connections between the hidden layer and the output layer play the role of weights of "linear" combinations of the "outputs" of SLPs, so that each output of the neural network depends on a linear combination of all the nodes in the hidden layer. In other words, each output of the

	Overall Performances		
Classifier Type	% error	Average cost	% false alarms
Fully connected MLP	1.55	0.029	3.57
Structured MLP	0.83	0.016	1.42

Table 3: Results obtained on the test set of two neural network architectures with 30 input nodes, 15 hidden nodes, and 5 output nodes. The hidden layer of the structured architecture is designed to process the three groups of features separately, while the hidden layer of the fully connected architecture is designed to process the entire feature set

	known attacks		unknown attacks	
Classifier Type	% error	Average cost	% error	Average cost
Fully connected MLP	0.64	0.011	12.00	0.192
Structured MLP	0.55	0.010	13.60	0.248

Table 4: Results obtained on the test set of the two neural network architectures with 30 input nodes. The table shows attack detection performances related to known attacks, i.e. attacks included in the training set, and unknown attacks, i.e. attacks that were not included in the training set.

network is a function of a linear combination of all the outputs of the SLPs. This network architecture has been designed for comparison purposes, as it exploits the same rationale of the multiple classifier approach, i.e. features belonging to different feature sets are processed separately by different nodes in the hidden layer. It is worth noting that similar approaches have been introduced in other application domains, such as multisensor remote sensing image classification [24].

The classification results on the test set are reported in Tables 3 and 4. The proposed structured architecture outperforms the fully connected architecture both in terms of the overall error and of the classification cost. The detailed analysis of the performances shows that the structured architecture exhibits higher performances in terms of false alarm rate and detection of known attack types, i.e. attack types included in the training set. On the other hand, the fully connected architecture, exhibits higher generalization capabilities, since it is capable of detecting a higher number of unknown attacks, i.e. attack types not included in the training set. These results confirm that there is a trade-off between false alarm rate and novel attack detection. It can be concluded that the structured architecture should be preferred as it exhibits a lower false alarm rate at the price of a slight decrease in attack

	Overall Performances		
Classifier Type	% error	Average cost	% false alarms
MLP - 4 intrinsic features	1.51	0.031	3.19
MLP - 7 content features	1.20	0.024	2.25
MLP - 19 traffic features	9.83	0.200	23.94

Table 5: Overall performances obtained on the test set of the three neural networks trained on distinct feature representations.

	known attacks		unknown attacks	
Classifier Type	% error	Average cost	% error	Average cost
MLP - 4 intrinsic features	0.73	0.017	16.00	0.320
MLP - 7 content features	0.82	0.017	14.40	0.280
MLP - 19 traffic features	2.11	0.047	31.20	0.880

Table 6: Results obtained on the test set of the three neural networks trained on distinct feature representations. The Table shows attack detection performances related to known and unknown attacks.

detection of novel attacks. In addition the structured architecture provides a higher percentage of detection of known attacks than the fully connected MLP.

In order to test the proposed multiple classifier approach, three neural networks have been trained on distinct feature representations, namely the 4 intrinsic features, the 7 content features, and the 19 traffic features. The three networks have the same fully connected architecture with 5 hidden neurons, 5 output neurons (the number of data classes), and a number of inputs equal to the number of features. The performances of these networks are reported in Tables 5 and 6. The overall performances of neural networks, except the network trained on traffic features, are quite similar to each other, the network trained on the content features providing the best results. This result indicates that the content feature set is well suited for the type of traffic at hand (ftp traffic), while the reverse is true as regards the traffic feature set. Table 6 reports performances of the three neural networks on known and unknown attacks. While performances of the network trained on the traffic features exhibit high error rates and high costs in agreement with its overall performances, the two networks trained on the intrinsic and content features show a similar behavior. In particular, they exhibit a good generalization ability with respect to unknown attacks. However, as can be expected, the best results in terms of attack detection are provided by the networks trained on the overall feature set.

	Overall Performances		
Combination Technique	% error	Average cost	% false alarms
Majority	0.89	0.018	1.29
Bayesian Average	0.87	0.017	1.33
Belief functions	0.82	0.014	0.87

Table 7: Overall performances obtained on the test set of the three methods used to combine neural networks trained on distinct feature representations.

	known attacks		unknown attacks	
Combination Technique	% error	Average cost	% error	Average cost
Majority	0.64	0.014	17.60	0.344
Bayesian Average	0.60	0.013	16.80	0.296
Belief functions	0.87	0.015	17.60	0.248

Table 8: Results obtained on the test set of the three methods used to combine results of the three neural networks trained on distinct feature representations. The Table shows attack detection performances related to known and unknown attacks.

The combination of the results of the above three networks trained on distinct feature representations are reported in Tables 7 and 8. It can be observed that the multiple classifier approach allows to improve the overall performances with respect to the individual classifiers, as well as to the neural networks trained on the 30-dimensional feature vector. In particular the combination by belief functions provides the best performances in terms of error rate, average classification cost, and false alarm rate. Good performances are also obtained with respect to the classification of known attacks, the two networks trained on the entire feature space performing slightly better. With respect to the ability of detecting unknown attacks, the best performances are attained by the neural networks trained on the 30-dimensional feature vector. It is worth noting that the analysis of classification costs, that take into account the severity of errors, shows that the differences in performances between combination methods and networks trained in the entire feature space are smaller. Again, these results are in agreement with the trade-off between novel attack detection capability and generation of large numbers of false alarms.

It can be concluded that the MCS paradigm provides an effective solution to the use of Pattern Recognition techniques in IDSs. In addition the MCS paradigm may also be used as a tool to address the open issue of data fusion for IDSs [19].

6 Conclusions

Current IDSs are designed to detect intrusions that exploit known vulnerabilities by coding exact sequences of events related to those intrusions. Attack signatures tailored to each particular attack instance are required in order to avoid the risk of unfeasible false alarm rates. Unfortunately the same vulnerability can often be exploited by minor changes in the attack sequence (e.g., modifying the transfer rate of malicious packets). As soon as an attack signature is developed and made publicly available, attackers slightly modify the attack sequence so that it does not match perfectly the attack signature stored in the IDSs. As a consequence these kinds of signatures cannot detect novel attacks or attack variations, thus requiring updating the IDSs.

Since it is not easy for human experts to code general attack signatures, researchers have recently proposed to use pattern recognition techniques to provide IDSs with generalization capabilities. In particular, pattern recognition tools are used to extract more complex and effective signatures than those hand-coded by human experts. To this end patterns have often been represented by a single feature vector made up of a set of features carrying information of a different kind. While the reported results show the effectiveness of these pattern recognition approaches, the generalization capabilities of pattern recognition tools may originate high false alarm rates, as it is very difficult to collect a representative set of normal traffic samples [6].

The proposed MCS approach based on distinct feature representations has been designed to provide a better trade-off between generalization capabilities and false alarm rate. It is worth noting that the proposed feature cathegorization is not unique, and may be expanded to include information from different sources. In addition, feature sets made up of the union of subsets extracted from the proposed feature categories may also be considered. The reported results showed the effectiveness of MCSs in providing more reliable results, as the final decision depends on the agreement among distinct classifiers. Other combination schemes should be devised to further improve the presented figures. In addition, more extensive testing is required to compare IDSs based on pattern recognition tools with traditional IDSs.

At present the following issues remain open for effective implementation of IDSs based on pattern recognition techniques:

- very little research has been carried out to produce a representative set of examples of normal traffic and attacks for training and testing

[4]. As a consequence pattern recognition techniques may suffer from insufficient training data;

- the intrusion detection problem formulated in terms of pattern recognition falls in the class of problems where data classes are very unbalanced, as attacks are very rare compared to normal traffic. The most suitable solution to this problem should be investigated further;
- as different classification errors exhibit different severities, misclassification costs may be introduced at training time to satisfy the requirement of the final user. Unfortunately, at present there is no generally agreed upon figure of merit that can be used to characterize the performance of IDSs [6].

In our opinion MCSs may represent a feasible solution to address the first two problems, as they are designed to exhibit higher robustness compared to uncertainties in training data. In addition this framework allows to address other advanced topics of IDS development such as the fusion of data from different sources.

References

[1] J. McHugh, A. Christie, and J. Allen, *Defending Yourself: The Role of Intrusion Detection Systems*, (IEEE Software, Sept./Oct. 2000) pp. 42-51.

[2] P.E. Proctor, *The Practical Intrusion Detection Handbook*, (Prentice Hall, 2001).

[3] D.E. Denning, *An Intrusion-Detection Model*, (IEEE Trans. on Software Engineering, SE-13(2), 1987) pp. 222-232.

[4] J. McHugh, *Testing Intrusion Detection Systems: A Critique of the 1998 and 1999 DARPA Intrusion Detection System Evaluations as Performed by Lincoln Laboratory*, (ACM Transactions on Information and System Security, 3(4), 2000) pp. 262-294.

[5] S. Northcutt, M. Cooper, M. Fearnow, K. Frederick, *Intrusion Signatures and Analysis*, (New Riders, Indianapolis, 2001).

[6] J. Allen, A. Christie, W. Fithen, J. McHugh, J. Pickel, E. Storner, *State of the Practice of Intrusion Detection Technologies*, (Tech. Rep. CMU/SEI-99-TR-028, 2000)

[7] R. Duda, P. Hart, D.G. Stork, *Pattern Classification*, (John Wiley & Sons, 2001)

[8] H. Debar, M. Becker, D. Siboni, *A Neural Network Component for an Intrusion Detection System*, (Proc. of the IEEE Symp. on Research in Security and Privacy, Oakland, CA, USA, 1992), pp. 240-250.

[9] J. Ryan, M.J. Lin, R. Miikkulainen, *Intrusion Detection with Neural Networks*, (in: Advances in Neural Information Processing Systems 10, M. Jordan et al., Eds., Cambridge, MA: MIT Press, 1998) pp. 943-949.

[10] J.M. Bonifacio et al., *Neural Networks Applied in Intrusion Detection Systems*, (Proc. of the IEEE World congress on Comp.Intell. (WCCI '98), 1998).

[11] A.K. Ghosh and A. Schwartzbard, *A Study in Using Neural Networks for Anomaly and Misuse Detection*, (Proc. of the USENIX Security Symposium, August 23-26, 1999, Washington, USA).

[12] J. Cannady, *An Adaptive Neural Network Approach to Intrusion Detection and Response*, (PhD Thesis, School of Comp. and Inf. Sci., Nova Southeastern University, 2000).

[13] R.P. Lippmann and R.K. Cunningham, *Improving Intrusion Detection Performance Using Keyword Selection and Neural Networks*, (Computer Networks, 34, 2000) pp. 597-603.

[14] S.C. Lee, D.V. Heinbuch, *Training a Neural-Network Based Intrusion Detector to Recognize Novel Attacks*, (IEEE Trans. on Systems, Man, and Cybernetics, Part A, 31, 2001), pp. 294-299.

[15] J. Kittler, M. Hatef, R.P.W. Duin, J. Matas, *On Combining Classifiers*, (IEEE Trans. on Pattern Analysis and Machine Intelligence, 20(3), 1998), pp. 226-229.

[16] J. Kittler and F. Roli (eds.), *Multiple Classifier Systems*, (LNCS 2096, Springer, 2001).

[17] C. Elkan, *Results of the KDD'99 Classifier Learning*, (ACM SIGKDD Explorations, 1, 2000), pp. 63-64.

[18] W. Lee and S.J. Stolfo, *A Framework for Constructing Features and Models for Intrusion Detection systems*, (ACM Trans. on Information and System Security, 3(4), 2000) pp.227-261.

[19] T. Bass, *Intrusion Detection Systems and Multisensor Data Fusion*, (Communications of the ACM, 43(4), 2000), pp. 99-105.

[20] S. Axelsson, *The Base-Rate Fallacy and the Difficulty of Intrusion Detection*, (ACM Trans. on Information and System Security, 3(3), 2000), pp. 186-205.

[21] L. Xu, A. Krzyzak and C.Y. Suen, *Methods for Combining Multiple Classifiers and Their Applications to Handwriting Recognition*, (IEEE Trans. Systems, Man and Cybernetics 22, 1992) pp. 418-435.

[22] http://www.ll.mit.edu/IST/ideval

[23] http://kdd.ics.uci.edu/databases/kddcup99/kddcup99.html

[24] F. Roli, *Multisensor Image Recognition by Neural Networks with Understandable Behaviour*, (International Journal of Pattern Recognition and Artificial Intelligence, 10, 1996) pp. 887-917.

Model-Based Pattern Recognition

Michal Haindl
Department of Pattern Recognition
Institute of Information Theory and Automation, Academy of Sciences CR
Prague, Czech Republic
E-mail: haindl@utia.cas.cz

Contents

D. Chen and X. Cheng (eds.),
Pattern Recognition and String Matching, 219–254.

1 Introduction

Recognition and processing of multi-dimensional data (or set of spatially related objects) is more accurate and efficient if we take into account all interdependencies between single objects. Objects to be processed like for example multi-spectral pixels in a digitized image, are often mutually dependent (e.g., correlated) with a dependency degree related to a distance between two objects in their corresponding data space. These relations can be incorporated into a pattern recognition process through appropriate multi-dimensional data model. If such a model is probabilistic we can use consistent Bayesian framework for solving many pattern recognition tasks. Data models are simultaneously useful to specify natural constraints and general assumptions about the physical world and a data capturing process hence they are essential in many data modelling or analytical procedures such as classification, segmentation, discontinuity detection, restoration, enhancement and scene analysis in general. Features derived from multi-dimensional data models are information preserving in the sense that they can be used to synthesize data spaces closely resembling original measurement data space as can be illustrated on the texture modelling application in the section 2.4.

Topological relations between objects are very often expressed by their indexation. Multi-dimensional data elements can be indexed (using multi-indices $r = \{r_1, r_2, \ldots\}$) on some regular finite or infinite lattices I, $r \in I$ (Fig.1) or even on completely irregularly distributed sites like vertices of some graph structure (Fig.2-left). Neighbourhood structures and indexation lattices can be even more complicated for higher dimensions and multiresolution data representations (Fig.2-right). Finite data lattices additionally require some assumption for their boundary conditions (e.g., free boundary conditions, toroidal boundary conditions, etc.) which can significantly influence the numerical efficiency of resulting algorithms (e.g., Markovian models). For example a typical digital image is represented by spectral variations defined over a finite rectangular lattice. In the case of usual colour image this vector has three components but multi-spectral images used in remote sensing applications can have tens of such components for every multi-

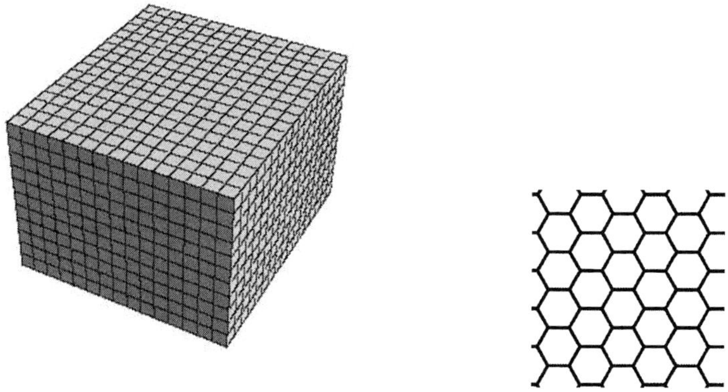

Figure 1: Regular 3D rectangular and 2D hexagonal lattices.

spectral pixel vector and the underlying index lattice is three-dimensional (Fig.1-left). Digital video data have the additional time component thus a four dimensional index lattice is needed. Such large amount of data is difficult to analyse, store or transport. Efficient processing of large data sets requires an underlying model explaining dominant statistical characteristics present in these data. Finally an image interpretation is an example of data lattice with irregularly distributed sites (Fig.2-left).

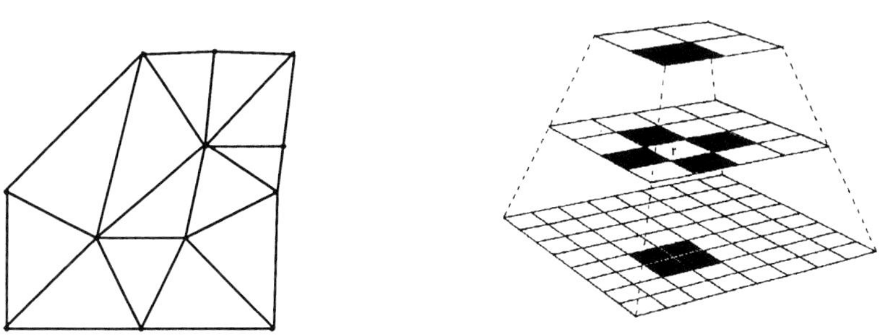

Figure 2: Delaunay neighbours for 2D irregularly distributed sites (left) and a regular multiple resolution 2D neighbourhood (right).

Mathematical multi-dimensional data models are useful for describing many of the multi-dimensional data types provided that we can assume some data homogeneity so some data characteristics are translation invariant. If this assumption does not hold, no advantage can be gained using data models for large data spaces like images. While the 1D models like time series are relatively well researched and they have rich application history in control theory, econometrics, medicine and many other recognition applications, multi-dimensional models are much less known and their applications are still limited. The reason is not only unsolved theory difficulties but mainly their huge computing power demands which prevented their wider use until recently. Several multi-dimensional data models may be thought of as a natural generalization of their 1D counterparts. However, in the nD domain, contrary to the 1D case, there is no natural order definition and hence terms like past or future are meaningless unless defined with respect to some specific and usually artificial order. Different order definitions in nD lead to different orthogonal decompositions. nD problems are also intrinsically different from 1D and consequently 1D models and techniques are not easily extended or transferred into the nD or even only to 2D world. For example the covariance matrix in 2D is in general difficult if not impossible to parameterize. Therefore a high quality goal in a nD space is better achieved by exploiting nD domains characteristics rather than imposing 1D ideas into nD. Dimensionality is another major difficulty. Many 1D calculation techniques can become intractable for small multi-dimensional data sets and for quite simple models. For example, even for a simple pixel classification into K classes of a small 512×512 image there are $K^{2^{18}}$ possible label configurations which prevents any exhaustive search based solutions.

This chapter explains general model-based approaches to several basic pattern recognition applications followed by a concise description of seven fundamental statistical multi-dimensional data model classes. For each model class a solution to parameter estimation and model data synthesis is outlined. Finally, the last section gives an overview of the strengths and weaknesses of studied multi-dimensional data model groups.

2 Pattern Recognition Modelling Applications

It is possible to divide data models applications into two broad categories: synthesis and analysis. The frequent synthesis application is the missing data reconstruction. Another common application is texture synthesis. Analytical applications include data classifications or segmentation, data space directionality analysis, compression and some others. The popular discontinuity detection and image restoration problems can be seen as classification problems with two and K (number of measurements quantization levels) classes, respectively. Recognition applications in this chapter are demonstrated on images because they are illustrative and monospectral images require relatively simple 2D models. However the model-based approaches are general and can be used for any multi-dimensional data recognition provided we are able to efficiently model the corresponding data spaces.

2.1 Contextual Classification

Classification solves the assignment of single objects (patterns) to one of several (K) prespecified categories

$$\Omega = \{\hat{\omega}_1, \ldots, \hat{\omega}_K\} \ .$$

The commonly used context-free Bayesian or other decision rules for object (e.g., multi-spectral image pixels, letters, phonemes, etc.) classification (classifiers) classify a single object using only this object's measurements (features) not using its possible mutual correlation with other objects in the segmented object space. It is well known that for example neighbouring image pixels are highly spatially correlated and similar relations hold also for other types of measurements. Such supervised (classification) or unsupervised (segmentation, clustering) algorithms neglect information from the classified object surroundings and thus inevitably deteriorate the classification error rate.

Contextual classification solves the problem of estimating labels in true unobservable object set ω given the observation data sets (e.g., an image) Y. Let $\omega = \{\omega_1, \ldots, \omega_N\}$ be the N-tuple random vector representing true labels $(\omega_i \in \hat{\Omega})$. Let vector $Y = \{Y_1, \ldots, Y_N\}$ represent the site observations in the same ordering as ω, Y_r is the d dimensional random vector representing measurements on the object at the site r. The standard context-free Bayes decision rule assigns Y_r to the class $\hat{\omega}_{r,i}$ subsequently for all $r \in I$ if

$$p(\omega_{r,i} \,|\, Y_r) = \max_{j=1,\ldots,K} \{p(\omega_{r,j} \,|\, Y_r)\} \ . \tag{1}$$

Using the usual notion of discrimination function $f(\omega_i)$, a decision rule can be formulated as follows:
Y_r is assigned to the class $\hat{\omega}_i$ if

$$f(\omega_{r,i}) = \max_{j=1,\ldots,K} \{f(\omega_{r,j})\} \ .$$

The optimal contextual Bayesian decision rule, which minimizes the average mean loss of single decision problems of an object (deterministic decision function, zero-one loss function) [26], should use all attainable information from data:

$$f(\omega_{r,i}) = p(\omega_{r,i} \,|\, Y) \ . \tag{2}$$

A solution of (2) is hardly a solvable task because of complicated posterior probability density and decision rule implementation. A practical classification scheme must be a compromise between information used and the computational complexity of an algorithm. The evaluation of the discrimination function (2) can be simplified by accepting some additional simplifying assumption. Usual Markovian simplification

$$p(\omega_{r,i} \,|\, Y) = p(\omega_{r,i} \,|\, Y_s \ \forall s \in I_r \cup \{r\}) \ ,$$

where $I_r \subset I$ denotes a neighbourhood of the site r such that $r \notin I_r$,

$$f(\omega_{r,i}) = p(\omega_{r,i} \,|\, Y_s \ \forall s \in I_r \cup \{r\}) \tag{3}$$

is substantiated from the fact that for many indexed object sets (e.g., images) the class conditional correlation between single object's measurements drops more rapidly with their distance than the same unconditional correlation. If we can also assume the conditional statistical independence of objects indexed in I_r

$$p(Y_s \ \forall s \in I_r \cup \{r\} \,|\, \omega_{r,i}) = \prod_{\forall \ s \in I_r \cup \{r\}} p(Y_s \,|\, \omega_{r,i}) \ ,$$

then we get further simplification:

$$f(\omega_{r,i}) = p(\omega_{r,i}) \prod_{\forall \ s \in I_r \cup \{r\}} p(Y_s \,|\, \omega_{r,i}) \ . \tag{4}$$

Combining both simplification, using the total probability theorem and possibly using also some other simplifying assumptions (e.g., knowledge of true classes of neighbouring objects) we can get the whole range [26] of possible contextual Bayesian classifiers. Statistical non Bayesian contextual classifiers can be derived similarly. The evaluation of statistical contextual classifiers discriminate functions requires appropriate simultaneous multidimensional probabilistic models. These can be models described in the subsequent section 3.

Classical image processing task of edge detection or more generally discontinuity detection in a multi-dimensional data space can be formulated also as a contextual classification problem with two site labels - edge, non-edge. Contextual techniques just described can be applied to it. The edge detection can be formulated as the estimation of an unobservable binary random field (ω) indexed on a dual lattice $\hat{I}$. The dual lattice is defined to have its sites placed midway between each neighbouring pair of I sites. The indirect measurement field for the corresponding edge sites are observable measurements Y on the lattice I. A possible maximum a posteriori (MAP) solution is to maximize

$$\max_{\omega} p(\omega \,|\, Y) \sim \max_{\omega} p(Y \,|\, \omega)\, p(\omega) \ .$$

If we assume ω to be Gibbs random field (20) and measurements Y conditionally independent on each other then

$$p(\omega \,|\, Y) \sim \prod_{r} p(Y_r \,|\, \omega_r)\, p(\omega)$$

is also a Gibbs random field with the corresponding energy function [44].

2.2 Segmentation

The goal of data space segmentation is the abstraction of complex data space into a description comprising a relatively small number of labeled regions (e.g., Fig.3). Every region is distinguished by homogeneous characteristics which are significantly different for each region to be discriminatory. The segmentation task can be formulated similarly as the subsequent restoration task to be a problem of recovering unobservable data from their degraded observable version. While for the restoration the goal is undegraded original data the segmentation goal is a segmentation map containing some class indicators.

Segmentation of a data space in the absence of a priori information is very difficult and still generally an unsolved task. The main problem is that the model and its parameters are unknown before segmentation; however, to effectively estimate model parameters the segmentation itself is needed. The spatial models impose prior constraints on acceptable labellings through a priori probabilities, so highly connected labellings are favourably weighted. A data segmentation can be based on a data redundancy measure. A good unsupervised classification should be highly redundant with the original data space. Alternatively the segmentation can be posed as data space estimation when the segmented map is to be recovered from a degraded data space (see section 2.3). Segmentation can be based on the MAP Bayesian estimator

$$\omega^* = \arg\max_{\omega} p(\omega \mid Y) = \arg\max_{\omega} \exp\left\{ \ln \frac{p(\omega \mid Y)}{p(\omega)} \right\} p(\omega) \tag{5}$$

where ω can be a discontinuity map or a segmentation map. In the case of the discontinuity map we cannot decide if unconnected distant regions share similar characteristics. The labeling field model can be assumed to be a Markov random field (MRF)

$$p(\omega) = \frac{1}{Z} \exp\{-Q(\omega)\} \ .$$

The MAP estimation minimizes the probability that any object in I will be misclassified and assigns the same cost to every incorrect segmentation regardless of the number of wrongly segmented objects. Alternatively the maximum marginal posterior probability

$$\omega^* = \{\omega_r^* : r \in I; \ \omega_r^* = \arg\max_{\omega_r \in \Omega} p(\omega_r \mid Y)\} \ ,$$

marginal posterior expectation

$$\omega^* = \sum_{\omega} \omega \, p(\omega \mid Y) = \{\omega_r^* : r \in I; \, \omega_r^* = \sum \omega_r \, p(\omega_r \mid Y)\} \ ,$$

clustering or some other criterion can be used.

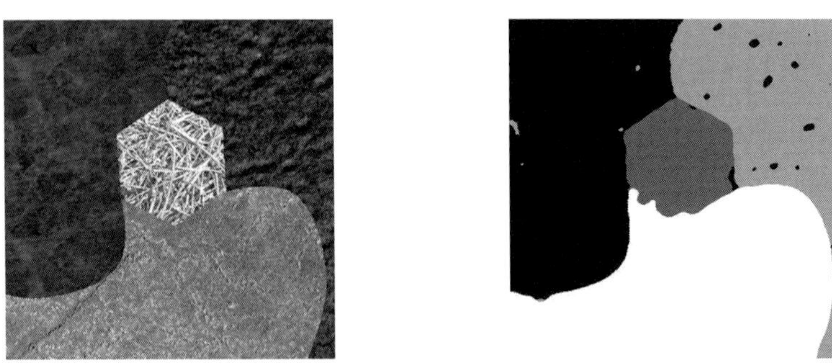

Figure 3: Colour textures mosaic and its segmentation based on a Markov random field model [31].

2.3 Restoration

Physical sensing systems and a recording media are imperfect and thus recorded data represent a degraded version of the original scene. Similarly such data are usually further corrupted during their processing, transmission or storage. The data restoration task is to recover an unobservable data given the observed corrupted data with respect to some statistical criterion. The restoration task can be even more complicated if for some data the corresponding observations are either missing (e.g., image scratch removal) or some data are missing and remaining data are corrupted. The simplest restoration method is to smooth the data with an isotropic linear or non-linear shift-invariant low-pass filter, these filters unfortunately tend to blur the location of boundaries. Model-based methods try to avoid this problem by an appropriate selection of a data model. The restoration application of the stochastic spatial process requires to describe spatial interactions through the choice of potential energy which depends on unknown parameters. This potential energy choice is difficult to substantiate and so is very often based on subjective considerations. Similarly also the parameters are sometimes chosen on an ad hoc basis.

Suppose ω represents a true but unobservable data space. The observable data are Y, a version of ω distorted by noise and/or some other form of distortion (e.g., blurring). Data restoration problem is to estimate an ideal distribution of data (e.g., radiant energy for images) from noise, sampling or blur corrupted observed values. The Bayesian MAP approach is again to maximize $p(\omega \mid Y)$, i.e.,

$$\max_{\omega} p(\omega \mid Y) \sim \max_{\omega} p(Y \mid \omega)\, p(\omega) \tag{6}$$

In contradistinction to segmentation ω is now a true measurements field (unobservable image). The oversmoothing can be diminished by introducing additional unobservable constraint process ϑ. The MAP task (6) is then replaced with

$$\max_{\omega,\vartheta} p(\omega, \vartheta \mid Y) \ . \tag{7}$$

Both MAP estimates (6), (7) can be used also for the missing data reconstruction problem with appropriately modified potential functions. The image degradation on Fig.4 is supposed to be approximated by the linear degradation model with additive noise:

Figure 4: Original, corrupted, and reconstructed Cymbidium image.

$$Y_r = \sum_{s \in I_r} f_s \omega_{r-s} + e_r$$

where f is a discrete representation of the unknown point-spread function and the degradation noise e is uncorrelated with the unobservable image. If the noisy image Y can be represented by a causal (2D or 3D) simultaneous autoregressive model (8), the fast recursive estimator of the unobservable (monospectral or colour) image in the form of corrupted image expectation $\omega = E\{Y\}$ can be found analytically [32].

2.4 Texture Modelling

Photo realism in virtual or mixed reality scenes cannot be accomplished without nature-like colour textures covering visualized scene objects. Similarly, realistic textures are used in computer games, CAD systems and some other computer graphics applications. There is no exact definition for texture since this notion is tied to the human semantic meaning, although some authors claim to give one (e.g., see [56] for a recent attempt to define texture). Texture is generally a visual property of a surface, representing the spatial information contained in object surfaces. A textured image can be understood to be a realization of random field and a texture modelling effort is simply to produce a visually indiscernible copy of a given image using modelling techniques.

A monospectral texture can be synthesized from a 2D model while the colour or multi-spectral texture requires a three dimensional model; however, if we are willing to sacrifice some spectral information, a 3D model can be approximated with a set of much simpler 2D models without com-

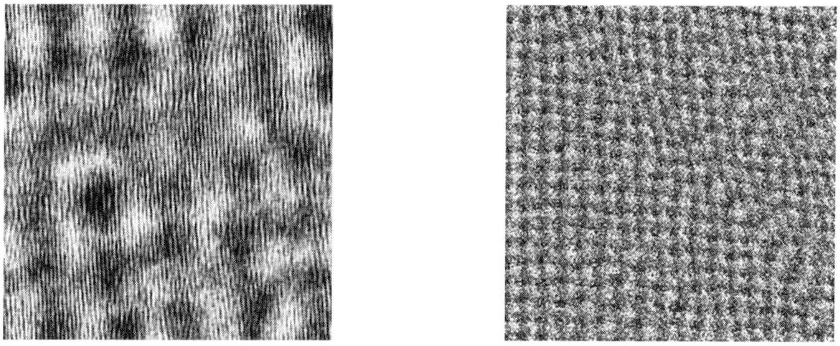

Figure 5: Synthesized wood and rattan colour textures.

promising its visual realism. Unfortunately there is no ideal texture model. Every model has its strong as well as weak sides. Among such possible models the Gaussian Markov random fields (22) are appropriate for texture modelling (Fig.5-left) not only because they do not suffer with some problems of alternative options (see [27], [33], [34] for details) but they are also easy to synthesize and still flexible enough to imitate a large set of natural and artificial textures. Similar observation holds for related simultaneous autoregressive models (8). Discrete mixtures models (27) can handle some regular textures (Fig.5-right) notoriously difficult for Markovian family models; however, they are less robust and rather time-consuming.

2.5 Compression

Multi-dimension data which can be generated from a model can be optimally compressed into a set of their model parameters irrespective of their size. The compression consists in such a case of the estimation of unknown model structure and parameters and the decompression is a corresponding model synthesis. For such data we get a compression ratio which cannot be reached using any other compression algorithm.

For example the 512×512 Gaussian Markov random fields (GMRF) realization on Fig.6 is characterized with four real number parameters. The JPEG compression with 75% quality has the compression ratio $2:1$, the compression ratio for 5% quality JPEG is still only $12:1$, while the MRF compression ratio is $16384:1$, respectively.

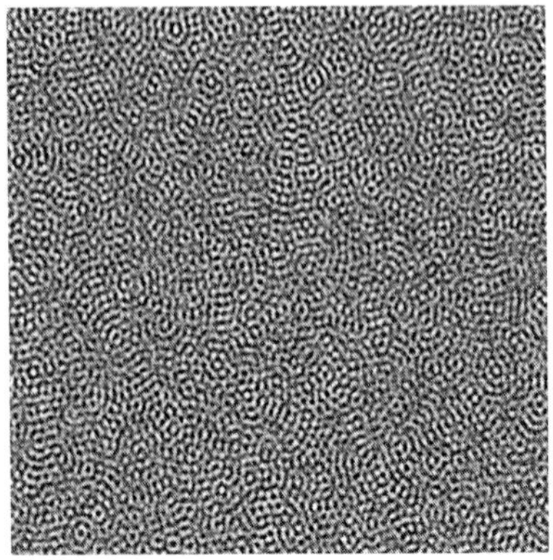

Figure 6: GMRF texture compression.

3 Multi-Dimensional Models

Several approaches to multi-dimensional data modelling exist, each of them has its advantages and also limitations. Existing models can be categorized using different criteria like deterministic and stochastic, causal and non causal, hierarchical and single-scale, syntactical and parametric, or another possible division is according to their primary modelling strength into low and high frequency subsets. We focus our attention on models with solved parameter estimation step, neglecting different ad hoc models or models with unknown parameter estimation which are useless for pattern recognition applications.

Several different data models exist and random field models are among the most powerful and flexible ones, because choosing different neighbour sets and different types of probability distribution for the random variables a variety of distinct data space types can be generated. It is possible to describe data spaces with sharp discontinuities of various orientation, patchy spaces or many other types of data configurations.

3.1 Random Mosaic Models

In the random mosaic models, the data region is first divided into convex polygonal cells, then every cell is independently assigned a class (feature vector, colour) ω_i according to a fixed set of probabilities $P_1, \ldots, P_K$. Let r, s be two points a distance d apart, $P(d)$ is the probability that both points are in the same cell, then [2]:

$$P(Y_r = \omega_j \mid Y_s = \omega_i) = P_j(1 - P(d)) ,$$
$$P(Y_r = \omega_i \mid Y_s = \omega_i) = P_i + (1 - P_i)\, P(d) .$$

Random mosaic models parameter estimation is based on model fitting techniques using known dependencies between the observed features (variogram, component width, perimeter, etc.) and the model parameters. If there is defined a distance between two classes (for example gray levels), we can modify the definition of the variogram ([52], [53]) of a random mosaic:

$$V(d) = E\{(Y_r - Y_{r+d})^2\} .$$

In a special case of constant measurement level in each cell and uncorrelated measurement levels in different cells the variogram can be simplified [52]

$$V(d) = 2\sigma^2(1 - P(d)) ,$$

where σ^2 is assumed to be the standard deviation of data space classes (measurements) and $P(d)$ is the probability that two points distance d apart are in the same cell. For the Poisson line model and the rotated checkerboard model we can compute for a given σ^2 the theoretical variogram as a function of the Poisson parameter τ or checkerboard square size b. Given a variogram of a real texture, these parameters can be obtained using fitting techniques.

3.1.1 Poisson Line Model

The texture area is divided into convex cells by a set of straight lines with random positions and orientations. Points (α, r) are generated [54] by a Poisson process of intensity τ/π, where $0 \leq \alpha < \pi$ and $-\infty < r < \infty$. Each line is defined as follows:

$$r = x \cos\alpha + y \sin\alpha.$$

The probability $P(d)$ can be found to be

$$P(d) = \exp\left\{-2\frac{\tau d}{\pi}\right\} ,$$

for some other statistical characteristics see [2], [54].

3.1.2 Rotated Checkerboard Model

The origin of a coordinate system as well as the orientation of axes is chosen with uniform probability (orientation in the interval $[0, \pi]$). Using the resulting coordinate grid, the plane is tessellated into square cells of side length b. Probability of the same class for two distant points [52]:

$$P(d) = \begin{cases} 1 - 4d/\pi b + d^2/\pi b^2, & d \leq b \\ 1 - 2/\pi - \frac{4}{\pi \cos(b/d)} - d^2/\pi b^2 + 4/\pi (d^2/b^2 - 1)^{1/2}, & b < d \leq b\sqrt{2} \\ 0, & d > b\sqrt{2} \end{cases}$$

A rotated hexagon model is analogous [53] to the checkerboard model, except that hexagons are used in place of squares. Another possibility is the rotated triangular model [2], where an equilateral triangular tessellation can be formed by connecting the growth centers of neighbouring cells in a hexagonal tessellation.

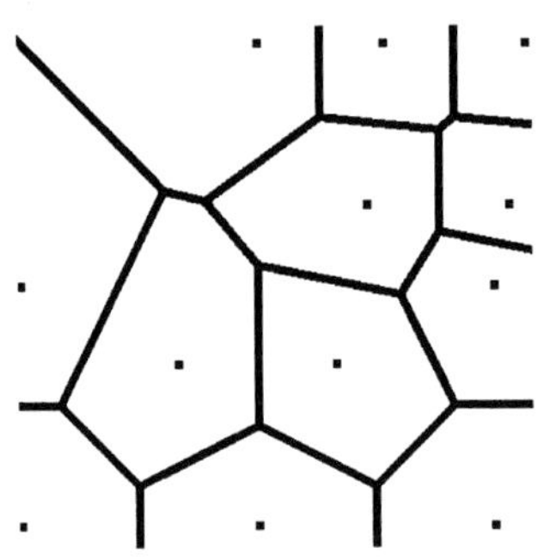

Figure 7: Voronoi 2D mosaic.

3.1.3 Voronoi Tessellation

The principle of this method (known also as the Dirichlet tessellation or the occupancy model) is to create a cell around each of the seed points (generated for example by a Poisson process). Each cell contains all points for which the corresponding seed point is the nearest one (see Fig.7). Voronoi tessellation shares similar disadvantages in a low natural objects modelling power with both previously mentioned methods, while the parameter estimation [54] is more difficult and even problematic on data with nonlinear or unclear borders between homogeneous areas.

If a data space can be divided into fixed-size cells, regularly replicated over a whole data space, we obtain a deterministic mosaic model (periodic tessellated model).

3.2 Syntactic Models

In the structural approach a data set is considered to be defined by subpatterns which occur repeatedly according to a set of well-defined placement rules within the overall pattern. A pattern is composed of fixed-sized windows as subpatterns. Each window is represented by a tree of which each node corresponds to a data element (e.g., pixel) in the window. A pattern primitive is the label representing the data element value (gray level of each pixel). A tree grammar is then used to characterize the windowed pattern. Some examples of this approach can be found in [20], [21]. In order to model the data set of interest, it is necessary to have the stochastic tree grammar actually inferred from the available data (image) samples. Such an inference procedure requires the inference of both the tree grammar and its production probabilities. Unfortunately, a general inference procedure for stochastic tree grammars does not exist and is still a subject of research [21]. Only some very special cases can be treated by existing inference algorithms, like for example [21]:

1. Windowed patterns are grouped into clusters using a similarity measure.

2. Each cluster consists of a finite set of trees having the same structure. A stochastic tree grammar is then inferred for each cluster.

3. The final texture grammar is the union of placement rules and all stochastic tree grammars.

Using such an algorithm it is still difficult to solve grammar inference for other then very simple structures. It is therefore difficult to model most realistic data structures, and the algorithm is computationally demanding due to employing large set searching procedures.

3.3 Fractal Models

Fractals are sets, whose Hausdorff-Besicovitch dimension, which in general is a real number, is strictly larger than their topological dimension [45], [4].

We limit ourselves to fractals with two additional properties, the similarity of each segment to all others and statistical invariance over wide transformation of scale. The primary control that fractal techniques provide over the resulting object is by the value of a single parameter determining the fractal dimension. The only examples of fractal objects being used to model natural phenomena are based on fractional Brownian functions. A random function Y_r is a fractional Brownian function (surface for 2D) if for all r and d:

$$P(\frac{Y_{r+d} - Y_r}{||d||^H} < t) = F(t) \ ,$$

where $F(t)$ is a cumulative distribution function. If Y_r is scalar, then the fractal dimension of Y_r is $D = 2 - H$. If $H = 0.5$ and $F(t)$ is the standardized Gaussian distribution, then Y_r is the classical Brownian function. Several useful properties of the fractal Brownian function were proven in [46]. The fractal dimension of a fractal Brownian function is invariant over transformation of scale. The fractal dimension of Brownian function can be measured directly or from Fourier power spectrum $P(z)$ of Y_r, as the spectral density of fractal Brownian function is proportional [46] to z^{-2H-1}. The fractal dimension can be estimated using methods [45] or [46] and fractals can be synthesized using the Fourier transformation [17], cylindrical integration, or midpoint displacement method.

3.4 Second-Order Statistical Models

This model generates a random field which has the second-order statistics similar to the given data we are trying to simulate. It is assumed, that second-order statistics are sufficient for data discrimination. Although this assumption is not generally valid (even not for visual data discrimination), such an approximation is still fairly accurate. The set of second-order joint density functions $p_{s,\bar{s}}(Y_1, Y_2)$, necessary for model analysis, can be estimated using occurrence matrices computed from a given data; the first order statistics are estimated from the corresponding histogram. The joint density function $p(Y_1, \ldots, Y_N)$ for homogeneous random fields can be found from known marginals $p(Y_s, Y_r)$ using linear programming techniques. For larger dimensionality problems the ART method [19] is better suited. A homogeneous random field with the second order spatial averages equal to those given can be synthesized using the method described in [5].

3.5 Autoregressive Models

Autoregressive or simultaneous autoregressive (AR) models from the class of simultaneous models have several favourable properties, which cause their wide usage in different applications. Simultaneous models are basically generalization of 1D time series models into nD. Such a generalization is not quite straightforward because unlike in 1D time series, where the existence of a preferred direction (time) is assumed, no such preferred ordering exists on a discrete n-dimensional lattice I. Other groups from the class of simultaneous models are simultaneous moving average (MA) and simultaneous autoregressive and moving average (ARMA) models. These other models have complicated parameter estimation. For example, estimation of ARMA model parameters is a nonlinear problem even in 1D, and some approximations have to be used. The advantage of AR models is their close relation with Markov random field models (MRF) as is demonstrated in Fig.8. For every AR model there exists a unique conditional MRF with an equivalent spectral density function, but the converse is not always true (except for the Gaussian case). The conditional MRF model is generally characterized by more parameters than the equivalent AR model (if it exists).

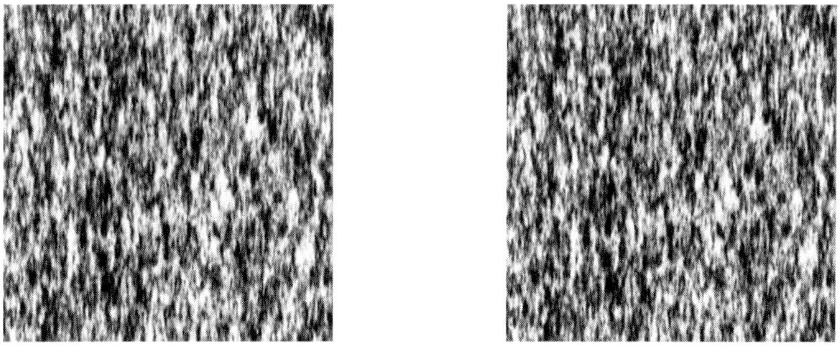

Figure 8: Equivalent realizations of 2D noncausal Gaussian AR (left) and GMRF random fields.

MA and ARMA models are not subsets of MRF models. The 2D AR model can be written as follows:

$$Y_r = \sum_{s \in I_r} a_s Y_{r-s} + e_r \ , \tag{8}$$

where e_r is a white Gaussian noise (or possibly a noise with some other distribution) with zero mean and a constant variance σ^2, and is uncorrelated with data from I_r. If I_r is a symmetric neighbourhood, we assume that symmetrically opposite neighbours have equal parameters. Otherwise, the parameters may not be identifiable [7]. We can rewrite AR model in the matrix form:

$$Y_r = \gamma X_r + e_r \ , \tag{9}$$

where

$$\gamma = [a_1, \dots, a_s] \ , \tag{10}$$

and X_r is a corresponding vector of Y_{r-s}. If we replace Y_r in (8) by $\tilde{Y}_r$ ($\tilde{Y}_{i,j} = Y_{i,j} - \mu$), the 2D joint probability density of process (8) on the $N_1 \times N_2$ lattice I can be written as

$$p(Y) = (2\pi\sigma^2)^{-\frac{N_1 N_2}{2}} |\Psi| \exp\left\{ -\frac{(Y-U)^T \Psi^T \Psi (Y-U)}{2\,\sigma^2} \right\} \ , \tag{11}$$

where Ψ is defined

$$\Psi = \begin{pmatrix} \Psi_0 & \Psi_1 & \dots & \Psi_{N_1-1} \\ \Psi_{N_1-1} & \Psi_0 & \dots & \Psi_{N_1-2} \\ \vdots & & \ddots & \vdots \\ \Psi_1 & & \dots & \Psi_0 \end{pmatrix}$$

and is for a torus lattice I a block circulant matrix, each block being again a circulant matrix and $E\{Y\} = U$. We can synthesize a random field realization described by AR from the equation

$$Y = \Psi^{-1}(E - B), \tag{12}$$

where the noise vector

$$E = [e_1, \dots, e_{N_1 N_2}]^T$$

has an arrangement corresponding with Y. B is a vector with border conditions and Ψ is an $N_1 N_2 \times N_1 N_2$ block Toeplitz type matrix of a_s coefficients with unit diagonal. If we assume zero boundary values $(B = 0)$, symmetric coefficients $(a_{r-s} = a_{r+s})$ and regularity of Ψ, then Ψ is regular, symmetric and positive definite. Under this condition the solution of (12) can be obtained iteratively by the conjugate gradient method [27], [13] (the Fletcher-Reeves version of the method). The AR model can be also synthesized using the Woods iterative synthesis [55] or using the efficient fast Fourier transformation (FFT).

The selection of an appropriate autoregressive model is important to obtain good results in modelling of a given random field. If the contextual neighbourhood is too small we can not capture all details of the random field. On the contrary, larger than necessary contextual neighbourhoods can introduce problems with numerical accuracy and can be too time consuming. The appropriate model can be chosen using pairwise hypothesis testing [7], [12], Akaike's information criterion (AIC) [37], or the Bayesian approach. The main disadvantage of the hypothesis testing approach is that the resulting decision rule is not transitive and also not consistent. The AIC method gives transitive decision rules but not consistent [38] ones. The Bayesian method selects a model $\mathcal{M}_i$ corresponding to $\max_j\{\, p(\mathcal{M}_j \,|\, Y)\,\}$ (or alternatively in the Fourier domain). Simultaneous conditional probability density can be evaluated analytically from

$$p(Y \,|\, \mathcal{M}_j) = \int\!\!\int p(Y \,|\, \gamma, \sigma^2)\, p(\gamma, \sigma^2 \,|\, \mathcal{M}_j)\, d\gamma d\sigma^2$$

only for causal models [28], otherwise some approximation (like [38]) is necessary:
choose the AR model k if $\quad k = \arg\min_j\{D_j\}$,
where

$$D_j = -\sum_{r\in I} \ln(1 - \gamma_k C_r + \gamma_k Q_r Q_r^T) + N_1 N_2 \ln \sigma_k^2 + m_k \ln(N_1 N_2) \ , \qquad (13)$$

γ_k, and σ_k^2 are iterative estimations of the corresponding parameters, C_r is equal to S_r except for the replacement of sin by cos, m_k is the number of neighbours in the k-th model, $Q_r = S_r S_r^T + C_r C_r^T$, and S_r is defined as follows:

$$S_r = \left[\, \sin(2\pi (N_1 N_2)^{-1/2} (r-1)^T s), \ldots \right]^T \qquad \forall s \in I_r \ . \qquad (14)$$

Bayesian parameter estimation for the AR model is complicated by a normalization factor which cannot be evaluated analytically for non-causal models and a numerical evaluation of this factor can be very difficult for larger set of parameters. The maximum likelihood (ML) estimates require numerical optimization methods due to the log-likelihood function, which is nonquadratic in γ. The ML estimate for σ^2 is (16). To avoid computationally expensive numerical methods, it is possible to use the least square method. If we define model prediction as

$$\tilde{Y}_r = E\{Y_r\} = \gamma X_r \ ,$$

then the non-consistent parameter estimators minimizing the variance of the prediction error are as follows:

$$\hat{\gamma} = E\{X_r X_r^T\}^{-1} E\{X_r Y_r\} = \left[\sum_{r\in I} X_r X_r^T\right]^{-1} \sum_{r\in I} X_r Y_r \ , \tag{15}$$

$$\hat{\sigma}^2 = \frac{1}{N_1 N_2} \sum_{r\in I} (Y_r - \hat{\gamma} X_r)^2 \ . \tag{16}$$

Parameters can be also alternatively estimated using the maximum pseudo-likelihood method or some approximation method like [38] which uses a quadratic approximation of the determinant Ψ.

Causal AR model are exceptional because all their parameters as well as the optimal model selection can be solved analytically under few reasonable additional assumptions [28], [29] using the Bayesian estimators (17)-(19):

$$\begin{aligned} D_j &= \ln\Gamma(\frac{\beta(r)-\eta+2}{2}) - \ln\Gamma(\frac{\beta(0)-\eta+2}{2}) - \frac{1}{2}\ln|V_{xx(r-1)}| \\ &\quad -\frac{\beta(r)-\eta+2}{2}\ln|\lambda_{(r-1)}| \ , \end{aligned} \tag{17}$$

$$\hat{\gamma}_{r-1}^T = V_{xx(r-1)}^{-1} V_{xy(r-1)} \ , \tag{18}$$

$$\hat{\sigma}_{r-1}^2 = \frac{\lambda_{(r-1)}}{\beta(r)} \ , \tag{19}$$

where

$$\begin{aligned} \lambda_{(r)} &= V_{yy(r)} - V_{xy(r)}^T V_{xx(r)}^{-1} V_{xy(r)} \ , \\ \beta(r) &= \beta(0) + r - 1 = \beta(r-1) + 1 \end{aligned}$$

$\eta = card\{I_r\}$ and $V_{yy(r)}, V_{xy(r)}, V_{xx(r)}$ are corresponding data gathering matrices defined as follows:

$$V_{vw(r-1)} = \sum_{k=1}^{r-1} V_k W_k^T + V_{vw(0)} \ .$$

The causal model can be synthesized directly from its defining equation (8). Similarly it is possible to find an analytical solution for the 3D causal AR model [30].

3.6 Markov Random Field Models

The Markov random field (MRF) is a family of random variables with a joint probability density on the set of all possible realizations Y of the lattice I, subject to following conditions:

$$p(Y) > 0, \qquad \forall Y,$$

(positivity condition), and

$$p(Y_r \,|\, Y_s \;\forall s \in I \setminus \{r\}) = p(Y_r \,|\, Y_s \;\forall s \in I_r) \ ,$$

(Markovianity in a strict sense, or local Markov property). We limit ourselves here to homogeneous and symmetric neighbourhood systems I_r. A hierarchy of MRF models can be defined using such a neighbourhood system, as in the following figure, where numbers on the place of 2D contextual neighbours indicate the order of the model (up to seventh order) relative to r:

$$\begin{array}{ccccccc} & & 7 & 6 & 7 & & \\ & 5 & 4 & 3 & 4 & 5 & \\ 7 & 4 & 2 & 1 & 2 & 4 & 7 \\ 6 & 3 & 1 & r & 1 & 3 & 6 \\ 7 & 4 & 2 & 1 & 2 & 4 & 7 \\ & 5 & 4 & 3 & 4 & 5 & \\ & & 7 & 6 & 7 & & \end{array} \qquad\qquad I_r^3 \quad \begin{array}{ccccc} & & \otimes & & \\ & \otimes & \otimes & \otimes & \\ \otimes & \otimes & r & \otimes & \otimes \\ & \otimes & \otimes & \otimes & \\ & & \otimes & & \end{array}$$

The neighbourhood systems I_r of the n-th order MRF model then contain all neighbours of r with position numbers $1, 2, \ldots, n$.

The 2D hierarchical neighbourhood system can be easily generalized into a nD hierarchical neighbourhood system as can be seen in the 3D example on Fig.9.

The Hammersley-Clifford theorem [7] states that Y is a MRF with strictly positive distribution and neighbourhood system I_r, if and only if the distribution of Y can be written as a Gibbs distribution

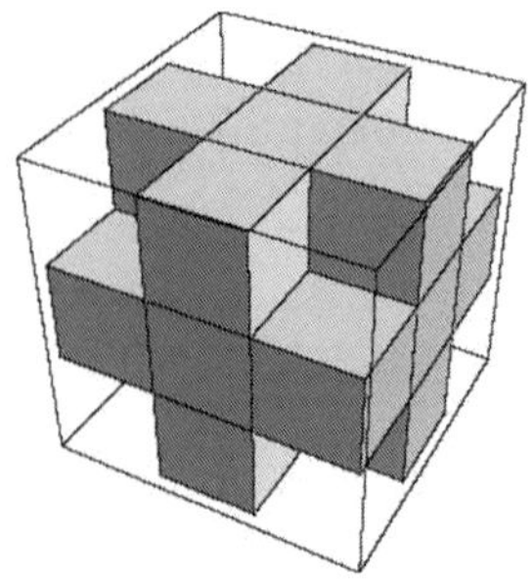

Figure 9: The second order 3D hierarchical neighbourhood.

$$p(Y) = \frac{\exp\{-Q(Y)\}}{Z} , \tag{20}$$

where the normalization constant is

$$Z = \sum_Y \exp\{-Q(Y)\} , \tag{21}$$

or

$$Z = \int_Y \exp\{-Q(Y)\}$$

in the case of a continuous random field and $Q(Y)$ is a Gibbs energy function. The normalization constant is with few exceptions intractable analytically as well as numerically. One of these exceptions is the Gaussian Markov random field (GMRF).

MRF parameter estimation is a complicated task even for single and simple MRF models. The complexity of estimation further increases for more than two-dimensional MRF models or if the neighbourhood system has to be simultaneously estimated. The complexity of posterior parameter distribution for most MRF models (the normalization constant Z is also a function of estimated parameters) and possible parameter priors means that analytical Bayesian estimation is rarely possible. Instead some Markov chain Monte Carlo method like the Gibbs sampler (section 3.6.2) or the Metropolis algorithm [43] (section 3.6.1) has to be used. For example the Metropolis algorithm can be applied as follows. Let $p(\acute{\theta}_i \mid \theta_i)$ be some parameter proposal distribution. A new i-th parameter value $\acute{\theta}_i$ is drawn from $p(\acute{\theta}_i \mid \theta_i)$. This new parameter value is accepted with probability

$$\min \left\{ 1, \frac{p(\acute{\theta} \mid Y)\, p(\acute{\theta}_i \mid \theta_i)}{p(\theta \mid Y)\, p(\theta_i \mid \acute{\theta}_i)} \right\} \, .$$

If the new value is rejected the previous value is retained. The maximum likelihood estimation (ML) maximizes the likelihood function (conditional probability or its logarithm):

$$\theta^* = \arg\max_{\theta} p(Y \mid \theta) \, .$$

The ML parameter estimation of the MRF model is again complicated by the difficulty associated with computing the normalization constant. Therefore it is necessary to use some approximation method, like for example the coding method [7] or the maximum pseudo-likelihood. To overcome the difficulties caused by mutual correlation of lattice pixels, the coding method labels lattice data elements in a way to obtain mutually independent site variables for a given contextual neighbourhood I_r. However the method has only partial utilization of the available data, different coding estimates for the same parameter can considerably differ and site labelling for larger dimensions and contextual neighbourhoods can be very complicated.

3.6.1 Metropolis Algorithm

Given the state Y^{t-1} of a MRF, another random configuration $\acute{Y}$ is chosen [43] such that $\acute{Y}_{(A)} = Y^{t-1}_{(A)}$, $\acute{Y}_A \neq Y^{t-1}_A$ where $A \subset I$ and $(A) = I \setminus A$. The transition kernel $\tilde{p}(Y^{t-1}_A \to \acute{Y}_A; Y^{t-1}_{(A)})$ is chosen to be symmetric in Y^{t-1}_A, $\acute{Y}_A$. The ratio

$$\alpha = \frac{p(\acute{Y})}{p(Y^{t-1})}$$

is computed. If $\alpha > 1$, or $Q(\acute{Y}) < Q(Y^{t-1})$ then $Y^t = \acute{Y}$, otherwise the transition is made with probability α. A variable ξ is selected from a standardized uniform distribution, if $\xi \leq \alpha$ then $Y^t = \acute{Y}$, otherwise $Y^t = Y^{t-1}$. Because a less favourable configuration is not automatically rejected the algorithm can escape from local minima.

A modification of the algorithm is the exchange algorithm [12], where $\acute{Y}$ is obtained from Y^{t-1} by exchanging values of two randomly chosen data elements. The exchange algorithm keeps the overall distribution of RF values fixed. The disadvantage of this method is its sensitivity to initial configuration [11]. In the single-flip [22] algorithm, $\acute{Y}$ is obtained from Y^t by changing the value of one randomly chosen data element $(card(A) = 1)$.

3.6.2 Gibbs Sampler

The Gibbs sampler [22] generates realizations from a given MRF using a relaxation technique similar to the Metropolis algorithm. The Gibbs sampler is a special case of the simulated annealing method with a fixed temperature or the Metropolis algorithm with zero rejection probability. The transition kernel $\tilde{p}(Y_A^{t-1} \rightarrow \acute{Y}_A; Y_{(A)}^{t-1})$ is independent of Y_A^{t-1}. Gibbs samplers usually require univariate updates $(card(A) = 1)$. The stationary configuration Y^0 is arbitrary. By repeatedly visiting all sites we always replace one data element with a value generated from the local characteristic of Gibbs distribution:

$$p(Y^t) = p(Y_r^t \,|\, Y_s^{t-1} \; \forall s \neq r)\, p(Y_s^{t-1} \; \forall s \neq r) \ ,$$

where

$$p(Y_r^t \,|\, Y_s^{t-1} \; \forall s \neq r) = \frac{\exp\{-Q_r^t(Y)\}}{\tilde{Z}} \ .$$

The convergence of the algorithm is assured by the relaxation theorem proven in [22] which states that if for each site from lattice the visit sequence contains this site infinitely often, then for every starting configuration Y^0 and every configuration Y,

$$\lim_{t \to \infty} p(Y^t = Y \,|\, Y^0) = \frac{\exp\{-Q(Y)\}}{Z} \ .$$

3.6.3 Gaussian Markov Random Field

If the local conditional density of the MRF model is Gaussian, we obtain the continuous GMRF (auto-normal) model:

$$p(Y_r \,|\, Y_s \; \forall s \in I_r) = \frac{1}{\sqrt{2\pi\,\sigma^2}} \exp\left\{-\frac{1}{2\sigma^2}\,(Y_r - \tilde{\mu}_r)^2\right\} \ ,$$

where the mean value is

$$E\{Y_r \,|\, Y_s \; \forall s \in I \setminus \{r\}\} = \tilde{\mu}_r = \mu_r + \sum_{s \in I_r} a_s (Y_s - \mu_s) \ ,$$

and the joint probability density (2D) function is

$$p(Y) = (2\pi\sigma^2)^{-\frac{N_1 N_2}{2}} |\Psi|^{\frac{1}{2}} \exp\left\{ -\frac{(Y-U)^T \Psi (Y-U)}{2\sigma^2} \right\} , \qquad (22)$$

where the $(N_1 N_2 \times 1)$ mean vector consists of single location means and the matrix Ψ has unit diagonal elements [7] and off diagonal elements are coefficients a_r all other entries are zero. Ψ is symmetric, but is also required to be positive definite. Coefficients of symmetric sites have to be equal, i.e., $a_{r-s} = a_{r+s}$. The covariance matrix is

$$\Sigma_Y = E\left\{ (Y-U)(Y-U)^T \right\} = \sigma^2 \Psi^{-1} .$$

GMRF (both finite-lattice I $(N_1 \times N_2)$ and infinite-lattice I) can be expressed as a stationary noncausal 2D autoregressive process described by the difference equation

$$Y_r = \mu_r + \sum_{s \in I_r} a_s (Y_s - \mu_s) + e_r ,$$

where $a_{r-t} = a_{r+t}$, μ_r is the mean value of Y_r, e_r is a stationary Gaussian noise with zero mean and autocorrelation given by

$$E\{e_r\, e_s\} = R_e(s) = \begin{cases} \sigma^2 & \text{if } s = r, \\ -\sigma^2 a_s & \text{if } s \in I_r, \\ 0 & \text{otherwise.} \end{cases}$$

An infinite lattice GMRF can be synthesized only by time consuming iterative methods. Finite toroidal lattice GMRF can be efficiently synthesized using inverse discrete Fourier transformation $Y = \mathcal{F}^{-1}\{\hat{Y}\} + U$, where U is the mean vector and $\hat{Y}$ is generated from the Gaussian generator $\mathcal{N}(0, N_1 N_2 S_{\tilde{Y}}(r))$ where $S_{\tilde{Y}}(i,j)$ is the power spectral density associated with $\tilde{Y}_r$. For GMRF models which have an equivalent AR model (e.g., Fig.8), we can use Woods iterative algorithm [55]. GMRF models which do not have an equivalent AR model, can be generated by modified Woods algorithm, but nothing is known about its convergence properties, because the convergence theorem [55] is not valid any more. For the optimal GMRF model selection we can use a modification of the Bayesian method (13), (14), where m_k is now a number of asymmetrical half neighbourhood parameters, and

$$D_j = -\sum_{r \in I} \ln(1 - 2\gamma_k C_r) + N_1 N_2 \ln \sigma_k^2 + m_k \ln(N_1 N_2) \ .$$

The GMRF normalization constant can be easily computed and the ML estimate maximizing the likelihood function (22) for $U = 0$ once the $\hat{\Psi}$ is found is:

$$\hat{\sigma}^2 = \frac{1}{N_1 N_2} \tilde{Y}^T \Psi \tilde{Y} \ . \tag{23}$$

To obtain $\hat{\Psi}$ estimation we have to substitute (23) into (22) and to solve $\max_\Psi \{p(Y)\}$ which yields

$$\min_{\Psi} \{ -\frac{1}{N_1 N_2} \ln |\Psi| + \ln(\tilde{Y}^T \Psi \tilde{Y}) \} \ .$$

This non-linear estimator requires a numerical solution. The pseudo likelihood estimator neglects mutual conditional dependence between data thus allows simple asymptotically normal but not efficient estimator:

$$\gamma_a = [a_{10}, a_{01}, \dots]^T = \left[\sum_{\forall r \in I} X_r^T X_r \right]^{-1} \sum_{\forall r \in I} X_r^T \tilde{Y}_r \ , \tag{24}$$

where

$$X_r = [\tilde{Y}_{r-s} \quad \forall s \in I_r] \ . \tag{25}$$

3.6.4 Binomial Markov Random Field

A discrete MRF example frequently used in modelling applications is the binomial Markov random field model. Let the random variable Y_r have K different values $\{0, \dots, K-1\}$, α is the probability of success in one trial, $K-1$ is the number of trials and the conditional probability of variable Y_r having value k to be conditionally binomial:

$$p(Y_r = k \mid Y_s \ \forall s \in I_r) = \binom{K-1}{k} \alpha^k (1-\alpha)^{K-1-k} \ ,$$

where

$$\alpha = \frac{\exp\{q_r\}}{1 + \exp\{q_r\}} \ ,$$

and q_r is a function of $\{Y_s \; \forall s \in I_r\}$ and parameters. For example, the model with the 2D first-order neighbourhood I_r^1 has the following form for q_r:

$$q_r = a_1 + a_2(Y_{r_1,r_2-1} + Y_{r_1,r_2+1}) + a_3(Y_{r_1-1,r_2} + Y_{r_1+1,r_2}) \; .$$

The mean value is

$$E\{Y_r = k \,|\, Y_s \; \forall s \in I_r\} = (K-1)\alpha$$

and the variance is $(K-1)\alpha(1-\alpha)$. Simulation of a binomial Markov random field (BMRF) can be done for example using the Metropolis algorithm (see section 3.6.1). BMRF parameters can be estimated using the coding technique. The log-likelihood function can be written in the following form:

$$\begin{aligned} \mathcal{L} &= \sum_{r=1}^{\tilde{N}_1 \tilde{N}_2} \log p(Y_r | Y_s \quad \forall s \in I_r) = \sum_{r=1}^{\tilde{N}_1 \tilde{N}_2} \left\{ \log \binom{K-1}{i} + \right. \\ & \left. i(q - \log(1+\exp\{q\})) + (K-1-i)\log(1 - \frac{\exp\{q\}}{1+\exp\{q\}}) \right\} \; , \end{aligned}$$

$$\frac{\delta \mathcal{L}}{\delta a_i} = \sum_{r=1}^{\tilde{N}_1 \tilde{N}_2} \left\{ i(\frac{\delta q}{\delta a_i} - \frac{\exp\{q\}}{1+\exp\{q\}} \frac{\delta q}{\delta a_i}) - (K-1-i)\frac{\exp\{q\}}{1+\exp\{q\}} \frac{\delta q}{\delta a_i} \right\} \; . \tag{26}$$

Maximalization of (26) is a demanding task using numerical methods. Another possibility is the maximum pseudo-likelihood method, then $\tilde{N}_1 = N_1$, $\tilde{N}_2 = N_2$.

3.7 Probabilistic Mixture Models

A multi-variate discrete mixtures model with components defined as products of univariate discrete probability distributions is defined

$$P(Y_{\{r\}}) = \sum_{m=1}^{M} P(Y_{\{r\}} \,|\, m) \, p(m) \tag{27}$$

where

$$Y_{\{r\}} = [Y_s \quad \forall s \in I_r \cup \{r\}] \; ,$$

$p(m)$ are probability weights, and the component distributions $P(\cdot\,|\,m)$ are factorizable, i.e., we can write

$$P(Y_{\{r\}}\,|\,m) = \prod_{s\in I_r\cup\{r\}} p_s(Y_s\,|\,m) \qquad Y_s \in Y_{\{r\}} \ , \tag{28}$$

$$P(Y_{\{r\}}) = \sum_{m=1}^{M} p(m) \prod_{s\in I_r\cup\{r\}} p_s(Y_s\,|\,m) \ , \tag{29}$$

$p_s(Y_s\,|\,m)$ are univariate discrete (component-specific) probability distributions. The conditional independence is not a restrictive assumption. It can be easily verified that, in discrete case, the class of finite mixtures is complete in the sense that any discrete probability distribution can be expressed in the form (29) for M sufficiently large. The unknown parameters of the approximating mixture can be estimated by means of the iterative EM algorithm [14], [24], which maximizes the likelihood function corresponding:

$$\mathcal{L} = \frac{1}{|\mathcal{S}|} \sum_{Y_{\{r\}}\in\mathcal{S}} \ln\left[\sum_{m=1}^{M} P(Y_{\{r\}}\,|\,m)\,p(m)\right] \tag{30}$$

The related iteration equations can be expressed [25] as follows:

$$q^{(t)}(m|\,Y_{\{r\}}) = \frac{P^{(t)}(Y_{\{r\}}\,|\,m)\,p^{(t)}(m)}{\sum_{j=1}^{M} P^{(t)}(Y_{\{r\}}\,|\,j)\,p^{(t)}(j)}, \tag{31}$$

$$p^{(t+1)}(m) = \frac{1}{|\mathcal{S}|} \sum_{Y_{\{r\}}\in\mathcal{S}} q^{(t)}(m|\,Y_{\{r\}}), \tag{32}$$

$$p_n^{(t+1)}(\xi\,|\,m) = \frac{1}{|\mathcal{S}|p^{(t+1)}(m)} \sum_{Y_{\{r\}}\in\mathcal{S}} \delta(\xi, Y_n)\,q^{(t)}(m\,|\,Y_{\{r\}}), \quad \xi \in \mathcal{K} \ , \tag{33}$$

where $\mathcal{K}$ is the set of possible data measurement values. The iteration scheme (31) – (33) has the monotonic property [24]:

$$\mathcal{L}^{(t+1)} \geq \mathcal{L}^{(t)}, \quad t = 0, 1, 2, \ldots$$

which implies the convergence of the sequence $\{\mathcal{L}^{(t)}\}_0^\infty$ to a stationary point of the EM algorithm. The advantageous property of the mixture model is easy computation of any univariate conditional distribution $p(Y_r\,|\,Y_{\{r\}})$ which subsequently serves for the corresponding random field synthesis. The implementation of the EM algorithm is simple but there are some well known computational problems, e.g., the proper choice of the number of components, the existence of local maxima of the likelihood function and the related problem of a proper choice of the initial parameter values. In some applications there might be problem to get sufficiently large learning set. Alternative probabilistic mixture models (e.g., Gaussian mixture model [50],[51]) can be similarly defined.

4 Summary

Several possible models exist for multi-dimensional data representation in model-based pattern recognition approaches, but not all of them are equally suitable for implementation especially in time-efficient or even real-time systems. The main disadvantage of random mosaic models is that they can model geometrical shapes in data spaces (straight lines, planes, rectangles, circles, spheres, cubes, etc.) which are not general enough for any data. For example they are not appropriate to model natural textures but can be used for certain man-made regular structures generation. Syntactic models at the present state of art suffer with unsolved grammar inference problems. Fractal models have the advantage that they can be approximated by a single fractal function over a range of scales, which potentially enables multi-scale data representation and recognition. Their second-order statistics are described by a single parameter. Such properties sound promising, but convincing studies are necessary to find how credible natural data modelled by fractals are, and in what range real fractals are scale invariant. Second order models were based on the assumption (later found to be wrong even for 2D image data) that the second order statistics are sufficient for a data space description to be visually indiscernible from original data. Although such an approximation is good enough for some data spaces, the synthesis step is done iteratively and is time consuming. Autoregressive models are highly flexible to model different data types; their advantage over equivalent Gaussian Markov random field models is their usually smaller number of parameters. On the contrary, the equivalent Gaussian Markov random field models are more general: for some GMRF models there does not exist a finite order AR model. Causal autoregressive models can be generated directly from the model defining equation and they have analytical ML or Bayesian parameter estimators, but in such a case we have to artificially limit spatial correlation and introduce causal direction, which is in most nD cases unnatural. Noncausal AR models on a toroidal lattice can be generated using the fast Fourier transformation. Another possibility is to use more complicated ARMA models, but their main problem is a highly nonlinear parameter estimation. Markov random fields are the most flexible (they can describe the most complex random fields from our surveyed models) from all described possibilities. Problems of general MRF are difficulties with evaluation of a normalizing constant, iterative synthesis, and non-linear parameter estimation. Markov random fields analysis as well as synthesis most often require extensive computation. An exception are the GMRF models, whose normalizing constant is easy to evaluate and if we assume the toroidal lattice, then we can use a rapid fast Fourier transformation based synthesis algorithm. The probabilistic mixture models have large data modelling potential and easy synthesis, however they require large learning sets for their reliable parameter estimation and they are computationally expensive.

The analytical or learning step of some pattern recognition applications (e.g., supervised classification) is not critical, because it can be often solved off-line. The supervised classifier has a database of different classes parameters and it is able to use them for fast object recognition. On the other hand, the recognition step is often done in real time, and it is therefore necessary to classify objects as quickly as possible. Iteration based algorithms like for example algorithms of the Metropolis type are not very handy, especially for larger data fields, because they are rather time consuming. Another disadvantage is that although these algorithms converge to a specified field, little is known about their convergence rate. Although time is often not so important for the analytical step of a classifier, ML parameter estimation needs numerical optimization for most of the AR and MRF models, so we have no guarantee to reach a global optimum. In our experimentation the MPL method (or Woods method) appears to be a rapid and sufficient alternative, at least for lower order models. The coding method of Besag uses only a fraction of the available data and offers no clear advantage over MPL. Similarly, the iterative estimation method [38] was slowly converging in several experiments.

The GMRF generator as well as the AR generator (in synthesis applications) based on the fast Fourier transformation, is especially advantageous because it needs only one inverse FFT to generate the desired random field, which is much faster than any other iterative generator. Another advantage of this method is the possibility to use a special FFT processor to build a real time GMRF or AR generator.

Comparing our different 2D simulations, we found that the convergence rate is different for each random field. While some models were already near their limiting random fields after 30 iterations, others needed 100 iteration steps and for others even this number was not sufficient. Hence it is difficult to find generally applicable stopping rules for iterative Markov chain Monte Carlo type of algorithms (e.g., simulated annealing, the Metropolis algorithm, Gibbs sampler, etc.).

Although model-based pattern recognition methods are universal, they are dependent on suitable multi-dimensional data models, but the knowledge about higher-dimensional models is still rather limited in comparison with the 1D models. The continuous progress in the random field theory together with the simultaneous improvement of computer technology is inevitably leading in the near future to many promising model-based pattern recognition applications as the core part of any machine perception system.

Acknowledgement

The support of the Grant Agency of the Czech Republic through grants 102/00/0030, 106/00/0030 and EC grant IST-2001-34744 RealReflect is gratefully acknowledged.

References

[1] N. Ahuja and A. Rosenfeld, Mosaic models for textures, *IEEE Trans. Pattern Anal. Mach. Int.* Vol.3 (1981) pp.1-11.

[2] N. Ahuja and B.J. Schachter, *Pattern models,* (J.Wiley, 1983).

[3] J. Bennett and A. Khotanzad, Maximum likelihood estimation methods for multispectral random field image models, *IEEE Trans. on Pattern Analysis and Machine Intelligence* Vol.21 No.3 (1999) pp.537-543.

[4] M. Barnsley, *Fractals Everywhere*, (Academic Press , 1988).

[5] C. Bennis and A. Gagalowicz, Hierarchical texture synthesis on 3-D surfaces, in W. Hansmann, F.R.A. Hopgood, and W. Strasser (eds.) *Eurographics '89* , (1989) pp.257-269.

[6] C. Bennis and A.Gagalowicz, 2-D Macroscopic Texture Synthesis , *Computer Graphics Forum* Vol. 8 (1989) pp.291–300.

[7] J. Besag, Spatial Interaction and the Statistical Analysis of Lattice Systems, *J. Royal Stat. Soc.* Vol. B–36 (1974) pp.192–236.

[8] J. Besag, On the Statistical Analysis of Dirty Pictures, *J. Royal Stat. Soc.* Vol. B–48 (1986) pp.259–302.

[9] C. Bouman and B. Liu, Multiple Resolution Segmentation of Textured Images, *IEEE Trans. Pattern Anal. Mach. Int.* Vol.13 (1991) pp.99–113.

[10] R. Chellappa, Two-dimensional discrete Gaussian Markov random field models for image processing, in L.N. Kanal and A. Rosenfeld (eds.) *Progress in Pattern Recognition 2*, (Elsevier, North-Holland, 1985).

[11] F.S. Cohen, Markov random fields for image modelling and analysis, in U.B. Desai (eds.) *Modeling and Application of Stochastic Processes*, (Kluwer Academic Publishers, Boston, 1986).

[12] G.R. Cross and A.K. Jain, Markov Random Field Texture Models, *IEEE Trans. Pattern Anal. Mach. Int.* Vol.5 (1983) pp.25–39.

[13] K. Deguchi and I. Morishita, Two-dimensional auto-regressive model for the representation of random image fields, *Proc. 6-th Int.Conf. on Pattern Recognition*, (Munich, 1982) pp.90–93.

[14] A. Dempster, N. Laird, and D. Rubin, Maximum likelihood from incomplete data via the EM algorithm, *Journal of the Royal Statistical Society, B*, Vol.39 (1977) pp.1–38.

[15] H. Derin and A.W. Cole, Segmentation of textured images using Gibbs random fields, *Comp. Graphics Image Proc.* Vol.35 (1986) pp.72–97.

[16] H. Derin and H. Elliot, Modeling and segmentation of noisy and textured images using Gibbs random fields, *IEEE Trans. Pattern Anal. Mach. Int.* Vol.9 (1987) pp.39–55.

[17] N. Dodd, Multispectral Texture Synthesis Using Fractal Concepts, *IEEE Trans. Pattern Anal. Mach. Int.*, Vol.9 (1987) pp.703–707.

[18] Z. Fan, F.S. Cohen, and M.A. Patel, Rotated and Scaled Textured Images using Markov Random Field Models, *IEEE Trans. Pattern Anal. Mach. Int.* Vol.13 (1991) pp.191–202.

[19] O.D. Faugeras and O.D. Garber, Algebraic reconstruction techniques for texture synthesis, *Proc. 5-th Int.Conf. on Pattern Recognition*, (Miami Beach, 1980) pp.792–782.

[20] K.S. Fu and S.Y. Lu, Computer generation of texture using a syntactic approach, *Computer Graphic*, Vol.12 (1978) pp.147–152.

[21] K. Fu, Syntactic Image Modelling Using Stochastic Tree Grammars , *Comp. Graphics Image Proc.* Vol.12 (1980) pp.136–152.

[22] S. Geman and D. Geman, Stochastic Relaxation , Gibbs Distributions and Bayesian Restoration of Images, *IEEE Trans. Pattern Anal. Mach. Int.*, Vol.6 (1984) pp.721–741.

[23] B. Gidas, A renormalization Group Approach to Image Processing problems, *IEEE Trans. Pattern Anal. Mach. Int.*, Vol.11 (1989) pp.164–180.

[24] J. Grim, On numerical evaluation of maximum likelihood estimates for finite mixtures of distributions, *Kybernetika*, Vol.18 (1982) pp.173–190.

[25] J. Grim and M. Haindl, Texture Modelling by Discrete Mixtures, accepted to *Computational Statistics and Data Analysis*, Elsevier (2002).

[26] M. Haindl, Contextual Classification, *Proc. Artificial intelligence application '90*, (Prague, 1990).

[27] M. Haindl, Texture Synthesis, *CWI Quarterly*, Vol. 4 (1991), pp. 305–331.

[28] M. Haindl and S. Šimberová, A Multispectral Image Line Reconstruction method, in P. Johansen and S. Olsen (eds.) *Theory & Applications of Image Analysis,* (World Scientific Publishing Co., Singapore, 1992).

[29] M. Haindl and S. Šimberová, A high - resolution radiospectrograph image reconstruction method, *Astronomy & Astrophysics*, Vol.115 (1996) pp.189-193.

[30] M. Haindl and S. Šimberová, A Scratch Removal Method, *Kybernetika*, Vol.34 (1998) pp.423-428.

[31] M. Haindl, Texture Segmentation Using Recursive Markov Random Field Parameter Estimation, in Bjarne K. Ersboll and Peter Johansen

(eds.) *Proceedings of the 11th Scandinavian Conference on Image Analysis*, (DSAGM, Kangerlussuaq, 1999) pp. 771-776.

[32] M. Haindl, Recursive Model-Based Image Restoration, in A. Sanfeliu, J.J. Villanueva, M. Vanrell, R. Alquezar, T. Huang, and J. Serra (eds.) *Proceedings of the 15th IAPR Int. Conf. on Pattern Recognition*, (Barcelona, IEEE Press, 2000) Vol. III, pp. 346 - 349.

[33] M. Haindl, Texture modelling, in B. Sanchez, J. M. Pineda, J. Wolfmann, Z. Bellahse, and F. Ferri (eds.) *Proceedings of the World Multiconference on Systemics, Cybernetics and Informatics*, (Orlando, International Institute of Informatics and Systemics, 2000), Vol. VII, pp. 634–639.

[34] M. Haindl and V. Havlíček, A multiresolution causal colour texture model, in F. J. Ferri, J. M. Inesta, A. Amin, and P. Pudil (eds.) *Advances in Pattern Recognition, Lecture Notes in Computer Science 1876*, (Berlin, Springer-Verlag, 2000) pp. 114 –122.

[35] J.M. Hammersley and D.C.Handscomb, Monte Carlo Methods, (Methuen, London, 1964).

[36] A. Jain, Advances in mathematical models for image processing, *Proc. IEEE* , Vol.69 (1981) pp.502–528.

[37] H. Kaneko and E. Yodogawa, A Markov Random Field Application to Texture Classification, *Proc. IEEE Conf. Pattern Recognition and Image Processing*, (Las Vegas, 1982).

[38] R.L. Kashyap and R. Chellappa, Estimation and Choice of Neighbors in Spatial-Interaction Models of Images, *IEEE Trans. Inf. Theory*, Vol.29 (1983) pp.60–72.

[39] R.L. Kashyap, Analysis and Synthesis of Image Patterns by Spatial Interaction Models, in L.N. Kanal and A.Rosenfeld (eds.) *Progress in Pattern Recognition 1*, (Elsevier, North-Holland, 1981).

[40] R. Kindermann and J.L. Snell, *Markov Random Fields and their Applications*, (American Mathematical Society , Providence, 1980).

[41] S.Y. Lu and K.S. Fu, A syntactic approach to texture analysis, *Comp. Graphics Image Proc.* Vol.7 (1978) pp.303–330.

[42] B.B. Mandelbrot, *The Fractal Geometry of Nature* , (Freeman , San Francisco, 1982).

[43] N. Metropolis, A.W. Rosenbluth, M.N. Rosenbluth, A.H. Teller, and E. Teller, Equations of state calculations by fast computing machines, *J. Chem. Phys.* Vol.21 (1953) pp. 1087-1091.

[44] S.G. Nadabar and A.K. Jain, Parameter Estimation in Markov Random Field Contextual Models Using Geometric Models of Objects, *IEEE Trans. Pattern Anal. Mach. Int.*, Vol. 18 (1996) pp. 326–329.

[45] S. Peleg, J.Naor, R.Hartley, and D.Avnir, Multiple resolution texture analysis and classification , *IEEE Trans. Pattern Anal. Mach. Int.*, Vol.6 (1984) pp.518–523.

[46] A.P. Pentland, Fractal-based description of natural scenes, *IEEE Trans. Pattern Anal. Mach. Int.*, Vol.6 (1984) pp.661–674.

[47] D.K. Pickard, A curious binary lattice process, *J. Appl.Probability* Vol.14 (1977) pp.717–731.

[48] Ch.J. Preston, *Gibbs states on countable sets*, (Cambridge Univ. Press, London, 1974).

[49] W. Qian and D.M.Titterington, Multidimensional Markov Chain Models for Image Textures , *J. Royal Stat. Soc.*, Vol.B-53 (1991) pp.661–674.

[50] S. Richardson and P.J. Green, On Bayesian Analysis of Mixtures with and Unknown Number of Components, *J. R. Statist. Soc.*, Vol. B-59 (1997) pp.731–792.

[51] S.J. Roberts, D. Husmeier, I. Rezek, and W. Penny, Bayesian Approaches to Gaussian Mixture Modeling, *IEEE Trans. Pattern Anal. Mach. Int.*, Vol. 20 (1998) pp.1133-1142.

[52] B.J. Schachter, A. Rosenfeld, and L.S. Davis, Random Mosaic Models for Textures, *IEEE Trans. Syst. Man Cyb.*, Vol.8 (1978) pp.694–702.

[53] B.J. Schachter and N. Ahuja, Random pattern generation processing, *Comp. Graphics Image Proc.* Vol.10 (1979) pp.95–114.

[54] D. Stoyan, W.S.Kendall, and J.Mecke, Stochastic geometry and its applications, (J.Wiley, 1989).

[55] J.W. Woods, Two-Dimensional Discrete Markovian Fields , *IEEE Trans. Inf. Theory*, Vol.18 (1972) pp.232–240.

[56] S.C. Zhu, X.W. Liu, and Y.N. Wu, Exploring Texture Ensembles by Efficient Markov Chain Monte Carlo - Toward a "Trichromacy" Theory of Texture, *IEEE Trans. on Pattern Analysis and Machine Intelligence* Vol.22 (2000) pp.554-569.

Structural Pattern Recognition in Graphs

Lawrence Holder
Department of Computer Science and Engineering
University of Texas at Arlington, Arlington, TX 76019
E-mail: holder@cse.uta.edu

Diane Cook
Department of Computer Science and Engineering
University of Texas at Arlington, Arlington, TX 76019
E-mail: holder@cse.uta.edu

Jesus Gonzalez
Instituto Nacional de Astrofisica Optica y Electronica Ciencias Computacionales
Luis Enrique Erro 1, Sta. Maria Tonantzintla, APDO. Postal 51 y 216, C.P. 72840, Puebla, Puebla. Mexico
E-mail: jagonzalez@inaoep.mx

Istvan Jonyer
Department of Computer Science and Engineering
University of Texas at Arlington, Arlington, TX 76019
E-mail: jonyer@cse.uta.edu

D. Chen and X. Cheng (eds.),
Pattern Recognition and String Matching, 255–279.

Contents

1 Introduction

Most pattern recognition approaches look for patterns in data represented as independent entities described by attributes. However, the relationships between entities are as important, if not more important, to the recognition of accurate and meaningful patterns. In this chapter we describe an approach to discovering patterns in relational data represented as a graph. Our approach is based on the minimum description length (MDL) principle [28], which measures how well various patterns compress the original database. This approach is implemented in the SUBDUE system. We begin with a discussion of related work. We then describe graph-based discovery, the main discovery algorithm, and the polynomially-constrained inexact graph matching algorithm at the heart of the discovery process. Next, we describe how this technique can also be used for clustering and concept learning. We illustrate the utility of the approach by applying the clustering and concept learning techniques to DNA and WWW data.

2 Related Work

Systems that deal with structural databases typically do not scale for large databases due to the increased combinatorics inherent in the richer representations. Inductive Logic Programming (ILP) systems can handle structural representations [15], but do not yet scale to large databases. Although ILP systems typically perform supervised learning, some systems (e.g., CLAUDIEN [27] and PROGOL [25]) attempt unsupervised discovery from structural data using techniques common in ILP systems. Our graph-based approach is constrained to run in polynomial time while still discovering relevant patterns [9, 11, 12, 10].

Other approaches to discovery in structural databases have been proposed [7, 20, 29, 33, 35]. Many of these approaches use a knowledge base of concepts to classify the structural data. These systems perform concept learning over examples and categorization of new data, and are typically not designed to perform unsupervised discovery. While the above methods process individual objects one at a time, our method is designed to process the entire structural database, consisting of many objects. Unlike many of the existing methods, SUBDUE can also discover knowledge in a supervised or unsupervised fashion.

The graph-based discovery approach relies on the ability to perform an inexact, or error-tolerant, graph isomorphism between two graphs. Our inexact graph match is based on a branch-and-bound search to find a minimal edit distance [4]. Such approaches have been refined [26, 23, 22], especially in their application to computer vision for matching artifacts of an image represented in graph form. Related work on computing an error-tolerant subgraph isomorphism for a set of graphs optimizes the graph algorithm by first matching common subgraphs of the set of graphs to be matched [24]. We also are interested in finding common subgraphs within a set of graphs, but also in a single graph, and our approach searches for such subgraphs motivated by the desire to find subgraphs that not only optimally compressing the graph, but also represent knowledge of interest to the user.

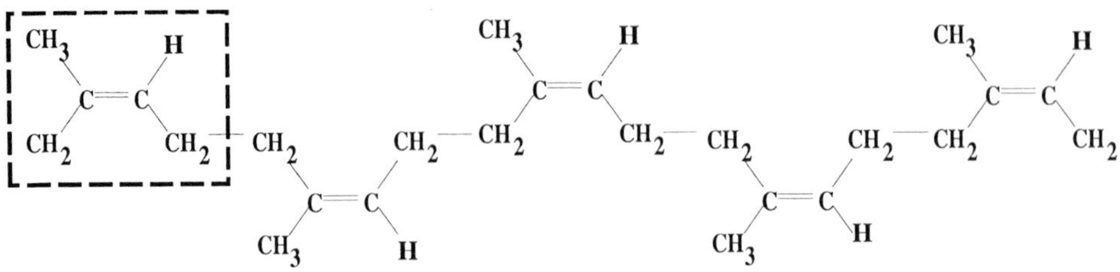

Figure 1: Natural rubber atomic structure.

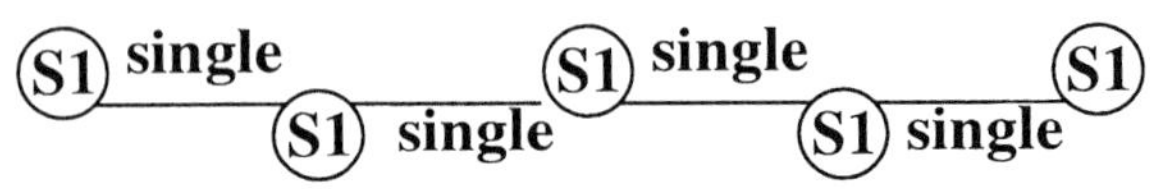

Figure 2: Rubber graph compressed using the discovered substructure.

3 Graph-Based Discovery

The graph-based discovery method discovers structural patterns, or substructures, in a graph representation of data guided by the MDL principle. Based on the MDL principle, the method discovers substructures that compress the original data and represent structural concepts in the data. Once a substructure is discovered, the substructure is used to simplify the data by replacing instances of the substructure with a pointer to the newly discovered substructure. The discovered substructures allow abstraction over detailed structures in the original data. Iteration of the substructure discovery and replacement process constructs a hierarchical description of the structural data.

3.1 Graph Representation

The substructure discovery approach represents structural data as a labeled graph. Objects in the data map to vertices or small subgraphs in the graph, and relationships between objects map to directed or undirected edges in the graph. A *substructure* is a connected subgraph within the graphical representation. This graphical representation serves as input to the substructure discovery algorithm. An *instance* of a substructure in a graph is a set of vertices and edges from the input graph that match, graph theoretically, to the graphical representation of the substructure.

Figure 1 shows a sample database that is input to SUBDUE, representing

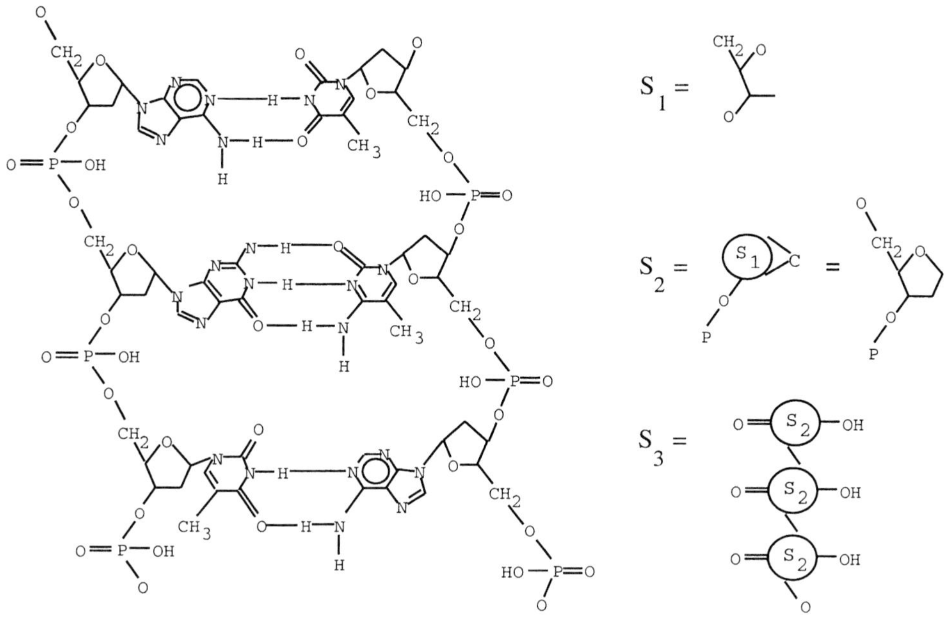

Figure 3: Sample results of Subdue on a partial DNA sequence.

the atomic structure of natural rubber. The input graph represents atoms as vertices, and single or double bonds as labeled undirected edges between the vertices. The highlighted substructure, which we label substructure S1, is selected as the best subgraph to describe the input database. The five instances of the substructure are each replaced by a single vertex representing the discovered concept. The graph that results from compressing the rubber database is shown in Figure 2. This discovery and compression process helps the user understand the database and exploit the knowledge content of the data encoded in the graph vertices and edges.

Figure 3 shows a sample input database containing a portion of a DNA sequence. In this case, atoms and small molecules in the sequence are represented with labeled vertices in the graph, and the single and double bonds between atoms are represented with labeled edges in the graph. SUBDUE discovers substructure S_1 from the input database. After compressing the original database using S_1, SUBDUE discovers substructure S_2, which when used to compress the database further allows SUBDUE to discover substructure S_3. Such repeated application of SUBDUE generates a hierarchical description of the structures in the database.

```
Subdue (Graph, Beam, Limit)
    queue Q = {v | v is a vertex in Graph having a unique label}
    bestSub = first substructure in Q
    repeat
        newQ = {}
        for each substructure S ∈ Q
            newSubs = Extend-Substructure (S, Graph)
                      in all possible ways
            Evaluate (newSubs)
            newQ = newQ ∪ newSubs mod Beam
            Limit = Limit - 1
        if best substructure in newQ better than bestSub
        then bestSub = best substructure in Q
        Q = newQ
    until Q is empty or Limit ≤ 0
    return bestSub
```

Figure 4: SUBDUE's discovery algorithm.

3.2 Discovery Algorithm

SUBDUE uses a variant of beam search [36, 16]. Beam search maintains a limited-length (i.e., *beam* length) queue of the best few patterns found so far. While limiting the queue length yields a suboptimal search, not limiting the queue length may result in an exponential-sized queue. The goal of the search is to find the substructure (pattern) that best compresses the description length of the input graph. A substructure in SUBDUE consists of a substructure definition and all its instances in the graph.

SUBDUE's discovery algorithm is shown in Figure 4 and is given the input graph, the beam length, and a limit on the total number of substructures considered by the algorithm. The initial state of the search is the set of substructures representing each uniquely labeled vertex and its instances. The only search operator is the *Extend-Substructure* operator. As its name suggests, *Extend-Substructure* extends the instances of a substructure in all possible ways by a single edge and a vertex, or by a single edge if both vertices are already in the substructure. The minimum description length (MDL) principle is used to evaluate the substructures.

The search progresses by applying the *Extend-Substructure* operator to each substructure in the current search frontier (queue), which is an ordered list of previously discovered substructures. The resulting frontier, however, does not contain all the substructures generated by the *Extend-Substructure* operator. The substructures are stored on a queue and are ordered based on their ability to compress the graph. The length of the queue is limited by the user-defined *Beam* parameter, where *mod Beam* means that only the best *Beam* substructures are kept on the queue after adding new substructures to the queue. The search terminates upon reaching a user-specified *Limit* on the number of substructures extended, or upon exhaustion of the search space. SUBDUE's run time is polynomial in *Beam* and *Limit*.

Once the search terminates and returns the best substructure, the graph can be compressed using this substructure. The compression procedure replaces all instances of the substructure in the input graph by a single vertex, which represents the substructure. Incoming and outgoing edges to and from the replaced substructure will point to, or originate from, the new vertex that represents the substructure. In our implementation, we do not maintain information on how vertices in each instance were connected to the rest of the graph, which results in lossy compression. Since the goal of substructure discovery is interpretation of the database, maintaining information to reverse the compression is of secondary importance.

The SUBDUE algorithm can be invoked again on this compressed graph. This procedure can be repeated a user-specified number of times, and is referred to as an iteration. The maximum number of iterations that can be performed on a graph cannot be predetermined; however, a graph that has been compressed into a single vertex cannot be compressed further.

To allow SUBDUE to discover substructures of particular interest to a scientist in a specific domain, the user can direct the search with expert-supplied background knowledge [12]. Background knowledge can take the form of known substructure models that may potentially appear in the database, or graph match rules to adjust the cost of each graph transformation. Unlike other existing approaches to graph-based discovery [8, 20, 30, 33, 37], SUBDUE is effective at finding interesting and repetitive substructures in any structural database with or without domain-specific guidance.

3.3 Minimum Description Length Principle

SUBDUE's search is guided by the minimum description length (MDL) principle developed by Rissanen [28]. According to the MDL heuristic, the best substructure is the one that minimizes the description length of the graph when compressed by the substructure. This compression is calculated as

$$Compression = \frac{DL(S) + DL(G \mid S)}{DL(G)}$$

where $DL(G)$ is the description length of the input graph, $DL(S)$ is the description length of the substructure, and $DL(G \mid S)$ is the description length of the input graph compressed by the substructure. The search algorithm attempts to maximize the value of the substructure, which is simply the inverse of the Compression. The description length of a graph is calculated here as the number of bits needed to encode an adjacency matrix representation of the graph. Additional details of the encoding scheme are reported in the literature [9].

3.4 Inexact Graph Match

Because instances of a substructure can appear in different forms throughout the database, an inexact graph match based on [4] is used to identify substructure instances. Performing an inexact match allows the discovered substructures to abstract away minor variations in the substructure instances. Subgraphs of the original database are considered to be instances of a substructure definition if the edit distance between the two graphs, or the cost of transforming the potential instance into a graph that is isomorphic with the substructure definition, does not exceed a user-defined value. Transformations between graphs can include addition or deletion of vertices, addition or deletion of edges, vertex label substitutions, and edge label substitutions. Each transformation is assigned a cost which can be customized by the user for a specific application.

Formally, an inexact graph match from graph g_1 to graph g_2 is a mapping $f : N_1 \rightarrow N_2 \cup \{\lambda\}$, where N_1 and N_2 are the sets of vertices of g_1 and g_2, respectively. A vertex $v \in N_1$ that is mapped to λ (i.e., $f(v) = \lambda$) is deleted. That is, the vertex has no corresponding vertex in g_2 and therefore maps to nothing (λ). Given a set of graph transformation costs, we define the cost of an inexact graph match $cost(f)$, as the sum of the cost of the individual transformations resulting from f, and we define $matchcost(g_1, g_2)$ as the value of the least-cost function that maps graph g_1 onto graph g_2. A SUBDUE parameter, the match threshold t, can be specified with a value between 0 and 1. A graph is considered to be an instance of a substructure if the matchcost of the two graphs is no more than t times the size of the

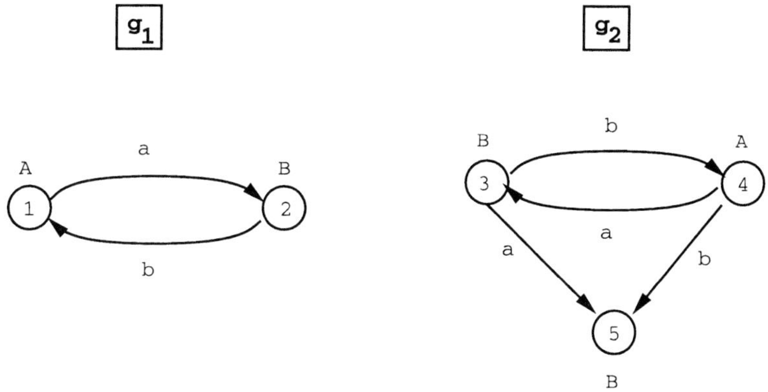

Figure 5: Two similar graphs g_1 and g_2.

larger graph. A threshold value of 0 constrains the algorithm to allow only exact matches, and a threshold value of 1 allows any two graphs to match.

Given g_1, g_2, and a set of distortion costs, the actual computation of $matchcost(g_1, g_2)$ can be determined using a tree search procedure. A state in the search tree corresponds to a partial match that maps a subset of the vertices of g_1 to a subset of the vertices in g_2. Initially, we start with an empty mapping at the root of the search tree. Expanding a state corresponds to adding a pair of vertices, one from g_1 and one from $g_2 \cup \{\lambda\}$, to the partial mapping constructed so far. A final state in the search tree is a match that maps all vertices of g_1 to g_2 or to λ. The complete search tree of the example in Figure 5 is shown in Figure 6. For this example we assign a value of 1 to each distortion cost. The numbers in circles in this figure represent the cost of a state. As we are eventually interested in the mapping with minimum cost, each state in the search tree is assigned the cost of the corresponding partial mapping. Thus the goal state to be found by our tree search procedure is the full mapping, or leaf node, with the minimum cost among all full mappings. From Figure 6 we conclude that the minimum cost inexact graph match of g_1 and g_2 is given by the mapping $f(1) = 4$, $f(2) = 3$. The cost of this mapping is 4. Given graphs g_1 with n vertices and g_2 with m vertices, $m \geq n$, the complexity of the full inexact graph match is $O(n^{m+1})$.

Because this routine is used throughout the discovery process, the complexity of the algorithm can significantly degrade the performance of the system. We offer a number of enhancements that improve the efficiency of

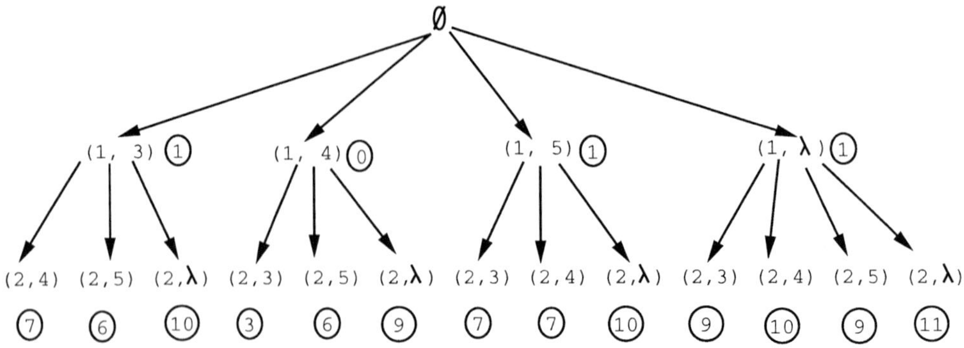

Figure 6: Search tree for computing matchcost(g_1,g_2) from Figure 5.

this algorithm. First, we apply a branch-and-bound search algorithm to the search space in which partial mappings are considered in nondecreasing order by match cost. The cost from the root to a given node is calculated as the cost of all distortions corresponding to the partial mapping for that node. Vertices from the matched graphs are considered in nonincreasing order by degree. Because branch-and-bound GUARANTEEs an optimal solution, the search ends when the first complete mapping (leaf node) is found.

Second, the user can limit the number of search nodes considered by the branch-and-bound procedure. The limit is defined as a function of n, the size of the larger graph. Once the number of nodes expanded in the search tree reaches the defined limit, the search switches to hill climbing or greedy searching using the cost of the partial mapping as the measure for choosing the best node at a given level.

Third, the match threshold itself can be used to prune portions of the search space. Because no graph transformations can be assigned a negative cost, as soon as a node is reached with a match cost greater than the allowed value, the subtree rooted at that node is removed from consideration.

Employing computational constraints such as a bound on the number of substructures considered (s) and the number of partial mappings considered during an inexact graph match (x), SUBDUE is constrained to run in polynomial time. The worst-case run time of the system is the product of the number of generated substructures, the number of instances of each substructure, and the number of partial mappings considered during graph match. This expression is equal to $(\sum_{i=1}^{s} i * ((v-1)-(i-1))) * (v(s-1)) * x$,

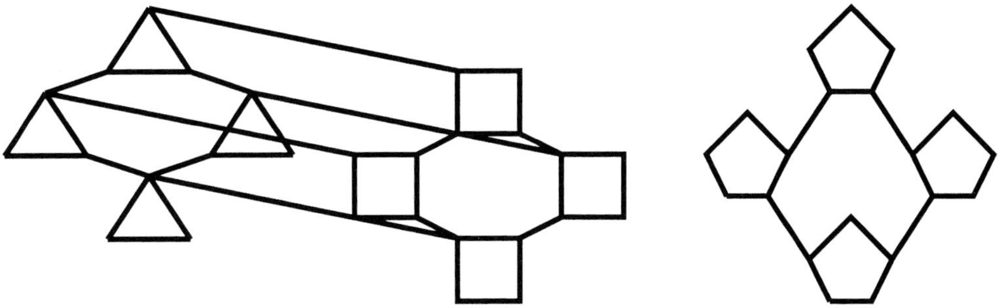

Figure 7: Sample geometric domain for clustering.

where v represents the number of vertices in the input graph.

4 Graph-Based Clustering

Cluster analysis has been studied and developed for model fitting, hypothesis generation and testing, data exploration and reduction, prediction based on groups, and finding true topologies [1]. The purpose of applying clustering to a database is to gain a better understanding of the data, in many cases by highlighting hierarchical topologies. An example of a hierarchical clustering is the classification of vehicles into groups such as cars, trucks, motorcycles, tricycles, and so on, which are then further subdivided into smaller groups based on observed traits. Although a number of relatively successful clustering systems have been constructed [6, 17, 33], few existing systems address the problem of clustering in discrete-valued, structural databases. Our approach centers on discrete-valued, structural databases that are represented as graphs.

One approach to structural clustering is to perform multiple iterations of SUBDUE, constraining the discovery algorithm to not reuse substructures discovered in previous iterations, until the whole graph is exhausted. Each iteration yields one cluster (substructure), which is used to compress the graph. This cluster can be inserted into a *classification lattice*. This lattice structure contrasts with previous research that generates classification trees, and is partially due to the fact that SUBDUE has the unique capability to represent more elaborate relationships in the data using graph structure.

4.1 Example

As an example, SUBDUE can be used to cluster the data shown in Figure 7. Visually, the input data consists of triangles, squares and pentagons, where vertices and edges in the figure map to vertices and edges in the input graph. SUBDUE first discovers the substructure describing the pentagon pattern in the input graph, which is inserted as a child cluster of the root node in the lattice. During the next two iterations, the square shape and triangle shape as discovered and inserted as children of the root. In the fourth iteration, SUBDUE returns the substructure describing two pentagon shapes connected by a single edge. This cluster is inserted into the classification lattice as the child of the cluster describing the pentagon, because the pentagon appears in the cluster definition. There are two links connecting this new cluster to its parent, because the parent cluster definition appears twice. In the last iteration, a substructure is discovered that contains a pair of squares connected by an edge, a pair of triangles connected by an edge, and an edge connecting the two pairs. This cluster is inserted as a child of two clusters from the first level of the lattice. The resulting lattice is depicted in Figure 8.

4.2 Application to DNA

We applied graph-based clustering to the DNA sequence data from Figure 3. The resulting lattice is shown in Figure 9. The lattice property is apparent in Figure 9, where the bottom-left nodes have multiple parents. This lattice describes 71% of the DNA sequence shown in Figure 3. As the lattice shows, smaller, more commonly occurring compounds are found first that compose the first level of the lattice. These account for more than 61% of the DNA sequence data. Subsequently identified clusters are based on these smaller clusters that are either combined with each other, or with other atoms or molecules to form a new cluster. The second level of the lattice extends the conceptual clustering description such that an additional 7% of the DNA is covered.

5 Graph-Based Concept Learning

We have extended our unsupervised discovery methods to perform supervised graph-based relational concept learning. Logic-based systems have dominated the area of relational concept learning, especially Inductive Logic Programming (ILP) systems. However, first-order logic can also be repre-

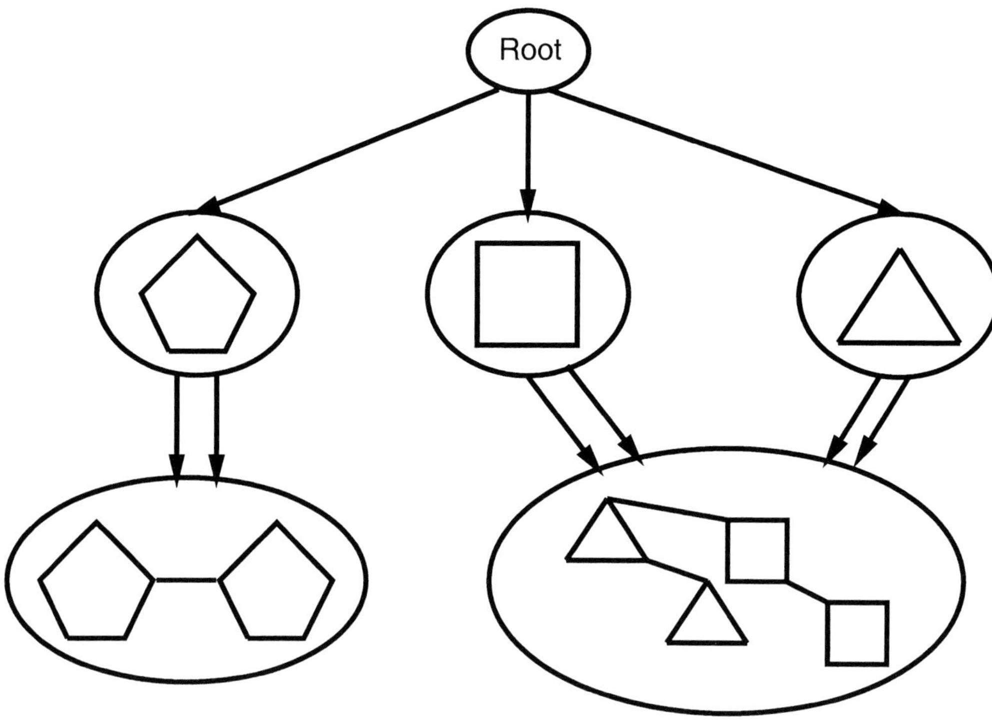

Figure 8: Clustering lattice for geometric domain in Figure 7.

sented as a graph, and in fact, first-order logic is a subset of what can be represented using graphs. Therefore, learning systems using graphical representations have the potential to learn richer concepts if they can handle the increased size of the hypothesis space.

Few general-purpose learning methods use a formal graph representation for knowledge, perhaps because of the arbitrary expressiveness of a graph and the inherent NP-hardness of typical graph-based learning routines. Graph-based learning methods for pattern recognition in chemical domains [2, 32, 34] have been successful due to the natural graphical description of chemical compounds. Some unsupervised learning methods (e.g., SUBDUE [9] and WARMR [14]) use a graph representation, but no domain-independent supervised concept learning systems use a graph representation (although there is some theoretical work on graph-based concept learning [21]).

The main challenge in adding concept-learning capabilities to SUBDUE

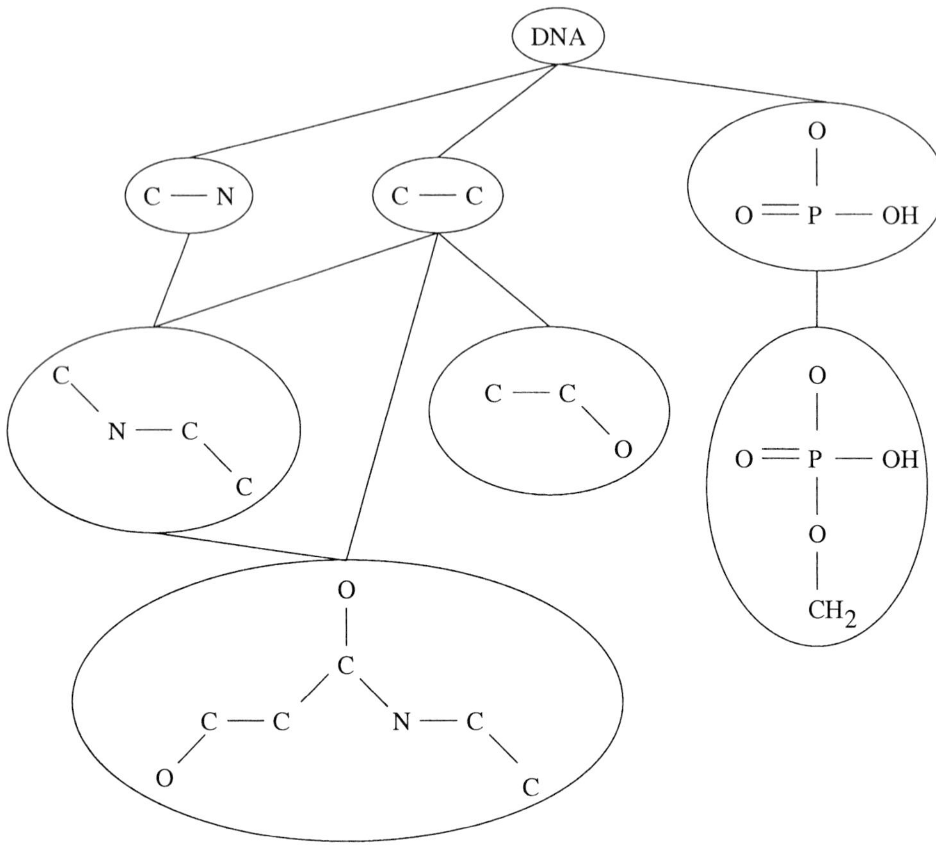

Figure 9: Clustering lattice for DNA domain.

is the inclusion of "negative" examples into the process. Substructures that describe the positive examples; but not negative examples, are likely to represent the target concept. Therefore, the SUBDUE concept learner (which we will refer to as SUBDUECL) accepts both positive and negative examples in graph format. Since SUBDUECL is an extension to SUBDUE, it uses SUBDUE's core functions to perform graph operations, but the learning process is different. SUBDUECL works as a supervised learner by differentiating positive and negative examples using a set-covering approach instead of graph compression. The hypothesis found by SUBDUECL consists of a set of disjunctions of conjunctions (substructures), i.e., the concept may contain several rules. SUBDUECL forms one of these conjunctions (rules) in each iteration. Positive example graphs that are described by the substructure found in a previous iteration are removed from the graph for subsequent iterations.

SUBDUECL uses an evaluation formula to give a value to all the generated substructures. This formula assigns a value to a substructure according to how well it describes the positive examples (or a subset of the positive examples) without describing the negative examples. Then, positive examples covered by the substructure increase the substructure value while negative examples decrease its value. In this formula the positive examples that are not covered and the negative examples covered by the substructure are considered errors, because the ideal substructure would be one covering all the positive examples without covering any negative example. Then, the substructure value is calculated as follows:

$$value = 1 - Error$$

where the error is calculated with respect to the positive and negative examples covered by the substructure using the following formula:

$$Error = \frac{\#PosEgsNotCovered + \#NegEgsCovered}{\#PosEgs + \#NegEgs}$$

$\#PosEgsNotCovered$ is the number of positive examples not covered by the substructure, and $\#NegEgsCovered$ is the number of negative examples covered by the substructure. $\#PosEgs$ is the number of positive examples remaining in the training set (remember that the positive examples that have already been covered in a previous iteration were removed from the training set), and $\#NegEgs$ is the total number of negative examples. This number does not change, because negative examples are not removed from the training set. Of two substructures with the same error, the substructure covering more positive examples is preferred.

In addition to replacing the compression-based evaluation measure with error-based measure mentioned above, the SUBDUECL algorithm differs from the original SUBDUE algorithm from Figure 4 in that SUBDUECL is called multiple times in a set-covering framework, each time adding a new substructure to the disjunctive hypothesis and removing covered positive examples. This process continues until either all positive examples are covered or no substructure exists discriminating the remaining positive examples from the negative examples (i.e., noise exists in the data).

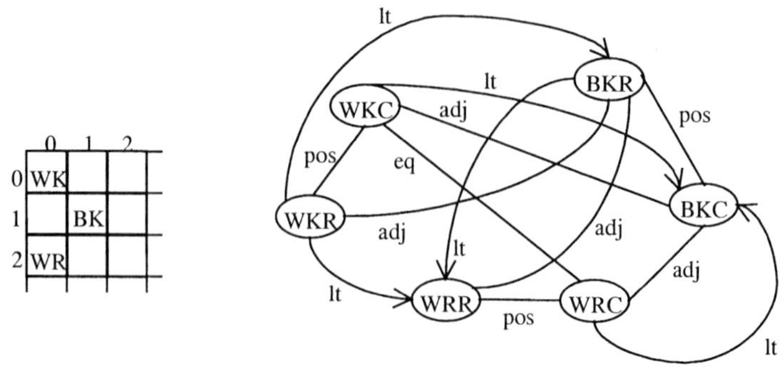

(a) (b)

Figure 10: An example from the chess domain: (a) Board configuration, (b) SUBDUECL's graphical representation of the example.

5.1 Example

As an example of SUBDUECL and its comparison to the ILP systems FOIL [5] and PROGOL [25], we present results from the chess endgame database available at the UCI Machine Learning Repository [3]. The chess domain consists of 20,000 examples of row-column positions for a white king, white rook and black king such that the black king is in check (positive) or not (negative). Therefore, if white's turn is next, then the positive examples are illegal configurations. The relational information in this domain consists of adjacency relations between board positions and less-than or equal relations between row and column numbers (0-7). Both FOIL and PROGOL extensionally define the three relations.

Figure 10b shows SUBDUECL's representation for the chess domain example in Figure 10a. Figure 10a shows a corner of the chess board and the positions of the white king (WK), white rook (WR), and black king (BK). SUBDUECL's representation in Figure 10b uses two vertices for the row and column for each piece: white king row (WKR), white king column (WKC), white rook row (WRR), white rook column (WRC), black king row (BKR), and black king column (BKC). The row and column vertices for each piece are connected by an undirected edge labeled "pos" representing the row/column position relationship. These row and column vertices are then connected by various numeric relationship edges, depending on whether they are less than (lt), equal to (eq), or adjacent (adj).

Due to computational constraints only a subset (5,000 examples) of the entire database was used for a 10-fold cross validation. This experiment involved partitioning the data into 10 subsets of 500 each. Ten trials are executed using one of the partitions as a test set evaluated on the patterns learned using the other 4,500 examples as training. The ten accuracy results are then averaged to arrive at a final accuracy value. The accuracy results are 99.74% for PROGOL, 99.34% for FOIL, and 99.74% for SUBDUECL. FOIL and SUBDUECL perform significantly better than PROGOL. In terms of number of rules, PROGOL learned 6 rules, FOIL learned 11 rules, and SUBDUECL learned 7 rules (substructures).

The rules learned by PROGOL and FOIL are shown below, where the arguments of the predicate illegal are illegal(WKC,WKR,WRC,WRR,BKC,BKR). Rule number 1 identifies the illegal board configurations where the black king and white king are in an adjacent row. Rule number two identifies the illegal board configurations where the white king and black king are in adjacent columns and rows. Rule three describes board configurations where the white king is in the same position as the white rook. Rule 4 says that board configurations where the white rook column is the same as the black king column are illegal. Rule 5 says that board configurations where the white rook row is the same as the black king row are illegal.

1. `illegal(A,B,C,D,A,E) :- adj(E,B).`
2. `illegal(A,B,C,D,E,F) :- adj(A,E), adj(B,F).`
3. `illegal(A,B,A,B,C,D).`
4. `illegal(A,B,C,D,C,E).`
5. `illegal(A,B,C,D,E,D).`

These rules say that a board configuration is illegal when:

1. The black king and white king are in an adjacent row.
2. The white king and black king are in adjacent columns and rows.
3. The white king is in the same position than the white rook.
4. The white rook column is the same as the black king column.
5. The white rook row is the same as the black king row.

The rules produced by FOIL are shown below.

```
1. illegal(A,B,C,D,E,D) :- lt(A,C).

2. illegal(A,B,C,D,C,F) :- lt(A,C).

3. illegal(A,B,C,D,E,F) :- adj(A,E), adj(B,F).

4. illegal(A,B,A,D,E,D).

5. illegal(A,B,C,D,C,F) :- lt(C,A).

6. illegal(A,B,A,B,E,F).

7. illegal(A,B,A,D,A,F) :- adj(D,F).

8. illegal(A,B,C,D,E,D) :- illegal(C,G,D,G,H,B).

9. illegal(A,B,A,D,A,F) :- adj(A,B), illegal(A,A,A,D,G,B).

10. illegal(A,B,C,D,E,D) :- illegal(A,C,B,G,E,A).

11. illegal(A,B,A,D,A,F) :- lt(D,B), illegal(A,F,A,F,G,A).
```

Rules 8-11 are recursive and not relevant to the domain. Rules 1-7 say that a board configuration is illegal when:

1. The white rook and the black king are in the same row, and the white king is in a lower column than the white rook.
2. The white rook and the black king are in the same column, and the white king is in a lower column than the white rook.
3. This rule is the same as rule 2 found by PROGOL.
4. The white king and the white rook are in the same column, and the white rook and the black king are in the same row.
5. The white rook and the black king are in the same column, and the white rook is in a lower column than the white king.
6. This rule is the same as rule 3 found by PROGOL.
7. The white king, the white rook and the black king are in the same column.

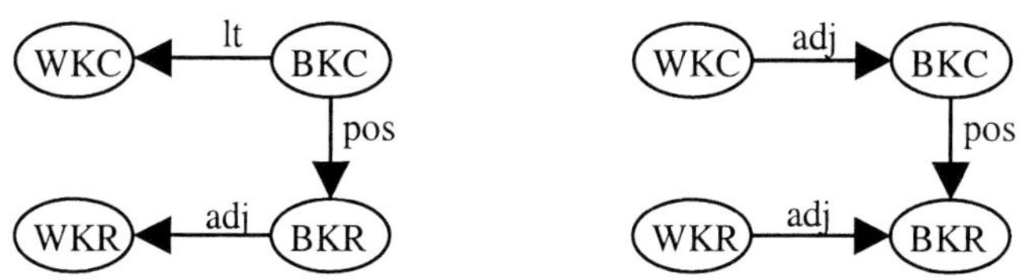

Figure 11: Two of seven substructures found by SUBDUECL in the chess domain.

SUBDUECL found the equivalent substructures for rules number 3, 4, and 5 of PROGOL. For rule number 2 found by PROGOL, SUBDUECL found two rules where the only difference between them is the direction of one edge. Figure 11 shows the last two of the seven substructures found by SUBDUECL in the chess domain. These two substructures found by SUBDUECL's are not directly equivalent to any of the rules found by PROGOL, but are similar to PROGOL's rule 1.

5.2 Application to the World-Wide Web

The Web domain consists of graphs created from web sites. We have three options for the information that these graphs contain: hyperlink structure, hyperlink structure + page's titles, and hyperlink structure + page's content. We used a Perl program to extract this information and convert it into its graph representation. The first experiment that we made for the Web domain consisted in differentiating the Web pages of professors and students. We made the professors' web pages our positive examples and the students web pages our negative examples. For this initial experiment we chose the web page structure + page's content option, because we wanted to give as much information as possible to SUBDUECL and observe its behavior. SUBDUECL was able to find a very small substructure that could differentiate the positive examples from the negative examples. This substructure is shown in Figure 12 and says that every professor has in their web page the word "box", but students do not have this word in their web pages. This is true because the word "box" is part of the address field of the professors' web pages.

For the second experiment in the web domain we used the hyperlink structure option. We chose this option, because we wanted to learn only structure without considering the web pages' content so that SUBDUECL did not learn a substructure with a word (or set of words) describing the

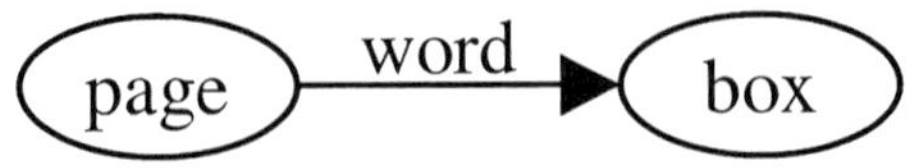

Figure 12: A substructure found in the web domain (Professors - Students).

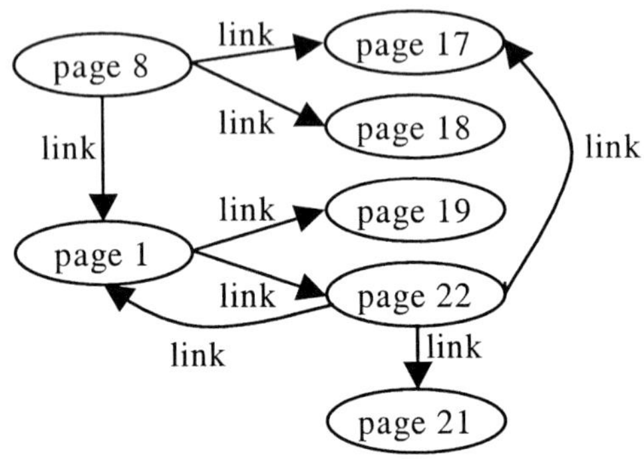

Figure 13: A substructure found in the web domain (Computer Stores - Professors).

domain as in the previous experiment. We chose the structure graphs of web sites of computer stores as our positive examples and the structure graphs of web sites of professors as our negative examples. Figure 13 shows a substructure found in this domain. This substructure covered 24 of a total of 29 positive examples without covering any of the negative examples. Although not all the examples covered by this substructure had the same interpretation, for most of them the following explanation applies: Node 8 represents a category of products linked to three subcategories of products represented by nodes 1, 17, and 18. Nodes 19 and 22 represent either more specific subcategories derived from the subcategory represented by node 1 or specific products. Node 21 represents a specialization of the subcategory represented by node 22.

The results in the web domain tell us that SUBDUECL is able to learn how to differentiate between the graph structure of different classes of web sites. Further experimentation has been done in both artificial and real domains to show that SUBDUECL is competitive with ILP systems [18].

6 Conclusions

Current databases are maintaining and increasing amount of structural information, and older databases can be augmented with such structural information (e.g., spatial and temporal relationships). There is a need for advanced pattern learning algorithms that can handle structural data. We described our graph-based discovery approach to this task, and its implementation in the SUBDUE system. Iterative executions of SUBDUE yield a hierarchical conceptual clustering of the data. Slight changes in the evaluation measure and top-level algorithm of SUBDUE yields SUBDUECL, a graph-based relational concept learner. These three functionalities (discovery, clustering, and concept learning) were illustrated using both artificial and real-world databases. Our conclusion is that the graph-based approach is both efficient and effective in identifying patterns in structural data. The approach is also competitive with older conceptual clustering techniques and inductive logic programming (ILP) approaches to relational concept learning.

We are pursuing several directions for enhancing the graph-based approach. Some of the enhancements are based on three advantages ILP systems have over SUBDUE. First, ILP systems can use variables to equate two arguments without specifying their value. Second, ILP systems can learn recursive concepts. Third, ILP systems are better at handling continuous values using inequalities and ranges. We are working on approaches to these three capabilities. We are working on introducing variables into a substructure definition based on the recognition of instances that have similar, but varying structure. We are moving to a graph grammar representation of substructures, clusters and concepts that will allow recursion. Lastly, we are investigating methods for learning ranges and distributions over continuous-valued labels in the graph. We believe these enhancements will give SUBDUE a clear advantage over ILP systems.

Finally, we will continue our empirical and theoretical analysis of SUBDUE. An increasing number of domains are becoming available that require structural pattern recognition to find desired patterns, e.g., domains in which relationships among a group of people might indicate suspicious behavior of the group. We have developed several parallel and distributed versions of SUBDUE to address very large databases, while maintaining the quality of the result [13]. We are also continuing our work in chemical domains (e.g., proteins, DNA, and carcinogens). We have begun a theoretical analysis of graph-based learning based on a probably-approximately correct

(PAC) analysis [19] of conceptual graphs [31] and an analysis of the Galois lattice [21].

Our graph-based approach to discovery, clustering and concept learning provides an efficient, effective, and uniform approach to the task of extracting knowledge from structural data. We will continue to investigate improvements to the algorithms and implementations, as well as apply our approach to various problems. Source code and data for the SUBDUE system and the aforementioned extensions are available at http://cygnus.uta.edu/subdue.

7 Acknowledgements

This work is supported by NSF grant ILI-9615272. This effort is also sponsored by the Defense Advanced Research Projects Agency (DARPA) and Air Force Research Laboratory, Air Force Materiel Command, USAF, under agreement number F30602-01-2-0570.

References

[1] G. H. Ball. Classification analysis. Technical Report SRI Project 5533, Stanford Research Institute, 1971.

[2] J. M. Barnard. Substructure searching methods: Old and new. *Journal of Chemical Information and Computing Sciences*, 33:532–538, 1993.

[3] C. L. Blake and C. J. Merz. UCI repository of machine learning databases, 1998.

[4] H. Bunke and B. T. Messmer. A new algorithm for efficient subgraph matching. In G. Vernazza, A. N. Vebetsanopoulos, and C. Braccini, editors, *Image Processing: Theory and Applications*, pages 303–307. Elsevier Science Publishers, 1993.

[5] R. M. Cameron-Jones and J. R. Quinlan. Efficient top-down induction of logic programs. *SIGART Bulletin*, 5(1):33–42, 1994.

[6] P. Cheeseman and J. Stutz. Bayesian classification (AutoClass): Theory and results. In U. M. Fayyad, G. Piatetsky-Shapiro, P. Smyth, and R. Uthurusamy, editors, *Advances in Knowledge Discovery and Data Mining*, chapter 6, pages 153–180. MIT Press, 1996.

[7] D. Conklin, S. Fortier, J. Glasgow, and F. Allen. Discovery of spatial concepts in crystallographic databases. In *Proceedings of the ML92 Workshop on Machine Discovery*, pages 111–116, 1992.

[8] D. Conklin and J. Glasgow. Spatial analogy and subsumption. In *Proceedings of the Ninth International Conference on Machine Learning*, pages 111–116, 1992.

[9] D. J. Cook and L. B. Holder. Substructure discovery using minimum description length and background knowledge. *Journal of Artificial Intelligence Research*, 1:231–255, 1994.

[10] D. J. Cook and L. B. Holder. Graph-based data mining. *IEEE Intelligent Systems*, 15(2):32–41, 2000.

[11] D. J. Cook, L. B. Holder, and S. Djoko. Knowledge discovery from structural data. *Journal of Intelligence and Information Sciences*, 5(3):229–245, 1995.

[12] D. J. Cook, L. B. Holder, and S. Djoko. Scalable discovery of informative structural concepts using domain knowledge. *IEEE Expert*, 11(5):59–68, 1996.

[13] D. J. Cook, L. B. Holder, G. Galal, and R. K. Maglothin. Approaches to parallel graph-based knowledge discovery. *Journal of Parallel and Distributed Computing*, 61(3):427–446, 2001.

[14] L. Dehaspe, H. Toivonen, and R. D. King. Finding frequent substructures in chemical compounds. In *Proceedings of the Fourth International Conference on Knowledge Discovery and Data Mining*, pages 30–36, 1998.

[15] S. Dzeroski. Inductive logic programming and knowledge discovery in databases. In U. M. Fayyad, G. Piatetsky-Shapiro, P. Smyth, and R. Uthurusamy, editors, *Advances in Knowledge Discovery and Data Mining*, chapter 5, pages 117–152. MIT Press, 1996.

[16] J. T. Favata. Offline general handwritten word recognition using an approximate beam matching algorithm. *IEEE Transactions on Pattern Analysis and Machine Intelligence*, 23(9), 2001.

[17] D. H. Fisher. Knowledge acquisition via incremental conceptual clustering. *Machine Learning*, 2(2):139–172, 1987.

[18] J. Gonzalez. *Empirical and Theoretical Analysis of Relational Concept Learning Using a Graph-Based Representation.* PhD thesis, Department of Computer Science and Engineering, University of Texas at Arlington, Arlington, TX, Aug. 2001.

[19] P. Jappy and R. Nock. Pac learning conceptual graphs. In *Proceedings of the Sixth International Conference on Conceptual Structures*, 1998.

[20] R. Levinson. A self-organizing retrieval system for graphs. In *Proceedings of the Fourth National Conference on Artificial Intelligence*, pages 203–206, 1984.

[21] M. Liquiere and J. Sallantin. Structural machine learning with galois lattice and graphs. In *Proceedings of the Fifteenth International Conference on Machine Learning*, pages 305–313, 1998.

[22] J. Llados, E. Marti, and J. J. Villanueva. Symbol recognition by error-tolerant subgraph matching between region adjacency graphs. *IEEE Transactions on Pattern Analysis and Machine Intelligence*, 23(10):1137–1143, 2001.

[23] B. Luo and E. R. Hancock. Structural graph matching using the em algorithm and singular value decomposition. *IEEE Transactions on Pattern Analysis and Machine Intelligence*, 23(10):1120–1136, 2001.

[24] B. T. Messmer and H. Bunke. A new algorithm for error-tolerant subgraph isomorphism. *IEEE Transactions on Pattern Analysis and Machine Intelligence*, 20(5):493–504, 1998.

[25] S. Muggleton. Inverse entailment and Progol. *New Generation Computing*, 13:245–286, 1995.

[26] R. Myers, R. C. Wilson, and E. R. Hancock. Bayesian graph edit distance. *IEEE Transactions on Pattern Analysis and Machine Intelligence*, 22(6):628–635, 2000.

[27] L. D. Raedt and M. Bruynooghe. A theory of clausal discovery. In *Proceedings of the Thirteenth International Joint Conference on Artificial Intelligence*, pages 1058–1063, 1993.

[28] J. Rissanen. *Stochastic Complexity in Statistical Inquiry.* World Scientific Publishing Company, 1989.

[29] J. Segen. Learning graph models of shape. In *Proceedings of the Fifth International Conference on Machine Learning*, pages 29–35, 1988.

[30] J. Segen. Graph clustering and model learning by data compression. In *Proceedings of the Seventh International Conference on Machine Learning*, pages 93–101, 1990.

[31] J. F. Sowa. *Conceptual Structures: Information in Mind and Machine.* Addison Wesley, 1984.

[32] S. Su, D. J. Cook, and L. B. Holder. Applications of knowledge discovery to molecular biology: Identifying structural regularities in proteins. In *Proceedings of the Pacific Symposium on Biocomputing*, pages 190–201, 1999.

[33] K. Thompson and P. Langley. Concept formation in structured domains. In D. H. Fisher and M. Pazzani, editors, *Concept Formation: Knowledge and Experience in Unsupervised Learning*, chapter 5. Morgan Kaufmann Publishers, 1991.

[34] J. T. L. Wang, B. A. Shapiro, D. Shasha, K. Zhang, and K. M. Currey. An algorithm for finding the largest approximately common substructures of two trees. *IEEE Transactions on Pattern Analysis and Machine Intelligence*, 20(8):889–895, 1998.

[35] P. H. Winston. Learning structural descriptions from examples. In P. H. Winston, editor, *The Psychology of Computer Vision*, pages 157–210. McGraw-Hill, 1975.

[36] P. H. Winston. *Artificial Intelligence.* Addison Wesley, 2nd edition, 1994.

[37] K. Yoshida, H. Motoda, and N. Indurkhya. Unifying learning methods by colored digraphs. In *Proceedings of the Learning and Knowledge Acquisition Workshop at IJCAI-93*, 1993.

Deriving Pseudo-Probabilities of Correctness Given Scores (DPPS)

Krassimir Ianakiev
Center of Excellence for Document Analysis and Recognition (CEDAR)
Department of Computer Science
State University of New York at Buffalo
UB Commons, 520 Lee Entrance, Suite 202
Amherst, New York 14228–2567
E-mail: ianakiev@cedar.buffalo.edu

Venu Govindaraju
Center of Excellence for Document Analysis and Recognition (CEDAR)
Department of Computer Science
State University of New York at Buffalo
UB Commons, 520 Lee Entrance, Suite 202
Amherst, New York 14228–2567
E-mail: govind@cedar.buffalo.edu

Contents

D. Chen and X. Cheng (eds.),
Pattern Recognition and String Matching, 281–294.

References

1 Introduction

Pattern recognition has been an active area of research for many years and fostered the development of many recognition techniques. Combining the results of these methods can improve the overall performance. Combining classifiers is particularly useful in the cases when the classifiers are very different (i.e use different features sets, training samples, cluster algorithms, measures, etc.). The outputs of such classifiers can be interpreted in many different ways: distance, probability, fuzzy measures and so forth. In addition, most of the known methods of combination assume that the scores returned by classifier are probabilities. A solution to this problems is to create a methodology that will transform the output scores of classifiers into probabilities. This chapter is an attempt to address these needs.

Our main goal is to describe how to apply the DPPS (Deriving Pseudo-Probabilities of Correctness Given Scores) model to recognition systems. Although a thorough theoretical investigation is beyond the scope of this section, we will demonstrate the effectiveness of the method by experimental results. The benefits of the application of DPPS is two-fold:
(i) improvement of the pattern recognition and
(ii) further applications of the derived pseudo-probabilities.
Experiments show an improvement in the range 0.56-1.5% in the decision making of a classifier when using the new pseudo-probabilities. We also get a uniform measure as a powerful tool for evaluating the performance of different recognition systems and interpreting their scores.

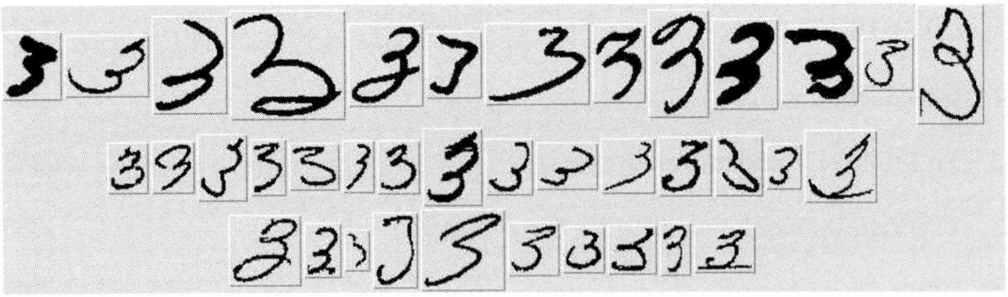

Figure 1: The Partition Set V_3^3 from **GSC**

A classifier of type III takes an unknown pattern for an input and outputs a decision vector of numerical scores, that classifier assigns to each of the classes. We will describe how DPPS transforms the output of the classifier into pseudo-probabilities of correctness, associated with each class. In order to completely separate the design of the classifier from the DPPS methodology, we will treat each classifier as a "black box", that transforms the input into a decision vector.

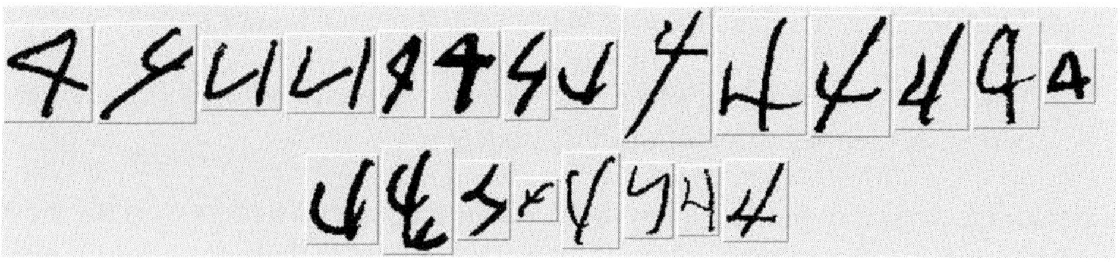

Figure 2: The Partition Set V_4^1 **CP** classifier

We will support the DPPS concepts by experiments using the following classifiers: **Polynomial** classifier ([4]), **Gradient** classifier ([3]), **GSC** classifier using k-nearest neighbor ([2]) and **CP**(a logistic regression combination of gradient and polynomial classifiers), developed at CEDAR. The **Polynomial** classifier is based on a polynomial discriminant algorithm trained to extract a relative weighting for each feature in each class. The **Gradient** classifier encodes local contour variations of a character image into a binary feature vector. **GSC** classifier uses 512 symbolic multi-resolutional features, which measure the image characteristics at local, intermediate, and global scales. The 192 **G**radient features measure edge curvature in a neighborhood of a pixel and provide information about stroke shape on a small scale. The 192 intermediate **S**tructural features measure short stroke types which span several pixels and give useful information about stroke trajectories. The 128 **C**oncavity features are used to detect stroke relationships which can span across the image. Features at all the 3 levels, **G, S,** and **C** are combined in a 512 binary feature vector and a weighted k-nearest neighbor algorithm is used for classification.

2 An Intuitive Problem Description

The following is an analogy that should clarify how the DPPS algorithm works. Teacher A and Teacher B want to evaluate the proficiency of students in Math. Both teachers give a question to n students $(S_1, \ldots S_n)$ and grade the students' responses. Teacher A examines the student's response and gives a score of 80% to the student (S_j) with the best answers (other students get lower scores). Teacher

B gives a score of 90% to the same student (S_j) which is also the highest score in his grading. Now observing the above, the following questions can be asked:

1. Is student S_j the most proficient in Math among the students examined?
2. Is the opinion of Teacher B about S_j's proficiency stronger than that of Teacher A, given that Teacher B gave the student a higher score?
3. When Teacher A (B) give the best student a score of 80% (90%), are they correct in their selection of the best student?

An intuitive answer to the first question would be that indeed student S_j is the most proficient among the students $S_1 \ldots S_n$, if he scores the highest consistently over many exams. This eliminates the possibility of chance occurrence. However, it must be ensured that all the exams administered are of equal difficulty.

An intuitive answer to the second question is that the information given is insufficient to make the determination. The grading policies of both teachers have to be studied over a large number of tests to qualify their grading behavior. In one teacher's mind 80% can mean more than what 90% means to the other teacher. The reason that comparison across teachers is difficult is because each teacher has a different notion of what a particular score means, i.e score are on different scales.

It is the third question we address here – in this case the set of all classes are the students, the classifier plays the teacher role and the input pattern is in the role of an exam. Our objective is to assess to what degree the classifier is correct in its ability to label unknown patterns and in particular, its ability to choose the best class. The degree of correctness over a large number of trials should provide the probability of correctness.

It is reasonable to assume that any class ω has the same chance of being the truth, i.e all classes are distributed uniformly. We circumvent the difficulty of computing $P(\omega|x)$ by falling back to the classic definition of probability. According to the definition probability is the frequency of outputs of an event of interest over many trials. However, given a pattern, a recognizer provides the same score for each class no matter how many times the recognition is performed. In other words, there is no randomness inherent in this process. So, one way to introduce randomness in this case is follows: instead of recognizing over and over the same pattern with the same result we generate many recognition trials with patterns "similar" to the input pattern. A set of "similar" patterns is generated in the re-training process by collecting the input pattern and its neighbors and then it is used to simulated the repeated recognition trials. In our analogy, this is the process of creating more exams for the students, where all the exams are of the same difficult level as the first one. Let us note that the "retraining" phase brings more knowledge and potential improved recognition rate.

Given an unknown input pattern X, we are interested in finding what is the chance (probabilistically) that a given class will be among the top choices (confusion set).

3 Theoretical Foundation of DPPS Methodology

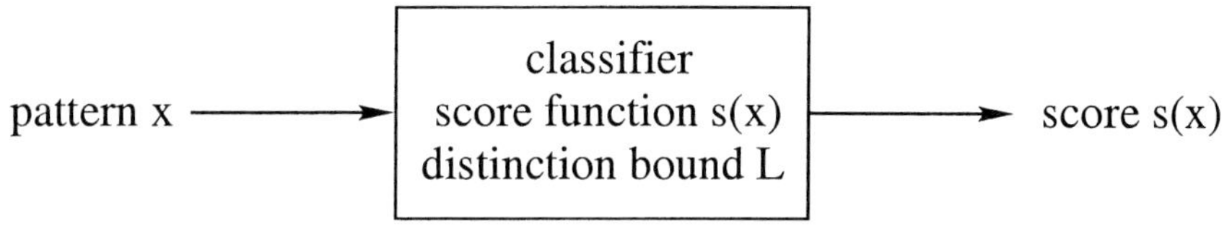

Figure 3: A model of an one dimensional classifier: if $s(x) < L$, then x belongs to class $\omega_1 \in N(a, \sigma^2)$, if $s(x) > L$, then x belongs to class $\omega_2 \in N(b, \sigma^2)$

Let us consider the following hypothetical example [1]. Suppose that a fish packing plant wants to automate the process of sorting incoming fish on a conveyor belt according to the species. As a pilot project it is decided to try to separate sea bass from salmon using optical sensing. A camera is set and it takes some sample images. The camera's signals are processed and a measurement value is returned, based on the classification made (i.e the returned score is compared to a threshold L).

This example describe a one dimensional classifier (see Figure 3). Let us assume that class ω_1 of salmon is normally distributed with mean a and deviation σ with respect to the scoring function. Let us also assume that class ω_2 of sea bass is normally distributed with mean b and deviation σ with respect to the scoring function. Also, without loss of generality, we can assume that $a < L < b$ (see Figure 4).

Theorem 3.1 *The classifier is theoretically optimal when* $L = \frac{a+b}{2}$

Proof: see [1].

Suppose that our classifier is not optimal, i.e. $L \neq \frac{a+b}{2}$. Our goal is to show that applying our methodology of computing pseudo-probabilities of correctness, we modify the classifier into the optimal one, i.e given unknown pattern x, if $s(x) < \frac{a+b}{2}$, then $x \in \omega_1$ and if $s(x) > \frac{a+b}{2}$, then $x \in \omega_2$.

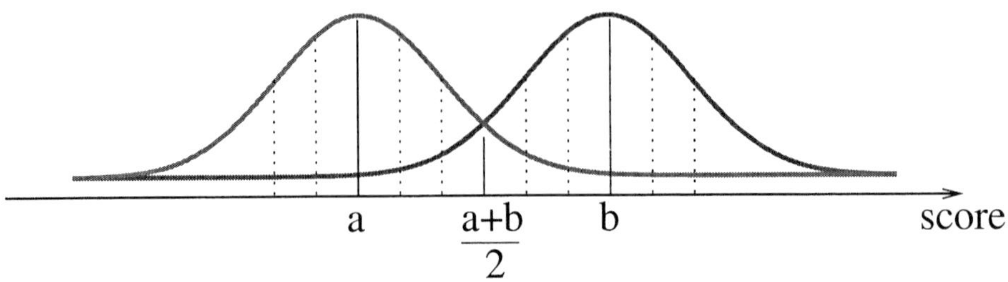

Figure 4: Distribution of salmon and sea bass on the scale of scoring function

Let V be the validation set with $2n$ samples, n samples from the first class ω_1 and other n samples from the second class ω_2. We will assume that n is large enough, so the set V represents both classes well.

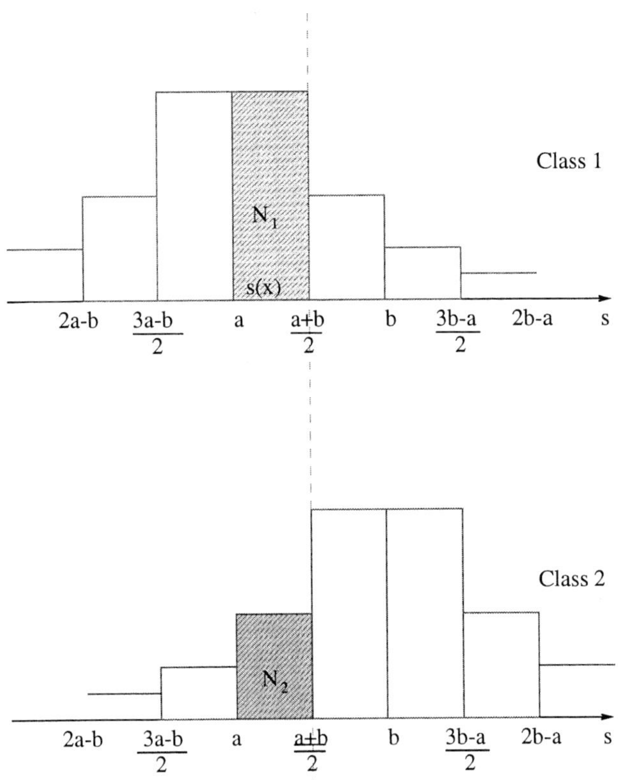

Figure 5:

Let

$$f(z,\mu,\sigma) = \frac{1}{\sqrt{2\pi\sigma^2}} e^{-\frac{(z-\mu)^2}{2\sigma^2}}$$

be the density function of a normal distribution with mean μ and deviation σ.

Let us consider the partition of the real line ($\mathcal{R}$) using the following points

$$\{\ldots, y_{-3} = 2a-b, y_{-2} = \frac{3a-b}{2}, y_{-1} = a, y_0 = \frac{a+b}{2},$$

$$y_1 = b, y_2 = \frac{3b-a}{2}, y_3 = 2b-a, \ldots\}$$

Let $X_{s(x)}$ denote the confusion set of an unknown pattern x, i.e the set of all samples from the validation set V that are similar to pattern x in view of the scoring function. More precisely, if $y_k < s(x) < y_{k+1}$ for the points y_k and y_{k+1} from the partition, then

$$X_{s(x)} = \{v \in V : y_k < s(v) < y_{k+1}\}.$$

Let N_i is the number of the images in $X_{s(x)}$, which belong to the class ω_i, $i = 1,2$. Then,

$$N_1 \approx n \int_{y_k}^{y_{k+1}} f(z,a,\sigma)\,dz$$

$$N_2 \approx n \int_{y_k}^{y_{k+1}} f(z,b,\sigma)\,dz.$$

Lets us make the following observation. Since $f(z,a,\sigma) > f(z,b,\sigma)$ for $z < \frac{a+b}{2}$ and $f(z,a,\sigma) < f(z,b,\sigma)$ for $z > \frac{a+b}{2}$, if $s(x) < \frac{a+b}{2}$, then it is obvious that the number N_1 of samples of class ω_1 in the confusion set $X_{s(x)}$ is greater than the number N_2 of samples of class ω_2 in the set $X_{s(x)}$ (see Figure 5). Then, the pseudo-probabilities of correctness are computed as follows:

$$\xi_i(x) = \frac{N_i}{|X_{s(x)}|} \qquad i = 1,2$$

implies that $\xi_1(x) > \xi_2(x)$ if $s(x) < \frac{a+b}{2}$ and $\xi_1(x) < \xi_2(x)$ if $s(x) > \frac{a+b}{2}$.

The same idea can be applied to the following model of a classifier (Figure 6) in order to improve the classifier performance.

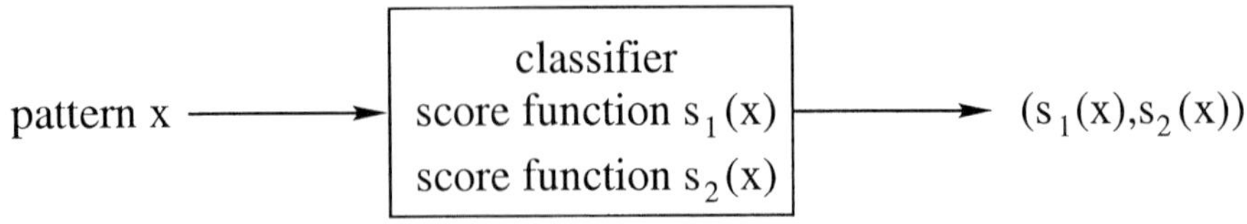

Figure 6: A model of a two dimensional classifier: if $s_2(x) < s_1(x)$, then x belongs to class ω_1, otherwise x belongs to class ω_2

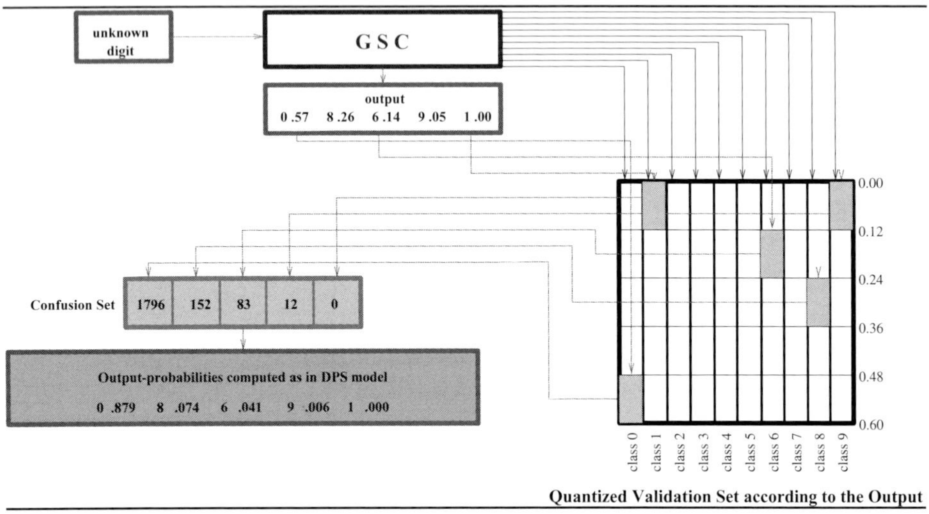

Figure 7: The DPPS Algorithm in this application

4 DPPS Algorithm

For simplicity, we assume that our classifier is the **GSC** digit recognizer [2]. It takes a digit image as an input and return a decision vector of the following format

$$(c_0\ s_0\ \ c_1\ s_1\ \ c_2\ s_2\ \ c_3\ s_3\ \ c_4\ s_4),$$

where c_0 is the first choice class with the highest score s_0, c_1 is the first choice class with highest score s_1 and so on. All scores are real numbers between 0 and 0.60.

We assume that all classes have the same chance of being the correct one. We use the set V of 20000 digit images as a validation set, that we use for re-training of the classifier. Each class is represent in V by 2000 digits. We propose four different techniques to quantize the validation set. They are described as follows:

1. Noisy class histogram (input partition)

In this case we quantize the set V with respect to the input. Then $V = \bigcup_{i=0}^{i=9} V_i$, where V_i is the subset of V containing all digits in V that belong to class ω_i. We divide each of the sets V_0, V_1, $\ldots V_9$ into q subsets $V_i = \bigcup_{j=0}^{j=q-1} V_i^j$ according to the score assigned by the recognizer to the top choice class. The set V_i^j contains all digits of class ω_i whose first choice score is in the interval $[0.60j/q, 0.60(j+1)/q)$.

2. Pure class histogram (input-output partition)

   In this case we quantize the validation set with respect to the input and output. We proceed in the same way as in the case of noise class histogram and then eliminate from the partition V_i^j all digits which are not recognized correctly, i.e the first choice class is different from the class the digit belongs to.

3. Mixed class histogram (output partition)

   In this case we quantize the validation set with respect to the output. Then $V = \bigcup_{i=0}^{i=9} V_i$, where V_i is the subset of V containing all digits in V that have class i as a first choice class in the decision vector, returned by the recognizer. We proceed in the same way as in the case of noise class histogram.

4. d-neighborhood

   For a given input digit image we use the entire validation set to form an appropriate d-neighborhood that contain all digits of the validation set that are "similar" to the input in distance topology, instead of using the one generated by the fixed quantization of the validation set. In other words, for given d, the d-neighborhood contains all samples from the validation set V, for which the following statement is true: The first choice class is among the classes in the decision vector and the difference of the corresponding scores is less than d. We will call that d-neighborhood a confusion set for the input. We also consider two possibilities-pure or mixed d-neighborhoods, which are described in the same way as in the case of pure and mixed histogram.

Let us note that all digits in each of the partition set V_i^j, $i, j = 0, 1 \ldots, 9$ are "similar" from the view point of the recognizer (Figure 1 and Figure 2).

method	test set 3-30398 digits	
	correctly recognized	% of correctly recognized
GSC	29532	97.15%
Noise	29548	97.20%
Pure	29583	97.32%
Mixed	29632	97.49%
Knn	29701	97.71%

Table 1: Comparison of GSC and DPPS using different techniques

Once we quantize the validation set, our next goal is to form the confusion set X_a for a given pattern a, that contains all "similar" samples from the viewpoint of the recognizer. In this particular application we have chosen $q = 5$, i.e. we have divided each of the sets $V_0, V_1, \ldots V_9$ into 5 parts. In case of noisy class histogram, V_i^j contains all digits in the validation set V of class ω_i, such that their top score in the decision vector, produced by the recognizer, is in the interval $[0.12j, 0.12(j+1))$. In case of pure class histogram, V_i^j contains all digits in the validation set V of class i, such that their first choice class is ω_i and the score in the decision vector, produced by the recognizer, associated with that class is in the interval $[0.12j, 0.12(j+1))$. In case of mixed class histogram, V_i^j contains all digits in the validation set V, such that their first choice class is ω_i and the score in the decision vector, produced by the recognizer, associated with that class is in the interval $[0.12j, 0.12(j+1))$.

In the first three cases (noise, pure and mixed class histograms) we proceed as follows. Let

$$(c_0\ s_0\ \ c_1\ s_1\ \ c_2\ s_2\ \ c_3\ s_3\ \ c_4\ s_4),$$

be decision vector associated with the digit a, with $0.12k_j \le s_j < 0.12(k_j + 1)$ for $j = 0,1,2,3,4$. Then $X_a = \bigcup_{j=0}^{j=4} V_{c_j}^{k_j}$. For the purpose of better illustrating how we construct the confusion set for the digit a, we will consider the following example The decision vector produced by the **GSC** recognizer on a unknown digit is

$$(8\ 0.289\ \ 3\ 0.083\ \ 0\ 0.078\ \ 5\ 0.039\ \ 2\ 0.005).$$

Then the confusion set

$$X = V_8^2 \bigcup V_3^0 \bigcup V_0^0 \bigcup V_5^0 \bigcup V_2^0.$$

class	size of the test set	GSC		DPPS	
		correctly recognized	% of correctly recognized	correctly recognized	% of correctly recognized
0	6986	6833	97.81	6852	98.08
1	6554	6496	99.12	6513	99.37
2	3954	3759	95.07	3764	95.19
3	2238	2170	96.96	2177	97.27
4	3191	3070	96.21	3075	96.36
5	1556	1497	96.21	1502	96.53
6	1474	1442	97.83	1446	98.10
7	1445	1391	96.26	1405	97.23
8	2049	1945	94.93	1958	95.58
9	951	929	97.69	943	99.15
total	30398	29532	97.15	29635	97.49

Table 2: Performance of GSC and mixed class histogram DPPS on different classes

class	size of the test set	GSC		DPPS	
		correctly recognized	% of correctly recognized	correctly recognized	% of correctly recognized
0	6986	6833	97.81	6867	98.30
1	6554	6496	99.12	6518	99.45
2	3954	3759	95.07	3770	95.35
3	2238	2170	96.96	2183	97.54
4	3191	3070	96.21	3082	96.58
5	1556	1497	96.21	1509	96.98
6	1474	1442	97.83	1453	98.58
7	1445	1391	96.26	1411	97.65
8	2049	1945	94.93	1963	95.80
9	951	929	97.69	945	99.36
total	30398	29532	97.15	29701	97.71

Table 3: Performance of GSC and d-neighborhood DPPS on different classes

In the case of d-neighborhood, we define the confusion set precisely: $X_a = \bigcup_{i=0}^{i=4} W_i^a$, where W_i^a contains all digits for which at least one of the following 5 statements is satisfied:

- the first choice class in the decision vector is c_0 and top choice score is in the interval $[s_0 - 0.06, s_0 + 0.06]$

class	size of the test set	Polynomial		DPPS	
		correctly recognized	% of correctly recognized	correctly recognized	% of correctly recognized
0	6986	6586	94.27	6613	94.66
1	6554	5841	89.12	5858	89.38
2	3954	3484	88.11	3538	89.50
3	2238	1858	83.02	1889	84.41
4	3191	3011	94.36	3034	95.08
5	1556	1428	91.77	1536	98.71
6	1474	1352	91.72	1382	93.76
7	1445	1344	93.01	1379	93.55
8	2049	1650	80.53	1768	86.29
9	951	807	84.86	829	87.17
total	30398	27361	90.00	27826	91.53

Table 4: Performance of Polynomial classifier and DPPS on different classes

- the first choice class in the decision vector is c_1 and top choice score is in the interval $[s_1 - 0.06, s_1 + 0.06]$
- the first choice class in the decision vector is c_2 and top choice score is in the interval $[s_2 - 0.06, s_2 + 0.06]$
- the first choice class in the decision vector is c_3 and top choice score is in the interval $[s_3 - 0.06, s_3 + 0.06]$
- the first choice class in the decision vector is c_4 and top choice score is in the interval $[s_4 - 0.06, s_4 + 0.06]$.

Here, $(c_0\, s_0\;\; c_1\, s_1\;\; c_2\, s_2\;\; c_3\, s_3\;\; c_4\, s_4)$, is the decision vector associated with the digit a.

Once we form the confusion set, then deriving the probability measure is very simple. If $|S|$ denotes the number of the elements (cardinality) in the set S, then the probability the unknown digit a being of class ω_i will be equal to the proportion of the number of elements of the set V_i, presented in the confusion set over total number of elements in the confusion set, i.e.

$$P(a \in \omega_i | V) = \frac{|V_i^{k_i}|}{|X_a|},$$

for $i = 0, 1, 2, 3, 4$

class	size of the test set	Gradient		DPPS	
		correctly recognized	% of correctly recognized	correctly recognized	% of correctly recognized
0	6986	6597	94.43	6626	94.85
1	6554	6107	93.18	6212	94.78
2	3954	3592	90.84	3613	91.38
3	2238	1989	88.87	2005	89.59
4	3191	3028	94.89	3030	94.95
5	1556	1441	92.61	1445	92.87
6	1474	1374	93.22	1378	93.49
7	1445	1299	89.90	1301	90.03
8	2049	1711	83.50	1824	89.02
9	951	863	90.75	895	94.11
total	30398	28001	92.11	28329	93.19

Table 5: Performance of Gradient classifier and DPPS on different classes

5 Experimental Results

In this section we give the experimental results from the performance of DPPS methodology in combinations with different classifiers. Table 1 contains the results from the DPPS method based on the scores given by the **GSC** recognizer. We have conducted experiments, involving all possible techniques described in previous section. Table 2 and Table 3 give more detailed comparison of the performances of **GSC** and DPPS model. Table 4 compares the performance of Polynomial classifier and DPPS model. Table 5 shows how DPPS compares to the gradient classifier and Table 6 shows the performance of enhanced CP classifier.

6 Some Final Remarks and Observations

First, we can see performance of the enhanced by the DPPS methodology classifier. The results also show that the extent of improvement depends on the distribution of the different classes.

Second, if the original recognition system performs more "poorly", DPPS improves the performance of the system by a greater degree.

Our hypothesis is that the performance of the algorithm will improve significantly provided the training set captures the properties of the recognition system, as well as optimal partition limits, needed for the quantization of the validation set. In some cases, it can be proved theoretically that proper validation set in combination with proper partition can improve the recognition system to the theoretically

class	size of the test set	CP classifier		DPPS	
		correctly recognized	% of correctly recognized	correctly recognized	% of correctly recognized
0	6986	6612	94.65	6672	95.51
1	6554	6372	97.22	6415	97.88
2	3954	3636	91.70	3661	92.59
3	2238	2094	93.57	2112	94.37
4	3191	3047	95.49	3088	96.77
5	1556	1467	94.28	1483	95.31
6	1474	1398	94.88	1421	96.40
7	1445	1301	90.03	1319	91.28
8	2049	1930	94.19	1947	95.02
9	951	909	95.58	918	96.53
total	30398	28766	94.63	29036	95.52

Table 6: Performance of CP classifier and DPPS on different classes

optimal.

DPPS modified classifier is almost as time consuming as the classifier itself in cases when d-neighborhood approach is not used. In the cases, when we choose to use d-neighborhood DPPS, the better correctness rate is a trade-off for increased classification time.

We have presented an algorithm for deriving probability values given recognition scores assigned to digits by a recognizer of type III. We have argued that these probability values are more informative and useful in a variety of applications including, but not limited to, multiple classifier methodologies.

References

[1] R. O. Duda and P. E. Hart and D. G. Stork, *Pattern Classification*, (2nd edition, 1997).

[2] J. T. Favata and G. Srikantan, A Multiple Feature/Resolution Approach to Handprinted Digit and Character Recognition, *International Journal of Imaging Systems and Technology*, (vol. 7, 1996) pp. 304–311.

[3] G. Srikantan, *Image sampling rate and image pattern recognition*, (Ph.D. Dissertation, SUNY at Buffalo, 1994).

[4] U. Srinivasan, *Polynomial discrimant method for handwritten digital recognition*, (Techical Report, SUNY at Buffalo, Sep. 1989).

Weighed Mean and Generalized Median of Strings

Xiaoyi Jiang
Department of Electrical Engineering and Computer Science
Technical University of Berlin
Franklinstrasse 28/29, D-10587 Berlin, Germany
E-mail: jiang@cs.tu-berlin.de

Horst Bunke
Department of Computer Science, University of Bern
Neubrückstrasse 10, CH-3012 Bern, Switzerland
E-mail: bunke@iam.unibe.ch

Contents

D. Chen and X. Cheng (eds.),
Pattern Recognition and String Matching, 295–314.

1 Introduction

Structural pattern recognition is characterized by the use of symbolic data structures such as strings, trees, and graphs for pattern representation. Consequently, the structural approach allows to model not only unary properties of patterns but also structural relationships between different patterns, or the parts of a pattern. However, compared to statistical pattern recognition, one of the shortcomings of the structural approach is the lack of a rich set of basic mathematical tools. As a matter of fact, the vast majority of all recognition procedures used in the structural domain are based on nearest neighbor classification using some distance function. By contrast, a large set of methods have become available in statistical pattern recognition, including various types of neural networks, decision theoretic methods, machine learning procedures, and clustering algorithms [13]. In this paper we put particular emphasis on novel work in the structural pattern recognition that aims at bridging the gap between statistical and structural pattern recognition in the sense that it may yield a basis for adapting various techniques from the statistical to the structural domain.

Two fundamental tools in dealing with feature vectors are given by the weighted mean of two vectors and the average, or mean, of a set of vectors. Consider two vectors, or equivalently points, in n-dimensional real space, $x, y \in \Re^n$. Their weighted mean is defined as a point z such that

$$z = \alpha x + (1-\alpha)y, \quad \alpha \in [0,1] \tag{1}$$

Clearly, if $\alpha = \frac{1}{2}$ then z is the (normal) mean of x and y. More generally, z is a point on the n-dimensional straight line segment that connect x and y, and the distance between z and both x and y is controlled via the parameter α. The weighted mean has a variety of applications, including self-organizing maps [20] for clustering and dimension-reduction purpose. The (normal) mean is useful in situations where we need to reduce a set of (similar) vectors to a single representative. The well-known k-means clustering algorithm [12] represents such a situation.

In the following we report our recent work in extending the concept of weighted mean and (normal) mean from vector space to the domain of strings. Section 2 describes an algorithm for computing the weighted mean of two strings and its application to the curve morphing problem. Section 3 is concerned with generalized median strings (the extension of mean to an arbitrary space is termed generalized median). The paper is then concluded by some discussions.

2 Weighted mean of two strings

The expression (1) cannot be directly translated into the domain of strings. But we notice that the weighted mean z of x and y satisfies:

$$\begin{aligned} d_E(z,x) &= \alpha \cdot d_E(x,y) \\ d_E(z,y) &= (1-\alpha) \cdot d_E(x,y) \end{aligned}$$

where d_E represents the Euclidean distance of two points. We have no difficulty to generalize this relationship to an arbitrary space U. Given a distance function $d(p,q)$ that measures the dissimilarity between two patterns $p, q \in U$, the weighted mean of p and q is a pattern $r \in U$ such that the properties

$$d(r,p) = \alpha \cdot d(p,q) \tag{2}$$
$$d(r,q) = (1-\alpha) \cdot d(p,q) \tag{3}$$

hold.

In dealing with strings we consider the set Σ^* of all strings over a finite alphabet Σ of symbols. The most popular distance function for strings is the Levenshtein edit distance. Given two words p and q over Σ, the Levenshtein edit distance $d(p,q)$ is defined in terms of elementary edit operations which are required to transform p into q. Usually, three different types of edit operations are considered, namely (1) substitution of a symbol $a \in p$ by a symbol $b \in q$, $a \neq b$, (2) insertion of a symbol $a \in \Sigma$ in q, and (3) deletion of a symbol $a \in p$. Symbolically, we write $a \to b$ for a substitution, $\epsilon \to a$ for an insertion, and $a \to \epsilon$ for a deletion. To model the fact that some edit operations may be more likely to occur than others, costs of edit operations, $c(a \to b)$, $c(\epsilon \to a)$, and $c(a \to \epsilon)$, are introduced. Let $s = l_1 l_2 \cdots l_k$ be a sequence of edit operations transforming p into q. We define the cost of this sequence by $c(s) = \sum_{i=1}^{k} c(l_i)$. The Levenshtein edit distance between p and q is given by

$$d(p,q) = \min\{c(s) \mid s \text{ is a sequence of edit operations transforming } p \text{ into } q\}.$$

The standard algorithm for computing $d(p,q)$ is based on dynamic programming [31]. It provides both the Levenshtein edit distance and the corresponding optimal edit sequence that transforms p into q with minimum cost. Using the Levenshtein edit distance, the weighted mean of two strings p and q is a string $r \in \Sigma^*$ that satisfies the relationship (2) and (3).

2.1 Computational procedure

Similarly to Wagner and Fischer [31], we assume that the cost function of any identical substitution $a \to a$, $a \in \Sigma$, is equal to zero and the cost function satisfies the triangular inequality, i.e.

$$c(a \to b) \ \leq \ c(a \to d) + c(d \to b)$$

for any three edit operations $a \to b$, $a \to d$, and $d \to b$ including insertions and deletions. We have the following result.

Theorem 2.1 *Let p and q be two strings, and S an optimal sequence of edit operations transforming p into q. For any subsequence S' of S with the corresponding cost $\beta = \sum_{e \in S'} c(e)$ we can construct a string r by applying S' to p. Then, we have $d(r, p) = \alpha \cdot d(p, q)$ and $d(r, q) = (1 - \alpha) \cdot d(p, q)$ with $\alpha = \frac{\beta}{d(p,q)}$.*

A proof of this theorem can be found in [5].

This theorem suggests a computational procedure as follows. To compute a weighted mean of two strings p and q, we first compute their edit distance $d(p, q)$. From the optimal sequence S of edit operations we select a subsequence of edit operations, S', with cost β and apply S' to generate a weighted mean r.

This basic computational procedure regards each edit operation of S as an atomic entity. The concept of edit operations can actually be extended to *decomposable* ones. We call an edit operation e decomposable into $e'(\neq e)$ and $e''(\neq e)$ if $e = e' \circ e''$ (concatenation of e' and e'', i.e. e' followed by e'') and $c(e) = c(e') + c(e'')$. Decomposable edit operations are particularly useful in situations where the symbols of Σ represent numerical quantities such as lengths and distances. As an example, let Σ be the set of all integers and the cost function be defined by:

- $c(a \to b) = |a - b|$ for $a, b \in \Sigma$
- $c(\epsilon \to a) = c(a \to \epsilon) = a$ for $a \in \Sigma$

Then, the edit operation $10 \to 8$ can decomposed into $10 \to 9$ and $9 \to 8$. Considering decomposable edit operations, Theorem 2.1 extends to (see [5] for a correctness proof):

Theorem 2.2 *Let p and q be two strings, and S an optimal sequence of edit operations transforming p into q. Furthermore, let $S' = \{e_1, e_2, \ldots, e_l\}$*

a subsequence of S, where each e_i is decomposable, i.e. $e_i = e'_e \circ e''_i$, $1 \leq i \leq l$, and $\beta = \sum_{i=1}^{l} c(e'_i)$ Then we can construct a string r by applying all operations e'_i to p such that $d(r,p) = \alpha \cdot d(p,q)$ and $d(r,q) = (1-\alpha) \cdot d(p,q)$ with $\alpha = \frac{\beta}{d(p,q)}$.

This theorem provides a computational procedure in a more general form.

Any string r generated by this general computational procedure is a weighted mean of p and q. Interestingly, the reverse is true as well. That is, any weighted mean of p and q can be actually generated by the procedure. Therefore, the computational procedure is both correct and complete. The readers are referred to [5] for a full discussion of this aspect.

2.2 Application to curve morphing

Shape morphing, also known as metamorphosis, shape interpolation, shape blending, and shape evolving, is to compute a continuous transformation from one shape (the *source*) to another (the *target*). It has widespread applications in various fields including computer graphics, industrial design, and robotics. A general discussion about curve morphing algorithms can be found in [8].

We apply the algorithm for computing the weighted mean of two strings to solve the curve morphing problem. This makes our approach very different from those known from the literature. The source curve C_1 and the target curve C_2 are represented by string p_1 and p_2, respectively. We require that the morphing process transform p_1 to p_2 though a series of strings (curves) $p(\alpha)$, $\alpha \in [0,1]$, such that $p(\alpha)$ is the weighted mean of p_1 and p_2 with parameter α. If we select k sample values $\frac{1}{k+1}, \frac{2}{k+1}, \ldots, \frac{k}{k+1}$ from the interval $[0,1]$, then this procedure will generate a total of k intermediate curves.

A curve is given by a sequence $C = (x_1, y_1) \ldots (x_n, y_n)$ of points in the xy-plane. For closed curves we assume that a starting point is specified in some way. To map such a curve into a string, we first sample the given data points such that the Euclidean distance between any pair of two consecutive points has a constant value Δ. That is, C is transformed into $\overline{C} = (\overline{x_1}, \overline{y_1}) \ldots (\overline{x_m}, \overline{y_m})$, where $d_E((\overline{x_i}, \overline{y_i}), (\overline{x_{i+1}}, \overline{y_{i+1}})) = \Delta$ for $i = 1, \ldots, m-1$. Then, a string $z_1 \ldots z_{m-1}$ is generated from $\overline{C}$ where z_i is the vector pointing from $(\overline{x_i}, \overline{y_i})$ to $(\overline{x_{i+1}}, \overline{y_{i+1}})$.

The edit distance $d(p_1, p_2)$ measuring the dissimilarity of two strings (curves) p_1 and p_2 is defined by the costs $c(a \to \epsilon) = c(\epsilon \to a) = |a| = \Delta$,

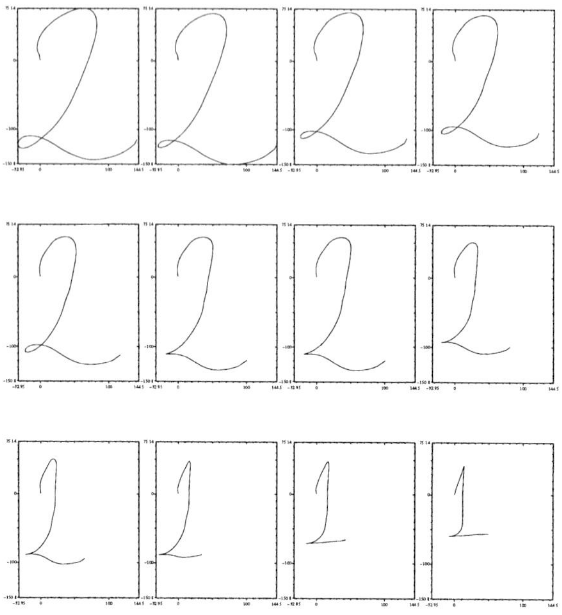

Figure 1: Morphing of two digits.

$c(a \rightarrow b) = |a - b|$. Notice that the minimum cost of a substitution is equal to zero (if and only if $a = b$), while the maximum cost is 2Δ. The latter case occurs if a and b are parallel and have opposite direction.

Figure 1 shows an example of curve morphing where a total of ten intermediate curves are generated and displayed in a top-down and left-right order. The source/target curves are online handwritten digits from a subset[1] of the UNIPEN database [10]. It can be clearly observed that with an increasing value of α, the curves represented by the intermediate strings become more and more similar to the target curve. Hence, the concept of weighted mean seems to coincide well with our intuitive notion of shape similarity.

Figure 2 illustrates the morphing process of two fish shapes from a large database[2]. Notice that both the source and the target curve are closed and need some special handling. First, we have to choose a starting point on each curve. A second problem is that the computed intermediate curves are

[1] Available at ftp://ftp.ics.uci.edu/pub/machine-learning-databases/pendigits/.
[2] Available at ftp://ftp.ee.surrey.ac.uk/pub/vision/misc/fish_contours.tar.Z.

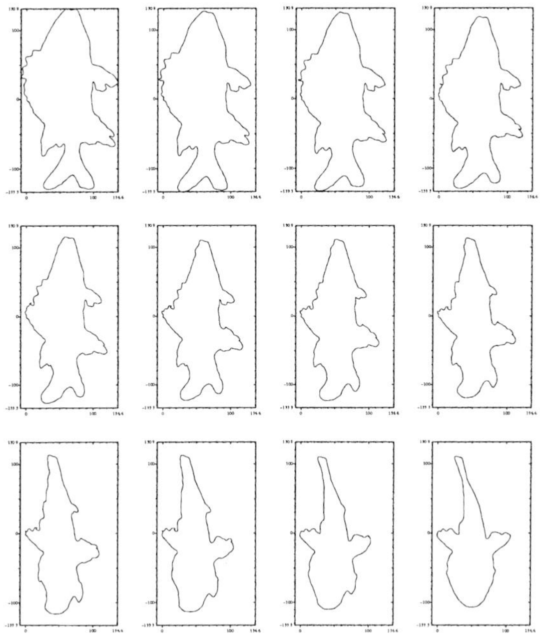

Figure 2: Morphing of two fish shapes.

not necessarily closed. Various solutions to these two problems are discussed in [16]. For the results in Figure 2 we simply manually specified the starting points. The following procedure was taken to ensure the closedness of the intermediate curves. Assume the amount $(-x_n, -y_n)$ is missing for the last point (x_n, y_n) of C to reach the origin. If we introduce a small correction $(-\frac{x_n}{n}, -\frac{y_n}{n})$ to each vector, the l-th point of C will take the new position $(x'_l, y'_l) = (x_l - l \cdot \frac{x_n}{n}, y_l - l \cdot \frac{y_n}{n})$. In this case the last point (x'_n, y'_n) corresponds to the origin.

Traditionally, multiscale curve representations are based on Gaussian smoothing [29]. Interestingly, curve morphing can be used for an alternative multiscale curve representation with the original curve as source and a circle as target [9]. Figure 3 shows such an example. In this experiment the radius of the circle is fixed to be $\sqrt{\frac{A}{\pi}}$ where A represents the area of the target shape, forcing the target circle to have the same area as the source curve.

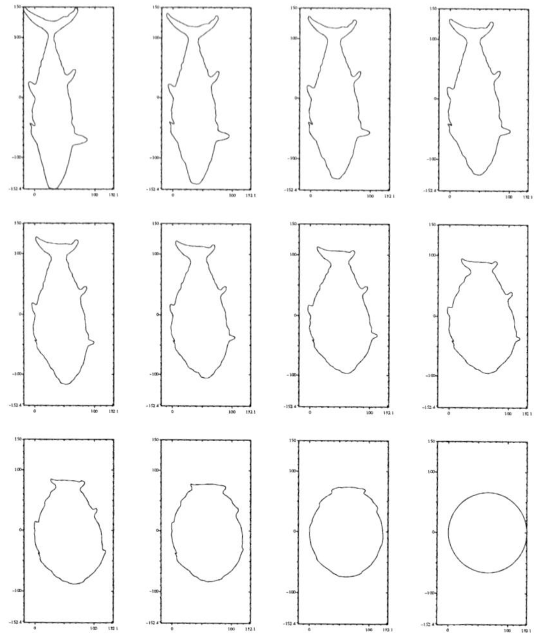

Figure 3: Multiscale representation of a fish shape.

2.3 Additional remarks

There are a number of well established procedures in statistical pattern recognition where a given pattern $x \in \Re^n$ is to be changed so as to make it more similar to another pattern $y \in \Re^n$. An example is a self-organizing feature map (SOM) [20], where the formula $x \leftarrow x + c(y - x)$ is applied. Originally, the SOM was proposed for feature vectors. Using the computational procedure for weighted mean string this updating formula can be adapted to strings. Hence, it opens up a way to implement the SOM in the domain of strings.

3 Generalized median of strings

The concept of average, or mean, is useful in various contexts. In sensor fusion, multisensory measurements of signals are averaged to produce the best estimate. Averaging the results of several classifiers is used in multiple classifier systems in order to achieve more reliable classification results. In clustering and machine learning, a typical task is to represent a set of (similar) objects by means of a single prototype.

In this section we explore the mean concept in the domain of strings. After an introduction of median strings we briefly discuss various methods for median string computation known from the literature. It turns out that the algorithm from the last section for determining the weighed mean of two strings can be used to compute median strings. It has the interesting property of being able to operate in a dynamic environment, where there is a steady arrival of new strings belonging to the considered set. We will describe this algorithm in some detail.

3.1 Median strings

The concept of mean can be defined in general form for an arbitrary representation space U. For this purpose we need a distance function $d(p,q)$ to measure the dissimilarity between any two patterns $p, q \in U$. Let S be a set of patterns from S. The essential information of S is captured by a pattern $\overline{p} \in U$ that minimizes the sum of distances to all patterns in S, i.e.

$$\overline{p} \;=\; \arg\min_{p \in U} \sum_{q \in S} d(p,q).$$

Pattern $\overline{p}$ is called a *generalized median* of S. If the search is constrained to the given set S, the resultant pattern

$$\hat{p} \;=\; \arg\min_{p \in S} \sum_{q \in S} d(p,q)$$

is called a *set median* of S. Notice that neither the generalized median nor the set median is necessarily unique. Because the set median is constrained to be equal to one of the given strings, the generalized median is the more general concept and therefore usually a better representation of the given patterns than the set median.

If U is the universe of real numbers and the distance function $d(p,q)$ is the absolute (squared) difference of p and q, then the generalized median simply corresponds to the scalar median (mean) known from statistics. Scalar median represents a powerful technique for image smoothing. Its extension to vector spaces [1, 2] provides a valuable image processing tool for multispectral/color images and optical flow.

In the domain of strings, several distance functions can be used to specify median strings. Besides the Levenshtein edit distance, other string distance functions are known from the literature, for instance, normalized edit distance [26], maximum posterior probability distance [19], feature distance [19], and others [7].

3.2 Computation of median strings

The computation of set median is a straightforward task. Independent of the underlying representation space we can always find the set median of N patterns by means of $\frac{1}{2}N(N-1)$ pairwise distance computations. (Notice, however, that each of these distance computations may have a high complexity.) The computational burden can be reduced if the distance function is a metric [18]. For non-metric distance functions an approximate set median search algorithm has been reported recently [27].

The generalized median provides a better representation of a set of patterns, but its computation is generally much more demanding than that of a set median. As a matter of fact, there is no general rule for computing the generalized median because any algorithm must be of constructive nature and the construction process crucially depends on the structure of the underlying patterns. The computation of generalized median string turns out to be NP-hard [11, 28]. Given N strings of length $O(n)$ each, an exact algorithm was proposed by Kruskal [21] based on an extension of the algorithm for string edit distance computation [31]. It needs both $O(n^N)$ time and space and thus suffers from a high complexity. Hence its applicability is restricted to rather small sets and short strings. If the sum of distances of the (unknown) generalized median string to all input strings is guaranteed to lie under some (relatively small) upper bound, then the computational expense can be reduced significantly, although remaining exponential [23]. In [24] heuristic information about the spatial location of the objects represented by the symbols in a string is used to speed up median string computation. A greedy algorithm is proposed in [6] that requires $O(n^2N|\Sigma|)$ time. Starting from an empty string, a new symbol is added in each iteration to the partial solution generated so far. All symbols from the alphabet are probed and that one is selected whose resultant new partial solution shows the smallest sum of distances. Recently, a slightly refined version of the greedy algorithm with the same time complexity was reported in [22]. The set median or the solution of the greedy algorithm can be used to start an iterative procedure that provides a further improved approximate generalized median string [19, 25]. It is based on a systematic perturbation of the current solution and is repeated until no more improvement is possible. This procedure needs $O(n^3 N|\Sigma|)$ time for each iteration. Finally, an approximate method is proposed in [15] based on genetic search.

3.3 Dynamic computation of generalized medians

In a dynamic context we are faced with the situation of steady arrival of new data items, represented by strings. At each point of time, t, the set S^t of existing strings is incremented by a new string, resulting in S^{t+1}, and its generalized median string should be computed. Doubtlessly, a trivial solution consists in applying any of the known approaches for median string computation to S^{t+1} [6, 15, 19, 21, 22, 23, 25]. By doing this, however, we compute the generalized median string of S^{t+1} from scratch without utilizing any knowledge about S^t, in particular its generalized median string. It seems that all known algorithms for computing generalized median strings are of such a static nature and thus not optimal in a dynamic context. In the following we describe a genuinely dynamic approach, in which the update scheme only considers the generalized median string of S^t together with the new data item, but not the individual members of S^t. In this manner the computational effort for each new string can be significantly reduced when compared to the use of a static algorithm. This dynamic algorithm is based on the procedure for computing the weighted mean of two strings from Section 2.

3.3.1 Algorithm

The inspiration for the algorithm comes from a fundamental fact in the n-dimensional real space. Under the distance function $d(p_i, p_j) = (p_i - p_j) \cdot (p_i - p_j)$, i.e. the squared Euclidean distance of p_i and p_j, the generalized median of a given set $S^t = \{p_1, p_2, \ldots, p_t\}$ of t points is the well-known mean:

$$\overline{p}^t = \frac{1}{t} \cdot \sum_{i=1}^{t} p_i.$$

When an additional point p_{t+1} is added to S^t, the resultant new set $S^{t+1} = S^t \cup \{p_{t+1}\}$ has the generalized median

$$\overline{p}^{t+1} = \frac{1}{t+1} \cdot \sum_{i=1}^{t+1} p_i$$

$$= \frac{t}{t+1} \cdot \overline{p}^t + \frac{1}{t+1} \cdot p_{t+1}$$

which is the weighted mean of $\overline{p}^t$ and p_{t+1} satisfying

$$\begin{aligned} d(\overline{p}^{t+1}, \overline{p}^t) &= \frac{1}{t+1} \cdot d(\overline{p}^t, p_{t+1}) \\ d(\overline{p}^{t+1}, p_{t+1}) &= \frac{t}{t+1} \cdot d(\overline{p}^t, p_{t+1}) \end{aligned}$$

Geometrically, $\overline{p}^{t+1}$ is a point on the straight line segment connecting $\overline{p}^t$ and p_{t+1} that has distances $\frac{1}{t+1} \cdot d(\overline{p}^t, p_{t+1})$ and $\frac{t}{t+1} \cdot d(\overline{p}^t, p_{t+1})$ to $\overline{p}^t$ and p_{t+1}, respectively.

On a heuristic basis we generalize the special case discussed above in the n-dimensional real space to an arbitrary space. Given a set $S^t = \{p_1, p_2, \ldots, p_t\}$ of t patterns and its generalized median $\overline{p}^t$, the generalized median of a new set $S^{t+1} = S^t \cup \{p_{t+1}\}$ is estimated by a weighted mean of $\overline{p}^t$ and p_{t+1}, i.e. by a pattern $\overline{p}^{t+1}$ such that

$$\begin{aligned} d(\overline{p}^{t+1}, \overline{p}^t) &= \alpha \cdot d(\overline{p}^t, p_{t+1}) \\ d(\overline{p}^{t+1}, p_{t+1}) &= (1-\alpha) \cdot d(\overline{p}^t, p_{t+1}) \end{aligned}$$

where $\alpha \in [0, 1]$. In the n-dimensional real space α takes the value $\frac{1}{t+1}$. For arbitrary spaces, however, we have no possibility to specify α in advance. Therefore, we resort to a search procedure. Remember that our goal is to find $\overline{p}^{t+1}$ that minimizes the expression

$$\mathrm{SOD}(\overline{p}^{t+1}) = \sum_{i=1}^{t+1} d(\overline{p}^{t+1}, p_i).$$

(where SOD stands for sum of distances). To determine the optimal α value we therefore go through a series of α values $0, \frac{1}{k}, \ldots, \frac{k-1}{k}, 1$ and choose that α value that results in the smallest $\mathrm{SOD}(\overline{p}^{t+1})$.

The procedure above is able to dynamically compute generalized medians in an arbitrary space. The only assumption is that we must have an algorithm for the computation of weighted mean in the particular space under consideration. In the domain of strings the algorithm described in Section 2 serves this purpose well. We apply the algorithm to $\overline{p}^t$ and p_{t+1} with some α probe value. Note that in general, the weighted mean of two strings is not unique. To take this fact into account we actually generate m different weighed means of $\overline{p}^t$ and p_{t+1} for each α probe value. Totally, we

---

Input: $\overline{p}^t$ (generalized median string of S^t)
p_{t+1} (new string)
Output: $\overline{p}^{t+1}$ (generalized median string of $S^{t+1} = S^t \cup \{p_{t+1}\}$)
Parameter: k, m

```
MinSOD = ∞;
for each α ∈ {0, 1/k, ..., (k−1)/k, 1} do {
  repeat m times {
    determine weighted mean p of p̄^t and p_{t+1} with α;
    if (SOD(p) < MinSOD) {
      p̄^{t+1} = p;
      MinSOD = SOD(p);
    }
  }
}
```

Figure 4: Algorithm for dynamically computing generalized median strings.

thus produce $(k+1)m$ candidates for $\overline{p}^{t+1}$. These candidates are evaluated by means of the function $\text{SOD}(\overline{p}^{t+1})$ and the updated generalized median string for the set S^{t+1} is chosen to be the one that minimizes $\text{SOD}(\overline{p}^{t+1})$. A summary of the resulting dynamic algorithm is given in Figure 4.

3.3.2 Experimental results

We report various experimental results that demonstrate the ability of the proposed algorithm to find reasonable approximate generalized median strings in a dynamic context. The proposed algorithm is also compared with the greedy algorithm [6] and set median.

We conducted an experimental evaluation on the online handwritten digits of the UNIPEN database [10] already used for curve morphing in Section 2.2. We took 10 samples of digit 1, 2 and 3 each, and 98 samples of digit 6; see Figure 5 for an illustration (for space reasons only ten samples of digit 6 are shown there). The results of the various algorithms are given in Figure 6, where we also list the value SOD and the computation time t in seconds. Note that the SOD values of digit 6 are substantially larger than

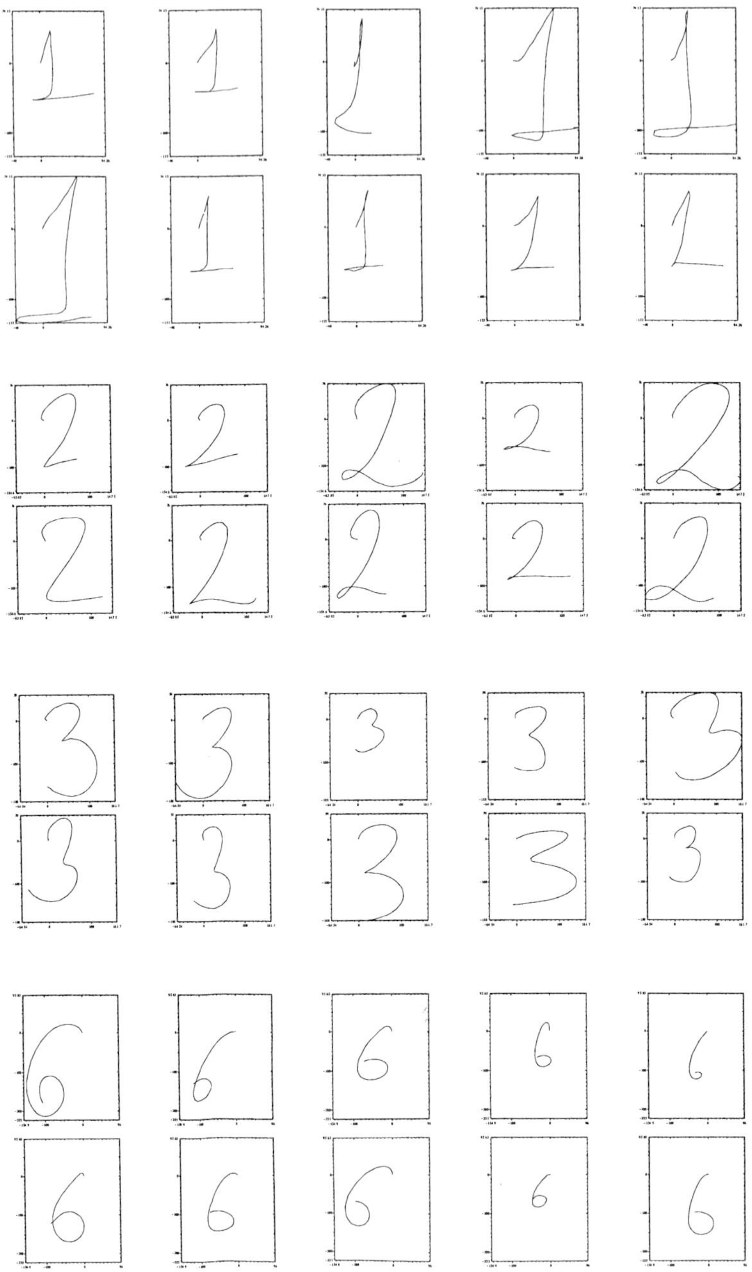

Figure 5: Samples of digits 1, 2, 3 and 6, respectively.

dynamic approach	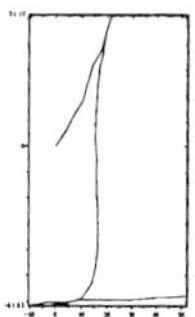 SOD = 1037 t = 13.5	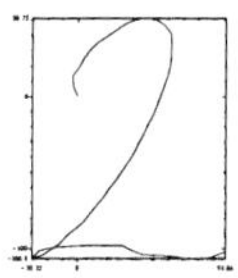 SOD = 1408 t = 26.8	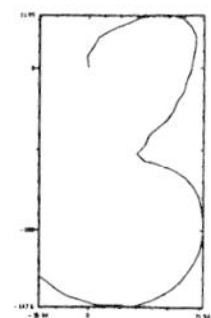 SOD = 1487 t = 22.8	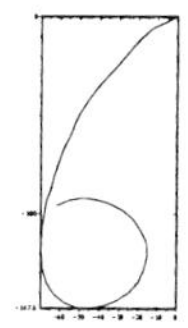 SOD = 11872 t = 1152.0
greedy	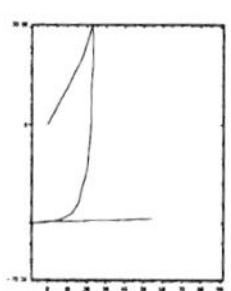 SOD = 1043 t = 7.8	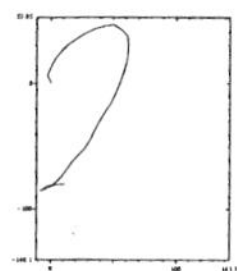 SOD = 2238 t = 18.0	 SOD = 2521 t = 11.5	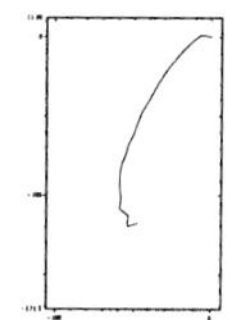 SOD = 20180 t = 586.7
set median	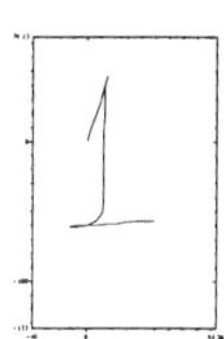 SOD = 1068 t = 0.3	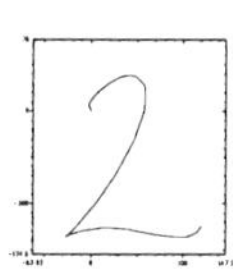 SOD = 1473 t = 0.7	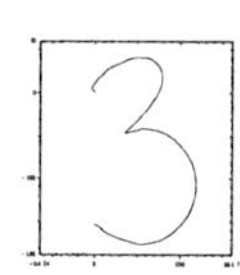 SOD = 1618 t = 0.6	 SOD = 12432 t = 47.0

Figure 6: Median computation of digits.

those of the other digits because of the definition of SOD as the sum, but not the average, of the distances to all input samples.

Based on subjective judgment, the medians computed by our algorithm represent reasonable shapes. On the other hand, except for digit 1, the greedy algorithm reveals some weakness. Looking at the median for digits 2, 3 and 6 it seems that the iterative process terminates too early, resulting in

	average SOD	std. dev.	minimum	maximum
digit 1	1040	8	1028	1056
digit 2	1408	16	1374	1455
digit 3	1478	15	1451	1514
digit 6	11785	37	11701	11895

Table 1: SOD for 50 tests with random order of samples.

a string (digit) much shorter than it should be. The reason lies in the simple termination criterion defined in [6]. It works well for the (short) words used there, but obviously encounters difficulties in dealing with longer strings occurring in this study. At first glance, our approach needs more computation time than the greedy algorithm. But one has to take into account that the recorded time is the total time of the dynamic process of adding one sample to the existing set each time, starting from a set consisting of the first sample. Therefore, totally 9 (97) generalized medians have effectively been computed for digits 1/2/3 (6). Taking the actual computation time for each update step of the whole dynamic process, which leads to a final set of all samples into account, our algorithm is attractive compared to static ones such as the greedy algorithm. If we use the static greedy algorithm to handle the same dynamic process, then the computation for digit 3, for instance, would require 46.1 seconds in contrast to 22.8 seconds of our dynamic approach. Considering the computational efficiency alone, set median is the best among the three methods. But the generalized median string is a more precise abstraction of the given set of strings with a smaller SOD value, which corresponds to the representation error.

Clearly, the performance of our dynamic algorithm depends on the order of input samples. The results shown in Figure 6 correspond to one particular order. To study the influence of the sample order on our algorithm, we generated 50 random orders for the samples of each digit and recorded the results; see Table 1. Interestingly, the results reported in Figure 6 represent the average performance of our algorithm very well. Even the worst case is still better than the set median.

An important application of generalized median strings is in the field of OCR [23, 24, 30]. Current OCR technology is far from being perfect. One possibility of performance enhancement lies in the combination of multiple classification results. It is well-known that OCR classifiers are sensitive to slight perturbations in their input. As a result, OCR error behavior can vary. The same page rescanned and re-OCR'ed, or OCR'ed by applying multiple classifiers to a single scan of source text may exhibit different sets of errors. Classifier combination aims at combining the outcome of multiple classifiers through voting procedures, with the goal of producing results better than those that could be achieved by any single classifier. Generalized median strings are very useful tools for such a voting procedure. We also evaluated the dynamic algorithm for computing generalized median strings on texts. For this application the performance is superior to that of the Greedy algorithm as well; see [17] for details.

3.4 Further remarks

As stated above, the generalized median string provides a means for multiple classifier combination in OCR. The same idea could be applied to speech recognition and speech reading (lipreading) as well. In addition the computational procedures for generalized median strings allow to extend clustering algorithms, such as k-means, from vectorial pattern representations into the domain of strings.

For the case $|S| = 2$, where the input set has only two strings p and q, there is a particular relationship between the generalized median and the weighted mean. Clearly, any weighted mean r of p and q minimizes the sum of distances to p and q, because $d(p,r) + d(q,r) = d(p,q)$ and there exists no r' such that $d(p,r) + d(q,r) < d(p,q)$ due to the triangular inequality. Hence, for the special case $|S| = 2$, any weighted mean is a generalized median. From a practical point of view, any procedure that computes a weighted mean can also be used for generalized median computation in this case. Regardless of the value of α, it will always return a generalized median.

4 Conclusions

In this paper we discussed the weighted mean and median of strings, and their computational procedures. These two concepts are interesting in their own right. But their combination will allow to extend clustering algorithms, such as k-means and SOM, from vectorial pattern representations into the domain of strings. This issue is currently under investigation.

From the general point of view, the procedures discussed in this paper can be regarded a contribution towards bringing the disciplines of statistical and structural pattern recognition closer together. In the field of statistical pattern recognition, a rich set of methods have become available during the last decades. But structural pattern recognition suffers from a limited repository of mathematical tools. In fact, most recognition procedures in the structural approach follow the nearest-neighbor paradigm, where an unknown pattern is matched against the full set of prototypes. The weighted mean string and the median string are an extension of the fundamental concepts of weighted mean and average in vector spaces. They enable to adapt clustering procedures originally developed in statistical pattern recognition to the domain of strings. Similar efforts have been undertaken in the domain of graphs [3, 14]. A general discussion on recent developments in both areas can be found in [4].

Acknowledgments

The authors would like to thank K. Abegglen, A. Kandel, and J. Crisik for their contribution to parts of the paper.

References

[1] J. Astola, P. Haavisto, and Y. Neuvo, Vector median filters, Proceedings of the IEEE, 78(4): 678–689, 1990.

[2] F. Bartolini, V. Cappellini, C. Colombo, and A. Mecocci, Enhancement of local optical flow techniques, Proc. of 4th Int. Workshop on Time Varying Image Processing and Moving Object Recognition, Florence, Italy, 1993.

[3] H. Bunke and S. Günter, Weighted mean of a pair of graphs, Computing, 67: 209–224, 2001.

[4] H. Bunke, Recent advances in structural pattern recognition with applications to visual form analysis, in C. Arcelli, L. Cordella, and G. Sanniti di Baja (eds.), *Visual Form 2001*, Springer-Verlag, 1–11, 2001.

[5] H. Bunke, X. Jiang, K. Abegglen, and A. Kandel, On the weighted mean of a pair of strings, Pattern Analysis and Applications, 2002. (to appear)

[6] F. Casacuberta and M.D. de Antoni, A greedy algorithm for computing approximate median strings, Proc. of National Symposium on Pattern Recognition and Image Analysis, 193–198, Barcelona, Spain, 1997.

[7] A.L.N. Fred and J.M.N. Leitão, A comparative study of string dissimilarity measures in structural clustering, Proc. of Int. Conf. on Document Analysis and Recognition, 385–394, 1998.

[8] J. Gomes, L. Darsa, B. Costa, and L. Velho, *Warping and Morphing of Graphical Objects*, Morgan Kaufmann Publishers, 1999.

[9] G. Gong and B. Parvin, A new regularized approach for contour morphing, Proc. of CVPR, 458–463, 2000.

[10] I. Guyon, L. Schomaker, R. Plamondon, M. Liberman, and S. Janet, UNIPEN project on on-line exchange and recognizer benchmarks, Proc. of ICPR, 29–33, 1994.

[11] C. de la Higuera and F. Casacuberta, Topology of strings: Median string is NP-complete, Theoretical Computer Science, 230(1/2): 39–48, 2000.

[12] A.K. Jain, M.N. Murty, and P.J. Flynn, Data clustering: A review, ACM Computing Surveys, 31(3): 264–323, 1999.

[13] A. Jain, R. Duin, and J. Mao, Statistical pattern recognition: A review, IEEE Trans. on PAMI, 22(1): 4–37, 2000.

[14] X. Jiang, A. Münger, and H. Bunke, On median graphs: Properties, algorithms, and applications, IEEE Trans. on PAMI, 23(10): 1144–1151, 2001.

[15] X. Jiang, K. Abegglen, and H. Bunke, Genetic computation of generalized median strings, 2001. (submitted for publication)

[16] X. Jiang, H. Bunke, K. Abegglen, and A. Kandel, Curve morphing by weighted mean of strings, Proc. of ICPR, 2002.

[17] X. Jiang, K. Abegglen, H. Bunke, and J. Csirik, Dynamic computation of generalized median strings, 2002. (submitted for publication)

[18] A. Juan and E. Vidal, Fast median search in metric spaces, in A. Amin and D. Dori (eds.), *Advances in Pattern Recognition*, Springer-Verlag, 905–912, 1998.

[19] T. Kohonen, Median strings, Pattern Recognition Letters, 3: 309–313, 1985.

[20] T. Kohonen, *Self-Organizing Map*, Springer-Verlag, 2001.

[21] J.B.Kruskal, An overview of sequence comparison: Time warps, string edits, and macromolecules, SIAM Reviews, 25(2): 201–237, 1983.

[22] F. Kruzslicz, Improved greedy algorithm for computing approximate median strings, Acta Cybernetica, 14: 331–339, 1999.

[23] D. Lopresti and J. Zhou, Using consensus sequence voting to correct OCR errors, Computer Vision and Image Understanding, 67(1): 39-47, 1997.

[24] V. Marti and H. Bunke, Use of positional information in sequence alignment for multiple classifier combination, in J. Kittler and F. Roli (Eds.), *Multiple Classifier Combination*, Springer-Verlag, 388–398, 2001.

[25] C.D. Martinez-Hinarejos, A. Juan, and F. Casacuberta, Use of median string for classification, Proc. of ICPR, 907–910, 2000.

[26] A. Marzal and E. Vidal, Computation of normalized edit distance and applications, IEEE Trans. on PAMI, 15(9): 926-932, 1993.

[27] L. Mico and J. Oncina, An approximate median search algorithm in non-metric spaces, Pattern Recognition Letters, 22(10): 1145–1151, 2001.

[28] J.S. Sim and K. Park, The consensus string problem for a metric is NP-complete, Journal of Discrete Algorithms, 2(1), 2001.

[29] H. Mokhatarian and A. Mackworth, A theory of multiscale curvature based shape representation of planar curves, IEEE Trans. on PAMI, 14: 789–805, 1992.

[30] S.V. Rice, J. Kanai, and T.A. Narker, An algorithm for matching OCR-generated text string, in H. Bunke, P. Wang, and H. Baird (eds.), *Document Image Analysis*, World Scientific, 263–272, 1994.

[31] R.A. Wagner and M.J. Fischer, The string-to-string correction problem, Journal of the ACM, 21(1): 168–173, 1974.

A Region-Based Algorithm for Classifier-Independent Feature Selection

Mineichi Kudo
Division of Systems and Information Engineering
Graduate School of Engineering
Hokkaido University, Sapporo 060-8628, JAPAN
E-mail: mine@main.eng.hokudai.ac.jp

Contents

1 Introduction

Feature selection is an important subject in pattern recognition. So far, many studies have been devoted to develop its methodology [1-8], and also some comparative studies on these algorithms [9, 10].

D. Chen and X. Cheng (eds.),
Pattern Recognition and String Matching, 315–340.

A feature selection algorithm can be specified by a quartet $(\mathcal{F}, c, s, g)$, where

$\mathcal{F} = \{f_1, f_2, \cdots, f_D\}$: an original feature set consisting of D features,
c: a criterion function of a given subset of $\mathcal{F}$,
s: a search procedure, or a selection procedure of a feature subset,
g: a goal of feature selection.

Let us investigate previous studies according to this notation. First, for c, the best is the Bayes discrimination rate, so that a criterion function is desirable as close as possible to it. From this point of view, many criterion functions, such as *Bhattacharyya distance*, *divergence* and *Chernoff bound* [11], have been proposed. These functions are based on the assumption that the probability density of a sample set or its functional form of the density is known. However, in general, such information is unknown, and criterion functions are therefore often calculated by regarding the density as a Gaussian. This assumption, however, has no reliable grounds, and in fact is not satisfied in many practical problems. Another approach for c is to estimate the discrimination rate of a discrimination rule directly from a training sample set. A *resubstitution method*, an *H-method* and a *U-method* (a *leaving-one-out method*) are such examples [12]. In particular, the leaving-one-out method is known to be able to utilize available samples effectively, but it requires designing classifiers N times, where N denotes the number of samples. This requires much computation cost even for not so large N.

Next, for s, an important fact is that the only way to select the best subset of d features from a set of D features is to explore all of their $\binom{D}{d}$ possible combinations [13, 14, 15]. Since this is usually not feasible, various suboptimal search procedures have been studied. The simplest one is a *forward (backward) sequential search* procedure, where the best (the worst) individual feature is chosen (removed) in the first round and then the best (the worst) pair including the best (the worst) individual feature is chosen (removed), and so on. The running time of this algorithm is $O(D^2)$. These algorithms have been developed to a *sequential forward floating search* and a *sequential backward floating search* procedures [4] with a little higher time complexity, which have been confirmed to show good performance in many cases [9, 10]. Another search procedure is a *branch and bound* algorithm [16, 17]. This algorithm enables us an exhaustive search under the assumption that a criterion function c is monotonic in inclusion relation with respect to feature subsets. This assumption is satisfied essentially only in the Bayes rule. Therefore, the procedure does not necessarily work well for practical classifiers except the Bayes rule. In addition, this procedure requires still a fairly large number of examinations of feature subsets. Recently, this procedure was greatly improved by Somol *et al.* [18], but it still requires much computation.

Last, as for g, there are two main goals: 1) to find a best feature subset effective for a specific classifier or 2) to find a universally effective feature subset independent of classifiers. The first goal is called *classifier-specific feature selection* and the second goal is called *classifier-independent feature selection*. The algorithms in [1, 3, 4, 19] belong to the former group and those in [2, 5, 7, 8] belong to the latter group. If we know in advance which classifier will be used after feature selection, the best performance of the classifier can be obtained by classifier-specific feature selection. However, it is generally impossible to know in advance which classifier should be used, and the latter goal is therefore more desirable. In this paper, we consider classifier-independent feature selection. In classifier-independent feature selection, the problem is to remove all *garbage features* that have no discriminative information, but it is very difficult to distinguish which features are informative and which features are garbage on the basis of a limited number of training samples. Moreover, to do classifier-independent feature selection requires estimation of the true densities of classes [5, 7] or estimation of the true class regions of classes [2, 8] as precisely as possible.

Classifier-independent feature selection also can be used for two-stage feature selection; a slightly larger feature subset is chosen so as not to include garbage features by classifier-independent feature selection, and then the best feature subset for a specific classifier is chosen from the feature subset chosen in the first stage by classifier-specific feature selection [20]. This two-stage feature selection is effective because classifier-independent feature selection algorithms are usually carried out faster than classifier-specific feature selection algorithms. In general, the latter needs to estimate the discrimination rates of a specified classifier on the basis of training samples by some re-sampling technique such as the leave-one-out or cross validation technique, whereas the former does not need such estimation.

In the following section, a method for constructing hyper-rectangles in order to approximate class regions is firstly described. Next, a structural index for measuring *separability* of classes is introduced. In the following section, a method for detecting garbage features and an algorithm for selecting the optimal feature subset are described. Finally, some experimental results are presented.

2 Subclass Method

The subclass method [21] is used for finding a set of hyper-rectangles in a class as the class region. Each hyper-rectangle is called a *subclass* and is defined on a discrete domain that is of equally spaced thresholds on each dimension. Let us introduce some notations. We assume that each sample takes D real values and is expressed by $x = \vec{x}D, x_i \in \mathbf{R}$. We quantize sample x to a discrete-value vector $\mathbf{x} = \vec{\mathbf{x}}D, \mathbf{x}_i \in \{0, 1, 2, \ldots, h+1\}$ by predetermined $h+2$ thresholds $\{\theta_i\}$ defined as

$$-\infty = \theta_0 < \theta_1 = \min_{\text{all training samples } x} x_i < \theta_2$$

$$< \cdots < \theta_h = \max_{\text{all training samples } x} x_i < \theta_{h+1} = +\infty$$

in each dimension i such that these thresholds satisfy $\theta_{i+1} - \theta_i = \text{const.}$ $(i = 1, \ldots, h-1)$. The individual hyper-rectangles are defined on this discrete domain $\{0, 1, \ldots, h, h+1\}^D$.

Let S^+ be a training sample set of a class under consideration and S^- be another training sample set of the other classes. Here, a sample x of S^+ is called a *positive sample* and a sample y of S^- is called a *negative sample*.

The aim is to find a family $\Omega(S^+, S^-)$ of subsets $X \subseteq S^+$ that satisfies

(1) (Exclusiveness) The hyper-rectangle spanned by X does not include any sample from S^-, and

(2) (Maximalness) X is maximal, in inclusion relation, in the family of all subsets that have the above-defined exclusiveness.

An example is shown in Fig. 1. Since this problem needs a combinational examination with respect to the training samples, a sub-optimal solution is usually obtained by a randomized algorithm [22]. The notation $\Omega(S^+, S^-; F)$ is used to clearly show in which feature space the subclasses are constructed. As shown in Fig. 1, a subclass is shown as a hyper-rectangle. Thus, a subclass has $2D$ sides corresponding to the dimensionality D.

3 Structural Index

In this section, the importance of a feature subset is evaluated on the basis of subclasses. The basic idea is very simple [2]. In general, with a large value of h, we can assume that every positive sample is included in one of all obtained subclasses. Thus, in most cases, a problem at hand is totally separable using

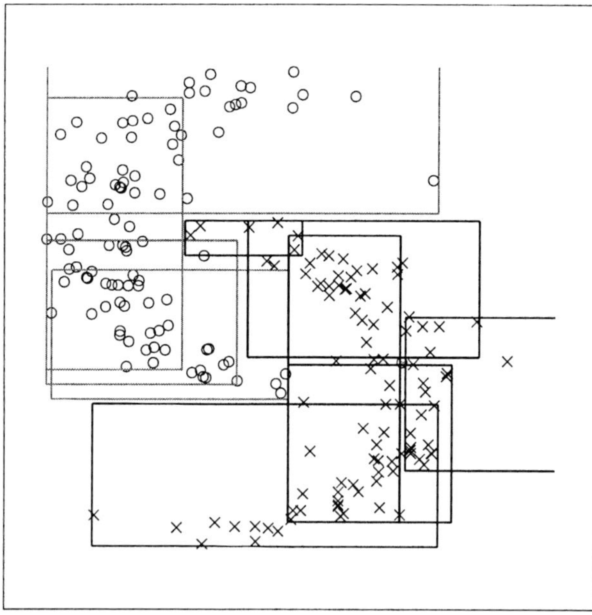

Figure 1: Example of subclasses.

subclasses. In that case, removal of an informative feature should give a great damage to the separability. In other words, if two sides of a subclass (a hyper-rectangle with $2D$ sides) are removed from the subclass in an axis corresponding to an informative feature, the modified subclass would include many negative samples. For example, in Fig. 1, if we remove all the vertical edges of the rectangles (subclasses) and extend all the horizontal edges to $\pm\infty$, only the top and bottom subclasses keep their exclusiveness and the others loose their exclusiveness, thus we know the first x-axis is important for discrimination.

Another idea is presented in our previous work [2] for determining the size of a feature subset to be selected. The idea is based on the fact that any garbage feature ideally cannot enlarge a subclass while keeping the exclusiveness. Conversely, removing garbage features never reduce subclasses. Thus, starting from the full feature set, removal of the worst feature can be stopped when the subclasses obtained with the remaining feature subset become much smaller than those with the feature subset before that removal.

In this paper, a sequential backward search is used as in [2]. Starting from the full feature set, at least one worst feature is removed at a time until a stop condition is satisfied. For the stop condition, we use a new structural index $I(\Omega; F)$ derived also from subclasses constructed in the feature space using F, where $I(\Omega; F)$ is calculated from a subclass family $\Omega(S^+, S^-; F)$ by

$$I(\Omega; F) = \frac{1}{|S^+|} \sum_{x \in S^+} |X_M(x)|/|S^+|, \text{ where}$$
$$|X_M(x)| = \max_{x \in X \in \Omega(S^+, S^-; F)} |X|.$$

That is, the maximal subclasses for individual positive samples are firstly collected, and then the index is calculated by summing up the sizes of those subclasses. It always holds that $0 \leq I(\Omega; F) \leq 1$. In addition, it should be noted that $I(\Omega; F)$ is monotonic in inclusion relation of feature subsets, that is, $I(\Omega; F) \leq I(\Omega; F')$ for $F \subseteq F'$. However, the randomized algorithm [22] may break the monotonicity a little, because it finds a subset $\hat{\Omega}$ of Ω stochastically. In our previous approach [2], we used the size of the maximum subclass only. This way may work in most cases but can be strongly affected by the randomness of the algorithm. While, $I(\Omega; F)$ as defined above is more robust against such a randomness because $I(\Omega; F)$ depends on many subclasses. Therefore, this index is used in this study.

In multi-class cases, $I(\Omega; F)$ is extended to

$$I(\Omega; F) = \sum_{\omega=1}^{C} P_\omega I_\omega(\Omega; F),$$

where C is the number of classes, P_ω is the *a priori* probability of class ω, and $I_\omega(\Omega; F)$ denotes $I(\Omega; F)$ for when class ω is a positive class.

Using this structural index $I(\Omega; F)$, the search is continued until a large degradation is found in $I(\Omega; F_i)$ for a sequence $\mathcal{F} = F_0 \supseteq F_1 \supseteq F_2 \supseteq \cdots$.

4 Automatic Determination of Size

In this section, we consider another stop condition for the search. First, we consider how well a class is separated from other classes with a feature subset. Let $\mathcal{F}$ be a full feature set and D be its size ($|\mathcal{F}| = D$). Also, let us denote a feature subset of size d by F_d, for instance, $F_D = \mathcal{F}$. In addition, let us assume that the subclasses are already obtained in the full feature space $\mathcal{F}$. That is, in what follows, unless stated otherwise, once the subclass are constructed in $\mathcal{F}$, they will never be reconstructed. Then we define *separability* $T(F)$ and *nonseparability* $T^c(F)$ of a feature subset $F \subseteq \mathcal{F}$ using these subclasses as

$$\begin{aligned} T(F) &= \{\text{the ratio of negative samples} \\ &\quad \text{excluded from subclasses when only } F \text{ is used}\} \\ T^c(F) &= \{\text{the ratio of negative samples} \\ &\quad \text{included in subclasses when only } F \text{ is used}\} \\ &= 1 - T(F). \end{aligned}$$

Here, $T(F)$ is calculated by counting the number of negative samples that are not included in the subclasses even after leaving all the sides corresponding to F and sweeping out the other sides in the subclasses. It holds that $T(\mathcal{F}) = 1$ and $T(\emptyset) = 0$.

In general, a feature subset F should be evaluated by 1) how well the class is separated from other classes when only F is used, and 2) how well the class is separated from other classes when F is removed. If F is valid solely, the value of $T(F)$ would be high, but may be low if F is valid only with some other features. However, even in the latter case, $T(F^c)$ would take a very low value, thus, $T^c(F^c)$ takes a high value, where F^c means the complementary set of F. This is because any feature subset that does not include F cannot show high separability. Especially, $T^c(F^c)$ is important for solving so-called *nesting problem* [13, 14, 15], i.e., the problem that the best single feature is not necessarily included in the best two features. In this paper, *separability measure* $V(F)$ is defined, by combining these two possibilities, as

$$V(F) = T(F) + T^c(F^c).$$

That is, both the separability of F and the nonseparability of its complement F^c are considered at the same time.

In multi-class cases, T, thus V too, is calculated as

$$T(F) = \sum_{\omega=1}^{C} P_\omega T_\omega(F), \quad V(F) = \sum_{\omega=1}^{C} P_\omega(T_\omega(F) + T^c_\omega(F^c)),$$

where C is the number of classes, P_ω is the *a priori* probability of class ω, and $T_\omega(F)$ is $T(F)$ for when class ω is under consideration. In what follows, we consider a class only in a two-class problem, but this argument is easily extended in this way without loss of generality.

Let us examine some properties about $V(F)$. First, since $0 \leq T(F), T(F^c) \leq 1$, we have

$$0 \leq V(F) = T(F) - T(F^c) + 1 \leq 2 \quad \text{(non-negativity)}.$$

Second,

$$V(F) + V(F^c) = (T(F) - T(F^c) + 1) + (T(F^c) - T(F) + 1) = 2. \quad (1)$$

Here, $V(\emptyset) = 0$ and $V(\mathcal{F}) = 2$ in the case where all positive samples are included in one of the subclasses. In addition, the following is obtained:

Property 4.1 *(Monotonicity)*

$$F \subseteq F' \Rightarrow V(F) \leq V(F').$$

(Proof)
This is derived from the monotonicity of T, that is $T(F) \leq T(F')$ for $F \subseteq F'$. This is because the addition of a feature f to F does not reduce the number of negative samples excluded from a subclass with F. Thus, from $T(F) \leq T(F')$ and $T(F^c) \geq T((F')^c)$, we have

$$V(F') - V(F) = (T(F') - T((F')^c) + 1) - (T(F) - T(F^c) + 1) \geq 0.$$

□

Next, let F_d^U and F_d^L be the best and the worst feature subset in V at size d. Then we have the following property.

Property 4.2

$$F_{D-d}^L = (F_d^U)^c, F_{D-d}^U = (F_d^L)^c.$$

(Proof)
Let us assume $F_{D-d}^L \neq (F_d^U)^c$. This means that $V(F_{D-d}^L) < V((F_d^U)^c)$. Then, from Eq. (1), $V((F_{D-d}^L)^c) > V(F_d^U)$. This is a contradiction. The second equality is derived in the same way. □

This property shows a *duality* of the best feature subsets and the worst subsets, that is, the best set of size d derives the worst set of size $D-d$ and vice versa.

Then, with notation $V^U(d) = V(F_d^U)$ and $V^L(d) = V(F_d^L)$, we have the following property.

Property 4.3 *(Monotonicity of the best and worst feature subsets)*

$$V^U(d-1) \leq V^U(d), \quad V^L(d-1) \leq V^L(d) \quad \textit{for} \quad 2 \leq d \leq D.$$

(Proof)
As for the best feature subsets $V^U(d)$, this is obvious because

$$V(F^U_{d-1}) \leq V(F^U_{d-1} + \{f(\notin F^U_{d-1})\}) \leq V(F^U_d).$$

As for the worst feature subsets $V^L(d)$, let us assume this is not true, that is, $V(F^L_d) < V(F^L_{d-1})$. Then, from the monotonicity of V, $V(F^L_d - \{f(\in F^L_d)\}) < V(F^L_d)$, and thus $V(F^L_d - \{f(\in F^L_d)\}) < V(F^L_{d-1})$. This is a contradiction.
□

Last, we define a measure $W(\cdot)$ to determine the size of a feature subset to be selected. We define $W(d)$ by

$$W(d) = V^U(d) - V^L(d).$$

This function can be rewritten as

$$W(d) = V(F^U_d) - V(F^L_d) = V(F^U_d) + V(F^U_{D-d}) - 2 = V^U(d) + V^U(D-d) - 2. \tag{2}$$

Then we have the following property.

Property 4.4 *(Symmetry)*

$$W(d) = W(D-d).$$

This means that it suffices to examine $d = 1, 2, \ldots, \lceil \frac{D}{2} \rceil$. Now, let us investigate how the value of $W(d)$ changes as the value of d increases one by one. First, from Eq. (2), we notice that

$$\begin{aligned} \Delta W(d) &= W(d) - W(d-1) \\ &= \left(V^U(d) + V^U(D-d)\right) - \left(V^U(d-1) + V^U(D-d+1)\right) \\ &= \left(V^U(d) - V^U(d-1)\right) - \left(V^U(D-d+1) - V^U(D-d)\right) \\ &= \Delta V^U(d) - \Delta V^U(D-d+1), \end{aligned} \tag{3}$$

where a notation $\Delta V^U(d) = V^U(d) - V^U(d-1)$ is used.

Next, let us investigate how garbage features affect the value of W. First, let us assume an ideal case in which all garbage features give no information on separability. Then we have the following basic theorem.

Theorem 4.5 *(An ideal case) If $d(< D)$ features have discriminative information and $D-d$ features are totally garbage, then*

$$W(1) \leq W(2) \leq \cdots \leq W(\min\{d, D-d\}) = \cdots = W(\lceil \frac{D}{2} \rceil).$$

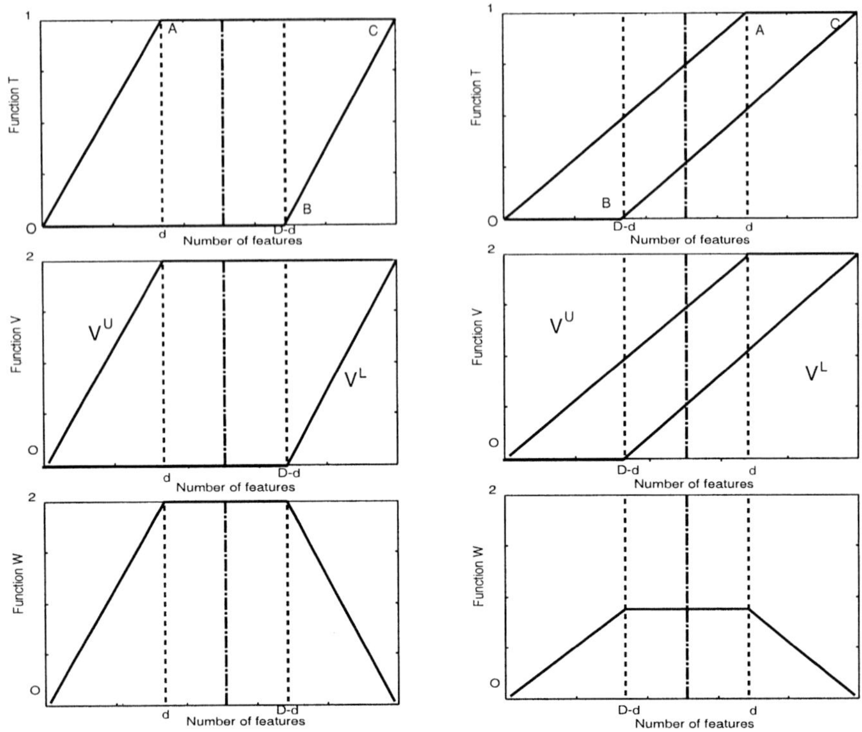

(a) case of $d \leq \lceil \frac{D}{2} \rceil$ (b) case of $d \geq \lceil \frac{D}{2} \rceil$

Figure 2: Two cases in proof.

(Proof)
Since all garbage features have no information from the assumption, we see

$$T(F_1^U) \leq T(F_2^U) \leq \cdots \leq T(F_d^U) = \cdots = T(F_D).$$

Let us discuss first the case of $d \leq \lceil \frac{D}{2} \rceil$ (Fig. 2(a)). If $e \leq d$, since $\Delta V^U(e) \geq 0$ and $\Delta V^U(D-e+1) = 0$, from Eq. (3), we have $\Delta W(e) \geq 0$. On the other hand, when $d < e \leq \lceil \frac{D}{2} \rceil$, we see that $\Delta V^U(e) = \Delta V^U(D-e+1) = 0$, thus $\Delta W(e) = 0$ too. Next, in the case of $d \geq \lceil \frac{D}{2} \rceil$ (Fig. 2(b)), in a similar discussion, we can see that $\Delta W(e) \geq 0$ for $e \leq D-d$ and $\Delta W(e) = 0$ for $D-d < e \leq \lceil \frac{D}{2} \rceil$. □

A more practical case is shown in Fig. 3. Here, d of D features are informative and the remaining $D-d$ features are garbage. In this situation, we have to cope with some *apparent discriminative information* found in garbage features. This is due to the finiteness of training samples. If only a

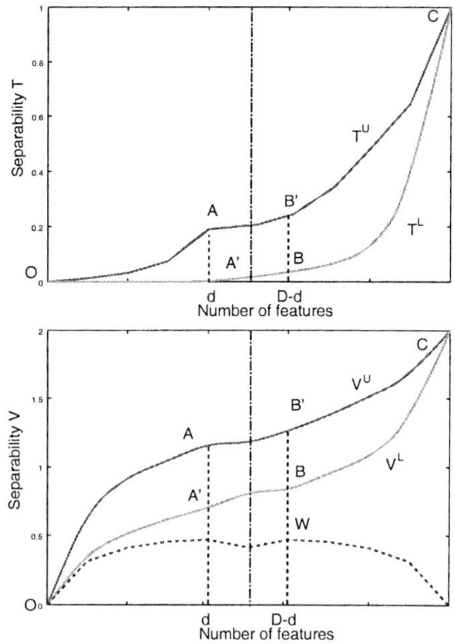

Figure 3: A general case.

limited number of training samples is available, purely garbage features can show a certain amount of discriminative information. We can easily imagine such a situation in a case in which a few points are placed randomly in a plane. In such a case, any partition into two groups of these points seems to be meaningful so that we regard the two features spanning the plane as being discriminative. Such an apparent discriminative information disappears as the number of training samples goes to infinity. In Fig. 3, point A shows the optimal feature subset including all d informative features and point B is its complement that includes only garbage features. In addition, a feature $f \in F_d^U - F_{d-1}^U$ should be removed from F_{D-d+1}^L to make F_{D-d}^L in most cases. Then, we have the following observations in separability T:

O1: The part O–A in the upper curve T^U may be convex upward and may also be concave upward. If informative features work individually and are added in the best order, this part would be convex. While, if those informative features work in combination, the curve would be concave. If each of the informative features has only a small amount of discriminative information, this curve would also be concave for the same reason as that for part A–C described below.

O2: In part B–C in T^L, which is the counter part of O–A, from point B the informative features are necessarily included in any feature subset. The lower curve would be concave, regardless of whether part O–A is concave or convex. This is because, if the curve O–A is convex upward, the informative features are added in the worst order in B–C and so naturally this curve would be concave. While, even if the curve O–A is concave upward, since such informative features work only in combination, this curve would also be a slightly concave.

O3: In part A–C in T^U, the upper curve would be concave in many cases. In this part, garbage features are added to informative features. Ideally, this curve should be flat, but in reality it increases due to apparent discriminative information of garbage features. In that case, a combination of garbage features dominates. Thus, this curve would be slightly concave upward.

O4: In part O–B in T^L, the lower curve would be straight or slightly concave for the same reason as that stated for part A–C. However, the concavity is less than that of A–C, because after all informative features are included in A–C, garbage features can easily increase the apparent separability in combination, while the addition of only garbage features in O–B do this totally at random.

Based on these observations in T, for the curves V^U and V^L, we can expect that part A–B' in V^U would be concave and part A'–B in V^L would be convex, thus,

$$W(1) \leq W(2) \leq \cdots \leq W(\min\{d, D-d\}) \geq \cdots \geq W(\lceil \frac{D}{2} \rceil)$$

when d of D features are informative. That is, a peak appears at size d or $D - d$ in the curve of $W(\cdot)$.

To confirm this, we conducted an experiment.

(Experiment)

In this experiment, one class consists of data drawn from a ten-dimensional Gaussian with mean $\mathbf{m_1} = \mathbf{0}$ and unit covariance matrix I, and another class consists of data drawn from a Gaussian with mean $\mathbf{m_2}$ and unit covariance matrix I. We dealt with two cases: 1) a low separability case with $\mathbf{m_2} = (1, 1, \ldots, 1, 0, 0, \ldots, 0)$ and 2) a high separability case with $\mathbf{m_2} = (2, 2, \ldots, 2, 0, 0, \ldots, 0)$. Here, the number of informative features is the number of 1's or 2's and is denoted by d. The number of training samples per class is 5,000, i.e., $N = 10,000$. Also, $D = 10$.

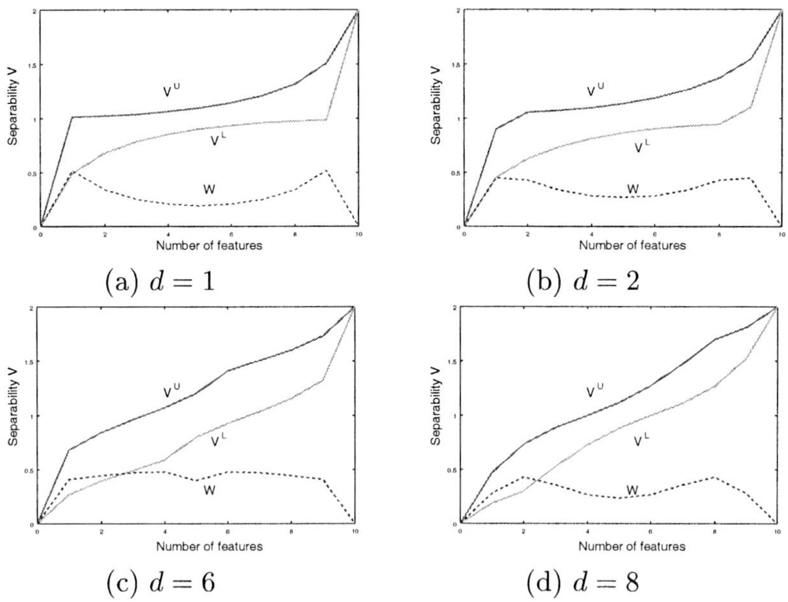

(a) $d = 1$ (b) $d = 2$

(c) $d = 6$ (d) $d = 8$

Figure 4: Results in a low separability case ($N = 10,000$).

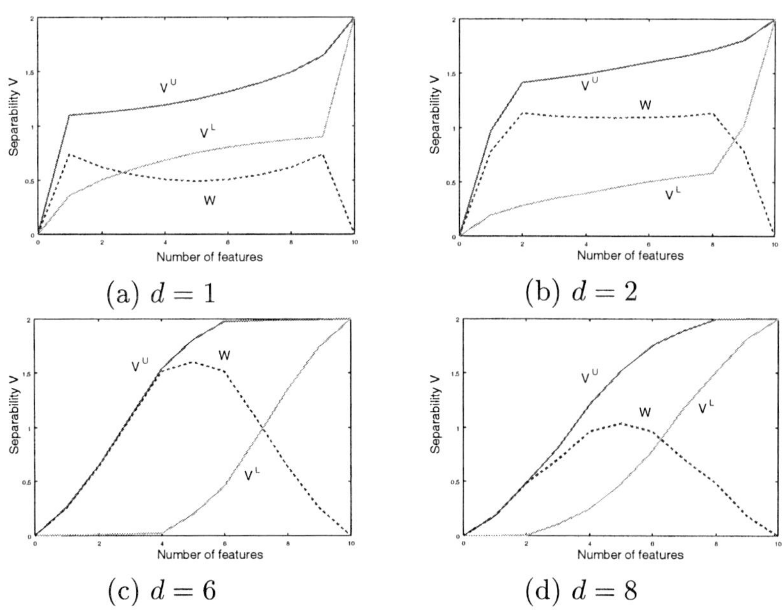

(a) $d = 1$ (b) $d = 2$

(c) $d = 6$ (d) $d = 8$

Figure 5: Results in a high separability case ($N = 10,000$).

The results are shown in Figs. 4 and 5. From the results, we notice that when a peak appears in function W, it correctly shows the optimal feature subset in most cases. However, in another similar experiment using a smaller number of training samples (the result is not shown in this paper), we also observed that a peak rarely appears and the reliability of the peak decreases either. A peak was also not found in the high separability case and when the number of informative features was larger than the number of garbage features as can be seen in Fig. 5 (c) and (d). Even if a peak exists, we must determine whether the peak is caused from an optimal feature subset of size d or that of size $D-d$, because in both cases the peak appears at the same point in range $[1, \lceil \frac{D}{2} \rceil]$. One possible way of doing this is to examine the value of V^U at the peak. If the optimal feature subset F has sufficient discriminative information and its complement does not, that is, $T(F) \geq 0.5$ and $T(F^c) < 0.5$, we have $V(F) = T(F) + T^c(F^c) > 0.5 + (1-0.5) \geq 1$. Hence, it may be possible to distinguish these two cases by examining whether $V^V(F_d) \geq 1$ or not. More reliable way is described in the next section.

Another observation is that, even in the case that no peak exists, the garbage features can be found as those with very small values in V^L (See Fig. 5 (c) and (d)). This situation arises when informative features have good discriminative information. In this case, it is an easy task to distinguish informative features from garbage features even when a peak does not appear.

5 Algorithm

Based on the discussion in the previous sections, an algorithm is proposed. The algorithm is shown in Figs. 6 and 7. This algorithm performs a sequential backward search and terminate it when the value of structural index I decreases largely. However, in each search step, we try to find a peak in the way described in Section 4, and if a peak is found then we terminate the search immediately.

In the main algorithm (Fig. 6), we change the search mode for peak detection at size L to either an exhaustive examination or a partial examination. This is because the cost of calculation of $V^U(d), V^L(d)$ $(d = 1, \ldots, \lceil \frac{D}{2} \rceil)$ is high even for not so large D. The size of the optimal feature subset is determined by investigating if a peak is found or not in the curve of W. If a peak is found at d in W, we have to make sure that it is caused from an optimal feature subset of size d or that of size $D-d$. We call a peak a

Algorithm

0. (Assumption and initialization)
   $\mathcal{F}$: original feature set
   L: level for changing search mode
   K: default level for search
   α: tolerance parameter for termination
   θ: cut-off value for garbage features

```
F ← 𝓕
F_old ← 𝓕
I_max ← I(F)
k ← K
while |F| ≠ ∅ do
        D ← |F|
        if D ≤ L then k ← ⌈D/2⌉
        Construct subclasses in F
        call Selection(F, k, D, θ) and
           receive the chosen subset F', the worst feature f_w,
           and a return value r.
        if r = 1 or r = 2 then /* positive or negative peak */
           output F' as the answer and terminate.
        else if r = 3 then /* garbage features */
              if I(F') < (1 − α)I_max then
                    F' ← F − {f_w} /* correction of decision */
              endif
        else if r = 0 then /* removal of the worst feature */
                    F' ← F − {f_w}
        endif
        if I(F') < (1 − α)I_max then
           output F_old as the answer and terminate.
        else
              if I(F') > I_max then I_max ← I(F') /* update of I_max */
              F_old ← F, F ← F'        /* update of F */
        endif
end
```

Figure 6: Algorithm

```
function Selection(F, k, D, θ)
{
        Examine all combinations of up to k features and
          calculate the values of V^U(j), V^L(j) and W(j) (j = 1,..., k).
        D_h ← ⌈D/2⌉
        if a peak is found at d < D_h in W then
          if k = D_h then /* exhaustive search */
(A1)        if V^L(d) ≥ (d/D_h) V^L(D_h) then
              return with F_d^U, F_1^L and r = 1 /* positive peak */
(B1)        else
              return with (F_d^L)^c, F_1^L and r = 2 /* negative peak */
            endif
          else /* partial search */

(A2)        if V^U(d) ≥ 1 then
              return with F_d^U, F_1^L and r = 1 /* positive peak */
(B2)        else
              return with (F_d^L)^c, F_1^L and r = 2 /* negative peak */
            endif
          endif
        else
(C)       if there exits e ≥ 1 such that V^L(e) < θ then
              return with (F_{e*}^L)^c, F_1^L and r = 3, /* garbage features */
              where e* = max{e | V^L(e) < θ}
(D)       else
              return with (F_1^L)^c, F_1^L and r = 0 /* removal of the worst feature */
          endif
        endif

}
```

Figure 7: Selection Procedure

positive peak if size d is correct and a *negative peak* if size $D - d$ is correct. The return value r from function *Selection* shows a positive peak for $r = 1$ and a negative peak for $r = 2$.

We terminate the procedure immediately if a peak is found (in the case of $r = 1$ or $r = 2$ in line 7). In addition, when some garbage features are found (in the case of $r = 3$ in line 9), we try to remove them all. However, even in the case, to confirm the correctness of this decision, we examine the value of I in lines 10–11. When there is no peak and no garbage features (in the case of $r = 0$ in line 13), we continue the procedure, removing the worst feature in line 14.

We terminate this algorithm when the value of structural index I gets down largely, except for termination by a peak. We stop the procedure when $I(F) < (1 - \alpha)I_{max}$ holds and output the feature subset obtained immediately one step before, where α is a predetermined *tolerance parameter* and I_{max} is the maximum value of I obtained so far and is initialized by $I(\mathcal{F})$.

The procedure for peak detection and for selection of a candidate feature subset is realized in function *Selection* (Fig. 7). The discussion in the previous section is reflected in this function. To distinguish a positive peak and a negative peak at d, we examine whether or not the value of $V^L(d)$ is over the value, at d, of a line connecting the origin and the value of $V^L(\lceil \frac{D}{2} \rceil)$ (Fig. 8). This line corresponds to a line connecting the values of V^U at the center $\lceil \frac{D}{2} \rceil$ and that at D (a dotted line in Fig. 8). This is based on the

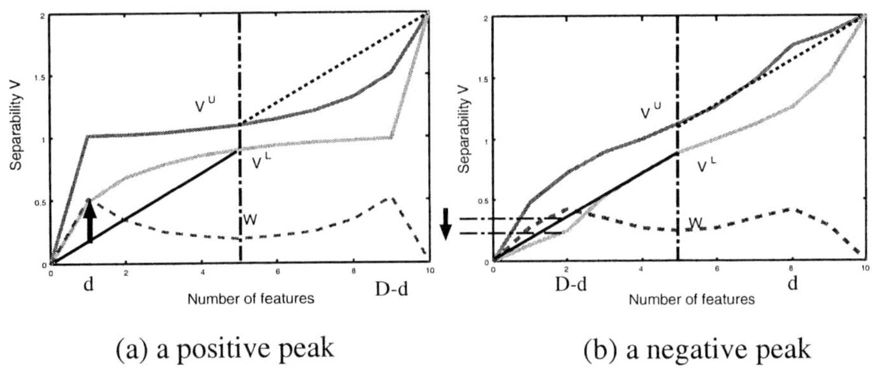

Figure 8: Discrimination of a positive peak and a negative peak.

fact that if the value of $V^U(d)$ at $d \geq \lceil \frac{D}{2} \rceil$ is a point of inflection in V^U, then it would appear above the line connecting the value of $V^U(\lceil \frac{D}{2} \rceil)$ and the value of $V^U(D)$ (Fig. 8(b)). If this is the case, such a peak is regarded as a negative peak (Fig. 8(b)), and otherwise such a peak is regarded as a positive peak (Fig. 8(a)). This judgment is realized in lines A_1 and B_1 of Fig. 7. However, this judgment requires the examination of all feature subsets at size $\lceil \frac{D}{2} \rceil$, which is usually not feasible. Therefore, instead lines A_2 and B_2 are used for a partial search in the case of $D < L$. In line A_2, a peak at d is regarded as a positive peak when $V^U(d) \geq 1$ and as a negative peak otherwise. In addition, we detect garbage features in line C. This is effective for removing many garbage features at once. It should be noted that subclasses are (re)constructed before calling *Selection*. In addition, it is also noted that the size D of the current feature set changes in the main loop of the algorithm and D is treated as the size of the full feature set in function *Selection*.

We carried out an experiment using the same data as those used in the experiment described in the previous section. The results are shown in Table 1. From this table, we can see that 1) a positive peak ($r = 1$) detects the optimal feature subsets almost correctly and 2) a negative peak ($r = 2$) also detects them but sometimes does not. The selection using positive peaks was more reliable than that using negative peaks. Overall, the selected subsets are sufficiently close to the optimal subsets, and the precision (P in Table 1) improves as the number of training samples increases. Here, in P, '+' means that some garbage features were added to all informative features and '-' means that only informative feature were chosen but not all. In addition, the number of construction of subclasses (M in Table 1) is much smaller than that needed in the simple sequential backward procedure. This is a great saving of computation time because the process of constructing the subclasses costs most. While, the procedure for detecting a peak does not require so much computation, since this procedure is done by bit-operations and by counting. As a whole, the complexity of this algorithm can be estimated as $O(D)$, ignoring the complexity for peak detection. Some incorrect estimations with a negative value of P obtained for rule $r = 0$ imply that the value of α is too large for those problems. This suggests that we should set this value in relation to the number of training samples. Although the values of I are not shown in Table 1, it was a surprising success that about half of the feature subsets F chosen by a peak did not satisfy $I(F) \geq (1 - \alpha)I_{max}$, that is, such feature subsets could not be found by the sequential backward search. The feature subsets obtained for $d = 1, 2, 4, 5$ and for $N = 1000$ in Table 1(a) are such examples.

6 Experiments

We applied our method to two synthetic datasets and one real dataset shown below. In the experiments, we used $\alpha = 0.05, \theta = 0.01, L = 10$ and $K = 3$. A value 3 of K means that we try to find a peak in the range of sizes $[1, 3]$ when $D > 10$. An exhaustive search is carried out for peak detection when $D \leq 10$. The results are shown in Tables 2,3 and 4 and in Fig. 9.

Table 1: Experimental results ($\alpha = 0.05, \theta = 0.01$). In the table, N is the number of training samples, d is the number of informative features, P is the difference in size of the chosen subset and the optimal subset, M is the number of construction of subclasses, and the column denoted by r's shows the decision rules in the used order.

(a) Low separability case

N	d	P	M	r's
200	1	0	1	1
	2	+2	1	1
	4	+3	5	0
	5	+4	3	0
	6	-1	3	0,2
	8	-2	3	0,2
	9	-1	4	0
1000	1	0	1	1
	2	0	1	1
	4	0	1	1
	5	0	3	0,2
	6	0	3	0,2
	8	0	1	2
	9	0	3	0
2000	1	0	1	1
	2	0	1	1
	4	0	1	1
	5	0	3	0,2
	6	0	1	2
	8	-1	1	2
	9	-2	1	2
4000	1	0	1	1
	2	-1	1	1
	4	+6	2	0
	5	0	3	0,2
	6	0	1	2
	8	0	1	2
	9	0	1	2

(a) High separability case

N	d	P	M	r
200	1	+8	1	2
	2	+4	3	3,0,2
	4	0	3	3,0,2
	5	0	4	3,0
	6	0	1	2
	8	0	1	2
	9	-1	1	2
1000	1	0	1	1
	2	+1	1	1
	4	+2	1	2
	5	0	4	3,0
	6	0	3	3,0
	8	-1	4	3,0
	9	-3	1	2
2000	1	0	1	1
	2	0	1	1
	4	-1	7	0,2
	5	-1	3	3,0
	6	-2	2	3,2
	8	-1	4	3,0
	9	-1	4	3,0
4000	1	0	1	1
	2	+1	1	1
	4	+2	1	2
	5	-1	2	3,2
	6	0	3	3,0
	8	-1	4	3,0
	9	-1	4	3,0

(Example 1) *Friedman*: A Friedman database [23].

This is a synthetic dataset. There are two classes of ω_1 and ω_2. The samples of ω_1 were generated according to a Gaussian with unit covariance matrix and mean zero. The samples of ω_2 surround those of ω_1 in the first four features that are distributed uniformly within a four-dimensional spherical slab centered at the origin with inner radius of 3.5 and outer radius of 4.0 in the samples of ω_2. The last six features of ω_2 are distributed as a Gaussian with unit covariance matrix and mean zero. In order to make the problem more realistic, the assumption of spherical symmetry is removed here by multiplying each feature by its feature number. That is, the first feature is multiplied by unity, the second by two, and so on. In this dataset, a combination of the first four features is important, although each of them does not show high separability. Each class has 1000 training samples.

(Example 2) *Waveform*: A waveform database taken from [24].

This is another synthetic dataset. In this database, there are three classes with 40 features with continuous values between 0 and 6. Each class is a random combination of two triangular waveforms with noise added. Each class is generated from a combination using two of three base waves. The samples were generated such that features $1, 2, \ldots, 21$ are necessary for class separation, whereas features $22, 23, \ldots, 40$ have random variables. The number of samples were 67 for class 1, 54 for class 2, and 79 for class 3, respectively.

(Example 3) *Sonar*: A sonar database taken from [24].

This is a natural dataset. The task is to discriminate between sonar signals bounced off a metal cylinder and those bounced off a roughly cylindrical rock using 60 features, each of which describes the energy within a particular frequency band, integrated over a certain period of time. The database consists of 111 patterns obtained by bouncing sonar signals off a metal cylinder at various angles and under various conditions and 97 patterns obtained from rocks under similar conditions.

In Table 2, within the range of sizes $7 - 10$, it is shown that only a worst feature was removed at one time, and at size 6, a negative peak was found

Table 2: Results of Friedman dataset ($\alpha = 0.05, \theta = 0.01$). Feature no. 1-4 are informative.

Step No.	Size	Chosen Feature Subset F	Decision Type r	$I(F)$		$(1-\alpha)I_{max}$
1	10	1,2,3,4,5,6,7,8,9,10	0	0.461		—
2	9	1,2,3,4,5,6,7,8,10	0	0.460	$\geq$	0.438
3	8	1,2,3,4,6,7,8,10	0	0.458	$\geq$	0.438
4	7	1,2,3,4,7,8,10	0	0.458	$\geq$	0.438
5	6	1,2,3,4,7,10	2	0.456	$\geq$	0.438
6	4	1,2,3,4	—	0.449	$\geq$	0.438

Table 3: Results of waveform dataset ($\alpha = 0.05, \theta = 0.01$). Feature no. 1-21 are informative.

Step No.	Size	Chosen Feature Subset F	Decision Type r	$I(F)$		$(1-\alpha)I_{max}$
1	40	1–40	0	0.864		—
2	39	1–25,27–40	0	0.859	$\geq$	0.821
		⋮				
20	21	3–20,22,24,33	0	0.821	$\geq$	0.821
21	20	3–20,22,33	—	0.806	$<$	0.821

and the algorithm was terminated. We expected to find a positive peak in the first round with $D = 10$, but it could not be found.

In Table 3, among the 21 informative features (nos. 1–21), only 18 features were chosen, and two garbage features were chosen in the final feature set of size 20. However, this chosen subset is close to the optimal feature subset, because feature nos. 1, 2, 20, and 21 have a small amount of discriminative information compared with that of the other features (nos. 3–19).

In Table 4, it can be seen that more than one worst feature were removed at one time in the first few steps. The algorithm was terminated when the value of the structural index I decreased to less than $(1-\alpha)I_{max}$ and we had a feature subset of size 35. To evaluate this result in detail, with $\alpha = 1.0$, we tried to choose a feature subset at alomost every size (Fig. 9). Here, to

Table 4: Results of sonar dataset ($\alpha = 0.05, \theta = 0.01$).

Step No.	Size	Chosen Feature Subset F	Decision Type r	$I(F)$		$(1-\alpha)I_{max}$
1	60	1–60	3	0.702		—
2	56	1,3–32,34–36,38–56,58–60	0	0.693	$\geq$	0.667
3	55	1,3–28, 30–32,34–36,38–56,58–60	3	0.693	$\geq$	0.667
4	53	1,3–22,24,25,27,28,30–32,34–36,38–56 58–60	0	0.689	$\geq$	0.667
		:				
22	35	1,3,5,6,9–13,15,17–19,21–25,28,30–32 34,36,38,39,43,45,47,50–52,54,55,58,60	0	0.669	$\geq$	0.668
23	34	1,3,5,6,9,11,1213,15,17–19,21–25,28,30–32 34,36,38,39,43,45,47,50–52,54,55,58,60	0	0.667	$<$	0.668

confirm the effectiveness of this algorithm for classifier-independent feature selection, we measured the recognition rates of the following seven classifiers: a linear classifier (LNR), a quadratic classifier (QDR), the nearest neighbor classifier (1NN), a decision tree (C4.5), the subclass classifier (SUB), a neural network (NRL) and a support vector machine (SVM). The recognition rates were calculated by the 10-fold cross validation technique. From Fig. 9, we see that at size 35 the chosen feature subset succeeded in improving the performance of all classifiers compared with the case in which all features were used. Some classifiers showed the best performance at a size less than 35, but the size depended on the classifier. In fact, it is difficult in the real world to judge if all and only garbage features are removed. Thus, in many cases, a feature subset that is slightly larger feature subset than the optimal feature subset consisting of all informative features should be accepted. In this sense, the result is satisfactory. It should also be noted that, in the experiment with $\alpha = 1.0$, after termination by a negative peak in size 5, the recognition rates of most classifiers greatly decreased.

7 Discussion

The proposed algorithm has two characteristics: 1) a structural index based on subclasses is used for measuring the sufficiency of a feature subset and used for terminating the search, and 2) the existence of a peak in a function W is used for finding the optimal feature subset. As for 2), although it cannot be expected that we always find a peak, once a peak is found, it is a strong cue to find the optimal feature subset. Conventional algorithms for feature selection have required either 1) the size of a feature subset to be selected or 2) a threshold to judge the sufficiency of a feature subset. From this point of view, the algorithm using a peak is superior to all those

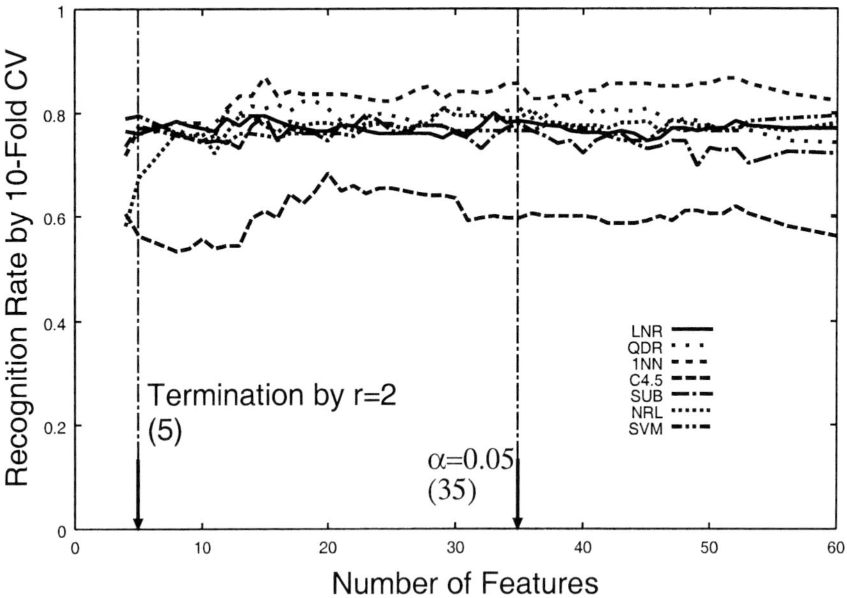

Figure 9: Discrimination rates of sonar dataset by six classifiers.

algorithms.

In the experiments, the termination by a peak was very effective and often was confidential in spite of the fact that the structural index showed a large degradation. It is, therefore, worth studying a way to make this judgment more reliable. To do this, we have to investigate the apparent discriminative information in detail and we need to analyze the behavior of the curve of W in terms of the number of samples and the dimensionality.

Another problem to be solved is the determination of a tolerance parameter α which is used for the case that the peak termination does not work. In this study, we used $\alpha = 0.05$. It is also desired to develop a method for determining this value automatically. To do this, we have to examine how this value is affected by the number of samples and also by the dimensionality.

8 Conclusion

A region-based algorithm for selecting $d(< D)$ features from all D features in the sense of classifier-independent feature selection has been proposed. This algorithm has the following merits: 1) it has a possibility to find the optimal feature subset automatically, 2) it exploits a sequential backward search but can remove more than one feature at a time, and 3) the evaluation of a feature subset is based on the estimated class regions in the feature subset and it can therefore be used for solving nesting problem. The second merit means that the algorithm runs in a linear order of the number of original features, although the algorithm includes a procedure whose complexity is higher but is ignorable in computational cost compared with that of the main procedure. This is important for large-scale problems.

Acknowlegment

This work was partly supported by Grant-in-Aid for Scientific Research (B), No. 60205101, from Japan Society for the Promotion of Science.

References

[1] J. Kittler, Feature set search algorithms. in C. H. Chen (ed.) *Pattern Recognition and Signal Processing*, (Sijthoff and Noordhoff, Alphen aan den Rijn, Netherlands, 1978) pp. 41–60.

[2] M. Kudo and M. Shimbo, Feature Selection Based on the Structural Indices of Categories, *Pattern Recognition* Vol. 26(1993) pp. 891–901.

[3] F. J. Ferri, P. Pudil, M. Hatef, and J. Kittler, Comparative study of techniques for large-scale feature selection, in E. S. Gelsema and L. N. Kanal (eds.) *Pattern Recognition in Practice IV*, (Elsevier Science B. V., 1994) pp. 403–413.

[4] P. Pudil, J. Novovičová, and J. Kittler, Floating Search Methods in Feature Selection, *Pattern Recognition Letters* Vol. 15(1994) pp. 1119–1125.

[5] J. Novovičová, P. Pudil, and J. Kittler, Divergence based feature selection for mulimodal class densities, *IEEE Trans. Pattern Anal. Machine Intell.* Vol. 18(1996) pp. 218–223.

[6] P. Pudil, Novovičová, and J. Kittler, Feature selection based on the approximation of class densities by finite mixtures of special type, *Pattern Recognition* Vol. 28(1995) pp. 1389–1397.

[7] H. J. Holz and M. H. Loew, Relative feature importance: A classifier-independent approach to feature selection. in E. S. Gelsema and L. N. Kanal (eds.) *Pattern Recognition in Practice IV*, (Amsterdam, Elsevier, 1994) pp. 473–487.

[8] M. Egmont-Petersen, W. R. M. Dassen and J. H. C. Reiber, Sequential Selection of Discrete Features for a Neural Networks - A Bayesian Approach to Building a Cascade, *Pattern Recognition Letters* Vol. 20(1999) pp. 1439–1448.

[9] A. Jain and D. Zongker, Feature Selection: Evaluation, Application, and Small Sample Performance, *IEEE Trans. Pattern Anal. Machine Intell.* Vol. 19(1997) pp. 153–157.

[10] M. Kudo and J. Sklansky, Comparison of Algorithms that Select Features for Pattern Classifiers, *Pattern Recognition* Vol. 33(2000) pp. 25–41.

[11] L. Kanal, Patterns in Pattern Recognition: 1968-1974, *IEEE Trans. Information Theory* Vol. 20(1974) pp. 697–722.

[12] G. T. Toussaint, Bibliography on Estimation of Misclassification, *IEEE Trans. Information Theory* Vol. 20(1974) pp. 472–479.

[13] T. M. Cover, The Best Two Independent Measurements Are Not the Two Best, *IEEE Trans. Information Theory* Vo. 4(1974) pp. 116–117.

[14] R. M. E. J. D. Elashoff and G. Goldman, On the Choice of Variables in Classification Problems with Dichotomous Variables. *Biometrika* Vol. 54(1967) pp. 668–670.

[15] G. T. Toussaint, Note on Optimal Selection of Independent Binary-Valued Features for Pattern Recognition *IEEE Trans. Information Theory* Vol. 17(1971), p. 618.

[16] P. M. Narendra and K. Fukunaga, A Branch and Bound Algorithm for Feature Subset Selection, *IEEE Transactions on Computers* Vol. 26(1977) pp. 917–922.

[17] B. Yu and B. Yuan, A More Efficient Branch and Bound Algorithm for Feature Selectionm *Pattern Recognition*, Vol. 26(1993) pp. 883–889.

[18] P. Somol, P. Pudil, and J. Grim, Branch & bound algorithm with partial prediction for use with recursive and non-recursive criterion forms. in

Advances in Pattern Recognition of Lecture Notes in Computer Science LNCS 2013 (Springer, 2001) pp. 230–239.

[19] P. Pudil, F. J. Ferri, J. Novovičová, and J. Kittler, Floating Search Methods for Feature Selection with Nonmonotonic Criterion Functions, in Proceedings of *12th International Conference on Pattern Recognition* (1994) pp. 279–283.

[20] M. Kudo and J. Sklansky, Classifier-Independent Feature Selection for Two-stage Feature Selection. in *Advances in Pattern Recognition of* Lecture Notes in Computer Science LNCS 1451 (Springer, 1998) pp. 548–554.

[21] M. Kudo and M. Shimbo, Optimal Subclasses with Dichotomous Variables for Feature Selection and Discrimination, *IEEE Trans. Systems, Man, and Cybern.*, Vol. 19(1989) pp. 1194–1199.

[22] M. Kudo, S. Yanagi and M. Shimbo, Construction of Class Regions by a Randomized Algorithm: A Randomized Subclass Method, *Pattern Recognition* Vol. 29(1996) pp. 581–588.

[23] J. H. Friedman, A Recursive Partitioning Decision Rule for Nonparametric Classification, *IEEE Transactions on Computers*, Vol. 26(1977) pp. 404–408.

[24] C. Blake and C. Merz, UCI Repository of machine learning databases [http://www.ics.uci.edu/~mlearn/MLRepository.html] (1998).

Inference of K-Piecewise Testable Tree Languages

Damián López, José Ruiz and Pedro García
Departmentamento de Sistemas Informáticos y Computación
Universidad Politécnica de Valencia, 46071 Valencia, SPAIN
E-mail: {dlopez,jruiz,pgarcia}@dsic.upv.es

Contents

1 Introduction

In the field of pattern recognition it is important to represent the objects as accurate as possible. The existence of primitives that can express different features of the objects help to obtain this accuracy.

The syntactic (or structural) representation of objects has been approached in different ways [1, 2]. There are primitives which have a high

D. Chen and X. Cheng (eds.),
Pattern Recognition and String Matching, 341–351.

representative power (web, plex, or graph grammars), but as the more representative power a primitive has, the more time it needs to operate, strings have been traditionally used to represent objects. Their expressive power is low, but the cost of operations on them, such as classification algorithms, is also low.

The development of efficient algorithms to manipulate primitives with greater representative power is a goal in the field of pattern recognition. In this way, algorithms that perform error correcting analysis for multidimensional languages [12, 9] permit the use of primitives such as trees in the modeling of the different classes that take place in a recognition task.

Due to the limits established by Gold in the field of grammatical inference [7], certain classes of string languages require the use of a structural sample to be inferred [18, 19, 15]. Although those algorithms could easily be adapted as to be considered tree language inference algorithms, this type of algorithms has been specifically considered in literature [5, 6, 13, 3, 11] and several of them have been implemented and used in practice (i.e. [10, 11]).

Another classical approach to string language inference consists in the use of only positive samples in the task of learning. Several important subfamilies of regular languages, like the k-Testable in the Strict Sense or the k-Piecewise Testable Languages, have been inferred in this way [4, 17]. Their common characteristic is that those languages can be defined by means of the segments (using consecutive symbols) or the subwords (using non consecutive symbols) that appear in their words. This features have proved to be fruitful in pattern recognition tasks [4, 16].

The aim of this work is to define a new family of tree languages that will be called k-Piecewise Testable Tree Languages ($k - PTTL$), where k is a strictly positive integer, and to propose an algorithm that, for a given value of k, infers the family of $k - PTTL$ from positive data in the limit.

The definition of $k - PTTL$ uses the concept of *tree subword*. Informally, a k-tree subword of a tree t is a tree of depth k, which is formed by *forks* of height one that appear in t and maintain the same node hierarchy. A language is $k - PTTL$ if it is saturated by a tree congruence in which two trees are equivalent if they have the same set of k-tree subwords.

The work is structured as follows, first we introduce some notation and the definitions that will be used along the work. Then, we define the new family of tree languages and establish some useful properties. Afterwards, we propose the inference algorithm and prove that it infers the desired family, the time complexity is also obtained. The conclusions and some lines of future work end this paper.

2 Notation and Definitions

Let $\mathcal{N}$ be the set of positive integers and let $(\mathcal{N}^*, \cdot)$ be the free monoid with the concatenation as the binary operation and λ as the identity. For $u, w \in \mathcal{N}^*$, we write $u \leq w$ if and only if there exists $v \in \mathcal{N}^*$ such that $w = u \cdot v$; $u < w$, if $u \leq w$ and $u \neq w$. The length of a word x is denoted by $|x|$.

A tree domain is a set $D \subseteq \mathcal{N}^*$ that fulfills the following conditions:

a) $v \in D$ and $u < v$ implies that $u \in D$.

b) If $u \cdot i \in D, i \in \mathcal{N}$ then $u \cdot j \in D \quad \forall j: \ 1 \leq j \leq i$.

Given an alphabet V, a *ranked alphabet* is a pair (V, r), where r is a finite relation in $(V \times \mathcal{N})$. Let $V_n = \{\sigma \in V | (\sigma, n) \in r\}$ and let V^T be the set of finite trees whose nodes are labelled with symbols in V. A tree can inductively be defined as follows:

$$\begin{array}{l} V_0 \subseteq V^T \\ \sigma(t_1, ..., t_n) \in V^T \quad \forall t_1, ..., t_n \in V^T, \quad \sigma \in V_n \end{array}$$

We define the *extended ranked alphabet* of V, denoted by V_e, as the pair (V, r_e), where $r_e = r \cup \{(a, 0) \mid \exists n: \ a \in V_n\}$.

Note that a tree $t \in V^T$ can be seen as a mapping $t: D \to V$, where D is the tree domain of t.

The size of a tree t, denoted by $\|t\|$, is inductively defined as follows:

$$\begin{array}{ll} \|a\| = 0 & \forall a \in V_0 \\ \|\sigma(t_1, ..., t_n)\| = 1 + \sum_{i=1...n} \|t_i\| & \forall t_1, ..., t_n \in V^T, \ \sigma \in V_n \end{array}$$

The depth of a tree t, denoted by *depth*(t), is defined as $Max(|x|)$, $x \in dom(t)$.

We denote by V^{T_k} the set of trees $V^{T_k} = \{t \mid t \in V^T \wedge depth(t) = k\}$.

The set of subtrees of t, $Sub(t)$, is defined as follows:

$$\begin{array}{ll} Sub(a) = \oslash & \forall a \in V_0 \\ Sub(\sigma(t_1, ..., t_n)) = \{t_1, ..., t_n\} \cup \bigcup_{i=1...n} Sub(t_i), & \\ \qquad \forall t_1, ..., t_n \in V^T, & \sigma \in V_n \end{array}$$

The subtree of $t \in V^T$ rooted at $x \in dom(t)$ is defined as:

$$t/x = \begin{cases} dom(t/x) = \{y \mid x \cdot y \in dom(t)\} \\ (t/x)(y) = t(x \cdot y) \end{cases}$$

The substitution of the subtree t/x with $s \in V^T$ is defined as follows:

$$t(x \leftarrow s)(y) = \begin{cases} t(y) \textbf{ if } |y| \leq |x|,\ y \neq x,\ y \in dom(t) \\ s(z) \textbf{ if } y = x \cdot z,\ z \in dom(s) \end{cases}$$

An equivalence relation $\equiv$ over V^T is *subtree invariant* whenever $t_1 \equiv t_2$ implies $t(x \leftarrow t_1) \equiv t(x \leftarrow t_2)$, $\forall t \in V^T$, $x \in dom(t)$. A tree set is regular if and only if is the union of some classes from a subtree invariant, finite index, equivalence relation.

A *deterministic tree automaton* is a four-tuple $A = (Q, V, \delta, F)$ where Q is a finite set of states; V is a ranked alphabet with $Q \cap V = \oslash$; $F \subseteq Q$ is the set of final states and $\delta = (\delta_0, ..., \delta_m)$ is a finite set of transitions defined as follows:

$$\begin{array}{ll} \delta_n : (V_n \times (Q \cup V_0)^n) \rightarrow Q & n = 1, ..., m \\ \delta_0(a) = a & \forall a \in V_0 \end{array}$$

δ can be extended to operate on trees in the following way:

$$\begin{array}{l} \delta : V^T \rightarrow Q \cup V_0 \\ \delta(\sigma(t_1, ..., t_n)) = \delta_n(\sigma, \delta(t_1), ..., \delta(t_n)) \textbf{ if } n > 0 \\ \delta(a) = a \end{array}$$

The language accepted by a tree automaton A is $L(A) = \{t \in V^T \mid \delta(t) \in F\}$.

Given a tree automaton $A = (Q, V, \delta, F)$ and a state $q \in Q$, we say that q is at level i if and only if $i = min\{depth(t) \mid t \in V^T, \delta(t) = q\}$.

3 Piecewise Testable Tree Languages

Definition 3.1 *Given $t, s \in V^T$, let $P^t(s)$ be defined as follows:*

$$\begin{array}{ll} P^t(a) = \{x \in dom(t) \mid t(x) = a\} & \forall a \in V \\ P^t(\sigma(t_1, ..., t_n)) = \{x \in dom(t) \mid \exists z_1, ...z_n & \\ \quad : \ x \cdot i \cdot z_i \in P^t(t_i)\} & \forall \sigma \in V_n, \\ & \forall t_1, ...t_n \in V^T \end{array}$$

Example 3.2 *Consider the tree t of the Figure 1, then:*

$$\begin{array}{rcl} P^t(b) & = & \{13, 1212, 1222\} \\ P^t(\sigma(a, b)) & = & \{121, 122\} \\ P^t(\sigma(a, \sigma(a, b), b)) & = & \{1\} \\ P^t(\sigma(\sigma(a, \sigma(a, b), b)), \sigma(c)) & = & \{\lambda\} \end{array}$$

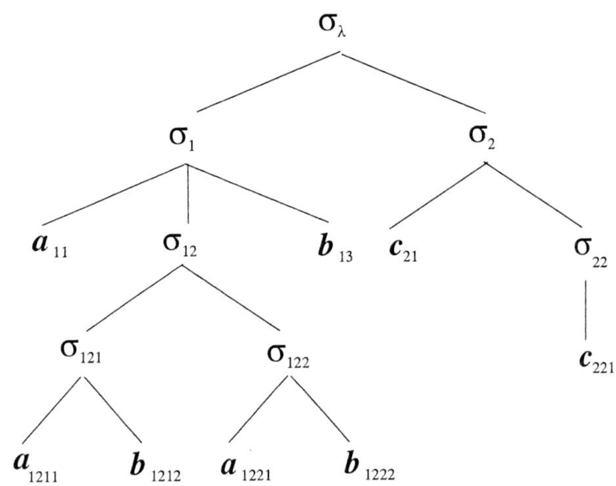

Figure 1: Tree example where the subscripts denote the domain of the node.

Definition 3.3 *Given a tree $t \in V^T$, the set $PSub(t)$ is defined as:*

$$PSub(t) = \{s \in V_e{}^T \mid P^t(s) \neq \oslash\}$$

and let also $kPSub(t)$ be defined as follows:

$$kPSub(t) = \begin{cases} t & \textbf{if } depth(t) < k \\ s \in V_e{}^T \mid s \in PSub(t) \quad depth(s) = k & \textbf{otherwise} \end{cases}$$

Example 3.4 *Given the tree of the Figure 1, the set $2PSub(t)$ and some members of $3PSub(t)$ are shown below:*

$$\begin{aligned} 2PSub(t) = \{&\sigma(\sigma(a,b),\sigma),\ \sigma(\sigma,\sigma(a,b)),\ \sigma(\sigma(a,b),\sigma(a,b)),\ \sigma(a,\sigma(\sigma,\sigma),b) \\ &\sigma(a,\sigma(a,b),b),\ \sigma(\sigma(a,\sigma,b),\sigma(c,\sigma)),\ \sigma(\sigma(a,\sigma,b),\sigma(c)) \\ &\sigma(\sigma(a,\sigma,b),\sigma),\ \sigma(\sigma(\sigma,\sigma),\sigma(c,\sigma)),\ \sigma(\sigma(a,b),\sigma(c,\sigma)) \\ &\sigma(\sigma(\sigma,\sigma),\sigma(c)),\ \sigma(\sigma(a,b),\sigma(c)),\ \sigma(\sigma(\sigma,\sigma),\sigma),\ \sigma(\sigma(a,b),\sigma) \\ &\sigma(\sigma,\sigma(c,\sigma)),\ \sigma(\sigma,\sigma(c)),\ \sigma(c,\sigma(c))\} \end{aligned}$$

$$\begin{aligned} 3PSub(t) = \{&\sigma(\sigma,\sigma(c,\sigma(c))),\ \sigma(\sigma(a,\sigma,b),\sigma(c,\sigma(c))),\ \sigma(\sigma(\sigma,\sigma),\sigma(c,\sigma(c))) \\ &\sigma(\sigma(a,\sigma(a,b),b),\sigma(c,\sigma(c))),\ \sigma(\sigma(\sigma(a,b),\sigma(a,b)),\sigma(c,\sigma(c))) \\ &\sigma(\sigma(\sigma(a,b),\sigma(a,b)),\sigma(c,\sigma)),\ \sigma(\sigma(\sigma(a,b),\sigma),\sigma(c,\sigma)) \\ &\sigma(\sigma(a,\sigma(\sigma,\sigma),b),\sigma),\ \sigma(\sigma(a,\sigma(a,b),b),\sigma) \\ &\sigma(a,\sigma(\sigma,\sigma(a,b)),b),\ ...\} \end{aligned}$$

As $depth(t) = 7$, the set $PSub(t)$ is the union of the sets $kPSub(t)$, $k : 0..7$

Definition 3.5 *Given $t, s \in V^T$, the equivalence relation $\equiv_k$ is:*

$$t \equiv_k s \iff kPSub(t) = kPSub(s)$$

The equivalence class of the tree t will be denoted by $[t]$. Clearly, the above equivalence relation is subtree invariant, and has the following properties:

1.- $\equiv_{k+1}$ refines $\equiv_k$.

2.- Given $t, s \in V^T$, $x \in dom(t)$, if $t(x \leftarrow s) \in [s]$, then $t(x \leftarrow (t(x \leftarrow ...t(x \leftarrow s)...))) \in [s]$.

3.- Any equivalence class $[t]$ is either a singleton or has infinite members.

Definition 3.6 *A tree language is said to be* k-Piecewise Testable *(*k-PTTL*) if and only if it is the union of some equivalence classes of $\equiv_k$. A language L is said to be* Piecewise Testable *if there exist a value of k for which L is k-Piecewise Testable.*

Given two trees $t, s \in V^T$, we say that $t \leq_k s$ if and only if $jPSub(t) \subseteq jPSub(s)$, $\forall j \leq k$. This relationship establishes a partial order in V^T, and it is easy to see that $t \in PSub(s)$ implies $t \leq_k s$.

When $\leq_k$ is extended to classes of equivalence, the quotient set $(V^T / \equiv_k, \leq_k)$ is a finite lattice with an absolute maximum (the class of trees $t \equiv_k [V^{T_k}]$), and an absolute minimum (the class of the empty tree). Considering the tree t and the order relation defined, each subtree $s = \sigma(s_1, s_2, ..., s_n)$ in a postorder traversal of the tree t, fulfills that $s_i \leq_k s$, $i : 1..n$.

From the above properties is easy to prove that:

1.- If L is a *k-PTTL*, then L is *j-PTTL* $\forall j \leq k$.

2.- If L is Piecewise Testable, then L is regular.

3.- Given a tree automaton A that accepts a *k-PTTL*, A has no transitions $\delta(\sigma, u_1, ..., q, ..., u_n) = p$ where p is at level i, q is at level j, and $i < j$.

4.- Given V a ranked alphabet, the class of *k-PTTL* has a finite number of languages, bounded by $|\mathcal{P}(\mathcal{P}(V_e{}^{T_k}))| \leq 2^{2^{|V_e|^{k \cdot bf}}}$ where bf is the maximum branching factor and where $\mathcal{P}$ denotes the power set.

3.1 Inference Algorithm

Considering the equivalence relation defined above, given a set of trees $\mathcal{T} = \{t_1, t_2, ..., t_k\}$, it is possible to obtain a *k-PTTL* $L_k(\mathcal{T})$ as it is shown below:

$$\forall t \in V^T \ t \in L_k(\mathcal{T}) \iff \exists s \in \mathcal{T} \ t \equiv_k s.$$

This definition has the following properties:

1.- $S \subseteq L_k(\mathcal{T})$.

2.- $L_k(\mathcal{T})$ is the smallest *k-PTTL* that contains S.

Proof. Let suppose that exist a *k-PTTL* L such that, $S \subseteq L \subset L_k(\mathcal{T})$. By definition 3.6, L must contain all the equivalence classes in S. Let $t \in V^T$, $t \in L_k(\mathcal{T}) - L$, then there exists no $s \in S$ such that $[s] = [t]$, which is a contradiction. □

3.- $L_k(\mathcal{T}') \subseteq L_k(\mathcal{T})$ whenever $\mathcal{T}' \subseteq \mathcal{T}$.

4.- $L_{k+1}(\mathcal{T}') \subseteq L_k(\mathcal{T})$.

5.- If $k > Max_{t \in \mathcal{T}}(depth(t))$ then $L_k(\mathcal{T}') = \mathcal{T}'$.

We proposose the algorithm depicted in Figure 2, which for a given positive sample $\mathcal{T}$ of the target language obtains a *k-PTTL* consistent with the sample that recognizes a subset of $L_k(\mathcal{T})$.

Example 3.7 *As an example of run, let* $k = 2$ *and consider the following set of trees* T *as input:*

$$\mathcal{T} = \left\{ \begin{array}{l} t_1 = \sigma(\sigma(a, \sigma(\sigma(\sigma(a), \sigma(a)), \sigma(a)), b), \sigma(c, \sigma(c, \sigma(c)))), \\ t_2 = \sigma(\sigma(\sigma(\sigma(\sigma(a), \sigma(a)), \sigma(a)), \sigma(a)), \sigma(c, \sigma(c, \sigma(c, \sigma(c))))), \\ t_3 = \sigma(\sigma(a, \sigma(\sigma(a), \sigma(a)), b), \sigma(c, \sigma(c, \sigma(c))))) \end{array} \right\}$$

The first iteration of the algorithm, using t_1 *as input, outputs the following automaton:*

$$\begin{array}{rlrl} \delta(\sigma, a) & = q_1 & \delta(\sigma, c) & = q_5 \\ \delta(\sigma, q_1, q_1) & = q_2 & \delta(\sigma, c, q_5) & = q_6 \\ \delta(\sigma, q_2, q_1) & = q_3 & \delta(\sigma, c, q_6) & = q_7 \\ \delta(\sigma, a, q_3, b) & = q_4 & \delta(\sigma, q_4, q_7) & = q_8 \in F \end{array}$$

Input :	*A tree set* $\mathcal{T} = \{t_1, t_2, ..., t_k\}$.
	A value k.
Output :	*A tree automaton* $A = (Q, V, \delta, F)$:
	$L(A) \subseteq L_k(\mathcal{T})$.
Method :	$Q = kPSub(Sub(\mathcal{T}))$
	$F = kPsub(\mathcal{T})$
	$\forall t \in \mathcal{T}$
	$\quad \forall s = \sigma(u_1, u_2, ..., u_p) \in Sub(t) \cup \{t\}$
	$\quad\quad \delta \supseteq \delta(\sigma, kPSub(u_1),$
	$\quad\quad\quad kPSub(u_2), ..., kPSub(u_p)) = kPSub(s)$
	$\quad$ **end**$\forall$
	end$\forall$
EndMethod.	

Figure 2: k-Piecewise Testable Tree Language Inference algorithm.

When t_2 is processed, the following productions are added to the output automaton:

$$\begin{aligned} \delta(\sigma, q_3, q_1) &= q_3 \\ \delta(\sigma, c, q_7) &= q_7 \\ \delta(\sigma, a, q_3, q_7) &= q_9 \in F \end{aligned}$$

Finally, after t_3 is processed, the following productions are added:

$$\begin{aligned} \delta(\sigma, a, q_2, b) &= q_{10} \\ \delta(\sigma, q_{10}, q_7) &= q_{11} \in F \end{aligned}$$

Theorem 3.8 *The Algorithm of figure 2 identifies the class of* k-PTTL *in the limit.*

Proof. The algorithm outputs an automaton that recognizes a superset of the input whenever a new structure $s = \sigma(s_1, ..., s_n)$, such that $\exists i : 1..n : \ s_i \equiv_k s$, is processed.

The finiteness of the lattice $(V^T / \equiv_k, \leq_k)$, and the ascending path the algorithm follows through the lattice assure the convergence. □

3.2 Time Complexity

In the process of inferring a *k-PTTL* the algorithm associates to every state of the automaton a set of piecewise subtrees. The number of possibly associated sets is bounded above by $|V_e|^{k \cdot bf}$, where bf, the maximum branching factor of the tree, is considered as a constant.

If we call t to the tree in $\mathcal{T}$ with the greatest amount of nodes, $kPSub(t) \leq \|t\|^{k \cdot bf}$. Since the processing of each $t \in \mathcal{T}$ implies the creation of $\|t\|$ transitions at most, if we consider k as a constant, as the number of different sets of piecewise subtrees is bounded, the time complexity of the algorithm could be expressed as $\mathcal{O}(p(\|t\|) \cdot n)$, where p is a polynomial and n denotes the number of samples in $\mathcal{T}$.

4 Conclusions and Future Work

The availability of powerful structural primitives to represent the objects is a key factor in pattern recognition. Multidimensional (tree) representations allow to model important features in disciplines like speech recognition and natural language processing. *Tree adjoining grammars* [8] have been used in this field, for instance to capture semantic caracteristics of verbs inside sentences [14].

In this work, the generalization of the notion of subword to tree structures allow us to characterize the class of *k-Piecewise Testable Tree Languages*. Together with the characterization, and the extension of several string languages properties to tree languages, a polynomial time inference algorithm is proposed which learns the class in the limit. The time complexity of the algorithm is also obtained.

This new family of formal languages allows the modelling of relations between components of the sentences that could be useful, not only in linguistics, but also in other pattern recognition tasks.

5 Acknowledgement

This work has been partially supported by the Spanish CICYT under contract TIC2000-1153.

References

[1] K. S. Fu *Syntactic pattern recognition and applications*, (Prentice-Hall, 1982).

[2] R. González and M. Thomason, *Syntactic Pattern Recognition. An Introduction*, (Addison-Wesley Publishing Company, 1978).

[3] H. Fernau, Learning tree languages from text, *Technical Report, Wilhelm-Schickard-Institut für Informatik*, Universität Tübingen, 2001.

[4] P.García and E. Vidal, Inference of k-testable languages in the strict sense and application to syntactic pattern recognition, *IEEE Transactions on Pattern Analysis and Machine Intelligence*, Vol.12 No.9 (1990) pp. 920-925.

[5] P. García, Learning k-testable tree sets from positive data, *Technical Report DSIC II/46/1993*, Departamento de Sistemas Informáticos y Computación, Universidad Politécnica de Valencia, 1993.

[6] P. García and J. Oncina, Inference of recognizable tree sets, *Technical Report DSIC II/47/1993*, Departamento de Sistemas Informáticos y Computación, Universidad Politécnica de Valencia, 1993.

[7] E. M. Gold, Language identification in the limit, *Information and Control* Vol.10 (1967) pp. 447-474.

[8] A. K. Joshi and Y. Schabes, Tree-Adjoining Grammars, in G. Rozenberg and A. Salomaa (eds.) *Handbook of formal Languages Vol.3*, (Springer, 1997) pp. 69-123.

[9] D. López, J. M. Sempere and P. García, Error correcting analysis for tree languages, *International Journal on Pattern Recognition and Artificial Intelligence*, Vol.14, No.3 (2000) pp. 357-368.

[10] D. López and I. Piñaga, Syntactic pattern recognition by error correcting analysis on tree automata, in F. Ferri, J. Iñesta, A. Amin and P. Pudil (eds.) *Advances in Pattern Recognition*, (Springer-Verlag, 2000) pp. 133-142.

[11] D. López and S. España, Error correcting tree language inference, *Pattern Recognition Letters*, Vol.23 No.(1-3) (2002) pp. 1-12.

[12] S. Y. Lu and K. S. Fu, Error-correcting tree automata for syntactic pattern recognition, *IEEE Transactions on Computers*, C-27 No.11 (1978) pp. 1040-1053.

[13] E. Mäkinen, On inferring linear single-tree languages, *Information Processing Letters* Vol.73 No.(1-2) (2000) pp. 1-3.

[14] M. Palmer, J. Rosenzweig and W. Schuler, Capturing motion verb generalizations in syncronous tree adjoining grammars, in P. Saint-Dizier (ed.) *Predicative forms in natural language and in lexical knowledge bases*, (Kluwer Academic Publishers, 1999) pp. 229-256.

[15] V. Radhakrisnan and G. Nagaraja, Inference of regular grammars via skeletons, *IEEE Trans. System, Man and Cybernetics.* SCM-17 (1987) pp. 982-992.

[16] J. Ruiz, Familias de lenguajes explorables: Inferencia inductiva y caracterizaciones algebraicas, *PhD. Thesis* (in spanish), Universidad Politécnica de Valencia, 1998.

[17] J. Ruiz and P. García, Learning k-Piecewise Testable Languages from Positive Data, in Miclet and de la Higuera (eds.) *Grammatical Inference: Learning Syntax from Sentences*, LCNS 1147 (1996)pp 203-210.

[18] Y. Sakakibara, Learning Context-Free Grammars from Structural Data in Polynomial Time, *Theoretical Computer Science* Vol.76 (1990) pp. 223-242.

[19] Y. Sakakibara, Efficient Learning of Context-Free Grammars from Positive Structural Examples, *Information and Computation* Vol.97 (1992) pp. 23-60.

Mining Partially Periodic Patterns With Unknown Periods From Event Stream

Sheng Ma
IBM T.J. Watson Research Center
Hawthorne, NY 10532
E-mail: shengma@us.ibm.com

Joseph L. Hellerstein
IBM T.J. Watson Research Center
Hawthorne, NY 10532
E-mail: jlh@us.ibm.com

Contents

D. Chen and X. Cheng (eds.),
Pattern Recognition and String Matching, 353–377.

1 Introduction

Temporal, periodic behavior is common in many application data [6][7][11] including web logs, stock data, alarms of telecommunications, and event logs of computer networks. This is because of routine behaviors of humane (e.g. morning rush every work day), seasonal effects (e.g. increase sales of stationary before a semester), or a consequence of routine tasks (e.g. rebooting print servers every morning or checking stock prices every 10 minutes). An example in computer system is that five repetitions every 10 minutes of a port-down event followed by a port-up event, which in turn is followed by a random gap until the next repetitions of these events. In fact, our study of event logs in production computer networks found that over 50% of the events can be explained by periodic temporal patterns.

Mining such patterns can often provide great value. Our experience in analyzing events of computer networks is that periodic patterns often lead to actionable insights. There are two reasons for this. First, a periodic pattern indicates something persistent and predictable. Thus, there is value in identifying and characterizing the periodicity. Second, the period itself often provides a signature of the underlying phenomena, thereby facilitating diagnosis. In either case, patterns with a very low support are of equally importance. For example, we found a one-day periodic pattern due to a periodic port-scan. Although this pattern only happens three times in a three-day log, it provides a strong indication of a security intrusion.

Unfortunately, existing work does not address key characteristics of these patterns:

1. Periodic behavior is not necessarily persistent. For example, in complex networks, periodic monitoring is initiated when an exception occurs (e.g., CPU utilization exceeds a threshold) and stops once the exceptional situation is no longer present. During the monitoring interval or "on" segment, the monitoring request and its response occur periodically. The "off" segment consists of a random gap in the periodicity until another exceptional situation initiates periodic monitoring. We refer this on-off windowed periodicity as partial periodicity. Although the fast Fournier transforms have been very successful in

dealing periodic behavior, it performs poorly on partial periodicity because of random off-segments.

2. Time information is critical for determining a periodic pattern. However, it is often imprecise due to lack of clock synchronization, the rounding issue, or network delays. Therefore, a time phase shift should be allowed.

3. Period lengths are not known in advance. Further, periods may span a wide range, from seconds to days. This implies the need of finding the periods instead of exhaustively searching all possible period lengths.

4. The number of occurrences of a periodic pattern typically depends on the period, and can vary drastically. For example, a pattern with a period of one day period has, at most, seven occurrences in a week, while one minute period may have as many as 1440 occurrences. Thus, support levels need to be adjusted by the period lengths.

5. Noise environment is inherent characteristics of many applications, especially computer or telecommunication networks. By noise, we mean that periodic events may be missing, and random events may occur in addition to periodic events.

This work develops algorithms for mining p-patterns taking in consideration of the above five issues. We begin with establishing a looser definition of periodicity to deal with item (1) and (2). In that, we introduce an on-off model to characterize partial periodicity, in which periodicity is confined in on-segments separated by off-segments. We then define the partially periodic temporal association to characterize a set of events that are partially periodic, which we refer to as a p-pattern. Our definition of p-patterns generalizes the partial periodicity defined in [6] by combining it with temporal associations (akin to episodes in [9]) and including a time tolerance to account for imperfections in the periodicities.

Finding p-patterns requires taking into account items (3) and (4) above. We structure this as two sub-tasks: (a) finding period lengths and (b) finding temporal associations. While the level-wise algorithm can be employed for (b), we develop an algorithm based on a chi-squared test for finding period lengths. Further, we develop two algorithms for discovering p-patterns based on whether periods or associations are discovered first. The performance trade-offs between these two approaches are studied. We conclude that the association-first algorithm has higher tolerance to noise while the period-first algorithm is more computationally efficient. We study effectiveness and efficiency of the algorithms using synthetic data. Further, we apply the developed algorithm to event logs obtained from two different networks: a middle-sized Intranet, and an outsourcing center. Many p-patterns are discovered in both logs, many of which led to diagnostic and corrective actions.

For related work, sequential mining[9][4][15][13][10] has been studied extensively, especially in domains such as event correlation in telecommunication networks [9], web log analysis[5][14], and transactions processing[6][13]. One theme here is discovering frequent temporal associations[9][13][15]; that is, finding a set of events that co-occur within a predefined time window. There has also been recent work in identifying periodic behaviors[6][7][11] closely related to our work. Ozden et. al.[11] study mining of cyclic association rules for full periodicities–patterns that are present at each cycle. As noted earlier, this is quite restrictive since periodicities may occur only intermittently. Han et. al.[6][7] study partially periodic patterns. While their concept of partially periodic patterns is in the same spirit as ours, their focus is symbol sequences, not time-based sequences. Further, they assume that period lengths are known in advance or that it is reasonable to employ an exhaustive search to find the periods. We make neither assumption. In addition, none of these studies consider the effect of noise–random occurrences of events of the same type as those in the periodic pattern. Nor do these studies address the problem of phase shifts or imperfections in the periodicity.

This sequel is organized as follows. Section 2 formulates the problem addressed. Section 3 presents our algorithm for finding possible periods. Section 4 describes our algorithms for finding p-patterns with unknown periods. Section 5 discusses empirical assessments of our algorithms. Our conclusions are contained in Section 6.

2 Definitions and Problem Statement

Fundamental to our work is the concept of an event. An event has two attributes: a type and an occurrence time.

Definition 1 *Let A be a set of event types. An event is a pair (a, t), where $a \in A$ is an event type and $t \in R$ is the occurrence time of the event. An* ***event sequence*** *S is an ordered collection of events, i.e. $\{(a_1, t_1), \ldots, (a_N, t_N)\}$, where $a_i \in A$ is the event type of the i-th event. t_i represents the occurrence time of the event, and $t_j \leq t_i$ for $1 \leq j \leq i \leq N$. Throughout, we use D to denote the set of events being mined.*

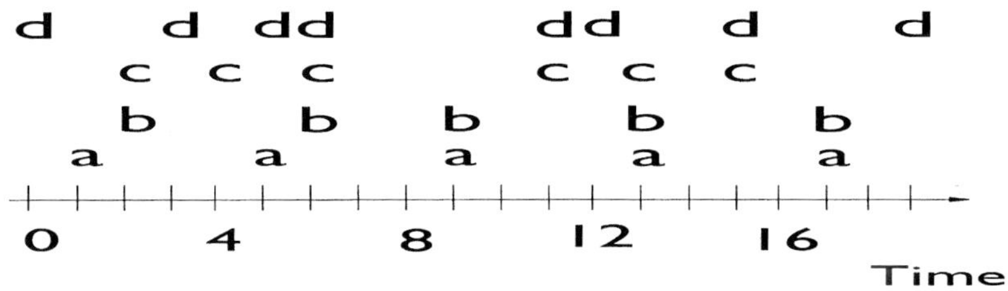

Figure 1: An illustrative event sequence. The event type of an event is labeled above its occurrence time.

Often, it is desirable to consider just the times associated with a sequence of events.

Definition 2 *A **point sequence** is an ordered collection of occurrence times.*

An event sequence can be viewed as a mixture of multiple point sequences of each event type. Figure 1 illustrates an event sequence, where the set of event types is $A = \{a, b, c, d\}$. The type of an event is labeled above its occurrence time. Figure 2 illustrates the point sequence of event type "d" $S_d^{'} = (0, 3, 5, 6, 11, 12, 15, 19)$. A point is plotted as "o" at its occurrence time on the time axis. We note that an event type may have a complex structure. For example, we can encode host name and alarm type into event type and so a "port down" alarm sent from host X has a different event type than a "port down" alarm sent from host Y.

Now, we define a periodic point sequence. We include the concept of tolerance to account for factors such as phase shifts and lack of clock synchronization.

Definition 3 $S' = (t_1, \cdots, t_N)$ *is a **periodic point sequence** with period* p *and time tolerance* δ *if a point occurs repeatedly every* $p \pm \delta$ *time units.*

For example in Figure 1, the point sequence of event type "a" is $S_a' = (1, 5, 9, 13, 17)$. Thus, event type "a" is periodic with period 4. "b" is also periodic, if $\delta \geq 1$.

The above definition of periodicity requires periodic behavior in the whole time duration. This is very stringent and cannot capture situations where events may be lost and a periodic behavior only persists in parts of the whole time duration. Thus, we consider a generalization. As illustrated

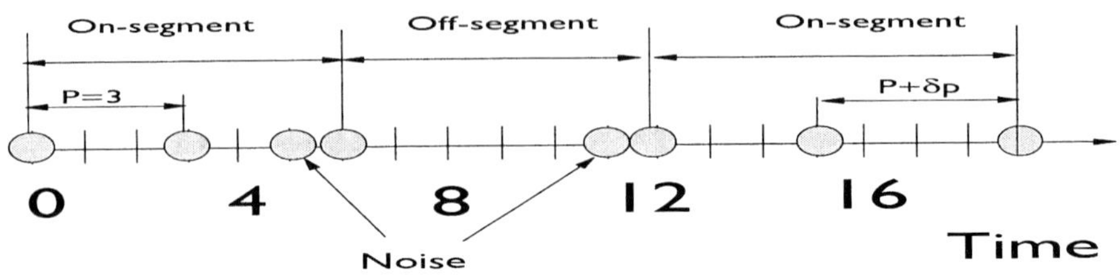

Figure 2: The point sequence for event type "d" in Figure 1. "o" represents a "d" event.

in Figure 2, we consider a point sequence consisting of on-off segments. During the on-segment, the point sequence is periodic with p. Then, there is a random gap, or off-segment, during which the point sequence is not periodic with p.

Definition 4 *A point sequence is* ***partially periodic*** *with period p and time tolerance δ, if points are periodic with p and δ during on-segments.*

Referring to the above example, "d" and "c" are partially periodic with periods 3 and 2, respectively, but not periodic. Figure 2 illustrates the point sequence for "d", and its "on" and "off" segments.

Thus far, we have described periodicities of point sequences. This can be generalized to a set of events that occur close in time.

Definition 5 *A set of event types $A_1 \subset A$ is a* ***partially periodic temporal association (p-pattern)*** *with parameters p, δ, w, and minsup, if the number of qualified instances of A_1 in D exceeds a support threshold minsup. A qualified instance $S_1 \subset D$ satisfies the following two conditions:*
(C1) The set of the event types of events in S_1 is A_1, and there is a t such that for all $e_i \in S_1$, $t \leq t_i \leq t + w$.
(C2) The point sequences for each event type in S_1 occur partially periodically with the parameters p and δ.

To illustrate this definition, we refer again to Figure 1. Let $w = 1$, $\delta = 1$, and $minsup = 2$. Then, $\{a, b\}$ is a p-pattern with length 2 and period 4.

Moreover, all non-null subsets of this pattern– $\{a\}, \{b\}$–are also p-patterns with length 2.

It can be easily verified that p-patterns are downward closed. That is, if A_1 is a p-pattern, then all non-null subsets of A_1 are also p-patterns. This means that a level-wise search can be used, thereby providing computational efficiencies.

2.1 Overview of Approach

Herein, we assume that δ, w, and $minsup$ are given. Thus, finding p-patterns require two sub-tasks: (1) finding (all) possible periods $\{p\}$, and (2) discovering temporal patterns A_1 with w.

We proceed as follows. Section 3 develops algorithms for finding possible periods. Section 4 develops algorithms to discover p-pattern with unknown periods by combining the two sub-tasks.

3 Finding Unknown Periods

This section develops an effective and efficient approach to finding periods of a point sequence. Effectiveness requires robustness to missing events and random occurrence of additional events as well as handling partially periodicity. Efficiency demands that the computation scale modestly as the number of events grows and the range of time-scales increases. Here, a point is a generic name for an instance of either an event type or a temporal association (i.e. a set of event types).

One approach to finding unknown periods is to use the fast Fourier transform (FFT), a well developed technique for identifying periodicities. There are two problems here. First, while the FFT is effective for finding a wide range of periodicities, it does not cope well with random off-segments in p-patterns. Further, the computational efficiency of FFT depends on the range of time scales. The computational complexity of FFT is $O(T \log T)$, where T is the number of time units. This is undesirable for sparse events with periodicities over a wide range of time-scales, which is often the case in network management. For example, although there may be hundred thousands of events in a month, there are over one billion milliseconds.

3.1 Chi-squared test for Finding Periods

We begin with some notation. Let $(t_1, \cdots, t_N)$ be a point sequence in a time window $[0, T]$, where N is the total number of points. We denote the i-th

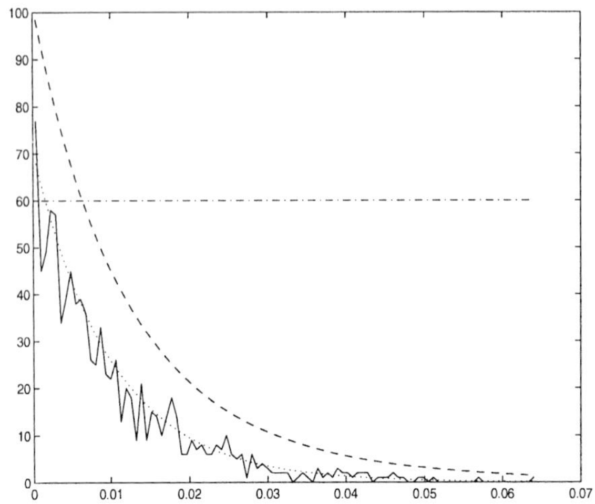

Figure 3: Distribution of inter-arrivals for a random point sequence. "-": empirical result. "..": theoretical distribution. "–": a threshold obtained by a chi-squared test. "-."': a constant threshold.

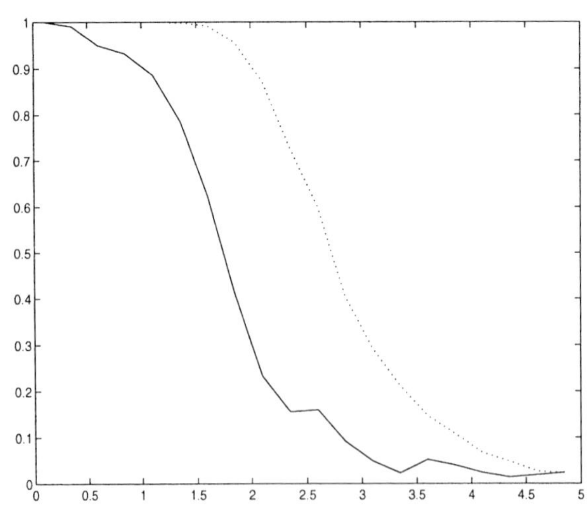

Figure 4: The effect of noise. y-axis: success percentage. x-axis: NSR ratio. "-": first-order inter-arrival. "..": second-order inter-arrival

inter-arrival time by $\tau_i = t_{i+1} - t_i$, where $1 \leq i \leq N-1$. We define two extreme point sequences: an ideal partially periodic sequence and a random sequence. An ideal partially periodic sequence is modeled by a sequence of on-off segments, in which a point re-occurs periodically during an "on" segment, and no point occurs during an "off" segment (e.g., event "a" in Figure 1). A random point sequence is an ordered collection of random points generated randomly and uniformly in $[0, T]$.

Now, we characterize inter-arrivals for an ideal partial periodic point sequence with parameters p and δ. Let $\iota_i = 1$ if both t_i and t_{i+1} are in the same on-segments; otherwise, $\iota_i = 0$. When $\iota_i = 1$, i.e. the i-th and $(i+1)$-th arrivals are in the same on-segment, $\tau_i = p + \mathbf{n}_i$, where $-\delta \leq \mathbf{n}_i \leq \delta$, characterizes a random phase shift. When $\iota_i = 0$, i.e. successive arrivals are not in the same on-segment, $\tau_i = \mathbf{r}_i$, a random variable characterizing an off-segment. Put these two cases together, we obtain

$$\tau_i = (\iota_i)(p + \mathbf{n}_i) + (1 - \iota_i)\mathbf{r_i}. \tag{1}$$

Now consider an arbitrary inter-arrival time τ and a fixed δ. Let C_τ be the total number of inter-arrivals in $[\tau - \delta, \tau + \delta]$. Intuitively, if τ is not equal to p, C_τ should be small; otherwise C_p should be large. Based on this observation, a naive algorithm for finding periods is to look for large values of C_τ. That is, an inter-arrival τ is declared to be a period, if

$$C_\tau > thres. \tag{2}$$

However, there is a problem. Intuitively, the number of partially periodic points with a large p is much smaller than that with a small p given the same on-segment length. Therefore, the naive algorithm favors small periods. To illustrate this, we randomly and uniformly generate 1000 points in the interval $[0, 100]$. The solid line of Figure 3 plots the distribution of inter-arrivals. A threshold used by the naive algorithm is illustrated by the dash-dotted line. The figure shows that the distribution of inter-arrivals decays as inter-arrivals increase. As the naive algorithm employs a fixed threshold regardless of period length, it tends to produce false positives for small periods and false negatives for large periods.

Clearly, we need a threshold that adjusts with the period under consideration. One approach is to use a chi-squared test. The strategy here is to compare C_τ with the number of inter-arrivals in $[\tau - \delta, \tau + \delta]$ that would be expected from a random sequence of inter-arrivals. The chi-squared statistics [8] is defined as

$$\chi^2_\tau = \frac{(C_\tau - NP_\tau)^2}{NP_\tau(1 - P_\tau)}, \tag{3}$$

where N is the total number of observations and P_τ is the probability of an inter-arrival falling in $[\tau - \delta, \tau + \delta]$ for a random sequence. Thus, NP_τ and $NP_\tau(1 - P_\tau)$ are the expected number of occurrences and its standard deviation, respectively. χ_τ^2 is the normalized deviation from expectation. Intuitively, the chi-squared statistic measures the degree of independence by comparing the observed occurrence with the expected occurrence under the independence assumption. A high χ_τ^2 value indicates that the number of inter-arrivals close to τ cannot be explained by randomness, which is a necessary condition for periodicity.

χ^2 is usually defined in terms of a confidence level. For example, 95% confidence level leads to $\chi^2 = 3.84$. Then, the above equation can be changed to

$$C'_\tau = \sqrt{3.84NP_\tau(1 - P_\tau)} + NP_\tau. \tag{4}$$

C'_τ can be used as a threshold to find possible periods. That is, we say that τ is a possible period, if

$$C_\tau > C'_\tau. \tag{5}$$

To compute P_τ , we note that a random event sequence approaches a Poisson arrival sequence[12] as the number of points increases. Further, it is well known that the inter-arrival times of a Poisson process are exponentially distributed. Using this, we obtain

$$P_\tau = \int_{p-\delta}^{p+\delta} \lambda \exp(-\lambda t) dt, \tag{6}$$

$$\approx 2\delta\lambda exp(-\lambda p), \tag{7}$$

where $\lambda = N/T$ is the mean arrival. The approximation is obtained based on the assumption that δ/p is small.

We conducted simulation experiments to assess the accuracy of this result. Figure 3 plots the results. The x-axis is the period (expected inter-arrival times), and the y-axis is the count of inter-arrival times for a period. Three curves are shown: (i) the empirical density of C_τ for a simulated exponential inter-arrival distribution (solid line), (ii) the expected value of C_τ in the simulated data (dotted line), and (iii) C'_p for 95% confidence level (dashed line). Observe that the χ^2 test results in a threshold that varies with period. Further observe that $C_\tau < C'_\tau$ in all cases, and hence there is no false positive.

3.2 Finding Periods in Noisy Data

As noise data is unavoidable in many applications, here we consider the ability of our test to detect partial periodicities in the presence of noise. By noise, we mean the addition of randomly occurring events in either an on-segment or an off-segment of the p-pattern.

This situation is best described with an example. Consider event "d" in Figure 2. This event has a period of 3 starting at time 0 and extending through time 18. However, there are also two noisy "d" events at times 5 and 11. Thus, the set of inter-arrival times are $\{3, 2, 1, 5, 1, 3, 3\}$. Clearly, noise events make it more difficult to detect the underlying period.

To gain more insight, we conducted a series of experiments. In each, there was a periodic point sequence with $p = 2$ in $[0, 200]$. We then gradually increase the number of noise points, which are generated uniformly and randomly. We quantify the noise by computing the ratio of the number of noise events to the number events in the partial periodicity. We refer to this as the **noise-to-signal ratio (NSR)**. For each NSR, 1000 runs were conducted. A run is successful if the period is correctly identified (i.e. above 95% confidence level), and no false period is found. Figure 4 plots the results. The x-axis is NSR, and the y-axis is the fraction of successful runs. The solid line displays the results for this experiment. Note that the success rate is above 90% if NSR does not exceed 1. However, as NSR increases above 1, the success rate degrades rapidly.

There is an intuitive explanation for the rapid drop-off in success rate as NSR increases beyond 1. As suggested by the example above, the performance of our test degrades rapidly when noise points lie between periodic events. With more noise events, this becomes increasingly likely. We note, however, that a noise point in an off-segment does less damage than one in an on-segment since the latter creates two inter-arrivals that do not have the period and deletes one that does. Thus, these experiments are in some sense a worst case since there is no off-segment.

Can we improve the performance of the above algorithm? Intuitively, we may increase the chance to detect the period if we take into consideration inter-arrivals between events separated by n other events. We refer to this as n-order inter-arrivals. Of course, this comes at a price—increased computational complexity. The dotted line in Figure 4 shows the performance of our test with second-order inter-arrivals. Note that the NSR value at which the success probability declines rapidly is close to 2, which is almost double the NSR value at which success probability declines rapidly when using first-order inter-arrivals.

3.3 Implementation Details

We now present an algorithm for finding periods based on the chi-squared test introduced earlier. Our discussion focuses on first-order inter-arrivals. The algorithm for higher-orders is essentially the same.

Let S be a point sequence. We assume that memory is larger than $|S|$ as we need the maximum of $|S|$ memory to hold counts in the worst case. This assumption is reasonable for a sequence of events of the same type since: (a) we only need to keep track of arrival instants and (b) this is a shorter sequence than raw data that contains a mixture of event types. When memory is problematic, indexing mechanism such as CF-tree[16] can be used to control the total number of buckets.

The algorithm takes as input the points s_i in S, the tolerance δ, and confident-level (e.g. 95%).

1. For $i = 2$ to $|S|$
   (a) $\tau = s_i.time - s_{i-1}.time$
   (b) If C_τ does not exist, then $C_\tau = 1$
   (c) Else, $C_\tau = C_\tau + 1$
2. Adjust Counts($\{C_\tau\}, \delta$) /*Adjust counts to deal with time tolerance*/
3. For each τ for which a C_τ exists
   (a) Compute threshold C'_τ based on Equation 5
   (b) If $C_\tau > C'_\tau$, then insert τ into the set of periods being output

This algorithm counts the occurrence of each inter-arrival time. Then, the computed counts are adjusted because of time tolerance δ by merging into a single group counts whose τ values are within δ of one another. Then, the test threshold is computed. If the threshold is exceeded by the test statistic, a possible period is found, and inserted into results. This procedure is repeated until all event types are processed.

It is easy to see that the above algorithm scales linearly with respect to the number of events. Further, the algorithm does not depend on the range of time scales of the periods.

4 Mining P-patterns With Unknown Periods

As we discussed, mining p-patterns with unknown periods is combination of two tasks: (a) finding periods, and (b) finding temporal association. In this section, we first overview algorithms for (b), i.e. mining associations. We then develop algorithms that combine the two tasks in different ways: the period-first algorithm and the association-first algorithm.

4.1 Overview of the Level-wise Algorithm

Many algorithms have been developed for mining associations efficiently ([2][3]). Our focus here is to illustrate basic idea, and then use the algorithm as a building block for mining p-patterns. Thus, we only discuss Apriori algorithm[2], and its variation for mining temporal associations ([9]). We also generalize it for mining p-patterns with a known period. The resulting algorithm also extends the algorithm developed by [7][6] so as to take into consideration of the time tolerance, and handle temporal data rather than sequence data as considered by the previous work. We will later use the last two algorithms (i.e. mining temporal associations, and mining p-patterns with a known periods) as the building blocks for developing algorithms for mining p-patterns with unknown periods.

Now, we reviews the general level-wise search, and its variations. Let L_k and C_k be the set of qualified patterns and candidate patterns at level k, respectively. Given events D and $C_1 = \{c \in 2^A \| c| = 1\}$, where A is the set of event types and C_1 is a set of patterns with length 1, the level-wise algorithm proceeds as follows.

(1) k =1;

(2) Count the occurrences in D for each pattern $v \in C_k$

(3) Compute the qualified candidate set: $L_k = \{v \in C_k | v.count > minsup\}$

(4) Compute the new candidate set C_{k+1} based on L_k

(5) if C_{k+1} is not empty, $k = k + 1$ and goto (2)

Step (3) computes the qualified candidate set L_k from C_k based on the supports. Step (4) computes a new candidate set C_{k+1} based on the previous qualified patterns in L_k. This is typically implemented by a join operation followed by pruning[1]. Step (2) counts occurrences of patterns in D. Step

(3) and (4) are similar for finding different patterns, while Step (2) varies based on the patterns to be found.

In mining associations, Step (2) augments the count of a pattern (called an item set) $v \in C_k$ by 1, if v is a subset of a transaction (see [2] for details). This can be generalized to an event sequence by introducing a time window to segment the temporal sequence into "transactions". In particular, Step (2) in mining temporal associations augments the count of a temporal association $v \in C_k$ (also called an Episode by [9]), if v is a subset of event types in a time window $[t-w, t]$[1]. Similarly, Step (2) in mining p-patterns with a known period augments the count of a p-pattern $v \in C_k$, if v satisfies two conditions: (1) v is a subset of event types in a time window $[t-w, t]$; and (2) v is a subset of those in the pervious window $[t-p-\delta'-w, t-p-\delta']$, where $0 \leq \delta' \leq \delta$ is corresponding to a possible time shift. Here, the first condition is required by Condition 1 of Definition 5. It is the condition to qualify an instance of a temporal association. The second condition is required by Condition 2 of Definition 5 for qualifying partial periodicity.

4.2 Discovering P-patterns with Unknown Periods

Now, we describe algorithms for mining p-patterns with unknown periods. Two steps are needed in achieving this goal: (1) finding the possible periodicities and (2) determining the temporal associations. These two steps can be combined in different orders resulting in two different algorithms: the period-first algorithm, and the association-first algorithm. The former does step (1) and then step (2), while the latter does step (2) and then step (1). A hybrid algorithm is also possible.

4.2.1 Period-first algorithm

This algorithm operates by first finding partial periodicities for each event type, and then constructing temporal associations for each period.

Let D be the set of events (sorted by time). The period-first algorithm is described as follows:

Period-first algorithm

Step 1: Find all possible periods

1 Group D by event type

2 Find possible periods for each event type

[1]Please refer [9] for subtle details.

Step 2: Find patterns

3 For each period p

1. Find related event types $A_p = \{a \in A | a \text{ has a period } p\}$
2. $C_1 = \{v \in 2^{A_p} || v| = 1\}$
3. Set support level $sup(p)$, and window size $w(p)$
4. Finding p-patterns with p

Step 1 finds the possible periods for individual events using the algorithm described in the preceding section. In particular, Line 1 groups events by type so that events in each group can be read out sequentially to find possible periods using algorithm discussed in Section 3.3. Alternatively, this step can be implemented in parallel to find periods for all event types simultaneously. However, tree indexing schemes, as those used in [16], are needed to handle a large amount of events.

Step 2 finds all patterns for each period. It first finds event types $A_p = \{a \in A | a \text{ has a period } p\}$. It then seeds initial candidate set C_1 and specifies the minimum support and window size. Last, the level-wise mining (Section 3.3) can be performed to find p-patterns with known p. Further, we note that Step 2 can be done in parallel for each period to gain further reductions in data scans.

The period-first algorithm uses the periods discovered to reduce the set of event types considered for temporal mining and hence significantly reduce the computational complexity of Step 2. Our experience has been that $|A_p|$ is much less than $|A|$ for two reasons. First, only periodic event types are included in A_p. Second, we only consider those event types with the period p. These result in a much smaller subset than the set of possible event types A.

Another advantage of the period-first algorithm is that the minimum support and the window size can be assigned separately for each period based on the period length and its support. This makes it possible to find a pattern with a very low support because of a large p[2], while avoiding drastic increase of the search space.

The computational complexity of Step 1 is determined by the group operation[3]. For Step 2, complexity is the sum of the complexity of finding temporal associations for each A_p.

---

[2]In our experience, we let minimum support be $minsup(p) = T/p\alpha$, and $w(p) = \beta p$, where $0 \leq \alpha, \beta << \leq 1$ are two user-controlled parameters.

[3]It is reasonable to assume that event logs are stored by time before mining.

There is, however, a disadvantage to this algorithm. If noise events are present, then our ability to identify the correct periods is reduced. Without the correct periods, Step 2 cannot output the correct p-patterns. In particular, as NSR increases, we expect the effectiveness of this algorithm to decrease.

4.2.2 Association-first Algorithm

The association-first starts by mining for temporal associations. The results of this step are then used to determine p-patterns.

How do we determine p-patterns from temporal associations? One approach is to check partial periodicity for each event type in a temporal association and then select only those associations whose event types occur in the same phases of the same period. This approach does not offer much benefit over the period-first algorithm.

An alternative approach is to find periodicities for the sequence of the instances of a temporal association. The algorithm, as we will further discuss, can operate at a very high NSR.

Association-first algorithm

Step 1 Find temporal patterns

1 Set support level and window size

2 $C_1 = A$

3 Run temporal mining

Step 2: Find periods of patterns

1. For each temporal association pattern
   (a) Find sequence of pattern occurrence times
   (b) Find periods for the sequence

Step 1 mines temporal associations. This can be done using the level-wise algorithm discussed in Section 4.1. Step 2 finds periods of temporal association patterns found in step 1. It is accomplished by two sub-steps. First, we identify a sequence of instances of a temporal association pattern. Doing so requires defining the time of each occurrence of a pattern. We use the time of the first event in each occurrence. Second, we employ the Chi-squared test discussed in Section 3 to find periods. As with the period-first algorithm, Step 2 can be implemented in parallel to find periods for multiple patterns simultaneously.

The association-first approach has a computational complexity comparable to that of temporal mining, which can be substantial for large patterns and low support levels. On the other hand, This algorithm has much better resistance to noise than the period-first algorithm for a p-pattern with at least 2 items. The reasoning is that when finding periods in Step 2b, difficulties arise if there is a random occurrence of a pattern between two periodic occurrences. However, the chances of a random occurrence of a temporal pattern should be much less than a random occurrence of an individual event, especially for larger patterns.

We note that there is a subtle consideration. If the window size (w) is large compared to the tolerance (δ) of a period, then we are actually finding a superset of the p-patterns. To understand this, we consider an example. Let $V = \{a, b, c\}$ be a temporal association. Suppose that we have the following occurrences:

$$\{(a, 1), (b, 3), (c, 5)\}$$

$$\{(b, 31), (c, 33), (a, 35)\}$$

$$\{(c, 61), (a, 63), (b, 65)\}$$

$$\{(b, 91), (a, 93), (c, 95)\}$$

Here, the inter-arrival times for a are $\{34, 28, 30\}$; for b, $\{28, 34, 28\}$; and for c, $\{28, 28, 34\}$. Let $\delta = 1$. The period-first algorithm finds that a is not partially periodic and in turn $\{a, b, c\}$ is not a p-pattern, while the aforementioned approach will claim it as a p-pattern because the occurrence time of a pattern is its first event.

4.2.3 Thoughts on a Hybrid Algorithm

Clearly, there is a trade-off between the period-first and association-first algorithms. The former provides efficiency while the latter provides robustness to noise.

Is it possible to trade robustness to noise for computational efficiency? One approach is to control the maximum size of the temporal patterns considered before checking for periodicities. Specifically, we can view period-first as looking for periodicities in temporal associations of size 1. Conversely, association-first looks for periodicities in temporal associations that are as large as possible. A hybrid algorithm could take as input a parameter

m that specifies the maximum size of a temporal association prior to checking for periodicities. Since the temporal association step largely determines computational complexity, we gain some control here. Also, larger temporal associations provide more noise robustness.

This algorithm is summarized as follows.

- Run level-wise algorithm to find all temporal patterns up to the m-th level
- Scan data to find an instance sequence for each m level pattern, and find possible periods
- Use period information to group m-level patterns
- For each pattern group run level-wise search

5 Experiments

This section assesses the algorithms for discovering p-patterns. These assessments consider both effectiveness and efficiency.

5.1 Synthetic Data

We compares the period-first and the association-first algorithms using synthetic data, in which true p-patterns are known in advance. Our focus here is to evaluate their performance in the presence of noise.

Each synthetic data consists of a mixture of random events and p-patterns, and is constructed by varying NSR, the ratio of the duration of on-segments to that of off-segments of a p-pattern, and the length of a p-pattern (i.e. the number of the event types of a p-pattern). We fix the following: ratio of off-segments to on-segments is 0.5; the number of p-patterns is 10; p-pattern length is 5. Further, the duration of an off-segment is exponentially distributed. This distribution is determined by the rate necessary to generate the number of p-patterns to achieve the NSR. Noise events are randomly and uniformly generated. The number of noise events is determined by NSR and the total number of events. Thus, a run is specified by the number of events and NSR.

We begin by assessing the scalability of the period-first and the association-first algorithms. NSR ratio is controlled between $0.5 \sim 1$. We vary the total

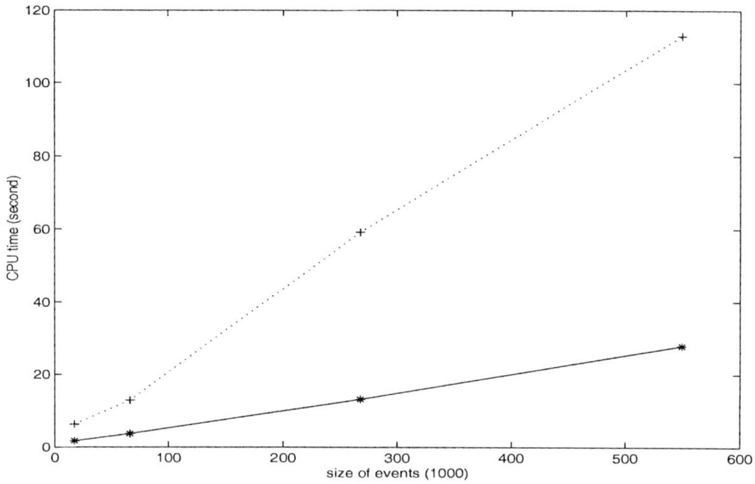

Figure 5: Average run time vs. the number of events

number of events. A run is effective if there is no false negative, i.e. all p-patterns are found. the minimum support is set[4] so that there is no false positive above level 2. Our experiment consists of 5 runs done with the same value of NSR and different random event streams. Figure 5 plots the average CPU time of the runs in each experiment against the total number of events (in thousands). We can see that the period-first algorithm is $3 \sim 5$ times faster than the association-first algorithm. This is not surprising since the period-first algorithm uses the information about the periods found in its first step to effectively constrain the search space of the level-wise algorithm.

Now we fix the number of events to be 100,000 and vary NSR. This provides an opportunity to see effectiveness of the algorithms. The successful rate is computed at the event type level. As shown in Table 2, both algorithms find all p-patterns at low NSRs. The period-first is considerably more efficient. As NSR increases, the effectiveness of the period-first degrades. This is a result of the period detection algorithm. As shown in Figure 3, for an $NSR > 1$ with first order inter-arrivals, the performance of this algorithm drops considerably. Even more extreme, at NSR in the range of 5 to 10, the period-first algorithm fails completely. Note that the CPU time drops for the period-first as NSR increases. This is because the period-first algorithm find less periods to conduct further searching. In contrast, association-first still finds 100% of the p-patterns.

[4]Our results are not sensitive to *minsup*. A typical *minsup* ranges from $0.1\% \sim 1\%$ of the total number of events.

One final comment is of interest here. This relates to the nature of the low effectiveness of the period-first algorithm if $NSR >> 1$. Under such circumstances, the problem is false negatives as a result of the period detection algorithm. That is, we fail to identify p-patterns that are present rather than falsely identify p-patterns that are not present.

	Effectiveness		Run Time(second)	
NSR	Period-first	Association-first	Period-first	Association-first
$0.5 \sim 1$	100%	100%	4.3	14.7
$1 \sim 2$	40%	100%	1.5	14.5
$5 \sim 10$	0%	100%	0.9	14.9

Table 1: Experimental results

5.2 Production Data

Now, we apply our algorithms to mine p-patterns in real data. Here, Our evaluation criteria are more subjective than the last section in that we must rely on the operations staff to detect whether we have false positives or false negatives.

Two data sets are considered. The first data set was collected from an Intranet containing hundreds of network elements (e.g., routers, hubs, and servers). The second data set was collected from an outsourcing center that supports multiple application servers across a large geographical region. Events in the second data set are mostly application-oriented (e.g. the cpu utilization of a server is above threshold), whereas those in the first data set are both network-oriented events (e.g. a link is down for a router) and application-oriented. An event in both data sets consists of three key attributes: host name, which is the source of the event; alarm type, which specifies what happened (e.g., a connection was lost, port up); and the time stamp of when the event occurred. In our preprocessing, we map each distinct pair of host and alarm type into a unique event type. The first data set contains over 10,000 events with around 400 event types over a three-day period. The second data set contains over 100,000 events with around 3000 event types over a two-week period.

Table 3 and Table 4 report our results of the two data sets by pattern size. Specifically, column 1 indexes the search level of the level-wise algorithm. Column 2 shows the size of C_k (i.e. the number of qualified patterns at level k) at each level. Column 3 indicates the number of large p-patterns. Here, large p-patterns refer to p-patterns that are not a subset of other p-patterns. Three more columns are used to indicate the diversity of p-patterns found: column 4 shows the number of different period lengths at each level; column 5 is the range (minimum to maximum) of the periods; and column 6 is the range (minimum to maximum) of the occurrence counts for the p-patterns at that level.

As the two tables show, many p-patterns are found. Even more striking, we found that over 50% events are partially periodic, and belong to an instance of at least one patterns. This indicates that our two data sets are highly periodic. Why are periodic behaviors so common here? Intuitively, two factors mainly contribute to this. The first relates to monitoring–when a managed element emits a high severity event, the management server often initiates periodic monitoring of key resources (e.g., router CPU utilization). The second consideration is a consequence of routine tasks, such as rebooting print servers every morning or backing up data every week.

One consequence of the above is that $NSR < 1$ for our data. This suggests that the period-first algorithm should be very effective in analyzing our data.

Further insights can be drawn from our results. First, not surprisingly, there are a large number of p-patterns consisting of a single or just two event types. However, there are also some fairly large patterns. Second, the range of periods is quite large–from 0 seconds to 1 day. So as the range of the number of occurrences. This suggests that FFT's sensitivity to the range of time scales would have very poor computational complexity. A further implication here is the importance of having algorithms that employ different minimum supports. For example, the daily patterns have a support no more than 3 in the three-day data, while no more than 14 in two-week data. Thus, daily patterns can only be discovered by very small minimum support.

We reviewed these patterns with the operations staff. It turned out that many of the p-patterns related to underlying problems. The p-patterns with period 1-day and lengths 10 and 11 were found to relate to a periodic port-scan, a possible indicator of a security intrusion. The pattern with period 5-minutes and length 1 resulted from an incorrect configuration of a hub. The pattern with period 60 seconds and 1 was caused by a router that did not get the most current routing information.

Level	Candi-date size	p-patterns	Different Periods	Min:Max Periods	Min:Max count
1	100	28	11	0:1-day	6 : 680
2	307	22	6	0:300	3:689
3	938	5	3	0:30	3:8
4	1917	1	1	4	3
5	3010	5	1	4	3
6	3525	3	1	4	3
7	3104	0	0		
8	2057	2	2	4:1-day	3
9	1017	0	0		
10	366	1	1	1 day	5
11	91	2	1	1 day	5
12	14	1	1	20	20
13	1	1	1	10	21

Table 2: Experimental results of the first data set

6 Conclusion

This paper addresses the discovery of partially periodic temporal associations (p-patterns), a pattern that is common in many applications. An example of a p-pattern in a computer networks is five repetitions every 30 seconds of a port-down event followed by a port-up event, which in turn is followed by a random gap until the next five repetitions of these events. Mining such patterns can provide great value. Unfortunately, existing work does not address key characteristics of these patterns, especially the presence of

Level	Candi-date size	p-patterns	Different Periods	Min:Max Periods	Min:Max count
1	1500	354	?	0:1-day	?
2	804	340	?	0:1-day	10 : 1890
3	781	103	?	0:2700	12 : 131
4	378	63	?	$4 : 1 - day$	11 : 160
5	148	24	?	$30\ 1 - day$	11 : 49
6	39	16	?	300	21 : 51
7	4	4	?	300	18 : 54

Table 3: Experimental results of the second data set

noise, phase shifts, the fact that periods may not be known in advance, and the need to have computationally efficient schemes for finding large patterns with low support.

We begin by defining partially periodic patterns (p-patterns) in a way that includes on-off segments, and phase shifts (via the tolerance parameter δ). Next, we construct an efficient algorithm for finding the period of a partially periodic pattern using a chi-squared test, and we study the performance of the proposed algorithm in the presence of noise. Further, we develop two algorithms for discovering p-patterns based on whether periods or associations are discovered first, and we study trade-offs between these approaches. One result is that the association-first algorithm has higher tolerance to noise while the period-first algorithm is more computationally efficient. In particular, using synthetic data we find that the period-first algorithm is three to five times faster than the association-first algorithm. On the other hand, the effectiveness (i.e., robustness to false negatives and false positives) of the period-first algorithm degrades rapidly if the noise-to-signal ratio exceeds 1. Using these insights, we apply the period-first algorithm to two types of event logs. Many p-patterns are discovered in the log, some of which led to diagnostic and corrective actions.

One area of future work is to explore hybrid algorithms that provide a way to control the trade-off between computational efficiency and effectiveness. Our current work suggests a couple of possibilities in this regard. One is to use the association-first approach but to limit the size of the temporal associations discovered before switching to the period first algorithm. Another strategy is to focus on the algorithm for finding unknown periods. We observe that by considering n-order inter-arrivals (the time between events separated by n other events), we can increase robustness to noise. Thus, we could employ the period-first algorithm but control the parameter n.

References

[1] R. Agrawal, T. Imielinski, and A. Swami. Mining association rules between sets of items in large databases. In *Proc. of VLDB*, pages 207–216, 1993.

[2] R. Agrawal and R. Srikant. Fast algorithms for mining association rules. In *Proc. of VLDB*, 1994.

[3] R. Agrawal and R. Srikant. Mining sequential patterns. In *Proc. of the 11th Int'l Conference on Data Engineering, Taipei, Taiwan*, 1995.

[4] C. Bettini, X. Wang, and S. Jajodia. Mining temporal relationships with multiple granularities in time sequences. *Data Engineering Bulletin*, 21:32–38, 1998.

[5] R. COOLEY, J. SRIVASTAVA, and B. MOBASHER. Web mining: Information and pattern discovery on the world wide web. In *Proceedings of the 9th IEEE International Conference on Tools with Artificial Intelligence (ICTAI'97)*, 1997.

[6] J. Han, G. Dong, and Y. Yin. Efficient mining of partially periodic patterns in time series database. In *Int. Conf. Data Engineering*, 1999.

[7] J. Han, W. Gong, and Y. Yin. Mining segment-wise periodic patterns in time-related database. In *Int'l Conf. on Knowledge Discovery and Data Mining*, 1998.

[8] H.O. Lancaster. *The Chi-squared distibution.* John Wiley & Sons, New York, 1969.

[9] H. Mannila, H. Toivonen, and A. Verkamo. Discovery of frequent episodes in event sequences. *Data Mining and Knowledge Discovery*, 1(3), 1997.

[10] T. Oates, M. Schmill, D. Jensen, and P. Cohen. A family of algorithms for finding temporal structure in data. In *6th Intl. Workshop on AI and Statistics*, 1997.

[11] B. Ozden, S. Ramaswamy, and A. Silberschatz. Cyclic association rules. In *Int. Conf. Data Engineering*, pages 412–421, 1998.

[12] S.M. Roos. *Introduction to probability and statistics for engineers and scientists.* John Wiley & Sons, New York, 1987.

[13] R. Srikant and R. Agrawal. Mining sequential patterns: Generalizations and performance improvements. In *Proc. of the Fifth Int'l Conference on Extending Database Technology (EDBT). Avignon, France.*, 1996.

[14] O. Zaane, M. Xin, and J. Han. Discovering web access patterns and trends by applying olap and data mining technology on web logs. In *Proc. Advances in Digital Libraries ADL'98*, pages 19–29, 1998.

[15] M. Zaki. Fast mining of sequential patterns in very large databases, 1997. Technical Report URCS TR 668, University of Rochester.

[16] T. Zhang, R. Ramakrishnan, and M. Livny. Birch: A new data clustering algorithm and its applications. *Data Mining and Knowledge Discovery*, pages 141–182, 1997.

Combination of Classifiers for Supervised Learning: A Survey

Sheng Ma
IBM T.J. Watson Research Center
Hawthorne, NY 10532
E-mail: `shengma@us.ibm.com`

Chuanyi Ji
Dept. of ECSE
Rensselaer Polytechnic Institute
Troy, NY 12180
E-mail: `chuanyi@ecse.rpi.edu`

Contents

D. Chen and X. Cheng (eds.),
Pattern Recognition and String Matching, 379–410.

1 Introduction

In many important application areas such as signal processing, pattern recognition, control and communication, nonlinear adaptive systems are needed to approximate underlying mappings through learning from examples. In order for approximations to be sufficiently accurate, a good performance is required for nonlinear adaptive systems. Meanwhile, many applications, especially those in emerging areas of wireless communication and networking [11][20][42][52][50], require the learning to be done in real-time in order to adapt to a rapidly changing environment. Other applications such as data mining and searching the web need to deal with very large data sets [19][36][55], and thus the learning time must scale nicely with respect to the size of data . Since the size of learning machines determines the memory required for implementation, a learning machine with a compact structure is also preferred. Therefore, a challenging problem is how to develop a learning system with a compact structure that can achieve a good performance, and be adapted in real time rapidly. The goal of this paper is to address the important issues of performance and real-time learning for nonlinear adaptive learning machines by first reviewing recent work in combinations of classifiers, and then detailing some of our work in this area.

In supervised learning, performance addresses the problem of how to develop a learning machine to achieve optimal performance on examples that are not included in a training set. Efficiency deals with the complexity of a learning machine in both space and training time. Specifically, the space-complexity of a classifier refers to its size, and the time-complexity characterizes the computational time needed to develop such a classifier. These three issues are interrelated.

The performance of a supervised learning system is characterized by its generalization error, which measures the distance between the output function of a trained model and an underlying target function. Most of the existing training methods for classifiers in supervised learning suffer from an intrinsic problem in pattern recognition: the bias and variance dilemma [25][21]. That is, if a classifier is too large, [1] it may overfit a particular training set and thereby fail to maintain small generalization error. A small classifier, however, may be insufficient to approximate an optimal solution. In addition, one algorithmic problem is how to find a good classifier, which can learn a complex optimization problem with possibly many local minima.

The size of a learning machine can be characterized by space-complexity, which is related to the number of free parameters (for instance, the number of weights of a neural-network classifier). A learning machine is considered to be efficient in space if its space-complexity scales as a polynomial function in terms of the dimension of feature vectors.[2] It has been found that compact classifiers like multiplayer feedforward neural networks are efficient approximators of a wide class of smooth functions. They possess a polynomial space-complexity [5]. Other learning machines, which consist of localized models such as nearest neighbor classifiers [13], Parzen windows

[1]For instance, feedforward neural networks as classifiers with too many neurons.

[2]Here the dimension of feature vectors is used as a measure of the complexity of a problem.

[18], and linear combinations of localized basis functions [43][5], may suffer from the so-called "curse of dimensionality." That is, their space-complexity scales exponentially with the dimension of the feature vectors [5][45][35].

Learning time can be characterized by the time-complexity as a scaling property with respect to the dimension of feature vectors. An adaptive learning system is considered efficient in time if the time complexity is polynomial. For example, training feedforward neural-network classifiers is slow, and in general, NP-complete [7][31]; while training the aforementioned localized classifiers can be done quickly. For example, nearest-neighbor classifiers simply remember all training samples, and therefore do not require training at all. This presents a dilemma existing between performance, space-, and time-complexity. These three fundamental issues – performance, efficiency, and space complexity – motivate the search for new solutions for adaptive learning machines, from training an individual model, to recent activities on combining well-trained models, and of many weak models.

Numerous methods have been developed in training an individual classifier. However, finding an optimal sized classifier is still difficult if not impossible. To address the performance issue, the combination of well-trained models has been proposed. The basic idea is to pick a slightly oversized model as a base for combination. Since the base model is oversized, it has a low bias but a high variance. The algorithm, therefore, relies on combinations to reduce the overall variance, and thus the generalization error.

Although the combination of well-trained models relaxes the need to optimize the size of a classifier, the learning time increases with the number of models combined. The combination of weak models is proposed to improve the efficiency of the combination of well-trained models while preserving the nice performance property. In particular, this approach uses the similar combination scheme but chooses a weak classifier as a base model. A weak classifier has a performance only slightly better than that due to random guessing, and thus has a large bias but a small variance. From the performance perspective, since different weak models are forced to learn different parts of a problem, combinations of many such weak models can reduce the overall bias and thus achieve good generalization. From the time-efficiency perspective, since the base model performs only slightly better than random guessing, it can be obtained efficiently. Furthermore, this approach uses an incremental learning procedure, and the whole training process can be very efficient.

The goal of this paper is to provide an overview of different approaches in combining weak classifiers. We would like to identify a general framework for various approaches with different grounds. Further, we would like to provide an overview of both theoretical and empirical results that demonstrate the advantages of combining weak classifiers in terms of both performance and efficiency.

The paper is organized as follows. Section 2 gives the background knowledge on performance and efficiency. Section 3 reviews the research on combining well-trained models to improve the performance. Section 4 discusses combinations of weak classifiers and how combined weak classifiers trade off among performance, time- and space-complexity. Section 5 discusses the empirical results of combining week classifiers. Section 6 discusses the theoretical results. Section 7 concludes the paper. As there is a large volume of the related work, we apologize for possible omissions.

2 Background: Performance and Efficiency

2.1 Notation

In a supervised learning environment, let $D = \{\mathbf{x}_n, t_n\}_{n=1}^N$ denote N training samples pairs, where $\mathbf{x}_n \in R^d$ is a d-dimensional feature vector of the n-th sample and t_n is the corresponding target. For simplicity, t_n is assumed to be a scalar in this paper. Furthermore, we assume that there is a function $f(\cdot)$ so that $t_n = f(\mathbf{x}_n)$.[3] For the function approximation problem, t_n is a real number. For classification, t_n indicates which class the n-th sample belongs to.

2.2 Performance

Let $f_D(\mathbf{x}; \mathbf{w})$ be a model with a set of parameters $\mathbf{w}$ and trained on training set D. The performance of $f_D(\mathbf{x}; \mathbf{w})$ can be measured in terms of the difference between a function $f(\mathbf{x})$ to be approximated and its approximation $f_D(\mathbf{x}; \mathbf{w})$ through the squared norm

$$\int \mid f(\mathbf{x}) - \mathbf{f_D}(\mathbf{x}; \mathbf{w}) \mid^2 \mathbf{p}(\mathbf{x})\mathbf{dx}, \tag{1}$$

where $p(\mathbf{x})$ is the probability density function of $\mathbf{x}$. If training samples are drawn randomly, the expected squared norm is

$$\mathbf{E}_D(\| f(\mathbf{x}) - \mathbf{f_D}(\mathbf{x}; \mathbf{w}) \|^2) = \mathbf{E_D}(\int \mid \mathbf{f}(\mathbf{x}) - \mathbf{f_D}(\mathbf{x}; \mathbf{w}) \mid^2 \mathbf{p}(\mathbf{x})\mathbf{dx}), \tag{2}$$

[3]For simplicity, this paper does not address possible noise terms.

where the expectation ($\mathbf{E}$) is done over all possible training sets D of the same size. The quantity, $\mathbf{E}_D(\| f(\mathbf{x}) - \mathbf{f_D}(\mathbf{x};\mathbf{w}) \|^2)$ can be considered as a measure of the performance. It is also called the generalization error of $f_D(\mathbf{x};\mathbf{w})$, since it measures the average performance of $f_D(\mathbf{x};\mathbf{w})$ in approximating the unknown function $f(\mathbf{x})$.

Pattern classification is considered as a special case of nonlinear regression and its generalization error can be defined as probability of incorrect classification, which is $\mathbf{E_D}(\Pr(\mathbf{f_D}(\mathbf{x};\mathbf{w}) \neq \mathbf{t}))$.

2.3 Performance Issues

Two important issues are related to the performance: the bias and variance dilemma, and the relation of which to possibly a complex objective function with many local minima.

The first issue is intrinsic, and is independent of algorithms used. The second issue is algorithm- and problem-dependent.

2.3.1 Bias and Variance Dilemma

The generalization error is affected by two factors: Bias and variance. Let $f(\mathbf{x};\hat{\mathbf{w}})$ be the best model in the model space. That is, $\hat{\mathbf{w}} = \arg\min_{\mathbf{w}} \int_{\mathbf{x}}(f(\mathbf{x}) - f(\mathbf{x};\mathbf{w}))^2\mathbf{p}(\mathbf{x})\mathbf{dx}$. Note that $\hat{\mathbf{w}}$ does not depend on the training data. The bias $Bias(\mathbf{w})$ and variance $Var(\mathbf{w})$ for a model can be defined as:

$$Bias(\mathbf{w}) = \mathbf{E}(\mathbf{f}(\mathbf{x}) - \mathbf{f}(\mathbf{x};\hat{\mathbf{w}}))^2, \tag{3}$$

where the expectation is on a randomly drawn $\mathbf{x}$, and

$$Var(\mathbf{w}) = \mathbf{E_D}(\mathbf{E}(\mathbf{f_D}(\mathbf{x};\mathbf{w}) - \mathbf{f}(\mathbf{x};\hat{\mathbf{w}}))^2). \tag{4}$$

The inner expectation is on $\mathbf{x}$, and the outer expectation is on randomly drawn training sets of the same size. Therefore, the generalization error can be decomposed into, and bounded by, a sum of the bias and variance [5][21][10]

$$\mathbf{E}_D(\| f(\mathbf{x}) - \mathbf{f_D}(\mathbf{x};\mathbf{w}) \|^2) \leq \mathbf{2}(\mathbf{Bias}(\mathbf{w}) + \mathbf{Var}(\mathbf{w})). \tag{5}$$

The first term of the right-hand side is the error due to the bias because of an inappropriate choice of the size of a class of models, when the number of training samples is assumed to be infinite. The latter is the error due to the variance because of the finite number of training samples[4].

[4]which is equivalent to the space-complexity versus the number of training samples.

For the special case when a class of models are chosen to be two-layer feed-forward neural-network classifiers with L sigmoid hidden units,[5] the bias and variance were bounded explicitly by Barron [5]. In particular, $O(\frac{1}{L})$ and $O(\frac{LdlogN}{N})$ are upper bounds for the bias and variance respectively, where $O(z)$ represents a quantity in the order of z. The fact that $O(\frac{1}{L})$ decreases with the number of hidden units suggests that a large neural network with many free parameters is less biased. As it has been shown by Barron [5] that feedforward neural networks are capable of approximating a wide class of smooth functions when the number of hidden units is sufficiently large, a larger set of neural networks (with a larger L) certainly contains a better approximation of $f(\mathbf{x})$ than a smaller set of neural networks (with L being smaller). That is why the bias is smaller when L is larger. However, when the number of training samples is finite, a network with an excessively large space-complexity will overfit the training set. This can be understood from the concept of information capacity of neural networks [1][14][6][28][38] in information theory. The information capacity is the maximum number of training samples that can be memorized by a neural network given a certain space-complexity. When the space-complexity exceeds the information capacity, learning machines are so large that they only memorize training samples. Since there are many such neural networks with the same space-complexity but different choices of weights which can memorize training samples, their (generalization) performance varies and thus leads to a large variance. That is, the average performance can decrease as L gets larger. This can be observed from the term $O(\frac{LdlogN}{N})$, which increases with respect to L. Therefore, a trade-off needs to be made between bias and variance.

2.3.2 Bias-variance and Local Minima

An issue which needs to be clarified is whether the bias and the variance are related to local minima, or the effectiveness of an algorithm at finding a good solution. In our view, the bias by definition only depends on the structure of a class of classifiers but not the choice of values of their parameters (such as weights of neural-net classifiers), and is therefore independent of any training algorithms. To understand how the variance relates to local minima, we may consider the sample variance. For any given (randomly drawn) training set D of size N, a classifier corresponding to $f_D(\mathbf{x}; \mathbf{w})$ can be found, which contributes to one sample in the sample variance. The true variance can be

[5] The number of free parameters is thus approximately $L(d+1)$.

estimated through a sufficiently large number of such samples. The definition given by Barron [5] assumed that each $f_D(\mathbf{x};\mathbf{w})$ corresponded to the global minimum of the error function, e.g. the best neural-network (classifier) that can be obtained given a training set. If the variance depended on a specific algorithm or applications, the theoretical result may lose its generality. This definition (given by Barron) simply suggests that the bias and variance dilemma is intrinsic for nonparametric regression (even if one had the most powerful algorithm to obtain the globally optimal classifier). Therefore, such a definition on the variance can also be considered to be algorithm independent.

In practice, for a given training set and a chosen error function, different algorithms result in different classifiers corresponding to different $f_D(\mathbf{x};\mathbf{w})$'s. If an algorithm is always better at finding a globally optimal solution, and another algorithm (e.g. a gradient-descent algorithm) gets stuck at a local minimum more often, then the former would have a smaller sample variance than the latter.

3 Combinations of Well-Trained Models

To alleviate the difficulties in finding an optimal structure of a single nonlinear classifier, methods are proposed to combine different models. The main idea is to train multiple models, such as neural networks, decision trees, or other classifiers individually, and then combine them as an ultimate model. The hope for using a combination to improve the generalization performance is that if individual models make classification errors differently, a combined model would be able to improve upon the performance of individuals.

Tremendous efforts have been made to investigate whether a combined model can indeed improve the performance, and how to combine models to achieve good performance [49][2]. Specifically, combinations of experts have shown to be able to work at least as well as the best expert in the pool of experts on predicting binary strings [12][37]. Combinations of neural networks [44][57] and regressors [9] have also been used for both classification and function approximation problems. Somewhat similar ideas were also used for collective agents like the classifier systems and genetic algorithms [22][26][27].

As a good review was offered by Dietterich [16] on this subject; in this section, we do not intend to survey all aspects of combinations of well-trained models. Rather, we focus on the general issues that are relevant to the next subject we discuss on combinations of weak classifiers.

3.1 Key Issues

A combined model can be represented in a simple form. Let $h_k(\mathbf{x})$'s be a set of total K models defined on a feature vector $\mathbf{x} \in \mathbf{R^d}$, where a $h_k(\mathbf{x})$ can be a neural-network classifier (or a single neuron). Let w_k be the weighting factor for the k-th model. The combination of these K models can be represented as

$$f_K(\mathbf{x}) = \sum_{\mathbf{k=1}}^{\mathbf{K}} \mathbf{w_k h_k}(\mathbf{x}). \tag{6}$$

$f_K(\mathbf{x})$ is essentially a linear combination of the so-called base models $h_k(\mathbf{x})$'s.

When a (two-class) classification is considered, the combined model becomes a combined classifier $C(\mathbf{x})$, where

$$C(\mathbf{x}) = \mathbf{I}(\mathbf{f_K}(\mathbf{x})). \tag{7}$$

$I(z)$ is an indicator function, where $I(z) = 1$ for $z > 0$, and $I(z) = 0$ otherwise. $C(\mathbf{x})$ is a label assigned to a feature vector $\mathbf{x}$ by a combined classifier. When $h_k(\mathbf{x})$'s are also classifiers, $C(\mathbf{x})$ is a combination of classifiers.

There are two key issues in combinations of models.

(1) How to obtain a set of base models, $\{h_k(\mathbf{x})\}_{\mathbf{k=1}}^{\mathbf{K}}$?

(2) Given a set of base models, how to choose an optimal set of weighting factors $\{w_k\}_{k=1}^{K}$ so that the generalization error of the combined model is minimized?

3.2 Obtaining Base Models

Numerous algorithms have been developed to explore the first question. These approaches are distinct in two ways. One type of approaches utilize base models with different architectures, and/or obtained through different algorithms. Another type of approaches use different ways to "perturb" the training process so that the base models obtained can have diverse error patterns. Having classifiers with diverse error patterns were shown to be crucial to effectively improving the performance through combinations of models [34][32]. This can be understood through two (extreme) examples. At one extreme, if these classifiers are identical, there will be no gain from any combination. At the other extreme, if these classifiers make independent errors with a probability less than 0.5, the overall number of errors made after a combination can decrease exponentially as the number of such classifiers increases [32]. However, the "diversity" is not easy to achieve because all models are trained to do essentially similar tasks, and thus more or less dependent. Therefore, a certain randomness should be introduced to "perturb" the learning procedure so that models can learn different parts of a problem, and thereby make errors as differently as possible. The common methods used to perturb training were summarized by Dietterich [17][16].

3.3 Combination

The other important issue for combination is how to determine the weighting factors. Intuitively, the outputs of base models $\{h_k(\mathbf{x})\}_{\mathbf{k=1}}^{\mathbf{K}}$ can be used to "train" the weighting factors $\{w_k\}_{k=1}^{K}$. However, the straightforward minimization approach often results in overfitting [41][39] and is therefore not applicable. This is because the base classifiers are highly correlated since they are designed to solve similar tasks. One simplest algorithm called majority vote uses the equal weighting for base models, i.e., $w_k = 1/K$ for all k. Clearly, this scheme is not optimal because it does not take full consideration of differences among base models. But majority vote has been used widely due to its simplicity [30][8][10]. More elaborate algorithms have been investigated for combination. Wolpert [57] proposed combining base models through minimizing the squared error using cross-validation. Breiman [9] later imposed proper constraints on the weights w_k's. Freund et al. [24][23] proposed using the confidence of a base model as the weight for that model. Principle component analysis [40] was also investigated to explore and thus discount correlation among individual base models.

3.4 Discussion

Both theoretical and empirical results [8][10] suggest that combinations of well-trained models can ease the bias-variance dilemma and improve the performance substantially. That is, we can select a base model with a relatively large size so that it has a small bias but a large variance. A combination scheme can be responsible for reducing the overall variance of a combined model [8][10][44][40]. Therefore, finding an optimal structure of a model is no longer important. Very often, however, the price paid for the gain in performance is a larger space-complexity. In addition, as several models are combined, each of which may take a long-time to train, an even longer training time results in for a combination. Since training time is critical for real-time applications and is more difficult to tackle than the space-complexity, a natural question to ask is whether combinations of models can be used to improve the time-complexity at a reasonable cost of the space-complexity. Combining *weak* models provides a promising answer to this question.

4 Performance and Efficiency: Combinations of Weak Classifiers

Although individual base models in a combination can be quite general, we focus on classifiers as base models in this section.

4.1 Three Approaches

There are three main approaches developed on combinations of weak classifiers: the boosting algorithm called Adaboost by Freund and Shapire [23][24], the stochastic discrimination (SD) by Kleinberg [32][33], and the combination of weak perceptrons (CWP) by Ji and Ma [15][29][30].

The concept of weak learning was first introduced by Kearns and Valiant as a theoretical question in the context of the probabilistic approximately correct (PAC) learning theory [6][54]. The question can be informally stated as " Does the existence of a weak learner imply the existence of an efficient strong learner?" Schapire [51] first proved the existence with a "yes" through a recursive and constructive approach. He showed that if a classification problem is solvable in the PAC framework, the problem can be solved through combinations of weak classifiers that can do a little better than random guessing. Later, Freund [23] showed in theory that a combination of weak classifiers through the simple majority vote can combine the weak classifiers into a strong classifier. The Adaboost [24] was further proposed as a practical algorithm for combination. Although this work [23][24] illustrated that weak classifiers could be combined to achieve what a strong classifier could do, it did not address whether combinations of weak classifiers had any advantage in performance or efficiency compared with either training an individual (strong) classifier or combining well-trained individual strong classifiers.

In an independent work, Kleinberg [32] proposed stochastic discrimination and showed that combining a large number of weak classifiers could improve the (training) performance monotonically. In addition, he also showed that the time-complexity of combined weak classifiers was polynomial. The theory, however, was built on an assumption that weak classifiers made independent classification errors. Such an assumption cannot be achieved in real situations.

Ji and Ma [30][29][15] proposed combinations of weak perceptrons (or neurons). Through experiments, this work showed that a very simple algorithm for generating and combining weak classifiers may achieve better efficiency and performance than training a strong classifier or combining well-trained classifiers. Such a simple procedure is to randomly generate weak perceptrons as weak classifiers and combine these weak perceptrons through a majority vote. Time- and space-complexity of a combined classifier was formally defined, and shown to be polynomial for a "special case" when the strength of weak classifiers was properly chosen. A trade-off was explicitly shown between time-complexity and space-complexity.

In the following, we further review these three algorithms and the corresponding results.

4.2 Weak Classifiers

4.2.1 Definition

The strength of a classifier $C(\mathbf{x})$ can be characterized by ν, the so-called weakness factor, where $\nu > 2$. Let $\frac{1}{2} - \frac{1}{\nu}$ be the required (generalization) error of classifier $C(\mathbf{x})$, i.e., $\Pr(C(\mathbf{x}) \neq \mathbf{t}) = \frac{1}{2} - \frac{1}{\nu}$. If $\nu >> 2$, the classifier is considered to be a weak classifier because it only performs a little better than random guessing. The larger the ν is, the weaker the weak classifier. The time- and space-complexity of $C(\mathbf{x})$ follows the general definition given in Section 2, but should take into account the weakness factor.

If the space- and time-complexity of a combined classifier scales polynomially with respect to both the dimension of feature vectors and parameters such as the weakness factor and a desired generalization error, the classifier is said to be efficient. Otherwise, if an exponential scaling is observed, they are considered to be inefficient.

A set of weak classifiers should satisfy the following two conditions: (1) each weak classifier should do better than random guessing, and (2) the set of classifiers should have enough computational power to learn a problem. The first condition ensures that each weak classifier possesses a minimum computational power. The second condition suggests that individual weak classifiers should learn different parts of a problem so that a collection of weak classifiers can learn an entire problem. If all the weak classifiers in a collection were to learn the same part of a problem, their combination would not do better than individual classifiers.

4.2.2 The Structure of Weak Classifiers

Both local and non-local classifiers have been proposed. Kleinberg et al. [33] proposed using a hyper-sphere as a weak classifier. Such a weak classifier dichotomizes all the samples that fall into the sphere as one class and those outside as the other class. This classifier is local because samples close to the center of a sphere are in the same class. This choice of weak classifiers, however, may suffer from the "curse of dimensionality" and thereby is difficult to handle high dimensional problems.

Kleinberg [33] and Freund et al. [24] independently proposed to use a half-space as a weak classifier. A half-space classifier classifies input features based on only one selected dimension (or feature) and ignores other dimensions. Therefore, the decision boundary of a half-space classifier is a hyperplane perpendicular to the selected dimension. Such a classifier is global and may be expected to have a nice (polynomial) scaling property with respect to the dimension of feature vectors, avoiding the curse of dimensionality for some cases. As shown by experimental results [24], the set of half-space weak classifiers is limited for representing an arbitrary decision boundary. As a result, the performance is not as good as that of combinations of stronger classifiers. To increase the representation power of the half-space classifiers, the union of two or more half spaces were also considered.

To generate a global classifier with a stronger representation power, Ji and Ma proposed using a perceptron (or a neuron) as a weak classifier [15][30]. Similar to a half-space classifier, the decision boundary of this classifier is also a hyperplane, and therefore global. However, the orientation of the hyperplane can be arbitrary. Therefore, a hyperplane classifier is much more flexible than a half-space classifier. In fact, the combination of these perceptrons (or neurons) forms a two-layer neural network that has been shown to be able to approximate arbitrary functions [5].

4.3 Algorithms for Combinations of Weak Classifiers

Once the structure of weak classifiers is determined, qualified weak classifiers can be generated and then combined.

4.3.1 Generation of a Qualified Weak Classifier

There are two fundamentally different approaches to generate a qualified weak classifier. One is to apply a training algorithm to find the "best" weak classifier. The other is to randomly generate a weak classifier from the set of all weak classifiers until a qualified classifier is obtained.

Freund et al. [24] used the first scheme, where a feature is first selected randomly and the best threshold is then obtained through exhaustive search. Both stochastic discrimination (SD) [33] and combinations of weak classifiers [30] generated a weak classifier through the second approach. The algorithm can be described as follows.

(1) Partition the training data into "cares" and "don't-cares."

(2) Randomly generate a candidate classifier from the classifier space,

(3) Test classification error rate of the candidate classifier on the "cares."

(4) If the error rate on the "cared" samples is less than a threshold $0.5 - 1/\nu$, then keep this classifier and go to step (1) to generate another weak classifier. Otherwise go to step (2), where ν is the weakness factor.[6]

In this algorithm, a random classifier is first chosen as a candidate classifier. The strength of this candidate classifier is then tested on a set of "cares," which refer to those training samples incorrectly classified. If the candidate passes the threshold, it is accepted and combined with previous qualified classifier. Otherwise, it is rejected and the procedure is repeated until a qualified candidate is found. Therefore, this approach of generating a qualified classifier can also be called trial-until-qualified (TUQ).

To generate a perceptron classifier randomly [30], the direction of a hyperplane is first generated randomly and uniformly. Then a feature vector is picked randomly to place the hyperplane in the feature space.

Three considerations motivate the choice of a randomized (TUQ) approach over the training

approach. First, if a qualified classifier is weak enough so that it can be obtained through only several trials, the TUQ may be computational cheaper than training a classifier. Second, the random sampling to obtain a weak classifier may reduce overfitting, which is the problem inherent associated with any training algorithm. Third, random sampling may facilitate theoretical analysis as we are to discuss later.

4.3.2 Combinations of Qualified Weak Classifiers

In general, the combination schemes developed to combine well-trained

models can be used to combine weak classifiers. Specifically, Freund [24] used the weighted majority vote and chose the confidence of the corresponding classifier as the weighting factors. Both combinations of weak perceptrons [30] and stochastic discrimination [33] used the the simple majority vote to combine weak classifiers.

[6] A typical value of ν used in [30] is between 50 and 200. A typical number of combined weak classifiers is from 500 to 5,000.

4.3.3 Adaptively Reweighing and Combining (ARC)

Several algorithms for combining either weak or strong models have been discussed. They are bagging [8], AdaBoost [23], stochastic discrimination [32][33], and a combination of weak hyperplanes [30][15]. Although these algorithms result from different theories and serve different purposes, they share common characteristics. Breiman [10] proposed an adaptively reweighing and combining (ARC) framework to capture the common characteristics in these algorithms, where ARC characterizes a combination algorithm as an iterative three-step procedure (assuming $k-1$ $(k \geq 1)$ classifiers have been obtained in the current combination):

(1) Resampling (or reweighing) the original training set to obtain the k-th training data set.

Bagging algorithm uniformly resamples the original training set. Boosting adaptively resamples the original training set. Stochastic discrimination and combinations of weak perceptrons resample uniformly on dynamically chosen "cares."

(2) Generating a qualified k-th classifier based on the k-th training data.

Bagging algorithm was designed to combine well-trained models. In practice, both neural networks and decision trees (CART) [8] have been used as base classifiers. Although Adaboost was originally designed to combine weak classifiers, it has been widely used to effectively boost the performance of strong classifiers [48][10]. SD and CWP all used the trial-until-qualified algorithm to generate randomly either a hyper-sphere or a perceptron but only keep those which exceed a certain performance threshold.

(3) Determining the weight for voting.

Adaboost uses the confidence over the performance of a model as the corresponding weight for voting. All other algorithms use the simple majority vote combination scheme.

5 Empirical results of combination of weak classifiers

In this section, empirical results are used to provide quantitative evaluation of combination of weak classifiers against other common approaches including combination of well-trained classifiers and single classifiers. In the latter cases, we use a neural network as a classifier.

Two synthetic problems are chosen to test the space- and time-complexity. Real applications from standard data bases are selected to compare the gen-

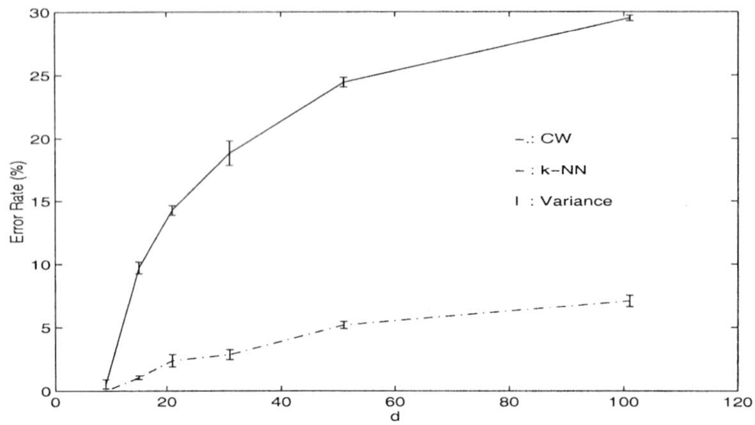

Figure 1: Performance on learning a perceptron

eralization performance of combinations of weak classifiers (CW) with that of other methods such as K-Nearest-Neighbor classifiers (K-NN)[7], artificial neural networks (ANN), combinations of neural networks (CNN), and stochastic discriminations (SD)[8].

5.1 Synthetic Problems

Two synthetic problems were designed to test the scaling properties on the performance of combined classifiers in terms of the dimension d of feature vectors.

5.1.1 A Perceptron

This problem is designed to learn an underlying perceptron with a d dimensional binary weight vector w_o which consists of all $(+1)$'s. Feature vectors x's are uniformly distributed in $\{-1, 1\}^d$. If an x satisfies $w_o^T x > 0$, x belongs to Class 1 with a label $t = 1$; otherwise, x belongs to Class 2 with a label $t = 0$. Our algorithm is used to learn a set of 2000 randomly-drawn samples when the dimension d of feature vectors is made to be larger and larger. That is, the size of the training set is made to be fixed, while the dimension of feature vectors is increasing. 10 different runs are conducted for each d, and the resulting classifiers are tested on another 4000 randomly-drawn

[7]The best result of different k is reported.

[8]We always test our method using exact data when compared with other methods. The details on the results due to other methods can be found in the related references.

samples. For comparison, the k-Nearest Neighbor classifiers also learn the same data sets. The resulting average generalization error as well as the standard deviations are given in Figure 1.

As shown in Figure 1, as d increases, the performance of k-NN decrease rapidly, thereby indicating the occurrence of the so-called curse-of-dimensionality as discussed in [45]. The degradation in performance is moderate with combinations of weak classifiers.

5.1.2 Two Overlapping Gaussians

To further test the scaling properties of combinations of weak classifiers, a non-linearly separable problem is chosen from a standard database called ELENA [3][4]. This database has been used by both neural network and machine learning communities to test and compare algorithms.

The problem is a two-class classification problem, where the distributions of samples in both classes are multi-variant Gaussians. Each dimension of the samples corresponds to an independent Gaussian random variable with zero mean, but samples in different classes have different standard deviations: 1 for samples in Class 1, and 2 for samples in Class 2. As shown in Figure 2 (for $d = 2$), there is a considerable amount of overlap between the samples in two classes, therefore the problem is non-linearly separable. The average generalization error and the standard deviations are given in Figure 3 for our algorithm based on 20 runs, and for other classifiers. The Bayes error is also given to show the theoretical limit [9].

Once again, the results show that the performance of kNN degrades very quickly. The performance of ANN is better than that of kNN but still deviates more and more from the Bayes error as d gets large. The combination of weak classifiers continues to follow the trend of the Bayes error.

5.2 Real Applications

5.2.1 Proben1 Data Sets

Three data sets, Card1, Diabetes1 and Gene1 were selected to test our algorithm from Proben1 databases which contain data sets from real applications

[9]The results of ANN and the Bayes error are from what reported in ELENA [4]. For ANN, a one hidden lay feedforward neural network with 10 or 20 hidden nodes was trained by Backpropagation algorithm.

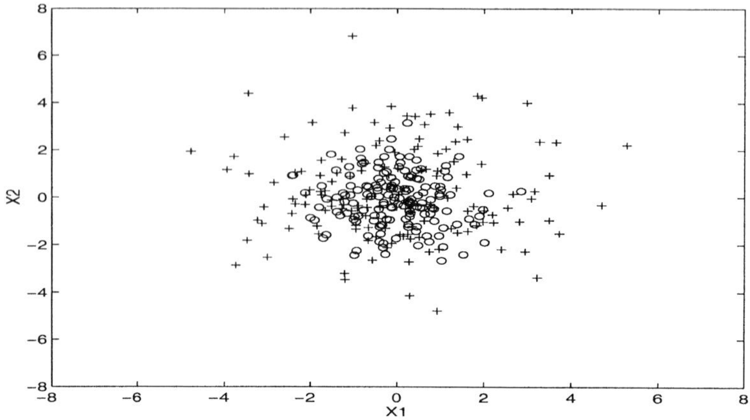

Figure 2: Two overlapping Gaussians

Algorithms	Card1	Diabetes1	Gene1
	(%) Error /σ	(%) Error /σ	(%) Error /σ
Combined Weak Classifiers	11.3/ 0.85	22.70 / 0.70	11.80 / 0.52
k Nearest Neighbor	15.67	25.8	22.87
Neural Networks	13.64/ 0.85	23.52/ 0.72	13.47/ 0.44
Combined Neural Networks	13.02/0.33	22.79 /0.57	12.08 / 0.23

Table 1: Performance on Card1, Diabetes1 and Gene1. σ: standard deviation

[46][10].

Card1 data set is for a problem on determining whether a credit-card application from a customer can be approved based on information given in 51-dimensional feature vectors. 345 out of 690 examples are used for training and the rest for testing. Diabetes1 data set is for determining whether diabetes is present based on 8-dimensional input patterns. 384 examples are used for training and the same number of samples for testing. Gene1 data set is for deciding whether a DNA sequence is from a donor, an acceptor or neither from 120 dimensional binary feature vectors. 1588 samples out of total of 3175 were used for training, and the rest for testing.

The average generalization error as well as the standard deviations are reported in Table 1. The results from combinations of weak classifiers are based on 25 runs. The results of neural networks and combinations of well-

[10] Available by anonymous ftp from ftp.ira.uka.de, as /pub/papers/techreports/1994/1994-21.ps.z.

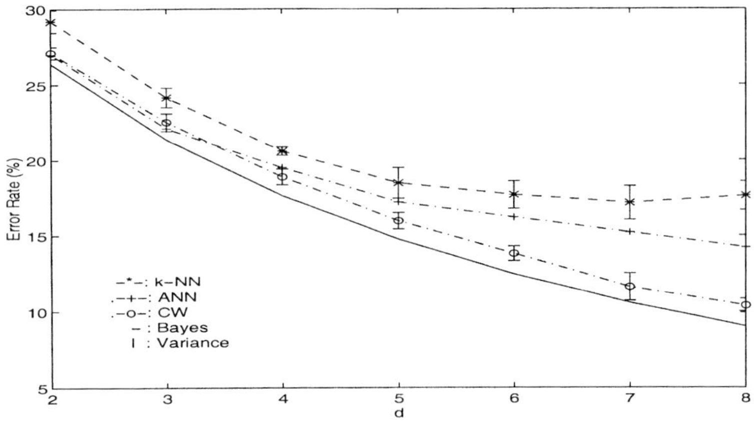

Figure 3: Performance versus the dimension of the feature vectors

trained neural networks are from [46][53] [11]. As demonstrated by the results, combinations of weak classifiers have been able to achieve the generalization performance comparable to or better than that of combinations of well-trained neural networks.

5.2.2 Hand-written Digit Recognition

Hand-written digit recognition is chosen to test our algorithm, since one of the previously developed method on combinations of weak classifiers (stochastic discrimination[32]) was applied to this problem. For the purpose of comparison, the same set of data as used in [32](from the NIST data base) is utilized to train and to test our algorithm. The data set contains 10000 digits written by different people. Each digit is represented by 16 by 16 black and white pixels. The first 4997 digits are used to form a training set, and the rest are for testing. Performance of our algorithm, k-NN, neural networks, and stochastic discriminations are given in Table 2. The results for our methods are based on 5 runs, while the results for the other methods are from [32].

The results show that the performance of our algorithm is slightly worse (by 0.3%) than that of stochastic discriminations, which uses a different method for multi-class classification by converting an M-class classification problem into $\frac{M(M-1)}{2}$ two-class classification problems[32].

[11] In [46][53], neural networks with many different architectures were trained by Back-propagation and several its variations. In [53], 3 to 7 well trained neural networks are combined by majority vote. Only the best reported results are listed in table 1.

Algorithms	(%) Error/σ
Combined Weak Classifiers	4.23 / 0.1
k Nearest Neighbor	4.84
Neural Networks	5.33
Stochastic Discriminations	3.92

Table 2: Performance on handwritten digit recognition.

Parameters	Perceptron	Gaussians	Card1	Diabetes1	Gene1	Digits
$1/2 + 1/\nu$	0.51	0.51	0.51	0.51	0.55	0.54
θ	0.51	0.51	0.51	0.54	0.54	0.53
2L+1	2000	2000	1000	1000	4000	20000
Average Tries	2~10	2	3	7	4	2

Table 3: Parameters used in our experiments.

5.3 Effects of Parameters

There are two parameters used in our algorithm: the threshold θ and the weakness factor ν. For most of the experiments described in this work, good performance can be obtained if θ is chosen to satisfy $\frac{1}{2} \leq \theta \leq \frac{1}{2} + \frac{1}{\nu}$. When the data is noisy, a slightly larger θ can be chosen to incorporate more samples into a set of cares. From our experience, the performance of a combined classifier is not very sensitive to the choice of θ as long as it is chosen within these guidelines.

The choice of weakness factor ν, strongly affects the size and training time of a combined classifier. Experiments on the problem of two 8-dimensional overlapping Gaussians given in Section 5.1.2 are done to test the effects of ν. The performance and the average training time (CPU-time on Sun Spac-10) of combined weak classifiers based on 10 runs are given for different ν's in Figures 4 and 5, respectively[12]. The results indicate as ν increases an individual weak classifier is obtained more quickly, but more weak classifiers are needed to achieve good performance.

A record of the parameters used in all the experiments on real applications are provided in Table 3. The average tries, which are the average number of times needed to sample the classifier space to obtain an acceptable weak classifier, are also given in the table to characterize the training time for these problems.

[12] θ is chosen to be $\frac{1}{2} + \frac{1}{\nu}$.

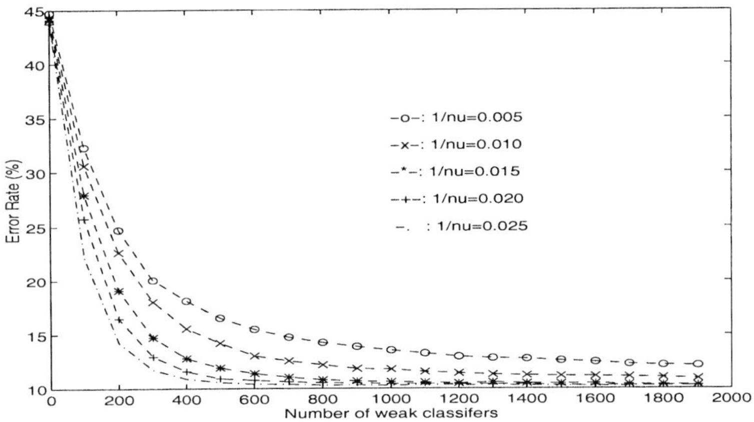

Figure 4: Performance versus the number of weak classifiers for different ν. nu: ν.

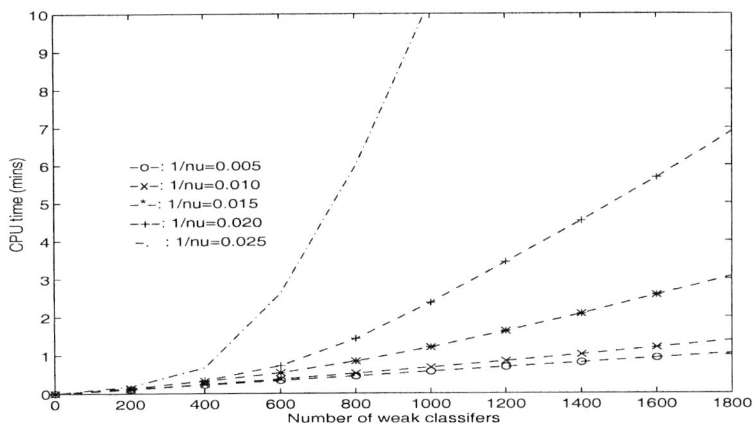

Figure 5: Training time versus the number of weak classifiers for different ν.

5.4 Training Time

To compare learning time with off-line BackPropagation (BP)[13], feedforward two layer neural network with 10 sigmoidal hidden units are trained by gradient-descent to learn the problem on the two 8-dimensional overlapping Gaussians. 2500 training samples are used. The performance versus

[13]Since our algorithm requires off-line learning, off-line BP is used to make comparison. However, it should be mentioned that on-line BP can be faster than the off-line BP.

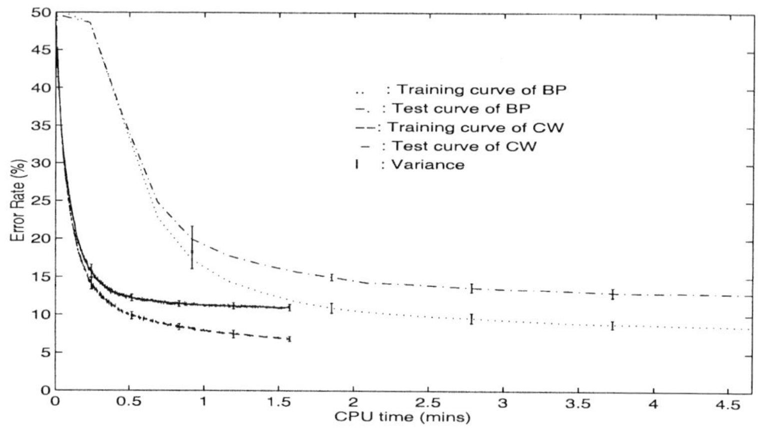

Figure 6: Performance versus CPU time

CPU time[14] are plotted for both our algorithm and BP in Figure 6. For our algorithm, 2000 weak classifiers are combined. For BP, 1000 epochs are used. The figure shows that our algorithm is much faster than the BP algorithm. Moreover, when several well-trained neural networks are combined to achieve a better performance, the cost on training time will be even higher. For instance, in Card1, Diabetes1 and Gene1 problems, seven well trained neural networks are combined to obtain an almost comparable performance to that of combinations of weak classifiers (see Table 1). Then the cost on training time would be 7 times that spent on training a single neural network. Therefore, compared to combinations of well-trained neural networks, combining weak classifiers is computationally much cheaper.

5.5 Discussions of Experimental Results

What have the experimental results shown us? First, we observe that our algorithm for combinations of weak classifiers has been able to achieve the best performance on all the synthetic and real problems except for hand-written digit recognition problem. More specifically, the performance of the combined weak classifiers is comparable or even better than combinations of well-trained classifiers, and out-performs individual neural network classifiers and k-Nearest Neighbor classifiers. In the meantime whereas the k-nearest neighbor classifiers suffer from the curse of dimensionality, a nice scaling property in terms of the dimension of feature vectors has been ob-

[14]Both algorithms are run on a Sun Sparc-10 sun workstation

served for combined weak classifiers.

Another important observation obtained from the experiments is that the weakness factor directly impacts the size of a combined classifier and the training time.

As the experimental results provide positive results, they also pose the following questions for us to answer:

(1) how to characterize the performance and efficiency of a combined classifier?

(2) how to choose the weakness factor ν?

Theoretical analysis will be carried out to provide answers to these questions. The analysis will be based on the same problem of learning a perceptron given in Section 5.1.1[15]. The insights gained can shed light on how to analyze our algorithm for more general problems in the future.

6 Theoretical results: Performance of Combinations of Weak Classifiers

Since the performance of Adaboost [24] when used to combine a large number of weak classifiers were found to be either comparable or somewhat inferior to combining strong classifiers, Adaboost algorithms have been used mostly to combine a small number of well-trained classifiers [24][10][48]. Through extensive experimental comparison s[10][48][24], Adaboost (adaptive boosting algorithm) for combining strong models has been shown to perform consistently better than training an individual model or a bagging combination algorithm.

Can combinations of weak classifiers do better in performance than training a strong classifier? Although the independent assumption used in the theory for stochastic discrimination is difficult to satisfy in reality, better performance has been obtained consistently when a large number of weak classifiers are combined and used in handwritten digit recognition [33]. In particular, the resulting combined classifiers were shown to be less sensitive to overfitting. This means the performance of a combined classifier was usually improved when more and more weak classifiers were combined. Similar to combinations of well-trained classifiers, this in fact shows that combinations of weak classifiers reduce the variance when more and more weak classifiers are combined.

[15] Since many methods with good performance and efficiency can be used to learn a binary perceptron[56], our intention is not to investigate the problem of how to learn a perceptron itself but to use it as an example for analyzing combinations of weak classifiers.

Combinations of weak perceptrons [30] were tested extensively on various synthetic and real data sets. The generalization performance of the combined weak perceptrons has been shown to be slightly better than combinations of well-trained classifiers, and outperforms individual neural network classifiers and k-nearest neighbor classifiers. Meanwhile, as will soon be shown, interesting phenomena have been observed that a trade-off can be made between performance and efficiency by combinations of weak perceptrons.

6.1 Efficiency of Combinations of Weak Classifiers

As the performance of combinations of weak classifiers is comparable to that of combinations of well-trained classifiers, what really are the advantages of using weak classifiers?

6.1.1 Stochastic Discrimination

Stochastic discrimination [32] first suggested that combinations of weak classifiers can be used to improve the training time. A theory was derived to show that when weak classifiers were assumed to make independent classification errors, the computational time for selecting a weak classifier was polynomial in terms of the dimension of feature vectors. As the space-complexity was also shown to be polynomial, the resulting time-complexity of a combined classifier was polynomial. Empirically, the training time needed to obtain a combination of weak classifiers was shown to be magnitudes faster than that of conventional training methods such as back-propagation.

Three factors were not included when the time-complexity was derived: the structure of weak classifiers, the weakness factor, and the statistical dependence among outputs of weak classifiers. As discussed in Section 5.2, the structure of weak classifiers relates directly to the space-complexity of a combined classifier, and thus determine its space efficiency. The weakness factor also affects the space-complexity of a combined classifier, since the weaker the weak classifiers are, the more weak classifiers may be needed to learn a problem, and the larger the space-complexity. This may in turn influence the time-complexity, since the time-complexity of a combined classifier can be regarded as the average training time needed to obtain a weak classifier multiplied by the space-complexity.[16]

[16] The number of weak classifiers by definition.

6.1.2 Combinations of Weak Perceptrons

To shed light on whether and why combinations of weak perceptrons are efficient, theoretical analysis was carried out on a simple example when combinations of weak classifiers are used to learn underlying perceptrons [30].

The generalization error $\Pr(C_{2L+1}(\mathbf{x})\mathbf{t} < \mathbf{0})$ of the combined classifier $C_{2L+1}(\mathbf{x})$ with $2L+1$ weak perceptrons was derived and bounded above by a quantity in the order of $\frac{\nu ln(2L+1)}{\sqrt{2L+1}}$, i.e.,

$$\Pr(C_{2L+1}(\mathbf{x})\mathbf{t} < \mathbf{0}) \leq \mathbf{O}(\frac{\nu \mathbf{ln(2L+1)}}{\sqrt{\mathbf{2L+1}}}), \tag{8}$$

where ν is the weakness factor. $O(z)$ stands for a quantity in the order of z.

Such a bound shows that the generalization error decreases at a polynomial rate in terms of the number of weak perceptrons. By setting the upper bound to be equal to ϵ_g, which is a bound on the desired generalization error, the space-complexity S of a combined classifier[17] can be obtained easily as

$$S \leq O(\frac{(\nu ln \nu)^2}{\epsilon_g^2}). \tag{9}$$

Therefore, S is polynomial in both the weakness of factor ν and ϵ_g.

The time-complexity T of a combined classifier is defined as the average number of samplings needed to obtain a combined classifier when a desired generalization error at most ϵ_g. Such a time-complexity was shown to satisfy

$$T = O(\frac{(\nu ln \nu)^2}{\epsilon_g^2}\sqrt{d}e^{\frac{d}{\nu^2}}), \tag{10}$$

for d and ν large but $\nu \ll d$.

Since the larger the weakness factor ν, the larger the space-complexity S, but the smaller the time-complexity T, a trade-off can be made between the space- and time-complexity by finding an optimal ν. Specifically, let $\frac{dT}{d\nu} = 0$ and assume d large, an optimal weakness factor ν_o can be obtained as $\nu_o = O(\sqrt{d})$. Therefore, when $\sqrt{\frac{d}{\ln d}} \leq O(\nu)$, the time complexity T is polynomial in the dimension d of feature vectors; otherwise, T is an exponential function of d. In the meantime, when this condition is satisfied, the space-complexity $S = O(\frac{dlnd}{\epsilon_g^2})$, is also polynomial in d. The existence of such a critical value for the weakness factor suggests that the polynomial time-complexity may

[17] S is the number of weak classifiers needed to achieve a certain generalization error.

be obtained at a cost of a larger size classifier compared to that of a well-trained classifier with a fixed structure. The cost, however, is theoretically tolerable, since it scales polynomially in the dimension d of feature vectors.

Discussion

Through analyzing the time-complexity, an intuitive explanation can be drawn on when and why combinations of randomly selected weak perceptrons are efficient. If weak classifiers are weak enough, i.e., the weakness factor ν is large enough, there will be many such weak classifiers. Therefore, the chance of getting a weak classifier is high at each sampling. That is, the number of times needed to sample the classifier space until a qualified weak classifier is accepted is small. As soon as weak classifiers are not too weak to destroy the polynomial space-complexity, the efficiency can be achieved both in time and space for combinations of weak perceptrons.

The theory provides an explicit relationship between performance and efficiency but is limited to a special case for learning a linear decision boundary. There are no similar results derived so far on nonlinear classification problems.

6.2 Open Issues

Due to the intrinsic difficulties of tackling performance and efficiency of nonlinear classifiers, many open issues need to be investigated on combinations of weak classifiers. Some of these open questions are

(1) how to show the optimality of a (randomized) algorithm for choosing and combining weak classifiers?

Assuming there were an infinite number of weak classifiers usable in a combination, this question essentially asks whether an algorithm is optimal in approximating a Bayes classifier. As an elegant analysis showed that the nearest-neighbor classifiers [13] are asymptotically Bayesian optimal, it remains open as to whether or not similar results could be obtained for a randomized algorithm using weak classifiers. The difficulty in obtaining such a result is that weak classifiers are statistically dependent.

(2) For what problems do a large number of weak classifiers exist to achieve a desired performance?

Even when an optimal algorithm is used to obtain a combination of weak classifiers, it is not clear whether a large enough number of weak classifiers exist so that the combined classifier can achieve a desired performance for a given structure (perceptron, for example) and a weakness factor. If the classification problem is too difficult, there may not exist any weak classifier, since a set of weak classifiers with the chosen structure may not have enough capacity [13][38][6][28] to do better than random guessing. (3) For what problems, do a large number of weak classifiers exist so that the time-complexity can be polynomial?

Not all problems can be solved in polynomial time (e.g. training non-linear neural classifiers is NP-complete). A polynomial complexity for randomly selecting a weak classifier is obtained based on the assumption that there exist a large number of weak classifiers. For example, if there exist at least a polynomial fraction of classifiers that can do better than random guessing, the average number of tries required to get one weak classifier is polynomial. Otherwise, choosing one acceptable weak classifier from an exponentially small fraction of all classifiers would take an exponential number of tries on the average. Since the number of acceptable weak classifiers depends on the problem, it is important to characterize problems for which a large number of weak classifiers do exist. The empirical results, however, are encouraging. That is, for many classification problems including those shown in this paper, a sufficiently large number of weak classifiers exist, and result in a powerful combined classifier.

7 Conclusion

We have reviewed several general techniques to improve efficiency and performance. In particular, we have discussed three different approaches , which improve the performance through making a trade-off between the bias and variance: (1) searching for an optimal structure for a single network. This was studied mainly for improving the performance of single model; (2) training several oversized models that have a low bias but a high variance, and then reducing the overall variance through combining these models. Combinations of well-trained models improve the performance through this approach; (3) training a large set of weak models that have a large bias but a small variance, and then reducing the overall bias and thus the generalization error by combining these weak models. Combinations of weak classifiers improve the performance through this scheme.

Combinations of weak classifiers, which use an incremental combination scheme and a randomized algorithm, have shown the potential to achieve time-efficiency as well as a good generalization performance. The cost is a polynomial space-complexity for benchmark problems, which is theoretically acceptable. Explicit relationships have been provided to illustrate the interrelation and the trade-off between performance and efficiency through combinations of weak classifiers. Randomized algorithms play an important role for combinations of (weak) classifiers.

Although much progress has been made in performance and efficiency of nonlinear adaptive systems, a lot of problems are still wide open for possible future research.

Acknowledgment

The authors would like to thank Leo Breiman and Andrew Barron for pointing out the issue of the existence of weak classifiers, and helpful discussions. The authors would like to thank T.K. Ho, Gene Kleinberg, George Nagy and Guohua Zhao for relevant references and helpful discussion. The support from the National Science Foundation (ECS-9312594 and (CAREER) IRI-95025 18) is gratefully acknowledged.

References

[1] Y.S. Abu-Mostafa, "Information theory, complexity and neural networks," *IEEE Communications Magazine*, vol.27, no.11, 25-28, Nov. 1989.

[2] J.A. and P.H. Swain. Consensus theoretic classification methods. *IEEE Transactions on Systems, Man, and Cybernetics*, 22:688-704, 1992.

[3] C. Aviles-Cruz, A. Guerin-Dugue, J.L. Voz and D. Van Cappel, "Deliverable R3-B1-P Task B1: Databases", Enhanced Learning for Evolutive Neural Architecture, ESPRIT-Basic Research Project Number 6891, June, 1995. Anonymous FTP: /pub/neural-nets/ELENA/databases/Benchmarks.ps.Z on ftp.dice.ucl.ac.be.

[4] F. Blayo, Y. Cheneval, J. Madrenas, M. Moreno and J.L. Voz, "Deliverable R3-B4-P Task B4: Benchmarks", Enhanced Learning for Evolutive Neural Architecture, ESPRIT-Basic Research Project Number 6891, June, 1995. Anonymous FTP: /pub/neural-nets/ELENA/databases/Benchmarks.ps.Z on ftp.dice.ucl.ac.be.

[5] A. Barron, "Universal Approximation Bounds for Artificial Neural Networks," *IEEE Trans. on Information Theory*, vol. IT-39, 930-944, 1993.

[6] E. Baum and D. Haussler,"What Size Net Gives Valid Generalization?" *Neural Computation*, 1(1), 151-160, 1989.

[7] A.L. Blum and R.L. Rivest, "Training A 3-Node Neural Network Is NP-Complete," *Neural Networks*, vol.5, 117-127, 1992.

[8] L. Breiman. Bagging predictors. *Machine Learning*, 24:132–140, 1996.

[9] L. Breiman. Stacked regressions. *Machine Learning*, 24:49–64, 1996.

[10] L. Breiman. Arcing classifiers. *Annals of Statistics*, 26:801–849, 1998.

[11] T. X Brown, H. Tong, S. Singh, "Optimizing admission control while ensuring quality of service in multimedia networks via reinforcement learning," to appear in Advances in Neural Information Processing Systems, ed. M. Kearns et al., MIT Press, 1999.

[12] N. Cesa-Bianchi, Yoav Freund, D.P. Helmbold, D. Haussler R.E. Shapire and M.K. Warmuth, " How to Use Expert Advice," *Proceeding of Foundation of Computer Science*, 382-391, 1993.

[13] T.M. Cover and P.E. Hart, "Nearest Neighbor Pattern Classification," *IEEE Trans. Inform. Theory,*, IT-13, 21-27, 1967.

[14] T.M. Cover, "Capacity Problems for Linear Machines," in *Pattern Recognition*, Thompson Book Co., 1968. ed. by L. Kanal.

[15] H.T. Demiral, S. Ma, and C. Ji. Combined power of weak classifiers. In *Proceeding of World Congress on Neural Networks*, pages 591–595, 1995.

[16] T.G. Dietterich. Machine-learning research. *AI Magazine*, 18:97–136, 1997.

[17] T.G. Dietterich. Discussion of Arcing classifiers. *Annals of Statistics*, 26:838–841, 1998.

[18] R.O. Duda and P.E. Hart, *Pattern Classification and Scene Analysis.* Wiley, New York, 1973.

[19] U. Fayyad, G. Piatetsky-Shapiro and P. Smyth, "From Data Mining to Knowledge Discovery: An Overview, " In *Advances in Knowledge Discovery and Data Mining*, Menlo Park, CA: The AAAI Press/The MIT Press, 1996.

[20] T.L . Fine, S. B. Wicker, T. Berger, and J. Halpern, "Sensor-Assisted ALOHA for Wireless Networks, " *Proc. 1998 IEEE International Symposium on Information Theory*, Aug., 1999.

[21] T.L. Fine, *Feedforward Neural Network Methodology* ,Springer-Verlag, 1999.

[22] S. Forest, " Emergent Computation: Self-Organizing, Collective, and Cooperative Phenomena in Natural and Artificial Computing Networks," *Physica D*, vol.42, 1-11, 1990.

[23] Y. Freund. Boosting a weak learning algorithm by majority. *Information and Computation*, 121:256–285, 1995.

[24] Y. Freund and R. Schapire. Experiments with a new boosting algorithm. In *Machine Learning: Proceedings of the Thirteenth International Conference*, pages 148–156, 1996.

[25] S. Geman, E. Bienenstock, and R. Doursat, "Neural Networks and the Bias/Variance Dilemma," *Neural Computation*, 4, 1-58, 1992.

[26] J.H. Holland, "A Mathematical Framework for Studying Learning Classifier Systems," *Physica*, vol. 22D, 307-317, 1986.

[27] J.H. Holland, "Genetic Algorithms," *Scientific American*, vol.267(1), 66-72, 1992.

[28] C. Ji and D. Psaltis, "Capacity of Two-Layer Networks with Binary Weights," *IEEE Trans. Information Theory*, Vol. 44, No.1, 256-268, Jan., 1998.

[29] C. Ji and S. Ma. Combined weak classifiers. In *NIPS*, pages 494–500, 1996.

[30] C. Ji and S. Ma. Combinations of weak classifiers. *IEEE Transactions on Neural Networks*, 8:32–42, 1997.

[31] S. Judd, *Neural Network Design and The Complexity of Learning*, MIT Press, 1990.

[32] E.M. Kleinberg, "Stochastic Discrimination," *Annals of Mathematics and Artificial Intelligence*, vol.1, 207-239, 1990.

[33] E.M. Kleinberg and T. Ho, "Pattern Recognition by Stochastic Modeling," *Proceeding of Third International Workshop on Frontiers in Handwriting Recognition*, 175-183, Buffalo, May 1993.

[34] A. Krogh and J. Vedelsby. Neural network ensembles, cross validation, and active learning. In *NIPS*, pages 231–238, 1995.

[35] S.R. Kulkarni, G. Lugosi, and S. S. Venkatesh, "Learning pattern classification-a survey," *IEEE Transactions on Information Theory*, Vol.44, No.6, 2178-2206, Oct. 1998.

[36] S. Lawrence, C.L. Giles, "Searching the Web: general and scientific information access," *IEEE Communications Magazine*, vol.37, no.1, 116-122, Jan. 1999.

[37] N. Little and M. Warmuth, "The Weighted Majority Algorithm," *The Third Workshop on Computational Learning Theory*, 1989.

[38] R.J . McEliece, E.C . Posner, E.R . Rodemich, S.S .Venkatesh, "The Capacity of the Hopfield Associative Memory," *IEEE Trans. Inform. Theory*, Vol. IT-33, No. 4, 461-482, July 1987.

[39] R. Meir. Bias, variance and the combination of least squares estimators. In *NIPS*, pages 295–302, 1995.

[40] C. Merz and M.J. Pazzani. Handling redundancy in ensembles of learned models sing principal components. In *Proceedings of the National Conference on Artificial Intelligence*, 1996.

[41] C. Merz and M.J. Pazzani. Combining neural network regression estimates with regularized linear weights. In *NIPS*, 1997.

[42] U. Mitra and H.V. Poor, "Neural Network Techniques for Adaptive Multiuser Demodulation," *IEEE Journal on Selected Areas in Communications*, vol.12, no.9, 1460-1470, Dec. 1994.

[43] J. Moody and C. Darken, "Fast Learning in Networks of Locally-Tuned Processing Units," *Neural Computation*, vol.1, 281-294, 1989.

[44] M.P. Perrone and L.N. Cooper, "When Networks Disagree: Ensemble Method for Neural Networks," Chap 10, *Artificial Neural Networks for Speech and Vision*, 1993.

[45] D. Psaltis, R.R. Snap and S.S. Venkatesh, "On The Finite Sample Performance of Nearest Neighbor Classifiers," *IEEE Trans. Inform. Theory*, IT-40, 820-837, May 1994.

[46] L. Prechelt, "PROBN1-A Set of Benchmarks and benchmarking rules for Neural Network Training Algorithms," *Technical Report* 21/94, *Fakultat fur Informatik, Universitat Karlsuhe*, D-76128 Karlsruhe, Germany, September 1994. Anonymous FTP:/pub/papers/techreports/1994/1994-21.ps.z on ftp.ira.uka.de.

[47] J.R. Quinlan. *C4.5: Programs for Machine Learning.* Morgan Kaufman, 1993.

[48] J.R. Quinlan. Bagging, boosting, and c4.5. In *Proceedings of the Thirteenth National Conference on Artificial Intelligence*, pages 725–730, 1996.

[49] S. Rangarajan, P. Jalote, and S.K. Tripathi. Capacity of voting systems. *IEEE Transactions on Software Engineering*, 19:698–705, 1993.

[50] I. Rish, M. Brodie and S. Ma. Accuracy versus efficiency in probabilistic diagnosis. *to be appeared in AAAI conference*, 2002.

[51] R. Schapire. The strength of weak learnability. *Machine Learning*, 5:197–227, 1990.

[52] M. Thottan and C. Ji, "Proactive Anomaly Detection Using Distributed Intelligent Agents," *Special Issue on Network Management-Today and Tomorrow, IEEE Net work, IEEE Network Magazine*, September 1998.

[53] K. Tumer and J. Ghosh. "Theoretical Foundations of linear and Order Statistics Combiners for Neural Pattern Classifiers." To be appeared in *IEEE Transactions on Neural Networks*, 1995.

[54] L.G. Valient, "A Theory of Learnable," *Communications of The ACM*, 27(11), 1134-1142, 1984.

[55] R. Vilalta, C. Apte, J.L. Hellerstein, S. Ma and S. Weiss, "Fault Prediction in Computer Networks," *IBM System Journal, Special Issue on AI*, August 2002.

[56] S. S. Venkatesh, "Directed Drift - A New Linear Threshold Algorithm for Learning Binary Weights Online," *J. Comput. Sys.* 46(2), 198-217, April, 1993.

[57] D. Wolpert, "Stacked Generalization," *Neural Networks*, vol. 5(2), 241-259, 1992.

Image Segmentation and Pattern Recognition: A Novel Concept, the Histogram of Connected Elements

Darío Maravall
Departamento de Inteligencia Artificial
Universidad Politécnica de Madrid
E-mail: dmaravall@fi.upm.es

Miguel Ángel Patricio
Departamento de Informática
Universidad Carlos III de Madrid
E-mail: mpatrici@inf.uc3m.es

Contents

D. Chen and X. Cheng (eds.),
Pattern Recognition and String Matching, 411–463.

1 Introduction

The segmentation process is a crucial step in any computer-based vision system or application, due to its inherent difficulty and the importance of its results, which are decisive for the global efficiency of the vision system. The objective of segmentation is to individualize any different regions present in any particular image. Our main concern in this chapter is to model the image segmentation process as a pattern recognition problem, which, as an important practical corollary, implies that any method or technique from the pattern recognition field can, in principle, be applied to solve the segmentation problem in any computer-based vision system or application.

To segment a digital image is to divide the image into individual, disjoint regions. In some cases, segmentation is equivalent to the complete recognition of the image, since the individualized regions can be directly classified -i.e. labeled- according to a previously established set of abstract patterns, without any further processing or automatic analysis of the segmented image. For the remainder of the cases, the segmented regions or objects have to be analyzed for recognition, usually by means of conventional pattern recognition techniques. Therefore, we can establish a preliminary classification of segmentation cases, into the following two groups.

1. **Direct segmentation**: the segmentation process simultaneously includes the complete recognition of the image.

2. **Indirect segmentation**: the segmented regions have to undergo further processing and analysis to be finally classified, either with metric, or statistical, or syntactic pattern recognition techniques.

The above distinction is created *de facto* by the actual objects present in a particular image. When the objects cannot be recognized by their form or shape, then the use of discriminant features associated with the sensory information stored in the image pixels is absolutely essential for their automatic recognition. On the contrary, for objects that can be recognized by their shape attributes, the segmentation process is only an intermediate stage, since its main objective is *simply* to differentiate, without recognizing, the pixels belonging to each object from the other pixels. Strictly speaking, however, both types of segmentation employ the same formal methods and techniques, all of which pertain to the pattern recognition field. What really happens is that, in one case, the segmentation process embraces the recognition process as well, whereas, in the other case, additional pattern recognition techniques must be applied for the complete analysis and final recognition of the digital image. In conclusion, pattern recognition techniques are used in any segmentation category, either for simultaneous pixel segmentation and classification as a pattern or class -the direct case- or for firstly differentiating each individual pixel and further recognizing the existing objects by means of their shape attributes.

2 Basic Definitions

According to the usual notation, a digital image is a bidimensional discrete function or signal $\{I(i,j)\}$:

$$\{I(i,j)\} \ / \ 0 \leq I \leq 2^q - 1; \ 0 \leq i \leq N-1; \ 0 \leq j \leq M-1 \tag{1}$$

where I is the intensity level associated with each pixel (i,j) or basic element of the image -the word pixel being the contraction of picture element-. The intrinsic nature of the function I is obviously sensory, although it can stem from several sensory sources, almost always belonging to the electromagnetic spectrum. The most frequent interpretation of I is light intensity, either in the visible or invisible range. Usually, I is the gray-level intensity provided by a black and white camera or the three basic red, green and blue light components obtained from a color camera. However, from a theoretical and general viewpoint, the physical nature of the function $I(i,j)$ is irrelevant as far as the segmentation and the recognition processes are concerned. In expression (1), q stands for the number of bits employed in the digitized image, which is usually $q = 8$, amounting to 256 intensity values: $0 \leq I \leq 255$. In the same expression, (i,j) stands for the coordinates -horizontal and vertical, respectively- of a generic pixel or, more precisely, the row and the column of the respective element –i.e. pixel- of the $\{I(i,j)\}$ matrix. Note that we make a distinction between the global digital image $\{I(i,j)\}$ and a particular pixel value $I(i,j)$.

The segmentation process of a generic image $\{I(i,j)\}$ produces its division into a set of disjoint regions $R_1, R_2, \ldots, R_S$ such that:

$$\begin{aligned} &(a) \quad \{I(i,j)\}_{NxM} = R = \bigcup_{i=1}^{S} R_i \\ &(b) \quad R_k \cap R_l = \emptyset\,, \quad \forall k \neq l \end{aligned} \tag{2}$$

The condition (a) means that every pixel $I(i,j)$ has been segmented. The corollary of condition (b) is the non-existence of overlapping regions. Obviously, it is impossible to guarantee for any specific digital image that the segmentation process will meet both restrictions. This, rather, should be interpreted as the ideal conditions and an optimum objective for any segmentation process.

3 Image Segmentation and the Pattern Recognition Paradigm

We use the pattern recognition paradigm to mean the basic idea of classifying objects into patterns or classes. The numerous existing methods and techniques in the field of pattern recognition can be divided into two main groups: (1) the numerical feature-based approach –also known as the metric or statistical approach- and (2) the syntactic approach.

The first approach is based on the conversion of the objects to be recognized into vectors of numerical features or discriminant variables, which can be classified by means of a vast body of mathematical methods (geometric, algebraic, statistical) and algorithmic methods (iterative procedures, neuro-fuzzy techniques etc.). The other approach exploits the qualitative structure of many physical objects and is based on the use of the theory of formal languages and automata theory. Although very efficient in the recognition of many physical objects, syntactic methods are seldom applied to the segmentation process, as they depend on the existence of regular structural relationships within the digital images at the pixel level, which is a rare occurrence in real-life digital images. For the remainder of the chapter, we shall focus on the numerical feature-based approach.

All the methods belonging to this approach revolve around the basic principle of converting the *objects* –in the broadest sense, which includes the whole digital image under analysis- into numerical vectors formed by the discriminant variables. As a result, the recognition process is a mapping between the numerical vectors and the set of all existing patterns or classes:

$$\forall \overrightarrow{x}? \quad \exists fd(\overrightarrow{x}) \ / \ \overrightarrow{x} \in \alpha_l \ ; \ l = 1, 2, \ldots S \tag{3}$$

Expression (3) symbolically summarizes the essence of the pattern recognition paradigm. Verbally, $\overrightarrow{x}$ in this expression stands for an unclassified vector -supposedly coming from a specific object- and the recognition involves applying what is known as a discriminant function $fd(.)$ to this vector. As a result, the vector $\overrightarrow{x}$ is classified as belonging to a particular class or pattern α_l.

The segmentation process can be modeled similarly. Indeed, given a generic pixel $I(i,j)$ –i.e. the object to be classified- the first step is to transform it into a numerical vector $\overrightarrow{x}(i,j)$. Afterwards, the segmentation process is aimed at labeling this pixel by assigning it to a pre-specified class, let's say α_l. For the direct segmentation case, this labeling process ends with the complete recognition of the whole image, provided that all the pixels have been labeled. In the indirect case, however, the labeled pixels have to be grouped into objects, which must then be recognized on an individual basis. Therefore, we can express the segmentation process as follows:

$$\forall (i,j) \rightarrow \overrightarrow{x}(i,j) \quad / \quad s(i,j) \ \triangleq \ \overrightarrow{x}(i,j) \in \alpha_l \ ; \ l = 1, 2, ..., S \tag{4}$$

which symbolizes that the segmentation process $s(i,j)$ is defined by the assignment of the respective pixel (i,j) to a specific class or pattern α_l.

As the feature-based pattern recognition paradigm and the segmentation process are very similar, it is feasible to apply the vast body of theoretical and applied resources available in this well-established field. In particular, from expressions (3) and (4), the segmentation process can be implemented as a discriminant function:

$$\forall \overrightarrow{x}(i,j) : s(i,j) \ \triangleq \ fd[\overrightarrow{x}(i,j)] \ / \ \overrightarrow{x}(i,j) \in \alpha_l \tag{5}$$

To explain these ideas, we develop below the foundations of the segmentation process as a strict pattern recognition problem by considering the most common case of grayscale digital images, where the pixel $I(i,j)$ is univariate –i.e. a scalar magnitude-. More specifically, $x(i,j) = I(i,j)$, where I stands for the grayscale intensity.

4 Segmentation of Grayscale Intensity Digital Images

As our main concern in this chapter is to discuss concepts and general principles, we have chosen a basic, but by no means trivial case, to provide a better introduction and understanding of the main ideas and principles of the segmentation process and its relationship to the pattern recognition field. Thus, let us suppose that we want to segment the text from the background in a digitized document. Therefore, our universe of discourse –i.e. the set of possible classes- is formed by α_1 -text- and α_2 -background-. Let us suppose also that the document has been digitized with 8 bits, which gives 256 possible grayscale levels for each pixel. In this specific case, segmentation involves classifying each pixel $I(i,j)$ either as text or as background. This process is known, in the jargon of automatic analysis of documents, as document binarization, since each pixel is labeled as 0 or as 1, depending on its final classification into either of the two classes. This is a typical case of indirect segmentation, as the segmented characters in the document have to undergo an additional recognition process based on discriminant features associated with their shapes. Figure 1 shows a digital document and its binarized image.

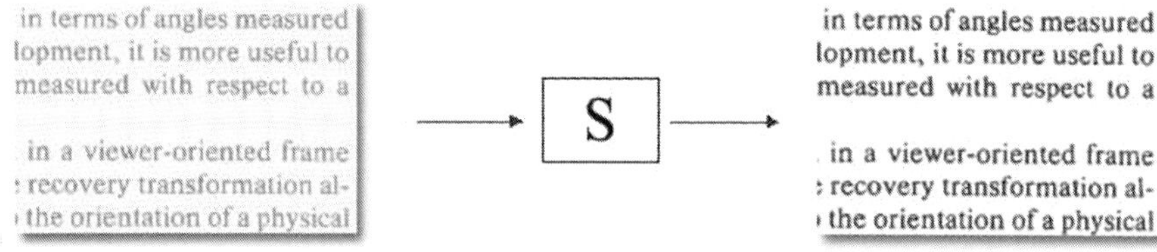

Figure 1: Digital document and the resulting image after its segmentation.

We are going to demonstrate that the segmentation process for this particular case can be formally modeled as a statistical pattern recognition problem. Indeed, let us take a look at the histogram of the digitized document of Figure 1 shown in Figure 2. This histogram is represented as a unidimensional discrete function $h(I_k)$, where the independent variable I_k stands for the range of grayscale intensity values, which, in this case, is $k = 0, 1, 2, \ldots, 255$, as the document has been digitized with 8 bits per pixel. With regard to the function $h(I_k)$ -the histogram itself- it represents the number of pixels with the generic grayscale level I_k . Therefore, the histogram gives the frequency of occurrence of the random event "grayscale intensity" appearing in the sampling space SS defined by all the pixels in the digital image; that is:

$$\begin{aligned} SS &\triangleq \{(i,j)\ ;\ \forall i,j\ /\ 0 \le i \le N-1\ ,\ 0 \le j \le M-1\} \\ h(I_k) &= card(i,j)\ /\ [I(i,j) = I_k]\ ,\ \forall i,j \end{aligned} \tag{6}$$

in which the possible values of the random event "$I(i,j) = I_k$" are I = 0,1,2, ..., 255.

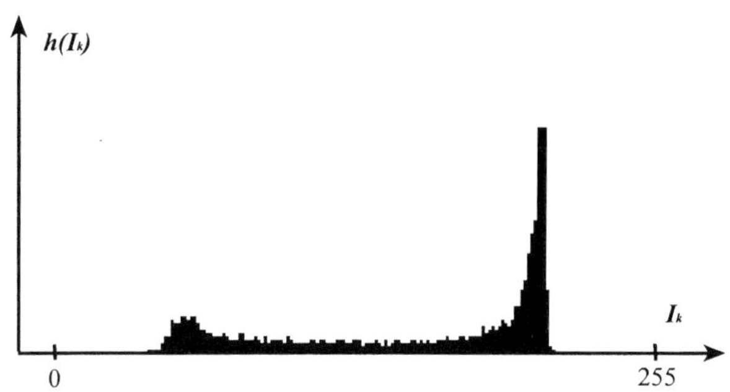

Figure 2: Histogram of the digitized document show in Figure 1.

Thus, the segmentation process can be modeled as a random process consisting of a sequential process in which all the pixels are classified according to their particular grayscale intensity. Having defined the segmentation operation in this way, the grayscale intensity histogram $\{h(I_k)\}$ can be approximated to the probability of occurrence of the event "$I(i,j) = x = I_k$", with $k = 0, 1, 2, \ldots, 255$. Therefore:

$$\lim_{\substack{N \to \infty \\ M \to \infty}} \{h(I_k)\}_{NxM} \approx p(x = I_k) \tag{7}$$

Bearing in mind that the random variable under consideration is the grayscale intensity –i.e. $x = I_k$ - , from now on we are going to generalize our notation and use the convention of denoting the random independent variable by x. As this random variable comes from two possible classes or patterns, α_1 and α_2, then we can express its probability density function, pdf, as:

$$p(x) = p(x/\alpha_1)p(\alpha_1) + p(x/\alpha_2)p(\alpha_2)$$
$$x = I_k \ , \ k = 0, 1, ..., 255 \tag{8}$$

where $p(x/\alpha_i)$ stands for the conditional probability of each class and $p(\alpha_i)$ the *a priori* probabilities. Given the event "$I(i,j) = x$", the optimum classification rule is based on the *a posteriori* probabilities:

$$p(\alpha_1/x) \underset{\alpha_2}{\overset{\alpha_1}{\gtrless}} p(\alpha_1/x) \tag{9}$$

The interpretation of this expression is straightforward, since x will be assigned to class α_1, if and only if its respective *a posteriori* pdf $p(\alpha_1/x)$ is greater than $p(\alpha_2/x)$. These *a posteriori* pdfs can be expressed as a function of the *a priori* pdfs:

$$p(\alpha_i/x) = \frac{p(x/\alpha_i)p(\alpha_i)}{p(x)} = \frac{p(x/\alpha_i)p(\alpha_i)}{p(x/\alpha_1)p(\alpha_1) + p(x/\alpha_2)p(\alpha_2)} \tag{10}$$
$$i = 1, 2$$

The use of the *a posteriori* pdfs leads to what are known as adaptive and learning methods, in which a set of labeled samples –the training set- is used for the iterative calculation of the discriminant functions, which are obtained as an approximation of the true *a posteriori* pdfs. On the other hand, the use of the *a priori* pdfs leads to well-established, conventional statistical estimation methods. As far as the segmentation of real-life digital images is concerned, the aprioristic approach is by far the most frequently used, for both methodological and practical reasons. The estimation of the *a priori* pdfs is quite straightforward and, combined with *a posteriori* or adaptive schemes, as will be shown later, produces quite acceptable results.

Therefore, as a first step, our main concern now is to estimate the *a priori* pdfs appearing in (7) that will be used to classify the image pixels, according to the inequality:

$$p(x/\alpha_1)p(\alpha_1) \underset{\alpha_2}{\overset{\alpha_1}{\gtrless}} p(x/\alpha_2)p(\alpha_2) \tag{11}$$

in which for the sake of giving a clear understanding of the concepts involved, we are still considering the two-class problem of segmenting the digital document in our example.

If all the elements appearing in expression (11) were known, then the optimum decision threshold, τ_0 , should be immediately obtained by making both terms in (11) equal:

$$p(x/\alpha_1)p(\alpha_1) = p(x/\alpha_2)p(\alpha_2) \quad \Rightarrow \quad x = \tau_0$$
$$x \in \alpha_1 \; iff \; x < \tau_0 \quad ; \quad x \in \alpha_2 \; iff \; x > \tau_0 \tag{12}$$

To explain these ideas, let us consider both classes α_1 and α_2 as Gaussians:

$$\alpha_1 \; : \; N(m_1, \sigma_1) \quad ; \quad \alpha_2 \; : \; N(m_2, \sigma_2) \tag{13}$$

The classification rule can be obtained straightforwardly as the following inequality:

$$\frac{(x-m_1)^2}{\sigma_1^2} \underset{\alpha_2}{\overset{\alpha_1}{\gtrless}} \frac{(x-m_2)^2}{\sigma_2^2} + 2\ln\frac{\sigma_2}{\sigma_1} + 2\ln\frac{p(\alpha_1)}{p(\alpha_2)} \tag{14}$$

that for the equality sign leads to the second degree polynomial:

$$\tau_0^2 + \alpha\tau_0 + \beta = 0$$
$$\alpha = \frac{2m_1\sigma_2^2 - 2m_2\sigma_1^2}{\sigma_1^2 - \sigma_2^2} \tag{15}$$
$$\beta = \frac{2\sigma_1^2\sigma_2^2}{\sigma_1^2 - \sigma_2^2}\ln\frac{p_1\sigma_2}{p_2\sigma_1} + \frac{m_2^2\sigma_1^2 - m_1^2\sigma_2^2}{\sigma_1^2 - \sigma_2^2}$$

where we have substituted the discriminant random variable by τ_0, which represents the optimum decision threshold, in the sense that it is the value that separates both classes with the minimum classification error -see expression (12)-.

Going back to our example, there are two main problems: (1) the restriction imposed on the grayscale intensity distribution of the text and the background implies that both must be Gaussians, and (2) provided that this restriction holds, the statistical parameters m_i, σ_i, $p(\alpha_i)$ with $i = 1, 2$ have to be estimated *a priori* and for ever. The Gaussian hypothesis restriction is less problematic, as it can be easily tested and changed for another statistical distribution in the design stage, if required by the specific distribution of the grayscale intensities in the classes at hand. The second restriction, however, is much more troublesome and has to be dealt with individually, usually by re-estimating the statistical parameters of the respective probability distributions after a preliminary segmentation process, by means of iterative or adaptive schemes.

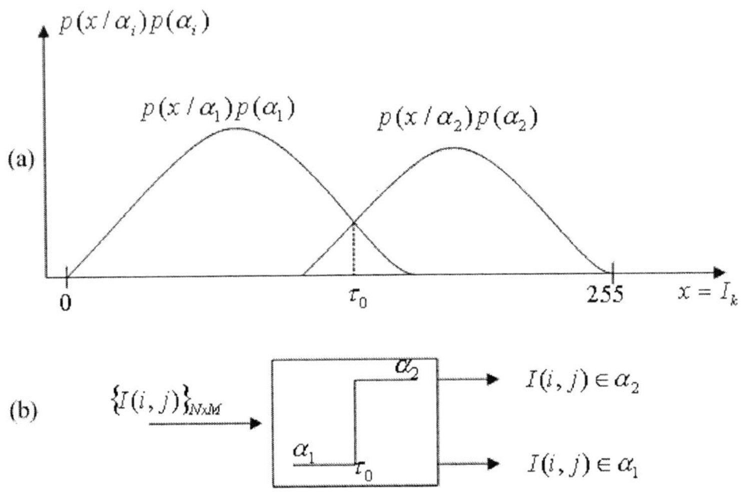

Figure 3: Symbolic representation of the segmentation process. In (a), the optimal threshold has been obtained, and in (b), this threshold has been applied to classify every pixel.

Let us suppose that we have successfully tested the Gaussian distribution hypothesis for both text and background pixels and have properly estimated their respective statistical parameters. Then, the segmentation process can be expressed compactly as follows:

$$\forall I(i,j) \;/\; I(i,j) = I_k \quad \Rightarrow \quad \begin{array}{l} (i,j) \in \alpha_1 \; iff \; I_k < \tau_0 \\ (i,j) \in \alpha_2 \; iff \; I_k > \tau_0 \end{array} \tag{16}$$

$$\forall (i,j) \;/\; 0 \leq i \leq N-1 \;;\; 0 \leq j \leq M-1$$

where the optimum threshold τ_0 has been computed according to expression (15) and with the respective estimated statistical parameters. This segmentation process has been symbolically depicted in Figure 3. Note that the histogram $\{h(x = I_k)\}$ has been ideally approximated to:

$$\{h(x = I_k)\} \approx p(x = I_k) = p(x = I_k/\alpha_1)p(\alpha_1) + p(x = I_k/\alpha_2)p(\alpha_2) \quad (17)$$

5 Adaptive Bayesian Segmentation

The main drawback of aprioristic recognition methods is that they are not robust with regard to the unavoidable changes that occur in the operational phase of any pattern recognition system or application, which is still more pronounced in the field of digital image analysis . Therefore, it is essential to make the segmentation process as flexible as possible, so that the automatic recognizer -i.e. the segmentation decision- adapts to the singularities of each particular digital image or, more specifically, to the histogram shape. In summary, the objective is to depart from the classification or segmentation given by the *a priori* model and then refine this initial segmentation by adapting the statistical parameters to the histogram of the digital image under scrutiny. Such a strategy falls within what are known as statistical or Bayesian adaptive segmentation methods.

Roughly speaking, the Bayesian adaptive or iterative segmentation process consists of a sequence of thresholds applied to the image histogram. This sequence starts with the threshold obtained from the a priori statistical parameters that appear in expression (15) and continues with a process that iteratively updates these parameters based on the image histogram. For our two-classes example, this updating process is given by the following expressions:

$$\hat{m}_1(l) = \frac{\sum_{k=0}^{\mu_0(l)} h(I_k)I_k}{\sum_{k=0}^{\mu_0(l)} h(I_k)}; \quad \hat{m}_2(l) = \frac{\sum_{k=\mu_0(l)+1}^{255} h(I_k)I_k}{\sum_{k=\mu_0(l)+1}^{255} h(I_k)}$$

$$\hat{\sigma}_1(l) = \frac{\sum_{k=0}^{\mu_0(l)} h(I_k)I_k^2}{\sum_{k=0}^{\mu_0(l)} h(I_k)} - \hat{m}_1^2(l)\ ; \quad \hat{\sigma}_2(l) = \frac{\sum_{k=\mu_0(l)+1}^{255} h(I_k)I_k^2}{\sum_{k=\mu_0(l)+1}^{255} h(I_k)} - \hat{m}_2^2(l) \quad (18)$$

$$\hat{p}_1(l) = \frac{1}{NxM}\sum_{k=0}^{\mu_0(l)} h(I_k)\ ; \quad \hat{p}_2(l) = 1 - \hat{p}_1(l) = \frac{1}{NxM}\sum_{k=\mu_0(l)+1}^{255} h(I_k)$$

which substituted in expression (15), give the optimum threshold $\tau_0(l+1)$. This process halts when the condition $\tau_0(l) = \tau_0(l+1)$ holds.

Having modeled the segmentation process as a conventional statistical pattern recognition problem, the practical and fundamental corollary is that any existing method or technique within the vast body of theoretical and applied results in the pattern recognition field can be, in principle, used for image segmentation. More specifically, we have modeled the segmentation of a digital image as a mixture of probability density functions that is materialized in the histogram. To understand this approach, we have to look at expressions (7) and (8), where the histogram has been approximated to an a posteriori total probability and, furthermore, we have hypothesized this total probability as being generated by several probability density functions -in our example, two Gaussian pdfs-. The mixture of Gaussians is a well-known and extensively studied topic in the pattern recognition discipline [1] - [5]. Consequently, the adaptive or iterative histogram-based segmentation process is a conventional mixture of Gaussians problem that can be tackled using an extensive battery of supervised and unsupervised techniques. Therefore it is no wonder that the related technical literature is quite voluminous [6] - [8].

Although iterative histogram-based segmentation can be considered a true mixture of Gaussians problem, it actually has very specific traits due to the peculiarities of digital images, the most outstanding being a consequence of the intrinsic spatial nature of digital images. For this reason, further refinements should be introduced in any iterative histogram-based Bayesian segmentation algorithm by exploiting the spatial information existing within any digital image. This is our next topic.

6 Contextual Segmentation.

The basic idea behind spatial or contextual refinement is to apply post-segmentation filtering, so that the segmentation results obtained in a generic neighborhood of each pixel -usually a nxn rectangle- determine the final decision about the classification of the pixel. A typical filtering operation is majority voting; i.e. the generic pixel in the middle of a given neighborhood is labeled -segmented- according to the majority label in the neighborhood. After every post-segmention filtering operation, a conventional iterative Bayesian segmentation algorithm is activated again, and this global process continues until the segmentation process finally converges.

Let us proceed with a more detailed study of the contextual segmentation process, again considering the document binarization example in order to attain a clearer understanding of the concepts involved. Thus, as a result of global Bayesian segmentation, all the pixels in the original image will have one of the two possible labels , α_1 or α_2. Remember that it is a two-class classification problem. Let us suppose that we use a $3x3$ neighborhood for local refinement, so that a generic pixel (i,j) is re-labeled according to the hegemonic label in its respective neighborhood. Once the majority post-segmentation filtering has been applied to the whole image, the statistical parameters of both classes are updated as follows:

$$\hat{m}_l = \frac{1}{N[\alpha_l]} \sum \sum_{\forall(i,j)\in\alpha_l} I(i,j) \; ; \quad \hat{\sigma}_l^2 = \frac{1}{N[\alpha_l]} \sum \sum_{\forall(i,j)\in\alpha_l} I^2(i,j) - \hat{m}_l^2$$
$$\hat{p}_l = \frac{N[\alpha_l]}{NxM} \; ; \quad l = 1,2 \tag{19}$$

and the new optimum threshold is updated according to expression (15). From a computational point of view, it should be noted that the statistical parameters are now obtained directly from the digital image, instead of being estimated from the histogram as before. Therefore, the computational load is increased by at least two orders of magnitude: from about 256 to NxM values. For this reason, it may sometimes be advisable to apply a thresholding operation based on the Euclidean distance:

$$\tau_0 = \frac{\hat{m}_1 + \hat{m}_2}{2} \tag{20}$$

which somewhat reduces the computational load associated with expression (19).

A contextual algorithm that has been quite widely used is the so-called Iterated Conditional Modes, ICM, [8] - [10]. The formal foundation of the ICM algorithm is relatively complex, so we merely describe its basic functioning, which involves computing both the two-class averages $\hat{m}_1$ and $\hat{m}_2$, and the between-class variance:

$$\hat{\sigma}^2 = \frac{1}{NxM} \sum_{i=0}^{N-1} \sum_{j=0}^{M-1} [I(i,j) - \hat{m}(i,j)]^2 = \frac{N_1\hat{\sigma}_1^2 + N_2\hat{\sigma}_2^2}{N_1 + N_2} \tag{21}$$

where N_1 and N_2 are the number of pixels that have been segmented as classes α_1 and α_2, respectively. The new labeling process is performed in accordance to the rule:

$$[I(i,j) - \hat{m}_1]^2 - \beta\hat{\sigma}^2 N(i,j) \underset{\alpha_2}{\overset{\alpha_1}{\lessgtr}} [I(i,j) - \hat{m}_2]^2 - \beta\hat{\sigma}^2 [8 - N(i,j)] \tag{22}$$

where $N(i,j)$ is the number of pixels within the 3x3 neighborhood that were classified as belonging to class α_1 and, therefore, $8 - N(i,j)$ is the number of pixels labeled as α_2. Obviously, the neighborhood shape is arbitrary and does not need to be of the nxn type. It can be shown that expression (22) is based on Markov random fields that we shall consider later [8]. Finally, it should be strengthened that global segmentation methods have a computational load that is proportional to the histogram range, whereas local methods -based on a further contextual or spatial refinement process- are proportional to the image range; i.e, to NxM. This serious disadvantage obliges the designers to make a careful analysis of the pros and cons before applying local post-segmentation refinement for any particular application. What is really interesting about spatial refinement or contextual post-segmentation is that, as opposed to purely histogram-based methods, which generally tend to destroy important spatial information inside a digital image, pixel classification not only depends on the strictly global statistical information provided by the histogram, but is also influenced by the local spatial distribution of the pixels themselves.

This line of thought has been very productive and the related technical literature is quite vast. In particular, attention has centered on two such methods: probabilistic and fuzzy relaxation algorithms, [12] - [13] and Markov random fields, [14] - [17]. We cannot go into further details here, but the relevance of combining the global segmentation process, based on statistical pattern recognition techniques, with the spatial information provided by the contextual refinement should be emphasized. In summary, these hybrid methods, which simultaneously exploit the global information condensed in the histogram and local information, can be expected to produce the best possible results in the segmentation process. Later on, we shall introduce a novel concept, the histogram of connected elements, which encompasses the global information of the histogram and the local information of the image itself. In this respect, it is a hybrid method too, although it has the advantage of integrating both kinds of information into just one element.

7 Multivariate and Multiclass Generalizations

So far, the formalization of segmentation as a statistical pattern recognition problem has been limited to the study of a two-class case -i.e. text and background in a digitized document- with a single discriminant variable: grayscale intensity. In this case, we have shown that the segmentation process is based on histogram thresholding, which is by far the most widely used technique in practice [18] – [21]. Thus, it is necessary to generalize our model in two directions: (a) multiple classes or patterns and (b) multiple discriminant features. The first generalization is immediate and trivial, as it does not require any conceptual change. The multivariate generalization is also straightforward, but it deserves a little attention as the change from a scalar variable x to a vector $\vec{x}$ involves somewhat more complex mathematics. Thus, the simple polynomial inequality in (14) or (15) is transformed into

$$(\vec{x}-\vec{m}_1)^T C_1^{-1}(\vec{x}-\vec{m}_1) \underset{\alpha_2}{\overset{\alpha_1}{\lesseqgtr}} (\vec{x}-\vec{m}_2)^T C_2^{-1}(\vec{x}-\vec{m}_2) + \ln\frac{|C_2|}{|C_1|} + 2\ln\frac{p(\alpha_1)}{p(\alpha_2)} \quad (23)$$

where C_1 and C_2 are the covariance matrices of classes α_1 and α_2. However, it should be noted that the univariate and the multivariate cases are conceptually identical and, for this reason, we shall not proceed further in this direction.

8 On the Concept of Discriminant Feature and its Generalization in Image Segmentation

It is really worthwhile and of practical relevance to carry out a deeper analysis of the discriminant feature concept within the field of image segmentation. More specifically, we are interested in seeing what kind of discriminant variables are relevant for image segmentation. We have already mentioned the grayscale intensity and the RGB components as the typical and by far most frequently used discriminant variables in image analysis and segmentation, as well as other signals of the electromagnetic spectrum, like UV, IR , radar signals and so on. These attributes are what we can call raw sensory information variables, in the sense that they are not the result of computer-based processing, but just readings from specific sensors.

Although it is quite unusual to adopt the following approach, we can extend the concept of discriminant feature or attribute from physical signals to mathematical or *abstract* variables, which are the result of computations on the above physical magnitudes. More specifically, we are referring to (1) linear and non-linear edge operators; (2) line and curve detectors and (3) texture operators.

The application of filters or operators to extract the borders existing in a digital image -which supposedly determine an object's contour- is one of the most commonly used operations in the field of image processing and analysis. The objective is to output a new image in which the borders are enhanced, allowing, in principle, more direct and efficient automatic analysis of the digital image. Unfortunately, this is not always the case, as shown in several examples given in Figure 4.

(a)

(b)

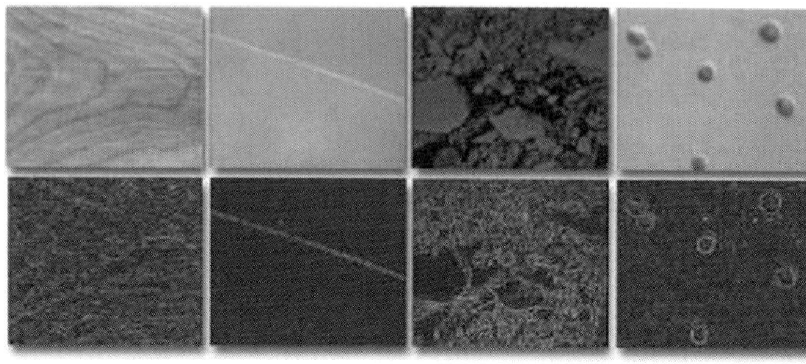

Figure 4: (a) Four distinct digital images. (b) Respective filtered images. Different edge operators have been applied in each case. Note that the automatic analysis of the transformed images appears to be much more problematic in some cases.

The technical literature about edge operators is literally overwhelming, and our interest here is exclusively to place them within the above theoretical framework of abstract attributes. Let us suppose that a certain edge extraction operator has been applied to a digital image $\{I(i,j)\}_{NxM}$ and, as a result, a new digital image $\{I_{bf}(i,j)\}_{NxM}$ has been produced, whose histogram is $h[I_{bf}(k)]$, $k = 0, 1, 2, \ldots 255$, where the subindexes stand for border filter. Then, we can apply a segmentation process equivalent to the ones described earlier except for the fact that it is based on a truly bimodal histogram, as there are only two classes -border and no-border-, thanks to the new discriminant feature. This idea is illustrated in Figure 5, where the application of a standard edge detector to a digital image makes possible a straightforward segmentation process.

By introducing the concept of *abstract* discriminant feature in our previous discussion, we have generalized the more intuitive and usual interpretation of the discriminant feature concept in the field of digital image segmentation, which is, basically, physical or sensory. Remember that in the subject of digital image processing, the basic information comes from the electromagnetic spectrum. Therefore, we can extend the somewhat rigid physical meaning of discriminant feature to the extremely flexible spaces

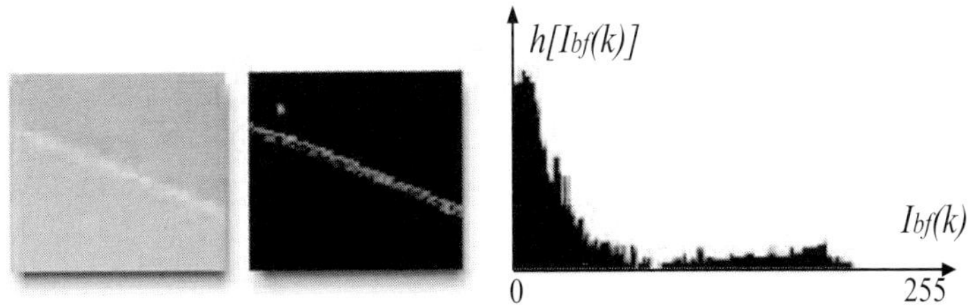

Figure 5: The original image, the image output after applying a gradient-based edge detector and its respective bimodal histogram.

formed by the abstract discriminant variables. This generalization will allow us to introduce a novel concept: histogram of connected elements. Indeed, the concept of texture itself belongs to this general category of abstract or processed discriminant variables. Following this line of thought, it is interesting to remark that other well-known operators aimed at the enhancement of spatial structures like lines and curves, can also be used as discriminant features. One of the best known and widely used of such operators is the Hough transform for the extraction of parameterized curves, although its computational load and its dependency on *a priori* knowledge of the curve parameters may preclude its application in some real-life situations.

However, there are some very simple operators that perform very efficiently, like the following operator for line and semilinear curve extraction, which the authors have successfully applied to the detection of very thin cracks in wood:

$$\frac{1}{7}\begin{bmatrix} 4 & 5 & 5 \\ -2 & 2 & 4 \\ -5 & -5 & -2 \end{bmatrix} \tag{24}$$

The excellent properties of this operator can be appreciated in Figure 6, which shows its application to two different images.

The application of the segmentation methods described above is straightforward when using the histograms of such filtered images. By way of an illustration, the iterative Bayesian method has been used in Figure 7 to segment the image appearing in Figure 6(d), where the objective is to segment the thin crack existing in the original wooden board. Note, however, that the final result is not entirely satisfactory, since part of the crack has been lost.

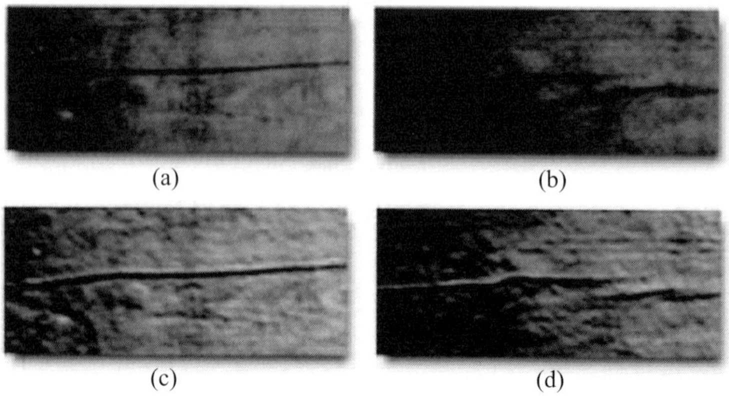

Figure 6: Examples of thin line or semilinear curve detection. (a) and (b) are the original images and (c) and (d) show their respective filtered images.

As a bridge between contextual or local segmentation and the novel concept of frequency histogram of connected elements, to be introduced later, we are now going to very briefly discuss two important groups of segmentation methods: (1) relaxation processes and (2) Markov random fields.

9 Relaxation Processes

As for any other local method, the basic idea is to label every pixel by taking into account the classification of the neighboring pixels. However, relaxation processes, as well as Markov random fields, have a rigorous and formal foundation, unlike some of the local methods discussed earlier.

When applying the relaxation process to the segmentation problem, the task of labeling -i.e. classifying- the generic pixel (i, j) as belonging to a certain class α_m can be expressed as

$$P_{\alpha_m}[(i,j)] \quad / \quad m = 1, 2, \dots S \; ; \; 0 \leq i \leq N-1 \; ; \; 0 \leq j \; M-1 \qquad (25)$$

Obviously, the probabilities are normalized to the unit:

$$\sum_{m=1}^{S} P_{\alpha_m}[(i,j)] = 1 \quad ; \quad \forall i, j \qquad (26)$$

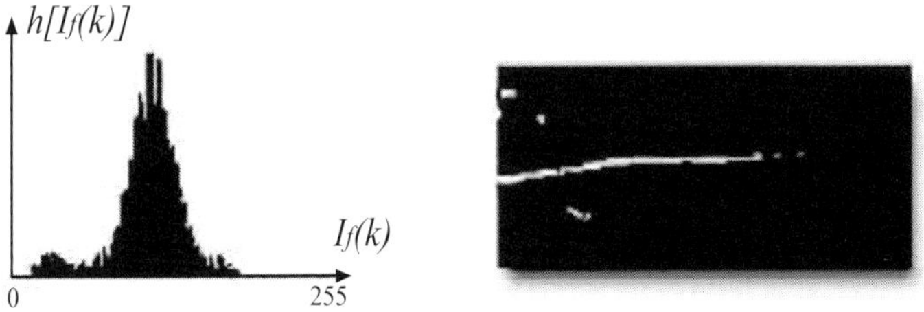

(a) (b)

Figure 7: (a) Histogram of the filtered image $\{I_f(i,j)\}$ obtained from the image in Figure 6(c). (b) Result of the segmented image, with part of the crack missing.

The next step is to introduce a compatibility function for the pixel neighborhood, with the intention of capturing the spatial information relevant to the segmentation process. First of all, a specific neighborhood spatial function $N[(i,j)]$ has to be defined and, afterwards, the respective compatibility function C can be expressed as follows:

$$C \triangleq \{C_{\alpha_m,\alpha_n}[(i,j),\ (k,l)]\} \quad / \quad 0 \leq m,n \leq S\ ;\ (k,l) \in N[(i,j)] \tag{27}$$

which represents the compatibility between the fact that pixel (i,j) *belongs to* class α_m and that the neighboring pixel (k,l) *belongs to* class α_n. Regarding the concept of neighborhood, let us consider a generic pixel (i,j), so that a neighborhood function $N[(i,j)]$ of (i,j) is defined as the set of pixels $\{(k,l)\} \in N[(i,j)]$ that meets a certain spatial constraint. For instance, four different neighborhood functions appear in Figure 8: (a) horizontal line neighborhood, (b) omnidirectional 3x3 neighborhood, (c) cross-like neighborhood and (d) diagonal neighborhood.

As regards the compatibility coefficients c_{α_m,α_n}, they can be considered as elements of a matrix defined by the following table:

(i,j) \ (k,l)	α_1	α_2	...	α_S
α_1	c_{α_1,α_1}	c_{α_1,α_2}	...	c_{α_1,α_S}
α_2	c_{α_2,α_1}	c_{α_2,α_2}	...	c_{α_2,α_S}
$\vdots$	$\vdots$	$\vdots$	$\vdots$	$\vdots$
α_S	c_{α_S,α_1}	c_{α_S,α_2}	...	c_{α_S,α_S}

$0 \leq i \leq N-1$
$0 \leq j \leq M-1$
$\forall (k,l) \in N[(i,j)]$

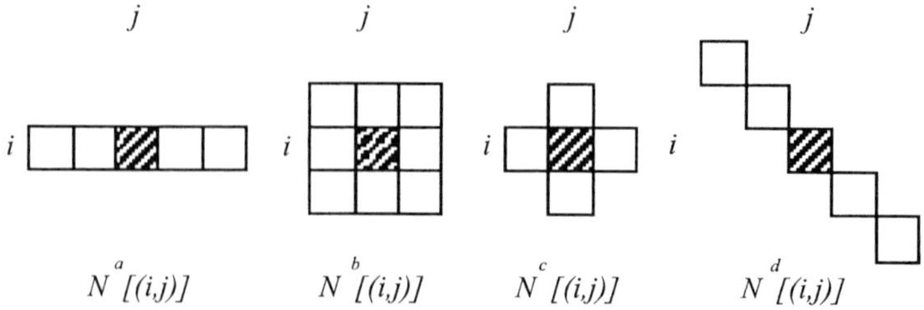

Figure 8: Examples of neighborhood functions.

where each row stands for the individual fact that the generic pixel (i, j) belongs to the respective class, while the columns indicates the label of the generic neighboring pixel (k, l), such that $(k, l) \in N[(i, j)]$.

It is straightforward to normalize the sum of each row's compatibility coefficients to the unit:

$$c_{\alpha_m}[(i,j)] = \sum_{n=1}^{S} c_{\alpha_m,\alpha_n}[(i,j),(k,l)] = 1 \tag{28}$$

since, given that pixel $(i, j) \in \alpha_m$, every neighboring pixel $(k, l) \in N[(i, j)]$ necessarily belongs to one of the existing classes.

Note that the words *belongs to* used in the paragraph just after expression (27) above are in italics, meaning that we are referring to true facts, although logically unknown to the classifier -i.e. to the segmentation process-. Thus, the compatibility functions are always fixed, aprioristic estimations made by the designer about *a priori* unknown, uncertain facts, like the real spatial distribution of the classes in a particular digital image. As a consequence, although the compatibility functions are essential for the success of the segmentation process, they have the stigma of being aprioristic and part of the rigid, *a priori* knowledge injected by the designer in the segmentation process. These are the facts, and nothing can be done in this respect. This is, undoubtedly, a serious drawback, since a poor design of the compatibility functions may significantly hinder the *a posteriori* adaptive segmentation process. Therefore, the design of the compatibility function is a crucial step to be carefully addressed.

To place this kind of spatial-based, iterative segmentation process within the theoretical and methodological framework of the statistical pattern recognition paradigm, we are now going to formalize the elements appearing in the definition of the compatibility functions -see expression (27)- as probability density functions (pdfs) and as conventional probabilities. As a matter of fact, the compatibility function in expression (27) can be immediately interpreted as a joint probability function:

$$c_{\alpha_m,\alpha_n}\left[(i,j),(k,l)\right] \triangleq p_{\alpha_m,\alpha_n}\left[(i,j),(k,l)\right] \equiv p\left[(i,j) \in \alpha_m \cap (k,l) \in \alpha_n\right] \tag{29}$$

$$0 \leq m,n \leq S\ ;\ \forall(k,l) \in N\left[(i,j)\right]$$

since the compatibility degree -*de facto*, the probability- of the fact that the pixel (i,j) and pixel (k,l) belongs simultaneously to classes α_m and α_n , respectively, coincides with the joint probability of the composed event. This joint pdf can be expressed as the product of a conditional pdf by an *a priori* probability:

$$\begin{aligned} c_{\alpha_m,\alpha_n}\left[(i,j),(k,l)\right] &\equiv p_{\alpha_m,\alpha_n}\left[(i,j),(k,l)\right] = \\ &= p_{\alpha_m/\alpha_n}\left[(i,j)/(k,l)\right] p_{\alpha_n}\left[(k,l)\right] \end{aligned} \tag{30}$$

Note that this notation is quite involved, and has been introduced for the sake of the clarity of our discussion. Obviously, our main concern now is to compute the probabilities associated with the classification of the generic pixel (i,j) into one of the possible patterns or classes. These classification probabilities can be expressed as follows:

$$\underset{m=1,2,\ldots S}{p_{\alpha_m}\left[(i,j)\right]} = \sum_{n=1}^{S} p_{\alpha_m/\alpha_n}\left[(i,j)/(k,l)\right] p_{\alpha_n}\left[(k,l)\right] \tag{31}$$

Since the generic pixel (i,j) possesses a set of neighbors -i.e. $(k,l) \in N[(i,j)]$- that do not necessarily have the same influence on it, we can define a weighted compatibility function as:

$$\begin{aligned} c_{\alpha_m}\left[(i,j)\right] &= \sum_{\forall(k,l)\in N[(i,j)]} \lambda(k,l) \sum_{n=1}^{S} p_{\alpha_m/\alpha_n}\left[(i,j)/(k,l)\right] p_{\alpha_n}\left[(k,l)\right] = \\ &= \sum_{\forall(k,l)\in N[(i,j)]} \lambda(k,l) \sum_{n=1}^{S} c_{\alpha_m,\alpha_n}\left[(i,j),(k,l)\right] \end{aligned} \tag{32}$$

where the index (k,l) scans all the pixels belonging to the neighborhood and $\lambda(k,l)$ is a neighbor's weight that quantifies the relative importance of each neighbor in the compatibility functions. The simplest choice is a constant λ equal to $1/card[N(i,j)]$.

Let us now describe the adaptive spatial segmentation process based on the above ideas. As a matter of fact, an iterative process for updating the probabilities of labeling a certain generic pixel (i,j) as belonging to one of the possible classes can be established as follows:

$$p_{\alpha_m}^{r+1}[(i,j)] = \frac{p_{\alpha_m}^{r}[(i,j)]\, c_{\alpha_m}^{r}[(i,j),(k,l)]}{\sum_{n=1}^{S} p_{\alpha_n}^{r}[(i,j)]\, c_{\alpha_n}^{r}[(i,j),(k,l)]} \quad ; \quad m = 1, 2, \ldots S \tag{33}$$

where the superindex r stands for the iteration order in the segmentation process. For the current iteration, the compatibility function is given by:

$$c_{\alpha_m}^{r}[(i,j),(k,l)] = \frac{1}{Card[N(i,j)]} \sum_{n=1}^{S} p_{\alpha_m/\alpha_n}[(i,j)/(k,l)]\, p_{\alpha_n}^{r}[(k,l)]$$
$$m = 1, 2, ...S \; ; \forall (k,l) \in N[(i,j)] \tag{34}$$

in which we have not considered distinct weights for the neighbors of the generic pixel (i,j). In expression (32), the aprioristic condition of the compatibility function can be explicitly stated as:

$$c_{\alpha_m}^{r}[(i,j),(k,l)] = \frac{1}{Card[N(i,j)]} \sum_{n=1}^{S} c_{\alpha_m/\alpha_n}[(i,j)/(k,l)]\, p_{\alpha_n}^{r}[(k,l)]$$
$$m = 1, 2, ...S \; ; \forall (k,l) \in N[(i,j)] \tag{35}$$

By way of an illustration, let us again consider the digital image in Figure 6(a), consisting of a wooden board with a very thin crack, which will also be used as a benchmark for this and other segmentation techniques later on. Obviously, the aim of segmentation is to detect this thin crack. Figure 6(c) shows the same wooden board after the application of a line enhancement operator. Clearly, there are two classes, α_1 = wood and α_2 = crack, in this case. As shown before, probabilistic relaxation algorithms are based on an iterative local refinement process starting from a global, histogram-based, segmentation. Figure 7(a) shows the histogram of the filtered wooden board image and Figure 7(b) the segmented image after applying a global iterative segmentation process. Note that the final segmentation is far from optimum, as there are crack losses and false alarms due to the fact that some pixels of sound wood were labeled as crack pixels. Let us see whether the application of a probabilistic relaxation algorithm improves these results.

The key element in any local adaptive segmentation technique -including relaxation processes- lies in the choice of the local distribution of the neighbor dependency. More specifically, as far as the relaxation process is concerned, the crucial point is the correct choice of the compatibility function. Since there are two classes α_1 and α_2 in our wooden board segmentation problem, the individual coefficients c_{α_1/α_1} , c_{α_1/α_2}, c_{α_2/α_1} and c_{α_2/α_2} have to be established. In fact, as the following relations hold:

$$c_{\alpha_m/\alpha_m}\left[(i,j),(k,l)\right] = 1 - c_{\alpha_m/\alpha_n}\left[(i,j),(k,l)\right]$$
$$m = 1,2 \; ; \; n = 1,2 \tag{36}$$

only two compatibility coefficients have to be introduced. A neighborhood spatial function has to be established to complete the definition of the compatibility functions. Thus, for our example, let us consider an omnidirectional 3x3 neighborhood function -see Figure 9-. Note also that because of the diagonal appearance of the crack inside this particular wooden board, such an omnidirectional neighborhood spatial restriction would not appear to be the best choice. Obviously, unless all the possible existing cracks in wooden boards had the same diagonal orientation, an omnidirectional spatial restriction is preferable, as it is the least *a priori* biased. In conclusion, we can establish the following compatibility coefficients:

$$\begin{array}{lll} c_{\alpha_1/\alpha_1}\left[(i,j),(k,l)\right] = 0.9 & \leftrightarrow & c_{\alpha_1/\alpha_2} = 0.1 \\ c_{\alpha_2/\alpha_1}\left[(i,j),(k,l)\right] = 0.2 & \leftrightarrow & c_{\alpha_2/\alpha_2} = 0.8 \end{array}$$
$$0 \le i \le N-1 \; ; \; 0 \le j \le M-1 \; ; \; \forall (k,l) \in N\left[(i,j)\right] \tag{37}$$

Note that the sound wood pixels have different compatibility coefficients from the crack pixels, due to the fact that cracks are thin and elongated, whereas the sound wood regions are much larger and their pixels are much more homogeneous. Figure 9 shows the respective neighborhood function and the compatibility coefficients.

It should be emphasized that the success of the relaxation process in the local, adaptive segmentation process depends basically on the neighborhood function $N[(i,j)]$ and, more specifically, on the table or matrix C that determines the compatibility coefficients. Figure 10 shows the segmentation result achieved with this type of probabilistic relaxation process on the same

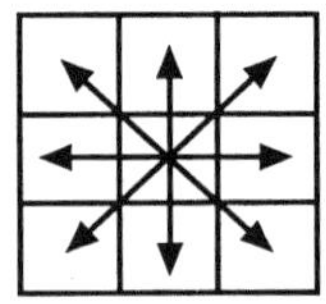

(i,j) \ (k,l)	α_1	α_2
α_1	0.9	0.1
α_2	0.2	0.8

Figure 9: The neighborhood spatial function and the compatibility matrix discussed in the text.

digital images as in Figures 6 and 7. Although crack detection has improved quite appreciably, there are still some losses and false alarms.

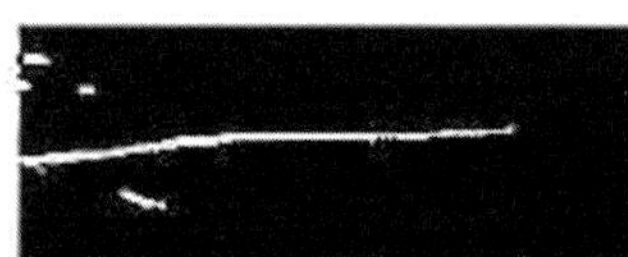

Figure 10: Segmentation results for the application of a relaxation process.

The relaxation process concept was introduced for the first time in the field of computer vision in [11] for the purpose of segmenting outdoor scenes and, since then, the related technical literature has grown constantly. For a good understanding of this approach, we recommend the textbook [12] and for a straightforward treatment of its application to image segmentation, the reference [13] is also highly recommendable.

10 Markov Random Fields

Relaxation processes -and, more specifically, compatibility functions- can be considered as a semiheuristic approach to the contextual segmentation process. A more rigorous approach for modeling the pixel spatial relationship in digital image segmentation is what are known as Markov random fields. The pioneering work on these methods started with [14] and since then this subject has received constant and considerable attention. Without going into the technicalities in too much detail, the fundamental idea behind this local segmentation paradigm is to establish a Markovian statistical law of the spatial distribution of the neighboring pixels that influences the classification or labeling of any generic pixel (i,j) in the image under scrutiny. Just like iterative relaxation algorithms, the iterative Markovian technique is founded upon the following same two strategic factors: (1) the neighborhood function and (2) the use of models –in this case, Markovian- of the distribution of the neighboring pixel discriminant features, that play a similar role to the compatibility functions in the relaxation algorithms. As a general remark, it can be said that although Markovian random models are a formal refinement on the semiheuristic relaxation processes, the truth is that –as is always the case when using formal models versus heuristic models- they are very sensitive to a correct or suitable choice of the statistical models for the neighboring pixels. For this reason, it is highly recommendable to carry out a careful analysis of the spatial distribution of the discriminant features to be used for image segmentation, before implementing the Markovian segmentation techniques. This ultimately leads to the problem of texture modeling. Unfortunately, this rigid aprioristic modeling of spatial statistical distributions -i.e. texture modeling- is what constitutes an important practical drawback, as the iterative segmentation process is meant to adapt itself to the specific spatial distribution of the particular digital image at hand, and any rigid, *a priori* restriction may hinder this adaptation process. In any case, it is the eternal dilemma between *a priori*, theoretical knowledge versus *a posteriori*, empirical evidence.

Just to give an idea of how the Markov random field segmentation methods operate, we shall very briefly describe one such method: Gibbs random fields [14], [16] and [17]. Let us suppose we have a certain neighborhood $N[(i,j)]$, so that a Gibbs random field is defined as follows:

$$p\left[s_m(i,j)/s\left\{N(i,j)\right\}\right] = \frac{1}{Z(i,j)} e^{-\frac{1}{T}\sum\limits_{N[(i,j)]} F\{N[(i,j)]\}} \qquad m = 1,2,...S \tag{38}$$

The first member in this expression represents a basic concept in Markov random fields; i.e. the probability of transition between the pixel states, where the concept of state is equivalent to a specific classification of the pixel into a class or pattern. Therefore, the interpretation of the first member is the probability of the generic pixel (i,j) being in state s_m -remember that there are S classes and, consequently, S states as well- provided that the pixels belonging to the neighborhood are in a certain state $s\{N[(i,j)]\}$, which is defined by the joint state of each neighboring pixel.

With regard to the second member of expression (38), it is based on Gibbs' statistical model of crystalline structures. Three elements appear in this model: $Z\{N[(i,j)]\}$, which is a normalization constant; T, a parameter of considerable importance, whose main objective is to adjust the rigidity of the neighboring pixels coupling -the greater T, the stronger the coupling- and, finally, $F\{N[(i,j)]\}$, which models the specific dependency among the pixel states within the neighborhood. This third element plays a very similar role to the compatibility functions in the relaxation algorithms. To give an idea of a possible dependency function, for a four-connected neighborhood, we have:

$$\begin{aligned} F\{N[(i,j)]\} = \; & s(i,j)\alpha + \beta[s(i-1,j) + s(i+1,j)] \\ & + \gamma[s(i,j-1) + s(i,j+1)] \end{aligned} \tag{39}$$

which depends on three free parameters that must be properly established. For an eight-connected neighborhood, the dependency function should have as many as ten parameters, which gives an idea of how complex this type of Markov models generally are.

After having selected the appropriate values of the elements appearing in (39), the iterative segmentation process proceeds as follows. Initially, the image pixels are classified by using any global segmentation process and, thereafter, a new labeling operation of all the pixels $\{(i,j)\}/0 \leq i \leq N-1; 0 \leq j \leq M-1$, is performed according to expressions (38) and (39), in which the pixel states -i.e. labels- are the result of the global segmentation. As a consequence, all the pixel labels have been updated, and a new iteration can be activated by again applying expressions (38) and (39) and so on. At each step, the pixels are classified as follows:

$$\begin{gathered} \max\left\{p_{\alpha_m}[(i,j)]\, p^r[s(i,j) = \alpha_m \,/\, s\{N[(i,j)]\}]\right\} \;\Leftrightarrow\; (i,j) \in \alpha_m \\ m = 1,2,...S \;;\; 0 \leq i \leq N-1 \;;\; 0 \leq j \leq M-1 \end{gathered} \tag{40}$$

where r stands for the iteration order and $p_{\alpha_m}[(i,j)]$ is the probability of the pixel (i,j) belonging to class α_m, as computed from the specific global segmentation algorithm applied previously. This iterative process ends when either it converges to a stable labeling or oscillates back and forth, in which case the process is halted.

Note that Gibbs Markov random fields have an additional design parameter, T, that is of the greatest importance. The symbol T directly refers to temperature, as it plays the same role as temperature in simulated annealing models, [24] and [25]. As mentioned above, high values of T are equivalent to weak dependency among the neighboring pixels and low values to strong dependency. As is well known, a correct choice of the parameter T is crucial in simulated annealing methods and, generally, an additional iterative loop is needed to converge to the optimum or near-optimum value of T. Consequently, the segmentation process based on Markov random fields and Gibbs statistical law tends to be extremely slow. If we also consider its sensitivity to a suitable model of the spatial dependency of the neighboring pixels, as determined by expressions (38) and (39), then we can easily understand the problems and difficulties arising in the application of these methods to real-life images.

As a final and general remark, it should be said that the methods described in this section -i.e. hybrid techniques based on the local segmentation performed by means of relaxation processes and Markov random fields or similar formal models of the local, spatial properties of pixels- are closely related to another relevant group of segmentation techniques; based on the automatic analysis of textures. Indeed, relaxation processes and Markov random fields are directly inspired by the concept of texture. Thus, it is a logical step to go on to analyze the basic functioning of texture analysis and recognition.

11 Segmentation Based on Texture Analysis and Recognition

The discriminant features based on the idea of texture belong to the second group of discriminant variables that we mentioned when discussing the generalization of the discriminant feature concept in image segmentation; i.e. they are abstract or, more precisely, transformed discriminant variables, as compared with raw sensory discriminant variables. However, as opposed to the usual spatial transformations and operators, texture-based features are intrinsically spatial -i.e. local or collective- and do not have an exclusively individual meaning. Indeed, although any spatial transformation - gradient, Laplacian and similar edge filters, line and curve detectors etc.- are the result of operations performed on the neighborhood of a pixel, the truth is that the segmentation process is ultimately performed on the transformed image and, more specifically, on its histogram. Therefore, only the pixel feature -i.e. the transformed value associated with each particular pixel- is considered in segmentation, without taking into account the collective distribution of the pixel feature in the neighborhood, except during the first stage when global segmentation is performed. Obviously, this global characteristic has a clear advantage, because any histogram-based segmentation algorithm is extremely fast. This computational advantage can only be exploited for individual discriminant features -as opposed to collective ones-. Unfortunately, at least from a purely computational standpoint, the concept of texture itself leads to discriminant features that are computed collectively.

On the other hand, hybrid segmentation methods combine the advantages typical of global histogram-based techniques and the specific qualities of a spatial refinement process, in which the information existing in the pixel neighborhood is also exploited. Precisely, we are going to show how the very idea of texture is directly connected with the local properties of the digital image, and, hence, the segmentation techniques based on textures are intrinsically local techniques, as the image is segmented using discriminant features -i.e. textures- computed in each pixel neighborhood. In the sequel, we shall present a brief review of the basic ideas and *modus operandi* of some texture-based segmentation techniques.

11.1 Basic Definitions

The texture concept, although essential and widely used in the automatic analysis of digital images, is hard to define objectively and formally, due to the inherent subjectivity of the idea of texture itself. Obviously, if the thing *per se* has a complex and rather polemic definition, its application to the segmentation problem gives rise to additional difficulties, due to the multiplicity of particular cases. Notwithstanding, we shall try to delimit the topic as much as possible. Let us introduce a formal definition of texture as a first and necessary step. Thus, initially, we define a texture as a regular pattern or structure, as regards both the physical space and the tonal or sensory attributes, where the latter are usually interpreted as light intensities or, generally, as belonging to the whole electromagnetic spectrum, obviously including the visible range. Therefore, three basic elements appear or coexist in the ontological definition of texture, which should be remarked upon: (1) a specific regular pattern or structure; (2) the physical space and (3) the sensory attributes or features. Thus, the texture concept implies the existence of a structure embodied in the physical space and based on sensory information. The coexistence of these three elements in a particular region of a digital image is the necessary, although not always the sufficient, condition for the existence of a possible texture.

Consequently, the automatic analysis of textures in digital images is a tricky task. To begin with, each of the three basic components are hard to define individually. What does regular pattern mean? Can they be mathematically modeled and how? If this is the case, is a deterministic or a statistical model the most appropriate? Similarly, how can a specific region in the physical space be established that embraces the texture itself? More specifically, what exact spatial resolution should be used? Just to focus on the last issue, it is well known that it is crucial for digital images to choose the right spatial resolution, as different kinds of textures can appear or just disappear depending on the resolution in question. Indeed, apart from being a strategic decision, the spatial resolution or observation window for texture analysis is extremely hard to optimize, as it can dramatically change within the same digital image. Additionally, it is related, somewhat, to what we can call the uncertainty principle in pattern recognition, which is, in turn, a particular case of the same principle in the general discipline of systems theory; i.e. the more rigid a model of the universe of discourse -here the textures- is, the less robust it will be against the real universe of application. Finally, what do we mean by a pattern based on the values taken by the sensory variables at hand? These questions are extremely involved and difficult to deal with individually and, considered as a whole, give rise to a really complex problem.

However, it is always possible to introduce some specific restrictions on the three above-mentioned elements and use the texture concept for practical matters in the automatic analysis and segmentation of digital images. For instance, it is very common to restrict the spatial domain of textures to some regular region -squares, circles, rectangles, etc.- and the sensory information to the grayscale intensity.

Roughly speaking, the spatial patterns or textures can be divided into two groups: (1) regular structures and (2) irregular structures. The first group is typical of artificial or man-made environments and objects, whereas the second group corresponds to the natural world. According to this distinction, the techniques for texture analysis and recognition can be classed as (1) syntactic methods and (2) statistical methods, respectively; although the latter can be also applied to regular structures. This division of texture analysis and recognition methods is equivalent to the well-known distinction in the discipline of pattern recognition between syntactic and statistical or metric methods.

In the automatic analysis and recognition of textures, the syntactic techniques have been less frequently used than the statistical techniques, among other reasons because the former are limited to regular textures, which are by far much less common than irregular textures. Even artificial objects, probably based on natural materials, have irregular textures, whose processing with syntactic techniques is less suitable than with statistical methods, as their texture patterns are seldom modeled by means of regular rules.

Therefore, we shall concentrate only on the statistical methods. Beforehand, we shall very briefly discuss a crucial issue in texture analysis, for both the syntactic and the statistical approaches, which is the problem of selecting the appropriate resolution of the observation window, also known as the region of interest (ROI), for texture analysis and recognition.

11.2 Selection of the Optimal Region of Interest for Texture Recognition

We are dealing, then, with a fundamental and strategic issue that radically influences the subsequent steps in the image segmentation process: the choice of the appropriate region or image window to search, detect and recognize any potential texture. An optimum window or ROI for texture recognition must meet two opposing conditions: (1) include the whole basic element of a single texture and (2) not include basic elements of more than one single texture. These two requisites are contradictory in the sense that the first condition tends to enlarge the image window, whereas the second one produces the opposite effect. If we add the fact that most textures have a random, impredictable spatial distribution very hard to model *a priori*, then it is easy to understand how difficult it is to choose the appropriate texture window in any practical situation. Indeed, it is an open research issue and also an unavoidable task that requires an extremely careful and well-designed solution for any computer vision application in which segmentation is performed by means of texture analysis and recognition.

Once the texture window has been established, the next step is to scan the entire digital image with the selected window. From a purely theoretical standpoint, the ideal scanning process should be from pixel to pixel, so that the segmentation process can be expressed as:

$$s\{I(i,j)\} \rightarrow \alpha_k \quad ; \quad 0 \leq i \leq N-1 \,;\, 0 \leq j \leq M-1 \,; \\ \alpha_k \in \{\alpha_1, \alpha_2, \ldots \alpha_T\} \tag{41}$$

In most cases, a post-segmentation process will be necessary in order to obtain a coherent regrouping of all the pixels into homogenous regions or individual objects. Post-segmentation techniques, like those previously described for the local refinement of the global, histogram-based, segmentation methods, can be applied to perform this regrouping. Obviously, the extremely high computational load is an important drawback of this theoretically optimum scanning process. At the other end of the scale is the scanning policy based on non-overlapping windows covering the whole digital image. In between these two extreme solutions, there are many possibilities for applying the texture window to search and classify the existing textures.

For non-optimal scanning processes, the final labeling of each window does not go down to the individual pixel, as it only segments the window as a whole. Depending on each particular case, it could be necessary to proceed with a finer segmentation within some of the segmented windows, giving rise to the well-known quad-tree structures.

11.3 Statistical Approach to Texture Analysis and Recognition

When speaking of a statistical approach, the point is that we are referring to the randomness of many textures and, hence, statistical methods are obviously the best candidates for the formal representation of such textures. Without trying to be exhaustive, we can group the statistical techniques as follows.

1. Spectral analysis, [6], [26] and [27].
2. First-order statistical parameters, [27] - [29].
3. Second-order statistics: co-occurrence matrices, [6], [27] and [30].
4. Miscellaneous: mathematical morphology, fractal descriptors, surface descriptors, contextual models including Markov fields etc. [7], [31] and [32].

Due to the variety and sometimes specificity of the techniques in group 4, we confine our discussion to the remaining groups. To begin with, the spectral techniques use the Fourier or similar transforms to compute the discriminant features. Let us consider a square texture window $\{I(i,j)\}_{nxn}$, such that $0 \leq i,j \leq n-1$, then its Fourier transform is:

$$I(u,v) = \frac{1}{N} \sum_{i=0}^{n-1} \sum_{j=0}^{n-1} I(i,j) e^{-\frac{2\pi j}{n}(iu+jv)} \tag{42}$$

$$0 \leq u,v \leq n-1$$

The power spectrum of the texture window is:

$$P(u,v) = \left\{ Re^2 \left[I(u,v) \right] + Im^2 \left[I(u,v) \right] \right\}^{1/2} \tag{43}$$

which is the Fourier transform of the autocorrelation function of the texture window, $\{I(i,j)\}_{nxn}$, $0 \leq i,j \leq n-1$:

$$R(\Delta i, \Delta j) = \frac{\sum_{i=0}^{n-1}\sum_{j=0}^{n-1} I(i,j)I(i+\Delta i, j+\Delta j)}{\sum_{i=0}^{n-1}\sum_{j=0}^{n-1} I^2(i,j)} \tag{44}$$

where Δi and Δj are pixel shifts within the texture window. Some of the properties of the autocorrelation function are of interest as far as texture discrimination is concerned. Thus, $R(\Delta i, \Delta j)$ has a maximum at $R(0,1)$, symmetrically decreasing with positive and negative values of Δi and Δj. An abrupt fall is typical of fine textures, while smoothness in the autocorrelation function denotes rough textures. Properties like the degree of symmetry, contrast and others can be used as discriminant variables as well, but, generally, the autocorrelation function cannot be considered as a good instrument for natural textures classification, [6] and [7]. On the contrary, the features associated with the Fourier transform of the autocorrelation function are of practical interest in texture discrimination, although they do not generally behave as efficiently as the features obtained from second-order statistics, [27], which, alongside first-order statistics are the most frequently used in practice. Computational reasons also play an important role in favor of what are know as spatial discriminant features -i.e. the first-order and the second-order statistics- versus the frequency-based discriminant attributes.

11.4 First-Order Spatial Statistical Parameters

First-order spatial statistics refer to two separate phenomena. Firstly, they are statistical parameters associated with the spatial distribution of a specific discriminant attribute of the digital image -or, rather, with a sub-image or texture window, to be more precise-. Secondly, they are absolute parameters -i.e. of the first-order type- and obtained from the values taken by the specific attribute at each pixel, considered as a unit and without considering the values of the remaining pixels. When, apart from the value of the discriminant feature at each pixel, the discriminant variable value at other pixels is also considered, then we have second-order statistics. *Mutatis mutandi*, third-order statistics are statistical parameters derived from the values of the discriminant attribute at three different pixels -i.e. the generic pixel (i,j) under consideration and another two-, and so on and so forth. Of course, these joint pixels are constrained to have a given spatial relationship. For reasons of both computational and discriminant efficiency, the technical literature has focused almost exclusively on first-order and second-order statistics, as they seem, generally, to discriminate most of the artificial and natural textures acceptably. Interestingly, Julesz's conjecture states that the human visual system is unable to distinguish among textures differing only in statistics greater than second-order, [33].

It is of interest to recall at this point that the attributes or features associated with a pixel can be diverse. Thus, the sensory features are attributes directly obtained from the physical sensors: grayscale intensity, RGB components intensity and other visible and invisible ranges of the electromagnetic spectrum. What we have previously called abstract discriminant features are attributes obtained as a result of the application of different mathematical operators and transformations: gradient, Laplacian, linear and nonlinear edge detectors, Hough transform, Fourier transform, wavelets transforms etc. For this kind of *abstract* discriminant attributes, it is possible to compute first-order statistics and, obviously, second-order statistics, too.

The element that concentrates the first-order global statistical information existing in the digital image is the image histogram, which, as we know, is the concrete realization of the probability density function of either the physical or abstract discriminant attribute at hand. Therefore, from the histogram, it is possible to compute a set of statistical parameters that are able to discriminate the region of interest of the image under analysis. For instance, the following statistics are commonly used:

$$\begin{aligned}
&Mean: && m = \sum_{k=0}^{p} x_k h(x_k) \\
&Variance: && \sigma^2 = \sum_{k=0}^{p} (x_k - m)^2 h(x_k) \\
&Skewness: && \mu_3 = \frac{1}{\sigma^3} \sum_{k=0}^{p} (x_k - m)^3 h(x_k) \\
&Kurtosis: && \mu_4 = \frac{1}{\sigma^4} \sum_{k=0}^{p} (x_k - m)^4 h(x_k)
\end{aligned} \tag{45}$$

Note, in expression (45), that we have introduced a generic discriminant feature x, instead of the usual grayscale intensity $I_k(i,j)$, $0 \leq k \leq 255$, whose value range is $0 \leq x \leq p$. This notation is meant to emphasize the key point that the pixel attributes for which we are computing the first-order statistics are not confined to the usual grayscale intensity or similar sensory features. Of course, $h(x_k)$ stands for the histogram of the discriminant attribute at hand.

From a discrimination point of view, the mean and variance provide global information about the texture or ROI window; i.e. its average value and its degree of randomness, whereas skewness informs about symmetry and kurtosis about marginal dispersion. As a general remark, the main advantage of first-order statistics is computational simplicity, although their capacity of discrimination for textures is not so attractive, except in some particular cases, [26] and [27].

A few references have appeared in the technical literature, [27] and [34], in which the statistics are obtained from the histogram of the distribution of the discriminant feature within a certain neighborhood of each generic pixel. This very interesting idea obviously leads to second-order statistics and, furthermore, is somewhat related to the novel concept of the histogram of connected elements, [35] and [36].

With regard to the above-mentioned concept of histogram of spatially distributed attributes, which can be properly called generalized histogram, one idea that immediately comes to mind is based on the difference of the grayscale level between the feature values at the generic pixel (i, j) and at the pixel determined by a pre-specified spatial displacement:

$$h(x, \vec{d}) \quad / \quad \begin{array}{l} x = |I(i,j) - I(i+\Delta i, j+\Delta j)| \\ \vec{d} \triangleq [\Delta i, \Delta j] \end{array} \tag{46}$$

Different generalized histograms are obtained by changing the discriminant attribute x and the displacement $\vec{d}$. Several discriminant statistics, [27], have been proposed for such generalized histograms. These include, for instance:

$$\begin{array}{ll} Mean: & m = \frac{1}{p} \sum\limits_{k=0}^{p} x_k h(y_k, \vec{d}) \\ Entropy: & e = -\sum\limits_{k=0}^{p} h(y_k, \vec{d}) \log \left[h(y_k, \vec{d}) \right] \\ Contrast: & c = \sum\limits_{k=0}^{p} x_k^2 h(y_k, \vec{d}) \end{array} \tag{47}$$

Although some other spatial predicates or restrictions, as defined by the different orientations of the generic pixel can be introduced in the generalized histogram -as well as other discriminant attributes- the truth is that this interesting concept has seldom been employed, as opposed to the extensive use of the conventional histogram and its related discriminant features.

11.5 Second-Order Statistics

Starting from the common background previously established in the introduction of first-order statistics, i.e. taking into account the fundamental concept of pixel attributes -remember that they can be either physical or abstract-, second-order statistical parameters are obtained when, apart from the individual values of these attributes at the generic pixel under consideration, the values of the features at *another* pixel are also considered. Obviously, this *other* pixel must be placed at a certain pre-specified position or, in other words, it must meet a spatial restriction or predicate with respect to the generic pixel. By changing this spatial predicate, we get a different set of second-order statistics. Thus, there are as many sets of second-order statistics as there are different pairs of pixels that can be formed with each generic pixel.

Although some pixel attributes other than the grayscale intensity have been proposed, the latter is by far the most commonly, if not the only one used in practice. Thus, focusing on the grayscale level as the only discriminant feature, the generalization of the histogram, which is the optimum information structure for the computation of first-order statistics, to the bidimensional case is what is known as co-occurrence matrix, which is a discrete bidimensional function. The co-occurrence matrix gives the number of times that each grayscale level appears in the digital image for every possible pair of pixels that can be formed considering a given pre-specified spatial predicate.

Let us suppose that we introduce a given spatial predicate consisting of the pair of pixels formed by a generic pixel (i, j) and its neighbor placed at a distance defined by a given vector $\vec{d} = (\Delta i, \Delta j)$. Then, the co-occurrence matrix will be:

$$\begin{gathered} MC\left[I_k, I_l, \vec{d}\right] = card\left\{[(i,j),(i+\Delta i, j+\Delta j)]\right\} \quad / \\ I(i,j) = I_k \,;\, I\left[(i+\Delta i, j+\Delta j)\right] = I_l \\ \\ \forall i,j \quad / \quad 0 \leq i \leq N-1 \,;\, 0 \leq j \leq M-1 \\ \forall k,l \quad / \quad 0 \leq k,l \leq 255 \end{gathered} \tag{48}$$

If all the elements of the co-occurrence matrix are divided by their total number of elements, we get the frequency histogram of the random event $\{I_k, I_l\} \cap \vec{d}$. In other words, we have a concrete realization of the total probability of such event:

$$h(I_k, I_l, \vec{d}) = \frac{MC(I_k, I_l, \vec{d})}{\sum\limits_{k=0}^{255}\sum\limits_{l=0}^{255} MC(I_k, I_l, \vec{d})} \tag{49}$$

Evidently, the co-occurrence matrix or bidimensional histogram conveys more information than the common unidimensional histogram, although it has several important drawbacks. These are; firstly, its computational complexity; secondly, its sensitivity to a proper choice of the spatial predicate and, finally, the so-called curse of dimensionality, which plagues many high-dimensional statistics in the field of pattern recognition [2].

In spite of the above-mentioned disadvantages, the co-occurrence matrix concept has been extensively used as it provides acceptable discrimination efficiency. Of course, the application of the co-occurrence matrix and the conventional histogram to the segmentation process still have the same basic problem arising with any segmentation method based on texture analysis and recognition, i.e. the suitable choice of the shape and size of the window for texture search, detection and discrimination. As mentioned above, too big a window may lead to poor results in the segmentation process, due to the mixture of several different textures, and too small a window can, apart from the extra computational load, produce the loss of the basic texture units. This is a fundamental issue and, as for any other segmentation method exploiting the local discriminant features existing in any specific digital image, the window needs to be carefully designed for each particular application.

12 A Novel Concept: the Frequency Histogram of Connected Elements

By way of a conclusion on texture-based segmentation, we can say that the statistical methods for texture analysis and recognition -by far the most widely studied and applied techniques in this field- can be divided into two main groups: (1) methods based on the unidimensional histogram and (2) methods based on bidimensional histograms, in which the idea of dimension is not associated with the discriminant variable or attribute, but with the notion of space. Thus, the unidimensional histogram is obtained from the values taken by the particular discriminant variable under consideration at each *individual* pixel, which is, obviously, a spatial predicate. Similarly, the bidimensional histogram is obtained from the values taken by the same specific attribute at each pair of pixels having a certain previously defined spatial restriction or predicate.

In other words, what differentiates unidimensional histograms from bidimensional histograms is not the dimension of the space of the possible discriminant variables -which is actually always unidimensional-, but the dimension of the respective spatial predicate. Therefore, the key point is the physical space of the pixels, rather than the space of the discriminant attributes.

Bidimensional histograms, computed from the co-occurrence matrices, provide useful information about the spatial distribution -about a specific two-point spatial relationship in actual fact- of the particular discriminant variable at hand, which is obviously relevant information as far as texture analysis and recognition are concerned. However, they have two serious drawbacks: their excessive computational load and, in particular, the curse of dimensionality that produces a considerable amount of irrelevant information and noise. On the contrary, unidimensional histograms, which are computationally speaking very attractive and even discriminant enough to efficiently solve the segmentation problem in many applications, do not provide as much discriminant information, in particular spatial and textural information, as bidimensional histograms do.

The authors have recently introduced a novel concept, [35] - [36], which intends to capture the strengths of both histograms. This concept, which we have called the *frequency histogram of connected elements* (FHCE), is a generalization of the unidimensional histogram and, at the same time, is related to the co-occurrence matrix and, hence, to the bidimensional histogram as well.

Being a conventional unidimensional histogram, the FHCE incorporates all the computational advantages, in terms of both simplicity and speed, inherent to histogram-based segmentation methods. Simultaneously, it includes information about the spatial distribution of the specific discriminant feature in the digital image, as bidimensional histograms do. The FHCE concept has an additional advantage in comparison to bidimensional histograms, as it is based on a much more powerful spatial function than the simple two-point relationships, typical of bidimensional histograms and co-occurrence matrices. This is the concept of *structuring element* or spatial predicate, which is somewhat, but not entirely, related to the structuring element concept used in morphological image processing [22]. Furthermore, there is yet another interesting advantage of the FHCE, as compared with the conventional unidimensional histogram. This is its flexibility with regard to the range of values of the discriminant variable, which is absolutely rigid in conventional unidimensional histograms. This means we can juggle with an interesting degree of freedom, what we have called the *connectivity level*, for texture analysis and recognition. Let us proceed now with a brief description of the theoretical foundation of the FHCE. But before doing so, we would like to remark that although this novel concept can be applied to any type of discriminant feature, either sensory or abstract, for the sake of clarity, we shall focus only on the typical grayscale intensity.

12.1 The Concept of Neighborhood

The neighborhood concept is central to the understanding of the FHCE concept. Let $\{I(i,j)\}_{NxM}$ be a digital image. If we denote the coordinates of a generic pixel as (i,j), the neighborhood of this pixel, $\mathcal{N}$, is defined as follows:

$$\mathcal{N} \triangleq \{\varphi_{i,j} \subset \{I(i,j)\}_{NxM}\} \tag{50}$$
$$\varphi_{i,j} = \{\forall (k,l) \ / \ D[(k,l),(i,j)] \text{ is true}\}$$

where D is a predicate defined by a distance-based condition. For instance, a valid definition of the neighborhood of a pixel $I(i,j)$ can be given by the set:

$$\varphi_{i,j}^{r,s} = \{\forall (k,l) \ / \ ||k-i|| \leq r \ and \ ||l-j|| \leq s\} \ ; \ r,s \in \mathrm{N} \tag{51}$$

which indicates that the neighborhood of the pixel is formed by a set of pixels whose distances are not greater than two integer values r and s, respectively. Figure 11 shows two instances of neighborhoods.

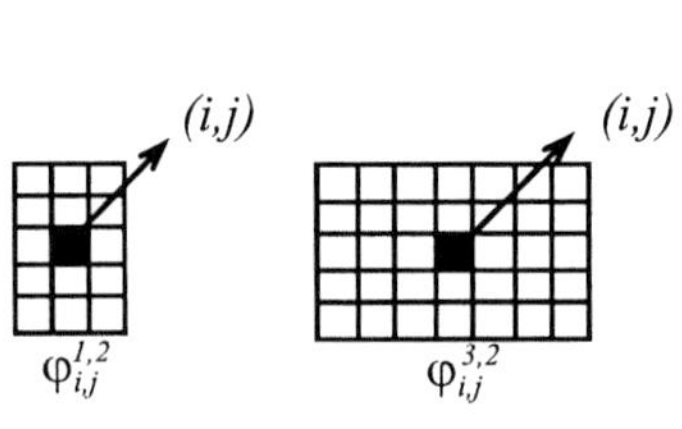

Figure 11: Two instances of neighborhood $(1,2)$ and $(3,2)$ of a generic pixel $I(i,j)$.

12.2 The Connected Element Concept

The concept of connected element is associated with the neighborhood. We mean by a connected element:

$$C_{i,j}(T) \triangleq \varphi_{i,j}^{r,s} \ / \ I(k,l) \subset [T-\varepsilon, T+\varepsilon] \ , \ \forall (k,l) \in \varphi_{i,j}^{r,s} \tag{52}$$

where I is the grayscale intensity or brightness of pixel (k,l). In other words, a connected element is any neighborhood unit such that its pixels have a grayscale level close to a given grayscale level T.

12.3 The Frequency Histogram of Connected Elements

The frequency histogram of connected elements (FHCE) is defined as:

$$\begin{aligned} H(T) &= \sum_{\forall (i,j) \in \{I\}} C_{i,j}(T) \\ 0 \leq T &\leq I_{max} - 1 \end{aligned} \tag{53}$$

That is to say, $H(T)$ approximates a density function for a random event occurring in a digital image $\{I(i,j)\}_{NxM}$. This event is related to the idea of connected element, which in turn is related to the pseudo-random structure of the grayscale intensity distribution of a particular texture. Our main purpose here is to demonstrate that this novel concept is an ideal tool for the automatic analysis of the textures appearing in a variety of digital images. From an intuitive standpoint, we can understand the information conveyed by the function $H(T)$: high values of $H(T)$ indicate that there is a sizeable number of connected elements in the image whose grayscale levels are homogeneous and close to level T.

Let us now show a simple case of building a FHCE and consider the toy image in Figure 12 that has a range of 0 to 7 grayscale levels.

4	6	1	0	0	3	3	7
6	7	1	0	1	3	7	5
3	5	2	1	2	7	5	6
6	6	0	2	2	5	4	7
3	3	1	1	2	4	7	5
4	4	0	0	2	4	5	6

Figure 12: Toy image for the computation of the FHCE.

In this simple, artificial image, there is a pixel ensemble appearing from the third to the fifth column, whose grayscale levels are close to 1. This ensemble or pixel grouping is clearly differentiated from the rest of the image, which possesses a definite random distribution of grayscale levels. First, we shall analyze the image grayscale histogram, shown in Figure 13.

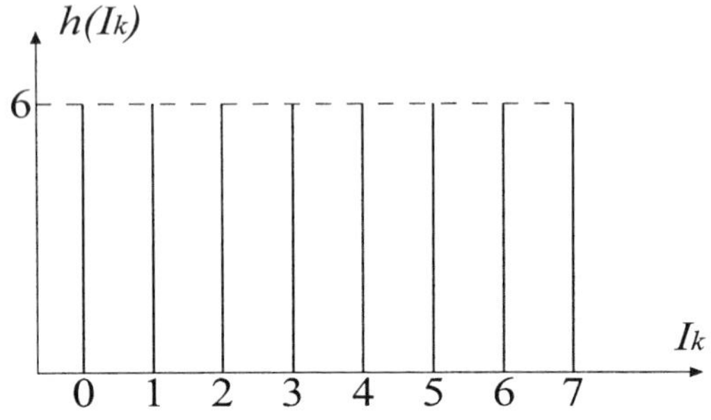

Figure 13: Grayscale intensity histogram of the image in Figure 12.

From the grayscale histogram, it is not possible to discover the above-mentioned spatial distribution of brightness around level 1. The only conclusion that can be inferred from the conventional histogram is the equiprobability of all the brightness levels. However, using the FHCE concept, the existing texture pattern can be discovered, as shown in the sequel. First, we shall build the FHCE taking the following steps.

Step 1. The first action is to define the shape of the neighborhood, which is a heuristic and domain-based process. In most cases, it is reasonably straightforward to choose the optimum dimension for the right neighborhood function. In this simple example, we find that the privileged direction of the two distinct texture regions in our image is vertical and, therefore, we should define a vertical neighborhood:

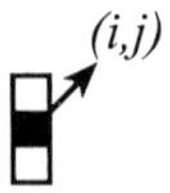

$\varphi_{i,j}^{0,1}$

Step 2. For this neighborhood, we choose the following connected element:

$$C_{i,j}(T) = \varphi_{i,j}^{0,1} \;/\; I(k,l) \subset [T-1, T+1] \;,\; \forall (k,l) \in \varphi_{i,j}^{0,1} \tag{54}$$

Step 3. Having completed the above steps, we can now compute the respective FHCE by just applying equation (53) to get the FHCE shown in Figure 14, where we have ignored the border pixels to avoid ambiguity.

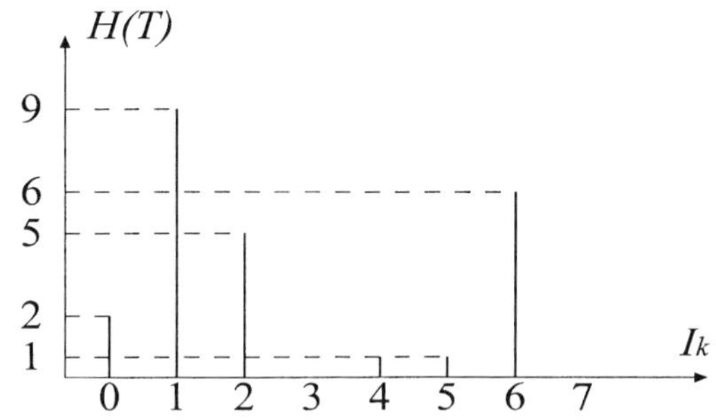

Figure 14: The FHCE of the image in Figure 12.

The pixels that are the centers of a connected element have been highlighted in Figure 15. Note that the grayscale levels of the pixels belonging to the neighborhoods defined must be 1,2 or 3; i.e. $[2-1, 2+1]$.

4	6	1	0	0	3	3	7
6	7	1	0	1	3	7	5
3	5	2	1	2	7	5	6
6	6	0	2	2	5	4	7
3	3	1	1	2	4	7	5
4	4	0	0	2	4	5	6

Figure 15: Pixels that are centers of a connected element have been highlighted.

We can now see the distinctive information provided by the FHCE as compared with the conventional grayscale histogram. This novel function $H(T)$ is bimodal for the toy image of Figure 12. It is, therefore, trivial to discriminate between the two regions of the image with different textures. Thus, if we choose level 3 as the threshold, the respective binary image of Figure 16 is perfect from the segmentation standpoint.

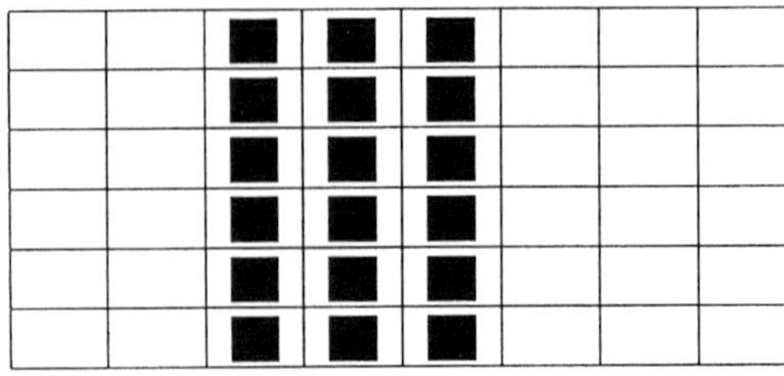

Figure 16: The result of binarizing the toy image with the threshold given by the FHCE.

Obviously, there is no universal connected element valid for any domain application. Thus, there is a critical and domain-dependent step in the FHCE design stage, which is responsible for the selection of the two parameters defining the optimum, connected element: the morphological component and the connectivity level.

12.4 Morphological components

We mean by morphological components those that determine the shape of the connected element; i.e., what we have called neighborhood. Obviously, these parameters are very dependent on the application domain, and final system efficiency will rely on the correct selection of the morphological components. At the beginning of this section, an instance of morphological component was used to build a pixel neighborhood with height $2 * s + 1$ and width $2 * r + 1$. Many different neighborhoods can be defined as shown in Figure 17.

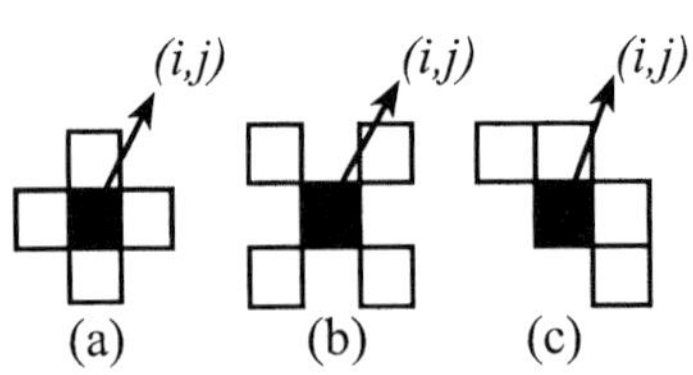

Figure 17: Some instances of morphological components.

The selection of the suitable morphological components for a particular application is not an easy task, and it requires a thorough empirical analysis. Even if specialized algorithms are used to choose the optimum set of morphological components, the final goal of the computer vision application at hand will determine the results. Thus, any knowledge we have about the application domain –i.e. our a priori knowledge of the digital images to be automatically analyzed- is crucial for the correct selection of the morphological components. As an illustration of this fact, just take a look at Figure 18, where different FHCEs have been computed and compared. These FHCEs have been obtained using different morphological components:

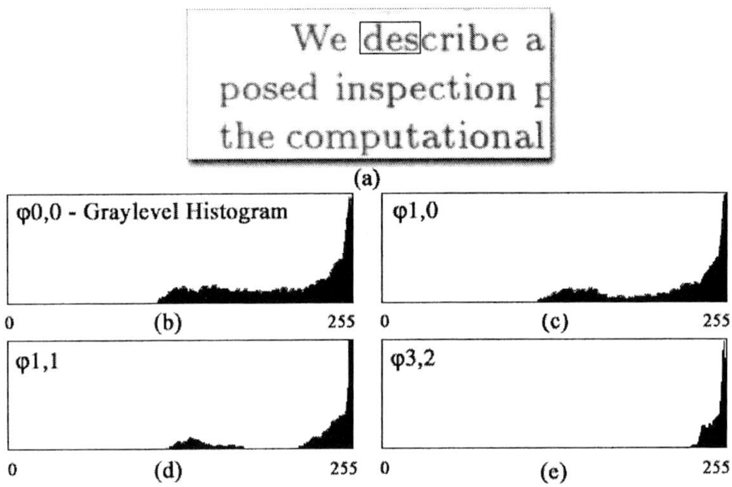

Figure 18: Portion of a digitized document and several FHCEs obtained using different morphological components.

From the FHCE shapes, it is apparent that the correct choice of the morphological components is vital for the final success of the automatic image analysis. Figure 18d is the optimum morphological component from the segmentation standpoint, as the two regions with distinct grayscale level distributions –i.e. text and background- are clearly distinguishable. On the contrary, some morphological components do not provide any information at all about such discrimination, in particular, the morphological components $\varphi^{0,0}$, $\varphi^{1,0}$ and $\varphi^{3,2}$.

12.5 Connectivity level

The connectivity level of a FHCE depends on the value ε that appears in the definition of connected element given by equation (52) and determines the restrictions that every pixel in a neighborhood must meet in order to form a connected element. As is the case for morphological components, the connectivity level is highly dependent on the application domain, and its correct choice is a matter of thorough experimentation. To illustrate this, let us compute several FHCEs with different connectivity levels. As before, we shall use a digitized document for this purpose -see Figure 19-.

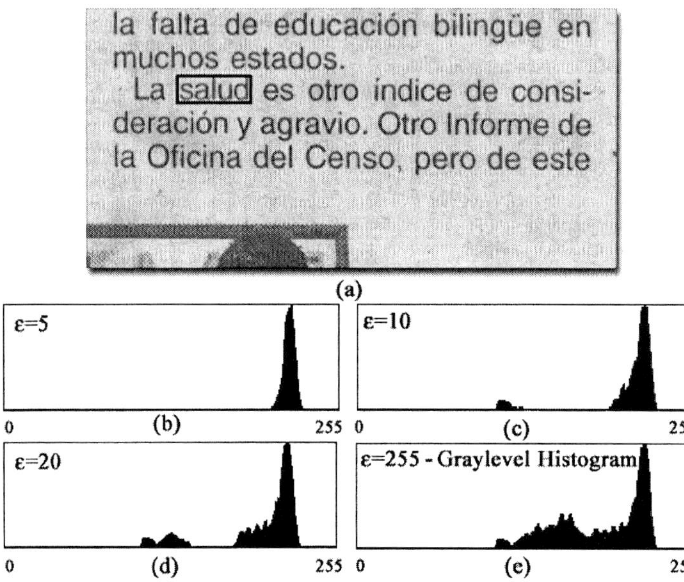

Figure 19: Digitized document and several FHCEs obtained with different levels of connectivity, decreasing from (b) to (e).

The connectivity level decreases from highest connectivity in Figure 19b to null connectivity in Figure 19e. Null connectivity means that every neighborhood defined by a morphological component is considered as a connected element, which in fact leads to the conventional grayscale level histogram. Looking at the different FHCEs in Figure 19 is clear that the choice of an optimum connectivity level is of vital importance. Thus, a clear bimodal distribution appears in Figure 19d, in which the first distribution -on the left-hand side- represents the connected elements formed by the text and the other distribution corresponds to the document background.

12.6 Computational issues

Through experimentation, we have found that the best results are obtained when applying a small window over the whole image. As happens with any other local operator, the key issue is to correctly select the size of the scanning window. For texture analysis, the basic idea is to apply a window whose size is big enough to capture the essential structure of any texture present in the image.

Another important point is the scanning process, i.e. the way the window is passed over the image. The common scanning process is to divide the whole image into a set of non-overlapping, adjacent cells or windows. This solution is very attractive from the computational point of view, but very inefficient for capturing the real textures occurring in the image. Thus, it is advisable to explore the image in more depth, where the extreme option is to apply the scanning window to every pixel in the image. This is extremely cumbersome for computational reasons, although optimum for automatic analysis of textures. The latter option bears a tremendous computational load, as the FHCE must be computed for each pixel as many times as there are possible values of T, while, at the same time, checking whether the neighborhood of the current pixel meets the condition of connected element. Remember that for a typical grayscale level image there are 256 different values of T.

In order to deal with real-time industrial applications, we have implemented an *ad hoc* procedure for obtaining the FHCE, which is based on computing, for each pixel, the maximum and the minimum grayscale level values in its neighborhood. The pixel neighborhood is classified as a connected element if the difference between the maximum and the minimum values divided by the dynamic range of the grayscale level histogram of the window centered on the current pixel is small. For instance:

$$\frac{i_{max} - i_{min}}{I_{max} - I_{min}} < \varepsilon \tag{55}$$

or

$$\left(1 - e^{-\beta\,|i_{max} - i_{min}|}\right) < \varepsilon \tag{56}$$

where

$$\frac{2}{I_{max} - I_{min}} \leq \beta \leq \frac{5}{I_{max} - I_{min}} \tag{57}$$

in which i_{max} and i_{min} are the maximum and minimum grayscale levels in the neighborhood, respectively, I_{max} and I_{min} are the maximum and the minimum brightness levels in the window, and ε is a very small numerical value – for instance, $\varepsilon = 0.1$-. If conditions (55) or (56) hold, then the respective neighborhood is labeled as a connected element, and a new event is added to the FHCE for $T = (i_{max}+i_{min})/2$. More sophisticated estimators of T may be attempted -for instance, the mean or the median grayscale levels of the neighborhood-, but there is a considerable additional computational load and only a slight improvement in the discrimination task.

12.7 A Comparative Study: Digital Angiograms Segmentation

In this section, we shall present the application of the FHCE to the segmentation of digital angiograms, comparing it with some other well-known and thoroughly tested methods. Although there are numerous methods available in the technical literature for angiogram segmentation [38] – [40], we believe that the application of the FHCE concept to this field is clearly beneficial, mainly for two reasons.

Efficiency. Most of the thresholding methods in angiogram segmentation are based on the grayscale level histogram. As will be shown later, the FHCE provides more reliable information for the thresholding problem.

Generalization. The FHCE is able to discriminate among multiple grayscale level distributions in a digital image. Although this section concentrates on the application of the FHCE to a two-class segmentation case - the vessels and the background-, the generalization for multiclass cases is straightforward.

An angiogram is a sequence of X-ray images obtained after injecting an intravascular contrast agent into the circulatory system. The automatic analysis of such images is not intended, under any circumstances, to supplant the knowledge, experience and analytical role of medical specialists. As there is no evidence of any such hypothetical supplanting in the foreseeable future, the main goal of automatic analysis -and, in particular, the segmentation process- of digital angiograms is to improve their quality, extract as much information as possible and help the medical expert with his/her diagnosis. Apart from the inherent difficulties of the automatic analysis and segmentation of digital angiograms, there is an additional problem caused by the considerable noise induced in the acquisition process, because of the dynamics involved -heart movements and contrast agent evolution-. We claim that our approach based on the FHCE concept is an attractive alternative for dealing with these problems, as compared with other existing methods, [38] - [40].

The main objective for the segmentation process in digital angiograms is to discriminate between the vessels and the background. By exploiting the empirical evidence that both classes or patterns have different grayscale intensity distributions, an automatic thresholding process can, in principle, be applied to get a first approximation of the real distribution of both classes in the particular image under analysis. Afterwards, a more specialized local segmentation process can be activated to take into account the spatial dependencies -i.e. the texture distributions- within the image.

Some segmentation methods, like majority vote filtering and iterated conditional modes (ICM) algorithm, can be used for the segmentation process. They are relatively simple and quite powerful segmentation procedures and are particularly well adapted to the kind of segmentation problems arising in digital medical images as considered here, where the objective is to discriminate between two different textures.

Both techniques perform extremely well for images with a clear difference in the grayscale intensity distributions of each pattern, as depicted in Figure 20, in which the vessels have been perfectly discriminated from the background.

However, there is a rather poor contrast between the vessels and the background in many regions of digital angiograms, due to the peculiarities of the acquisition process. For these regions, the results obtained with the two above adaptive Bayesian methods are clearly far from optimal, as illustrated in Figure 21, where the vascular structures have not been adequately discriminated from the background. As will be shown in the sequel, the FHCE method performs very well in segmenting the vascular structures in these low-contrast regions of the digital angiogram.

To begin with, as is the case with any other local segmentation method, the key issue when applying the FHCE approach is to correctly select the size of the scanning window. In texture analysis, the basic idea is to apply a window whose size is big enough to capture the essential structure of any texture present in the image. In our particular application to the segmentation of digital angiograms, after exhaustive experimentation, we have found that a window of 40x30 pixels seems to be optimum in most cases. The scanning process is another important point. As an excellent compromise between computation times and discrimination efficiency, we have applied a scanning process where two consecutive windows are half-overlapping.

The next step is to select the two basic FHCE variables: (a) the morphological component and (b) the connectivity level. After some experimen-

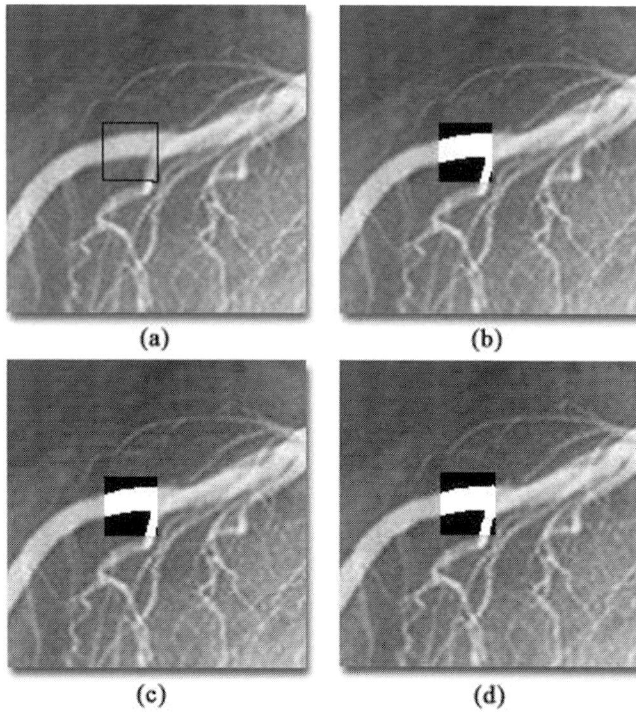

Figure 20: In (a), the region of interest to be segmented in the original image is denoted by a rectangle. In (b), (c) and (d), the majority vote filter, the ICM algorithm and the FHCE segmentation results, respectively, are shown.

tation, the optimum *morphological component* happens to be the vertical neighborhood $\varphi_{i,j}^{0,1}$, and the *connectivity level* holds the following grayscale distribution:

$$\frac{i_{max} - i_{min}}{I_{max} - I_{min}} < 0.1 \tag{58}$$

where i_{max} and i_{min} are the maximum and the minimum grayscale values in the neighborhood, respectively, and I_{max} and I_{min} are the maximum and the minimum grayscale values in the whole image.

For these values, the automatic thresholding of the respective FHCE is straightforward, and the vascular structures or blood vessels are efficiently segmented from the background. Figure 22 shows the results obtained with the FHCE method for the two angiograms of Figure 21. In particular, note

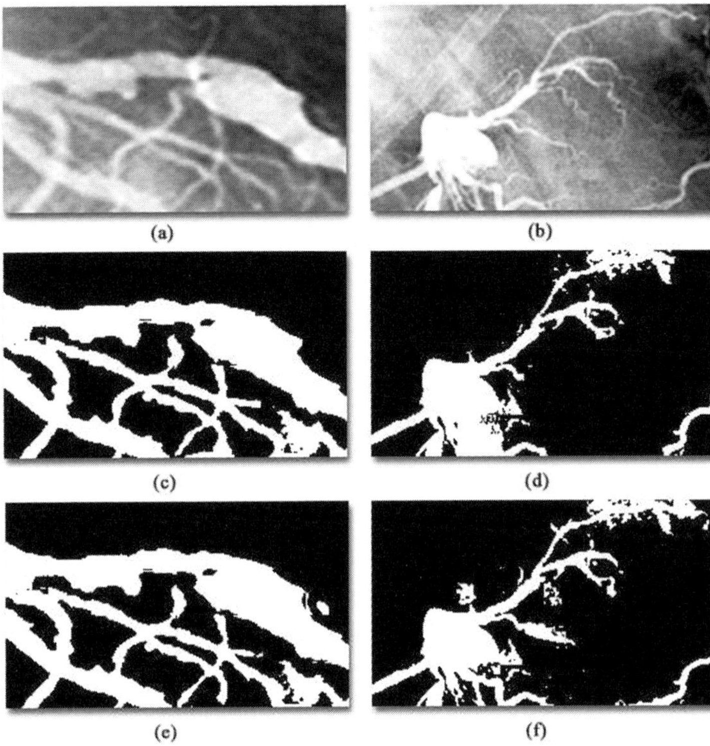

Figure 21: In (a) and (b), two low-contrast angiograms are shown. In (c)-(d) and (e)-(f), the majority filter and the ICM algorithm segmentation results, respectively, are shown for each individual angiogram. Note the poor results, as the blood vessels are not properly discriminated.

the excellent segmentation of the vascular structures in the low-contrast regions of the angiograms, as compared with Bayesian methods.

12.8 Concluding Remarks

This chapter has reviewed some of the main issues related to the segmentation problem in the automatic analysis and recognition of digital images. As a first step, the image segmentation process was modeled as a typical pattern recognition problem, which means that, in principle, the vast body of theoretical, methodological and practical results available in this discipline can be applied. However, due to the peculiarities and, in particular, to the intrinsic spatial nature of digital images, these pattern recognition

Figure 22: Results of the segmentation process applying the FHCE method for the two angiograms of Figure 21. Note the efficiency in segmenting the low-contrast regions of the angiograms.

techniques have to be combined with specific digital image processing techniques. As far as the segmentation problem is concerned, the key point is to exploit the spatial information existing in any particular digital image. Thus, the remainder of the chapter is devoted to a critical review, under the umbrella of the pattern recognition paradigm, of some of the most popular and widely tested segmentation methods. Finally, the chapter introduces a novel concept recently proposed by the authors, called histogram of connected elements, which is a generalization of the conventional histogram in two directions: the discriminant feature space and the physical space. Comparisons carried out with other well-established segmentation methods have shown the advantages and the practical interest of this novel concept for the segmentation of textured digital images.

References

[1] A.K. Jain, R.P.W. Duin, J. Mao, Statistical Pattern Recognition: A Review, *IEEE Transactions on Pattern Analysis and Machine Intelligence*, Vol. 22 No. 1 (2000) pp. 4-38.

[2] R.O. Duda, P.E. Hart, D.G. Stork, *Pattern Classification*, Second Edition (John Wiley and Sons, Inc., New York, 2001).

[3] D. Chen, X. Cheng, A Simple Implementation of the Stochastic Discrimination for Pattern Recognition, in F.J. Ferri, J.M. Inesta, A. Amin, P. Pudil (eds.) *Advances in Pattern Recognition*, LNCS 1876 (Springer, Berlin, 2000) pp. 882-887.

[4] T. Hastie, R. Tibshirani, J. Friedman, *The Elements of Statistical Learning*, (Springer, New York, 2001).

[5] K. Fukunaga, Statistical Pattern Recognition, in C.H. Chen, L.F. Pau and P.S.P. Wang (eds.) *Handbook of Pattern Recognition and Computer Vision*, (Singapore, World Scientific, 1993) pp. 33-60.

[6] R.M. Haralick, L.G. Shapiro, *Computer and Robot Vision*, Vol I (Addison-Wesley, Reading, Mass., 1992).

[7] M. Sonka, V. Hlavac, R. Boyle, *Image Processing, Analysis and Machine Vision*, Second Edition (PWS Publishing, Pacific Grove, Ca., 1999).

[8] C.A. Glasbey, G.W. Horgan, *Image Analysis for the Biological Sciences*, (John Wiley and Sons, New York, 1995).

[9] J. Besag, On the Statistical Analysis of Dirty Pictures (with discussions), *Journal of the Royal Statistical Society*, Series B 48 (1986) pp. 259-302.

[10] J. Besag, Towards Bayesian Image Analysis, *Journal of Applied Statistics*, Vol. 16 No. 3 (1989) pp. 295-406.

[11] A. Rosenfeld, R.A. Hummel, S.W. Zucker, Scene Labelling by Relaxation Operations, *IEEE Transactions on Systems, Man, and Cybernetics*, Vol. 6 No. 6 (1976) pp. 420-433.

[12] A. Rosenfeld, C. Kak, *Digital Picture Processing*, Second Edition, Vol. 2 (Academic Press, 1982).

[13] J.A. Richards, X. Jia, *Remote Sensing Digital Image Analysis*, Third Edition (Springer, 1999).

[14] S. Geman, D. Geman, Stochastic Relaxation, Gibbs distributions and the Bayesian Restoration of Images, *IEEE Transactions on Pattern Analysis and Machine Intelligence*, Vol. 6 No. 6 (1984) pp. 722-741.

[15] R. Chelappa, A. Jain, *Markov Random Fields. Theory and Applications*, (Academic Press, 1993).

[16] S.Z. Li, *Markov Random Fields Modeling in Image Analysis*, (Springer, 2001).

[17] G.L. Gimelfarb, *Image Textures and Gibbs Random Fields*, (Kluwer Academic Publishers, Dordrecht, 1999).

[18] P.K. Sahoo, S. Soltani, A.K.C. Wong, SURVEY: A Survey of Thresholding Techniques, *Computer Vision, Graphics, and Image Processing*, Vol. 41 (1988) pp. 233-260.

[19] J.R. Parker, *Algorithms for Images Processing and Computer Vision*, (John Wiley and Sons, 1997).

[20] M. Seul, L. O'Gorman, M.J. Sammon, *Practical Algorithms for Image Analysis*, (Cambridge University Press, 2000).

[21] B. Batchelor, F. Waltz, *Intelligent Machine Vision: Techniques, Implementations and Applications*, (Springer, 2001).

[22] S.J. Sangwine, R.E.N. Horne (eds.), *The Colour Image Processing Handbook*, (Kluwer Academic Publishers, Boston, 1998).

[23] K. Abend, T.J. Harley, L.N. Canal, Classification of Binary Random Patterns Images, *IEEE Transactions on Information Theory*, Vol. 11 (1965) pp. 538-544.

[24] Ch. Therrien, An Estimation-Theoretic Approach to Terrain Image Segmentation, *Computer Vision, Graphics, and Image Processing*, Vol. 23 (1983) pp. 313-326.

[25] Ch. Therrien, *Decision, Estimation and Classification*, (John Wiley and Sons, New York, 1989).

[26] R.W. Conners, C.A. Harlow, A Theoretical Comparison of Texture Algorithms, *IEEE Transactions on Pattern Analysis and Machine Intelligence*, Vol. 2 No. 3 (1980) pp. 204-222.

[27] M.D. Levine, *Vision in Man and Machine*, (Mc Graw-Hill, New York, 1985).

[28] M.A. Patricio, D. Maravall, Segmentation of Text and Graphics/Images using the Gray-Level Histogram Fourier Transform, in F.J. Ferri, J.M. Iesta, A. Amin, P. Pudil (eds.) *Advances in Pattern Recognition*, LNCS 1876 (Springer, Berlin, 2000) pp. 757-766.

[29] W.K. Pratt, *Digital Image Processing*, Third Edition (John Wiley and Sons, New York, 2001).

[30] C.C. Gotlieb, H.E. Kreyszig, Texture Descriptors Based on Coocurrence Matrices, *Computer Vision, Graphics, and Image Processing*, Vol. 51 No. 1 (1990) pp. 70-86.

[31] M. Tuceryan, A.J. Kain, Texture Analysis, in C.H. Chen, L.F. Pau and P.S.P. Wang (eds.) *Handbook of Pattern Recognition and Computer Vision*, (Singapore, World Scientific, 1993) pp. 235-276.

[32] R. Chelappa, R.L. Kashyap, B.S. Manjunath, Model-Based Texture Segmentation and Classification, in C.H. Chen, L.F. Pau and P.S.P. Wang (eds.) *Handbook of Pattern Recognition and Computer Vision*, (Singapore, World Scientific, 1993) pp. 277-310.

[33] B. Julesz, Textons, the Elements of Texture Perception and their Interations, *Nature*, Vol.290, March 12 (1981) pp. 84-92.

[34] J.S. Weszka, C.K. Dyer, A. Rosenfeld, A Comparative Study of Texture Measure for Terrain Classification, *IEEE Transactions on Systems, Man, and Cybernetics* Vol.6 No.4 (1976) pp. 269-285.

[35] M.A. Patricio and D. Maravall, A Novel Concept: the Connected Elements Histogram and its Application to Document Binarization, *In Proc. of the IX Spanish Symposium on Pattern Recognition and Image Analysis*, Vol. I (2001) pp. 43-48.

[36] M.A. Patricio and D. Maravall, Wood Texture Analysis by Combining the Connected Elements Histogram and Artificial Neural Networks, in J. Mira, A. Prieto (eds) *Bio-inspired Applications of Connectionism*, LNCS 2085, (Springer, Berlin, 2001) pp. 160-167.

[37] J. Serra, P. Soille, *Mathematical Morphology and its Application to Image Processing*, (Kluwer Academic Publishers, Dordrecht, 1994).

[38] P.H. Eichel, E.J. Delp, K. Koral, A.J. Buda, Method for a Fully Automatic Definition of Coronary Arterial Edges from Cineangiograms, *IEEE Transactions on Medical Imaging*, Vol. 7 No. 4 (1988) pp. 313-320.

[39] R. Nekovei, T. Sun, Back-Propagation Network and its Configuration for Blood Vessel Detection in Angiograms, *IEEE Transactions on Neural Networks*, Vol.6 No.1 (1995) pp. 64-72.

[40] R. Poli, G. Balli, An Algorithm for Real-Time Vessel Enhancement and Detection, *Computer Methods and Programs in Biomedicine*, Vol. 52 (1997) pp. 1-22.

Prototype Extraction for k-NN Classifiers using Median Strings[1]

Carlos D. Martínez-Hinarejos, Alfons Juan, Francisco Casacuberta
Departament de Sistemes Informàtics i Computació
Institut Tecnològic d'Informàtica
Universitat Politècnica de València, Camí de Vera, s/n, 46071, València, Spain
E-mail: {cmartine,ajuan,fcn}@iti.upv.es

Contents

1 Introduction

The k-Nearest Neighbour rule is one of the most popular non-parametric classification techniques in Pattern Recognition. This technique requires

[1]This work was partially supported by the Spanish MCT under projects TIC2000-1703-C03-01 and TIC2000-1599-C02-01.

D. Chen and X. Cheng (eds.),
Pattern Recognition and String Matching, 465–476.

a set of good prototypes to represent pattern classes. One possibility is to define the given training set as the set of prototypes. Obviously, this approach presents a high computational cost if the training set is large. Alternatively, clustering techniques allow for the description of a training corpus in terms of clusters. A cluster is formed by patterns with certain simmilarities [1]. These clusters can be represented by a set of prototypes. The selection of adequate prototypes is one of the most important problems in Pattern Recognition.

The usual prototype selection techniques are editing techniques and condensing techniques. Editing techniques reduce the prototypes set by eliminating the prototypes which are bad classified using the k-NN rule; this technique provides an error reduction (the outlier prototypes, which are an error source, are eliminated) and a time complexity reduction (the total number of prototypes is reduced). The most popular editing algorithms are the Wilson algorithm [2] and the multiedit algorithm [3]. Condensing techniques reduce the prototypes set by keeping only the prototypes which allow all the prototypes from the original set to be correctly classified using the k-NN rule. These techniques are usually based on the *consistency* property [4].

An alternative prototype selection technique can be based on extracting the most centered prototype for each class, or on building a new prototype which is the most centered point in the representation space. Data points are usually feature vectors and an appropriate prototype for each cluster can be computed from the mean vector of the cluster. However, there are many applications where strings (sequences of discrete symbols) are more appropriate to represent data [5] (character contours, vector quantization in speech decoding, chromosomes, etc.). The optimal prototype for a cluster of strings is the *(generalized) median string*, which minimizes the sum of distances to each string in the cluster.

In general, the problem of finding the median string of a given set of strings is a **NP**-Hard problem [6], and only approximations to the median string could be achieved in a reasonable time. Several heuristic approximations to median string have been proposed in the last few years. One of the most popular heuristics is using the *set median string* as an approximation to the median string. The search of the set median string is constrained to the given input set, and is a polynomial problem. However, the set median string cannot be a good approximation to the median string in many cases (for example, as in the extreme case of a set of two strings).

Another heuristic approach was proposed in [7] for computing the median string by systematic perturbation of the set median string of the set; this approach is not time-consuming if the median is in the vicinity of the set median. Other approaches were proposed in [8, 9], and a method for more complex structures was proposed in [10].

A new greedy simple algorithm was proposed in [11] to efficiently compute a good approximation to the median string of a set. An improvement on this algorithm was proposed in [12], using iterative refinement of an initial solution. A new definition of median string was also proposed in [13] and was applied to Nearest Neighbour classifiers. The preliminary results obtained using a chromosome corpus allowed us to show that such proposals were appropriate for computing a good approximation to the median string in small clusters. The application of median strings to large clusters is presented in this work, using the same chromosome corpus which was used in previous experiments.

This chapter is organized as follows: Section 2 defines the median and set median strings and describes the methods used to obtain their approximations. Section 3 describes the corpus used and the prototype extraction. Section 4 describes the experiments carried out. Section 5 presents conclusions and some interesting open questions.

2 Median string

In order to define the concepts of median string and set median string, it is necessary to previously define the concept of distance between two strings. Given an alphabet Σ (which is a finite set of symbols), a string over Σ, s, is defined as a finite sequence from the symbols of Σ. We denote by $|s|$ the length (number of symbols) of s. The free monoid, Σ^*, over Σ, is the set of all the strings over Σ.

Given two strings s and t from Σ^*, the most common definition of distance between s and t is the *edit distance* (also known as *Levenshtein distance*) [14]. This distance computes the optimal sequence of editing operations (insertion, deletions and substitutions) needed to transform string s into string t. This distance can be computed in polynomial time [14].

The main problem with the traditional edit distance is that it depends on the length of the strings being compared. Therefore, a new definition called *normalized edit distance* was proposed in [15] , which is length-independent and can also be computed in polynomial time in the average case.

The distance value is usually defined using a set of weights for each basic edit operation, in an attempt to achieve the lowest value of the sum of the weights of the edit operations used. The weight of the substitution of the symbol a by the symbol b is denoted by $\gamma(a, b)$; the weight of the deletion of the symbol a is denoted by $\gamma(\lambda, a)$, and the weight of the insertion of the symbol a by $\gamma(a, \lambda)$. Note that not all possible edit weights define a true *distance* between two strings. To compute the Levenshtein distance, $\gamma(a,b) = 1$ if $a \neq b$, $\gamma(a,a) = 0$, $\gamma(\lambda, a) = 0$ and $\gamma(a, \lambda) = 0$.

Once we have defined the concept of distance between two strings, we can define the concepts of median string and set median string. Let Σ be an alphabet. Given a finite set of string $S \subset \Sigma^*$, the *median string* of S (m_S) is given by:

$$m_S = \operatorname*{argmin}_{t \in \Sigma^*} \sum_{r \in S} d(t,r) \tag{1}$$

where d is the distance used.

However, in degenerate cases, this definition does not provide adequate prototypes (for example, in a set with two strings, either of these strings satisfies the definition). Therefore, using the similarity between the concept of median string and the concept of mean vector for Euclidean spaces, the following definition can be used:

$$m_S = \operatorname*{argmin}_{t \in \Sigma^*} \sum_{r \in S} (d(t,r))^2 \tag{2}$$

The definition of the set median string is very similar to the one given in Equation 1, but is constrained to the given set S. Therefore, the *set median* of the set S (sm_S) is given by:

$$sm_S = \operatorname*{argmin}_{t \in S} \sum_{r \in S} d(t,r) \tag{3}$$

Obtaining the set median string of a set S has a time complexity of $O(l_S^2 \cdot |S|^2)$, where $l_S = \max_{s \in S} |s|$ and $|S|$ is the number of strings in S. Although obtaining the set median string is less time-consuming than obtaining any approximation to the median string, the given prototypes can be worse than the ones obtained by the approximations. Nevertheless, the set median string can be used as an initial solution to the algorithm which gives the approximation to the median string.

Using the set median as initialization is not the only alternative. We have used a greedy technique to compute another initial string. This greedy technique generates a sequence of prefixes of such an initial string, choosing each prefix from a set of hypotheses. This set of hypoteses is built by adding one symbol from Σ to the previous prefix computed. The new prefix is the one that achieves the least edit distance to the prefix set of S. The stopping criteria is based on the accumulated distance to the set S (the process stops when this accumulated distance increases with respect to the previous iteration).

Therefore, this initialization process can be described using the following algorithm, which is fully described in [11]:

1. Let M^0 be the current hypothesis (initially, the empty string); let sc be the score (initially, $sc = \infty$)
2. For each symbol $b \in \Sigma$,
   - Make $M_b^0 = M^0 \cdot b$
   - Compute $d_b = \sum_{t \in S} \min_{p \in \{q \mid q \text{ is a prefix of } t\}} d(M_b^0, p)$
3. Let $a \in \Sigma$ be the symbol such that d_a is the minimum of all d_b, $\forall b \in \Sigma$
4. Make $sc_{new} = \sum_{t \in S} d(M_a^0, t)$
5. If $sc_{new} < sc$ then
   - Make $M^0 = M_a^0$ and $sc = sc_{new}$
   - Go to step 2
6. Else stop and return M^0

The time complexity of this algorithm is $O(l_S^2 \cdot |S| \cdot |\Sigma|)$.

The obtained initial string can be improved by successive refinements using an iterative process. In this process, the edition operations (insertion, deletion and substitution) are applied to each position on the current string, in an attempt to reduce the accumulated distance to S with respect to the current string. The process is repeated until there is no improvement. More specifically:

For each position i in the current approximated median string M

1. Build alternatives

   Substitution: Make $M_{sub} = M$. For each symbol $a \in \Sigma$

   - Make M'_{sub} the result string of substituting the ith symbol of M by symbol a.
   - If the accumulated distance of M'_{sub} to S is lower than the accumulated distance from M_{sub} to S, then make $M_{sub} = M'_{sub}$.

Deletion: Make M_{del} the result string of deleting the ith symbol of M.

Insertion: Make $M_{ins} = s$. For each symbol $a \in \Sigma$

- Make M'_{ins} the result of adding a at position i of M.
- If the accumulated distance from M'_{ins} to S is lower than the accumulated distance from M_{ins} to S, then make $M_{ins} = M'_{ins}$.

2. Choose an alternative: from the set $\{M, M_{sub}, M_{del}, M_{ins}\}$, take the string M' with the least accumulated distance to S. Make $M = M'$.

M is the returned value at the end of the process. The time complexity of this algorithm is $O(l_S^3 \cdot |S| \cdot |\Sigma|)$ for each iteration. The complexity of obtaining this approximation is clearly greater than the one to obtain the set median string (because usually $l \cdot |\Sigma| > |S|$ and more than one iteration is needed to converge). Thus, this process is more time-consuming than the process of obtaining the set median string.

3 Experimental Framework

3.1 Chromo Corpus

The data used in the experiments was extracted from a database of approximately 7,000 chromosomes that were classified by cytogenetic experts [17]. Each digitized chromosome image was automatically transformed into a string by using the procedure illustrated in Fig. 1. This procedure starts by obtaining an idealized, one-dimensional density profile that emphasizes the band pattern along the chromosome. The idealized profile is then mapped into a string composed of symbols from the alphabet $\{1, 2, 3, 4, 5, 6\}$. Then, this string is difference coded to represent signed differences of successive symbols, using the alphabet $\Sigma = \{e, d, c, b, a, =, A, B, C, D, E\}$ ("=" for a difference of 0; "A" for +1; "a" for -1; etc.). See [18] for a detailed description of this procedure.

The chromosome dataset comprises 200 string samples for each of the 22 non-sex chromosome types, i.e. a total of 4400 samples. An additional piece of information contained in this dataset is the location of the *centromere* in each chromosome string. However, this position is difficult to accurately determine in a fully automatic way and, thus, we have decided not to use it in this work.

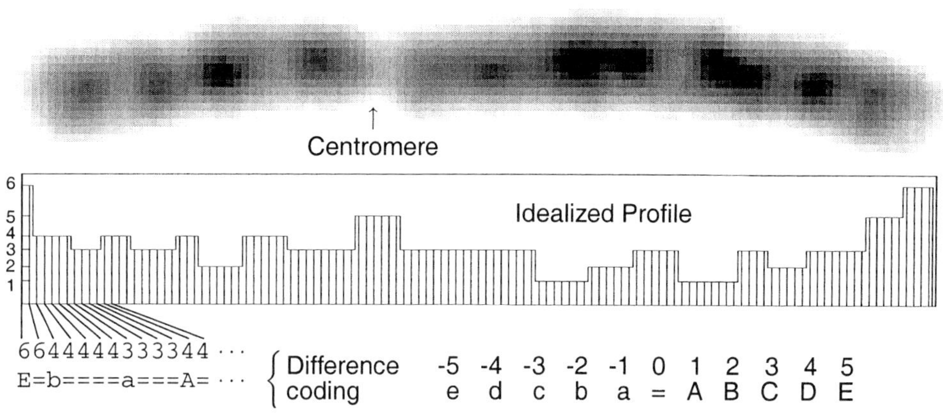

Figure 1: Chromosome preprocessing. Each digitized chromosome image is transformed into a string by first computing an idealized, one-dimensional density profile and then difference-coding successive density values.

3.2 Median String Extraction

In the prototype extraction, the *normalized edit distance* [16] was used. We used this distance with the matrix of editing weights shown in Table 1. This table shows that the substitution of any symbol in Σ by other symbol in Σ is defined as its absolute difference between the signed differences. The values associated to the insertion and deletion weights are the same because of the symmetrical interpretations of insertion and deletion operations. These values are chosen so that any substitution involving $s \in \Sigma$ has a lower cost than the equivalent deletion-insertion/insertion-deletion pair.

This matrix of editing weights gives a better tuning of the distance function than the matrix traditionally used [16], although the only changes are the values for the insertion and deletion weights of the symbols in $\{c, b, a, =, A, B, C\}$ (in previous works, these weights were always 5). The adequacy of these weights to this corpus has been demonstrated by the classification errors obtained using a 2-fold cross-validation, where both sets are chosen to have 100 samples from each of the 22 non-sex chromosome types (i.e., 2,200 samples in each set). The results obtained are compared with several other techniques in Table 2), using $k = 12$ (the value that achieves the best classification results). These results show the excellent performance of the k-Nearest Neighbour classifier.

Table 1: Matrix of editing weights for substitutions, insertions (right column) and deletions (bottom row).

γ	e	d	c	b	a	=	A	B	C	D	E	λ
e	0	1	2	3	4	5	6	7	8	9	10	5
d	1	0	1	2	3	4	5	6	7	8	9	5
c	2	1	0	1	2	3	4	5	6	7	8	4
b	3	2	1	0	1	2	3	4	5	6	7	4
a	4	3	2	1	0	1	2	3	4	5	6	3
=	5	4	3	2	1	0	1	2	3	4	5	3
A	6	5	4	3	2	1	0	1	2	3	4	3
B	7	6	5	4	3	2	1	0	1	2	3	4
C	8	7	6	5	4	3	2	1	0	1	2	4
D	9	8	7	6	5	4	3	2	1	0	1	5
E	10	9	8	7	6	5	4	3	2	1	0	5
λ	5	5	4	4	3	3	3	4	4	5	5	∞

Table 2: Error rates and their corresponding 95% confidence intervals for several classification techniques tested on the chromosome database under the same conditions.

Classification technique	Error %	95% conf.
k-NN (Normalized Distance)	3.9	$[3.3, 4.5]$
Constrained Markov Nets [19]	7.3	$[6.5, 8.1]$
ECGI algorithm [20]	7.5	$[6.7, 8.3]$
Multilayer Perceptron [20]	9.1	$[8.3, 10.0]$
Hidden Markov Models [21]	9.3	$[8.5, 10.2]$

4 Experiments and Results

The experiments are focused to demonstrate the better performance of the median string approximations when using k-NN classifiers [1] compared to the performance of the set median when prototypes are extracted from big clusters (for example, more than 10 strings per cluster). The results for prototypes extracted from small clusters were reported in [22]. Note that the first result reported in Table 2 corresponds to the extreme case (not in the goal of the experiments presented in this chapter) when each cluster is composed by only one training sample (the smallest cluster). In this case, each class is represented by all the training samples of the class.

The same 2-fold cross-validation division on the chromo corpus (as in

Section 3.2) was performed. Each class of chromosomes was divided into several clusters, from 1 (i.e., no clustering) up to 9 clusters, in order to get several prototypes for each class. The set median and four different approximations to the median string were extracted from each cluster. These approximations were obtained using the two initial strings (set median and greedy median string) and the two definitions of distance.

After obtaining the prototypes, several classification experiments were performed using a classical k-NN classifier [1] , using k values of 1, 3, 6, 9, 12 and 15. We present some error-classification results in Figure 2. This figure shows only classification results for $k \in \{1, 3, 12\}$ and for only one type of approximated median string (with set median initialization and optimization for the definition given in Equation 2). We only report results with one kind of approximated median string due to the insignificant differences in classification results given by the four kinds of approximated median strings obtained.

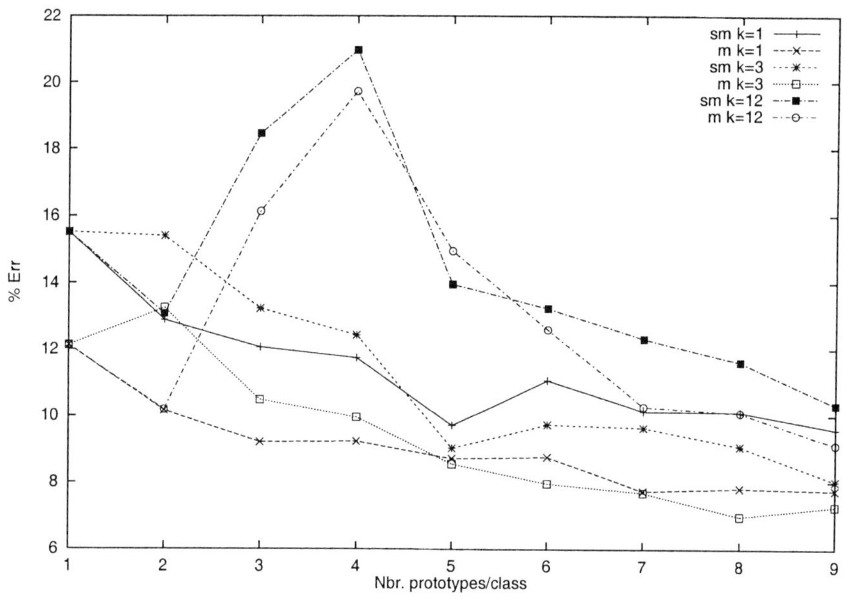

Figure 2: k-NN classification error rates for $k \in \{1, 3, 12\}$ as a function of the number of clusters/prototypes (from 1 to 9 per class), for set median (sm) and approximated median string (m).

Figure 2 shows two main properties. On the one hand, you can see that the greater the k value, the worse the error rates are. This fact is justified by the low number of prototypes used (there are not enough prototypes to achieve good results when k is greater than a moderate value). On the other hand, you can see that the median string achieves better classification results than the set median string (except in one case, when there are 5 prototypes per class and $k = 12$). The relative improvement in the classification results varies irregularly depending on the k value, but it is approximately 20 % in most cases (specially for a low number of prototypes).

5 Conclusions and Open Questions

The results of the experiments carried out demonstrate that the proposed approximations to the median string are better prototypes than the set median for big clusters when using a k-NN classifier. These results also show us the importance of choosing a good k value which depends on the number of prototypes obtained per class. The experiments also showed that there are no significant differences between the severals approximations proposed, although we do not show results with the other approximations for the sake of simplicity.

The most important question is to extend the generality of these results, that is, use approximated median strings in other classification tasks in order to confirm the improvement with respect to set median and other classification methods. Other problems to be worked on are reducing the time complexity of the approximation methods (for example, using beam-search in the distance calculation), applying these methods to cyclic strings, comparing them with other approximation methods and adapting these methods to sequences of vectors.

References

[1] R. O. Duda, P. Hart, and D. G. Stork, *Pattern Classification*, (John Wiley, 2001).

[2] D. L. Wilson, Asymptotic properties of nearest neighbour rules using edited data, *IEEE Transactions on Systems, Man and Cybernetics*, 2 (1972) pp. 408–421.

[3] P. Devijver and J. Kittler, *Pattern Recognition: A Statistical Approach.* Prentice Hall, Inc., London, 1982.

[4] P. E. Hart, The condensed nearest neighbour rule, *IEEE Transactions on Information Theory*, 14 (1968), pp. 515–516.

[5] Fu, K. S., 1982. Syntactic Pattern Recognition. Prentice-Hall.

[6] C. de la Higuera and F. Casacuberta, The topology of strings: two np-complete problems, *Theoretical Computer Science* 230 (2000) pp. 39–48.

[7] T. Kohonen, Median strings, *Pattern Recognition Letters* 3 (1985) pp. 309–313.

[8] F. Kruzslicz, A greedy algorithm to look for median strings, in: *Abstracts of the Conference on PhD Students in Computer Science*, (Institute of informatics of the József Attila University, 1988).

[9] I. Fischer and A. Zell, String averages and self-organizing maps for strings, in: *Proceeding of the Second ICSC Symposium on Neural Computation*, (2000) pp. 208–215.

[10] X. Jiang, A. Münger, and H. Bunke, On Median Graphs: Properties, Algorithms, and Applications, *IEEE Transactions on Pattern Analysis and Machine Intelligence* 23 (10) (2001) pp. 1144–1151.

[11] F. Casacuberta and M. de Antonio, A greedy algorithm for computing approximate median strings, in: *Proceedings of the VII Simposium Nacional de Reconocimiento de Formas y Análisis de Imágenes*, (1997) pp. 193–198.

[12] C. D. Martínez, A. Juan and F. Casacuberta, Use of Median String for Classification, in: *Proceedings of the 15th International Conference on Pattern Recognition*, (Vol. 2, Barcelona, Spain, 2000) pp. 907–910.

[13] C. Martínez, A. Juan and F. Casacuberta, Improving classification using median string and nn rules, in: *Proceedings of IX Simposium Nacional de Reconocimiento de Formas y Análisis de Imágenes*, (2001) pp. 391–394.

[14] R. Wagner and M. Fisher, The string-to-string correction problem. *Journal of the ACM* 21 (1974) pp. 168–178.

[15] E. Vidal, A. Marzal and P. Aibar, Fast computation of normalized edit distances, *IEEE Transactions on Pattern Analysis and Machine Intelligence* 17 (9) (1995) pp. 899–902.

[16] A. Juan, and E. Vidal, On the Use of Edit Distances and an Efficient k-NN Search Technique (k-AESA) for Fast and Accurate String Classification, in: *Proceedings of the 15th International Conference on Pattern Recognition*, (Vol. 2, Barcelona, Spain, 2000) pp. 680–683.

[17] C. Lundsteen, J. Philip and E. Granum, Quantitative Analysis of 6895 Digitized Trypsin G-banded Human Metaphase Chromosomes, *Clinical Genetics* 18 (1980) pp. 355–370.

[18] E. Granum and M. Thomason, Automatically Inferred Markov Network Models for Classification of Chromosomal Band Pattern Structures, *Cytometry* 11 (1990) pp. 26–39.

[19] J. Gregor and M. G. Thomason, A Disagreement Count Scheme for Inference of Constrained Markov Networks, in: L. Miclet and C. de la Higuera (Eds.), *Grammatical Inference: Learning Syntax from Sentences*, (Vol. 1147 of Lecture Notes in Computer Science, Springer, 1996) pp. 168–178.

[20] E. Vidal, and M. J. Castro, Classification of Banded Chromosomes using Error-Correcting Grammatical Inference (ECGI) and Multilayer Perceptron (MLP), in: *Proceedings of the VII Simposium Nacional de Reconocimiento de Formas y Análisis de Imágenes*, (Vol. 1, Bellaterra, Spain, 1997) pp. 31–36.

[21] E. Vidal, M. J. Castro and J. A. Sánchez, Classification of Banded Chromosomes, (tech. rep., DSIC, Universidad Politécnica de Valencia, Spain) 1997.

[22] C. D. Martínez-Hinarejos, A. Juan, F. Casacuberta, Median String for k-Nearest Neighbour classification, *Pattern Recognition Letters*, acepted for revision.

Cyclic String Matching: Efficient Exact and Approximate Algorithms

Andrés Marzal[1], Ramón Mollineda[2], Guillermo Peris[1], Enrique Vidal[2]

[1]*Departamento de Lenguajes y Sistemas Informáticos*
Universidad Jaume I, Castellón, Spain

[2]*Instituto Tecnológico de Informática*
Universidad Politécnica de Valencia, Valencia, Spain

E-mail: amarzal@lsi.uji.es, rmollin@iti.upv.es, peris@lsi.uji.es, evidal@iti.upv.es.

Contents

D. Chen and X. Cheng (eds.),
Pattern Recognition and String Matching, 477–497.

1 Introduction

Many pattern recognition problems require the evaluation of a measure of similarity between two strings of symbols. Applications range from molecular biology to speech and image analysis [13].

String dissimilarities can be properly measured in terms of elementary differences along an adequate *correspondence* or *alignment* between the strings compared. Such differences are the *substitution* of one symbol in x by its corresponding one in y, the *deletion* of symbols in x, and *insertions* of new symbols in y, where x and y are the two strings compared. These elementary differences can be seen as active *edit operations* that would transform x into y. Levenshtein defined a primary concept of *edit distance* between two strings x and y as the smallest number of *edit operations* needed to change x into y [6]. A widely adopted generalization of the Levenshtein metric is to weight the edit operations. The edit distance is reformulated as the minimum weighted sequence of edit operations that changes x into y. An efficient dynamic programming procedure for performing this optimization has been presented in many independent publications, being [15] the most cited among them. This procedure, that will be called Wagner and Fischer's algorithm, runs in $\mathcal{O}(mn)$ time, where m and n represent the lengths of x and y, respectively.

A common extension arises from considering a string x equivalent to any possible rotated version of itself. This relation is formally stated as an equivalence class, which is known as *cyclic string*. It can be useful,

This work has been partially supported by the Spanish CICYT under grant TIC2000-1599-CO2-01.

for example, in digital image analysis, where strings can represent closed contours of 2D objects without a definite starting point, or in bacterial and mitochondrial DNA analysis, which is typically circular [4].

In such cases, the measure of similarity between two cyclic strings, the so called *cyclic edit distance*, is defined as the minimum weighted sequence of edit operations that changes some rotation of x into some rotation of y [7]. A more feasible and equivalent statement is also given in [7] as the minimum weighted sequence needed to transform some rotation of x into y. This exhaustive approach takes $\mathcal{O}(m^2n)$ time.

Maes [7] proposed an efficient Divide and Conquer algorithm that runs in $\mathcal{O}(mn \log m)$ time. A later approach for cyclic strings comparison was introduced by Gregor and Thomason [3]. It is also $\mathcal{O}(m^2n)$, but it is more efficient in certain cases [8]. A more recent Branch and Bound algorithm has been presented by Marzal and Barrachina [8], which has a worst time complexity in $\mathcal{O}(mn \log m)$, but whose average computing time was shown to be significantly lower than Maes' algorithm, since it is based on a partial exploration of the same search space.

Another way to handle this problem is to sacrifice the solution optimality while reducing computational cost. Bunke and Bühler [1] proposed a $\mathcal{O}(mn)$ suboptimal method to approximate the cyclic edit distance which was successfully used for 2D shape recognition. Based on this technique, Mollineda, Vidal, and Casacuberta [11] introduced two new, more accurate, approximate algorithms with the same $\mathcal{O}(mn)$ computing time.

This paper reviews Maes' algorithm, its Branch and Bound speed-up and approximate computation approaches. The Branch and Bound methods are also extended to deal with arbitrary edit operations weights. The following sections include an analysis of previous methods to show the relationships among them attending to both accuracy (for approximate techniques) and computing time. Experiments with synthetic data and with data from real applications show both the efficiency gain of Branch and Bound and approximate methods with respect to Maes, and the quality of the results obtained with the suboptimal techniques.

2 Edit Distance

Let Σ be an alphabet of symbols and let Σ^* be the set of all finite-length strings over Σ. Let λ denote the empty string. An *edit operation* is an ordered pair $(a, b) \in (\Sigma \cup \{\lambda\}) \times (\Sigma \cup \{\lambda\}) - \{(\lambda, \lambda)\}$ denoted by $a \to b$. There are four edit operations: *matching* ($a \to a$), *substitution* ($a \to b$, for

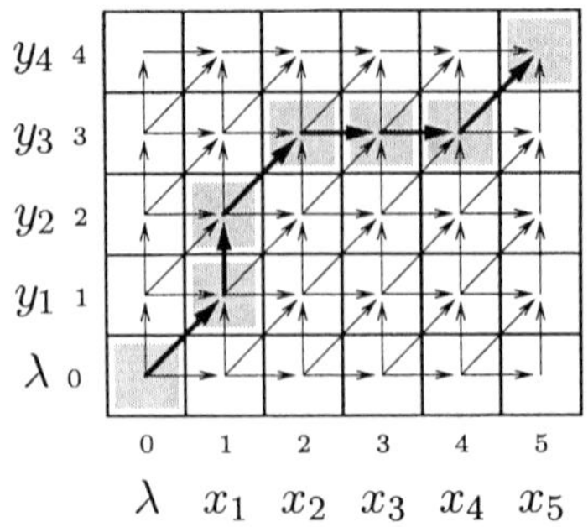

edit operations	**corresponding arcs**
$x_1 \to y_1$	$[(0,0),(1,1)]$
$\lambda \to y_2$	$[(1,1),(2,1)]$
$x_2 \to y_3$	$[(2,1),(3,2)]$
$x_3 \to \lambda$	$[(3,2),(3,3)]$
$x_4 \to \lambda$	$[(3,3),(3,4)]$
$x_5 \to y_4$	$[(3,4),(4,5)]$

Figure 1: Edit graph for two strings x and y. Paths must start at node $(0,0)$ and end at node $(5,4)$. A (non necessarily optimal) path is highlighted on the graph and its associated sequence of edit operations is shown on the right.

$a \neq b$), *deletion* ($a \to \lambda$) and *insertion* ($\lambda \to a$).

A function $\gamma : (\Sigma \cup \{\lambda\}) \times (\Sigma \cup \{\lambda\}) \to \mathbb{R}^{\geq 0}$ assigns a cost to each edit operation. For the sake of notation simplicity and without loss of generality, we will define γ in terms of three constants γ_S, γ_D, and $\gamma_I \in \mathbb{R}^{>0}$: insertion costs $\gamma(\lambda \to a) = \gamma_I$, deletion costs $\gamma(a \to \lambda) = \gamma_D$, substitution costs $\gamma(a \to b) = \gamma_S$, and matching costs $\gamma(a \to a) = 0$.

An *edit sequence* is a sequence of edit operations, $E = e_1 e_2 \dots e_k$. Its cost is the sum of the costs of the individual edit operations, i.e., $\gamma(E) = \sum_{i=1}^{k} \gamma(e_i)$. The *edit distance* d between strings x and y is defined as

$$d(x,y) = \min\{\gamma(E) \mid E \text{ is an edit sequence that changes } x \text{ into } y\}. \quad (1)$$

The so-called *edit graph* [15] of x and y is a matrix of $(m+1) \times (n+1)$ nodes whose arcs represent edit operations: horizontal arcs correspond to deletions, vertical arcs to insertions, and diagonal arcs to substitutions and matchings. Figure 1 shows an example of edit graph. The computation of a minimum cost edit sequence is reduced to finding an optimal path from node $(0,0)$ to node (m,n), which can be performed in $\mathcal{O}(mn)$ time by Dynamic Programming.

3 Cyclic Strings

Given a string $x = x_1 x_2 \dots x_m \in \Sigma^*$, its *cyclic shift* is defined as $\sigma(x) = x_2 x_3 \dots x_m x_1$. Any *rotation* of x can be obtained by repeatedly applying σ. The application of k cyclic shifts to x will be noted $\sigma^k(x)$. A *cyclic string* $[x]$ is an equivalence class defined by an *equivalence relation* $\equiv$ in Σ^* such that, given two strings x and $\bar{x} \in \Sigma^*$,

$$x \equiv \bar{x} \Leftrightarrow \bar{x} = \sigma^k(x) \quad \text{for some } k \in \{0, 1, \dots, m-1\}. \tag{2}$$

The edit distance between two cyclic strings $[x]$ and $[y]$, is defined as

$$D([x],[y]) = \min\{d(\bar{x}, \bar{y}) \mid x \equiv \bar{x}, y \equiv \bar{y}\}. \tag{3}$$

From now on, this measure will be called *cyclic edit distance.*

The previous definition can be naturally extended to introduce the edit distance between a *cyclic* string $[x]$ and a *linear* string y by

$$D([x], y) = \min\{d(\bar{x}, y) \mid x \equiv \bar{x}\}. \tag{4}$$

4 Computation of the Cyclic Edit Distance

A simple way to compute the edit distance between two cyclic strings can be derived from the following lemma:

Lemma 4.1 (Maes) *Given two cyclic strings $[x]$ and $[y]$, $D([x],[y]) = D([x], y)$.*

The proof of this lemma is trivial. It leads straightaway to an exhaustive procedure to compute $D([x],[y])$ by comparing the m possible rotations of x with y in $\mathcal{O}(m^2 n)$ time.

All the methods that we will study work on an *extended edit graph* associated to the concatenation of x with itself, $x \cdot x$, and y (see Figure 2). In the extended edit graph, which has $(2m+1) \times (n+1)$ nodes, paths can depart from any node $(i, 0)$ where $0 \leq i \leq m$ and must arrive to $(i+m, n)$. Paths correspond to alignments between y and substrings of $x \cdot x$ of length m.

4.1 Maes' algorithm: an exact computation procedure

Maes' algorithm [7] is based on the following theorem:

Theorem 4.1 (Maes) *Let $P(k)$ be the path associated to an optimal edit sequence between $\sigma^k(x)$ and y on the extended edit graph of $x \cdot x$ and y. Given three integers $0 \leq j < k < l \leq m$, if the optimal edit paths $P(j)$ and $P(l)$ do not cross each other, then there is an optimal edit path $P(k)$ which does not cross either $P(j)$ or $P(l)$.*

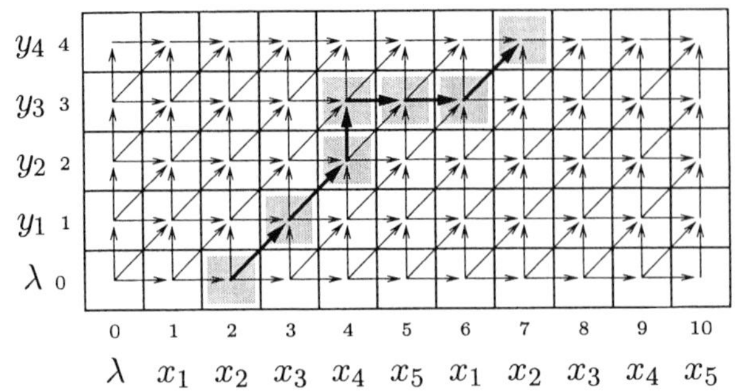

Figure 2: Extended edit graph. Paths can start at any node $(i, 0)$, for $0 \leq i \leq m$, and must end at $(i + m, n)$. A valid path defining a sequence of edit operations between $x_3x_4x_5x_1x_2$ and y has been highlighted.

The algorithm first computes $P(0)$ and $P(m)$ (which is $P(0)$ shifted m positions to the right on the extended edit graph) by means of Wagner and Fischer's algorithm. Then, it computes $P(\lceil m/2 \rceil)$ by visiting only the nodes between $P(0)$ and $P(m)$ on the extended edit graph. The new path allows to compute $P(\lceil m/4 \rceil)$ by visiting only the nodes between $P(0)$ and $P(\lceil m/2 \rceil)$, and $P(\lceil 3m/4 \rceil)$ by visiting only the nodes between $P(\lceil m/2 \rceil)$ and $P(m)$; and so on. Once $P(0)$, $P(1)$, ..., $P(m)$ are known, the cyclic edit distance is computed as the weight of the best path. Figure 3 depicts this Divide and Conquer procedure. It can be easily seen that this procedure exhaustively explores a binary tree with $\mathcal{O}(\log m)$ levels. All the computations performed at each level are $\mathcal{O}(mn)$ and, therefore, the global time complexity is $\mathcal{O}(mn \log m)$.

4.2 Bunke and Bühler algorithm (BB): an approximate computation procedure

The Bunke and Bühler algorithm (BB) [1] estimates $D([x], y)$ by searching in the extended edit graph the optimal path departing from any node $(i, 0)$, for $0 \leq i \leq m$, and arriving to any node (j, n), for $m \leq j \leq 2m$, without any control over the length of the substring of $x{\cdot}x$ aligned with y.

The BB estimation of $D([x], y)$, that will be denoted as $D_{BB}(x, y)$, is performed by initializing all starting nodes with 0 and computing the minimum value of any optimal path arriving at a final node. Therefore, the BB can be computed in $\mathcal{O}(mn)$ time complexity.

It was shown in references [14] and [1] that, given two strings x and y, the BB algorithm computes $D_{BB}(x, y) = d(x', y)$, where x' is the substring of $x{\cdot}x$

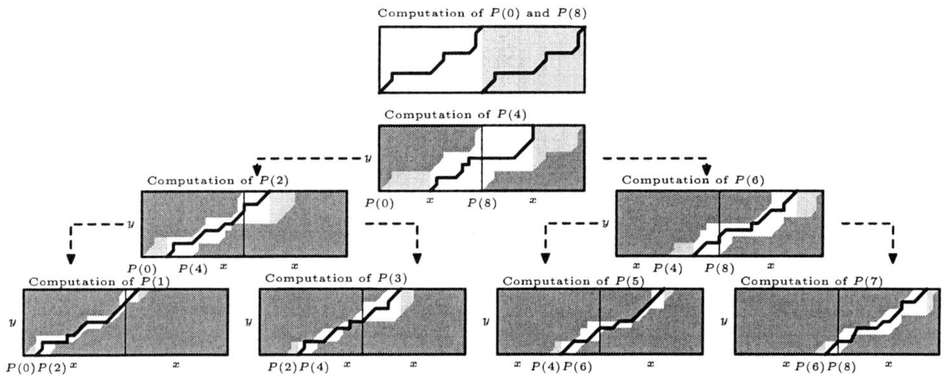

Figure 3: Search space for the computation of the cyclic edit distance by Maes' algorithm for $m = 8$. Each frame represents an extended edit graph. Optimal paths are drawn with thick lines and never cross already known optimal paths. Shaded regions in each frame are not visited when computing optimal paths.

that most resembles y. The corresponding edit path can be considered as a *pseudo-alignment* between $[x]$ and y. The following lemma is straightforward from the above discussion:

Lemma 4.2 *The BB estimation is a lower bound of the exact cyclic edit distance:* $D_{BB}(x, y) \leq D([x], y)$

There are significant differences between estimations given by $D_{BB}(x, y)$ and $D_{BB}(y, x)$. Let us assume that $m > n$. When $D_{BB}(x, y)$ is computed, m', the length of the substring of $x{\cdot}x$ most similar to y, is expected to be closer to n than to m, which is the "ideal" length. Thus, the quality of the estimation may inversely depend on the difference $m - n$. On the other hand, if $D_{BB}(y, x)$ is computed, n', the length of the substring of $y{\cdot}y$ aligned with x, will tend to be closer to both m and n. Consequently, the estimation given by $D_{BB}(y, x)$ is expected to be more accurate than the one obtained by $D_{BB}(x, y)$ when $m > n$. Exhaustive experiments with both synthetic and real data, clearly confirmed this expectation [10].

5 Speeding Up the Exact Computation of Cyclic Edit Distance: a Branch-and-Bound Approach

Maes' algorithm explores all the states in a search space (see Figure 3). The Branch and Bound algorithm introduced by Marzal, Barrachina, and Peris

(MB) in [8] and [12] explores only those states that may lead to the optimal path in the extended edit graph (Figure 4).

Formally, a state in the search space is a pair (j, l) where $0 \leq j < l \leq m$ and $l - j > 1$, and it represents the set of rotations $\{\sigma^k(x) : j < k < l\}$. We are interested in an optimal rotation included in a state, that is, a $\sigma^{k^*}(x)$ such that $d(\sigma^{k^*}(x), y) \leq \min_{j<k<l} d(\sigma^k(x), y)$. The algorithm starts by computing $d^* = d(x, y)$ and initializing the set of live search states, S, to $\{(0, m)\}$. Every state (j, l) in S is scored by $g(j, l)$, a lower bound on $\min_{j<k<l} d(\sigma^k(x), y)$. The algorithm iteratively performs the following steps until the set of live states is empty:

1. the state (j, l) in S with the lowest value of $g(j, l)$ is selected;
2. the edit distance $d(\sigma^k(x), y)$ and the edit path $P(k)$ for $k = \lceil (j - l)/2 \rceil$ are computed by visiting only nodes between $P(j)$ and $P(l)$ in the extended edit graph (Theorem 4.1); then, if $d(\sigma^k(x), y) < d^*$, the value of d^* is updated;
3. two new states, (j, k) and (k, l), are created and inserted in S.

The algorithm implicitly prunes the live states set by not inserting new states in S if their scores are not lower than d^*, and by finishing when the lower bound of the best state in S is greater or equal than d^*. The set S can be implemented as a *min-heap* ordered by the value of g.

When S is empty or d^* is lower or equal than the best value of g on any state of S, the algorithm ends returning d^* as the value of $D([x], [y])$.

The MB algorithm requires defining g, the lower bound function. If $g(j, l)$ is defined as 0 for all $0 \leq j < l \leq m$ (a trivial lower bound), the algorithm is equivalent to Maes' algorithm. In the following theorem, a lower bound is proposed:

Theorem 5.1 *For all k such that $j < k < l$,*

$$g(j, l) = \max\left(0, \frac{d(\sigma^j(x), y) + d(\sigma^l(x), y) - (\gamma_I + \gamma_D)(l - j)}{2}\right) \leq d(\sigma^k(x), y).$$

(The proof can be found in [9].) This lower bound can be easily extended to arbitrary γ functions by replacing γ_I and γ_D by the maximum value of any insertion or deletion of a symbol in Σ, respectively.

In the worst case, the Branch and Bound algorithm will explore the full search space, thus leading to the cost of Maes' algorithm plus the overhead introduced by the computation of the lower bound function and the management of S. The lower bound, g, can be evaluated in $\mathcal{O}(1)$ and it is computed

function $\mathrm{MB}(x : \Sigma^*, y : \Sigma^*) : \mathbb{N}$ // x is a string of length m
aux $g : \mathbb{N} \times \mathbb{N} \to \mathbb{R}$ // *such that* $g(j,l) \leq \min_{j<k<l} d(\sigma^k(x), y)$
var $S : 2^{\mathbb{N}\times\mathbb{N}}$ // *set of live states*
P : array $[0..m]$ of paths
begin
$d^* := d(x, y)$
Let $P[0]$ be the optimal path associated to $d(x, y)$.
Let $P[m]$ be $P[0]$ shifted m units to the right.
$S := \{(0, m)\}$
while $S \neq \emptyset$ **and** $d^* > \min_{(j,l)\in S} g(j,l)$
$(j,l) := \arg\min_{(j',l')\in S} g(j',l')$
$S := S - \{(j,l)\}$
$k := \lceil (j+l)/2 \rceil$
$d^* := \min(d^*, d(\sigma^k(x), y))$ (between $P[j]$ and $P[l]$)
Let $P[k]$ be the optimal path underlying $d(\sigma^k(x), y)$
if $k - j > 1$ **and** $g(j,k) < d^*$ **then** $S := S \cup \{(j,k)\}$
if $l - k > 1$ **and** $g(k,l) < d^*$ **then** $S := S \cup \{(k,l)\}$
return d^*
end

Figure 4: The MB algorithm, a Branch and Bound procedure to compute the exact cyclic edit distance.

at most m times. The cost introduced by the management of the *min-heap* S is $\mathcal{O}(m \log m)$: at most m states will be inserted into and extracted from S. Therefore, the worst case time complexity of this algorithm is $\mathcal{O}(mn \log m)$.

It should be noted that, in practice, the algorithm can run much faster [8]. For instance, it is expected to perform better on close (similar) cyclic strings than on distant ones.

6 Improving the Accuracy of Approximate Solutions

6.1 Completing BB *pseudo-alignments*

An extension to the original BB algorithm (EBB) was reported in [11]. Let m' be the length of the longest substring of $x{\cdot}x$ that can be transformed into y with cost $D_{BB}(x, y)$. The EBB algorithm adds a $m - m'$ deletions at the end of the obtained pseudo-alignment, thus completing it. In this way, a *suboptimal* but complete edit sequence between a rotation of x and y is built by the EBB algorithm. The cost of the complete edit sequence obtained by

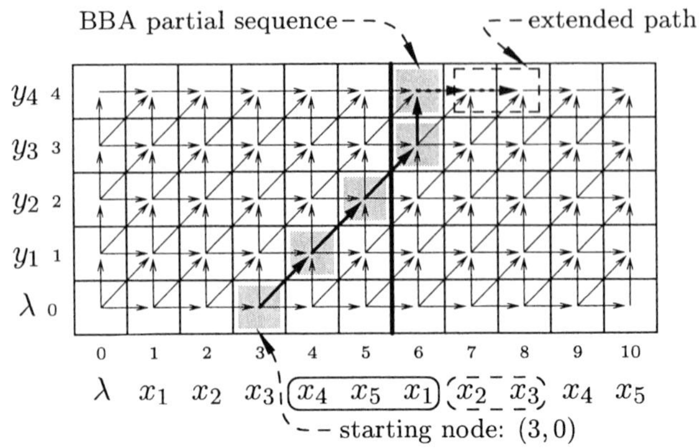

Figure 5: EBB procedure. BB transforms y into the substring $x_4x_5x_1$. The symbols x_2 and x_3 are not considered by the path. EBB adds $x_2 \rightarrow \lambda$ and $x_3 \rightarrow \lambda$ to the sequence of edit operations so that $x_4x_5x_1x_2x_3$, a rotated version of x, is transformed into y.

the EBB procedure will be denoted by $D_{EBB}(x, y)$.

In the example of Figure 5, the string y has been aligned with the substring $x_4x_5x_1$ via the partial edit sequence $x_4 \rightarrow y_1$, $x_5 \rightarrow y_2$, $x_1 \rightarrow y_3$, $\lambda \rightarrow y_4$. Two symbols of x (x_2 and x_3) have not been aligned with y. By adding the deletions of these unaligned symbols to the partial edit sequence, a complete one is built.

The EBB algorithm stores the column number of the starting node of the optimal path arriving at every node in the extended edit graph. At every final node, the length of the $x \cdot x$ substring underlying its arriving optimal path is computed by subtracting the stored initial column number from the current column number. Before extending any partial path, this procedure is also used to keep the length of the substrings of $x \cdot x$ aligned with y lower than or equal to m.

Lemma 6.1 *The EBB estimation is an upper bound of the cyclic edit distance:* $D_{EBB}(x, y) \geq D([x], [y])$.

The proof of this lemma is based on the fact that the EBB algorithm builds full, suboptimal edit sequences between some members of $[x]$ and y, while $D([x], y)$ is the cost of the minimum cost edit sequence between y and any $\bar{x} \in [x]$.

The estimation $D_{EBB}(x, y)$ can be computed in $\mathcal{O}(mn)$ time, the same time complexity of the BB algorithm.

6.2 A weighted combination of the lower and upper-bounded solutions

Lemmas 4.2 and 6.1 give lower and upper bounds of the cyclic edit distance, i.e., $D_{BB}(x,y) \leq D([x],y) \leq D_{EBB}(x,y)$. A different approximate solution was also introduced in [11], as a weighted average between the lower estimate, $D_{BB}(x,y)$, and the upper one, $D_{EBB}(x,y)$:

$$D_W(x,y) = \alpha \cdot D_{BB}(x,y) + (1-\alpha) \cdot D_{EBB}(x,y). \tag{5}$$

The coefficient (weight) α can be computed by minimizing the sum of squared relative errors of the weighted solutions with respect to the exact distances on a training set T of string pairs of the problem at hand. The approximation error function,

$$E_W = \sum_{(x,y)\in T} \left(\frac{D([x],y) - D_W(x,y)}{D([x],y)} \right)^2, \tag{6}$$

can be minimized for a value of α such that $dE_W/d\alpha = 0$, that is,

$$\alpha = \frac{\sum_{(x,y)\in T} (D_{EBB}(x,y) - D_{BB}(x,y)) \cdot \left(\frac{D_{EBB}(x,y)}{D([x],y)} - 1 \right)}{\sum_{(x,y)\in T} \frac{(D_{EBB}(x,y) - D_{BB}(x,y))^2}{D([x],y)}}. \tag{7}$$

This *weighted* average between BB and EBB will be referred to as WeBB, and its value will be denoted by $D_W(x,y)$. Since both bounds can be computed simultaneously in $\mathcal{O}(mn)$ time, their combined approximation keeps the same asymptotic cost.

7 Experimental Analysis

The experiments are intended to show the efficiency gain of the MB algorithm with respect to Maes' algorithm and to prove the effectiveness of the approximate techniques (BB, EBB, and WeBB) regarding both accuracy and computing time.

A number of experiments were conducted on synthetic and on real data. A first set of experiments was performed on synthetic strings in order to compare time and accuracy indexes as a function of different parameters: cyclic string length, dissimilarity degree between cyclic strings, and alphabet size. In a second set, approximate and exact methods were applied to real-world classification tasks. Besides comparing efficiency and accuracy, classification performance was studied.

In order to measure accuracy in approximate methods, we computed the approximation error E_M as the *Mean Squared Relative Error* (MSRE), for M in $\{D_{BB}, D_{EBB}, D_W\}$,

$$E_M = \frac{\sum_{(x,y)\in T} \left(\frac{D([x],y) - M(x,y)}{D([x],y)} \right)^2}{|T|}, \tag{8}$$

where T is a test set of string pairs.

The computing time of the implemented algorithms is presented as a ratio of the time needed by Wagner and Fischer's algorithm (WF) to compute the edit distance between the same (non-cyclic) strings. For the sake of simplicity and to allow a better comparison amoung the different methods, all the experiments were run with the Levenshtein cyclic and non-cyclic edit distances ($\gamma_I = \gamma_D = \gamma_S = 1$).

7.1 Experiments with synthetic strings

Each experiment with synthetic cyclic strings consists of two series of 100 cyclic strings that were generated according to the following procedure: first, a "seed" cyclic string of a given length was randomly drawn from an alphabet; then, 99 new strings were obtained by distorting the seed with a given number of edit operations; finally, all the strings were randomly rotated. One of the data sets was used to estimate the weight α used in D_W and the other was used to measure the accuracy and running time on the $\binom{100}{2} = 4950$ cyclic string pairs.

7.1.1 Cyclic strings with different dissimilarity degree

In order to assess how much the MB efficiency depends on the degree of similarity between the compared cyclic strings, two experiments were performed. Two groups of cyclic strings sets were generated with growing dissimilarity degree, one for short cyclic strings (40 symbols long) and other for long ones (320 symbols long). The seed for each data set was drawn from an alphabet composed of 8 symbols. The rest of the cyclic strings was obtained by editing the seed with n random edit operations, where $n \in \{1, 2, 4, 8, 16, 32\}$ for the short cyclic strings sets and $n \in \{1, 2, 4, 8, 16, 32, 64, 128, 256\}$ for the long cyclic strings sets.

Running times for all the methods are shown in Figure 6. As expected, the MB algorithm performs much better for close cyclic strings that for distant ones. The time needed by EBB and WeBB was close to that required by BB.

The MSRE of approximate techniques is shown in Figure 7. The method WeBB is the most accurate and both BB and EBB obtain similar results, being BB slightly better in most cases. The estimation is worse for distant cyclic strings.

In (5), α weights the importance of BB and EBB in WeBB. Therefore, the value of α can be interpreted as an index of relative accuracy of BB and EBB with respect to cyclic strings features (length, distance, and alphabet size). For example, values lower than 0.5 represent closer estimates from EBB, while values greater than 0.5 mean that BB is a better approximation than EBB. WeBB always takes advantage of the most accurate bound and reduces the influence of the other. The values obtained for α are $\{0.59, 0.54, 0.47, 0.59, 0.46, 0.34\}$ for short cyclic strings and $\{0.53, 0.64, 0.82, 0.66, 0.43, 0.56, 0.56, 0.62, 0.62\}$ for long ones.

7.1.2 Cyclic strings with different lengths

In this experiment we study efficiency and accuracy for growing cyclic string length. Since there are significant differences of MB running time for close and distant cyclic strings, both cases are separately considered.

The experiment was run on 6 training and test sets of cyclic strings from an 8-symbols alphabet. The length of the seed cyclic string in each set was 10, 20, 40, 80, 160, and 320, respectively. The distortion applied to these cyclic strings consisted of length/10 edit operations for the close cyclic strings sets and length/2 for the distant ones.

The running time of all the techniques is shown in Figure 8. For close cyclic strings, the time required by the exact MB procedure was significantly lower than the one needed by Maes' algorithm. Approximate methods are insensitive to the length of the cyclic strings and faster than the exact methods.

Figure 9 illustrates the approximation error for the suboptimal techniques BB, EBB, and WeBB. A very important error reduction was achieved by WeBB with respect to BB, specially for longer and distant strings. For close strings, all the methods were very accurate, being WeBB clearly better.

The optimal values for α were $\{0.50, 0.68, 0.48, 0.61, 0.45, 0.57\}$ (close cyclic strings) and $\{0.54, 0.41, 0.44, 0.51, 0.59, 0.70\}$ (distant cyclic strings). In the case of distant cyclic strings, all the values were lower than 0.5, assessing the fact that EBB makes notably better estimates than BB (see Figure 9).

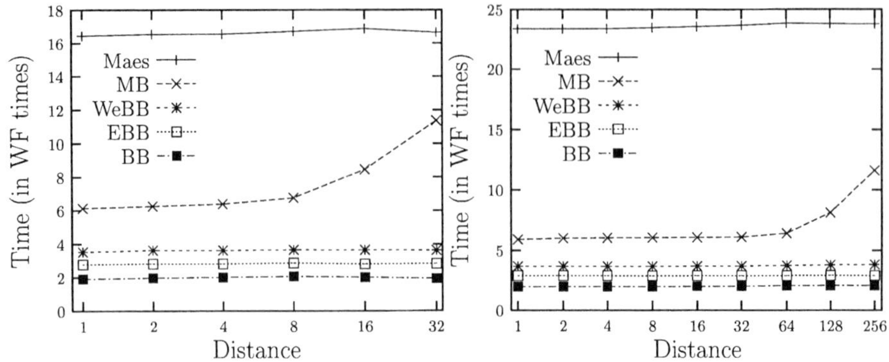

Figure 6: Time (ratio of the time needed by the WF algorithm) as a function of distance among cyclic strings with an 8 symbols alphabet for seeds of length 40 (left) and 320 (right).

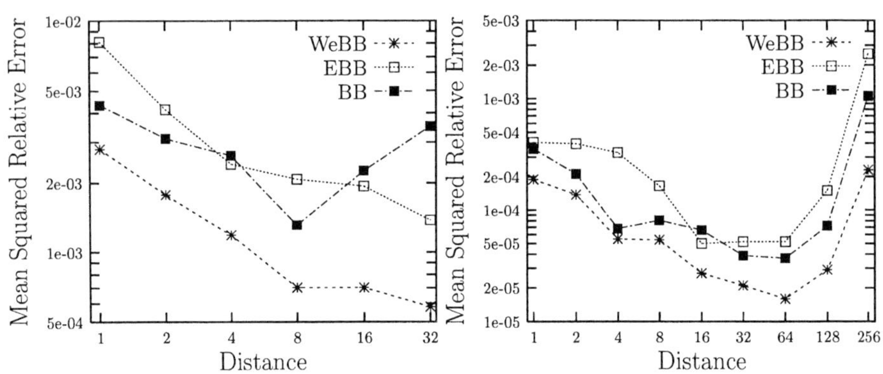

Figure 7: Approximation error of suboptimal techniques (BB, EBB, WeBB) as a function of distance among strings using an alphabet with 8 symbols, considering short strings of length 40 (left) and long strings of length 320 (right).

7.1.3 Cyclic strings from alphabets of different sizes

This section analyzes the behavior of the algorithms for cyclic strings drawn from different alphabet sizes. Once again, we separately study close and distant cyclic strings.

The alphabet size took values in the set $\{2, 4, 8, 16, 32\}$. The length of the seed cyclic string in each data set was 160. The close cyclic strings were obtained by applying 16 edit operations to their respective seeds and distant ones were obtained by applying 80 edit operations.

The values obtained for α in WeBB are $\{0.56, 0.41, 0.45, 0.40, 0.35\}$ for short cyclic strings and $\{0.45, 0.55, 0.59, 0.57, 0.54\}$ for long ones.

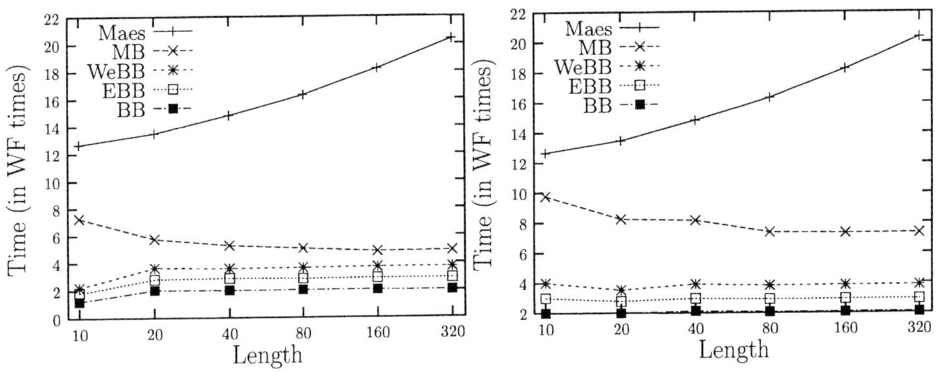

Figure 8: Time (ratio of WF's running time) as a function of cyclic strings length with an 8 symbols alphabet for close (left) and distant (right) cyclic strings.

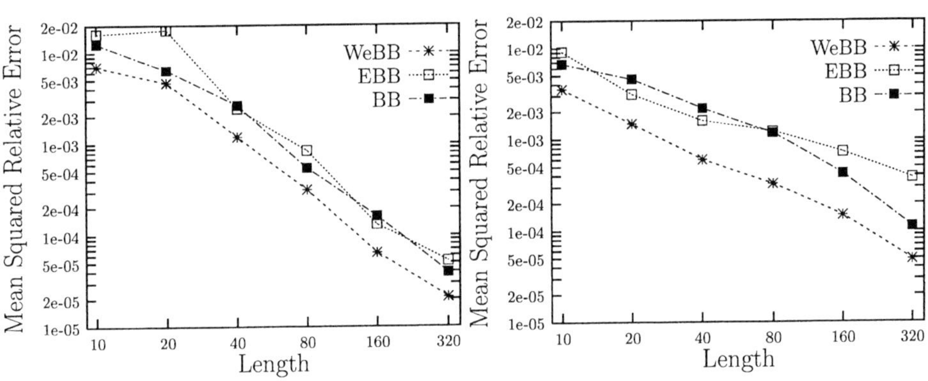

Figure 9: Approximation error of approximate techniques (BB, EBB, WeBB) as a function of cyclic string length using an alphabet with 8 symbols, for close (left) and distant cyclic strings (right).

The results of this section confirm the main trends already observed in the other experiments. Once again, the exact method MB was notably faster than Maes (see Figure 10) and WeBB was largely the best suboptimal technique (see Figure 11).

7.2 Experiments with real-world data

In order to study efficiency and accuracy on real-world data and to show the interest of the cyclic edit distance computation on pattern recognition problems, we performed experiments with two different 2D shapes classification tasks.

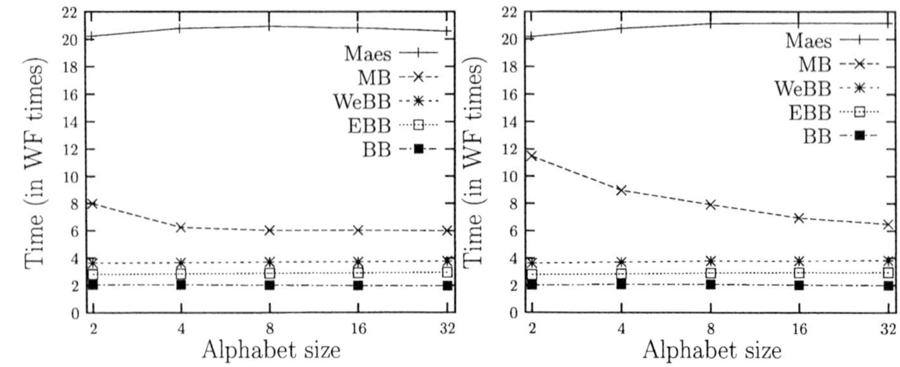

Figure 10: Time (in WF execution times) as a function of alphabet size among strings with an 8 symbols alphabet for close (left) and distant (right) cyclic strings.

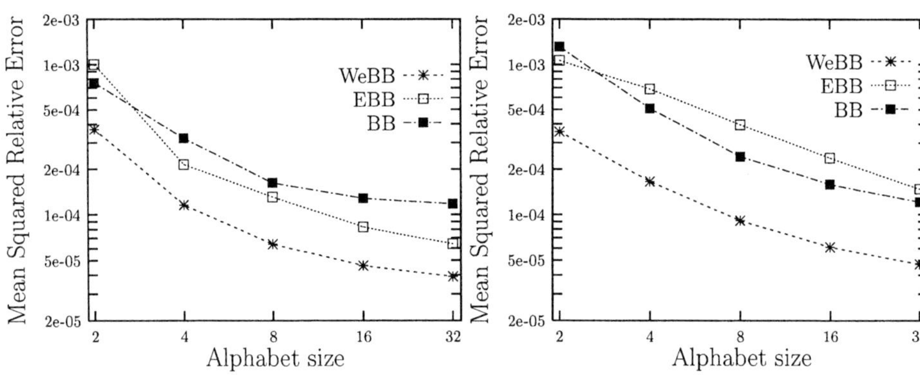

Figure 11: Approximation error of suboptimal techniques (BB, EBB, WeBB) on test sets as a function of the alphabet size, considering close strings (left) and distant strings (right).

7.2.1 Handwritten digits image data

A data set containing 1000 handwritten digit images (100 instances of each of the 10 digits) was used. The outer contour of every digit (see examples in Figure 12) was represented by using the standard {4,8}-directions chain-encoding procedures [2]. The average length of chain-code strings of the two resulting sets was about 170 and 146 symbols for 4-directions and 8-directions, respectively.

The two sets of chain-code contours were divided, preserving class distributions, into two subsets of 500 strings each, one for training and one for test. For time and accuracy experiments, training sets were used to estimate the values of α needed by WeBB, being 0.19 and 0.26 for 4-codes and

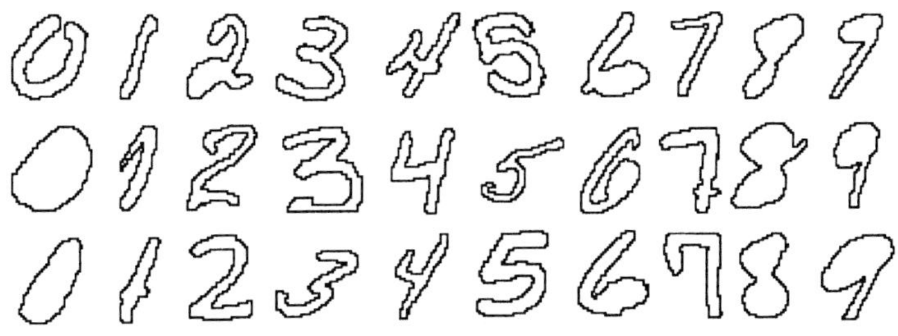

Figure 12: Sample of 4-direction contours of handwritten digits.

Table 1: Approximation error of the non-cyclic edit distance and the suboptimal techniques BB, EBB, WeBB with respect to exact cyclic edit distance, for chain-code representations of handwritten digits.

Chain-code	non-cyclic	BB	EBB	WeBB
4 directions	0.0140	0.0033	0.0005	0.0003
8 directions	0.0078	0.0020	0.0005	0.0002

Table 2: Running time (as a ratio of WF running time) to approximate (by suboptimal methods BB, EBB, WeBB) and to exactly compute (by the exact methods MB and Maes) a cyclic edit distance on the chain-code representations of handwritten digits.

Chain-code	BB	EBB	WeBB	MB	Maes
4 directions	2.04	2.93	3.72	12.92	20.79
8 directions	2.03	2.97	3.77	13.07	20.44

8-codes string sets, respectively. The accuracy and time results are shown in Table 1 and Table 2. They agree with the ones computed on synthetic strings: MB was significantly faster than Maes and WeBB was much more accurate than BB.

Apart from comparing efficiency and accuracy, a real classification task was performed. Each test set was classified with the k-Nearest Neighbor (k-NN) rule. Classification results for different values of k are shown in Figure 13. The suboptimal techniques presented in this paper (WeBB and EBB) achieved better classification performance than BB and the non-cyclic edit distance (WF) and were always comparable to the exact algorithm MB.

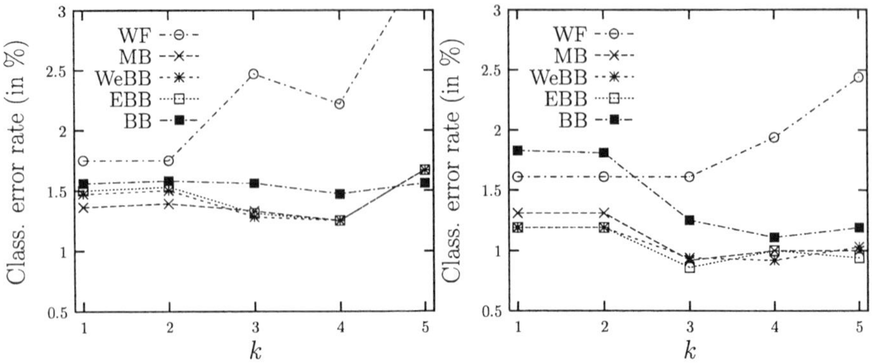

Figure 13: k-NN classification error rate for the different methods in the digits recognition task.

Figure 14: Sample of chicken pieces from the 5 classes.

7.2.2 Chicken pieces image data

This data set consists of 446 binary images, each one containing the silhouette of a chicken piece [11]. Each piece belongs to one of five categories: wing (117 samples), back (76), drumstick (96), thigh and back (61), and breast (96) (see Figure 14). Pieces were placed without considering orientation. All silhouettes were clipped and scaled into 64×64 pixels images. A standard 4-direction chain-encoding procedure [2] was applied and the resulting chain-codes were re-encoded into a rotation-invariant representation, where new codes specify relative change of angles [5]. The average length of the resulting chain-codes is 225.

From these 446 chain-code strings, two (training and test) subsets of 100 samples were randomly selected. The corresponding training set of

Table 3: Approximation error of the non-cyclic edit distance and the suboptimal techniques BB, EBB, WeBB with respect to exact cyclic edit distance in the chicken pieces task.

BB	EBB	WeBB
0.0031	0.0010	0.0003

Table 4: Running time (as a ratio of WF running time) of BB, EBB, WeBB, MB and Maes algorithms on the chain-code representations of chicken pieces.

BB	EBB	WeBB	MB	Maes
1.7	2.4	3.0	12.2	21.9

Table 5: Classification error (in %) for the non-cyclic edit distance, all the approximate methods and the exact cyclic edit distance (MB) in the chichen pieces classification task.

non-cyclic	BB	EBB	WeBB	MB
32.5	24.6	23.5	21.9	22.6

string pairs was used to estimate the parameter α. The minimum MSRE was achieved for $\alpha = 0.265$. The approximation errors of the suboptimal techniques are shown in Table 3 and the running time (relative to WF) of all the methods is shown in Table 4. Accuracy and efficiency results are coherent with the experiments on synthetic data.

A classification experiment was also carried out. Classification error rate was estimated through a "leaving one out" scheme with the 1-Nearest Neighbor rule using the 446 samples. To estimate the value of α, the data set was partitioned into 4 subsets keeping the frequencies among classes. From each subset i in $\{0, 1, 2, 3\}$, a related set of all different non-ordered pairs of strings was built and used to compute the value of its corresponding α_i. To classify a contour in the subset i the coefficient $\alpha_{(i+1)\mathrm{mod}4}$ was used. In this way, each cyclic pattern did not take part in the computation of the coefficient α used in its classification. The resulting error rates are shown in Table 5. The WeBB was clearly better than BB and as good as exact methods.

8 Conclusions

This paper addresses the problem of computing the edit distance between cyclic strings. A survey of known techniques is presented along with new approaches to efficiently compute the exact and approximate values of this dissimilarity measure.

The MB algorithm, a Branch and Bound procedure, computes the exact cyclic edit distance significantly faster (up to four times) than Maes' algorithm.

The suboptimal methods presented are based on the Bunke and Bühler algorithm (BB), which computes a lower bound of the cyclic edit distance. A first extension to BB (EBB) enlarges partial alignments built by BB to complete edit sequences and obtains an upper estimation of the cyclic edit distance. WeBB, a new, more accurate estimation, is based on minimizing the squared relative error of a weighted combination of BB and EBB. Both EBB and WeBB keep the asymptotic cost of BB.

Experiments on both synthetic and real data, showed that WeBB was the best suboptimal technique at large, achieving much more accurate approximations than BB and EBB. WeBB takes advantage of the most accurate bound between BB and EBB, and reduces the influence of the other.

Classification experiments have also been carried out in which the cyclic edit distance is shown to be a better dissimilarity measure than the (non-cyclic) edit distance for 2D shape recognition. WeBB constitutes a good trade-off between accuracy and computing time and it seems to be a very attractive approach to deal with classification tasks.

References

[1] H. Bunke and U. Bühler, Applications of Approximate String Matching to 2D Shape Recognition, *Pattern Recognition* Vol. 26 No. 12 (1993) pp. 1797–1812.

[2] H. Freeman, Computer Processing of Line Drawing Images, *Computer Surveys*, Vol. 6 (1974) pp. 57–98.

[3] J. Gregor and M. G. Thomason, Dynamic Programming Alignment of Sequences Representing Cyclic Patterns, *IEEE Transactions on Pattern Analysis and Machine Intelligence* Vol. 15 No. 2 (1993) pp. 129–135.

[4] D. Gusfield, *Algorithms on Strings, Trees, and Sequences: Computer Science and Computational Biology*, (Cambridge University Press, 1997).

[5] R. C. Gonzalez and R. E. Woods, *Digital Image Processing*, (Addison-Wesley, 1992).

[6] V. I. Levenshtein, Binary Codes Capable of Correcting Deletions, Insertions and Reversals, *Cybernetics and Control Theory* Vol. 10 No. 8 (1966) pp. 707–710.

[7] M. Maes, On a Cyclic String-to-String Correction Problem, *Information Processing Letter* Vol. 35 No. 2 (1990) pp. 73–78.

[8] A. Marzal and S. Barrachina, Speeding Up the Computation of the Edit Distance for Cyclic Strings, *Proceedings of 15th International Conference on Pattern Recognition* Vol. 2 (Barcelona, Spain, IEEE Computer Society, 2000) pp. 895–898.

[9] A. Marzal and G. Peris, Dependencia del cálculo de la distancia de edición cíclica con las funciones de coste, *Internal Report DLSI 01-02/01*, Universitat Jaume I, Spain. (In Spanish.)

[10] R. Mollineda, *Técnicas de Agrupamiento Jerárquico para la Selección de Prototipos y Clasificación basadas en Distancias. Uso en Cadenas Cíclicas*, Ph.D. Dissertation. Universidad Politécnica de Valencia, Spain, 2001. (In Spanish.)

[11] R. A. Mollineda, E. Vidal and F. Casacuberta, Efficient Techniques for a Very Accurate Measurement of Dissimilarities between Cyclic Patterns, *Lecture Notes in Comp. Science* Vol. 1876 (Springer, 2000) pp. 337–346.

[12] G. Peris and A. Marzal, Fast Computation of Cyclic Edit Distances: Dependence on the Cost Functions, *Pattern Recognition and Image Analysis. Proceedings of the IX Spanish Symposium on Pattern Recognition and Image Analysis.* Vol. 1, Universitat Jaume I, Spain (2001) pp. 121–126.

[13] D. Sankoff and J. B. Kruskal, *Time Warps, String Edits, and Macromolecules: the Theory and Practice of Sequence Comparisson*, (Addison-Wesley, 1983).

[14] P. H. Sellers, The Theory and Computation of Evolutionary Distances: Pattern Recognition, *Journal of Algorithms* Vol. 1 (1980) pp. 359–373.

[15] R. A. Wagner and M. J. Fischer, The String-to-String Correction Problem, *Journal of ACM* Vol. 21 No. 1 (1974) pp. 168–173.

Homogeneity, Autocorrelation and Anisotropy in Patterns

Andrés Molina
Departamento de Informática
Universidad de Jaén. Avd. Madrid 35, 23071 Jaén. Spain
E-mail: molina@ujaen.es

Contents

1 Introduction and Previous Works

The interpretation of the spatial distribution of a phenomenon can only be done by evaluating both the global (large scale trend or values in each point in the space) and local scale effects due to the interaction of each point with

D. Chen and X. Cheng (eds.),
Pattern Recognition and String Matching, 499–516.

its neighboring points [7]. The absence of these local effects make the values of the phenomenon vary depending on the place. In other words, the values observed in a window change systematically, hence, there does not exist any spatial dependence among values and the process is heterogeneous or non-stationary. On the contrary, if the existence of local effects is detected, the process is spatially homogeneous or stationary. Spatial dependence is a particular case of homogeneity. In images whose elements show spatial correlation, it is a fact that the existence of a concrete value of the phenomenon makes it more likely that this value should occur in near places. The existing statistics for determining the existence of spatial dependence among elements of an image are very varied [1][12], and they include non-spatial techniques, such as ANOVA, error terms in spatial autoregresive systems, Moran's I, Geary's C, Getis's G*, variograms, correlograms....[6]. All of them determine the existence or absence of autocorrelation in an image, but they do not provide spatial information about the direction in which it is manifested, that is, the directional trend of the variation of parameters defining the thematic characteristics of the image. In this paper we propose a method, based on first and second-order bivariate circular statistics, that will allow us to study the spatial variability of the phenomenon, to establish the existence of spatial dependence in an image and to calculate the direction in which this appears, by using a parametric procedure based on the standard and confidence ellipses for normal samples. A study about the flow of seawater is presented as an example of the potential use of the proposed method.

2 Difference vectors Matrix

Let $W_{n,n}\,(d, w_{c,c})$ be the $n \times n$ matrix that keeps the values of a window obtained from an image whose central element (element that has d rows both up and down and d columns both right and left) is w_{cc}. From $W_{n,n}$ the difference vectors matrix $V_{n,n}$ or matrix of first order vectors is obtained. For this purpose, every element $w_{i,j}$ is compared with the element which is diametrically opposed to it; if $w_{i,j}$ is greater than $w_{2k+2-i,2k+2-k}$, we will assign the difference between these two values $(w_{i,j} - w_{2k+2-i,2k+2-k})$ to $v_{i,j}$ and zero value to $v_{2k+2-i,2k+2-k}$. If $w_{i,j}$ is not greater than $w_{2k+2-i,2k+2-k}$, the assignment will the inverse:

FOR $(k = 1..d)$
  FOR $(i = 1..2k+1)$ $(j = 1..2k+1)$
      IF $w_{i,j} > w_{2k+2-i,2k+2-k}$

THEN $v_{i,j} \leftarrow (w_{i,j} - w_{2k+2-i,2k+2-k})$
$v_{2k+2-i,2k+2-k} \leftarrow 0$
ELSE $v_{i,j} \leftarrow 0$
$v_{2k+2-i,2k+2-k} \leftarrow (w_{i,j} - w_{2k+2-i,2k+2-k})$

The element $v_{i,j}$ of the matrix $V_{n,n}$ represents a vector with length $r = v_{i,j}$ and angle $\Phi = \arctan\left[\frac{j-(k+1)}{-(i-(k+1))}\right]$, which can be interpreted as a variation of the characteristic with intensity $v_{i,j}$ in direction Φ . From the $V_{n,n}$ and taking values $d = 1, 2, 3, ..., (n-1)/2$, we obtain $V_{3,3}(1), V_{5,5}(2), ..., V_{n,n}((n-1)/2)$, submatrix of $V_{n,n}$ whose central element is $v_{(2d+1)/2,(2d+1)/2}$. Each of the matrixes $V_{(2d+1)/2,(2d+1)/2}(d)$ will be formed by the elements $v_{i,j}$ whose distance to the central element is less or equal to d. Given $V(d)$, its mean variation vector $\overline{m}_d$ (of length r_d and mean angle $\overline{\Phi}_d$) is calculated as follows [3]:

$$N = \sum_{i=1}^{2d+1} \sum_{j=1}^{2d+1} b_{i,j}(d)\, v_{i,j} \quad \textbf{(1)}$$

$$\overline{x} = \frac{1}{N} \sum_{i=1}^{2d+1} \sum_{j=1}^{2d+1} b_{i,j} v_{i,j} \cos\left[\arctan \frac{j-(d+1)}{(-i-(d+1))}\right] \quad \textbf{(2)}$$

$$\overline{y} = \frac{1}{N} \sum_{i=1}^{2d+1} \sum_{j=1}^{2d+1} b_{i,j} v_{i,j} \sin\left[\arctan \frac{j-(d+1)}{(-i-(d+1))}\right] \quad \textbf{(3)}$$

$$r = \sqrt{\overline{x}^2 + \overline{y}^2}; \qquad \overline{\Phi}_d = \left\{ \begin{array}{ll} \arctan\left[\frac{\overline{y}}{\overline{x}}\right] & if\ \overline{x} > 0 \\ 180 + \arctan\left[\frac{\overline{y}}{\overline{x}}\right] & if\ \overline{x} < 0 \end{array} \right\} \quad \textbf{(4)}$$

$$b_{i,j}(d) = \left\{ \begin{array}{ll} 1 & if |i-d| = 0 \quad or \quad |j-d| = 0 \\ 0 & Otherwise \end{array} \right\} \quad \textbf{(5)}$$

If the mean variation vectors $[d = 1..8]$ are calculated from the images in Figure 1, we will obtain the values shown in Tables 1 and 2 respectively.

d	1	2	3	4	5	6	7	8
r_d	0.79	0.78	0.77	0.66	0.61	0.66	0.63	0.44
$\overline{\Phi}_d$	292	290	281	270	253	237	225	222

Table 1: Mean variation vectors calculated from Fig.1 (Left).

The analysis of spatial dependence at local level is carried out by exploring the spatial covariance structure of the pattern. For that, the most

d	1	2	3	4	5	6	7	8
r_d	0.35	0.23	0.06	0.10	0.03	0.07	0.06	0.26
$\overline{\Phi}_d$	143	28	93	231	158	245	19	256

Table 2: Mean variation vectors calculated from Fig.1 (Right).

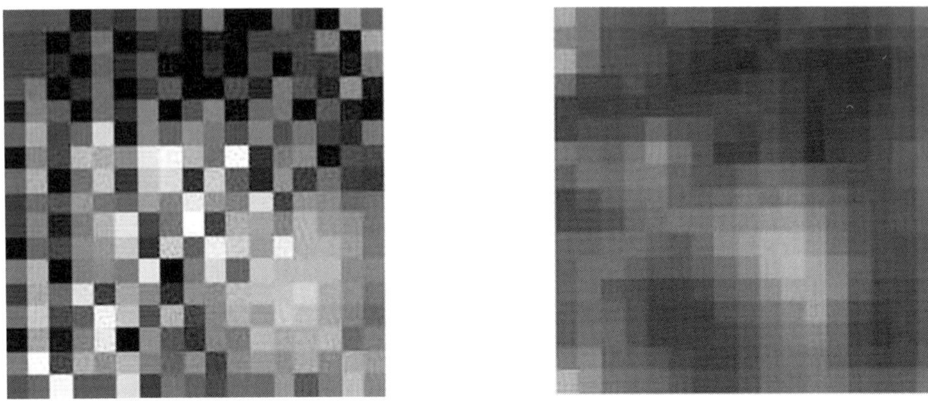

Figure 1: Patterns with Moran's I= 0.34 (Left) and 0.82 (Right).

commonly technique used is the autocorrelation test based on any of the statistics designed for it, such Moran's I, Geary's C, Getis' G* or Lagrange multiplier. Moran'I [4] is the most widely used one. For a matrix, these statistics at k lag are estimated as

$$I^{(k)} = \frac{n\sum_{i=1}^{n}\sum_{j=1}^{n}\delta_{i,j}^{(k)}(z_i-\overline{z})(z_j-\overline{z})}{\sum_{i=1}^{n}(z_i-\overline{z})^2\sum_{i=1}^{n}\sum_{j=1,j\neq i}^{n}\delta_{i,j}^{(k)}} \quad \textbf{(6)}$$

The expected value and variance are

$$E(I) = -\frac{1}{n-1} \qquad Var(I) = \frac{n^2 S_1 - nS_2 + 3S_0^2}{S_0^2(n^2-1)} \quad \textbf{(7)}$$

where

$$S_0 = \frac{1}{2}\sum_{i=1}^{n}\sum_{j=1}^{n}\left(\delta_{i,j}^{(k)} + \delta_{j,i}^{(k)}\right)^2 \quad S_1 = \sum_{i=1}^{n}\left(\sum_{i=1}^{n}\delta_{i,j}^{(k)} + \sum_{j=1}^{n}\delta_{j,i}^{(k)}\right)^2 \quad \textbf{(8)}$$

Here, $\delta_{i,j}^{(k)}$ denotes the $k-$spatial connection between the i-th and j-th cells. Although the correlation coefficients are usually restricted to (-1,1) range, this is not Moran's I case. In order to force the (-1,1) range, I must be divided by a correction factor c:

$$c^{(k)} = \frac{n}{\sum_{i=1}^{n}\sum_{j=1,j\neq i}^{n}\delta_{i,j}^{(k)}}\left(\frac{\sum_{i=1}^{n}\left(\sum_{j=1}^{n}\delta_{i,j}^{(k)}(z_i-\overline{z})\right)^2}{\sum_{i=1}^{n}(z_i-\overline{z})^2}\right) \quad \textbf{(9)}$$

A larger positive value is associated to a clustered pattern. On the contrary, a negative value different from zero implies a scattered pattern. Values next to zero represent lack of spatial dependence and are consequently associated to random spatial patterns. By observing Eq.(6), we can see that it is not a sound measure for patterns whose cells have the same value, since we find indeterminations when these expressions $\frac{0}{0}$ obtain. If we calculate the first-lag autocorrelation of Figures 1 by using Moran's Index (Eq.6) and by examining the eight neighboring cells connected to each cell, we obtain the values shown in Table 3. We can observe that the left pattern in Fig. 1 presents a certain level of heterogeneity in the spatial distribution of the values. This fact is confirmed by a low index of autocorrelation (0.34). The right pattern, on the contrary, shows an index near to one (0.82), which indicates a high spatial dependence.

	Fig.1 (Left)	Fig1 (Right)
Number of cells included	289	289
Mean of cells included	91.82	67.02
Standard deviation (σ) of cell values	3.74	1.80
Moran's I	0.34	0.82

Table 3: Autocorrelation parameters of Fig. 1.

3 Bivariate and Second-Order Analysis

It is evident that, from the observations made for a concrete spatial lag, we cannot draw a conclusion about the behavior of values in the image for other lags, no matter how many data are available and how sophisticated the analysis may be. For this reason, the statistical analysis must be performed in two steps, called first and second-order analysis or first and second stage of analysis [5]:

- For each k-lag $[k = 1..d]$ we reduce the variation vectors by calculating the mean variation vector $\overline{m}_k$.

- We combine the mean variation vectors $\overline{m}_1$, $\overline{m}_2$, ..., $\overline{m}_k$ of the above step and test their significance. Only then can we draw a statistical inference about the directional behavior of data.

The vector $\overline{m}_d$ is described by an angle $\overline{\Phi}_d$ and a module r_d; in other words, both the angles and the amplitude of its module have to be considered. Under this condition, the pairs $\left(r_1, \overline{\Phi}_1\right)$, $\left(r_2, \overline{\Phi}_2\right)$, ...,$\left(r_d, \overline{\Phi}_d\right)$ become bivariate and second-order sample, and their treatment is considered as second-order analysis [2].

3.1 Standard Ellipse

Among the tools used for second-order analysis, we find the standard ellipse [3], which serves exclusively for descriptive purposes. The tips of the vectors of a second-order sample form a scatter diagram of data with standard deviations in the X and Y directions and a certain trend upward or downward. The standard ellipse describes this behavior in a condensed form: assuming normality, roughly 40% of the data points fall inside the ellipse and 60% outside. The parent population need not be normal, although it is desirable that it does not deviate too much from normality. In order to draw the standard ellipse two means, two standard deviations and a correlation coefficient are required:

$$x_i = r_i \cos \overline{\Phi}_i \qquad y_i = r_i \sin \overline{\Phi}_i \quad \textbf{(10)}$$

$$\overline{x} = \tfrac{1}{d} \sum_i^d x_i \qquad \overline{y} = \tfrac{1}{d} \sum_i^d y_i \quad \textbf{(11)}$$

$$s_1 = \tfrac{1}{d-1} \sum_i^d \left(x_i - \overline{x}\right)^2 \qquad s_2 = \tfrac{1}{d-1} \sum_i^d \left(y_i - \overline{y}\right)^2 \quad \textbf{(12)}$$

$$Cov\left(x, y\right) = \tfrac{1}{d-1} \sum_i^d \left(x_i - \overline{x}\right)\left(y_i - \overline{y}\right) \qquad r = Corr\left(x, y\right) \tfrac{Cov(x,y)}{s_1 s_2} \quad \textbf{(13)}$$

The equation of the standard ellipse is

$$A\left(x - \overline{x}\right)^2 + 2B\left(x - \overline{x}\right)\left(y - \overline{y}\right) + C\left(y - \overline{y}\right)^2 = D \quad \textbf{(14)}$$

where

$$A = s_1^2 \qquad B = -r s_1 s_2 \qquad C = s_2^2 \qquad D = \left(1 - r^2\right) s_1^2 s_2^2 \quad \textbf{(15)}$$

The midpoint of the ellipse is at $(\overline{x}, \overline{y})$; the semi-axes a and b $(a < b)$ are

$$a = \left[\frac{2D}{A+C-R}\right]^{\frac{1}{2}} \qquad b = \left[\frac{2D}{A+C+R}\right]^{\frac{1}{2}} \quad \textbf{(16)}$$

where

$$R = \left[(A-C)^2 + 4B^2\right]^{\frac{1}{2}} \quad \textbf{(17)}$$

The sample shows the maximum variability in a direction of angle θ. This is the angle by which the major axis is inclined versus the X axis $(-90 < \theta < 90)$.

$$\theta = \arctan\left[\frac{2D}{A+C+R}\right] \quad \textbf{(18)}$$

The values of the parameters described, calculated from Table 1 and Table 2, are the ones shown respectively in Tables 4 and 5. Fig. 2 and 3 show the representation of the ellipses.

$\overline{x}$	$\overline{y}$	s_1	s_2	r	A	B	C	D	a	b	θ
0.073	-0.368	0.189	0.370	-0.062	0.137	0.004	0.036	0.005	0.370	0.189	272

Table 4: Values of the standard ellipse calculated from vectors of Table 1.

$\overline{x}$	$\overline{y}$	s_1	s_2	r	A	B	C	D	a	b	θ
-0.076	-0.596	0.295	0.164	-0.860	0.027	0.042	0.087	0.0006	0.103	0.075	332

Table 5: Values of the standard ellipse calculated from vectors of Table 2.

As commented above, a standard ellipse is useful for descriptive purposes solely. Nevertheless, from it, we can estimate whether it is reasonable to assume normality in the parent population. If the mean vector angles are uniformly spaced around the coordinates origin, we can consider that the population from which the sample is drawn does not differ from randomness or one-sideness, avoiding normality. When this occurs, the origin falls into the ellipse. For this reason we can consider, in an approximate way, that the parent population is normal if the standard ellipse contains the origin of the coordinates.

3.2 Confidence Ellipse

By the standard ellipse, we have described the spatial behaviour of the mean variation vectors quantifying its directional trend by calculating θ. Nevertheless, we must determine if the directional trend is caused by random

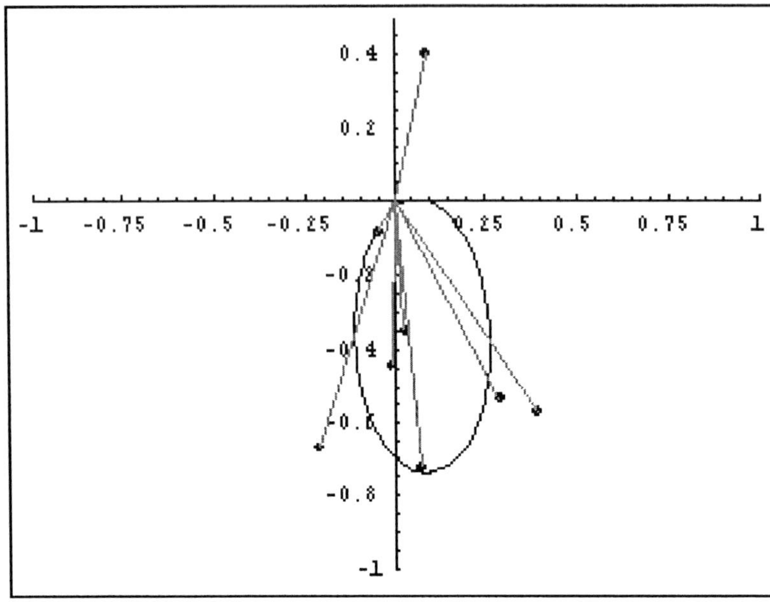

Figure 2: Standard ellipse depicted from parameters contained in Table 4.

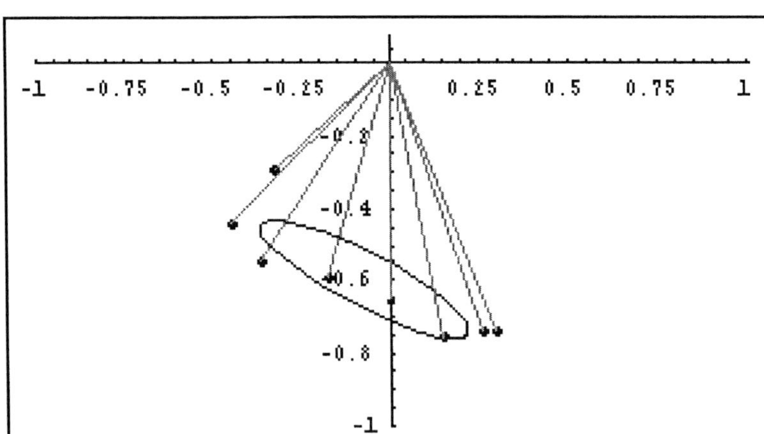

Figure 3: Standard ellipse depicted from parameters contained in Table 5.

fluctuations of the vectors or whether, on the contrary, this is caused by the existence of directedness or anisotropic variability [10]. For testing directedness, the confidence ellipse is used. Confidence ellipse includes a region in the XY-plane that covers the unknown population centre with a preassigned probability $Q = 1 - \alpha$, therefore being a tool for statistical inference. Assuming normality, this region has the shape of an ellipse which has the same centre as the standard ellipse and the same inclination (θ) versus the major axis. The problem of determining the existence of directedness is solved by generating the confidence ellipse and testing whether the origin is in its interior. If it is not the case, the population centre cannot coincide with the origin, and $(\overline{x}, \overline{y})$ is significantly different from it, so that it may be concluded that the mean vectors as a group are oriented in the direction ψ

$$\psi = \arctan \left[\frac{\overline{y}}{\overline{x}} \right] \quad \textbf{(19)}$$

For the confidence ellipse, the coefficients A, B and C in Eq.(14) have the same values for standard ellipse in Eqs.(15). The coefficient D being

$$D = \left(1 - r^2\right) s_1^2 s_2^2 n^{-1} T^2 \quad \textbf{(20)}$$

where

$$T^2 = 2 \tfrac{n-1}{n-2} F_{2,n-2}\left(\alpha\right) \quad \textbf{(21)}$$

Here, $F_{2,n-2}\left(\alpha\right)$ denotes the critical F value with 2 and $n - 2$ degree of freedom and significance level α. Confidence ellipse has the same center and the same θ value as the standard, since it is independent from the variable D, as Eqs.(11)(18) reveal. In both ellipses, the principal axes coincide. Only the semi-axes are variable. Let the semi-axes have the special parameter $D = 1$. Then we obtain from Eqs.(16) for arbitrary values $D \neq 0$

$$a = a_1 D^{\frac{1}{2}} \qquad\qquad b = b_1 D^{\frac{1}{2}} \quad \textbf{(22)}$$

Since a and b are proportional to $D^{\frac{1}{2}}$, we obtain

$$a = a_s n^{\frac{1}{2}} T \qquad\qquad b = b_s n^{\frac{1}{2}} T \quad \textbf{(23)}$$

where a_s and b_s are the semi-axes of the standard ellipse. From Eq.(14) it is deduced that the points which are inside the region limited by the confidence ellipse fulfil the inequality

$$A\left(x - \overline{x}\right)^2 + 2B\left(x - \overline{x}\right)\left(y - \overline{y}\right) + C\left(y - \overline{y}\right)^2 < D \quad \textbf{(24)}$$

Fig. 1	Left ($\alpha = 0.05$)	Right ($\alpha = 0.1$)	Right ($\alpha = 0.05$)
T^2	8.07	11.99	11.99
a	0.37	0.45	0.40
b	0.19	0.23	0.09

Table 6: Parameters of confidence ellipse calculated from Table 2

If the origin falls within the ellipse, the population centre could coincide with the origin and the sample is not directed. In this case, inequality (24) is fulfilled with the special values $x = 0$ and $y = 0$. From Eqs.(20)(24), the condition for the existence of directedness with a level of significance is

$$T^2 < T^2(\alpha) \quad \textbf{(25)}$$

where

$$T^2(\alpha) = \frac{n}{1-r^2}\left[\frac{\overline{x}^2}{s_1^2} - \frac{2r\overline{x}\overline{y}}{s_1 s_2} + \frac{\overline{y}^2}{s_2^2}\right] \quad \textbf{(26)}$$

Analysing the standard ellipse (Fig. 4) obtained from Fig. 1 (right), we observe that the assumption of normality is reasonable, so that a confidence ellipse is applicable to it. We want to determine the confidence ellipse at a level of significance $\alpha = 5\%$ or a confidence coefficient of $Q = 95\%$ from Fig.1 (right). As a first step we calculate T^2 by using Eq.(21). From a table of the F-distribution we read $F_{2,6} = 5.14$. Hence $T^2 = 11.99$. From Eqs.(20)(23) we obtain D, a and b parameters shown in Table 6. Our confidence ellipse (dashed curve depicted in fig.4) is somewhat bigger than the standard ellipse. Since the confidence ellipse does not contain the origin in its interior, $(\overline{x}, \overline{y})$ differs significantly from the origin. Therefore, the mean vectors are oriented as a group, where the direction at $\psi = -97.25$.

In Fig.2 we observe that, although the standard ellipse does not contain the origin, it is very close to it. Consequently, the assumption of normality must be taken with certain reservations. Fig.5 shows the confidence ellipses with errors $\alpha = 0.05$ and $\alpha = 0.1$ drawn from the parameters of Tables 5 and 6. The inside ellipse does not contain the coordinates' origin, while the exterior one does, assuming the existence of directedness ($\psi = -77.76$) with a confidence $Q = 90\%$, but not with $Q = 95\%$.

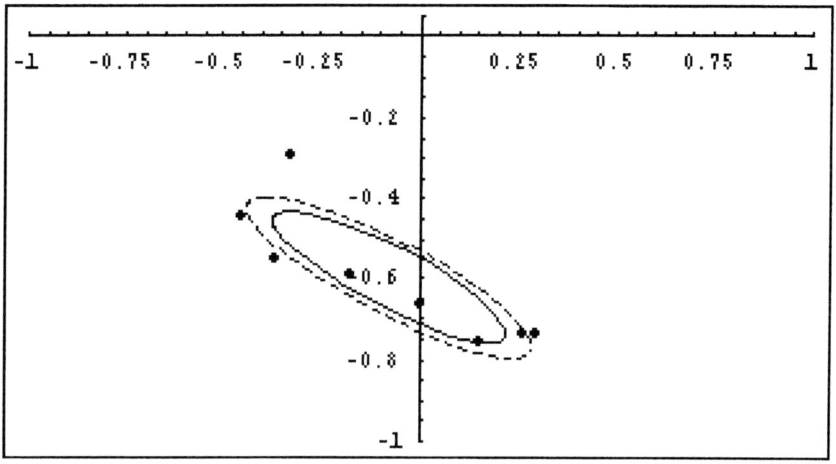

Figure 4: Confidence (dashed curve) and standard ellipses depicted from values of Tables 4 and 6.

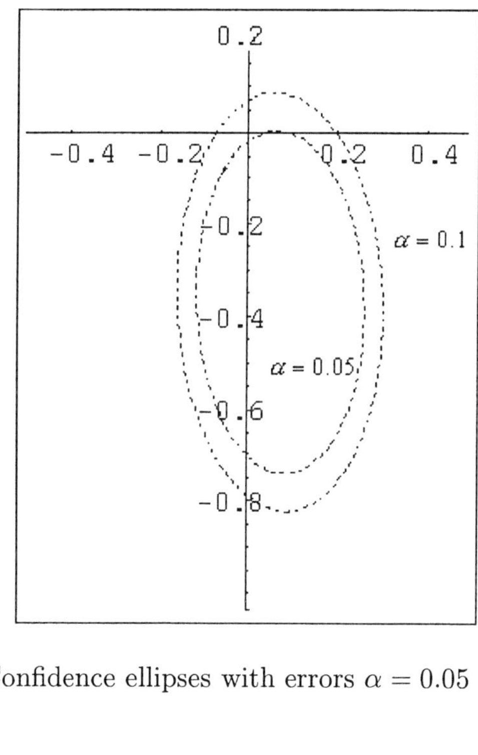

Figure 5: Confidence ellipses with errors $\alpha = 0.05$ and $\alpha = 0.1$.

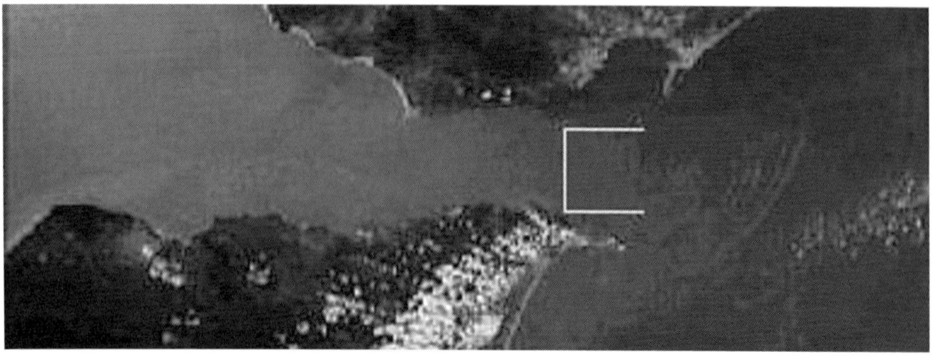

Figure 6: Strait of Gibraltar, seen from STS-58 flight (STS or Space Transportation System is the NASA's name for the overall shuttle program).

4 Graphical Model of Flow in Seawater

Fig.6 is a photograph [9] of the Strait of Gibraltar, as seen from the STS-58 flight. It belongs to the digital images collection of the NASA's Johnson Space Center [8]. The Johnson Space Center provides all catalogued shuttle mission Earth-looking imagery from the first shuttle mission, STS-1, through STS-76 [11]. The Bay of Cádiz on the southwest coast of Spain, the Rock of Gibraltar and the Moroccan coast are visible in it. A swift current flowing through the strait produces complicated patterns in the water surface due to sunlight reflected on it. In order to study the flow direction, we select the window of 37 by 37 pixels (Fig 7) marked in the image. This pattern shows the values of light reflected on the 0-255 grey scale, the central sub-pattern being the window of 27 by 27 pixels in the dashed box. For each location of the central sub-pattern, we obtain the $27 \times 27 = 729$ matrixes and from them, the difference vector matrixes will be calculated. Then, we will test the existence of anisotropy and quantify the spatial trend of the pattern in the case of directedness statistical evidence. Anisotropy is a property of stationary processes. This implies that its study can only be carried out under the assumption of stationarity in both global and local effects. For this reason, before the directedness analysis, the homogeneity level of the pattern must be checked, because if it showed spatial independence at global or local level, the interpretation of the results obtained in such analysis would not make any sense.

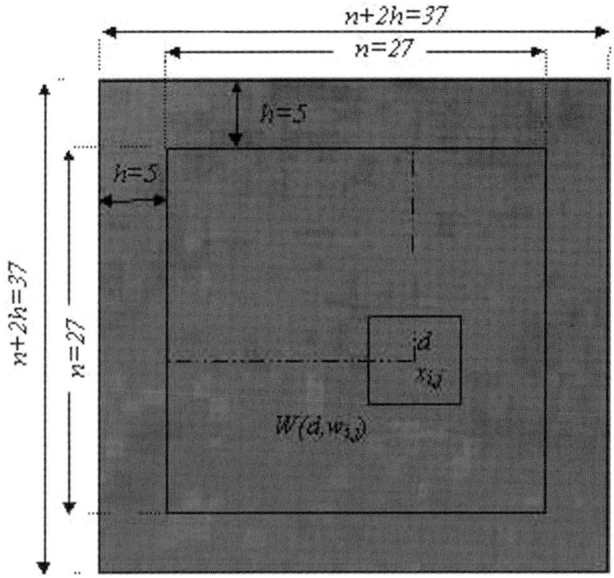

Figure 7: Window of 37 by 37 pixels marked in Fig.6. The central sub-pattern selected is the squared window of 27 by 27 pixels.

4.1 Analysing Global Homogeneity.

Among the techniques used for the study of homogeneity in local effects, the standard one-way analysis of variance (ANOVA) is the most simple one. A basic assumption underlying the analysis of variance is that of equality of variances. If this is not performed, ANOVA does not provide a valid test of the equality of pattern means. Assuming normality, in the population from which the patterns have been drawn, the pattern variances should be distributed as Chi-Square with (11×11)-1=120 degrees of freedom, the lower and upper limits of 95%-confidence intervals being 95.70 and 146.57 respectively. Values lower than the inferior limit denote patterns with low variance, and consequently with a strong spatial dependence, while values higher than the upper limit indicate a high variability or heterogeneity. Statistics have been calculated by estimating the variance value of the population (324.96) as the variance of a large sample of pixels drawn from the whole image, all of them being lower than 95.70. That is why stationarity in global effects can be assumed.

Parameter	Estimate	Standard Error	Confidence Interval
β_1	-0.1204	0.0300	$(-0.2158, -0.0250)$
β_2	0.0262	0.0090	$(-0.0025, 0.0550)$

Table 7: Fit parameters of the expected indexes by a function $\widehat{I} = \beta_1 + \beta_2 h$

4.2 Testing Local Spatial Dependence.

A common practice for exploring spatial dependence is to estimate the spatial correlation at different lags for the construction of a correlogram where $I^{(k)}$ is plotted against the k lag. The observation of this correlogram, and especially its peaks, allows us to obtain information about the level and spatial trend of autocorrelation and the lags on which this reaches its maximum and minimum values. Another method to analyze the variation of the spatial dependence level of a pattern is to calculate $I^{(1)}$ from the matrixes $\{W_{2h+1,2h+1}(d, x_{i,j}), d = 1..5\}$ defined in Section 2. For each location $x_{i,j}$, we will obtain 5 indexes $\left\{I^{(1)}(d, x_{i,j}), d = 1..5\right\}$ of autocorrelation. Each of these groups of 5 indexes is fitted, after removing the detected indeterminations, by a linear function $\widehat{I} = \beta_1 + \beta_2 d$. From Eq.(7) the expected indexes $\{E(I(d)), d = 1..5\} = \left\{\frac{-1}{8}, \frac{-1}{24} \frac{-1}{48} \frac{-1}{80} \frac{-1}{120}\right\}$ are obtained. Table 7 shows the parameters of the linear fit of these values and their confidence intervals calculated with a signification $\alpha = 5\%$. Among the 729 cells, only 15 of them (Fig.8) can be considered spatially independent, since they present slope values out of the confidence interval.

4.3 Analysing Directedness.

As was said in Section 3, the first step when analysing directedness is testing if there is statistical evidence of an anisotropic behaviour. For each cell $x_{i,j}$ of the central sub-pattern, we calculate $\{V_{2h+1,2h+1}(d, x_{i,j}), d = 1..5\}$and its mean variation vectors $\{\overline{m}_d, d = 1..5\}$ by applying Eqs.(10-13). After obtaining the mean variation vectors of each cell of the central sub-pattern, we determine the parameters of the confidence ellipse that allow us to obtain the statistic T^2 (Eq.(26)) and we will apply condition (Eq.(25)) for testing the anisotropy. All confidence ellipses have been calculated with 5 vectors, so that the critical value $T^2(0.05)$ (Eq.(21)) for all cells is 25.47. The direction in which the anisotropy appears in each location is given by ψ (Eq.(19)), and it is shown graphically in Fig.9. The existence of homogeneity, dependence and anisotropy, can be seen in Fig. 8. The symbols have

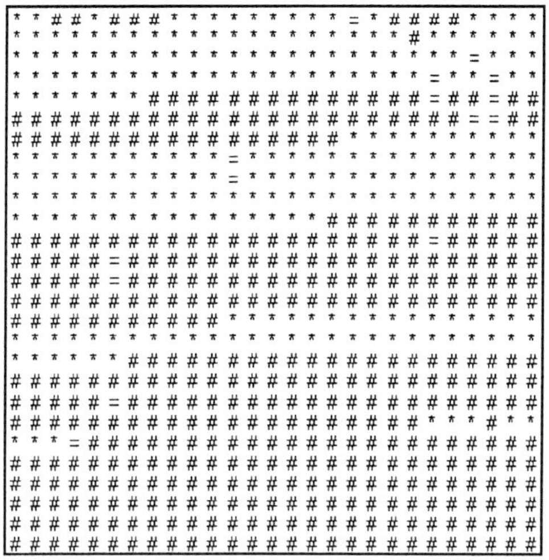

Figure 8: Homogeneity, dependence and anisotropy of each cell bellonging to central sub-pattern of Fig.7.

the following interpretation: = Homogeneity in global effects and lack of spatial dependence in local ones; * Homogeneity in global effects and spatial dependence in local ones with isotropic behaviour; # Homogeneity in global effects and spatial dependence in local ones with anisotropic behaviour. By analysing this figure, we can observe that only 15 of the 729 cells present homogeneity at a global level and lack of spatial dependence in local effects, and although 7 of them are located on the upper right corner of the pattern, they do not form a clearly defined cluster. The rest of the cells present global homogeneity and local spatial dependence. As for directedness, we can observe the existence of a cluster of anisotropic cells that are located around the lower two thirds of the pattern. The upper third is distributed among three well defined clusters, two of them with isotropic cells and another one with directed cells. The existence of such a widespread stationarity at local and global levels, leads us to think that the degree of anisotropy of each cell will intensify if we increase the number of mean variation vectors used for the calculation of the parameters that define confidence ellipses, as the directional effects are emphasized, in contrast to random fluctuations that keep constant. This assertion is confirmed if we observe Fig.10 since the number of cells without statistical evidence of anisotropy decreases as value d increases. Inspecting the results of the ψ directions obtained, it is seen that the process variations in the horizontal direction are different from the

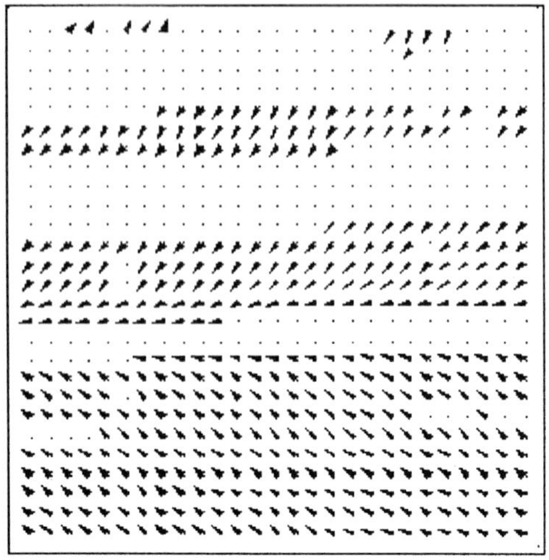

Figure 9: Direction ψ in which anisotropy appears in each cell of the central sub-pattern calculated with mean variation vectors. The points represent isotropic cells.

ones in the vertical direction, since there is a clear anisotropic trend in east-west direction. This is due to the tide that produces a flow of Atlantic water through the Strait of Gibraltar into the Mediterranean Sea. This flow generates internal waves that, in the photograph, are rendered into the different intensities of light as it is shown in Fig.6 and 7.

5 Concluding Remarks.

Previous methods about analysis of directional dependence in patterns are restricted to a visual examination of the phenomenon and the behaviour of the variograms in different directions. The intuition and "savoir faire" of the researcher assess the existence of anisotropy and quantify the direction in which it appears. The procedure described in the paper, on the contrary, allows us to explore the directional variability of the pattern and the normality of parent population from which the sample has been drawn. Besides, it is possible to test the existence of directedness and to calculate the angle in which the anisotropy is revealed. For this purpose, we have used two tools (standard and confidence ellipses) based on first and second-order bivariate circular statistics.

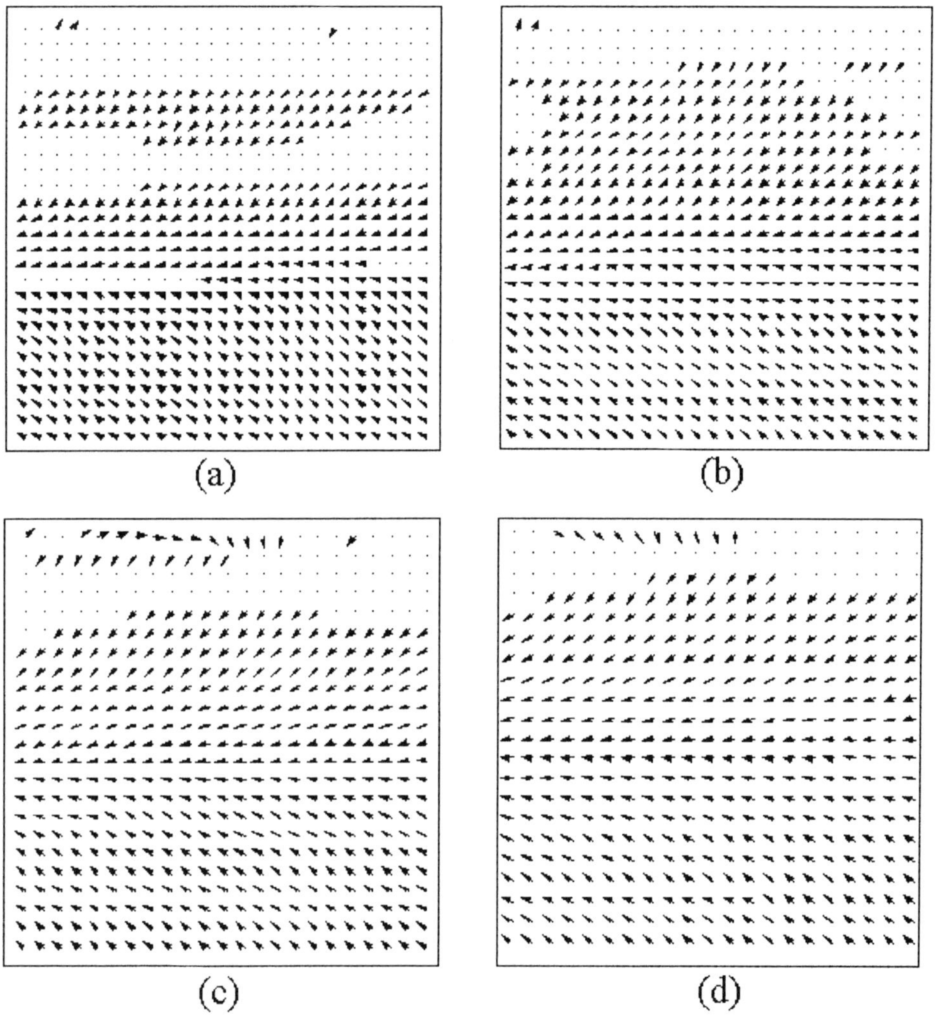

Figure 10: Direction ψ in which anisotropy appears in each cell of the central sub-pattern calculated with (a) $d = 7$. (b) $d = 8$. (c) $d = 9$. (d) $d = 10$.

Acknowledgements

This work has been funded by grants number TIC2001-2099-C03-03 and REN2001-3890-C02-02 from MCYT (Ministry of Science and Technology of the Spanish Government), and European Union FEDER program.

References

[1] T.C. Bailey and A.C. Gattrel, *Interactive Spatial Data Analysis*, (Edinburgh, Longman, 1995).

[2] E. Batschelet, Second-Order Statistical Analysis of Directions, in K. Schmidt-Koening and W.T. Keeton (eds.), *Animal, Migration, Navigation and Homing*, (Berlin, Springer-Verlag, 1978) pp.3-24.

[3] E. Batschelet, *Circular Statistics in Biology*, (London, Academic Press, 1981).

[4] Y.H. Chow, Spatial Patterns and Spatial Autocorrelation, in A.U. Frank and W. Kuhn, (eds.) *Spatial Information Theory. Lectures Notes in Computer Science (Vol. 988)* (Berlin, Springer-Verlag (1995) pp. 365-376.

[5] D.R. Cox and D.V. Hinkley, *Theoretical Statistics*, (London, Chapman and Hall, 1974).

[6] N. Cressie, *Statistics for Spatial Data.* (Chichester, John Wiley, 1993)

[7] A. Getis, Homogeneity and Proximal Databases, in: S. Fotheringham and P. Rogerson (eds.) *Spatial Analysis and GIS.*, (London, Taylor&Francis, 1994) pp. 105-120.

[8] *Johnson Space Center Home Page.*

http://www.jsc.nasa.gov/.

[9] *JSC Digital Image Collection - STS058-73-009.*

http://images.jsc.nasa.gov/images/pao/STS58/10083794.htm.

[10] A. Molina and F.R.Feito, Design of Anisotropic Functions Within Analysis of Data, in F.E.Wollter and N.M. Patrikalakis (eds.) *Proc. IEEE Computer Graphics International '98* (1998) pp. 726-729.

[11] *STS Mission History and Manifest.*

http://lpsweb.ksc.nasa.gov/EXODUS/Forms/sts.shtml.

[12] H. Wackernagel, *Multivariate Geostatistics*, (Berlin,Springer-Verlag, Berlin, 1995).

Robust Structural Indexing through Quasi-Invariant Shape Signatures and Feature Generation

Hirobumi Nishida
Ricoh Software Research Center
1-1-17 Koishikawa, Bunkyo-ku, Tokyo 112-0002, Japan
E-mail: hn@src.ricoh.co.jp

Contents

D. Chen and X. Cheng (eds.),
Pattern Recognition and String Matching, 517–544.

1 Introduction

Information retrieval from large image databases is an essential functionality for the reuse, manipulation, and editing of multimedia documents [9, 24]. Images have some components in terms of representation, such as color, texture, and shape [5]. Color and texture are mathematically and physically tractable, and their properties and variations can be represented in well-structured forms in terms of some statistical methods [27]. On the other hand, shape is another essential component, but shape analysis and representation are still difficult research subjects in spite of intensive research carried out for decades. In particular, shapes observed in natural scenes are often occluded, corrupted by noise, and partially visible.

It is an important problem to retrieve efficiently model shapes that have parts similar to the query shape presented to the image database [14]. A number of methods have been proposed for matching and recognition of partially visible shapes. Many approaches consider certain types of invariant shape features or signatures along boundary contours, employing dynamic programming techniques for finding optimal subparts in terms of similarity of the shape signatures [1, 2, 6, 25, 30] or Hough voting techniques for identifying locations of subparts [29]. For instance, the *arclength versus turning angle graph* [29, 30] is invariant to the Euclidean transformation group composed of rotation and translation. Furthermore, invariant features with respect to the similarity transformation group include the shape descriptor called sphericity [1] and the total curvature function [25]. However, these methods are inefficient when the number of models is large, because a significant amount of computations are required for finding out optimal subparts with respect to a single model and the computation time is proportional to the number of models. Indexing techniques [8, 11, 12] are potential approaches to improving efficiency, and geometric hashing approaches [3, 13, 28] have been proposed to recognize partially visible shapes. The geometric hashing is composed of the computation of some invariant features with respect to a specific transformation group and the voting for the transformation space. Its main drawback is that it is sensitive to noise and shape deformations, and its performance in terms of classification accuracy is degraded drastically even due to small changes of shapes [7].

Efficient and robust information retrieval from large image databases by shape [14] is a challenging problem, and shape retrieval from image databases has been studied recently for improving efficiency and robustness [4, 16, 17]. In particular, the problem is intractable when the shape is partially visible. Stein and Medioni [28] try to improve the robustness against noise and local shape deformations by extracting shape features from several versions of piecewise linear approximations of boundary segments. Mokhtarian [15] applies the curvature scale space approach along with an object indexing techniques in order to improve the robustness and efficiency.

Efficiency and robustness are important, but sometimes incompatible criteria for performance evaluation. The improvement of robustness implies that the scheme for classification and retrieval should tolerate certain types of variations and deformations for images. Obviously, it may lead to inefficiency if some brute-force methods are employed such as a generate-and-test strategy by generating various images with a number of different parameters. A key to achieving both efficiency and robustness is through a compact and well- structured representation of images that tolerate variations and deformations. Furthermore, an idea to achieving both efficiency and robustness is to handle deformations in the *feature domain* by generating features that can be extracted from deformed patterns. *Feature generation models* can be obtained through an analysis of feature transformations due to some particular types of shape deformation. Robustness can be improved by incorporating generated features into the indexing process. It has been confirmed that efficiency of classification and robustness against noise and local shape transformations can be improved at the same time by this feature indexing approach incorporating shape feature generation models [22]. It is now an attractive and interesting subject to apply this approach to retrieval of model shapes that have parts similar to the query shape presented to the image database.

In this paper, based on the structural feature indexing with feature generation models, an efficient, robust method is presented for retrieval of model shapes that have parts similar to the query shape presented to the image database. The design of algorithms and data structures proceeds in the following steps:

1. Representation: Based on convex/concave structures incorporating quantized-directional features along curves, a compact shape representation with simple, efficient computation is explored so that the curves can be described by a few components with rich features. In

order to derive high-level models for coping with noise and local shape deformations, the representation should hold some rigorous properties.

2. Quasi-invariant shape signature: The shape signature is a key feature used in the indexing. In order to retrieve images for a query given as a partially visible shape, the shape signature should tolerate rotation, scaling, and translation, and should also be rich enough to characterize the shape. We define the quasi-invariant shape signature in terms of the shape representation based on convex/concave structures incorporating quantized-directional features.

3. Feature transformations: Features based on convex/concave structures are transformed by noise, scales of observations, and local shape deformations. Therefore, to cope with such deformations, an analysis of feature transformations is carried out with respect to some particular types of shape deformations, leading to transformation rules of structural features composed of a small number of distinct cases. We extend the analysis given by Nishida [22] to the new shape signature introduced in this work. Features that can be extracted from deformed patterns caused by noise and local shape deformations are generated by applying the transformation rules to the features extracted from the image patterns.

4. Model database organization by structural indexing: For efficient manipulation of a large number of models, a large table is constructed for a model set by assigning a table address to a feature extracted from the image patterns and storing a list of the model identifiers with the corresponding feature as a table item. To cope with feature transformations caused by noise, scales of observation, and local shape deformations, the generated features by the transformation rules are also used for the model database organization with structural indexing.

5. Retrieval by voting for models: In the retrieval, from the features extracted from the query shape, features are also generated by the transformation rules. Model identifier lists are retrieved from the tables by using the addresses computed from the generated features, and the transformation parameters are computed between the model shape and the query shape. Voting boxes are prepared for the transformation space associated with each model, and voting is carried out for the transformation parameters computed for each model on the lists.

The query shape is classified efficiently by selecting out some models according to some scores based on the number of votes.

This paper is organized as follows: In Section 2, a structural representation of curves by quasi-convex/concave features along with quantized-directional features [18, 19] is outlined. In Section 3, based on the shape representation outlined in Section 2, we describe the quasi-invariant shape signature, the model database organization through feature indexing, and the shape retrieval through voting. In Section 4, the transformation rules of shape signatures are introduced to generate features that can be extracted from deformed patterns caused by noise and local shape deformations. The proposed algorithm is described in Section 5 based on the shape signature, the feature indexing, and the voting along with the transformation rules of the shape signature. In Section 6, the proposed method is demonstrated with a large image database obtained from real data, and is validated by systematically designed experiments with a large number of synthetic data. Section 7 concludes this paper.

2 Shape Representation

The structural representation of curves [18, 19] is outlined in this section, based on quasi- convex/concave structures incorporating $2N$ quantized-directional features (N is a natural number). As shown in Fig. 1(a), the curve is first approximated by a series of line segments. On a 2-D plane, we introduce N-axes together with $2N$ quantized-direction codes. For instance, when $N = 4$, eight quantized-directions are defined along with the four axes as shown in Fig. 1(b). Based on these N-axes together with $2N$ quantized-direction codes, the analysis is carried out hierarchically.

A curve is decomposed into *sub-segments* at extremal points along each of the N-axes. Fig. 1(c) illustrates the decomposition of a contour shown in Fig. 1(a) into sub-segments when $N = 4$. For adjacent sub-segments a and b, suppose that we turn counterclockwise when traversing them from a to b, and the joint of a and b is an extremal point along the axes toward the directions $(j, j+1\,(mod\ 2N), \ldots, k)$. Then, we write the concatenation of these two sub-segments as $a \xrightarrow{j,k} b$. For instance, the joint of sub-segments H and G in Fig. 1(c) is an extremal point along the three axes toward the directions 3, 4, and 5. Therefore, the concatenation of H and G is written as $H \xrightarrow{3,5} G$. In this way, we obtain the following concatenations for the sub-segments illustrated in Fig 1(c).

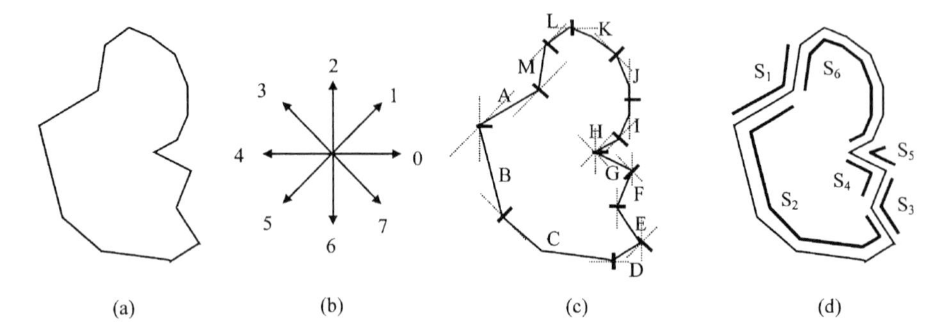

Figure 1: (a) A closed contour with a polygonal approximation, (b) quantized-directional codes when $N = 4$, (c) sub-segments when $N = 4$, (d) segments when $N = 4$.

$$L \xrightarrow{3,3} M,\ K \xrightarrow{2,2} L,\ J \xrightarrow{1,1} K,\ I \xrightarrow{0,0} J,$$
$$H \xrightarrow{7,7} I,\ H \xrightarrow{3,5} G,\ F \xrightarrow{0,1} G,\ F \xrightarrow{4,4} E,$$
$$D \xrightarrow{7,0} E,\ C \xrightarrow{6,6} D,\ B \xrightarrow{5,5} C,\ A \xrightarrow{3,4} B,\ A \xrightarrow{7,7} M.$$

By linking local features around joints of adjacent sub-segments, some sequences of the following form can be constructed:

$$a_0 \xrightarrow{j(1,0),j(1,1)} a_1 \xrightarrow{j(2,0),j(2,1)} \cdots \xrightarrow{j(n,0),j(n,1)} a_n, \tag{1}$$

where $j(i, 0)$ and $j(i, 1)$ are direction codes assigned to the concatenation of a_{i-1} and a_i. A part of the contour corresponding to a sequence of this form is called a *segment*. Furthermore, the starting point of the segment is defined as the end point of a_0, and the ending point is as the end point of a_n. When a segment is traversed from its starting point to its ending point, one turns counterclockwise around any joints of sub-segments. The following segments, as shown in Fig. 1(d), are generated from the 13 sub-segments shown in Fig. 1(c):

$$S_1 : A \xrightarrow{7,7} M,\ S_2 : A \xrightarrow{3,4} B \xrightarrow{5,5} C \xrightarrow{6,6} D \xrightarrow{7,0} E,$$
$$S_3 : F \xrightarrow{4,4} E,\ S_4 : F \xrightarrow{0,1} G,\ S_5 : H \xrightarrow{3,5} G,$$
$$S_6 : H \xrightarrow{7,7} I \xrightarrow{0,0} J \xrightarrow{1,1} K \xrightarrow{2,2} L \xrightarrow{3,3} M.$$

A segment is characterized by a pair of integers $\langle r, d\rangle$, *characteristic numbers*, representing the angular span of the segment and the direction of the first sub-segment:

$$r = \sum_{i=1}^{n} (j(i,1) - j(i,0))_{mod\ 2N} + \sum_{i=1}^{n-1} (j(i+1,0) - j(i,1))_{mod\ 2N} + 2 \quad (2)$$

$$d = j(1,0) \quad (3)$$

If

$$(j(i,1) - j(i,0))_{mod\ 2N} = 1 \quad \text{for } i = 1, 2, \ldots, n,$$

and

$$(j(i+1,0) - j(i,1))_{mod\ 2N} = 1 \quad \text{for } i = 1, \ldots, n-1,$$

the angular span r coincides with the number of sub-segments in the sequence. The characteristic numbers are given by $\langle 2,7\rangle$, $\langle 7,3\rangle$, $\langle 2,4\rangle$, $\langle 3,0\rangle$, $\langle 4,3\rangle$, and $\langle 6,7\rangle$, respectively, for the six segments shown in Fig. 1(d). The reader is referred to Nishida and Mori [23] for the intuitive idea of the characteristic numbers.

Adjacent segments are connected by sharing the first sub-segments or last ones of the corresponding sequences. These two types of connection are denoted by $S \overset{h}{-} T$ and $S \overset{t}{-} T$, respectively, for two adjacent segments S and T. For instance, connections are denoted by

$$S_1 \overset{h}{-} S_2 \overset{t}{-} S_3 \overset{h}{-} S_4 \overset{t}{-} S_5 \overset{h}{-} S_6 \overset{t}{-} S_1$$

for the six segments shown in Fig. 1(d).

3 Quasi-Invariant Shape Signature, Indexing, and Voting

Based on the shape representation outlined in Section 2, we describe the shape signature, the model database organization through feature indexing, and the shape retrieval through voting. For the model database organization, we assume that each model shape is presented to the system as line drawings or boundary contours of objects. In the shape retrieval, we assume that line drawings or parts of some model shape can be given as a query to the system.

3.1 Quasi-Invariant Shape Signatures

In order to retrieve images for a query given as a partially visible shapes, the shape signature is required to tolerate rotation, scaling, and translation. Therefore, features depending on orientation, size, and location cannot be employed as shape signatures. Based on the characteristic numbers and connections of segments extracted from model shapes or query shapes, the shape signature is constructed to satisfy this requirement. We assume that a series of n segments S_i $(i = 1, 2, \ldots, n)$ have been extracted with characteristic numbers $\langle r_i, d_i \rangle$ and lengths l_i. The angular span r_i does not depend on orientation, size, or location. (We will later cope with certain problems caused by quantizing the direction.) Furthermore, the lack of information due to dropping orientation, size, and location can be compensated by employing a triplet of the angular spans of two consecutive segments and their length ratio as the shape signature. From two consecutive segments S_i and S_{i+1} connected as $S_i \overset{c}{-} S_{i+1}$, $c \in \{h, t\}$, the quasi-invariant shape signature is constructed as follows:

$$\left(r_i, r_{i+1}, c, \left\lfloor Q \cdot \frac{l_{i+1}}{l_i + l_{i+1}} \right\rfloor \right),$$

where Q is the number of quantization levels for length-ratio parameters $l_{i+1} / (l_i + l_{i+1})$.

3.2 Indexing

From each model shape, shape signatures are extracted from all pairs of consecutive segments. There are a large variety of curve segments that have the identical shape signature. Therefore, shapes cannot be classified by the shape signatures alone, and some additional information is needed to reduce the candidate model sets for the query shape. The expressive power of the shape signature can be enriched with some simple, additional geometrical parameters. This integration of structural features with simple geometrical information has been quite successful in that complex deformed patterns can be classified with a few features [20]. Furthermore, the geometrical parameters are also exploited in generating hypotheses about the pose of the query shape with respect to the model shape in consideration. Since it is essential to estimate the pose in shape classification, transformation parameters need to be computed in the voting process. This mechanism is commonly incorporated into most object recognition systems based on the structural indexing approach [13, 28].

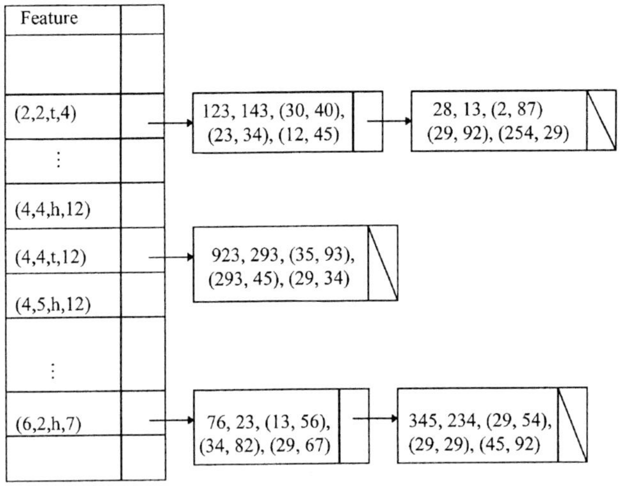

Figure 2: Model database organization by structural indexing. Each table item stores a list whose element is composed of the model identifier, length, location of the center of gravity, and locations of the two end points of the curve segment.

Based on the above discussion, a large table, as illustrated in Fig. 2, is constructed for a model set by assigning a table address to a shape signature and storing there a list whose item is composed of the following elements:

1. the model identifier that has the corresponding shape signature
2. shape parameters of the curve segment corresponding to the shape signature, namely length, location of the center of gravity, and locations of the two endpoints, computed on the model shape.

The elements of type (2) are also used for generating hypotheses about the geometrical transformation applied to the query shape with respect to the model shape in consideration.

3.3 Voting

Classification of the query shape is carried out by voting for the transformation space associated with each model. For each model, voting boxes are prepared for the quantized transformation space $(\sigma, \theta, x_T, y_T)$, where σ is the scaling factor, θ is the rotation angle, and (x_T, y_T) is the translation vector. Shape signatures are extracted from the curve segment given as a query to the shape database. For each extracted shape signature, model identifiers and shape parameters are retrieved from the table by computing the table address. By comparing the shape parameters of the extracted shape signature with the registered parameters, the transformation parameters $(\sigma, \theta, x_T, y_T)$ can be computed for each model and the voting box corresponding to the transformation parameters associated with the model is incremented by one. In the implementation, transformation parameters σ and θ are computed from the line segment connecting the two endpoints, and (x_T, y_T) is computed from the location of the center of gravity. Furthermore, the scaling factor σ is normalized into the interval [0,2] by mapping $\sigma \to 2 - 1/\sigma$ if $\sigma > 1$.

4 Feature Generation Models

Shape signatures extracted from the curve are sensitive to noise and local shape deformations, and therefore, the correct model does not necessarily receive many votes as expected for the ideal case. Furthermore, when only one sample pattern is available for each class, techniques of statistical or inductive learning from training data cannot be employed for obtaining *a priori* knowledge and feature distributions of deformed patterns. To cope with these problems, we analyze the feature transformations caused by some particular types of shape deformations, constructing feature transformation rules (see Fig. 3). Based on the rules, we generate segment features that can be extracted from deformed patterns caused by noise and local shape deformations. In both processes of model database organization and classification, the generated features by the transformation rules are used for structural indexing and voting, as well as the features actually extracted from curves. The following three types of feature transformations are considered in this work:

1. Change of convex/concave structures caused by perturbations along normal directions on the curve and scales of observation, along with transformations of characteristic numbers (the angular span of the segment and the direction of the first sub-segment).

2. Transformation of characteristic numbers caused by small rotations.

3. Transformation of length-ratio parameters caused by noise and local shape deformations.

We describe these three types of transformation in the rest of this section.

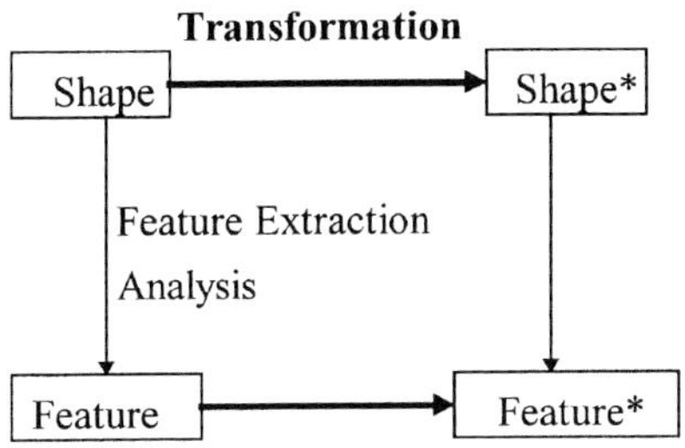

Figure 3: When shapes are deformed by some transformations, features are also transformed accordingly.

4.1 Transformation of Convex/Concave Structures

The convex/concave structures along the curve are changed by noise and local deformations, and also depend on scales of observations. For instance, two parts of curves shown in Fig. 4(a) are similar to one another in terms of global scales, but their structural features are different. When $N = 4$, the curve shown on left is composed of three segments connected as $S_1 \overset{t}{-} S_2 \overset{h}{-} S_3$ with characteristic numbers $\langle 6,6\rangle$, $\langle 2,6\rangle$, and $\langle 3,2\rangle$, whereas the one shown on right is composed of five segments connected as $S'_1 \overset{t}{-} S'_2 \overset{h}{-} S'_3 \overset{t}{-} S'_4 \overset{h}{-} S'_5$ with characteristic numbers $\langle 6,6\rangle$, $\langle 2,6\rangle$, $\langle 2,2\rangle$, $\langle 2,6\rangle$, and $\langle 3,2\rangle$. To cope with such deformations, structural features on the two curves are edited so that their features can become similar to one another. For instance, the structural features illustrated in Fig. 4(a) can be edited by merging the two segment blocks $\{S_1, S_2, S_3\}$ and $\{S'_1, S'_2, S'_3, S'_4, S'_5\}$ to two segments S and S' as shown Fig. 4(b).

In general, rules can be introduced for generating characteristic numbers from a segment block (a set of consecutive segments).

Rule 4.1 *[22] Let $\langle r_i, d_i\rangle$ be the characteristic number of the segment s_i. From a segment block, a characteristic number is generated according to the following rules:*

1. *From a segment block $\left\{ s_i \,\middle|\, i = 0, 1, \ldots, n-1;\ s_0 \overset{h}{-} s_1 \overset{t}{-} \cdots \overset{t}{-} s_{n-1} \right\}$, where n is odd, a characteristic number*

$$\left\langle \sum_{i=0}^{n-1} (-1)^i r_i, d_{n-1} \right\rangle$$

is generated if $r_{2i-2} - r_{2i-1} + r_{2i} \geq 2$ *for* $i = 1, 2, \ldots, \lfloor n/2 \rfloor$ *and* $\sum_{i=0}^{n-1}(-1)^i r_i \geq 2$.

2. *From a segment block* $\left\{s_i \;\middle|\; i = 0, 1, \ldots, n-1;\ s_0 \overset{t}{-} s_1 \overset{h}{-} \cdots \overset{h}{-} s_{n-1}\right\}$, *where* n *is odd, a characteristic number*

$$\left\langle \sum_{i=0}^{n-1}(-1)^i r_i, d_0 \right\rangle$$

is generated if $r_{2i-2} - r_{2i-1} + r_{2i} \geq 2$ *for* $i = 1, 2, \ldots, \lfloor n/2 \rfloor$ *and* $\sum_{i=0}^{n-1}(-1)^i r_i \geq 2$.

These rules can be introduced from some mathematical properties mentioned in Nishida [18, 19]. Their theoretical justification and motivation are given in detail in Nishida [22]. In the structural indexing and voting processes, for an integer M specifying the maximum number of segments to be merged, shape signatures are generated by applying Rule 4.2 described below to segment blocks.

Rule 4.2 *Let* $\langle r_i, d_i \rangle$ *be the characteristic number of the segment* s_i, *and* $\lambda(s_j s_{j+1} \cdots s_{j+k-1})$ *be length of the curve composed of* k *consecutive segments* $s_j s_{j+1} \cdots s_{j+k-1}$. *From a segment block, a shape signature is generated according to the following rules:*

1. *From a segment block*

$$\begin{aligned}\Sigma(s_j, m, n) \;=\; & \Big\{s_{j+k} \;\Big|\; k = 0, 1, \ldots, m-1, m, \ldots, m+n-1; \\ & s_j \overset{h}{-} s_{j+1} \overset{t}{-} \cdots \overset{t}{-} s_{j+m-1} \overset{h}{-} s_{j+m} \overset{t}{-} \cdots \overset{h}{-} s_{j+m+n-1}\Big\},\end{aligned}$$

where m *and* n *are odd such that* $1 \leq m, n \leq M$, *a shape signature*

$$\left(\sum_{k=0}^{m-1}(-1)^k r_{j+k}, \sum_{k=0}^{n-1}(-1)^k r_{j+m+k}, h, \left\lfloor Q \cdot \frac{\lambda(s_j s_{j+1} \cdots s_{j+m-1})}{\lambda(s_j s_{j+1} \cdots s_{j+m-1} s_{j+m} \cdots s_{j+m+n-1})} \right\rfloor \right)$$

is generated if $\sum_{k=0}^{m-1}(-1)^k r_{j+k} \geq 2$, $\sum_{k=0}^{n-1}(-1)^k r_{j+m+k} \geq 2$, $r_{j+2k-2} - r_{j+2k-1} + r_{j+2k} \geq 2$ *for* $k = 1, \ldots, (m-1)/2$, *and* $r_{j+m+2k-2} - r_{j+m+2k-1} + r_{j+m+2k} \geq 2$ *for* $k = 1, \ldots, (n-1)/2$.

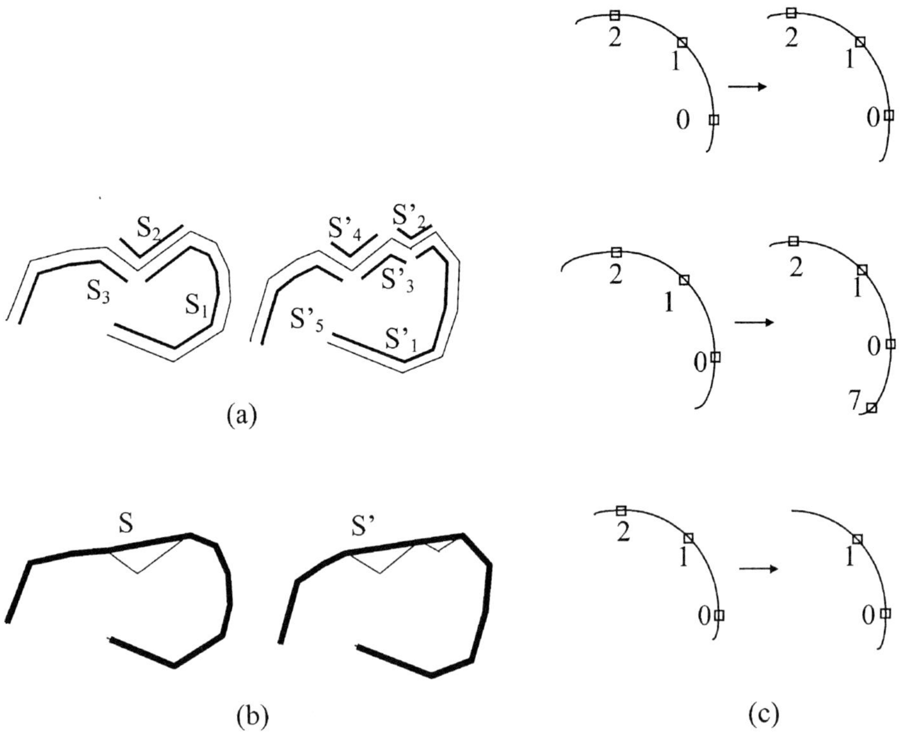

Figure 4: (a) Part of curves similar to one another in terms of global scales, (b) editing structural features by merging segment blocks, (c) transformations of characteristic numbers of segments by small rotations.

2. From a segment block

$$\Sigma(s_j, m, n) = \left\{ s_{j+k} \,\middle|\, k = 0, 1, \ldots, m-1, m, \ldots, m+n-1; \right.$$
$$\left. s_j \overset{t}{-} s_{j+1} \overset{h}{-} \cdots \overset{h}{-} s_{j+m-1} \overset{t}{-} s_{j+m} \overset{h}{-} \cdots \overset{t}{-} s_{j+m+n-1} \right\},$$

where m and n are odd such that $1 \leq m, n \leq M$, a shape signature

$$\left(\sum_{k=0}^{m-1} (-1)^k r_{j+k}, \sum_{k=0}^{n-1} (-1)^k r_{j+m+k}, t, \right.$$

$$\left\lfloor Q \cdot \frac{\lambda(s_j s_{j+1} \cdots s_{j+m-1})}{\lambda(s_j s_{j+1} \cdots s_{j+m-1} s_{j+m} \cdots s_{j+m+n-1})} \right\rfloor \Big)$$

is generated if $\sum_{k=0}^{m-1}(-1)^k r_{j+k} \geq 2$, $\sum_{k=0}^{n-1}(-1)^k r_{j+m+k} \geq 2$, $r_{j+2k-2} - r_{j+2k-1} + r_{j+2k} \geq 2$ *for* $k = 1, \ldots, (m-1)/2$, *and* $r_{j+m+2k-2} - r_{j+m+2k-1} + r_{j+m+2k} \geq 2$ *for* $k = 1, \ldots, (n-1)/2$.

For instance, when $N = 4$ and $M = 3$, from the six segments illustrated in Fig. 1(d) with characteristic numbers $\langle 2, 7\rangle$, $\langle 7, 3\rangle$, $\langle 2, 4\rangle$, $\langle 3, 0\rangle$, $\langle 4, 3\rangle$, and $\langle 6, 7\rangle$, the following shape signatures (length-ratio omitted) are generated by Rule 4.2: (2, 7, h), (2, 8, h), (7, 2, t), (7, 3, t), (8, 4, t), (2, 3, h), (2, 5, h), (3, 6, h), (3, 11, h), (3, 4, t), (5, 2, t), (4, 6, h), (4, 11, h), (6, 2, t), (11, 2, t), (11, 3, t). In total, at most $n \cdot \lceil M/2 \rceil^2$ shape signatures are generated from n segments.

4.2 Transformation of Characteristic Numbers by Small Rotations

The characteristic number $\langle r, d\rangle$ ($r \geq 2$) can be transformed by rotating the shape. Rules can be introduced for generating characteristic numbers by rotating the shape slightly (see Fig. 4(c)).

Property 4.1 *[22] By rotating the segment by angle* Ψ ($-\pi/N \leq \Psi \leq \pi/N$)*, the characteristic number* $\langle r, d\rangle$ *can be transformed into one of the following: (1)* $\langle r, d\rangle$*, (2)* $\langle r+1, d-1\rangle$*, (3)* $\langle r+1, d\rangle$*, (4)* $\langle r-1, d\rangle$ ($r \geq 3$)*, (5)* $\langle r-1, d+1\rangle$ ($r \geq 3$)*.*

Rule 4.3 *When the curve composed of the two consecutive segments* S_1 *and* S_2 *with characteristic numbers* $\langle r_1, d_1\rangle$ *and* $\langle r_2, d_2\rangle$ *is rotated by angle* Ψ ($-\pi/N \leq \Psi\pi/N$)*, the angular spans* r_1 *and* r_2 *can be transformed into one of the 9 cases: (1)* (r_1, r_2) *, (2)* $(r_1, r_2 - 1)$*, (3)* $(r_1, r_2 + 1)$*, (4)* $(r_1 - 1, r_2)$*, (5)* $(r_1 - 1, r_2 - 1)$*, (6)* $(r_1 - 1, r_2 + 1)$*, (7)* $(r_1 + 1, r_2)$*, (8)* $(r_1 + 1, r_2 - 1)$*, (9)* $(r_1 + 1, r_2 + 1)$*. Note that the cases (4-6) are applicable only if* $r_1 \geq 3$*, and that the cases (2), (5), and (8) are applicable only if* $r_2 \geq 3$.

The detailed derivation of this rule is given in Appendix. For instance, when $N = 4$ and $M = 3$, the 16 shape signatures have been generated by Rule 4.2 from the six segments illustrated in Fig. 1(d). Then, by applying Rule 4.3 to these generated ones, 120 shape signatures, in total, are further generated .

4.3 Transformation of Length-Ratio Parameters

The length-ratio parameters of segments are also changed by local deformations and noise. In the structural indexing and voting processes, the parameter is quantized into L intervals (L is a positive integer), and therefore, we need to take into consideration quantization errors of the parameters along with local shape deformations and noise. We introduce rules for generating quantized values of the length-ratio parameters $p \equiv l_j/(l_j + l_{j+1})$.

Rule 4.4 *Let α be a parameter $0 \leq \alpha \leq 1$.*

1. *If $i \leq pQ \leq i+\alpha/2$ $(0 < i < Q)$, then integers i and $i-1$ are generated as quantized values of p.*

2. *If $i+1-\alpha/2 \leq pQ < i+1$ $(0 \leq i < Q-1)$, then integers i and $i+1$ are generated as quantized values of p.*

3. *Otherwise, integer $\lfloor Qp \rfloor$ is generated as quantized value of p.*

If we assume that value of parameter p is distributed uniformly over the interval [0,1], the average number of the generated shape signatures by Rule 3 is given by

$$2\alpha + (1-\alpha) = 1 + \alpha.$$

5 Algorithm

In the model database organization step by structural indexing, shape signatures are generated from each model shape by Rules 4.2–4.4, and the model identifier with the shape parameters is appended to the list stored at the table address corresponding to each generated shape signature. For each model i $(i = 1, 2, \ldots, n)$, let c_i be the number of shape signatures generated by Rules 4.2 and 4.3. For instance, $c_i = 120$ for the contour shown in Fig. 1(a) when $N = 4$ and $M = 3$. In the classification and retrieval by voting for models and the transformation space, from shape signatures extracted from the query shape, shape signatures are generated by Rules 4.2 and 4.3. Model identifier lists are retrieved from the tables by using the addresses computed from the generated shape signatures, and the transformation parameters are computed for each model on the lists. The voting box is incremented by one for the model and the computed transformation parameters. Let v_i $(i = 1, 2, \ldots, n)$ be the maximum votes among the voting boxes associated with the model i. The query shape is classified by selecting out some models according to the descendant order of v_i/c_i.

The details are described for the indexing and voting processes in the rest of this section.

5.1 Indexing

In the model database organization, model shapes are, in general, given as a set of line drawings or boundary contours of objects. Suppose that for the model shape under consideration, segments S_i $(i = 0, 1, \ldots, n-1)$ with characteristic numbers $\langle r_i, d_i \rangle$ have been extracted such that

$$S_0 \overset{t}{-} S_1 \overset{h}{-} \cdots \overset{t}{-} S_{n-1}.$$

For each $i = 0, 1, \ldots, n-1$, shape signatures are generated by applying Rule 4.2 to segment blocks starting from S_i:

$$\Sigma\left(S_i, k, l\right) = \left\{ S_{i+j} \,\middle|\, j = 0, 1, \ldots, k-1, k, \ldots, k+l-1 \right\},$$

where k and l are odd such that $1 \leq k, l \leq M$. Then, Rules 4.3 and 4.4 are applied to the generated shape signatures. A table address is computed from each of the generated shape signatures, and a list item composed of the model identifier and the shape parameters is appended to the list stored there, as illustrated in Fig. 2. If we assume that the length-ratio parameter is uniformly distributed over the interval [0,1], the average number of shape signatures generated from the model shape composed of n segments is given by

$$O\left(n\left(1+\alpha\right) \cdot \lceil M/2 \rceil^2\right).$$

5.2 Voting

In the classification by voting, voting boxes are prepared for tuples of (Model-ID, σ, θ, x_T, y_T). Query shapes are, in general, given as a set of line drawings or parts of some model shape. Suppose that for the query shape, a series of m segments S_i $(i = 0, 1, \ldots, m-1)$, with characteristic numbers $\langle r_i, d_i \rangle$ have been extracted. For each $i = 0, 1, \ldots, m-1$, shape signatures are generated by applying Rule 4.2 to segment blocks starting from S_i:

$$\Sigma\left(S_i, k, l\right) = \left\{ S_{i+j} \,\middle|\, j = 0, 1, \ldots, k-1, k, \ldots, k+l-1 \right\},$$

where k and l are odd such that $1 \leq k, l \leq M$ and $i + k + l - 1 < m$. Then, Rule 4.3 is applied to the generated shape signatures. In total, $O\left(n \cdot \lfloor M/2 \rfloor\right)^2$ shape signatures are generated from the query shape. A table address is computed from each of the generated shape signatures, and a list stored there is retrieved from the table. For each list item composed of the model identifier and the shape parameters, the transformation parameters $(\sigma, \theta, x_T, y_T)$ are computed, and the voting box for (Model-ID, σ, θ, x_T, y_T) is incremented by one.

6 Shape Retrieval System

A prototype system for shape retrieval from image databases was developed based on the proposed method for structural feature indexing with feature generation models. Some experimental trials were carried out to validate the effectiveness of the proposed method for shape retrieval.

6.1 Outline of the system

In the model database organization by the structural indexing, a feature-model table as shown in Fig. 2 is constructed by computing shape features in a coordinate system specific to each model shape. The parameters used in the shape feature computation and the feature indexing are set as follows:

1. Parameters for Feature Computation
   - Number of quantized directions: 32 ($N = 16$)
   - Error tolerance for piecewise linear approximation: 2% of the longer side of the bounding box (the upright rectangle just enclosing the shape).
2. Parameter for Feature Indexing
   - Maximum number of segments to be merged: $M = 9$
   - Number of quantized levels of the length-ratio parameters of segments: $Q = 25$
   - Perturbation of the length-ratio parameters of segments: $\alpha = 0.1$
3. Quantization of Transformation Parameters
   - Scaling factor σ: 7 intervals
   - Rotation angle θ: 32 intervals

- Translation vector (x_T, y_T): 2×2 intervals

The shape feature database was constructed for the image data set publicly available through the www site

http://www.ee.surrey.ac.uk/Research/VSSP/imagedb/demo.html

from the VSSP Center of the University of Surrey, UK [16, 17]. The data set is composed of 1100 boundary contours of marine creature images, originally scanned from some printed books. Some experimental results for shape retrieval are presented in Figs. 5 and 6. In each figure, the query shape is presented at the top, with model shapes arranged from top-left in the descending order of scores v_i/c_i assigned to the models.

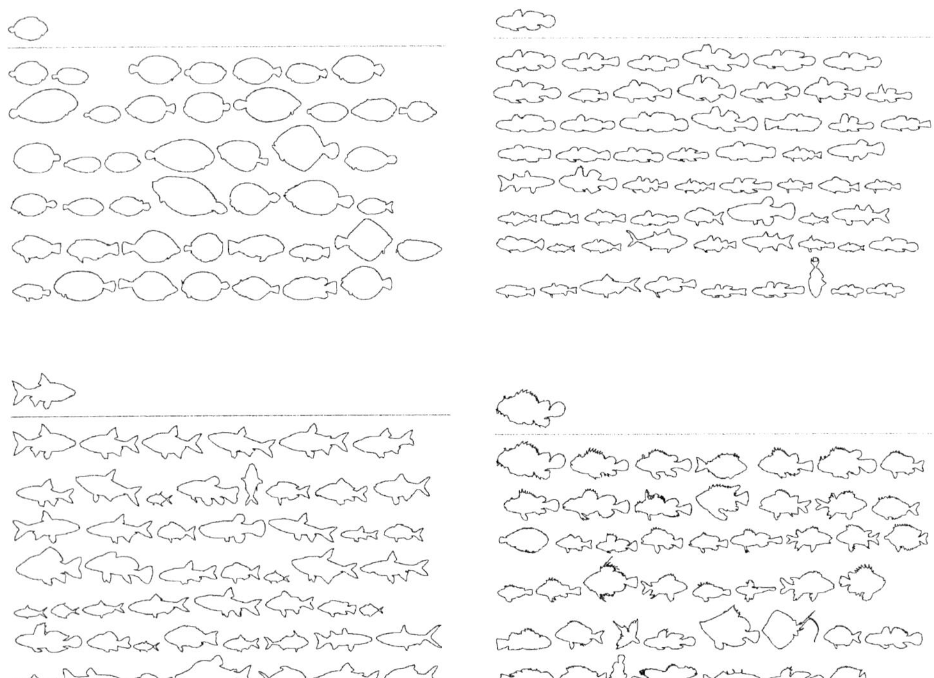

Figure 5: Example of shape retrieval from the shape database of marine creature images. The query shape is presented at the top, and retrieved shapes are arranged in the descending order of the scores from the top-left.

The system was also tested with the shape database composed of 400 boundary contours of plant leaves [10]. Some experimental results for shape retrieval are presented in Fig. 7.

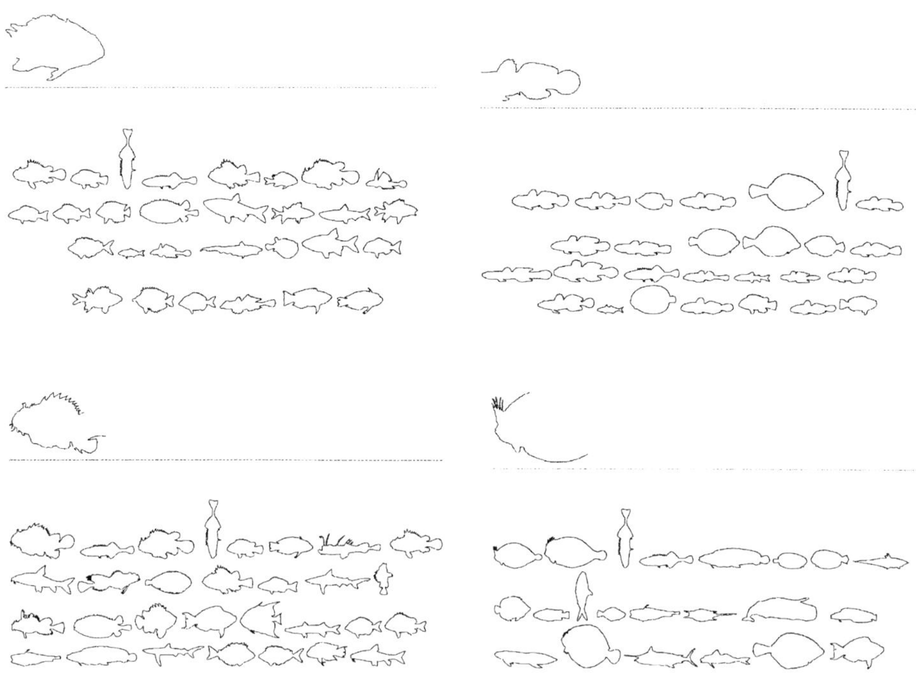

Figure 6: Example of shape retrieval from the shape database of marine creature images. The query shape is presented at the top, and retrieved shapes are arranged in the descending order of the scores from the top-left.

6.2 Quantitative Evaluation

In this section, the proposed algorithm is evaluated quantitatively in terms of the robustness against noise and shape deformations, based on the systematically designed, controlled experiments with a large number of synthetic data. The experimental design is composed of the generation of model patterns and their deformations to be used as testing samples. A modification of the midpoint displacement algorithm based on fractional Brownian motion is employed together with affine transformations for the model pattern generation [21]. From each model pattern expressed as a polygon, a number of testing samples are generated by applying arbitrary rotations and random perturbations along the curves. The deformation process for a model pattern is composed of the following two steps:

1. For a vertex P of the polygon, let A and B be its two adjacent vertices, and M be the midpoint of line segment AB. For a given parameter β,

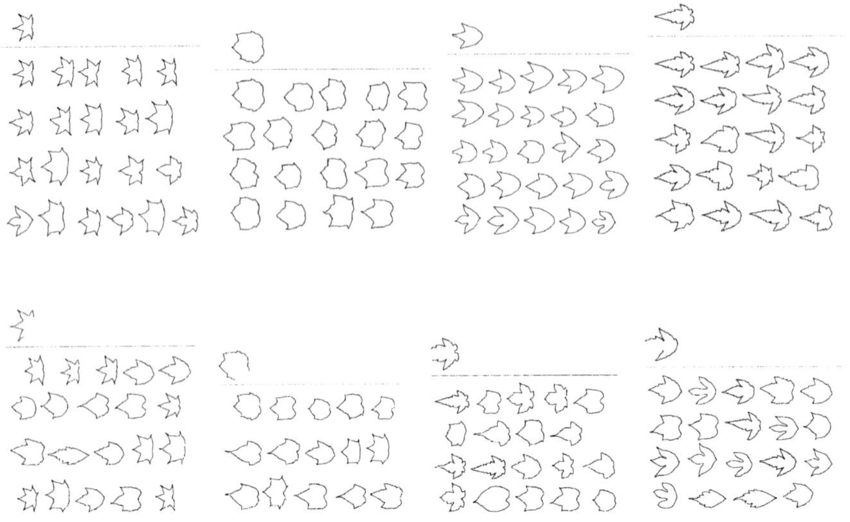

Figure 7: Example of shape retrieval from the shape database of plant leaf images.

move point P by $r(-\beta, \beta) \cdot \overline{PM}$ along the line passing through P and M. This operation is applied to all points on the polygon.

2. After the operation (1), the polygon is rotated arbitrarily.

We examined the probability that the correct model is included in top t% choices for various values of β when curves composed of $r\%$ portions of a model shape are given as queries. For given values of r and β, a sub-contour of $r\%$ of length is randomly extracted from the model shape, and then, it is deformed by the deformation process as described above. Some examples of query shapes are presented in Fig. 8 for some specific values of β and r.

The main contribution of this work is to incorporate the shape feature generation into the structural indexing for coping with shape deformations and feature transformations. Therefore, the performance was compared with a naive method extracting features from several versions of piecewise linear approximations of the curve with a variety of error tolerances for approximations.

We carried out several experimental trials by changing the number of models from 200 to 500, examining the classification accuracy in terms of the deformed portions of model shapes given as queries to image databases. In the implementation of the naive method, by changing the error tolerance

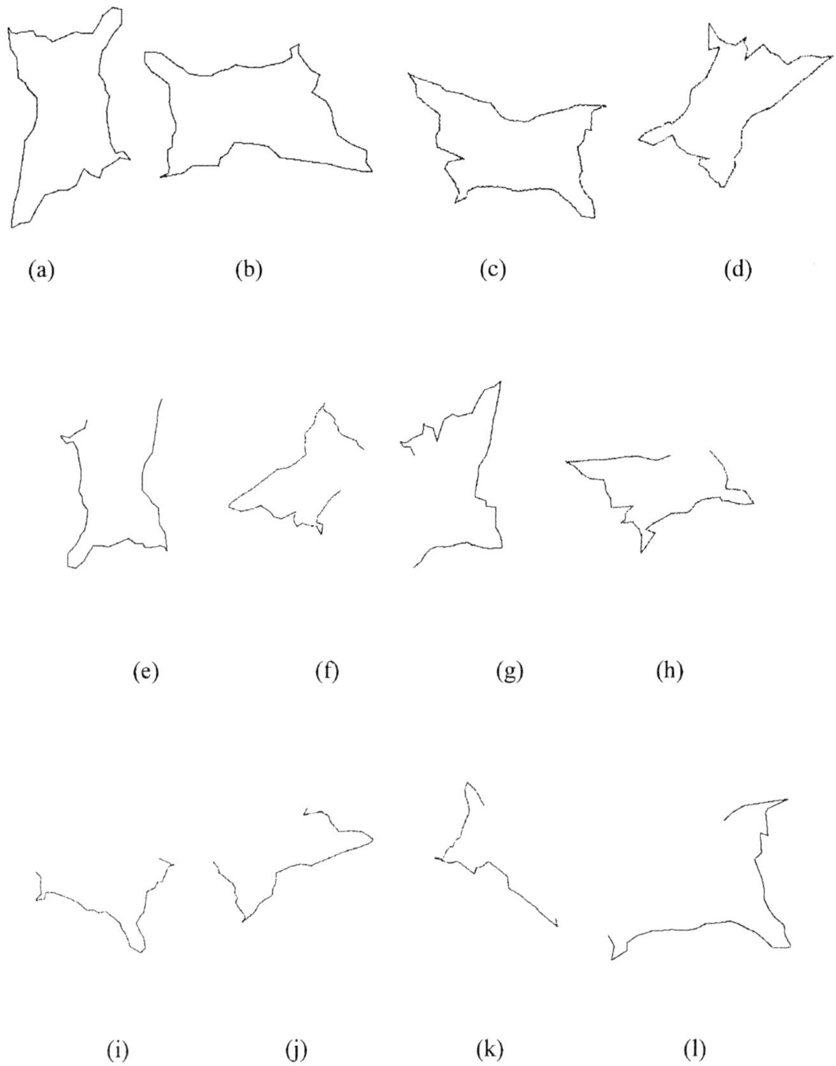

Figure 8: Examples of synthetic query shapes for some values for β and r. (a–d) $r = 100\%$, (e–h) $r = 80\%$, (i–l) $r = 60\%$; (a, e, i) $\beta = 0.0$, (b, f, j) $\beta = 0.5$, (c, g, k) $\beta = 1.0$, (d, h, l) $\beta = 1.5$.

with Ramer's method [26] from 1% to 20%, with a step of 1%, of the widest side of the bounding box of the curve, twenty versions of approximations were created for each model shape and the query shape.

Table 1: Average classification rates (%) for top 1%, 2%, 3%, 4%, 5%, and 10% choices when $\beta \in [0.0, 0.5]$ in terms of the portion of model shapes (r%) presented as queries, in comparison with the naive method.

Portion	Method	1%	2%	3%	4%	5%	10%
100%	proposed	100.0	100.0	100.0	100.0	100.0	100.0
	naive	92.8	96.8	98.3	98.6	98.8	99.7
90%	proposed	100.0	100.0	100.0	100.0	100.0	100.0
	naive	92.1	96.1	97.6	98.2	98.7	99.6
80%	proposed	99.6	99.7	99.7	99.7	99.8	99.9
	naive	86.3	92.3	94.8	96.1	97.3	98.9
70%	proposed	97.8	98.6	99.1	99.3	99.4	99.6
	naive	75.8	86.0	89.2	91.3	93.0	96.2
60%	proposed	92.4	95.0	95.9	96.5	96.7	98.0
	naive	61.6	73.1	79.0	82.5	84.7	91.6
50%	proposed	77.0	83.5	86.1	87.3	88.7	91.3
	naive	49.5	61.6	67.7	71.3	75.0	85.1

Table 2: Average classification rates (%) for top 1%, 2%, 3%, 4%, 5%, and 10% choices when $\beta \in [0.5, 1.0]$ in terms of the portion of model shapes (r%) presented as queries, in comparison with the naive method.

Portion	Method	1%	2%	3%	4%	5%	10%
100%	proposed	99.8	99.8	99.8	99.9	99.9	99.9
	naive	80.6	87.4	89.9	92.1	92.8	96.5
90%	proposed	99.2	99.6	99.7	99.7	99.7	99.8
	naive	76.8	84.6	88.7	90.5	91.6	95.8
80%	proposed	96.0	97.8	98.2	98.6	99.1	99.5
	naive	69.5	79.8	83.9	86.4	88.1	93.1
70%	proposed	87.2	91.7	93.7	94.5	95.4	97.4
	naive	52.2	64.8	70.8	75.1	77.9	86.5
60%	proposed	75.1	82.7	85.7	87.9	89.0	93.4
	naive	39.5	51.2	58.3	63.3	66.8	78.1
50%	proposed	54.4	64.7	70.0	72.5	74.6	82.3
	naive	27.7	37.4	43.9	48.5	52.3	66.8

Table 3: Average classification rates (%) for top 1%, 2%, 3%, 4%, 5%, and 10% choices when $\beta \in [1.0, 1.5]$ in terms of the portion of model shapes (r%) presented as queries, in comparison with the naive method.

Portion	Method	1%	2%	3%	4%	5%	10%
100%	proposed	91.1	94.3	96.2	96.7	97.3	98.2
	naive	55.1	65.8	71.0	74.2	76.2	84.2
90%	proposed	84.3	89.4	91.2	92.2	93.1	96.3
	naive	53.6	63.8	68.5	71.8	74.2	83.2
80%	proposed	68.4	75.9	79.8	82.5	84.9	90.1
	naive	40.7	52.0	57.7	62.4	66.0	76.5
70%	proposed	51.9	61.6	67.0	71.1	74.1	81.0
	naive	29.0	39.5	46.7	51.2	54.4	67.0
60%	proposed	36.6	46.0	51.7	56.2	59.0	68.5
	naive	17.5	27.3	33.3	37.7	41.3	55.5
50%	proposed	21.3	27.7	32.6	36.5	39.6	51.2
	naive	12.1	17.9	22.9	26.8	30.7	43.7

Tables 1–3 present the average classification rates for top 1%, 2%, 3%, 4%, 5%, and top 10% choices when $\beta \in [0.0, 0.5]$, $\beta \in [0.5, 1.0]$, and $\beta \in [1.0, 1.5]$. For instance, when a curve segment composed of 70% portions of a model shape subject to the deformation process with $\beta \in [0.5, 1.0]$ is given as a query to shape databases of 500 models, the correct models are included in top 20 choices (4%) with probability 94.5% for proposed algorithm and with probability 75.1% for the naive method.

Furthermore, average processing time for a single shape query is presented in Table 4 in terms of the size of image databases, in comparison with the naive method. The statistics were calculated on implementations with C programming language on Sun Sparc Ultra 2 running under Sun OS version 5.5.1. Clearly, significant improvements of robustness against noise and local shape deformations can be observed for the proposed algorithm in terms of classification accuracy without a significant degradation of efficiency. Through the experiments, the effectiveness has been verified through the experiments for the shape signature and the shape feature generation models.

Table 4: Average processing time (sec/query) of a single shape query ($r = 100\%$) for the proposed algorithm in terms of the size of image databases (number of models), in comparison with the naive method.

♯Models	200	300	400	500
Proposed Method	2.2	3.4	4.3	5.5
Naive Method	1.3	2.1	2.9	3.9

7 Conclusion

Efficient and robust information retrieval from large image databases is an essential functionality for the reuse, manipulation, and editing of multimedia documents. Structural feature indexing is a potential approach to efficient shape retrieval from large image databases, but the indexing is sensitive to noise, scales of observation, and local shape deformations. It has now been confirmed that efficiency of classification and robustness against noise and local shape transformations can be improved at the same time by the feature indexing approach incorporating shape feature generation techniques [22]. In this paper, based on this approach, an efficient, robust method has been presented for retrieval of model shapes that have parts similar to the query shape presented to the image database. The effectiveness has been confirmed by experimental trials with a large database of boundary contours obtained from real images and has been validated by systematically designed experiments with a large number of synthetic data.

A Derivation of Rule 4.3

We describe the derivation of Rule 4.3 in the appendix.

Property A.1 *[18, 19] Let S and T be segments with characteristic numbers $\langle r_S, d_S \rangle$ and $\langle r_T, d_T \rangle$.*

1. *If $S \underline{h} T$, then $d_S \equiv d_T + N \pmod{2N}$.*

2. *If $S \underline{t} T$, then $r_S + d_S \equiv r_T + d_T + N \pmod{2N}$.*

Based on Properties 4.1 and A.1, the following properties can be derived.

Property A.2 *Suppose that two segments* S_1 *and* S_2 *with characteristic numbers* $\langle r_1, d_1 \rangle$ *and* $\langle r_2, d_2 \rangle$ *are connected as* $S_1 \overset{h}{-} S_2$. *When the curve composed of the two segments is rotated by angle* Ψ $(-\pi/N \leq \Psi \leq \pi/N)$, *the characteristic numbers of* S_1 *and* S_2 *can be transformed into one of the 11 cases: (1)* $\langle r_1, d_1 \rangle$-$\langle r_2, d_2 \rangle$, *(2)* $\langle r_1, d_1 \rangle$-$\langle r_2+1, d_2 \rangle$, *(3)* $\langle r_1, d_1 \rangle$-$\langle r_2-1, d_2 \rangle$, *(4)* $\langle r_1+1, d_1 \rangle$-$\langle r_2, d_2 \rangle$, *(5)* $\langle r_1+1, d_1 \rangle$-$\langle r_2+1, d_2 \rangle$, *(6)* $\langle r_1+1, d_1 \rangle$-$\langle r_2-1, d_2 \rangle$, *(7)* $\langle r_1+1, d_1-1 \rangle$-$\langle r_2+1, d_2-1 \rangle$, *(8)* $\langle r_1-1, d_1 \rangle$-$\langle r_2, d_2 \rangle$, *(9)* $\langle r_1-1, d_1 \rangle$-$\langle r_2+1, d_2 \rangle$, *(10)* $\langle r_1-1, d_1 \rangle$-$\langle r_2-1, d_2 \rangle$, *(11)* $\langle r_1-1, d_1+1 \rangle$-$\langle r_2-1, d_2+1 \rangle$.

Property A.3 *Suppose that two segments* S_1 *and* S_2 *with characteristic numbers* $\langle r_1, d_1 \rangle$ *and* $\langle r_2, d_2 \rangle$ *are connected as* $S_1 \overset{t}{-} S_2$. *When the curve composed of the two segments is rotated by angle* Ψ $(-\pi/N \leq \Psi \leq \pi/N)$, *the characteristic numbers of* S_1 *and* S_2 *can be transformed into one of the 11 cases: (1)* $\langle r_1, d_1 \rangle$-$\langle r_2, d_2 \rangle$, *(2)* $\langle r_1, d_1 \rangle$-$\langle r_2+1, d_2-1 \rangle$, *(3)* $\langle r_1, d_1 \rangle$-$\langle r_2-1, d_2+1 \rangle$, *(4)* $\langle r_1+1, d_1-1 \rangle$-$\langle r_2, d_2 \rangle$, *(5)* $\langle r_1+1, d_1-1 \rangle$-$\langle r_2+1, d_2-1 \rangle$, *(6)* $\langle r_1+1, d_1-1 \rangle$-$\langle r_2-1, d_2+1 \rangle$, *(7)* $\langle r_1+1, d_1 \rangle$-$\langle r_2+1, d_2 \rangle$, *(8)* $\langle r_1-1, d_1+1 \rangle$-$\langle r_2, d_2 \rangle$, *(9)* $\langle r_1-1, d_1+1 \rangle$-$\langle r_2+1, d_2-1 \rangle$, *(10)* $\langle r_1-1, d_1+1 \rangle$-$\langle r_2-1, d_2+1 \rangle$, *(11)* $\langle r_1-1, d_1 \rangle$-$\langle r_2-1, d_2 \rangle$.

Then, Rule 4.3 can be introduced from Properties A.2 and A.3 in a straightforward way.

References

[1] N. Ansari and E.J. Delp, Partial shape recognition: a landmark-based approach, *IEEE Transactions on Pattern Analysis and Machine Intelligence*, vol. 12, no. 5 (1990), pp. 470–483.

[2] H. Bunke and U. Buhler, Applications of approximate string matching to 2D shape recognition, *Pattern Recognition*, vol. 26, no. 12 (1993), pp. 1797–1812.

[3] A. Califano and R. Mohan, Multidimensional indexing for recognizing visual shapes, *IEEE Transactions on Pattern Analysis and Machine Intelligence*, vol. 16, no. 6 (1994), pp. 373– 392.

[4] A. Del Bimbo and P. Pala, Image indexing using shape-based visual features, in *Proceedings of the 13th International Conference on Pattern Recognition* (Vienna, Austria, August 1996), vol. C, pp. 351–355.

[5] M. Flickner, H. Sawhney, W. Niblack, J. Ashley, Q. Huang, B. Dom, M. Gorkani, J. Hafner, D. Lee, D. Petkovic, D. Steele, P. Yanker, Query by image and video content: the QBIC System, *Computer*, vol. 28, no. 9 (1995), 23–32.

[6] J.W. Gorman, O.R. Mitchell, and F.P. Kuhl, Partial shape recognition using dynamic programming, *IEEE Transactions on Pattern Analysis and Machine Intelligence*, vol. 10, no. 2 (1988), pp. 257–266.

[7] W.E.L. Grimson and D.P. Huttenlocher, On the sensitivity of geometric hashing, in *Proceedings of the Third International Conference on Computer Vision* (Osaka, Japan, December 1990), pp. 334–338.

[8] W.I. Grosky and R. Mehrotra, Index-based object recognition in pictorial data management, *Computer Vision, Graphics, and Image Processing*, vol. 52 (1990), pp. 416–436.

[9] V.N. Gudivana and V.V. Raghavan, Content-based image retrieval systems, *Computer*, vol. 28, no. 9 (1995), pp. 18–21.

[10] C. Im, H. Nishida, and T.L. Kunii, Recognizing plant species by leaf shapes - a case study of the Acer family, in *Proceedings of the 14th International Conference on Pattern Recognition* (Brisbane, Australia, August 17–20, 1998), pp. 1171–1173.

[11] A. Kalvin, E. Schonberg, J.T. Schwartz and M. Sharir, Two-dimensional, model-based, boundary matching using footprints, *International Journal of Robotics Research*, vol. 5, no. 4 (1986), pp. 38–55.

[12] T.F. Knoll and R.C. Jain, Recognizing partially visible objects using feature indexed hypotheses, *IEEE Journal of Robotics and Automation*, vol. 2 (1986), pp. 3–13.

[13] Y. Lamdan and H. Wolfson, Geometric hashing: a general and efficient model-based recognition scheme, in *Proceedings of the Second International Conference on Computer Vision* (Tampa, FL, December 5-8, 1988), pp. 238–249.

[14] R. Mehrotra and J.E. Gary, Similar-shape retrieval in shape data management, *Computer*, vol. 28, no. 9 (1995), pp. 57–62.

[15] F. Mokhtarian, Silhouette-based occluded object recognition through curvature scale space, *Machine Vision and Applications*, vol. 10, no. 3 (1997), pp. 87–97.

[16] F. Mokhtarian, S. Abbasi, and J. Kittler, Efficient and robust retrieval by shape content through curvature scale space, in *Proceedings of the First International Workshop on Image Database and Multimedia Search* (Amsterdam, The Netherlands, August 1996) pp. 35–42.

[17] F. Mokhtarian, S. Abbasi, and J. Kittler, Robust and efficient shape indexing through curvature scale space, in *Proceedings of the Sixth British Machine Vision Conference, BMVC '96* (Edinburgh, UK, September 1996), pp. 53–62.

[18] H. Nishida, Curve description based on directional features and quasi-convexity/concavity, *Pattern Recognition*, vol. 28, no. 7 (1995), pp. 1045–1051.

[19] H. Nishida, A structural model of curve deformation by discontinuous transformations, *Graphical Models and Image Processing*, vol. 58, no. 2 (1996), pp. 164–179.

[20] H. Nishida, Shape recognition by integrating structural descriptions and geometrical/statistical transforms, *Computer Vision and Image Understanding*, vol. 64, no. 2 (1996), pp. 248–262.

[21] H. Nishida, Matching and recognition of deformed closed contours based on structural transformation models, *Pattern Recognition*, vol. 31, no. 10 (1998), pp. 1557–1571.

[22] H. Nishida, Structural shape indexing with feature generation models, *Computer Vision and Image Understanding*, vol. 73, no. 1 (1999), pp. 121–136.

[23] H. Nishida and S. Mori, Algebraic description of curve structure, *IEEE Transactions on Pattern Analysis and Machine Intelligence*, vol. 14, no. 5 (1992), pp. 516–533.

[24] A. Pentland, R.W. Picard, and S. Sclaroff, Photobook: content-based manipulation of image databases, *International Journal of Computer Vision*, vol. 18, no. 3 (1996), pp. 233–254.

[25] A. Pikaz and I. Dinstein, Matching of partially occluded planar curves, *Pattern Recognition*, vol. 28, no. 2 (1995), pp. 199–209.

[26] U. Ramer, An iterative procedure for the polygonal approximation of plane curves, *Computer Graphics and Image Processing*, vol. 1 (1972), pp. 244–256.

[27] E. Saber, A.M. Tekalp, R. Eschbach, K. Knox, Automatic image annotation using adaptive color classification, *Graphical Models and Image Processing*, vol. 58, no. 2 (1996), pp. 115–126.

[28] F. Stein and G. Medioni, Structural indexing: efficient 2-D object recognition, *IEEE Transactions on Pattern Analysis and Machine Intelligence*, vol. 14, no. 12 (1992), pp. 1198–1204.

[29] J.L. Turney, T.N. Mudge, and R.A. Volz, Recognizing partially occluded parts, *IEEE Transactions on Pattern Analysis and Machine Intelligence*, vol. 7, no. 4 (1985), pp. 410–422.

[30] H.J. Wolfson, On curve matching, *IEEE Transactions on Pattern Analysis and Machine Intelligence*, vol. 12, no. 5 (1990), pp. 483–489.

Energy Minimisation Methods for Static and Dynamic Curve Matching

Edgard Nyssen, Bart Truyen, Hichem Sahli
Electronics Department
Vrije Universiteit Brussel, Pleinlaan 2, B-1050 Brussels, Belgium
E-mail: ehnyssen@etro.vub.ac.be

Contents

D. Chen and X. Cheng (eds.),
Pattern Recognition and String Matching, 545–563.

1 Introduction

Curve matching is an important problem in pattern recognition with a variety of applications including model based recognition. In these applications the two matched curves are usually very similar. An application of curve matching to model based recognition involves typically a decision whether a model curve and an image curve are the same, up to some scaling or a transformation and some permitted level of noise. Most of the work on curve matching relates directly to contour matching, where one should further make a distinction between dense matching and feature matching. The latter approach is based on a set of features, calculated for both contours. In that case, the distance of the contours in feature space is used as matching criterion. Dense matching is usually formulated as a parameterisation problem, with some cost function to be minimised. The cost might be defined as the elastic energy needed to transform one curve to the other [1, 2, 3], but other alternatives exist [4, 5, 6].

In this paper, the dense matching approach is considered. We introduce recently developed generic methods for curve matching. These methods can be used as a preprocessing step for problems of feature extraction and classification in image analysis. They are typically useful for handwritten text analysis and the quantitative analysis of image features. Two approaches are considered here. The first approach uses a description of statically recorded analogous curves, and is typically used to match contours, extracted from images, using an edge finding method. The methodology can be used in applications were velocity vectors of moving contours must be estimated accurately from image series [7]. The second approach uses description of dynamically recorded analogous curves and is suited for problems where the dynamic features of the curves are relevant for the analysis. A typical application example of the latter technique is handwritten signature verification.

The paper is organised as follows. Section 2 introduces the theoretical basis of the methods for static and dynamic curve matching, described in this paper. Simultaneously, it reviews earlier work on curve matching by Duncan [8] and Cohen [2, 9]. Section 3 is a detailed description of the Static Curve Matching methodology and discusses its theoretical background (3.1), the numerical implementation of the solution methods (3.2), a specific hierarchical approach which improves the convergence behaviour of the algorithms (3.3), and a technique for the calculation of curvature involving smoothing by Gaussian filtering (3.4). An application example in cardiac Magnetic Resonance (MR) imaging has been added to the discussion (3.5). Section 4 explains how the previously introduced techniques can be adapted to perform Dynamic Curve Matching (4.1). A practical example illustrates how these can be applied in the analysis of handwritten texts and more precisely, for handwritten signature verification (4.2).

2 The Methods of Duncan and Cohen

First, we review the methods of Duncan [8] and Cohen [2, 9].

2.1 Mathematical Formulation

Let $C_{\mathbf{P}}$ and $C_{\mathbf{Q}}$ be two continuous and smooth contours of an image sequence; $C_{\mathbf{Q}}$ is obtained by a (non rigid) transformation of the contour $C_{\mathbf{P}}$. The contours $C_{\mathbf{P}}$ and $C_{\mathbf{Q}}$ are parametrised by s and s', the length along the two contours. For the coordinates of points $\mathbf{P}$ on contour $C_{\mathbf{P}}$, we have parametrised expressions $x_{\mathbf{P}}(s)$ and $y_{\mathbf{P}}(s)$ and similarly for the coordinates of points $\mathbf{Q}$ on contour $C_{\mathbf{Q}}$, $x_{\mathbf{Q}}(s')$ and $y_{\mathbf{Q}}(s')$. The starting points for s and s' are chosen manually, preferably in corresponding low curvature regions of the contours in order to reduce the sensitivity of s and s' for errors in the manually indicated Cartesian co-ordinates of the starting points. The problem is to determine, for each parameter value s on $C_{\mathbf{P}}$, a corresponding value s' on $C_{\mathbf{Q}}$. In order to do this, we must define a deformation measure which locally compares the neighbourhoods of s and s'. We assume that the curvatures $\kappa_{\mathbf{P}}$ and $\kappa_{\mathbf{Q}}$ remain almost invariant in these neighbourhoods. This allows one to introduce a local bending energy measure of the form

$$E_{\text{curve}} = \int_{\delta s} \left(\kappa_{\mathbf{Q}}(s') - \kappa_{\mathbf{P}}(s)\right)^2 \mathrm{d}s \ , \tag{1}$$

where δs is measured over $C_{\mathbf{P}}$.

We also wish the displacement vector field to vary smoothly along the contours. Duncan [8] uses a smoothness term similar to the one of Hildreth [10]. The two term fitting criterion is given by

$$E = E_{\text{curve}} + \lambda E_{\text{smooth}} \ , \tag{2}$$

where

$$E_{\text{smooth}} = \oint_{C_{\mathbf{P}}} \left\| \frac{\partial \left(\mathbf{Q}(s') - \mathbf{P}(s)\right)}{\partial s} \right\|^2 \mathrm{d}s \ . \tag{3}$$

E_{smooth} measures the variation of the displacement vector $\mathbf{PQ}$ along the curve $C_{\mathbf{P}}$, $\|\cdot\|$ denotes the norm associated to the scalar product $\langle \cdot, \cdot \rangle$ in the space $\Re^2$, and λ is a relative weighting factor (a high λ value means heavy smoothing).

Duncan [8] finds a displacement field by direct minimisation of a discrete form of equation (2). The resulting displacement vectors in this approach may, however, map points not belonging to the two contours. This problem was solved by Cohen [2]. His mathematical model can be summarised as follows.

Given two contours $C_\mathbf{P}$ and $C_\mathbf{Q}$, parametrised by $s \in [0,1]$ and $s' \in [0,\alpha]$ (where α is the length of the curve $C_\mathbf{Q}$), we have to determine a function $\mathrm{f} \in G = \{\mathrm{g} \mid \mathrm{g} : [0,1] \to [0,\alpha]\}$; $s \to s'$ satisfying

$$\mathrm{f}(0) = 0 \text{ and } \mathrm{f}(1) = \alpha \tag{4}$$

and

$$\mathrm{f} = \arg\min_{\mathrm{g}\in G} E(\mathrm{g}) \ , \tag{5}$$

where E is defined in equation (2).

Cohen obtains the function f, which satisfies equation (5) and conditions (4), by a variational method. This method finds a local minimum of the functional $E(\mathrm{f})$ as the solution of the Euler-Lagrange equation $\nabla E(\mathrm{f}) = 0$:

$$\mathrm{f}''\|\mathbf{Q}'(\mathrm{f})\|^2 + \kappa_\mathbf{P}\langle \mathbf{N}_\mathbf{P}, \mathbf{Q}'(\mathrm{f})\rangle + \frac{1}{\lambda}[\kappa_\mathbf{P} - \kappa_\mathbf{Q}(\mathrm{f})]\kappa'_\mathbf{Q}(\mathrm{f}) = 0 \ , \tag{6}$$

with boundary conditions (4). In equation (6), $\mathbf{Q}$ is a parameterisation of the contour $C_\mathbf{Q}$, $\mathbf{Q}'(\mathrm{f})$ is the tangent vector of $C_\mathbf{Q}$, $\mathbf{N}_\mathbf{P}$ is the normal vector to the contour $C_\mathbf{P}$, and $\kappa'_\mathbf{Q}$ is the derivative of the curvature of the curve $C_\mathbf{Q}$. The solution of equation (6) with boundary conditions (4), critically depends on the appropriate choice of the initial estimation $\mathrm{f}(s)|_{t=0} = \mathrm{f}_0(s)$ in the associated evolution equation:

$$\begin{cases} \frac{\partial \mathrm{f}(s)}{\partial t} + \mathrm{f}''(s)\|\mathbf{Q}'(\mathrm{f}(s))\|^2 + \kappa_\mathbf{P}\langle \mathbf{N}_\mathbf{P}, \mathbf{Q}'(\mathrm{f})\rangle + \frac{1}{\lambda}[\kappa_\mathbf{P} - \kappa_\mathbf{Q}(\mathrm{f})]\kappa'_\mathbf{Q}(\mathrm{f}) = 0 \ , \\ \mathrm{f}(s)|_{t=0} = \mathrm{f}_0(s) \ . \end{cases} \tag{7}$$

To avoid local minima in $E(\mathrm{f})$, $\mathrm{f}_0(s)$ should already be a good approximation of the real solution. Cohen [2] solves (7) with the finite element method.

2.2 Initial estimation of the solution

The most critical aspect of the solution of (7) consists of finding a satisfactory initial tentative solution $\mathrm{f}_0(s)$.

The method consist of the following steps:

- The interval $[0, 1]$, is subdivided at node points $s_i \in [0, 1]$, $i = 0, \ldots, N+1$. The aim is to search for a set of of points $\mathbf{Q}(s'_i)$ on the curve $C_{\mathbf{Q}}$, corresponding to the points $\mathbf{P}(s_i)$ on the curve $C_{\mathbf{P}}$, where the correspondence is given by the function $\mathrm{f}_0(s_i) = s'_i$.

- A search area S_i on the curve $C_{\mathbf{Q}}$ is associated to each node s_i. This search area is a set of points, consisting of point s''_i, which is the nearest distance point to s_i on $C_{\mathbf{Q}}$, and $(N_{\text{search}} - 1)/2$ points at each side of s''_i. In order to guarantee a plausible tentative solution $\mathrm{f}_0(s)$, s''_i is required to satisfy the constraint given by: $s''_i \approx s_i/\alpha$.

- Among the candidates in S_i, the point s'_i is chosen, which minimises the bending energy, given by (1).

3 Static Curve Matching

3.1 Mathematical Formulation

Equation (7) is complicated and difficult to solve, even with a finite element method. Here, we present a simplified equation, without losing the bending energy and smoothness requirements expressed in (2).

Equation (3) can be rewritten as:

$$E_{\text{smooth}} = \oint_{C_{\mathbf{P}}} \left\| \lim_{\Delta s \to 0} \left(\frac{\Delta \mathbf{Q}(s')}{\Delta s} - \frac{\Delta \mathbf{P}(s)}{\Delta s} \right) \right\|^2 \mathrm{d}s \; .$$

According to Schwartz' inequality [11], we have

$$\left\| \lim_{\Delta s \to 0} \left(\frac{\Delta \mathbf{Q}(s')}{\Delta s} - \frac{\Delta \mathbf{P}(s)}{\Delta s} \right) \right\|^2 \geq \left(\lim_{\Delta s \to 0} \left(\left\| \frac{\Delta \mathbf{Q}(s')}{\Delta s} \right\| - \left\| \frac{\Delta \mathbf{P}(s)}{\Delta s} \right\| \right) \right)^2 \; . \quad (8)$$

In the limit, the arc length can be approximated as:

$$\lim_{\Delta s' \to 0} \Delta s' = \|\Delta \mathbf{Q}(s')\| \; ,$$

$$\lim_{\Delta s \to 0} \Delta s = \|\Delta \mathbf{P}(s)\| \; .$$

In our problem, $s \to s'$ is a continuous one to one mapping, so

$$\lim_{\Delta s' \to 0} \Delta s' = \lim_{\Delta s \to 0} \Delta s'(s) \; .$$

Substituting this relation in equation (8), we get

$$\left\| \lim_{\Delta s \to 0} \left(\frac{\Delta \mathbf{Q}(s')}{\Delta s} - \frac{\Delta \mathbf{P}(s)}{\Delta s} \right) \right\|^2 \geq \left| \lim_{\Delta s \to 0} \left(\frac{\Delta s'}{\Delta s} - \frac{\Delta s}{\Delta s} \right) \right|^2 \equiv \left| \frac{\partial (\mathrm{f}(s) - s)}{\partial s} \right|^2$$

or

$$E_{\mathrm{smooth}} = \oint_{C_{\mathbf{P}}} \left\| \frac{\partial \left(\mathbf{Q}(s') - \mathbf{P}(s) \right)}{\partial s} \right\|^2 \mathrm{d}s \geq \oint_{C_{\mathbf{P}}} \left| \frac{\partial (\mathrm{f}(s) - s)}{\partial s} \right|^2 \mathrm{d}s \,, \tag{9}$$

where $(\mathrm{f}(s) - s)$ is the displacement due to the deformation. Equation (9) establishes that the newly introduced smoothness term

$$E_{\mathrm{smooth}} = \oint_{C_{\mathbf{P}}} \left| \frac{\partial (\mathrm{f}(s) - s)}{\partial s} \right|^2 \mathrm{d}s \tag{10}$$

is uniformly bounded by the original smoothness energy equation (3) as defined in Cohen [2].

Substituting the bending energy expression (1) and the smoothing energy expression (10) in (2), leads to a simplified solution of the Euler-Lagrange equation:

$$\begin{cases} \mathrm{f}'' + \frac{1}{\lambda}[\kappa_{\mathbf{P}}(s) - \kappa_{\mathbf{Q}}(\mathrm{f}(s))]\kappa'_{\mathbf{Q}}(\mathrm{f}) = 0 \,, \\ \text{boundary conditions (4)} \,. \end{cases} \tag{11}$$

Equation (11) describes a 2-point non-linear boundary value problem [12, 13].

3.2 Numerical Solution of the Non-Linear Boundary Value Problem

Equation (11) describes a physical stationary problem. Such problems can be solved either by discretising the differential equation in space and applying a time-independent numerical scheme, or by introducing time dependency and an associated mixed space-time discretisation for which we only retain the time-asymptotic, steady state numerical solution. For non-linear problems, like the one we are facing in equation (11), the second approach is to be preferred [14], because it allows to solve the set of non-linear equations by a linear iterative process which is chosen to guarantee convergence. This leads to the following parabolic differential equation:

$$\frac{\partial \mathrm{f}}{\partial t} = \frac{\partial^2 \mathrm{f}}{\partial s^2} - b \equiv \mathrm{H}(s) \,, \tag{12}$$

with

$$b = -\frac{1}{\lambda} \left[\kappa_{\mathbf{P}}(s) - \kappa_{\mathbf{Q}}(\mathrm{f}(s)) \right] \frac{\partial \kappa_{\mathbf{Q}}(\mathrm{f}(s))}{\partial s} \,.$$

Finite difference approximation. A finite difference approximation of equation (11) is formulated, based on splitting the domain $[0,1]$ of s at node points s_i, $i \in \{0,1,2,\ldots,N+1\}$, into intervals of equal length h. This procedure yields the following numerical tractable expression:

$$\frac{w_{i-1}-2w_i+w_{i+1}}{h^2}=b_i\ ,\ \forall i \in \{1,2,\ldots,N\}\ ;$$

w_i and b_i are the values of $\mathrm{f}(s)$ and $\mathrm{b}(s)$, respectively at the node points s_i. The boundary conditions are

$$w_0=0\ ,\ w_{N+1}=\alpha\ ,$$

and the following $N\times N$ non-linear system is obtained:

$$\begin{aligned}
-2w_1+w_2 &= h^2 b_1(s_1,w_1)\ ,\\
w_1-2w_2+w_3 &= h^2 b_2(s_2,w_2)\ ,\\
&\cdots\\
w_{N-2}-2w_{N-1}+w_N &= h^2 b_{N-1}(s_{N-1},w_{N-1})\ ,\\
w_{N-1}-2w_N &= h^2 b_N(s_N,w_N)-\alpha\ ,
\end{aligned}$$

or, in matrix notation:

$$\mathbf{A}\mathbf{w}=h^2\mathbf{b}(\mathbf{s},\mathbf{w})\ . \tag{13}$$

Equation (12) now can be written as

$$\frac{\partial \mathbf{w}}{\partial t}=\mathbf{H}=h^{-2}\mathbf{A}\mathbf{w}-\mathbf{b}(\mathbf{s},\mathbf{w})\ ,$$

where

$$\mathbf{A}=\begin{bmatrix}
-2 & 1 & 0 & 0 & 0 & \cdots & 0\\
1 & -2 & 1 & 0 & 0 & \cdots & 0\\
\cdots & \cdots & \cdots & \cdots & \cdots & \cdots & \cdots\\
0 & 0 & \cdots & \cdots & 1 & -2 & 1\\
0 & 0 & \cdots & \cdots & 0 & 1 & -2
\end{bmatrix},$$

$$\mathbf{w}=[w_1\, w_2\, \cdots\, w_{N-1}\, w_N]^{\tau}\ ,$$

$$\mathbf{b}(\mathbf{s},\mathbf{w})=$$
$$\left[b_1(s_1,w_1)\, b_2(s_2,w_2)\ \cdots\ b_{N-1}(s_{N-1},w_{N-1})\, b_N(s_N,w_N)+\alpha h^{-2}\right]^{\tau}\ ,$$

and where τ denotes the matrix transpose.

Predictor-corrector scheme. Since both $\mathbf{w}$ and $\mathbf{b}$ are unknown and mutually dependent, two interleaved iterative procedures may be considered.

1. in the *predictor* step, a first guess $\bar{\mathbf{w}}^{(k+1)}$ of $\mathbf{w}^{(k+1)}$ is obtained as:

$$\bar{\mathbf{w}}^{(k+1)} = \mathbf{w}^{(k)} + \Delta\tau\mathbf{H}^{(k)} ,$$

   with $\Delta\tau$ the discretisation step.

2. in the *corrector* step, $\mathbf{H} \equiv h^{-2}\mathbf{A}\mathbf{w} - \mathbf{b}(\mathbf{s}, \mathbf{w})$ and $\mathbf{w}^{(k+1)}$ are calculated according to the following Beam and Warming scheme [14]:

$$(1+\xi)\mathbf{w}^{(k+1)} - (1+2\xi)\mathbf{w}^{(k)} = \Delta\tau[\theta\bar{\mathbf{H}}^{(k+1)} + (1-\theta)\mathbf{H}^{(k)}] - \xi\mathbf{w}^{(k-1)},$$

   with $\bar{\mathbf{H}}^{(k+1)} = \mathbf{H}(\bar{\mathbf{w}}^{(k+1)})$.

Setting $\theta = 1$ and $\xi = 0$ leads to the following simple and stable iteration scheme [14]:

$$\begin{aligned} \bar{\mathbf{w}}^{(k+1)} &= \mathbf{w}^{(k)} + \Delta\tau\mathbf{H}^{(k)} , \\ \mathbf{w}^{(k+1)} &= \mathbf{w}^{(k)} + \Delta\tau\bar{\mathbf{H}}^{(k+1)} . \end{aligned} \tag{14}$$

The sequence (14) is first order accurate in time and second order in space. The stop criterion is chosen as:

$$d^{(k)} = \frac{1}{N}\sum_{i=1}^{N} |H_i^{(k)}| < \eta .$$

Ensuring convergence of the iteration scheme. In order to ensure convergence, a heuristically derived alternative scheme is adopted in the curve matching method. From equation (13), one can obtain $\bar{\mathbf{w}}^{(k+1)}$ directly from

$$\bar{\mathbf{w}}^{(k+1)} = h^2\mathbf{A}^{-1}\mathbf{b}(\mathbf{w}^{(k)}) . \tag{15}$$

Since $\mathbf{A}$ is a symmetric positive definite matrix, its inverse always exists. We define the following corrector step, which is found to be appropriate for ensuring convergent behaviour:

$$\mathbf{w}^{(k+1)} = (1-\beta)\mathbf{w}^{(k)} + \beta\bar{\mathbf{w}}^{(k+1)} , \tag{16}$$

where $0 \leq \beta \leq 1$. It can be shown that equation (15) is locally stable if β is carefully chosen. For a more detailed discussion of convergence, we refer to [15].

3.3 Hierarchical Approach

The definition of the initial estimation f_0 is known to have a significant effect on the convergence of the curve matching algorithm. Anticipating, a hierarchical approach has been designed to reduce this dependency. In the iterative approximation method defined by (15), and (16), the solution is obtained by updating a tentative solution sequentially through a local relaxation process. Here, we consider the application of this procedure, after smoothing $C_{\mathbf{P}}$ and $C_{\mathbf{Q}}$. This is achieved by the convolution of $x_{\mathbf{P}}(s)$, $y_{\mathbf{P}}(s)$, $x_{\mathbf{Q}}(s')$, and $y_{\mathbf{Q}}(s')$ with a Gaussian kernel. The degree of smoothness is determined by the Gaussian parameter σ. By smoothing $C_{\mathbf{P}}$ and $C_{\mathbf{Q}}$, the difference of the two curvatures $\kappa_{\mathbf{P}}$ and $\kappa_{\mathbf{Q}}$, as well as the first and second derivatives of $\kappa_{\mathbf{Q}}$, are reduced. A fast local matching is achieved by applying the result obtained at a certain degree of smoothing as a pre-estimate for the iteration at a lower degree of smoothing. Therefore, instead of having to find a good initial estimate f_0 for the fitting of the original curves with full detail, it is now sufficient to define f_0 at the highest degree of smoothing, which is a much easier task. An example of the procedure is given in Fig. 1. As an alternative to the multi-smoothing strategy, one might envisage gradually reducing λ. A first objection against this approach is that this scheme needs much more steps to get a good matching than the multi-smoothing scheme. A second problem, encountered when gradually decreasing the λ value, is concerned with reaching the global minimum of the fitting criterion (2). Minimising (2) can be regarded as a Lagrange multiplier problem for which it is only guaranteed to reach the global minimum if λ is progressively increased, starting from a sufficiently small value. In the multi-smoothing scheme based on changing σ, the noise in the curvature data is suppressed because the curvatures are calculated on smoothed data; this extra noise suppression is not achieved by a scheme which gradually reduces λ. Moreover, for multimodal optimisation problems, such as those considered in this paper, the multi-smoothing approach is appropriate to remove local suboptimal minima in the fitting criterion.

3.4 Some Considerations Regarding the Computation of Curvature

Parametric Representation. We define $\mathrm{x}(s)$ and $\mathrm{y}(s)$, where $0 < s < L$, and L being the length of the contour. In practice, the calculations are based on digitised images, yielding non-uniformly sampled contours. This

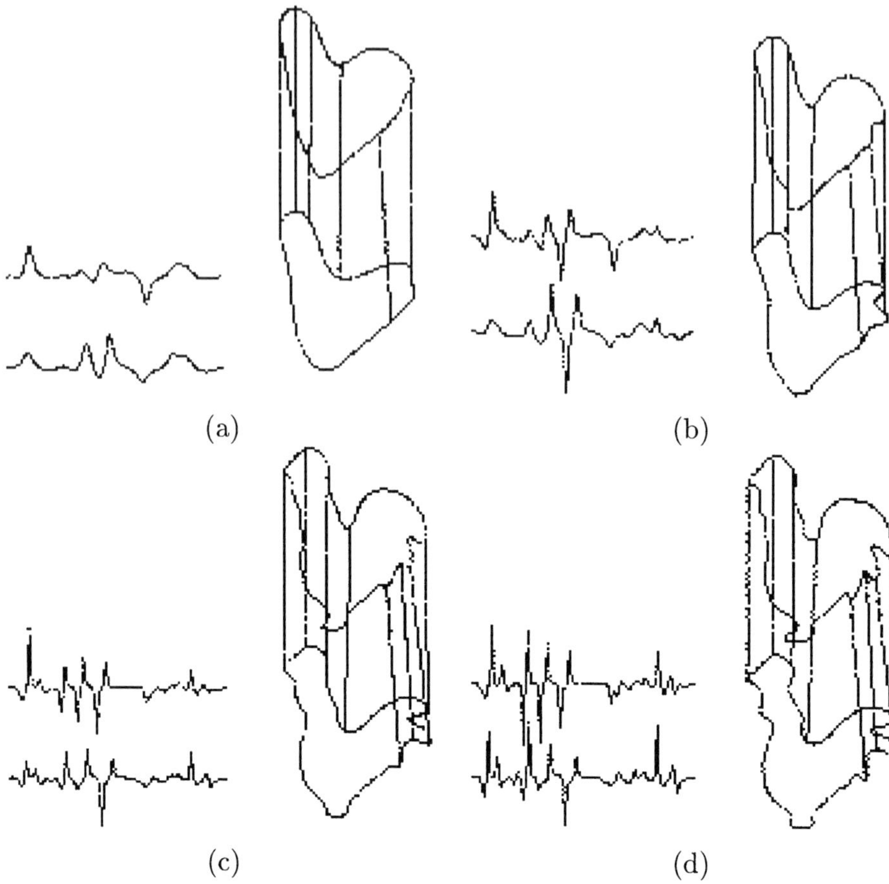

Figure 1: Example of a four step matching of contours. (a) First step: matching at coarsest scale $\sigma_c = 8$. (b) Second step: matching at scale $\sigma_2 = 6.5$ starting the iteration from f_0 at scale $\sigma_c = 8$. (c) Third step: matching at scale $\sigma_3 = 5.375$, starting from f_0 at scale $\sigma_2 = 6.5$. (d) Fourth step: matching at finest scale $\sigma_f = 2$ starting from f_0 at scale $\sigma_3 = 5.375$.

problem necessitates resampling of the contours using interpolation. In our calculations, three point Lagrange interpolation is applied. Define

$$\begin{aligned} \dot{\mathrm{x}}(s) &= \mathrm{d}x/\mathrm{d}s\ , \quad & \dot{\mathrm{y}}(s) &= \mathrm{d}y/\mathrm{d}s\ , \\ \ddot{\mathrm{x}}(s) &= \mathrm{d}^2x/\mathrm{d}s^2\ , \quad & \ddot{\mathrm{y}}(s) &= \mathrm{d}^2y/\mathrm{d}s^2\ . \end{aligned}$$

A curvature is calculated, using the following formula:

$$\kappa(s) = \frac{\dot{\mathrm{x}}(s)\ddot{\mathrm{y}}(s) - \ddot{\mathrm{x}}(s)\dot{\mathrm{y}}(s)}{(\dot{\mathrm{x}}(s)^2 + \dot{\mathrm{y}}(s)^2)^{3/2}} \ . \tag{17}$$

Gaussian Filter. Discrete contours are inherently noisy. Abrupt changes in the contours are amplified by differentiation and give rise to severe distortion in the curvature values. This implies that a certain degree of smoothness must be imposed to the discrete curve before differentiation. According to Marr [16], we calculate $\dot{\mathrm{x}}(s)$ and $\ddot{\mathrm{x}}(s)$ by convoluting $\mathrm{x}(s)$ with the first and the second derivative of the Gaussian. A similar procedure is applied on $\mathrm{y}(s)$. Note that this procedure can be applied to optimise the matching performance, by adopting a multi-smoothing scheme. Gaussian filtering is preferred because of its attractive scaling properties for planar curves [17]. The main advantage of Gaussian filtering is that no new inflection points are created in the contours, when the degree of smoothing is increased [18].

With $*$ denoting the convolution operator, we define

$$\begin{aligned} \dot{\mathrm{X}}(s,\sigma) &= \mathrm{x}(s) * \dot{\mathrm{g}}(s,\sigma) \ , \quad \dot{\mathrm{Y}}(s,\sigma) = \mathrm{y}(s) * \dot{\mathrm{g}}(s,\sigma) \ , \\ \ddot{\mathrm{X}}(s,\sigma) &= \mathrm{x}(s) * \ddot{\mathrm{g}}(s,\sigma) \ , \quad \ddot{\mathrm{Y}}(s,\sigma) = \mathrm{y}(s) * \ddot{\mathrm{g}}(s,\sigma) \ , \end{aligned}$$

where

$$\begin{aligned} \mathrm{g}(s,\sigma) &= \frac{1}{\sigma\sqrt{2\pi}} \exp\left(-\frac{s^2}{2\sigma^2}\right) , \\ \dot{\mathrm{g}}(s,\sigma) &= \frac{\partial}{\partial s}\mathrm{g}(s,\sigma) = \frac{-s}{\sigma^3\sqrt{2\pi}} \exp\left(-\frac{s^2}{2\sigma^2}\right) , \\ \ddot{\mathrm{g}}(s,\sigma) &= \frac{\partial^2}{\partial s^2}\mathrm{g}(s,\sigma) = \left(\frac{s^2}{\sigma^5\sqrt{2\pi}} - \frac{1}{\sigma^3\sqrt{2\pi}}\right) \exp\left(-\frac{s^2}{2\sigma^2}\right) , \end{aligned}$$

are the Gaussian function, its first and its second derivative respectively. Rewriting equation (17), we get a formula for the discrete curvature, parametrised by the Gaussian scale constant σ:

$$\kappa(s,\sigma) = \frac{\dot{\mathrm{X}}(s,\sigma)\ddot{\mathrm{Y}}(s,\sigma) - \ddot{\mathrm{X}}(s,\sigma)\dot{\mathrm{Y}}(s,\sigma)}{(\dot{\mathrm{X}}^2(s,\sigma) + \dot{\mathrm{Y}}^2(s,\sigma))^{3/2}} \ .$$

3.5 Application

One interesting application of static curve matching is the analysis of Magnetic Resonance (MR) images of the heart [15]. These images represent the heart muscle at different moments of the cardiac cycle, in a specific slice. When looking at such image series, the cardiologist is mainly interested in the local displacement of the heart muscle at different time instances, especially in the ventricular region of the heart. Abnormal dynamic behaviour of the heart muscle may indeed reveal a pathologic condition. Fig. 2 shows a time series of cardiac MR images superimposed with the contours of the inner ventricular wall. The contours are obtained using edge detection.

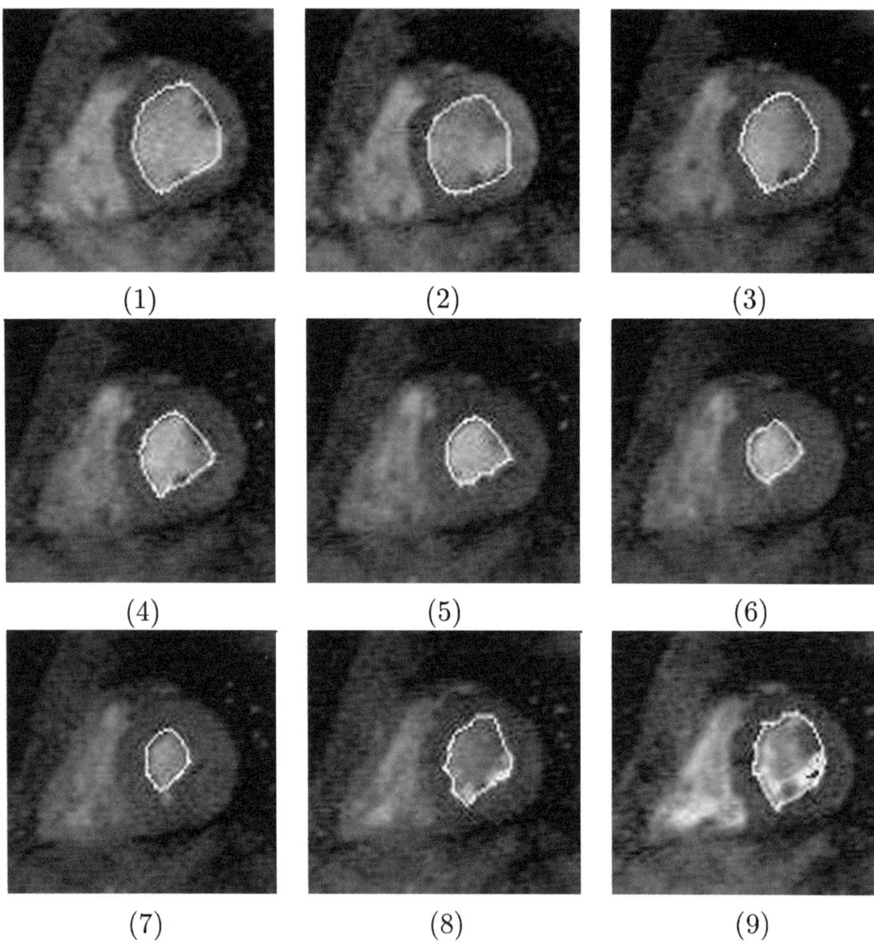

Figure 2: A set of spin echo MR heart images with superimposed LV endocardial contours (short axis orientation).

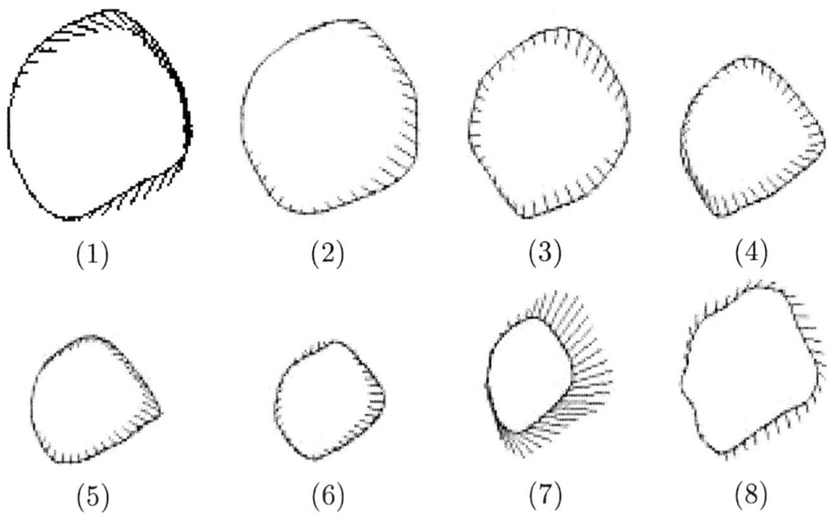

Figure 3: Displacement field of the successive contours.

Curve matching has been applied to successive pairs of contours in order to establish a displacement field, revealing the details of the muscular movements. This displacement field is shown in Fig. 3. Note that we apply the methodology to a time series of images. Each curve is however extracted from an image corresponding to one time instance and for that reason, we are dealing here with *static* curve matching. In dynamic curve matching, the curves themselves are generated by a temporal process, and the matching criteria take into account their dynamic features.

4 Dynamic Curve Matching

4.1 Theoretical and Practical Approach

In some curve matching applications, the dynamic characteristics of the curves play an important role. E.g., in signature verification, two signatures which appear to be similar on the basis of static comparison criteria, may reveal forgery when the dynamic characteristics of the curves are compared.

One approach to dynamic curve matching is based on a parametric description of the curves $(x(t), y(t))$, where the time t plays a similar role as the length s along the curve in section 2 [19]. Relevant local dynamic characteristics of the curves are the curvilinear velocity $\mathrm{v_c}(t)$, defined as

$$\mathrm{v_c}(t) = |\mathbf{v}(t)| = \sqrt{\mathrm{v}_x^2(t) + \mathrm{v}_y^2(t)} \ ,$$

and the angular velocity $\mathrm{v_a}(t)$, defined as

$$\mathrm{v_a}(t) = \dot{\Theta}(t) = \tan^{-1}\frac{\mathrm{v}_y(t+1)}{\mathrm{v}_x(t+1)} - \tan^{-1}\frac{\mathrm{v}_y(t)}{\mathrm{v}_x(t)} \ .$$

This is a basis for a reformulation of the fitting criterion (2). The following expressions for the local bending energy term E_{curve}, and the smoothness term E_{smooth} are used, in analogy to the ones given by equations (1) and (10):

$$E_{\text{curve}} = \int_{\delta t} (\mathrm{v}_Q(t') - \mathrm{v}_P(t))^2 \, \mathrm{d}t \ ,$$

where v is either $\mathrm{v_c}$ or $\mathrm{v_a}$, and

$$E_{\text{smooth}} = \oint_{C_P} \left| \frac{\partial(\mathrm{f}(t) - t)}{\partial t} \right|^2 \mathrm{d}t \ .$$

The approach to this problem boils down to the solution of the Euler-Lagrange equation (11), where κ is substituted by v and s by t.

4.2 Applications

In [19] is described how the technique can be applied as a basis for the comparison of handwritten signatures. During the experiments, these signatures were sampled on a digital pen tablet device, at a fixed sample rate (100 points/second), in order to obtain the position of the pen tip as a function of time. Since signatures are usually complex curves, the hierarchical approach for curve matching is not used. Instead, the solution methodology uses a plausible initial matching guess, based on the segmentation of the signatures at well defined landmarks. These landmarks are chosen among the "corner" points on the signatures, where usually the curvilinear velocity $\mathrm{v_c}(t)$ is minimum, or the angular velocity $\mathrm{v_a}(t)$ is maximum. For more details about the proposed procedure for extracting, selecting and matching the corner points of two signature, which is rather elaborate, we refer to the paper [19]. Figure 4 shows the result of the abovementioned corner extraction and selection procedure.

An initial guess of the matching of both curves is obtained from the matching of the landmarks and linear interpolation of point positions for

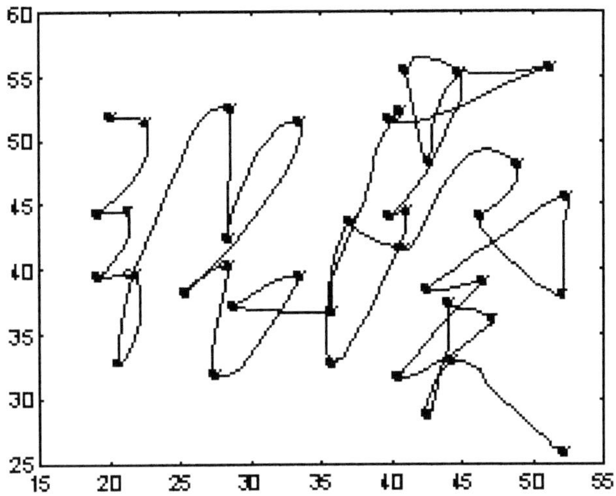

Figure 4: Sample of the result of corner extraction and selection, applied to a signature in Chinese handwriting.

matching point pairs between the landmarks. Figure 5 illustrates the results of matching two signature curves, using the dynamic curve matching technique described in 4.1, using the curvilinear velocity information.

In [20], a three-stage online signature verification system is described, based on the abovementioned curve matching techniques. In this system, a signature is regarded as a pattern which has to be classified as either genuine or forgery, based on a number of dissimilarity measures. The classification criterion uses a training set of genuine reference signatures. The three verification stages are applied in sequence to the signature under test, which is only accepted as genuine if it passes the three stages.

- the first stage is based on global features of the signatures — the duration of signing and the curve length — and the dissimilarity measure is the Mahalanobis distance between the tested feature vector and the mean feature vector of the training set;

- the second stage uses a combination of criteria based on features which are derived from the corners, extracted from the tested signature and the reference signatures;

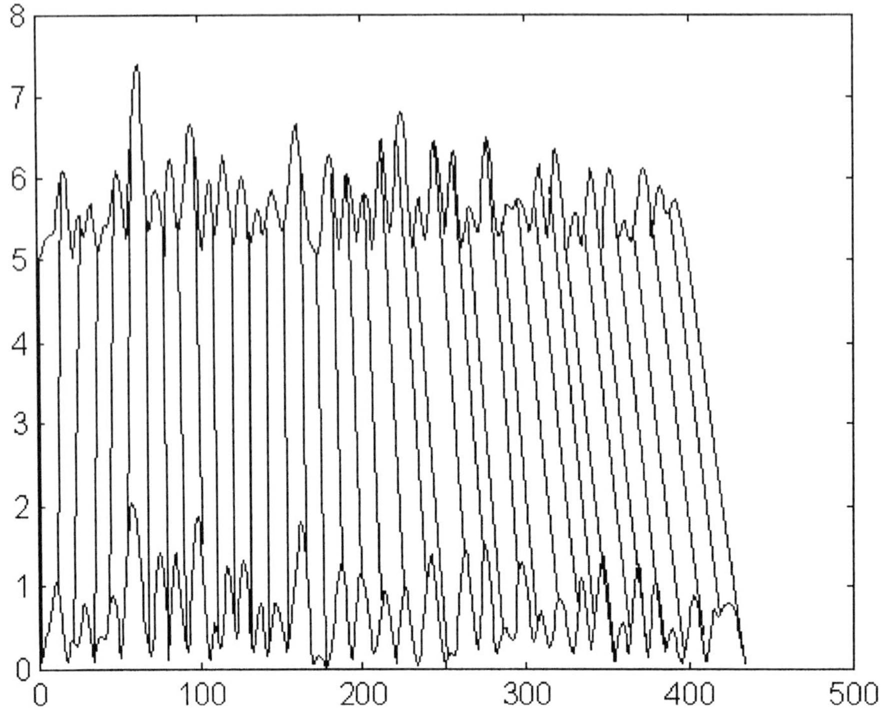

Figure 5: Sample of the point-to-point matching of curvilinear velocity curves, applied to corresponding signatures in Chinese handwriting.

- the third stage implements a function-based method, which uses the dynamic curve matching techniques, described here, to calculate the matching energy and a measure for the extent of the parts of the tested signature where the writing speed in terms of curvilinear velocity is out of range.

In the experiments, 306 genuine signatures and 302 forgeries were collected for 19 persons. For each of them, at least 15 genuine signatures and 15 forgeries were recorded. Eight genuine signatures per subject were used for training the system. For the test set, consisting of the remaining 154 genuine signatures and the 302 forgeries, the whole three-stage verification system showed a false acceptance ratio of 0 and a false rejection ratio of only

5.8%, thus fitting the requirements of a high security level.

5 Acknowledgements

Except for the work of Duncan and Cohen, the curve matching methods presented here were developed at the Electronics Department of the VUB. Several scientific collaborators of the ones who participated in these developments are no longer members of our research group. We would like to take the opportunity to thank these people, and particularly Yang Xin and Ioannis Pratikakis, for their contribution to the development of the static curve matching methodology and its application to the analysis of heart and brain MR images [15]. Likewise we would like to thank Kui Zhang, who adapted the static technique to dynamic curve matching problems and used the technique for online signature verification [19, 20]. We finally would like to thank Rudi Deklerck for providing us some supplementary documentation.

References

[1] A. Brint and M. Brady. Stereo matching of curves. *Image and vision computing*, 8:50–56, 1990.

[2] I. Cohen, N. Ayache, and P. Sulger. Tracking points on deformable curves. In *Proc. Second European Conference on Computer Vision 1992*, May 1992.

[3] R. Basri, L. Costa, D. Geiger, and D. Jacobs. Determining the similarity of deformable shapes. In *Proc. IEEE Workshop on Physics Based Modeling in Computer Vision*, pages 135–143, 1995.

[4] E. Arkin, L. Paul Chew, H. Huttenlocher, K. Kedem, and J. Mitchel. An efficiently computable metric for comparing polygonal shapes. *PAMI*, 13:209–216, 1991.

[5] A. Del Bimbo and P. Pala. Visual image retrieval by elastic matching of user sketches. *PAMI*, 19:121–132, 1997.

[6] D. Sankoff, J.B. Kruskal, and Day Whe. Time warps, string edits, and macromolecules - the theory and practice of sequence comparison. *Journal of Classification*, 1(2-3):281–285, 1984.

[7] J. S. Park and J. H. Han. Contour matching: a curvature-based approach. *Image and Vision Computing*, 16:181–189, 1998.

[8] J. S. Duncan, R. Owen, P. Anandan, L. Staib, T. McCauley, A. Salazar, and F. Lee. Shape-based tracking of left ventricular wall motion. In *Proc. Computers in Cardiology 1990*, pages 23–26. IEEE Computer Society, Sep. 1990.

[9] I. Cohen, N. Ayache, and P. Sulger. Tracking points on deformable objects using curvature information. In Ajit Singh, Dmitry Goldgof, and Demetri Terzopoulos, editors, *Deformable Models in Medical Image Analysis*, pages 306–314. IEEE Computer Society, 1998.

[10] E. C. Hildreth. *The Measurement of Visual Motion.* The MIT Press, Cambridge, Massachussetts, 1984.

[11] A. Jennings and J. J. Mckeown. *Matrix Computation.* Wiley, second edition, 1992.

[12] R.L. Burden and J.D. Faires. *Numerical Analysis.* PWS-KENT Publishing Company, Boston, 1989.

[13] H.B. Keller. *Numerical Methods for Two-Point Boundary-Value Problems.* Blaisdell, Waltham, Massachussetts, 1968.

[14] C. Hirsch. *Numerical Computation of Internal and External Flows*, volume 1. John Wiley & Sons, 1989.

[15] Yang Xin, Bart Truyen, Ioannis Pratikakis, and Jan Cornelis. Hierarchical contour matching in medical images. *Image and Vision Computing Journal*, 14(6):417–433, Jun. 1996.

[16] D. Marr and T. A. Poggio. A computational theory of human stereo vision. In *Proc. Royal Soc. London*, volume B204, pages 301–328. John Wiley & Sons, 1979.

[17] A. L. Yuille and T. A. Poggio. Scaling theorems for zero crossings. *IEEE Trans. Pattern Anal. Mach. Intell.*, PAMI-8(1):15–25, Jan. 1986.

[18] A. Witkin. Scale-space filtering. In *Proc. Proc. 8th Int. Joint Conf. Artif. Intell.*, pages 1019–1022, Karlsruhe, Germany, 1983.

[19] Kui Zhang, Ioannis Pratikakis, Jan Cornelis, and Edgard Nyssen. Using landmarks to establish a point-to-point correspondence between signatures. *Pattern Analysis and Applications*, 3:69–75, Mar. 2000.

[20] Kui Zhang, Edgard Nyssen, and Hichem Sahli. A multi-stage online signature verification system. *Pattern Analysis and Applications*, 2002. in publication: PAA ref. 2104.

Recent Feature Selection Methods in Statistical Pattern Recognition

Pavel Pudil, Jana Novovičová, Petr Somol[1]
Department of Pattern Recognition
Institute of Information Theory and Automation, Academy of Sciences of the Czech Republic, 182 08 Prague 8, Czech Republic
E-mail: `pudil@utia.cas.cz`, `novovic@utia.cas.cz`, `somol@utia.cas.cz`

Contents

[1]The authors are also with the Faculty of Management, University of Economics. J. Novovičová is also with the Faculty of Transportation Sciences, Czech Technical University.

D. Chen and X. Cheng (eds.),
Pattern Recognition and String Matching, 565–615.

1 Introduction

Pattern recognition can be with certain simplification characterized as a classification problem combined with dimensionality reduction of pattern feature vectors which serve as the input to the pattern classifier. This reduction is achieved by extracting or selecting a subset of features which optimizes an adopted criterion.

The chapter is devoted to feature selection problem in statistical pattern recognition, as this is the area to which our own contribution has been acknowledged by the pattern recognition community.

2 Dimensionality Reduction

We shall use the term "pattern" to denote the D-dimensional data vector $\mathbf{x} = (x_1, \ldots, x_D)^T$ of measurements, the components of which are the measurements of the features of the entity or object. Following the statistical approach to pattern recognition, we assume that a pattern $\mathbf{x}$ is to be classified into one of a finite set of C different classes $\Omega = \{\omega_1, \omega_2, \cdots, \omega_C\}$. A pattern $\mathbf{x}$ belonging to class ω_i is viewed as an observation of a random vector $\mathbf{X}$ drawn randomly according to the known class-conditional probability density function $p(\mathbf{x}|\omega_i)$ and the respective *a priori* probability $P(\omega_i)$.

One of the fundamental problems in statistical pattern recognition is representing patterns in the reduced number of dimensions. In the majority of practical cases, the dimensionality of the pattern descriptor space can be rather high. It is the consequence of the fact that in the design phase it is extremely difficult or practically impossible to evaluate directly the "usefulness" of particular descriptors or input variables. Thus it is desirable and important to include into the original set all the "reasonable" descriptors the designer can think of and to reduce the set later on. The reason is that no subsequent mathematical processing can add information missing in the originally designed measurement set.

The aim of dimensionality reduction can be stated as the task to determine a set of new d features of the input feature set of D features (if possible $d \ll D$), so as to maximize (or minimize) an adopted criterion. The set of d features can be either formed by a subset of the original set of D features or alternatively derived from the original features by a certain transformation.

Though there are numerous techniques to achieve dimensionality reduction, they can be divided basically either according to their aim or according to the adopted strategy - in both cases into two categories:

- According to the aim:
  1. *dimensionality reduction for optimal data representation*
  2. *dimensionality reduction for classification.*

While the first aims to preserve the topological structure of data in a lower-dimensional subspace as much as possible, the second one aims to enhance in a feature subspace the discriminatory power of the selected subset.

- According to the adopted strategy:
  1. *feature selection* (subset selection)
  2. *feature extraction* (feature transformation).

The first strategy is to select the hopefully best subset of the input feature set. The second strategy is to find a transformation from the D features to a lower dimensional space. New features are linear or nonlinear combinations of the original features.

The choice between feature selection (FS) and feature extraction (FE) depends on the application domain and the specific training data which is available.

FS leads to savings in measurements cost since some of the features are discarded and the selected features retain their original physical interpretation. FS methods will be more suitable in cases when the potential user puts emphasis on preserving the interpretability of original data and prefers decision-making on the basis of meaningful features. For example, most problems of computer-assisted medical decision-making belong to this class of tasks.

On the other hand, transformed features generated by feature extraction may provide a better discriminative ability than the best subset of given features, but these new features may not have a clear physical meaning. In the sequel we shall concentrate on FS problem only.

3 Feature Selection

Given a set of D features, X_D, let us denote $\mathcal{X}_d$ the set of all possible subsets of size d, where d represents the desired number of features. Let J be some criterion function. Without any loss of generality, let us consider a higher value of J to indicate a better feature subset. Then the feature selection problem can be formulated as follows: find the subset $\tilde{X}_d$ for which

$$J(\tilde{X}_d) = \max_{X \in \mathcal{X}_d} J(X).$$

One important property of FS criterion is the *monotonicity property*: Given two subsets of the feature set X_D, A and B such that

$$A \subset B \Rightarrow J(A) < J(B). \tag{1}$$

That is, evaluating the feature selection criterion on a subset of features of a given set yields a smaller value of the feature selection criterion.

Assuming that a suitable criterion function has been chosen to evaluate the effectiveness of feature subsets, feature selection is reduced to a search problem that detects an optimal feature subset based on the selected measure.

The basic approach is to build up a subset of required number of features incrementally starting with the empty set - *bottom-up* approach or to start with the complete set of features and remove redundant features until d features retain - a *top-down* approach.

If X_k represents a set of k features then in the bottom-up approach, the best set at a given iteration, $\tilde{X}_k$, is the set for which the feature selection criterion has its maximum value

$$J(\tilde{X}_k) = \max_{X \in \mathcal{X}_k} J(X). \tag{2}$$

For the set $\tilde{X}_k$, $k = 1, \ldots, D-1$ satisfying (2), the term *superset of size k* will be used. The set $\mathcal{X}_k$ of all sets of features at a given step is determined from the set at the previous iteration. It means that the set of features at one stage of an iterative procedure is used as a starting point to find the set at the next stage.

If $\bar{X}_k$ represents a set of $D-k$ features after removing k features from the set X_D then in the top-down approach, the best set at a given iteration, $\tilde{\bar{X}}_k$, is the set for which the feature selection criterion has its maximum value over the set of all subsets of $D-k$ features. For the set $\tilde{\bar{X}}_k$, $k = 1, \ldots, D-1$ the term *superset of size $D-k$* will be used.

3.1 Search Methods for Feature Selection

The methods for FS in statistical pattern recognition can be split into two basic strategies for feature subset selection:

1. *Optimal methods*: These include e.g. exhaustive search methods which are feasible for only small size problems and accelerated search methods (as Branch and Bound algorithm).
2. *Suboptimal methods*: These methods essentially tradeoff the optimality of the selected subset for computational efficiency. These include e.g. Best Individual Features, Sequential Forward and Backward Selection, Plus l-Take Away r, their generalized versions, and genetic algorithms.

Although the exhaustive search is a sufficient procedure to guarantee the optimality of a solution, in many realistic problems it is computationally prohibitive. In practice, therefore, one has to rely on computationally feasible procedures which avoid the exhaustive search but may yield suboptimal results. A comprehensive list of suboptimal procedures can be found e.g. in the monography of Devijver and Kittler [2]. A very useful taxonomy of FS methods current available in pattern recognition has been recently published in the papers by Jain and Zongker [7], Kudo and Sklansky [8].

The well known *Branch and Bound* (BB) algorithm guarantees to select an optimal feature subset of size d without involving explicit evaluation of all the possible combinations of d measurements. However, the algorithm is applicable only under the assumption that the feature selection criterion used satisfies the monotonicity property. This assumption precludes the use of the error rate as the criterion. Moreover, the search algorithm becomes computationally prohibitive for problems of high dimensionality. In order to overcome this problem, Siedlecki and Sklansky [20] explored the idea of approximate monotonicity together with other optimization methods like genetic algorithms and Monte Carlo approaches to develop search algorithms which works well even in case when the FS criterion is not monotonic.

Our own research and experience with FS has led us to the conclusion that *there exists no unique optimal approach* to the problem. Some approaches are more suitable under certain conditions, and different approaches are more appropriate under other conditions. These conditions are related to the level of our *knowledge of the problem*, thus we can talk about "knowledge-based FS". Hence we have attempted to extend the "battery" of available tools by developing several new methods, each of them suitable for different conditions, trying to cover the majority of situations which can be encountered in practice. These methods are introduced in the sequel.

3.2 Levels of Problem Knowledge

There are perhaps two basic classes of situations with respect to *a priori* knowledge of the underlying probability structures:

1. *Some a priori knowledge is available*
   It is at least known that probability density functions (pdfs) are unimodal. In these cases, a probabilistic distance measure (like Mahalanobis, Bhattachaarya, etc.) may be appropriate as the evaluation criterion. As pointed out by Siedlecki and Sklansky [20] the error rate is even better (provided it can be reasonably computed). The criterion used here to select the subset is of course tacitly based also on certain assumptions about the underlying probability structures.

   New Tool:
   For this type of situations, in the category of optimal methods we have developed a family of Fast Branch and Bound algorithms, utilizing the prediction mechanisms and described in Section 4. In the category of suboptimal methods we have developed a family of Floating Search algorithms which yield close to optimal solution, are computationally effective

(facilitating FS in high dimensional problems) and do not require the fulfillment of monotonicity property. In a recent comparative study of currently available subset search strategies carried out by Jain and Zongker [7] were the Floating Search algorithms evaluated as the most powerful ones. Since then we have extended the family by more sophisticated methods, namely Adaptive Floating Search and Oscillating Search which are able to find even better solution though at the expense of higher computational time. They will be mentioned in Section 5.

2. *No a priori knowledge is available*
   We cannot even assume that pdfs are unimodal or suspect they are multimodal. The only source of available information is provided by the training data. Feature selection in such a case becomes a very challenging problem. The early solutions of this problem suffer from serious shortcomings.

   New Tool:
   For these situations we have developed two new approaches aimed to cope reasonably well in such circumstances, conceptually very different from those mentioned above. They are based on approximating unknown conditional pdfs by finite mixtures of a special type and are discussed in Section (6). Our first paper describing one of the approaches has been awarded the British Machine Vision Conference (BMVC'93) Science Prize for the best theoretical result [17].

4 Recent Optimal Search Methods

The problem of optimal feature selection (or more generally of subset selection) is difficult especially because of its time complexity. All known optimal search algorithms have an exponential nature. The only alternative to the exhaustive search is the *Branch and Bound* algorithm [4] and ancestor algorithms based on a similar principle. All BB algorithms rely on the monotonicity property of the FS criterion. By a straightforward application of this property many feature subset evaluations may be omitted.

Before discussing more advanced algorithms, let us briefly summarize the essential BB principle. The algorithm constructs a search tree where the root represents the set of all D features, X_D and leaves represent target subsets of d features. While tracking the tree down to leaves the algorithm successively removes single features from the current set of "candidates" ($\bar{\mathcal{X}}_k$ in k-th level). The algorithm keeps the information about both the till-now

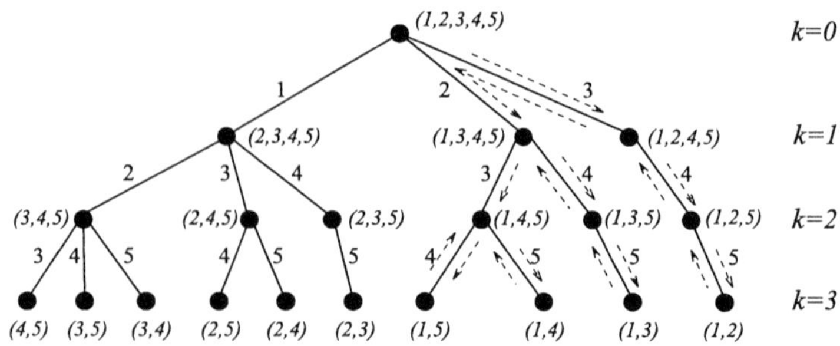

Figure 1: *Example of a "branch and bound" problem solution, where $d = 2$ features are to be selected from a set of $D = 5$ features. Dashed arrows illustrate the way of tracking the search tree.*

best subset of size d and the corresponding criterion value, denoted as the *bound*. Anytime the criterion value in some internal node is found to be lower than the current *bound*, due to condition (1) the whole sub-tree may be cut-off and many computations may be omitted. The course of the BB algorithm is illustrated in Fig. 1. This scheme in its simplest form is known as the "Basic" BB algorithm. For details see [2, 4].

4.1 Branch and Bound Properties

When compared to the exhaustive search, every BB algorithm requires additional computations. Not only the target subsets of d features $\bar{X}_{D-d}$, but also their supersets $\bar{X}_{D-d-j}$, $j = 1, 2, \cdots, D-d$ have to be evaluated. The BB principle does not guarantee that enough sub-trees will be cut-off to keep the total number of criterion computations lower than in exhaustive search.

To reduce the amount of criterion computations an additional heuristic has been introduced in a more powerful "Improved" BB algorithm [2, 4]. Let the *criterion value decrease* be the difference between the current criterion value and the value after removal of a particular feature. Let *bad* features be those features, removal of which from the current candidate set causes only a slight *criterion value decrease*. Let *good* features be those ones, whose removal causes a significant *criterion value decrease*. (Exact definition of 'slight' and 'significant' is not needed for the explanation). The "Improved" BB attempts to order search tree branches so that the tree would become asymmetrical with its dense (left) part containing mostly *good* features and its sparse (right) part containing mostly *bad* ones. Tracking first the right tree part would result in faster *bound* increase and usually in more effective sub-tree cut-offs (of denser tree parts) in later algorithm stages.

However, even the "Improved" BB cannot avoid weak search performance in certain situations. It may result from simple facts that nearer to the root: a) criterion value computation is usually slower (evaluated feature subsets are larger), and b) sub-tree cut-offs are less frequent (higher criterion values may be expected for larger subsets, what reduces the chance of the criterion value to remain under the *bound*, which is updated in leaves). The BB algorithm usually spends most of time by tedious, less promising evaluation of tree nodes near the root. This effect is most apparent for $d \ll D$. Moreover, the "Improved" BB needs a considerable number of additional computations to find its required node ordering.

Currently the most effective way of further improving the Branch and Bound performance is based on prediction mechanisms that replace a significant number of slow criterion computations by means of fast value prediction. The prediction based algorithms are described in following sections.

4.2 Prediction Mechanism Based Branch and Bound

The *Fast Branch and Bound* (FBB) [22] algorithm aims to reduce the number of criterion function computations in internal search tree nodes. A simplified algorithm description is as follows: it attempts to utilize the knowledge of past feature-dependent *criterion value decreases* for future prediction of criterion values without the need of real computation. Prediction is allowed under certain circumstances only, e.g. not in leaves. Both the really computed and predicted criterion values are treated in the same way, i.e. for ordering of node descendants during the tree construction phase.

If the predicted criterion value remains significantly higher than the current *bound*, we may expect that even the actual value would not be lower and the corresponding sub-tree could not be cut-off. In this situation the algorithm continues to construct the consecutive tree level. However, if the predicted value comes close to the *bound* (and therefore there arises a chance that the real value is lower than the *bound*), the real criterion value must be computed. Only if real criterion values are lower than the current *bound*, sub-trees may be cut-off. Note that this prediction scheme does not affect the optimality of obtained results. The FBB algorithm course remains similar to that of the "Improved" BB, possible sub-tree cut-offs are allowed according to real criterion values only. Possible inaccurate predictions may result in nothing worse than constructing sub-trees, which would have been

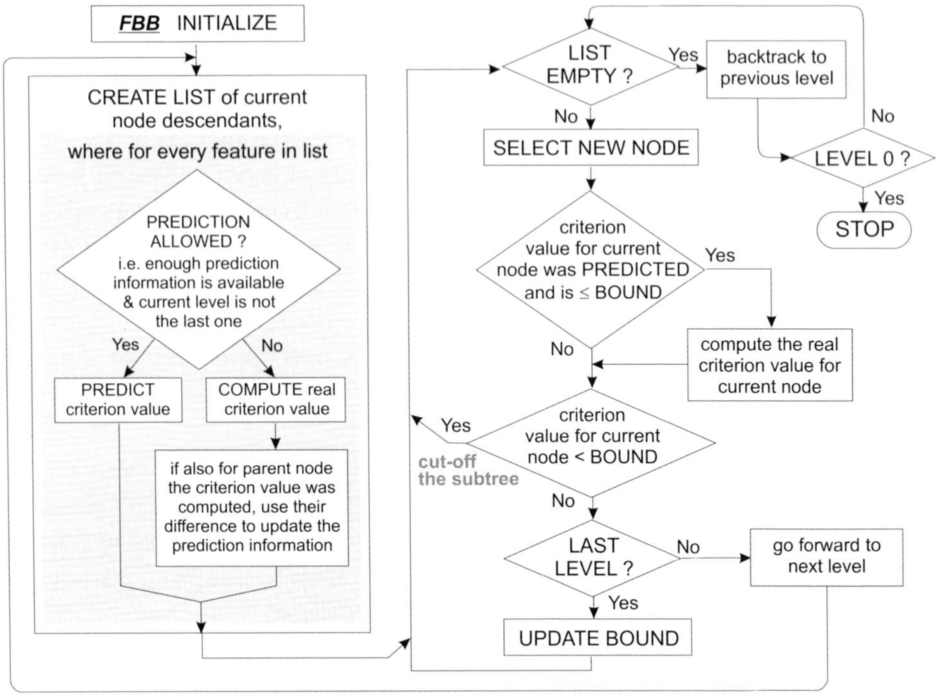

Figure 2: *A simplified diagram of the FBB algorithm*

pruned out by means of classical BB algorithms. However, this situation is usually strongly outweighed by criterion computation savings in other internal nodes, especially near the root, where criterion computation tends to be slower.

The prediction mechanism utilizes the information about the averaged criterion value change caused by removing a particular feature. Obviously, the following trade-off arises: we want to use prediction as frequently as possible to save time, on the other hand we have to compute the actual criterion values to precise (or at least to estimate) the information needed for the prediction mechanism. Different features appear in different search tree construction stages, therefore we need to collect the prediction information consecutively and separately for every feature.

Technically the algorithm uses a *vector of feature contributions to the criterion value* (denoted **A**) for storing the information about estimated *criterion value decrease* for each feature. Also, an information about the nature of processed criterion values must be stored in a *vector of value types*.

This is used to: a) ensure that prediction information in **A** is updated using real criterion value differences only and b) to check whether a sub-tree can be cut-off (this may follow only from real criterion values descending below the current *bound*). A simplified flowchart of the FBB algorithm is in Fig. 2. General algorithm properties are discussed in Section 4.4. The algorithm and its behaviour is discussed in following section or in [22].

4.2.1 Fast Branch and Bound Algorithm Formal Description

The algorithm description is based on notion from the monography [2]. We will use following symbols:

Constants:

D – number of all features,

d – required number of selected features,

$\delta \geq 1$ – *minimum required number of feature contribution evaluations*, this integer constant may be used for delaying the prediction mechanism to prevent prediction in initial algorithm stages when the collected prediction information may be insufficient yet,

$\gamma \geq 0$ – "optimism" constant, may be used to shift the algorithm functionality towards the exhaustive search (γ close to 0) or towards the classical non-predictive BB (with increasing γ), the reference value is $\gamma = 1$,

Other symbols:

Y – set of all D features,

$J(.)$ – criterion function,

k – tree level ($k = 0$ denotes the root),

$\bar{\chi}_k = \{\xi_j \mid j = 1, 2, \cdots, D - k\}$ – current "candidate" feature subset in k-th tree level,

q_k – number of current node descendants (in consecutive tree level),

$\mathcal{Q}_k = \{Q_{k,1}, Q_{k,2}, \ldots, Q_{k,q_k}\}$ – ordered set of features assigned to edges leading to the current node descendants (note that "candidate" subsets $\bar{\chi}_{k+1}$ corresponding to current node descendants are fully determined by features $Q_{k,i}$ for $i = 1, \cdots q_k$),

$\mathbf{J}_k = [J_{k,1}, J_{k,2}, \ldots, J_{k,q_k}]^{\mathrm{T}}$ – vector of criterion values corresponding to the current node descendants in consecutive tree level ($J_{k,i} = J(\bar{\chi}_k \setminus \{Q_{k,i}\})$ for $i = 1, \cdots, q_k$),

$\mathbf{T}_k = [T_{k,1}, T_{k,2}, \ldots, T_{k,q_k}]^{\mathrm{T}}$, $T_{k,i} \in \{\text{"}C''\text{, "}P''\}$ for $i = 1, \cdots, q_k$ – criterion value type vector (records the type of corresponding $J_{k,i}$ values),

$\Psi = \{\psi_j \mid j = 1, 2, \cdots, r\}$ – control set of r features being currently

available for search-tree construction, i.e. for building consecutive descendant vector $\mathcal{Q}_k$; the Ψ set serves for maintaining the search tree topology,
$\mathcal{X} = \{x_j \mid j = 1, 2, \cdots, d\}$ – current best subset of d features,
X^* – current *bound* (criterion value corresponding to $\mathcal{X}$),
$\mathbf{A} = [A_1, A_2, \ldots, A_D]^{\mathrm{T}}$ – *vect. of feature contributions to the crit. value,*
$\mathbf{S} = [S_1, S_2, \ldots, S_D]^{\mathrm{T}}$ – *counter vector* ($\mathbf{S}$ with $\mathbf{A}$ serve for prediction),
$\mathbf{V} = [v_1, v_2, \ldots, v_{q_k}]^{\mathrm{T}}$ – temporary vector for sorting purposes only,

REMARK 1. It is necessary to store all values q_j, ordered sets $\mathcal{Q}_j$ and vectors $\mathbf{J}_j$ and $\mathbf{T}_j$ for $j = 0, \cdots, k$ during the algorithm run to allow backtracking.

The algorithm is to be initialized as follows:
$k = 0$ (starting in the root),
$\bar{\chi}_0 = Y$, $\Psi = Y$, $r = D$,
X^* – lowest possible value (computer dependent)
$S_i = 0$ for all $i = 1, \cdots, D$
$\delta = 5$, $\gamma = 1.1$ (recommended default values)

FBB algorithm

Whenever the algorithm removes some feature y_i from the current "candidate" subset and computes the corresponding real criterion value $J(\bar{\chi}_k \setminus \{y_i\})$ in k-th tree level, and if also the predecessor value $J(\bar{\chi}_k) \equiv J(\bar{\chi}_{k-1} \setminus \{y_j\})$ (after previous removal of some feature y_j) had been computed (as indicated by $T_{k-1,y_j} =$"C"), then update the prediction information as follows:

$$A_{y_i} = \frac{A_{y_i} \cdot S_{y_i} + J_{k-1,y_j} - J(\bar{\chi}_k \setminus \{y_i\})}{S_{y_i} + 1}$$

and let

$$S_{y_i} = S_{y_i} + 1$$

Step 1: *Select descendants of the current node to form the consecutive tree level:* first set their number to $q_k = r - (D - d - k - 1)$. Construct an ordered set $\mathcal{Q}_k$ and vectors $\mathbf{J}_k$ and $\mathbf{T}_k$ specifying the current node descendants in the following way: for every feature $\psi_j \in \Psi, j = 1, \cdots, r$ if $k + 1 < D - d$ (nodes are not leaves) and $S_{\psi_j} > \delta$ (prediction allowed), let

$$v_j = J_{k-1,q_{k-1}} - A_{\psi_j}$$

i.e. predict by subtracting the appropriate prediction value based on ψ_j feature from the criterion value obtained in the parent node, otherwise the value must be computed, i.e. let

$$v_j = J(\bar{\chi}_k \setminus \{\psi_j\}).$$

After obtaining all v_j values sort them in the ascending order

$$v_{j_1} \leq v_{j_2} \leq \cdots \leq v_{j_r}$$

and for $i = 1, \cdots, q_k$ let
$Q_{k,i} = \psi_{j_i}$
$J_{k,i} = v_{j_i}$ if v_{j_i} records a computed value
$J_{k,i} = J_{k-1,q_{k-1}} - \gamma \cdot A_{\psi_{j_i}}$ if v_{j_i} records a predicted value
$T_{k,i} =$"C" if v_{j_i} records a computed value
$T_{k,i} =$"P" if v_{j_i} records a predicted value
To avoid future duplicate testing, features ψ_{j_i} in $\mathcal{Q}_k$ cannot be used for the construction of consecutive tree levels, so let $\Psi = \Psi \setminus \mathcal{Q}_k$ and $r = r - q_k$.

Step 2: *Test the right-most descendant node (connected by the Q_{k,q_k}-edge):* if $q_k = 0$, all descendants were tested, go to Step 4 (backtracking). If $T_{k,q_k} =$"P" and $J_{k,q_k} < X^*$, compute the real value $J_{k,q_k} = J(\bar{\chi}_k \setminus \{Q_{k,q_k}\})$ and mark $T_{k,q_k} =$"C". If $T_{k,q_k} =$"C" and $J_{k,q_k} < X^*$, then go to Step 3. Else let $\bar{\chi}_{k+1} = \bar{\chi}_k \setminus \{Q_{k,q_k}\}$. If $k + 1 = D - d$, then you have reached a leaf, go to Step 5. Otherwise go to next level: let $k = k + 1$ and go to Step 1.

Step 3: *Descendant node connected by the Q_{k,q_k}-edge (and its possible sub-tree) may be cut-off:* return feature Q_{k,q_k} to the set of features available for the tree construction, i.e. let $\Psi = \Psi \cup \{Q_{k,q_k}\}$ and $r = r + 1$, $\mathcal{Q}_k = \mathcal{Q}_k \setminus \{Q_{k,q_k}\}$ and $q_k = q_k - 1$ and continue with its left neighbour; go to Step 2.

Step 4: *Backtracking:* Let $k = k - 1$. If $k = -1$, then the complete tree had been searched through; stop the algorithm. Otherwise return feature Q_{k,q_k} to the set of "candidates": let $\bar{\chi}_k = \bar{\chi}_{k+1} \cup \{Q_{k,q_k}\}$ and go to Step 3.

Step 5: *Update the* bound *value:* Let $X^* = J_{k,q_k}$. Store the currently best feature subset $\mathcal{X} = \bar{\chi}_{k+1}$ and go to Step 2.

REMARK 2. In Step 1 for $k = 0$ the term $J_{-1,q_{-1}}$ denotes the criterion value for the set of all features, $J(Y)$.

4.3 Improving the "Improved" Algorithm

The FBB algorithm often enables the fastest Branch and Bound search currently possible, however it exhibits some drawbacks. Let us address the two most important ones – unusability with recursive criterion forms and theoretical impossibility to exclude prediction mechanism failure (and consecutive slow down). We show in the following that due to flexibility of the prediction mechanism concept it is possible to define a slightly different algorithm resolving (to a certain extent) some of FBB drawbacks.

Let us focus on the "Improved" BB [4] heuristics for ordering of internal tree node descendants. Assume every BB constructs a search tree with a given topology (e.g. the "minimum solution tree" [27]). It is apparent that given the topology, different feature assignments to the tree edges may be defined. The "Improved" BB attempts to utilize the topology and tree node ordering for increasing the overall performance (as noted in section 4.1). However, the ordering heuristic requires a significant number of additional computations. Let us illustrate this drawback in Fig. 1 – when constructing the first level (i.e. specifying the ordering of root descendants), the algorithm evaluates the *criterion value decrease* for every available feature (all 5 features), although only 3 features are to be ordered and assigned to first level edges. The problem is that without evaluating all 5 features the algorithm cannot decide which 3 to choose.

Suppose our intention is to find the same (or very similar) ordering of tree nodes as given by means of the "Improved" BB algorithm with reduced number of criterion evaluations. To achieve this goal we utilize the prediction mechanism as defined for purposes of the FBB algorithm. The new algorithm will construct the consecutive tree levels in several phases. First the *criterion value decrease* will be *predicted* for every feature being currently available for tree construction. The features will be sorted descending according to the *predicted criterion value decreases*. Then, the required number of features (beginning from the feature with highest *predicted criterion value decrease*) will be taken to form the consecutive tree level. Real criterion values therefore need to be computed for the level-forming features only (e.g. 3 instead of 5 in the above example). In all other aspects the algorithm course remains equal to the "Improved" BB. The new algorithm based on this idea is called *Branch and Bound with Partial Prediction* (BBPP). A simplified flowchart of the BBPP algorithm is in Fig. 3. For a more detailed discussion of the algorithm and its behaviour see [24].

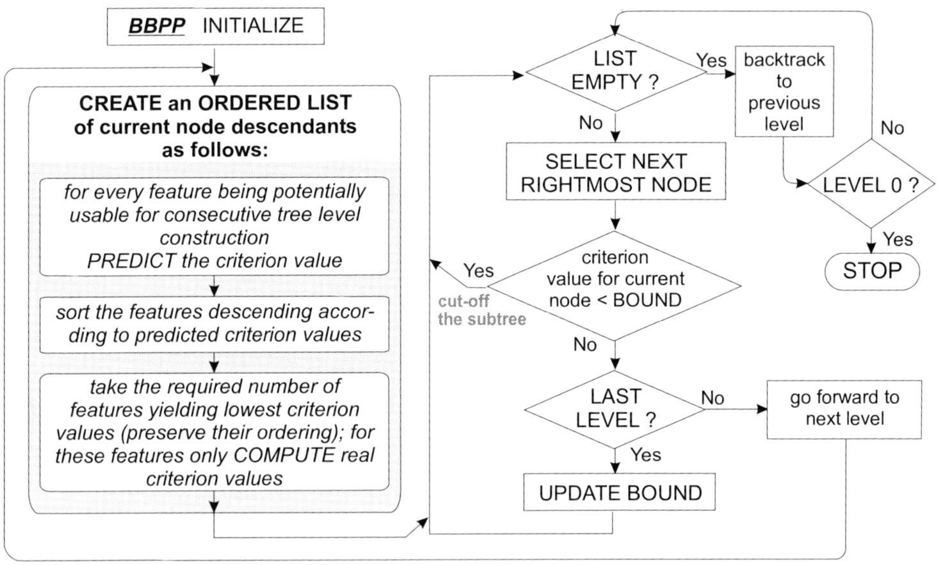

Figure 3: *Simplified diagram of the BBPP algorithm*

4.4 Predictive Branch and Bound Properties

The predictive algorithms (FBB, BBPP) may be expected to be most effective when the individual feature contributions to the criterion value do not change strongly in relation to different subsets. Practical tests on real data exhibit this property in the majority of cases. Both the FBB and BBPP showed to be effective even in cases, when due to difficult statistical dependencies individual feature contributions failed to remain stable.

When compared to classical BB algorithms the predictive algorithms always spend additional time for maintaining the prediction mechanism. However, this additional time showed not to be important, especially when compared to time savings arising from the pruned criterion computations.

The FBB prediction mechanism may fail to gain computational time under certain circumstances. The FBB is not suitable for use with recursive criteria, where calculating $J(\bar{X}_k)$ value requires the knowledge of previous value $J(\bar{X}_{k-1})$. The use of prediction in FBB leads to performing the criterion calculations not level by level but irregularly in different stages. This makes it impossible to calculate successive criterion values recursively. In general, the FBB algorithm shows to be most advantageous when used with computationally demanding non-recursive criterion functions with higher than linear computational complexity (in terms of the feature set size).

When compared to the FBB algorithm, the BBPP may be expected to be more robust. A potential failure of the prediction mechanism would have only indirect influence on the overall algorithm performance. A potentially wrong ordering of internal tree nodes (i.e. assigning of features to edges) would eventually decrease the efficiency of sub-tree cut-offs, however, the basic advantage over the "Improved" BB algorithm – reducing the number of additional computations – remains always preserved.

4.5 Experimental Results of Optimal Search Methods

The algorithms have been tested on a number of different data sets. Here we present representative results computed on 30-dimensional mammogram data (2 classes – 357 benign and 212 malignant samples) obtained from Wisconsin Diagnostic Breast Center via the UCI repository - ftp.ics.uci.edu. We used both the recursive and non-recursive Bhattacharyya distance as the criterion function. Performance of different methods is illustrated in Fig. 4 and Fig. 5 by a graph of total computational time and a graph of criterion evaluation numbers. The graph of criterion values is not included, because all the methods yield the same optimum values.

We compare all the results especially with the results of the "Improved" BB algorithm [2, 4], because this algorithm is generally accepted to be the most effective optimal subset search strategy. In case of non-recursive criterion functions the comparison includes the recent FBB algorithm, which cannot be used with recursive functions. Note that in case of non-recursive criterion computations we implemented all BB algorithms so as they construct the "minimum solution tree" [27].

Although the FBB algorithm usually finds the optimum after the smallest number of computations, its principle prevents it to be used with recursive criterion functions. The graphs in Fig. 4 and Fig. 5 illustrate that the BBPP operates faster than the "Improved" BB regardless the fact whether used with non-recursive or recursive criterion forms. According to expectation, the new algorithm, when used with recursive criterion forms, brings a less significant speedup. The computational complexity of recursive criterion forms is usually significantly lower than in non-recursive case, although the computational time spent by the prediction mechanism remains the same.

REMARK 3. When used with a recursive criterion form, no BB algorithm may utilize the "minimum solution tree" [27] due to the necessity to preserve criterion value computation sequence. The minimum solution tree assumes

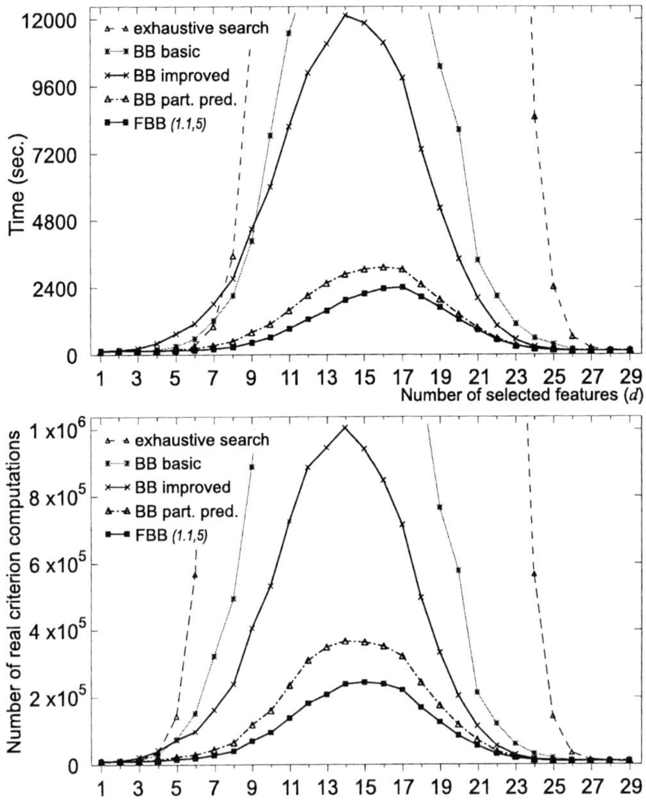

Figure 4: *Optimal subset search methods performance when maximizing the* non-recursive *Bhattacharyya distance on 30-dimensional data (Wisconsin Diagnostic Breast Center). Results computed on a Pentium II-350 MHz computer.*

shortening of straight paths to leaves, what breaks the criterion computation sequence. Because of this reason numbers of computations differ in recursive and non-recursive case.

Both for the FBB and BBPP a slight shift of their graphs to the right may be observed when compared to the "Improved" BB algorithm. The prediction mechanism based algorithm acceleration relates to the number of criterion evaluation savings in internal search tree nodes, therefore with decreasing d the search tree depth increases and allows more effective operation of the prediction mechanism.

The majority of experiments produced results similar to those in Fig. 4

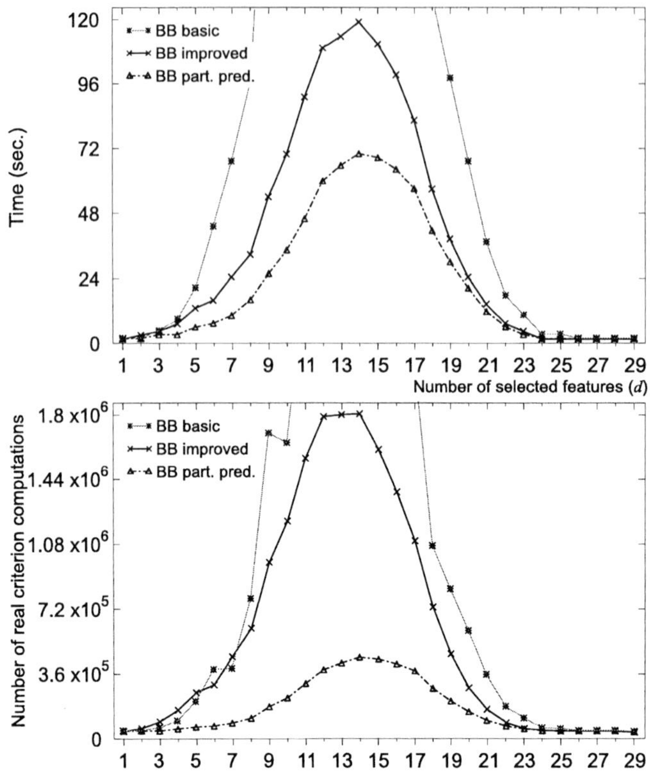

Figure 5: *Optimal subset search methods performance when maximizing the* **recursive** *Bhattacharyya distance on 30-dimensional data (Wisconsin Diagnostic Breast Center). Results computed on a Pentium II-350 MHz computer.*

and Fig. 5. In one isolated worst case the speed of the BBPP used with recursive criterion form remained comparable to the speed of "Improved" BB. Theoretically we can not exclude the prediction mechanism failure – if the individual feature contributions to criterion value were unstable, i.e. changed too often and too strongly, either the FBB or the BBPP operation could become comparable with the "Basic BB" algorithm. However, such a situation seems highly improbable and we have not met it in our experiments.

Apart from measuring computational times and numbers of computations we can compare the algorithms also by observing the behaviour of their search tree construction. Table 1 illustrates difference in effectiveness

Numbers of sub-tree cut-offs in different Branch & Bound algorithms search tree levels

selecting 7 features out of 30 (Wisconsin Diagnostic Breast Center mammogram data), non-recursive Bhattacharyya distance

k =	1	2	3	4	5	6	7	8	9	10	.....	18	19	20	21	22
Basic BB	0	0	0	0	0	0	0	136	2228	44399	.....	83211	34697	8982	2160	391
Improved BB	0	0	0	0	0	717637	501701	518751	165407	73087	.....	6	0	0	3	0
BBPP	0	0	0	0	0	728096	508671	486499	165792	78540	.....	42	25	4	3	2
FBB (1.1, 5)	0	0	0	0	0	160104	820656	571660	291146	97440	.....	42	15	8	0	2

Table 1: *Comparison of sub-tree cut-off numbers performed in different BB algorithms ($k = 0$ represents the root, $k = 23$ represents the leaves).*

Branch & Bound algorithms search tree tracking characteristics

selecting 7 features out of 30 (Wisconsin Diagnostic Breast Center mammogram data), non-recursive Bhattacharyya distance

Algorithm	Total number of leaves	Computations saved in leaves	Computations saved by prediction	Full Criterion evaluations	Computational time
Basic BB	2035799	1898137	0	333378	00:18:04
Improved BB	2035799	2026177	0	163444	00:28:18
BBPP	2035799	2024761	125700	37689	00:03:30
FBB (1.1,5)	2035799	2024117	181467	19338	00:01:18

Table 2: *Comparison of numbers of feature configurations tested in different BB algorithms.*

of different BB algorithm search tree construction. Numbers of sub-tree cut-offs in different tree levels are shown, while k values close to 0 represent cut-offs closer to the root and values closer to 23 represent cut-offs closer to the leaves. The distribution of values is important for comparison purposes more than the exact values themselves. More effective algorithms may be expected to cut-off more sub-trees closer to the root, what results in more significant decrease of the number of criterion evaluations to be done. As expected, the "Improved BB" orders the tree in the best way, together with the "BB with Partial Prediction". The FBB prediction mechanism allows slightly worse tree ordering, however it compensates this drawback by reducing dramatically the number of computations that could not be avoided in other algorithms. Table 2 then illustrates the differences in the numbers of criterion evaluations performed by different BB algorithms.

4.6 Summary of Recent Optimal Methods

The only optimal subset search method usable with non-monotonic criteria is the trivial exhaustive (full) search. However, because of exponential nature of the search problem, alternative methods are often needed. The simplest in a series of Branch and Bound algorithms, the "Basic" BB, is overperformed by other BB variants. The "Improved" BB operates usually several times faster than the "Basic" BB. Recent prediction based algorithms ("Fast BB" and "BB with Partial Prediction") usually outperform the "Improved" BB by factor 3–20, although such behavior cannot be always guaranteed.

The described algorithms behavior has been verified during a number of experiments on different datasets. All the algorithms have been tested e.g. on a 32-dimensional breast cancer dataset, 20-dimensional marble stone dataset, 60-dimensional sonar data (from computational time reasons for limited subset sizes only), etc. Description of some of our experiments can be found in [22, 24].

Even with most advanced search algorithms the optimal search problem remains often prohibitive. All the known algorithms have an exponential nature. Their performance is usually incomparably better than the performance of the simple exhaustive search, however their performance is not guaranteed. Moreover, for most of problems the performance of optimal subset selection algorithms cannot be even predicted and therefore their application may have to be preceded by preliminary testing and adjusting.

In general we recommend the BBPP algorithm if there are no particular reasons for using another one. Its prediction mechanism allows significant time savings when compared to "Basic" or "Improved" Branch and Bound algorithms. The algorithm has been experimentally proved to be robust and well suitable for use with different criterion functions, both in recursive and non-recursive form.

5 Recent Suboptimal Search Methods

Despite the advances in optimal search methods, for larger size problems we have to resort still to suboptimal methods. For this class of methods we have developed recently a family of new algorithms presented in this section. They are:

1. *sequential floating search methods*
2. *adaptive floating search methods*
3. *oscillating search methods*

Earlier methods suffered from so called nesting of feature subsets which significantly deteriorated the performance of algorithms. The first attempt to overcome this problem was to employ either the plus l- take away r (denoted (l, r)) or generalized (l, r) algorithms which involve successive augmentation and depletion process. Consequently, the resulting dimensionality in respective stages of both algorithms is fixed depending on the prespecified values of l and r. Unfortunately, there is no theoretical way of predicting the values of l and r to achieve the best feature set. Alternatively, instead of fixing these values, there is a couple of more sophisticated procedures of which these values are flexibly changing so as to approximate the optimal solution as much as possible. Although both of these methods switch between including and excluding features, they are recognized as two different algorithms according to the dominant direction of the search. The search in the forward direction is known as the *sequential forward floating selection* (SFFS), while in the opposite direction is known as the *sequential backward floating selection* (SBFS) [15]. They are known as floating methods because the resulting dimensionality in respective stages of the algorithm is not changing monotonously but is actually "floating" up and down.

5.1 Sequential Floating Search

The sequential forward floating selection procedure consists of applying after each forward step a number of backward steps as long as the resulting subsets are better than previously evaluated ones at that level. Consequently, there are no backward steps at all if the performance cannot be improved. The same applies for the sequential backward floating selection procedure. Both algorithms allow a 'self-controlled backtracking' so they can eventually find good solutions by adjusting the trade-off between forward and backward steps dynamically. It is possible to say that, in a certain way, they compute only what they need without any parameter setting.

Before describing the SFFS and SBFS algorithms formally, the following definitions have to be introduced.

Let $X_k = \{x_{i_1}, \ldots, x_{i_k}\}$ be the set of k features selected from the original set of D possible features, $X_D = \{x_i \mid i = 1, \ldots, D\}$, where $\{i_1, \ldots, i_k\}$ is a subset of $\{1, \ldots, D\}$. Then $X_D - X_k = \{x_{j_1}, \ldots, x_{j_{D-k}}\}$ is the set of features not yet selected, $\{j_1, \ldots, j_{D-k}\}$ is the complement of $\{i_1, \ldots, i_k\}$ in $\{1, \ldots, D\}$. We say that the feature x_{i_r} from the set X_k is the *best* (*worst*) feature *in the set* X_k if

$$J(X_k - x_{i_r}) = \min_{s=1,\ldots,k} J(X_k - x_{i_s}) \quad \left(J(X_k - x_{i_r}) = \max_{s=1,\ldots,k} J(X_k - x_{i_s}) \right).$$

We say that the feature x_{j_p} from the set $X_D - X_k$ is the *best* (*worst*) feature *with respect to the set* X_k if

$$J(X_k+x_{j_p}) = \max_{q=1,\ldots,D-k} J(X_k+x_{j_q}) \quad \left(J(X_k + x_{j_p}) = \min_{q=1,\ldots,D-k} J(X_k + x_{j_q})\right).$$

5.1.1 Sequential Forward Floating Search Algorithm

SFFS procedure is basically a bottom-up search procedure which adds (includes) new features by means of applying the basic sequential forward selection (SFS) procedure starting from the current feature set, followed by a series of successive conditional removing (exclusion) of the worst feature in the newly updated set provided a further improvement can be made to the previous sets. The SFFS method can be described algorithmically as follows:

SFFS algorithm

Suppose k features have already been selected from the complete set of features X_D to form set X_k with the corresponding value $J(X_k)$ of the criterion function J. In addition, the values of $J_i = J(X_i)$ for all preceding supersets of size $i = 1, 2, \cdots, k-1$ are known and stored.

Step 1: *(Inclusion)* Using the basic SFS method, select from the set of available features, $X_D - X_k$ the best feature with respect to the set X_k, say x^+, and add it to the current set X_k to form new feature set X_{k+1}; i.e.

$$x^+ := \arg \max_{x \in X_D - X_k} J(X_k + x), \quad X_{k+1} := X_k + x^+.$$

Step 2: *(Conditional exclusion)* Find the worst feature in the set X_{k+1}, say x^-, i.e.

$$x^- = \arg \max_{x \in X_{k+1}} J(X_{k+1} - x).$$

If the feature x^- is the same as x^+, set $J_{k+1} = J(X_{k+1})$, $k = k+1$ and go to Step 1; otherwise remove this feature from the set X_{k+1} to form a new feature set $X_k^{'}$, i.e.

$$X_k^{'} = X_{k+1} - x^-.$$

Note that now $J(X_k^{'}) > J(X_k) = J_k$. If $k = 2$, then set $X_k = X_k^{'}$ and $J_k = J(X_k^{'})$ and go to Step 1, otherwise go to Step 3.

Step 3: *(Continuation of conditional exclusion)* Continue removing the features from the set $X_k^{'}$ to form reduced sets $X_{k-1}^{'}$ while

$$J(X_{k-1}^{'}) > J_{k-1}.$$

Set $k = k - 1$ or $k = 2$; then continue with Step 1.

The algorithm is initialized by setting $k = 0$ and $X_0 = \emptyset$, and the SFS method is used until a feature set of size 2 is obtained. Then the algorithm continues with Step 1.

A simplified flowchart of the SFFS algorithm is given in Fig. 6.

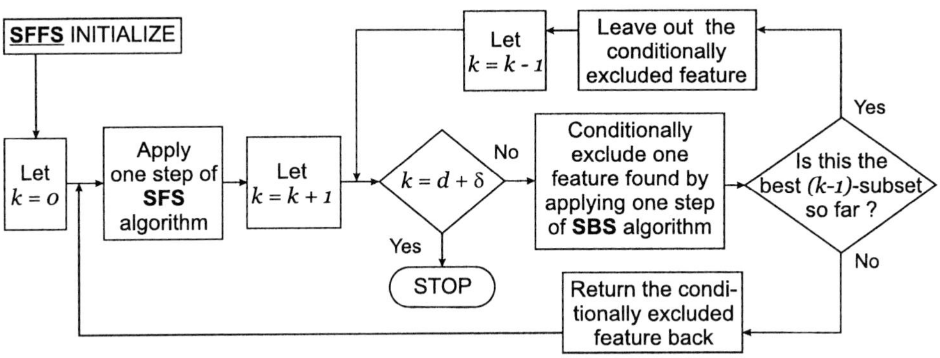

Figure 6: *Flow chart of SFFS algorithm*

The terminating condition $k = d + \delta$ in the flowchart Fig. 6 means that in order to fully utilize the properties of SFFS search we should not stop the algorithm immediately after it reaches for the first time the dimensionality d. By leaving it to float up and back a bit further, the potential of the algorithm is better utilized and a subset of size d outperforming the first one achieved is usually found. In practice we can let the algorithm either go up to the original dimensionality D, or if it is too large, then the value of δ can be determined heuristically (e.g. according to the value of maximum number of backtracking steps prior to reaching d first time).

5.1.2 Sequential Backward Floating Search Algorithm

SBFS procedure is a top-down search procedure which excludes features by means of applying the basic sequential backward selection (SBS) procedure starting from the current feature set and followed by a series of successive conditional adding of the best significant feature from the available features if an improvement can be made to the previous sets. SBFS is the backward counterpart of the SFFS algorithm obtained in a straightforward manner by substituting adding by removing and the initialization in the previous description. The SBFS method can be described algorithmically in a similar way to the SFFS method as follows:

SBFS algorithm

Suppose k features have already been removed from the complete set of features $\bar{X}_0 = X_D$ to form a feature set $\bar{X}_k$ with the corresponding criterion function $J(\bar{X}_k)$. Furthermore, the values of all supersets $\bar{X}_i,\ i = 0, 1, \ldots, k-1$, are known and stored.

Step 1: *(Exclusion)* Use the basic SBS method to remove the worst feature, say x^-, from the current set $\bar{X}_k$ to form a reduced set $\bar{X}_{k+1}$, i.e.

$$x^- = \arg\max_{x \in \bar{X}_k} J(\bar{X}_k - x), \quad \bar{X}_{k+1} = \bar{X}_k - x^-.$$

Step2: *(Conditional adding)* Find among the already removed features the best feature with respect to the set $\bar{X}_{k+1}$, say x^+, i.e.

$$x^+ = \arg\max_{x \in X_D - \bar{X}_{k+1}} J(\bar{X}_k + x).$$

If x^+ is the same as x^- then set $k = k + 1$ and go to Step 1; otherwise add x^+ to the set $\bar{X}_{k+1}$ to form a new set $\bar{X}'_k$, i.e.

$$\bar{X}'_k = \bar{X}_{k+1} + x^+.$$

Note that now $J(\bar{X}'_k) > J(\bar{X}_k)$.

Step 3: *(Continuation of conditional inclusion)* Continue adding features to the set $\bar{X}'_k$ to form the new enlarged set $\bar{X}'_{k-1}$, while

$$J(\bar{X}'_{k-1}) > J(\bar{X}'_k).$$

Set $k = k - 1$ or $k = 2$; then continue with the Step 1.

The algorithm is initialized by setting $k = 0$ and $\bar{X}_0 = X_D$ and the SBS method is used until a feature set of size $D - 2$ is obtained (it means until 2 worst features are removed). Then the algorithm continues with Step 1.

5.2 Adaptive Floating Search

Though the floating search algorithms have been found very successful and evaluated by independent comparative studies as the currently most efficient algorithms, even their performance can be still somewhat improved, though at the expense of the increased computational time. In this section the generalization of "classical" floating search algorithms, called "adaptive floating search" (AFS) algorithms is presented.

For the sake of simplicity, let us denote original floating search methods (SFFS and SBFS) together as "classical floating search" methods and denote them by CFS. CFS methods use only single feature adding or removing, respectively, in the course of the algorithm.

Our new search strategy aims to utilize the best of both generalized strategies and classical floating strategies. The course of search is similar to that of CFS, but the individual search steps are generalized. However, the new algorithm is by no means just a generalized version of CFS.

The basic generalization of CFS would be to replace the simple SFS or SBS steps inside the CFS procedure by their generalized versions GSFS(o) or GSBS(o), respectively. Unfortunately for a potential user, it is generally not known which value of o to choose to get the best results. Moreover, the choice of the value of o, too high for the particular problem, results in unnecessary increase of computing time.

As opposed to the above mentioned generalized methods, the AFS method does not need the user-specified level of generalization. This level (value of o) is determined dynamically in the course of search according to the current situation so as to achieve better results. In this sense, AFS brings about a similar improvement in comparison with the generalized search strategies as the CFS search brought about in comparison with simple "non-generalized" strategies (CFS introduced dynamical "floating" of adding / removing steps).

Because of the time-exacting character of generalized steps (especially when used in high dimensional problems) we introduced a user defined parametric limit r_{max}, restricting the maximum generalization level which the method can use. The current generalization level, which changes in the course of search, is denoted by o. The AFS is called "adaptive" because of its ability to adjust the limit under which the actual generalization level can be automatically set. Simply said, the nearer the current subset size (k) is to the final one (d), the higher is the generalization limit. This characteristic aims to save computing time by limiting the generalization levels while the current subset is still far from the desired one. Therefore, we introduce the variable r representing the actual generalization limit for a given dimension.

To summarize, r_{max} is a user-specified absolute generalization limit, r is the actual generalization limit determined adaptively by the algorithm for the current subset ($r \leq r_{max}$ always holds), o is the current generalization level depending on the current situation ($o \leq r$ always holds).

REMARK 4. For $r_{max} = 1$ the AFS is identical to classical floating search.

Adaptive determination of r is done as follows: at the beginning of every forward or backward algorithm phase, respectively:

1. if $|k - d| < b$, let $r = r_{max}$
2. else if $|k - d| < b + r_{max}$, let $r = r_{max} + b - |k - d|$
3. else let $r = 1$

Here b denotes the neighbourhood of the final dimension, where the highest generalization levels are allowed. Basically it is possible to set $b = 0$.

Thus, in the generalized course of the AFS algorithm $o = 1$ is used in usual algorithm stages (e.g., in the beginning). Only special algorithm stages (when conditional forward, resp. backward steps brought no improvement) allow increasing of o and, therefore, a more detailed search.

By setting r_{max} or b to higher values, the user has a possibility to let the algorithm perform more thorough search with better chances to find the optimal solution, of course at the expense of longer computation time. The setting of these two parameters is not so critical as setting the generalization level in classical GSFS(o), resp GSBS(o) and (l, r). Increasing r_{max} or b does not lead to a different search, but to a more detailed search.

REMARK 5. Every AFS algorithm run includes steps identical with CFS ones. Also for this reason we expect the AFS algorithm to find equal or better solutions than CFS. The computer time needed for AFS is expected to be substantially longer (due to the generalization) than for CFS. However, if we constructed the generalized floating search in the simple way, it would consume incomparably more time.

CFS were found to occasionally prefer worse working subsets in the course of search.

To describe such a case, let us remind the principle of CFS (specifically SFFS) first:

Step 1. Add the *best* feature to the current subset of size k. Set $k = k + 1$.

Step 2. Conditionally remove the *worst* feature from the current subset.

Step 3. If the current subset is the so-far best subset of size $k - 1$, set $k = k - 1$ and go to Step 2. Else return the conditionally removed feature and go to Step 1.

Note that backward steps are conditional. Only backward steps bringing improvement are allowed. On the other hand, forward steps cannot be conditional. If they were conditional, the algorithm could theoretically fall into an infinite cycle (repeated by conditional adding and removing a feature). Because of their unconditionality, the forward steps can lead to finding worse subset, than was the so-far best one of a given dimension. A less prospective "search branch" is thus uselessly followed.

Remedy of this problem that we have adopted is simple. If the forward step found a subset worse than the best one known so-far, the current one is forgotten and the so-far best one becomes the current one. Note that this "violent" swapping of current subset cannot lead to infinite cycling as finding of a worse subset by the forward step must have been preceded by finding a better subset in some lower dimension.

A simplified flowchart of the ASFFS algorithm is in Fig. 7. For formal description of the adaptive floating algorithms see [21].

5.3 Oscillating Search

In this section a new suboptimal subset search method for feature selection is introduced. As opposed to other till now known subset selection methods the *oscillating search* (OS) is not dependent on prespecified direction of search (forward or backward). The generality of oscillating search concept allowed us to define several different algorithms suitable for different purposes. We can specify the need to obtain good results in very short time, or let the algorithm search more thoroughly to obtain near-optimum results. In many cases the oscillating search over-performed all the other tested methods. The oscillating search may be restricted by a preset time-limit, what makes it usable in real-time systems.

Note that most of known suboptimal search strategies are based on step-wise adding of features to initially empty feature set, or step-wise removing features from the initial set of all features, Y. One of search directions, *forward* or *backward*, is usually preferred, depending on several factors, the expected difference between the original and the final required size being the most important one. Regardless of the direction, it is apparent, that all these algorithms spend a lot of time testing feature subsets having cardinalities far distant from the required size d.

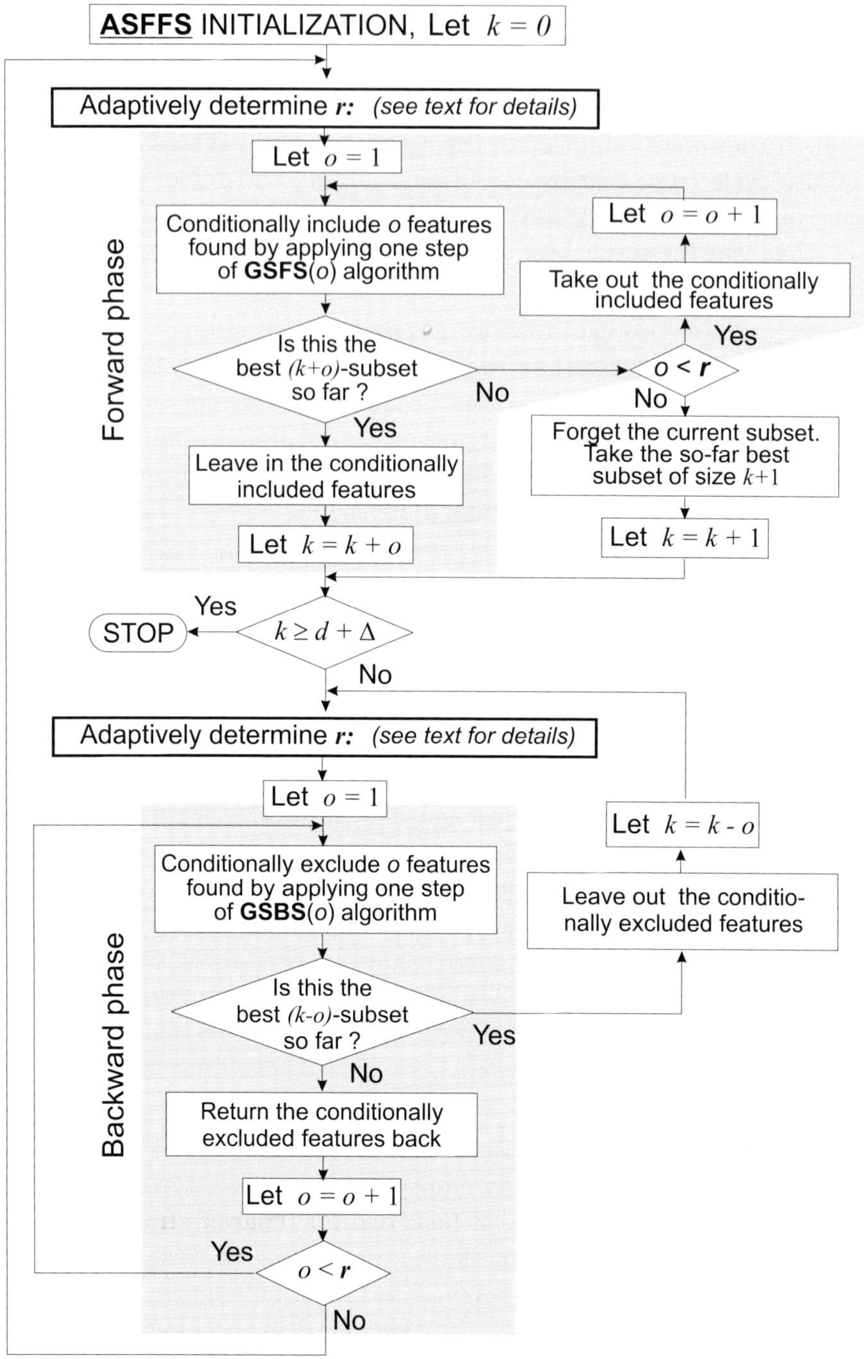

Figure 7: *Simplified flowchart of ASFFS algorithm*

5.3.1 Oscillating Search Algorithms

Unlike other methods, the OS is based on repeated modification of the current subset X_d of d features. This is achieved by alternating the *down-* and *up-swings*. The *down-swing* removes o worst features from the current set X_d to obtain a new set X_{d-o} at first, then adds o best ones to X_{d-o} to obtain a new current set X_d. The *up-swing* adds o "good" features to the current set X_d to obtain a new set X_{d+o} at first, then removes o "bad" ones from X_{d+o} to obtain a new current set X_d again. Let us denote two successive opposite swings as an *oscillation cycle.* Using this notion, the oscillating search consists of repeating oscillation cycles. The value of o will be denoted *oscillation cycle depth* and should be set to 1 initially. If the last oscillation cycle did not find better subset X_d of d features, the algorithm increases the oscillation cycle depth by letting $o = o + 1$. Whenever any swing finds better subset X_d of d features, the depth value o is restored to 1. The algorithm stops, when the value of o exceeds the user-specified *limit* Δ. The course of oscillating search is illustrated in Fig. 8.

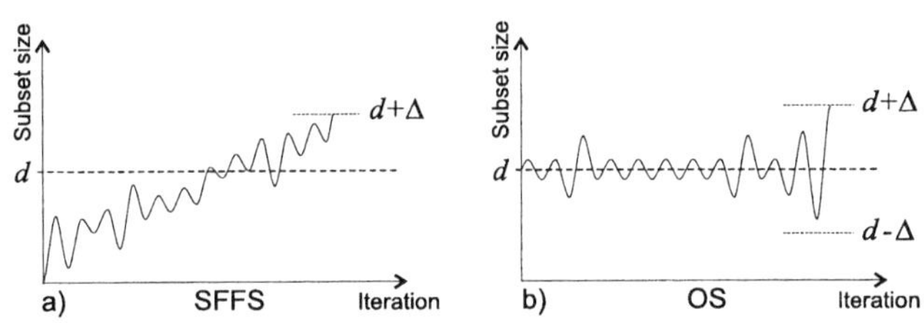

Figure 8: *Graphs demonstrate the course of search algorithms: a) sequential floating forward selection, b) oscillating search.*

Every oscillation algorithm assumes the existence of some initial set of d features. Obtaining such an initial set will be denoted as an *initialization.* Oscillating algorithms may be initialized in different ways; the simplest ways are random selection or the SFS procedure. From this point of view the oscillating search may serve as a mechanism for tuning solutions obtained in another way.

Whenever a feature o-tuple is to be added (or removed) from the current subset in the till now known methods, one of two approaches is usually utilized: the *generalized* adding (or removing) finds the optimum o-tuple by means of exhaustive search (e.g. in GSFS, GSBS) or the *successive adding (or removing) single features o-times* (e.g. in basic (l, r)), which may fail to find the optimum o-tuple, but is significantly faster.

In fact, we may consider finding feature o-tuples to be equivalent to the feature selection problem, though at a "second level". Therefore, we may use any search strategy for finding feature o-tuples. Because of proved effectiveness of *floating search* strategies we adopted the floating search principle as the third approach to adding (or removing) feature o-tuples in oscillating search. Note that in such a way one basic idea has resulted in defining a couple of new algorithms, as shown in the sequel.

For the sake of simplicity, let us denote the adding of feature o-tuple by ADD(o) and the removing of feature o-tuple by REMOVE(o). Finally, we introduce three versions of oscillating algorithm:

1. *Sequential oscillating search*: ADD(o) represents a sequention of o successive SFS steps (see [2]), REMOVE(o) represents a sequention of o successive SBS steps.

2. *Floating oscillating search*: ADD(o) represents adding of o features by means of the SFFS procedure (see [15]), REMOVE(o) represents removing of o features by means of the SFBS procedure.

3. *Generalized oscillating search*: ADD(o) represents adding of o features by means of the GSFS(o) procedure (see [2]), REMOVE(o) represents removing of o features by means of the GSBS(o) procedure.

Oscillating search algorithm

REMARK 6. c serves as a swing counter.

Step 1: (*Initialization*) By means of the SFS procedure (or otherwise) find the initial set X_d of d features. Let $c = 0$. Let $o = 1$.

Step 2: (*Down-swing*) By means of the REMOVE(o) step remove the worst feature o-tuple from X_d to form a new set X_{d-o}. (* If the $J(X_{d-o})$ value is not the so-far best one among subsets having size $d - o$, stop the down-swing and go to Step 3. *)
By means of the ADD(o) step add the best feature o-tuple from $Y \setminus X_{d-o}$ to X_{d-o} to form a new subset X'_d. If the $J(X'_d)$ value is the so-far best one among subsets having required size d, let $X_d = X'_d$, $c = 0$ and $o = 1$ and go to Step 4.

Step 3: (*Last swing did not find better solution*) Let $c = c + 1$. If $c = 2$, then none of previous two swings has found better solution; extend the search by letting $o = o + 1$. If $o > \Delta$, stop the algorithm, otherwise let $c = 0$.

Step 4: (*Up-swing*) By means of the ADD(o) step add the best feature o-tuple from $X_D \setminus X_d$ to X_d to form a new set X_{d+o}. *(** If the $J(X_{d+o})$ value is not the so-far best one among subsets having size $d+o$, stop the up-swing and go to Step 5. **)*
By means of the REMOVE(o) step remove the worst feature o-tuple from X_{d+o} to form a new set X'_d. If the $J(X'_d)$ value is the so-far best one among subsets having required size d, let $X_d = X'_d$, $c = 0$ and $o = 1$ and go to Step 2.

Step 5: (*Last swing did not find better solution*) Let $c = c + 1$. If $c = 2$, then none of previous two swings has found better solution; extend the search by letting $o = o + 1$. If $o > \Delta$, stop the algorithm, otherwise let $c = 0$ and go to Step 2.

REMARK 7. Parts of code enclosed in *(** and **)* brackets may be omitted to obtain a bit slower, more thorough procedure.

The algorithm is described also by a float chart in Fig. 9. The three introduced algorithm versions differ in their effectiveness and time complexity. The *generalized oscillating search* gives the best results, but its use is restricted due to the time complexity of generalized steps (especially for higher Δ). The *floating oscillating search* is suitable for finding solutions as close to optimum as possible in a reasonable time even in high dimensional problems. The *sequential oscillating search* is the fastest but possibly the least effective algorithm version with respect to approaching the optimal solution.

5.3.2 Oscillating Search Properties

All oscillating algorithms may be characterized as algorithms independent on the direction of search (neither bottom-up nor top-down). Every algorithm cycle attempts to improve the current subset of size d. Therefore OS algorithms do not loose time by evaluating subsets of distant cardinalities. All OS versions overcome effectively the "nesting problem" which SFS and SBS suffer from.

The fastest improvement of the current subset may be expected in initial phases of the algorithm, because of the low initial cycle depth. Later, when the current feature subset evolves closer to optimum, low-depth cycles fail to improve the subset and therefore the algorithm increases the depth (o) value. Though this will increase the chance to get closer to optimum, the trade-off

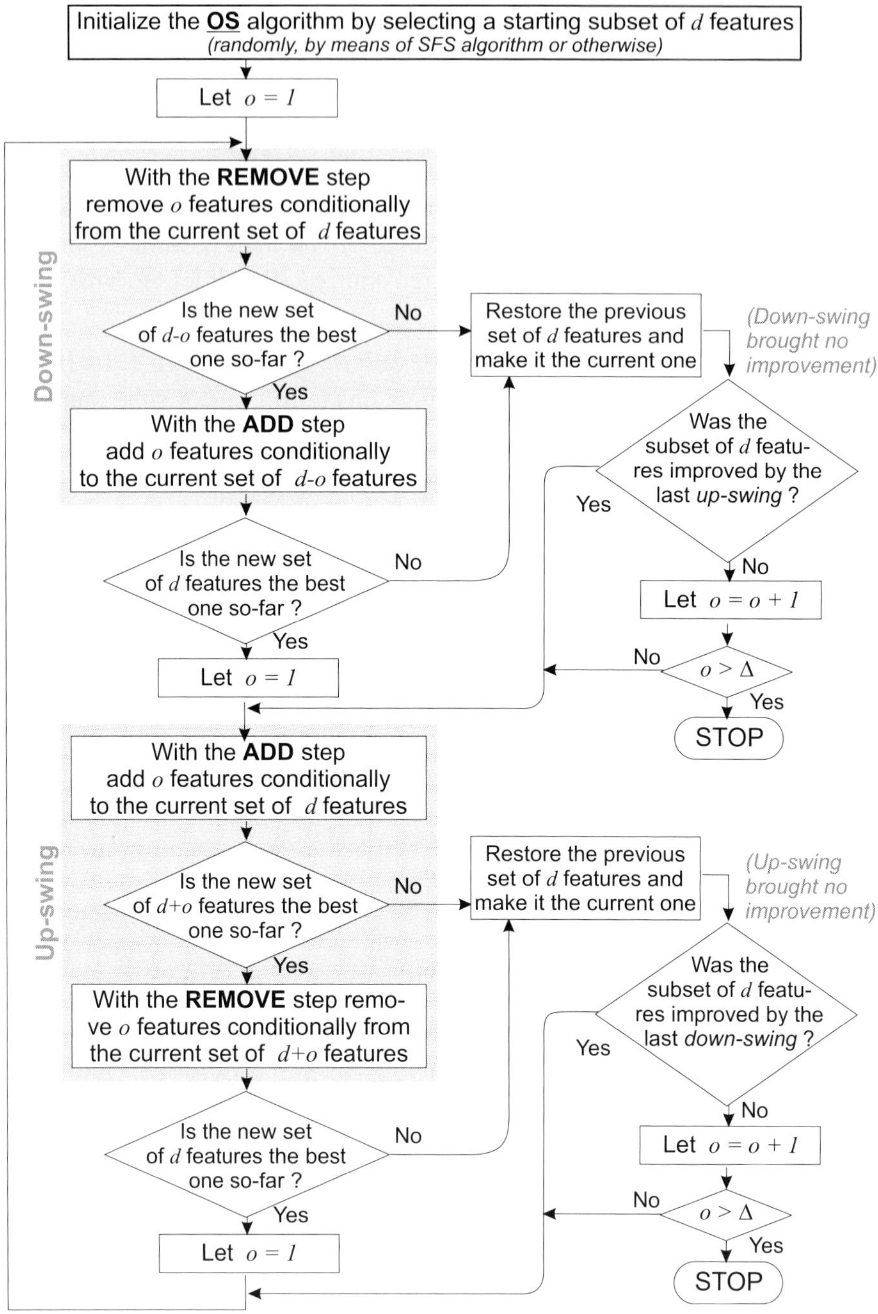

Figure 9: *Simplified Oscillating Search algorithm flowchart. Remark: meaning of the ADD and REMOVE symbols is discussed in text.*

between finding a better solution and the computational time clearly exists. It has been found out that it is often possible to find out very good solution after a short time corresponding to the first few low-depth cycles. This algorithm behavior is advantageous, because it is possible to obtain usually very good solutions after very short time. The oscillating search principle proves to be useful for different purposes:

- It may be looked upon as a universal tuning mechanism, being able to improve solutions obtained in any other way. In this sense the oscillating search is usable in addition to any subset search method.

- Let us denote the sequential OS with $\Delta = 1$, initialized by SFS, the *basic oscillating search* - BOS. This simple algorithm is suitable for problems being currently solved by SFS or SBS methods. Both the time and implementation complexity of the BOS is comparable to the complexity of these simple methods. However, the BOS algorithm is much more effective.

- The floating OS finds usually one of the best achievable suboptimal solutions. When used for solving high dimensional problems, the floating OS finds usually better solutions, than both classical and adaptive floating strategies [15, 21].

- The randomly initialized sequential OS is a very fast heuristic, that often finds results not achievable in other ways. However, because of its randomized nature, we may not expect it to find good results each time.

- Repeating of oscillation cycles allows to stop the algorithm after specified amount of time and still to obtain a usable solution. Because of this property the OS is suitable also for use in real-time systems.

Suboptimal search algorithms are often criticized because of their dependence on user parameters. The OS procedure assumes user-setting of the *oscillation cycle depth* Δ. Setting higher Δ value should result in more thorough search. Note, that we may consider the maximum meaningful Δ value, which is $max\{d, D-d\}$. Based on the knowledge of this maximum, we may allow the user to specify Δ value relatively, i.e. independently of current problem properties (d and D values). Maximum performance would be therefore achieved by utilizing OS (100%), although lower values (20 – 30%) are more suitable for most tasks, especially when computational time is to be restricted.

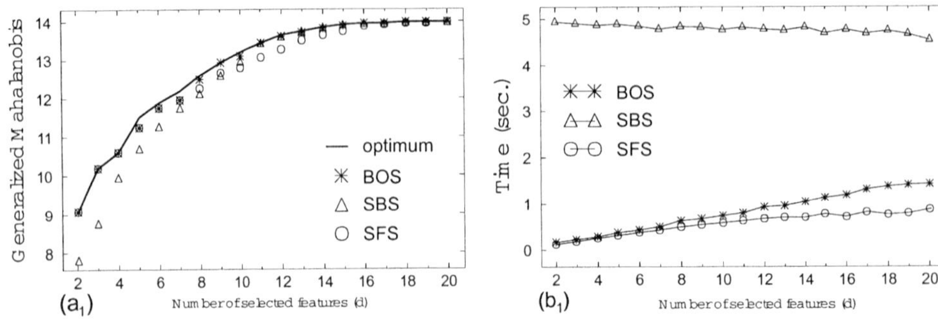

Figure 10: *Comparison of simple search strategies; 30-dimensional data (Wisconsin Diagnostic Breast Center).*

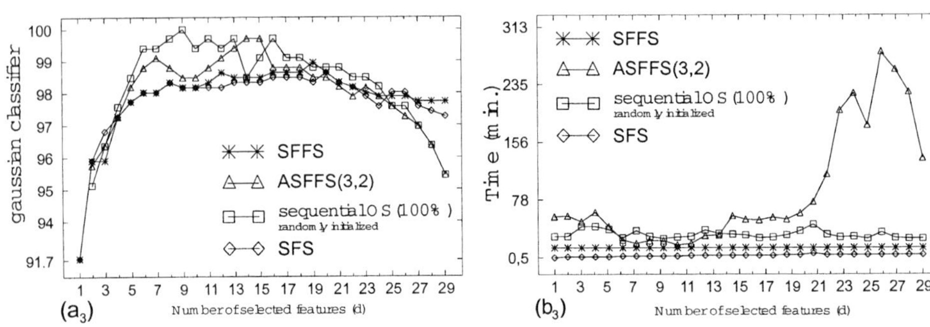

Figure 11: *Comparison of search strategies on high dimensional problem; 65-dimensional data (mammogram, PRIM lab.,University of California, Irvine).*

5.4 Experimental Results of Suboptimal Search Methods

The described suboptimal sequential search methods have been tested on a large number of different problems. We demonstrate their performance on 2-class, 30-dimensional mammogram data from Wisconsin Diagnostic Breast Center (obtained from ftp.ics.uci.edu) and 2-class, 65-dimensional mammogram data from PRIM Lab., University of California, Irvine.

Graphs in Fig. 10 compare the basic oscillating search with simple sequential selection methods. The BOS algorithm finds better solutions, though its computational time remains comparable with the SFS. Therefore, we find the BOS algorithm to be for simple tasks more efficient. Fig. 11 demonstrates suitability of the OS search for solving high dimensional problems. The floating OS finds usually the best solutions, but its computational

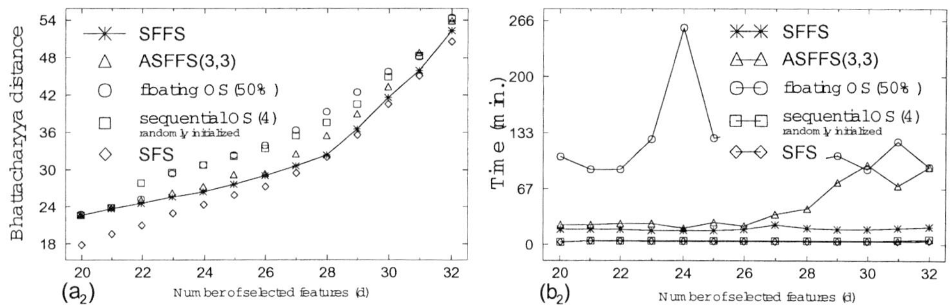

Figure 12: *Comparison of search strategies on 30-dimensional data (W.D.B.C.).*

time may be rather high. Note the effectiveness of the randomly initialized sequential OS. Fig. 12 also illustrates the potential of randomly initialized search. For classification purposes, the data set was divided into two equally-sized halves for training and testing. Because of randomized nature, finding best solutions cannot be always guaranteed.

5.5 Summary of Recent Suboptimal Methods

A battery of feature selection strategies has been extended by several new methods. They are the Floating Search methods, evaluated recently as the most effective feature selection methods, and furthermore, Adaptive Floating Search and Oscillating Search methods which have all been presented in this section.

The experimental comparison of various search techniques currently available has also been carried out. The Floating Search algorithms performance has been found superior to all earlier methods as demonstrated by experiments on various types of data [15] and also as evaluated by independent comparative studies by Jain and Zongker [7].

The performance of adaptive floating search has been compared with that of "classical" floating search on a number of real data. Some experimental results on two real data sets (60 and 65-dimensional), presented above, can also be found in [21]. These results demonstrate that ASFFS yielded even better results than classical SFFS (of course at the expense of longer computation time due to a more thorough search).

The oscillating search methods were also tested on a large number of different problems. Some experimental results on two real data sets (one of them 65-dimensional), presented above, can also be found in [23]. Generally we can state that the basic oscillating search finds better solutions than the simple sequential selection methods, though its computational time remains comparable with the SFS. Therefore, we find the BOS algorithm to be for simple tasks more efficient. The floating OS finds usually the best solutions, but its time requirements may be rather high.

Generally, though of heuristic nature, Floating Search methods provide either the optimal or a close to optimal solution, but also require much less computational time than the Branch and Bound method and most other currently used suboptimal strategies. The computational efficiency allows the use of floating search even for large scale FS problems. To test the different algorithms in a large search space we also considered a document recognition problem. It involved discrimination between correct and defective records of banking documents consisting of 360 optical measurements. The Floating Search methods outperformed in this problem the traditional suboptimal search methods and also yielded better results than the genetic algorithms, particularly for higher dimensions [3].

It should be stressed that, as opposed to the Branch and Bound method, the Floating Search methods (also ASFS and OS) are also tolerant to deviations from monotonic behaviour of the feature selection criterion function. It makes them particularly suited in conjunction with nonmonotonic FS criterion like the error rate of the classifier which according to a number of researchers seem to be the only legitimate criterion for feature subset evaluation.

However, when neither underlying probabilistic structures or even their functional form are known, the use of probabilistic distance measures as evaluation criterion may result in selecting features corresponding to rather poor classifier performance. For these case we have developed a completely new conceptually different approach, based on the so called approximation model, which is presented in the next section.

As for the Adaptive Floating Search methods, owing to a more thorough search than "classical" floating search, they have a potential of finding a solution even closer to the optimal one. The trade-off between the quality of solution and the computational time can be controlled by user's setting of certain parameters.

Finally, the Oscillating Search can be utilized for tuning up the solution found by any of the previous methods. The approach can be used both in cases when the quality of solution is our priority and in cases when on the other hand the priority is given to the speed of finding a reasonable solution.

6 Feature Selection by Using Mixture Model

The problem of feature selection becomes particularly difficult in the situation where the multivariate class-conditional probability density functions (pdfs) and perhaps also the class prior probabilities are either partially or completely unknown.

Mixture models, as the methods of density estimation, provide more flexible models than the simple models. The reason behind this is that mixture models are able to represent arbitrarily complex pdfs. This makes mixtures also suited for representing complex class-conditional densities in supervised learning (classification) (see e.g. [6, 9, 13, 19, 26] and references therein).

We have developed the approach to feature selection problem based on the approximating the unknown class pdfs by modified finite mixture of product parameterized components (see Pudil *et al.* [17, 16] and Novovičová *et al.* [11, 12]). In terms of the required computer storage it is considerably more efficient than nonparametric pdf estimation methods.

The method is suitable especially if the underlying distributions of some or all given classes are multimodal or they conceal the existence of subpopulations.

Our approach to FS presented in this section can serve the multiple goal of:

- learning the structure of multivariate distributions,
- identifying the most important feature variables and thus facilitating the dimensionality reduction,
- deriving automatically a decision rule based on the selected features for both dichotomic and multiclass problems.

As far as the FS problem is concerned, there are two different methods with respect to the criterion employed. Though both methods exploit a common basic approximation model, the way of selecting features is completely different. The same holds for their purpose and optimal applicability. Since it is not possible to present a detailed formalized description of the approach in this section, the reader is referred to respective original sources. However, we attempt to provide at least an outline of the common statistical model used for both the methods and then the respective methods of FS.

6.1 Modified Mixture Model with Product Components

Consider the problem of classifying an input pattern described by feature vector $\mathbf{x}$ into one of C classes. The Bayes rule for minimum error is: Assign input pattern $\mathbf{x}$ to class ω_i if

$$p(\mathbf{x}|\omega_i)P(\omega_i) > p(\mathbf{x}|\omega_j)P(\omega_j) \;\; j = 1, ..., C, \;\; j \neq i. \tag{3}$$

Although the a priori probabilities are often known or easily estimated from the training data, the pdfs $p(\mathbf{x}|\omega_i)$ are seldom specified in practice. So, it is necessary to estimate them from training data. Since we expect multimodal distributions, we seek a class of approximating distributions which are computationally tractable. One approach is based on mixture representation for pdf $p(\mathbf{x}|\omega)$ of the form:

$$p(\mathbf{x}|\theta_\omega) = \sum_{m=1}^{M_\omega} \alpha_m p_m(\mathbf{x}|\theta_m^\omega) = \sum_{m=1}^{M_\omega} \alpha_m \prod_{i=1}^{D} p_{mi}(x_i|\theta_{mi}^\omega), \tag{4}$$

in which each of the mixture components $p_m(\mathbf{x}|\theta_m^\omega)$ is chosen to be a product of marginal pdfs (or to be factorial over the individual feature variables). Therefore the mixture has the specific feature that component densities are independence models. Here M_ω denotes the number of mixture components and $p_m(\mathbf{x}|\theta_m^\omega)$ denotes the density of $\mathbf{x}$ in the ω-th class within any of the components. The α_m^ω are the relative weights for each mixture component m ($\sum_{m=1}^{M_\omega} \alpha_m^\omega = 1,\ \alpha_m^\omega > 0$ for $1 \leq m \leq M_\omega$); $\theta_\omega = \{\alpha_1^\omega, \ldots, \alpha_M^\omega, \theta_1^\omega, \ldots, \theta_{M_\omega}^\omega\}$ denotes the set of parameters for the overall mixture model with M_ω components. The number of components is assumed to be known or given.

The component densities may be of different parametric forms. It is usually assumed that all the components are of the same parametric form, for example the multivariate Gaussian.

Making substitution

$$p_m(\mathbf{x}|\theta_m^\omega) = g_0(\mathbf{x}|\theta_0) g(\mathbf{x}|\theta_m^\omega, \theta_0, \phi), \tag{5}$$

in (4), the following modification of the model (4) for ωth class-conditional pdf of $\mathbf{x}$ is introduced in the considered approach presented in [16, 11]:

$$p(\mathbf{x}|\alpha_\omega, \theta_\omega, \theta_0, \phi) = \sum_{m=1}^{M_\omega} \alpha_m^\omega g_0(\mathbf{x}|\theta_0) g(\mathbf{x}|\theta_m^\omega, \theta_0, \phi). \tag{6}$$

Each component density $g(\mathbf{x}|\theta_m^\omega, \theta_0, \phi)$ includes additional binary parameters $\phi_i \in \{0,1\}$

$$g_m(\mathbf{x}|\theta_m^\omega, \theta_0, \phi) = \prod_{i=1}^{D} \left[\frac{f_i(x_i|\theta_{mi}^\omega)}{f_i(x_i|\theta_{0i})} \right]^{\phi_i} \tag{7}$$

$$\theta_m^\omega = (\theta_{m1}^\omega, \theta_{m2}^\omega, \dots \theta_{mD}^\omega), \quad \phi = (\phi_1, \phi_2, \dots, \phi_D) \in \{0,1\}^D$$

and

$$g_0(\mathbf{x}|\theta_0) = \prod_{i=1}^{D} f_i(x_i|\theta_{0i}), \quad \theta_0 = (\theta_{01}, \theta_{02}, \dots, \theta_{0D}) \tag{8}$$

is a nonzero 'background' pdf common to all classes defined as a product of marginal densities.

The function g is actually defined on a subspace $\mathcal{X}_{(l)} \subset \mathcal{R}^l$: $\mathcal{X}_{(l)} = \mathcal{X}_{i_1} \times \mathcal{X}_{i_2} \times \ldots \times \mathcal{X}_{i_l}$, $\mathcal{X}_{i_k} \subset \mathcal{R}$, $\{i_1, \dots, i_k\}$ is the subset of $\{1, \dots, D\}$, $(k = 1, \dots, l)$ specified by nonzero binary parameters ϕ_{i_k}. The univariate functions f_i is assumed to be from a parametric family of probability density functions parameterized by θ.

The proposed model is based on the idea to posit a common 'background' density for all classes and to express each class pdf as a mixture of a product of this 'background' density with a class-specific function defined on a subspace of the feature vector space $\mathcal{X}$. This subspace is chosen by means of the parameters ϕ_i and the same subspace of $\mathcal{X}$ for each component density is used in all classes, which is distinction from the model proposed by Grim in [5]. The introduced model (6) is formally the same as the model (4) with the difference, that any specific univariate function $f_i(x_i|\theta_{mi}^\omega)$ is substituted by the 'background' density $f_i(x_i|\theta_{0i})$ whenever ϕ_i is zero. In this way the binary parameters ϕ_i can be looked upon as *control variables* due to the fact that the complexity and the structure of the mixture (6) can be controlled by means of that parameters.

For any choice of ϕ_i the finite mixture (6) can be rewritten by using (7) and (8) as

$$p(\mathbf{x}|\alpha_\omega, \theta_\omega, \theta_0, \Phi) = \sum_{m=1}^{M_\omega} \alpha_m^\omega \prod_{i=1}^{D} [f_i(x_i|\theta_{0i})^{1-\phi_i} f_i(x_i|\theta_{mi}^\omega)^{\phi_i}] \tag{9}$$

$$\alpha_\omega = (\alpha_1^\omega, \alpha_2^\omega, \dots, \alpha_{M_\omega}^\omega), \quad \theta_\omega = (\theta_1^\omega, \theta_2^\omega, \dots, \theta_{M_\omega}^\omega).$$

Setting some $\phi_i = 1$, we replace the function $f_i(x_i|\theta_{0i})$ in the product in (9) by $f_i(x_i|\theta_{mi}^\omega)$ and introduce a new independent parameter θ_{mi}^ω in the mixture (9). The actual number of involved parameters we specify by the condition

$\sum_{i=1}^{D} \phi_i = \gamma, \quad 1 \leq \gamma \leq D$. The modified mixture model (9) reduces the computational complexity of the corresponding Bayes decision rule. It is easy to see that the background density g_0 given by (8) may be reduced in the Bayes formula for a posteriori density of $\mathbf{x}$ and the same holds for the inequality (3) in the Bayes decision rule.

Using model (9), we can represent pattern by vector $\mathbf{x}_{(d)} = (x_{i_1}, ..., x_{i_d})^T$ which is a subvector of dimension d of $\mathbf{x}$, where $d < D$, $\{i_1, \ldots, i_d\}$ is a subset of the set $\{1, \ldots, D\}$. Therefore, a new pattern may be classify according to the *pseudo-Bayes decision rule:*
Decide that new pattern is from class ω_l if

$$P(\omega_l) \sum_{m=1}^{M_{\omega_l}} \alpha_m^{\omega_l} \prod_{k=1}^{d} f_{i_k}(x_{i_k}|\theta_{mi_k}^{\omega_l}) \geq P(\omega_j) \sum_{m=1}^{M_{\omega_j}} \alpha_m^{\omega_j} \prod_{k=1}^{d} f_{i_k}(x_{i_k}|\theta_{mi_k}^{\omega_j}) \tag{10}$$

$$j = 1, \ldots, C, \quad 1 \leq i_k \leq D.$$

We use the term 'pseudo-Bayes' as the rule is applied in a lower-dimensional subspace corresponding to the selected subvector $\mathbf{x}_{(d)}$. The pseudo-Bayes rule facilitates decision-making in a subspace of dimensionality d, while the remaining $D - d$ unused features form the background distribution.

6.2 Estimation of Model Parameters

Consider the ωth training set is $\mathcal{T}_\omega$ and T_ω is the number of training samples from the ωth class with unknown pdf $p(\mathbf{x}|\omega)$. It is assumed that $p(\mathbf{x}|\omega)$ may be approximated with the sufficient accuracy by a set of pdfs (6).

In order to estimate the unknown parameters of the model (6) we maximize the log-likelihood function for the training samples defined as

$$L(\alpha, \theta, \theta_0, \phi) = \sum_{\omega \in \Omega} \frac{P(\omega)}{T_\omega} \sum_{\mathbf{x} \in \mathcal{T}_\omega} \log g_0(\mathbf{x}|\theta_0) \sum_{m=1}^{M_\omega} \alpha_m^\omega g(\mathbf{x}|\theta_m^\omega, \theta_0, \phi) \tag{11}$$

by means of EM (expectation-maximization) algorithm. The two steps of this algorithm are specified as follows:

E-step: Given a parameters $(\alpha, \theta, \theta_0, \phi)$ compute for $m = 1, ..., M_\omega, \mathbf{x} \in \mathcal{T}_\omega$

$$w(m|\mathbf{x}, \omega) = \frac{\alpha_m^\omega g(\mathbf{x}|\theta_m^\omega, \theta_0, \Phi)}{\sum_{j=1}^{M_\omega} \alpha_j^\omega g(\mathbf{x}|\theta_j^\omega, \theta_0, \phi)}. \tag{12}$$

Note that each weight $w(m|\mathbf{x},\omega)$ is the posterior probability that observation $\mathbf{x}$ falls in mth component given that it is observed to fall into class ω.

M-step : Under fixed weights (12) compute the new values α, θ

$$\hat{\alpha}_m^\omega = \frac{1}{T_\omega} \sum_{\mathbf{x}\in\mathcal{T}_\omega} p(m|\mathbf{x},\omega), \quad \sum_{m=1}^{M_\omega} \hat{\alpha}_m^\omega = 1 \tag{13}$$

$$p(\mathbf{x}|m,\omega) = \frac{w(m|\mathbf{x},\omega)}{\sum_{\mathbf{y}\in\mathcal{T}_\omega} w(m|\mathbf{y},\omega)}, \tag{14}$$

$$\hat{\theta}_{mi}^\omega \quad = \quad \arg\max_\theta \{ \sum_{\mathbf{x}\in\mathcal{T}_\omega} p(\mathbf{x}|m,\omega) \log f_i(x_i|\theta)\}. \tag{15}$$

Estimation of the binary parameters $\phi_i,\ i = 1,\ldots,D$ is included in the M-step of the EM algorithm and it depends on different criteria for considering features (see [16, 12]).

6.3 Feature Selection for Approximation

Two methods have been derived depending on different criteria for features evaluation.

The approximation method selects a feature subset of d features $X_d = \{x_{i_k} \mid k = 1,\ldots,d\}$ from the set X_D of D possible features representing the pattern, $d < D$, by choosing that vector $\phi_d = (\phi_{i_1},...,\phi_{i_d})$ for which the best approximation is obtained. The criterion we use for measuring the error resulting from approximating the true pdf $p(\mathbf{x}|\omega)$ by $p(\mathbf{x}|\alpha_\omega,\theta_\omega,\theta_0,\phi)$ for all $\omega \in \Omega$ is a mixture, in the true proportions $P(\omega_1),\ldots,P(\omega_C)$, of the Kullback-Leibler distances between the true and the postulated class conditional probability density functions of $\mathbf{x}$.

The approximation method selects features with respect to their approximation or representation quality, which may not in particular cases coincide with their "discriminative" quality. Consequently, the method is particularly convenient for the problems of multivariate data representation in a lower-dimensional space or pattern interpretation. Is is also applicable to multiclass problems. For the cases when the discrimination between classes is the primary goal, the following "divergence" method has been developed.

6.4 Feature Selection for Discrimination

In order to select those features that are most useful in describing differences between two possible classes, we have developed another method [11, 12] for feature selection. Similarly to the "approximation" method it utilizes the same general model for approximating unknown class conditional pdfs by finite mixtures of parameterized densities (6). However, in this case the Kullback's J-divergence [1] in terms of a posteriori probabilities (or equivalently the Kullback-Leibler measures of discriminatory information between two classes mixed in the proportions in which the classes truly occur) is used as the appropriate evaluation criterion. The goal of the method is to maximize the divergence discrimination, hence the name of 'divergence' method.

The proposed approach is especially suitable for multimodal data and is restricted to two classes.

6.5 Properties of Approximation and Divergence Methods

These methods share two interesting characteristics:

1. Owing to the convenient form of postulated model, a feature subset's contribution to the chosen criterion is a sum of individual contributions, which can be assessed independently for each feature (see [16, 11, 12]).
2. The methods therefore require only the ranking of a sum of individual contributions without any search procedure to obtain a required subset of d features.

An important characteristic of our approach is that it effectively partitions the set X_D of all D features into two disjunct subsets X_d and $X_D - X_d$, where the joint distribution of the features from $X_D - X_d$ is common to all the classes and constitutes the background distribution, as opposed to the features forming X_d, which are significant for discriminating the classes. The joint distribution of these features constitutes the "specific" distribution defined in (8). According to these features alone, a new pattern $\mathbf{x}$ is classified into one of C classes.

It has been proved that just under this partition of the feature set X_D the following holds:

1. The "approximation" method minimizes the Kullback-Leibler distance between the true and the postulated class conditional pdfs.
2. The "divergence" method maximizes the Kullback J-divergence.

Both the methods also yield automatically a *pseudo-Bayes plug-in rule* based on the selected d features and the approximations of the true conditional pdfs: Decide that a new pattern is from class ω_l if

$$P(\omega_l) \sum_{m=1}^{M_\omega} \hat{\alpha}_m^{\omega_l} \prod_{k=1}^{d} f_{i_k}(x_{i_k}|\hat{\theta}_{mi_k}^{\omega_l}) = \max_{j=1,\dots,C}\{P(\omega_j) \sum_{m=1}^{M_\omega} \hat{\alpha}_m^{\omega_j} \prod_{k=1}^{d} f_{i_k}(x_{i_k}|\hat{\theta}_{mi_k}^{\omega_j})\}. \tag{16}$$

It means that a new feature vector is classified into one of C classes according to only d selected features $x_{i_1}, \dots, x_{i_d}$.

The proposed approach to feature selection based on approximating the unknown class probability density functions by a finite mixtures of parameterized densities of a special type is somewhat more realistic than the other parametric approaches. It is particularly useful for the case of multimodal distributions when other feature selection methods based on distance measures (e.g. Mahalanobis distance, Bhattacharyya distance), would totally fail to provide reasonable results as has been shown in [16] and [11]. The reason is that the use of these measures in their simple analytic form incorporates the additional information that pattern vectors are multivariate Gaussian, which may not be true.

Two proposed methods yield the feature subset of required size, minimizing or maximizing the used criterion, without involving any search procedure. Furthermore, they provide a pseudo-Bayes plug-in rule employing the selected features. Consequently, the problems of feature selection and classifier design are solved simultaneously.

6.6 Experimental Results of Approximation Method

Experiments on various sets of simulated and real-world data have been carried out, though only the results on texture data are demonstrated here. In all the experiments the equal *a priori* class probabilities were considered and separate training and test sets were used.

The error rate of the pseudo-Bayes classifier (16) has been compared with the error rate of the Bayes classifier using the features selected by the Branch and Bound algorithm with the Bhattacharyya distance for multivariate Gaussian distribution as the feature selection criterion.

Texture Data

A number of different images have been tested but two colour images from a project at the University of Surrey were specifically chosen since they are not well separated in the measurement space. They correspond to "rosa baveno" and "bianca castilla" types of marble.

Each colour image was divided into two halves, the top half was used for training and the bottom half was used for classification propose. The size of each image was 256×256 from which sample sub-images were selected with window of size 100×100 randomly placed. Then a 26-dimensional feature vector was extracted from each sub-image where the first 8 features are texture features and the remaining 18 features are colour features. The texture features were derived from the discrete cosine transform (DCT) filter of size 3×3. The colour features were gathered from the 3-dimensional histogram model of the colour texture from which the statistical description in the form of energy, entropy, local homogeneity, inertia, mean and variance were used. The sample size for both training and test sets were 1000. Because of the high dimensionality of the original measurements of this particular experiment, when the distributions are assumed to be Gaussian, the sequential forward floating selection (SFFS) method (see [15]) is used to select features. The use of the Branch and Bound algorithm is not feasible since it requires a large memory and long computational time.

The results obtained with the approximation method of subset selection for mixtures consisting of 2, 3 and 4 components are shown in the following Table 3. The quality of the selected subset of features has been tested by the classification error on the independent test set (not used for training). The above discussed pseudo-Bayes decision rule derived simultaneously by the method and based only on the selected features has been used. The estimate of the probability of error for the subset of i selected features is denoted $\mathrm{Pe}(X_i)$.

In order to compare our approach with a "conventional" one, a multivariate Gaussian classifier was used as an alternative. The features have been selected in this case, using the Bhattacharyya distance for the subset evaluation, by means of the optimal Branch and Bound (BB) algorithm that finds the optimal solution to maximize the criterion. Please, note that though for this another approach $\widetilde{X}_i$ is also a subset of the size i, it is generally different from the subset X_i selected by the approximation method.

From Table 3, it is obvious that the Gaussian classifier with features selected by BB algorithm gives rather big errors as apparently the assumption that the distributions are unimodal Gaussian is not appropriate. In contrast, when a mixture of normal densities is used, a much smaller error rate is obtained. A substantial dimensionality reduction can be made without significantly deteriorating the classifier performance. In other words,

METHOD	Estimate of probability of classification error			
1. Pseudo-Bayes classifier based on Approx. method	Pe(X_6)	Pe(X_{10})	Pe(X_{14})	Pe(X_{26})
Mixture of 2 components	0.100	0.064	0.030	0.030
Mixture of 3 components	0.065	0.028	0.016	0.007
Mixture of 4 components	0.055	0.012	0.007	0.007
2. Gaussian classifier with features selected by BB	Pe($\widetilde{X}_6$)	Pe($\widetilde{X}_{10}$)	Pe($\widetilde{X}_{14}$)	Pe($\widetilde{X}_{26}$)
	0.235	0.287	0.169	0.169

Table 3: *Classifier performance on different feature subset size of image data*

redundant features have been detected especially with the mixture of four components. This can be clearly seen for the subsets of 10 and 14 features.

6.7 Experimental Results of Divergence Method

The divergence method was tested on synthetic and real data though only the results achieved on real data speech recognition are described in the sequel. Its performance is compared with that of the ordinary multivariate normal model and in the case of speech data also with that of the "approximation model" [14, 16]. Owing to different assumptions about the underlying probability distributions, the comparison has been made using the misclassification rate obtained from both types of classifiers, i.e. classifier based on the mixture of normal distribution (the pseudo-Bayes classifier) and the classifier based on multivariate normal distribution (the Gausian classifier). Separate training and test sets were used in all experiments. In addition, the *a priori* probabilities in all experiments were taken to be equal for all classes. Note that when the unknown distribution is assumed to be unimodal multivariate normal, the best feature subset of size d is selected from the set of D available measurements using the branch and bound algorithm with the Bhattacharyya distance as the criterion function, unless stated otherwise.

Speech Data

The data used to train, test and compare the proposed method was a set of 1418 pattern vectors of the utterances "YES" and "NO" spoken over the public switched telephone network. Each 15-dimensional feature vector contained 5 segments of 3 features derived by low order linear prediction analysis. From this set, 798 samples were used for training and the remaining 620 samples to form the test set. Both sets contain nearly equal number of samples for each pattern class. The data was supplied by British Telecom. The results of the experiment using 2, 3, 4 and 5 mixture components are shown in Fig. 13.

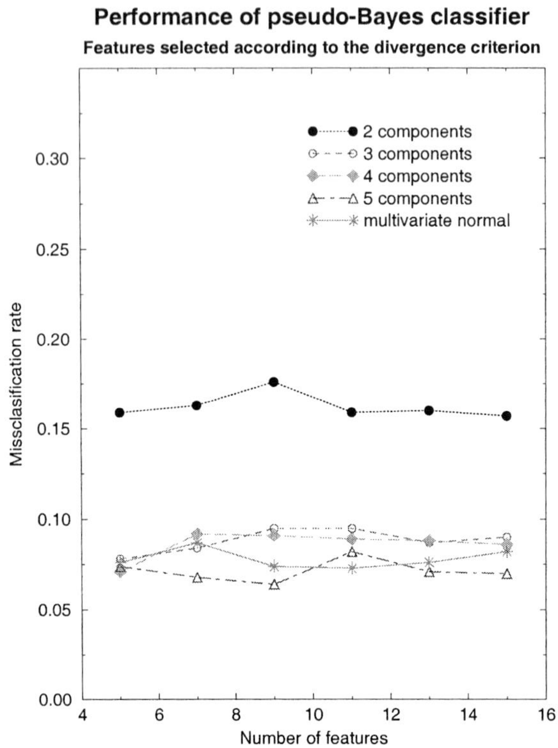

Figure 13: *Performance of pseudo-Bayes classifier*

As reported already in [17], a detailed analysis has shown that the data lies in narrow regions along parallel hyperplanes, i.e. the data are highly correlated and unimodal. It is therefore pertinent to use the Gaussian classifier in this case. To approximate these distributions by a mixture of independent Gaussian distributions properly would require a higher number of mixture components. However, owing to a small sample size of the training set, this is not possible. For this reason the feature selection method based on pdf approximation [17] failed to provide good results in this case. The comparison with results reported in [14, 16] show that the new method based on divergence outperforms in this case the method based on approximation. Moreover, though the data are very favourable for the Gaussian classifier, the divergence method yields comparable results and in the case of 5 components even better ones. A higher number of mixture components is likely to perform better because of the larger flexibility in fitting multimodal pdfs with different widths of the individual modes. However, the training set has to be of reasonable size compared to the number of parameters of the mixture components to be estimated.

7 Summary

This chapter has been devoted to the problem of feature selection in statistical pattern recognition which is considered to be one of two key stages in the whole process of pattern recognition. The other one is classification itself, which has to be usually preceded by feature selection resulting in dimensionality reduction of feature vectors. The main contributions of the research reported in this chapter can be summarized as follows:

- The current state of art in reduction of dimensionality for decision problems of classification type has been critically analyzed. A number of feature subset search strategies has been reviewed, analyzed and compared. Following the analysis of their respective advantages and shortcomings, the conditions under which certain strategies are more pertinent than others have been established. Moreover, the concept of *a priori* knowledge based approach to feature selection has been presented.

- Based on a detailed study of *the Branch and Bound* (BB) algorithm principle, new algorithms for optimal subset search have been defined. Their prediction mechanism allows significant time savings when compared to the "Basic" or "Improved" Branch and Bound algorithms. The algorithms were experimentally proved to be robust and well suitable for use with different criterion functions, both in recursive and non-recursive form. The first one, denoted *Fast Branch and Bound* (FBB) algorithm, aims to reduce the number of criterion function computations in internal search tree nodes. The algorithm requires the criterion function to fulfill the *monotonicity condition* just like the BB algorithm. Although the FBB algorithm requires usually several times less criterion computations than any other BB algorithm, its suitability for many practical problems is limited if the recursive criterion forms are to be used. To resolve this limitation we have designed a new, more universal algorithm, denoted *Branch and Bound Algorithm with Partial Prediction.* As opposed to the Fast Branch and Bound algorithm, the new algorithm is well suitable for use with recursive criterion forms. Though the new algorithm does not operate as effectively as the FBB algorithm, it is able to find the optimum significantly faster than any other Branch and Bound algorithm. Though the algorithm has been described in the context of feature selection, its usability may be without doubt much broader, similarly to the known "classical" Branch and Bound algorithm.

- Floating Search algorithms which yield close to optimal solution, are computationally effective (facilitating FS in high dimensional problems) and do not require the fulfillment of monotonicity condition have been described. In a recent comparative study of currently available subset search strategies carried out by A. Jain and D. Zongker the Floating Search algorithms were evaluated as the most powerful ones. The subsets found by Floating Search are practically identical with those found by a computationally tedious "exhaustive" combinatorial search. Comparison with the Branch and Bound method showed that Floating Search is much faster. The paper [15] is widely cited in the literature.

- Further extensions of floating search principle have lead to the design of adaptive floating search algorithms for feature selection. They are able to achieve the results even closer to the optimal ones than "classical" floating search, though usually at the expense of prolonged computational time. The oscillating search algorithm is another extension of the battery of available tools. It is particularly convenient for "tuning up" the solution found by other methods.

- Two feature selection methods suitable for cases, when no *a priori* information on underlying probability structures is known, have been developed and successfully employed. They are based on approximating the unknown class conditional densities by finite mixtures of the densities of a special type. One of them - the *approximation method* is primarily designed for feature selection aimed at optimal data representation, the other one - *divergence method* is designed for classification - in order to enhance the discriminative ability of features.

Future Development

Though it is certainly too premature to claim a real development of the knowledge guided approach to subset selection, the presented methods can serve as a starting point to achieve such an ambitious goal. To implement it, we are developing an integrated environment which would incorporate a whole family of methods together with a sort of expert system that should guide a user according to his goals and the degree of the problem knowledge he has to the semi-automatic choice of the most suitable method. The core of the system is represented by the software package FS Toolbox (Feature Selection Toolbox) for solving feature selection tasks, the first version of which has already been completed [25, 18].

Acknowledgements

The work has been supported by grants of the Czech Ministry of Education CEZ:J18/98:31600001, the Grant Agency of the Czech Republic No 402/01/0981 and the Complex research project of the Academy of Sciences of the Czech Republic K1019101.

References

[1] D.E. Boekee and J.C.A. van der Lubbe, Some aspects of error bounds in feature selection, *Pattern Recognition* Vol.11 (1979) pp. 353–360.

[2] P.A. Devijver and J. Kittler, *Pattern Recognition: A Statistical Approach*, (Prentice-Hall, 1982)

[3] F.J. Ferri, P. Pudil, M. Hatef, and J. Kittler, Comparative study of techniques for large-scale feature selection, in E.S. Gelsema and L.N. Kanal (eds.) *Pattern Recognition in Practice IV* (Elsevier Science B.V., 1994) pp. 403-413.

[4] K. Fukunaga, *Introduction to Statistical Pattern Recognition*, (Academic Press, Inc., 1990).

[5] J. Grim, Multivariate statistical pattern recognition with nonreduced dimensionality, *Kybernetika* Vol.22 No.2 (1986) pp.142–157.

[6] T. Hastie and R. Tibshirani, Discriminant analysis by Gaussian mixtures, *Journal Royal Statist. Soc. B* Vol. 58 (1996) pp. 155-176.

[7] A.K. Jain and D. Zongker, Feature selection: Evaluation, application and small sample performance, *IEEE Transactions on Pattern Analysis and Machine Intelligence* Vol.19 No.2 (1997) pp. 153-158.

[8] M. Kudo and J. Sklansky, Comparison of algorithms that select features for pattern classifiers, *Pattern Recognition* Vol.33 No.1 (2000) pp. 25-41.

[9] G.J. McLachlan, *Discriminant Analysis and Statistical Pattern Recognition*, (New York, John Wiley&Sons, 1992).

[10] P.M. Narendra and K. Fukunaga, A branch and bound algorithm for feature subset selection, *IEEE Transactions on Computers* C-26 (1977) pp. 917-922.

[11] J. Novovičová, P. Pudil, and J. Kittler, Divergence based feature selection for multimodal class densities, *IEEE Transactions on Pattern Analysis and Machine Intelligence* Vol.18 No.2 (1996) pp. 218-223.

[12] J. Novovičová and P. Pudil, Feature selection and classification by modified model with latent structure, in: M. Karný, K. Warwick, and V. Kurková (eds.) *Dealing With Complexity: Neural Network Approach* (Springer Verlag, 1997) pp. 126-140.

[13] H.Ch. Palm, A new methods for generating statistical classifiers assuming linear mixtures of Gaussian densities, in: *Proceedings of the 12th International Conference on Pattern Recognition*, Jerusalem (1994) pp. 483-486.

[14] P. Pudil, J. Novovičová, N. Choakjarernwanit, and J. Kittler, Selection of features optimizing a finite mixture approximation of nonparametric class densities, University of Surrey, Technical Report, VSSP-TR-3/92, November 1992.

[15] P. Pudil, J. Novovičová, and J. Kittler, Floating search methods in feature selection, *Pattern Recognition Letters* Vol.15 No.11 (1994) pp. 1119-1125.

[16] P. Pudil, J. Novovičová, and J. Kittler, Feature selection based on approximation of class densities by finite mixtures of special type, *Pattern Recognition* Vol.28 (1995) pp. 1389-1398.

[17] P. Pudil, J. Novovičová, and J. Kittler, Simultaneous learning of decision rules and important attributes for classification problems in image analysis, *Image and Vision Computing* Vol.12 (1994) pp. 193-198.

[18] P. Pudil, J. Novovičová, and P. Somol, Feature selection toolbox software package, *Pattern Recognition Letters* Vol.23 No.4 (2002) pp. 487-492.

[19] B. Ripley, *Pattern Recognition and Neural Networks*, (Cambridge, Mass., Cambridge University Press, 1996).

[20] W. Siedlecki and J. Sklansky, On automatic feature selection, *International Journal of Pattern Recognition and Artificial Intelligence* Vol.2 No.2 (1988) pp. 197-220.

[21] P. Somol, P. Pudil, J. Novovičová, and P. Paclík, Adaptive floating search methods in feature selection, *Pattern Recognition Letters* Vol.20 No.11,12,13 (1999) pp. 1157-1163.

[22] P. Somol, P. Pudil, F.J. Ferri, and J. Kittler, Fast branch & bound algorithm in feature selection, in *Proceedings of the 4th World Multiconference on Systemics, Cybernetics and Informatics SCI 2000*, Orlando, Florida, Vol VII Part 1 (2000) pp. 646-651.

[23] P. Somol and P. Pudil, Oscillating search algorithms for feature selection, in Proceedings of the 15th IAPR International Conference on Pattern Recognition, (Barcelona, 2000) pp. 406-409.

[24] P. Somol, P. Pudil, and J. Grim, Branch & bound algorithm with partial prediction for use with recursive and non-recursive criterion forms, in *Lecture Notes in Computer Science LNCS 2013* (Springer 2001) pp. 230-239.

[25] P. Somol and P. Pudil, Feature selection toolbox as a multi-purpose tool in pattern recognition, in A. Fred and A.K. Jain (eds.) *Pattern Recognition in Information Systems Proceedings of the 1st International Workshop on Pattern Recognition in Information Systems* (Setbal, ICEIS Press, 2001) pp. 91-102.

[26] A. Webb, *Statistical Pattern Recognition*, (London, Arnold, 1999).

[27] B. Yu and B. Yuan, A more efficient branch and bound algorithm for feature selection, *Pattern Recognition* Vol.26 (1993) pp. 883-889.

Fast Image Segmentation under Noise

RosaMaria Romano
Istituto per le Applicazioni del Calcolo
Consiglio Nazionale delle Ricerche, Rome, Italy
E-mail: romano@iac.rm.cnr.it

Domenico Vitulano
Istituto per le Applicazioni del Calcolo
Consiglio Nazionale delle Ricerche, Rome, Italy
E-mail: vitulano@iac.rm.cnr.it

Contents

D. Chen and X. Cheng (eds.),
Pattern Recognition and String Matching, 617–646.

References

1 Introduction

Image segmentation is one of the most interesting and difficult topics within image processing and a big effort has been devoted to it by many researchers. Its goal consists of determining the image components according to some criteria, in agreement with the human eye working [3, 4, 5, 26]. This is a very difficult task but, at the same time, necessary. Identification of object boundaries in a given scene becomes more and more important within computer vision. In fact, image segmentation constitutes a starting point for a wide class of higher level processes such as object recognition, image classification and so on, in which inaccurate edge detection invalidates the results of a successive computation.

The large number of approaches proposed for trying to solve the image segmentation problem led the scientific community to propose the following classification [21, 39]:

- models designed for detecting image discontinuities. They are based on detecting edges, i.e. sharp changes of image intensity;
- models oriented to detect uniform regions according to one or more criteria. Edges will correspond to the boundaries (or contours) of these regions.

Moreover, the increasing development of this field on the basis of more complicated approaches, gave rise to alternative and more detailed classifications. In other words, though *rough*, these classifications show an increasing difficulty in embracing all different approaches in literature. For example in [54] Zhu and Yuille proposed the following taxonomy:

- *Local filtering.* This approach is based on a local information of a given scene and then has the drawback of producing not closed contours.
- *Region growing and merging.* These approaches consist of splitting an image in different and non overlapping regions, using a given criterion of homogeneity. They may produce irregular edges.

- *Snake and Balloon.* Techniques belonging to this class are based on using curves moving on the image (usually called *active contours*) by means of forces based on local information and acting on the normal of the curve points. The minimization of a suitable energy should lead these curves to lie on the right position. A smooth contour, along with problems tied to the aforementioned minimization, such as necessity of good initial conditions, constitute the main disadvantages;

- *Global optimization.* Unlike the three classes mentioned above, this approach is based on using a global energy whose minimum consists of image contours. It theoretically leads to a better performance because it uses the global information of the considered image; on the other hand it suffers from all the difficulties derived from the minimization of the above energy.

An immediate criticism may be that there are too many classifications and each is based on a different point of view. However, the previously mentioned taxonomy allows us to understand the complexity and the diversity of this field. It is then clear that an exhaustive presentation cannot be dealt with. In this chapter we will focus on image segmentation based on *variational models.* Even though we are focusing on a particular class of approaches a certain degree of generality will be required.

2 Image Segmentation under Noise

Variational approaches are tied to image segmentation under noise. The latter is a well known problem in the image processing field and can be formally written as (see [21] p. 268 and [46]):

$$f = hu + \eta \tag{1}$$

where $f : \Omega \subseteq R^2 \to R$ is the noisy image, u the noiseless one (to be determined) while h and η are two kinds of distortion. We outline that this is an ill-posed problem. Strictly speaking, a problem is ill-posed in the sense of Hadamard [15], if its solution does not satisfy one or more of the following constraints:

- existence;

- uniqueness;

- stability.

In the following we will assume that $h = I$, where I is the identity operator and η is an additive noise with zero-mean gaussian distribution. In other words we will consider our data without blur usually due to different reasons such as camera motion during image acquisition, atmospheric turbulence and so on. This will seem an over simplification but it is not so [21]. The problem remains consistent and a lot of papers have been devoted to it.

Different approaches have been used in literature to face this problem and the most widely known are:

- *Wavelet-based approaches.* They are based on the Wavelet Transform computation, which is a linear step, followed by non-linear operations on its coefficients. There are two main approaches.
  - In the first one, Mallat tries to detect good discontinuities, discarding the spurious ones due to noise, by computing local maxima of the WT's decomposition coefficients[28]. This approach allows us to achieve not only restoration, i.e. the recovered function u in (1), but also edge detection (i.e. u's discontinuities).
  - In the second one, WT's coefficients are suitably thresholded; the resulting signal is then processed by the Inverse Wavelet Transform to obtain the de-noised signal. In this case only restoration is achieved and, if the threshold is suitably selected, the result has been shown to be optimal in the min-max sense (see chap. 10 of [29]). Segmentation may be performed as further step.
- *Variational based approaches.* In this case restoration and segmentation are dealt with in a single step by minimizing a suitable energy, as will be more clear in the following.

The remainder of the chapter will focus on the second class of approaches.

3 Segmentation as Weak Continuity Problem

Traditionally, *edge detection* has been seen as computation of singularity points of the gradient of the image function. In particular, if we denote our image by $f : \Omega \subseteq R^2 \rightarrow R$, its singularities will correspond to maxima for its (suitably computed) first derivatives and zero crossing for its second derivatives. Some well known examples, widely used to this task, were Sobel, Prewitt and Roberts filters as gradient operators; the Laplacian one was used as second derivative [21].

However, in the early Eighties a great conceptual step forward was made by understanding that many image processing problems can be seen as inverse problems. In particular edge detection and then derivative operation can be considered as a solution of the inverse (and then ill-posed) problem:

$$f(x) = \int_{-\infty}^{x} g(x)dx. \tag{2}$$

Moreover, since such an operation is not robust under noise, edge detection problem can be stated as:

$$\underbrace{||\int_{-\infty}^{x} g(x)dx - f(x)||^2}_{\text{faithfulness to the data}} + \lambda \underbrace{||T[g(x)]||}_{\text{regularizing term}} \tag{3}$$

where $||\cdot||$ is a norm and $T[\cdot]$ is a regularizing functional. The first term of (3) represents the faithfulness to the data while the second one is regularizing. The parameter λ balances these two requirements.

Thus, the solution is a constrained minimum of the aforementioned problem and then is a *variational approach.*

Variational approaches look to be the most powerful tool for both restoration and detection because they allow us to deal with noise and, at the same time, they preserve discontinuities characterizing the image information.

The first functional which was able to achieve both restoration and detection was the Mumford-Shah one [35]:

$$E(u, K) = \int_{\Omega} (u(x) - f(x))^2 \, dx + w_1 \int_{\Omega \setminus K} \nabla u \cdot \nabla u \, dx + w_2 \int_K ds, \tag{4}$$

where u is the first unknown of the problem and represents the restored image, while K is the set of discontinuities of the image and then the second unknown of the problem. It is easily noticed that the functional in (4) is very close to the form in (3). In fact, the first term of (4) represents again faithfulness of u to data (i.e. to the input image f); the second term is regularizing since producing a selective smoothing. In other words, smoothing is allowed only where discontinuities (i.e. K's points) are not.

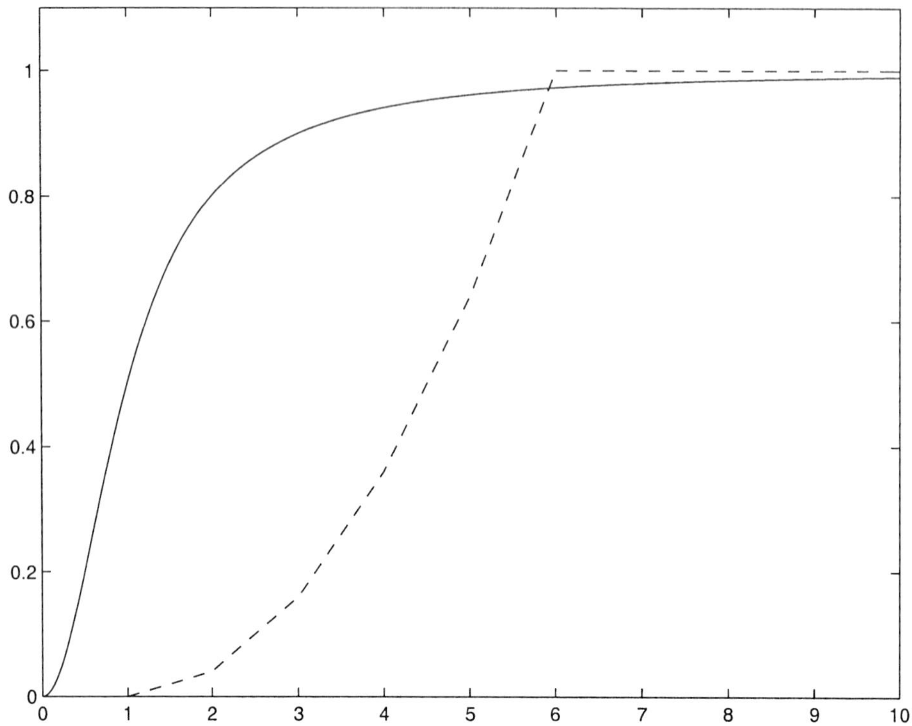

Figure 1: Two examples of non convex regularizing functions: a) the truncated parabola of the Mumford-Shah functional (dashed line), b) Geman-McClure one (solid line).

Finally the third term provides a penalization on the length of discontinuity curves. w_1 and w_2 are weight factors providing a good compromise among these different components.

Problems related to both edge detection and restoration are sometimes called *edge-preserving regularization* or *weak continuity*. In particular, the acronym *weak continuity* is due to the fact that the Mumford-Shah functional behavior may be seen as a *weak string* (in 1-D) or *weak membrane* (in 2-D); weakness consists of breaking these elastic frameworks in correspondence to the discontinuities of the input signal (1-D) or image (2-D) [6].

The resulting functional is not convex and Blake and Zissermann showed that second and third term are equivalent to a regularizing function whose shape is a truncated parabola [6], as shown in Fig. 1. They proposed a clever algorithm to compute the minimum: *GNC* (Graduated Non Convexity).

It is well known that non-convexity leads to the use of more complicated algorithms because of the presence of many local minima.

GNC consists of building a sequence of functionals where the first one is convex, while the last one is the *original.* The intermediate functionals are built by a gradual deconvexification of the convex functional till converging to the last one (true functional). The minimum, i.e. the solution of the first (convex) functional is easily achieved and constitutes the starting point for the minimization of the second functional. The process is iterated for all functionals of the aforementioned sequence. This algorithm is a valid alternative to *Simulated Annealing* since, though it does not guarantee the absolute minimum computation (but good enough), it is very efficacious.

In one sense the Geman-McClure functional [13, 14] can be seen as an evolution of the discrete approximation by a truncated parabola of the Mumford-Shah funtional. Regularizing term is plotted in Fig. 1. It can be written as follows:

$$E(u) \quad = \quad \underbrace{\int_\Omega (u(x) - f(x))^2 \quad dx}_{\text{faithfulness to the data}} \quad + \quad \underbrace{w^2 \int_\Omega \varphi(|\nabla(u)|) \quad dx}_{\text{regularizing term}} \tag{5}$$

where $|\nabla(u)|$ is the modulus of the gradient of u, w is a weight factor balancing these two components, while:

$$\varphi(t) = \frac{t^2}{1+t^2}. \tag{6}$$

The functional of Geman and McClure can also be seen as an evolution of previous approaches oriented to the only restoration. In fact, many choices for the function φ had previously been proposed in literature and a useful review is in [46].

For instance, in the standard regularization of Tikhonov [47], a possible choice is $\varphi(t) = t^2$, restricting the class of admissible functions to a Sobolev space containing smooth functions. Such a choice entails a global smoothing and then discontinuities representing the information contained in a pictorial scene are destroyed. With regard to total variation, where $\varphi(t) = t$ [9], resulting edge preserving is not so effective.

When the function φ results *half-quadratic*, (5) yields only restoration: non-convexity allows us to obtain an effective regularization and edge detection.

It is obvious that dealing with edge detection and restoration increases the complexity of the problem more so than dealing with the only restoration one. In fact, it can be shown [9, 23, 46] that a regularizing function, that is convex and linear at infinity, guarantees a unique solution in a Bounded Variation functions space. Nevertheless the use of a non convex φ allows us to get a selective smoothing preserving the information contained in the *good* discontinuities, but it also represents an ill-posed problem: the existence of a minimum is guaranteed only in the discrete version of the functional, while stability and uniqueness are still an open problem.

The Geman-McClure functional is non-convex and can be minimized by GNC as the Mumford-Shah one [10]. However, it can be also minimized by a gradient descent method on its Euler-Lagrange equations as well as with the following (dynamic) scheme:

$$\begin{cases} \frac{\partial u(t,x)}{\partial t} = [f(x) - u(t,x)] + w \quad \text{div}\{\frac{\varphi'[|\nabla u(t,x)|]}{2|\nabla u(t,x)|} \nabla u(t,x)\} \\ \frac{\partial u}{\partial n}\Big|_{\partial\Omega} = 0 \end{cases} \tag{7}$$

where t is a parameter to achieve minimization while n is a vector normal to the boundary $\partial\Omega$ of the image domain Ω.

The computational aspect, again, helps us to understand the successive evolution: the Perona-Malik approach based on *heat equation*. In fact, in [36], the problem is tied to the anisotropic diffusion that allows us to recover edges of the original image without blurring them. In fact, this model prefers to produce a smoothing inside regions more than their boundaries. The problem is then to determine the solution of the following equations:

$$\begin{cases} \frac{\partial u(t,x)}{\partial t} = \text{div}[c|\nabla u(t,x)|)\nabla u(t,x)] \\ \frac{\partial u}{\partial n}\Big|_{\partial\Omega} = 0 \\ u(0,x) = f(x). \end{cases} \tag{8}$$

where $c|\nabla u(t,x)|$ is the conduction coefficient while t is now a scale parameter.

It is easy to see that the above equation is similar to (7) except for the term relative to faithfulness to data.

The main drawback of this approach is tied to the *termination time*. In fact, since the problem has an only trivial solution [53], the time chosen for stopping the energy minimization process is fundamental to achieving a good performance. This problem is usually solved by heuristic methods.

Recently, Liu, Wang and Ramirez proposed an interesting approach [25] to achieving better results than in [36]. It consists of a two step model where the first phase is based on a non-linear smoothing inside an *a priori* fixed number of windows centered in each pixel. The smoothing operation is performed only if the probability of *smoothness* is high for all windows, otherwise it is performed in only one direction — if close to an edge. In the second step a simple Sobel filter is performed on the resulting image. The major drawback is a long computing time and poor results on textures.

We outline that the path drawn till now is just one possible way to read the effort made by the scientific community by means of a variational approach. It is evident that for each of the previously mentioned steps, there are many subtle and deep aspects that allowed researchers to take one way rather than others. For instance, any new step forward made in this field, may be read from a statistical as well as a deterministic point of view. These two approaches, though seemingly in competition, have helped each other to achieve better results. This is to say that a *complete history* is much more complicated and does not constitute the topic of this chapter.

4 The Concept of Texture

It is evident that for real pictorial scenes, only using edge detection is not enough for a complete understanding of the image under study. Textures can be observed in a lot of objects of our ordinary life such as grass, wood and so on. Nevertheless, although the concept of texture is of great interest in many fields such as remote sensing, materials study, medicine and so on, there is no precise definition for it yet. This difficulty stems from the fact that textures are easily discriminated by the human eye but difficult to be formally defined. This aspect was extensively studied in the early part of this century by Gestalt psychologists [51] and successively by many other researchers in neurophysiology as well as experts in computer science. For instance, in [45] Tamura et al. tried to understand the features corresponding to the human visual perception. They listed the following: *coarseness, contrast, directionality, line-likeness, regularity* and *roughness*. Other studies have found that Gabor filters seemed to follow human visual system at early stages [8, 22].

The definition of texture is based on the concept of *textel*, i.e. texture element. Textures are seen as repetition of this basic element, which in turn is composed of image pixels whose placement is random and/or periodic. It is then evident that natural textures result nearly random while the synthetic ones deterministic.

A more detailed analysis of the aforementioned definition of texture leads to defining three basic components of textures: *grey level primitives*, i.e. textels, characterizing texture grain usually called coarseness, *spatial primitives*, i.e. displacement rules of textels and *resolution* at which texture is acquired.

With regard to the latter it is worth outlining that it has a fundamental role in texture analysis since many textures, though different, may result equal at a given scale level. It is well known Marr's sentence: "*textures live at different scale levels*" [30]. The concept of scale generalizes, in a sense, concepts of micro and macro-structure of textures. Some authors are used to outlining this point, since many practical problems are dealt with a specific scale level in order to extract interesting features.

In the following we will deal with textures analysis. Such approaches can be split in two wide classes (for the sake of brevity, the authors have selected some basic references, without considering evolutions):

- *Statistical Approaches.* Approaches belonging to this class are mainly based on grey level features of a texture more than its spatial ones. They are usually statistical. Approaches belonging to this class are based on **Autocorrelation function** ([18] etc.), **Edge density**([19, 42] etc.), **Image transforms** ([48] etc.), **Co-occurrence matrix** ([16] etc.), **Histogram features** ([21] etc.), **Gabor filters**([11] etc.), **Probabilistic spatial dependence** ([31, 38] etc.) and so on.

- *Structural Approaches.* In this case the spatial placement of textels is mainly taken into account. Approaches based on **Edge density** ([43] etc.), **Relative extrema** ([33] etc.), **Mosaic models** ([2] etc.), **Syntactic models** ([27] etc.), **Fractal dimension** ([1] etc.) and so on, belong to this class.

The above review could seem not complete but, as for edge detection, the large literature relative to texture analysis does not allow us to give here a deep and exhaustive review. Moreover, many approaches result combinations of two or more simpler approaches, or little variants of them. Some basic reviews can be found in [12, 16, 49].

5 Variational Approaches for Texture Segmentation

After the previous review about some variational approaches to texture segmentation, it is clear that it is not easy to get these two fields to meet. In fact, both variational formulation for edge detection and texture analysis constitute two wide and complicated fields of image processing with many open problems. Thus, although an extensive and deep study on edge detection, a little has been done for texture segmentation through variational frameworks. On the other hand the non precise definition of texture seems not suitable for a strongly theoretical and precise formulation as the variational one. Nevertheless some attempts have been made by the scientific community as the following ones.

In [20], Koepfler, Lopez and Morel proposed a very fast algorithm with a pyramidal architecture, able to compute a hierarchy of segmentations. Starting from the hypothesis that textures have to be analyzed at different scale levels, a particular case of the Mumford-Shah functional is used, where regularizing term is not. The reader may consider the energy in (4) where w_1 tends to infinity. Thus w_2 becomes the parameter relative to the scale level. Moreover, the model is vectorial since performed on the output of a pre-defined number of filters corresponding to selective features. The model consists of performing segmentations at different scale levels, i.e. at different values of w_2. For each step a merging operation is performed on two different regions O_i, O_j if the following energy is decreasing:

$$E(K \backslash \partial(O_i, O_j)) - E(K) = \frac{|O_i||O_j|}{|O_i| + |O_j|} ||u_i - u_j||^2 - w_2 \int_{\partial(O_i,O_j)} ds. \quad (9)$$

In this case the energy $E(K)$ only depends on K, since for each region the value of the recovered function u is constant. The corresponding algorithm results very fast, i.e. with an *O(n ln n)* complexity where n is the number of pixels of the input image.

In [24] Lee, Mumford and Yuille combine the classical variational approach with a Gabor-wavelet based representation (*Coupled-Membrane model*). In fact, it is well known that such wavelets give a representation very similar to the visual cortex one, resulting very efficient in segmenting textures. The energy to minimize is then:

$$E(u, K) = \int_{\Omega} (u(x, y, \sigma, \theta) - Wf(x, y, \sigma, \theta))^2 \, dx \, dy \, dlog(\sigma) \, d\theta \; +$$

$$\int_{\Omega\setminus K}(w_\sigma\frac{\partial u}{\partial log\sigma})^2+(w_\theta\frac{\partial u}{\partial\theta})^2+(w_1\frac{\partial u}{\partial x})^2+(w_1\frac{\partial u}{\partial y})^2\,dx\,dy\,dlog(\sigma)\,d\theta\;+$$

$$w_2\int_K ds,(10)$$

where the differences with (4) are evident. The input data is now $Wf(x,y,\sigma,\theta)$, i.e. the wavelet transform of the image f. The whole problem is then four-dimensional (Ω is four dimensional now), with two spatial and two frequency variables. In particular the two frequency variables σ and θ represent respectively dilation and rotation of a given filter.

The main drawback of this approach is a very high computational effort due to four-dimensional coupled nature of the proposed formulation.

The most famous approach is the *Region Competition* one proposed by Zhu and Yuille [54]. For the sake of brevity we will not give the complete and very elegant mathematical formulation of this approach here. Nevertheless the underlying idea is very simple and, at the same time, attractive. The initial condition consists of putting a curve in the domain of the input image. On each point of this curve two opposite forces, due to the local statistics of the image of both sides of the curve itself, will act. Thus, there is a motion of the curve over the domain till equilibrium. The whole energy is designed to make curves to stay on boundaries of homogeneous regions. Again, homogeneity will depend on selected statistics. The final result will be achieved putting many curves over the domain and minimizing the corresponding energy.

This model is very attractive since deals with many open problems in image segmentation such as *shape* and *size* of the regions where to compute local statistics moving the aforementioned curves. We will see that also in the alternative approach for textures segmentation that will be presented later, we will have the same problems. Coming back to the *region competition* approach, its main drawback is the *a priori* knowledge of the number of different regions of the input image (*seeds* problem). More precisely, if one does not put a sufficient number of curves, some regions will not be detected. On the other hand, putting a number of curves greater than the true number of regions will lead to an additional computational effort.

To the best of the authors' knowledge, all proposed approaches missed a detailed analysis of the textures behavior under noise. In other words, there is no study geared to understanding how texture structure can change while corrupted by noise. The reader can easily understand that noise makes changes in textels, but in general, in the structure of the texture at any scale level. Thus variational frameworks for textures segmentation have to be designed to be invariant under noise or, at least, to minimize noise effects. To this aim, in the following of the chapter *VRENS* (Variational Representation for an Efficient under Noise Segmentation) will be presented, as first attempt to understand what may be the most suitable features to segment textures under noise [40, 41].

6 A Variational Representation to Segment Images under Noise

Let us suppose that an image showing a real scene is to be segmented. There will be some *quite flat* regions, i.e. regions that can be easily dealt with by a usual edge detector, and other texture ones. Then, a functional like the Mumford-Shah one will be able to detect the discontinuities with a given threshold (tied to w_2 tuning) and at a given scale level (tied to w_1 threshold). It is worth outlining that discontinuities of the continuous case are equivalent to finite differences with an amplitude greater than a given threshold in the discrete case. These edge points are very important since they are relative to objects in the pictorial scene and, above all, they are not due to noise. In the following of the section we will see how to build measures based on them and their behavior for increasing noise.

The result of the Mumford-Shah functional consists of both the recovered image u and the set K of *edge points* at a given scale level. In the following we will assume that the domain Ω is discretized by a suitable mesh and then also u and K. These latter allow us to introduce the following characteristic function $\chi : \Omega \to R$:

$$\chi(x,y) = \begin{cases} 1 & \text{if } (x,y) \in K \\ 0 & \text{otherwise.} \end{cases} \tag{11}$$

It is then possible to define its *colored* version:

$$\tilde{\chi}(x,y) = u(x,y)\chi(x,y) \qquad x,y \in \Omega. \tag{12}$$

$\tilde{\chi}(x,y)$ is a representation (still an image) of the original image where only the edge points contribute to information.

Moreover, let W_1 and W_2 be two regions of Ω. In the following we will call them *windows* because we are interested in looking at image information inside of them. Be W_1 and W_2 such that $\partial W_1 \bigcap \partial W_2 = \gamma \neq \emptyset$, where with

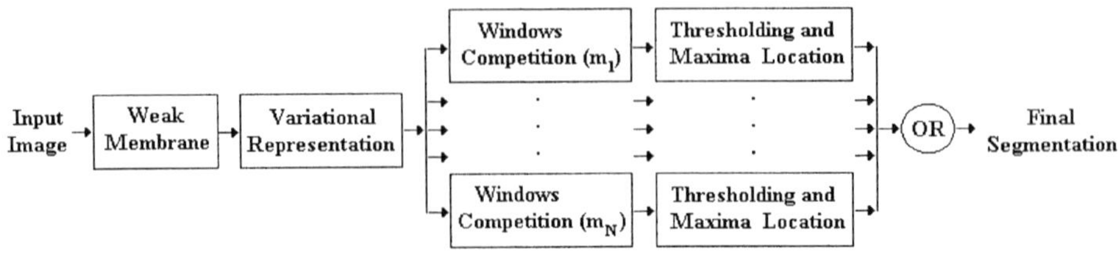

Figure 2: *VRENS'* block scheme.

∂W_i we indicate the boundary of the i-th window. Thus, γ is the contact curve between these two adjacent windows whose interiors intersection is empty.

Starting from a vector $\mathbf{m}$ consisting of N statistics to be determined, we can then introduce N channels. For each channel we can compute:

$$S_i(x,y) = |\, m_i[\tilde{\chi}(x,y)]_{W_1} - m_i[\tilde{\chi}(x,y))]_{W_2}\,| \qquad i = 1,..,N \tag{13}$$

where $m_i[\tilde{\chi}(x,y)]_{W_i}$ is the statistic m_i computed on $\tilde{\chi}$ points belonging to the window W_i. In other words we can imagine that for each point (x,y) of Ω we put two *touching* windows as in Fig. 6. Then we can compute $S_i(x,y)$ as in (13).

It is easy to see that in correspondence to *edge points*, S_i will assume maxima values while in uniform regions will be zero or close to it. We stress the fact that we have also textured regions so that edge points will also be inside of the regions of the image. Then, after a threshold operation, performed on the resulting S_i, oriented to eliminate spurious responses (due to noise), its local maxima locations are determined. It is clear that we will have a threshold t_i for each one of the N channels. At this stage we obtain N (one for each channel) binary images whose ones will be in correspondence to the maxima of the relative S_i. Finally, in order to obtain the final segmentation an *OR* operation is performed on them.

The block scheme of the proposed technique is shown in Fig. 2, where S_i computation phase has been called *windows competition* because of its analogy with the region competition approach.

This simple framework, as we will see, is very simple and effective under noise, but at same time, allows us to face some common problems in image segmentation:

- How many and which statistics have to be selected for segmentation. Now the problem is to select features showing a good robustness to noise;
- What size the regions W_1, W_2 must have. In other words, in order to determine whether a point is an edge or not, how local has the information to be?
- What has the shape of windows W_1 and W_2 to be?

6.1 Statistics Selection

With regard to the first problem, i.e. which statistics have to be selected, we have got to take into account the trade-off between a low computational cost and good segmentation performances. In other words we are interested in looking for the least number of statistics (and then of channels) allowing us to achieve good results under noise.

We outline the requirement *under noise*, since more complex statistics could be selected without this constraint, and then a more accurate segmentation can be achieved — for a review about segmentation without noise see also [37, 39].

In presence of noise the problem of segmentation becomes different. As previously mentioned, the more noise amplitude increases the more the structure of textures changes. Hence, we have to select few features but able to describe textures under noise.

We will consider here *mean* and *edges density*. In particular this latter was known since the early Seventies [43].

It is interesting to outline the *double meaning* of *edge density*: it may be seen as a structural feature as well as a statistical one. Structural since there is the hypothesis that two or more textures can be discriminated by such a feature only if they show a good spatial regularity of their textels placement. On the other hand, its statistical meaning gives an interesting information relative to textures coarseness that, as we have already seen, is one of the most important criteria of human visual system (see the aforementioned features proposed by Tamura et al.). In addition, we outline that the combination of *mean* and *edge density* is very similar to another combination well known by segmentation experts: *mean* and *variance* of image grey levels, widely used in many important frameworks as region competition [54].

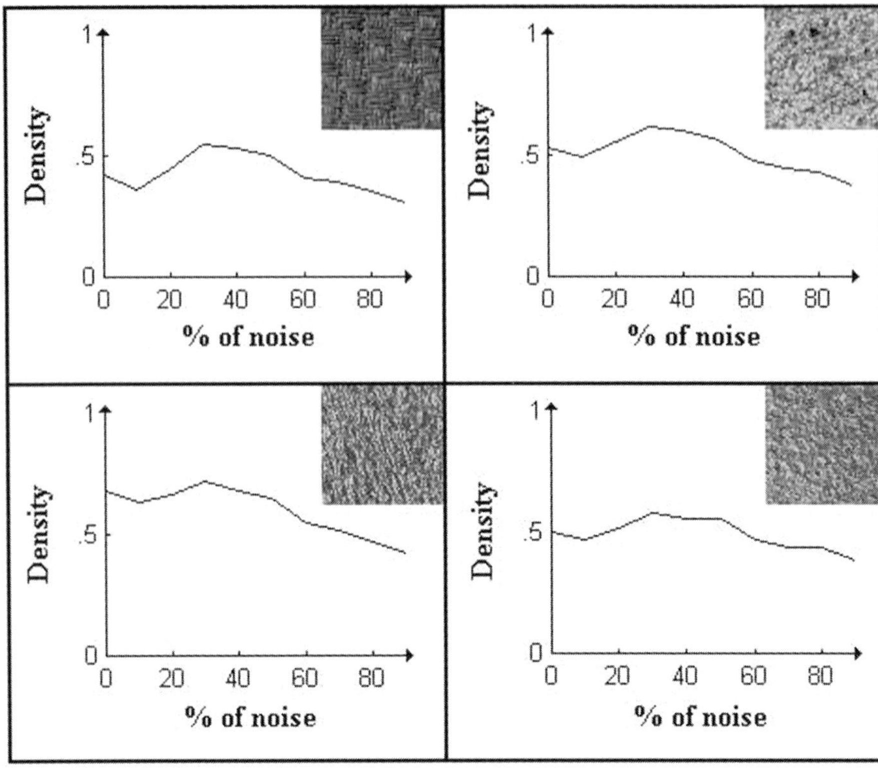

Figure 3: Brodatz textures and their edge density for increasing gaussian noise. Edges have been obtained by the Mumford-Shah functional with $w_1 = 2$ (fixed scale level) and increasing values of w_2. It can be noticed that the *edge density* curves show a very similar behavior and do not cross each other. It follows that a model based on *edge density* difference can achieve a good segmentation till 90% on the above textures composition.

Summing up, *mean* and *edge density* correspond respectively to color and coarseness of image regions, according to human eye working, and can be defined as follows:

$$\mu(W) = \frac{1}{|W|} \sum_{i,j \in W} \tilde{\chi}(i,j) \qquad \rho(W) = \frac{1}{|W|} \sum_{i,j \in W} \chi(i,j) \tag{14}$$

where W is the window we are considering over the image and $|W|$ its area.

As matter of fact our mean is different from the mean computed in the

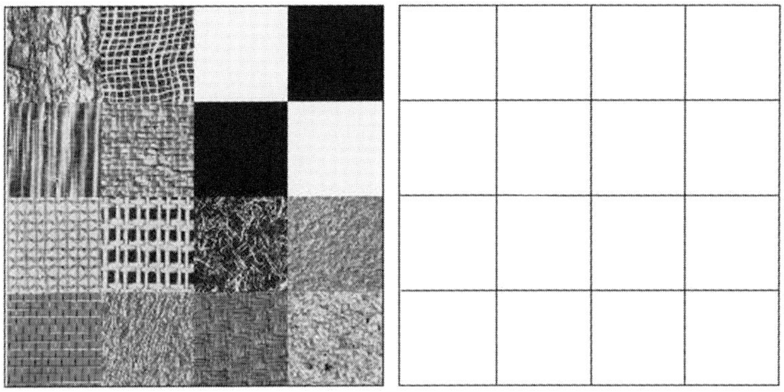

a) b)

Figure 4: 256x256x8 bits image (chessboard) composed of Brodatz textures a) and its perfect segmentation obtained using *mean* and *edge density* through the technique in [40] b). We outline that for all manual compositions showed in the following we have used the Brodatz textures [7] though other test textures databases may be used [32, 34].

classical sense. In fact, from the above definition it is computed considering only edge points. Nevertheless, if we take into account that the human eye works a low-pass filter, they are not so different.

6.2 Competing Windows Size

The aim of this section is to determine what should be the size of the two competing windows W_1 and W_2. In fact, it is possible to show that they must be wide enough by means of simple considerations . These windows contain all available information we have about the points belonging to γ, i.e. the curve between them.

In the next sections some results about this point will be discussed.

6.2.1 *Mean* Behavior under Noise

In order to simplify the formalism, we study the *mean* behavior under noise in the 1-D case. For images the results can be easily extended. Let us consider x_i $0 \le i \le M$ a constant signal whose value is equal to μ with an added noise having normal distribution $N(0, \sigma^2)$. The probability that there is a discontinuity, i.e. a value between two consecutive samples greater than a given threshold $h_0 \in R$, can be indicated with $P(|x_{i-1} - x_i| > h_0)$ where $1 \leq i \leq N$.

Since

$$|x_{i-1} - x_i| > h_0 \equiv \begin{cases} x_i > x_{i-1} + h_0 \\ x_i < x_{i-1} - h_0 \end{cases}$$

and considering x_i, x_{i-1} independent, with a suitable quantization we can assume

$$\sum_{\mu-3\sigma}^{\mu+3\sigma} \frac{1}{2\pi\sigma^2} e^{-\frac{(x_{i-1}-\mu)^2+(x_i-\mu)^2}{2\sigma^2}} \simeq 1 \tag{15}$$

(discarding .3% of the whole probability). If $h_0 < 6\sigma$ we obtain:

$$\sum_{\mu-3\sigma}^{\mu+3\sigma} \Big(\sum_{max(x_{i-1}+h_0,\mu-3\sigma)}^{\mu+3\sigma} + \sum_{\mu-3\sigma}^{min(x_{i-1}+h_0,\mu+3\sigma)} \Big) \frac{1}{2\pi\sigma^2} e^{-\frac{(x_{i-1}-\mu)^2+(x_i-\mu)^2}{2\sigma^2}}. \tag{16}$$

Now, let us suppose that we have to segment a piecewice constant signal assuming values μ_1 and μ_2 with added gaussian noise ($N(0, \sigma)$).

Considering $\tilde{\mu_1}$ and $\tilde{\mu_2}$ the means computed as in 14, at a given scale level and for wide windows, they will approximate μ_1 and μ_2.

If we want n edge points (the least number required to determine a region), known the previously mentioned probability P, the region size should be $|W| \geq \frac{n}{P}$. In other words we need competing windows W_1 and W_2 to be greater than a given size[52].

Since two different regions cannot be detected when $\tilde{\mu}_1 \simeq \tilde{\mu}_2$, selecting a high value of h_0 (with respect to μ_1, μ_2 and σ), we detect few edge points (the queues of the gaussian distribution) with small oscillations on the resulting mean (good detection). This entails wide distances between two consecutive edge points and so a bad localization. On the other hand, with little values of h_0 we have many edge points (good localization) but high oscillations of the observed mean and then a low robustness to the noise. We have to get a trade-off between these two different factors.

While for the mean we have found well known results tied to the *uncertainty principle* [52], as regards the *edge density* the situation is more interesting as shown in the next section.

6.2.2 *Edge Density* Behavior under Noise

Let us see what happens when the *edge density* of textures is considered. In the following we will use qualitative considerations avoiding a strong formalism.

Be

$$f(x, y) = u(x, y) + \eta(x, y) \qquad \forall x, y \in \Omega, \tag{17}$$

where u is the noiseless image while η is noise with distribution $N(0, \sigma)$.

Starting from the Mumford-Shah functional, we obtain N_i edge points for a given scale level and a threshold h_0 (tied to w_2 [6]). In order to see how the *edge density* behaves we have got to study the two following cases: low or high amplitude noise with respect to the image information.

This difference is important since showing different behaviors.

- In the first case we can notice that the edge points of the noiseless image trivially have a very high probability of remaining edges in the noisy image, since this latter has a little distortion. Textels are not strongly modified and coarseness, i.e. *edge density*, does not basically change.

- In the second case the situation is more complicated. In fact, considering a noiseless texture t, its density is

  $$\rho = \frac{ep}{|W|} \tag{18}$$

  where ep are the edge points of the noiseless image and $|W|$ is the size of the window we are studying. When we add a high amplitude noise, ρ changes as follows:

  $$\tilde{\rho} = \frac{\tilde{ep} + sp}{|W|}, \tag{19}$$

  where $\tilde{ep}$ are the new edge points, i.e. under noise, and such that $\tilde{ep} \leq ep$. In fact some discontinuities may disappear, making the effect we will call *missing good edges*. Other discontinuities may be emphasized but do not change ρ. At the same time there are some spurious edges sp due to noise and whose locations are where ep are not. The latter effect will be called *new spurious edges*.

For two windows W_1 and W_2 wide enough, the two effects: *missing good edges* and *new spurious edges* are similar and opposite in both the windows. In other words the observed *edge density* is quite constant as shown in Fig. 3. The wider W_1 and W_2 are the more realistic this situation is. In other words ,wide windows, i.e. many pixels, make this behavior more and more true. The sum of these two effects is exploited by *VRENS* to achieve good results under a strong noise, since for two generic textures the final result seems to only depend on the difference of their noiseless *edge densities*.

In a preliminary paper we achieved a perfect segmentation on the composition shown in Fig. 4 till 90 % of additive (zero-mean) gaussian noise[40]. In that case we used a modified Split and Merge [17] that was equivalent to know which and where the edge points were. Nonetheless that paper stressed the potentialities of two simple features as *mean* and *edge density*, along with the fact that Split and Merge technique is not able to segment generic shapes as shown in Fig. 5. In fact, in this case small square blocks close to the edges are required against the aforementioned considerations relative to *mean* and *edge density* behavior.

In Fig. 4 the image regions were squared and then we knew in advance their shape. Usually it is not true and then next section will deal with the problem of optimal shape for competing regions.

6.3 Windows Shape

We have seen that a point of an image is an edge point if it belongs to the boundary between two different regions, which are homogeneous under some criteria. The choice of the shape of the windows in which to compute the aforementioned statistics, and then of the contact curve, is strongly dependent on the edges to be detected. Usually, without additional *a priori* hypotheses, it is assumed that the shape of the image regions is unknown. Moreover, in order to obtain accurate edges the contact curve γ should be minimized. In [54] (p. 893) Zhu and Yuille proposed *elliptical windows with their major axes parallel to the boundary.* This was basically due to some requirements relative to the functional they used. Usually, the simplest shape that satisfies the previous constraints is the circle. In the following we will consider diamonds (see Fig. 6), for computational reasons. We stress the fact that this is an open problem too, as outlined in [54]. In fact, the shape should take into account not only efficaciousness of edge detection but

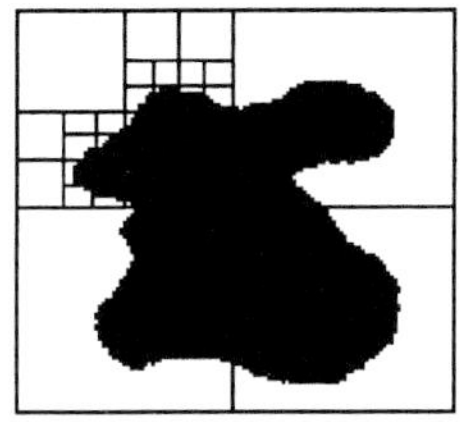

Figure 5: Quadtree technique applied to a generic shape developed here only for the topmost and leftmost square. As it can be noticed it requires small squares yielding wrong results under noise (see text for details).

also textels size and shape, as well as noise amplitude that modifies these two latter features. Moreover, it should also consider the minimization of wrong results whenever a window lies on two or more regions of the image. In fact, in this case, the resulting statistic value ($S_i(x, y)$ in equation 13) will correspond to the one of more than one region of the image. To the best of the authors' knowledge, window shape is selected taking into account the *a priori* regularity of the resulting edges, more than the other aforementioned aspects, since they are not *a priori* known on real images. But such a choice strongly contributes to the final result.

In *VRENS* edge points will be detected along horizontal, vertical and the two main diagonal directions. In particular, in order to optimize the segmentation, for a given statistic m_k let us consider the pixels of i-th row in $\tilde{\chi}(i, j)$ (horizontal direction from left to right). The contact point between W_1 and W_2 will assume locations of pixels belonging to i-th row to compute $S_k(i, j)$. If $S_k(i, j) < t_k$, we have a uniform region. Then the new W_1 will become the old W_2 while the new W_2 will have its leftmost vertex coinciding with new W_1's rightmost one. In other words we skip all the pixels of the old W_1 diameter instead of computing $S_k(i, j)$ for all the pixels of i-th raw. This leads to a computational effort saving. It turns out that the diameter d of the diamonds is to be tuned considering the distance among the edges of the image. This information usually is not known *a priori*; then d will have to be tuned heuristically.

7 Some Results

VRENS has been tested on many images. In this section some results, oriented to outline some interesting aspects along with the potentialities of

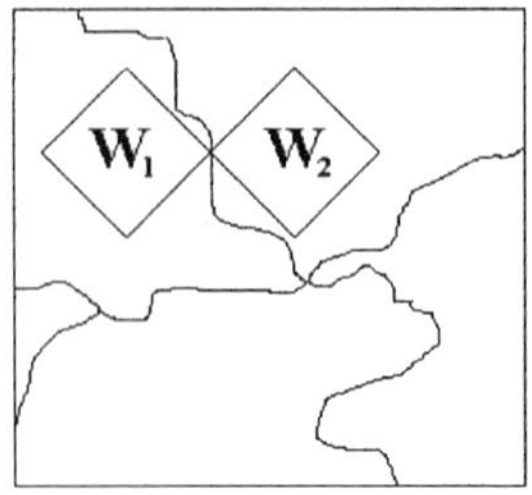

Figure 6: A diamond as shape for the two test windows W_1 and W_2. γ is composed of an only point.

the proposed approach, will be shown.

Before showing them, we outline that for all simulations the Mumford-Shah functional, minimized by GNC, has been used. The two used statistics are *mean* and *edge density* whose thresholds will be denoted by t_μ and t_ρ respectively, while d is the diameter of the windows.

The first example shown in Fig. 7, is relative to the test image contained in [25]. This is useful to compare our technique with other well known algorithms of edge detection. In particular, VRENS seems to be as good as Perona-Malik model, achieving comparable results — see [25] for a visual comparison with the following models: i) SUSAN [44], ii) Perona-Malik [36], iii) Weickert [50], iiii) Liu-Wang-Ramirez [25]. Thresholds values at 10% of noise are: $w_1 = 2$, $w_2 = .05$, $t_\mu = 1$, $t_\rho = 6.5$ and $d = 9$, while at 30% of noise: $w_1 = .01$, $w_2 = .01$, $t_\mu = 1$, $t_\rho = 7$ and $d = 13$. Though its performances are satisfying, we outline that *VRENS* is not as good as Liu, Wang and Ramirez approach [25]. Of course, the latter has been designed to be particularly efficacious in detecting smooth regions under noise. In fact the maximum likelihood method works very well because the only (gaussian) distribution is due to the additive noise. Some difficulties occur when there are the textures distributions too.

As regards images with textures the situation is strongly different. We have considered the composition presented in [25] and shown in Fig. 8. In this case, VRENS works noticeably better, achieving good results till 90% of noise, whereas textures look to lose their intrinsic structure. In this case our statistics show a strong robustness to the noise following the theoretical considerations of the previous sections. At 10% of noise, thresholds values are: $w_1 = 2$, $w_2 = 1$, $t_\mu = .3$, $t_\rho = 20$ and $d = 31$, while at 90% of noise: $w_1 = .5$, $w_2 = .5$, $t_\mu = .8$, $t_\rho = 9.3$ and $d = 31$.

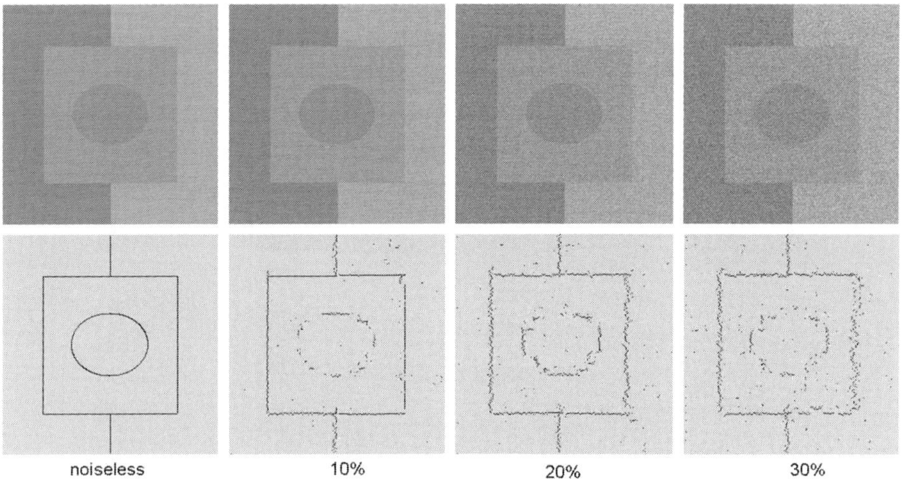

Figure 7: *VRENS*'s segmentation on the test image proposed in [25] for increasing additive (zero-mean) gaussian noise.

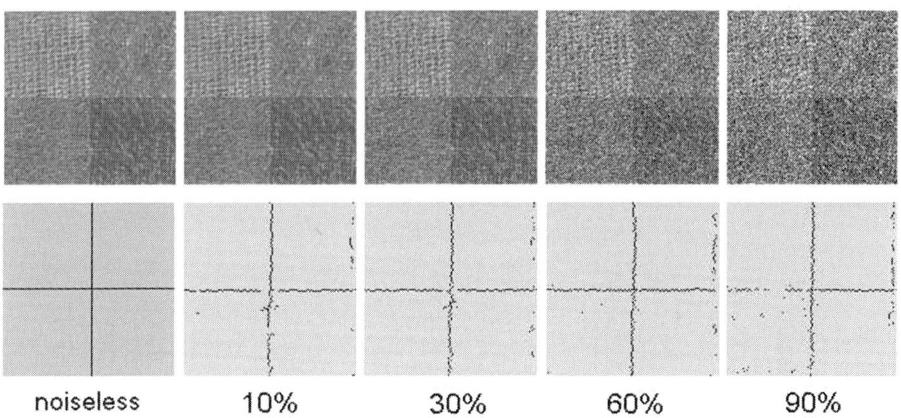

Figure 8: *VRENS*'s results on the textures composition proposed in [25]. Our approach shows a strong robustness in agreement with theoretical considerations made in Section 6.2.2.

The last example consists of an image showing a real scene (Fig. 9). In this case, with regard to the noiseless image, the imperfect segmentation is due to the irregular texture of the leopard's skin — for instance in correspondence of the snout or the neck. Nonetheless, though more detailed

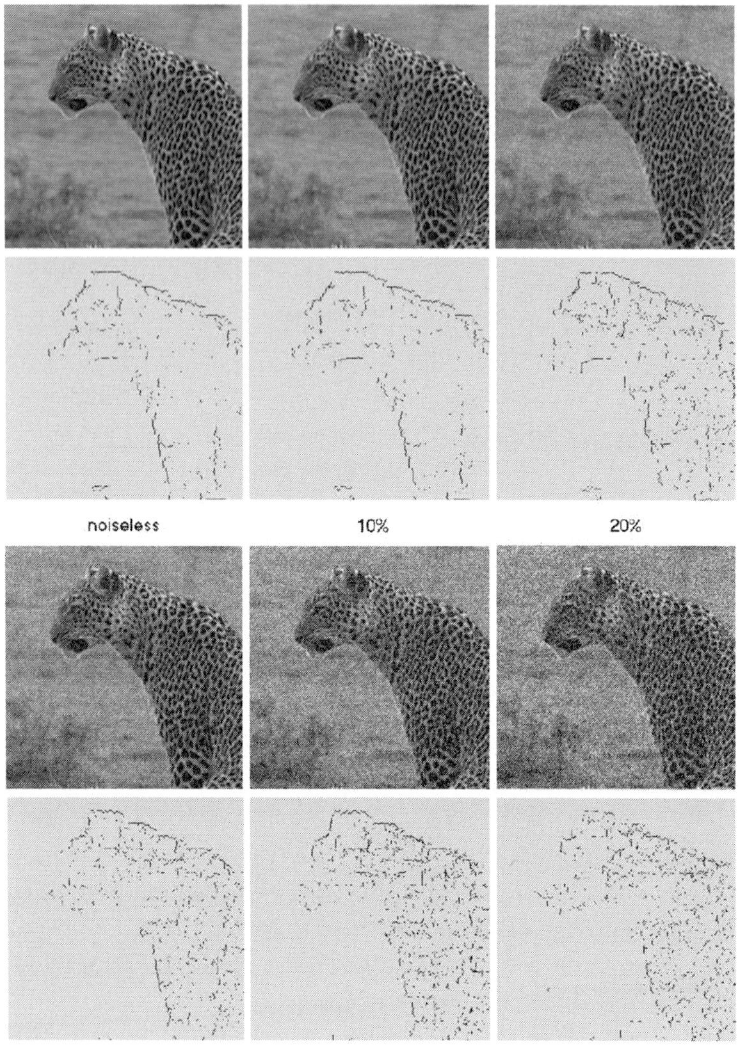

Figure 9: A real image showing a leopard on the grass with increasing gaussian additive noise. The coarse segmentation oriented to discriminate between the leopard's skin and the background does not drastically change under strong noise.

segmentations are also possible, this coarse one has been selected in order to discriminate between the leopard and the background. For increasing noise, it can be seen that the problems are only inside the leopard skin where there are different but very close textures — in terms of *edge density*. On the contrary, leopard and grass keep on being different and then they are discriminated under a strong noise, in agreement with the theoretical results of Section 6.2.2.

Finally, VRENS is very effective if we consider that, for example, for the image at 30% of noise in Fig. 8, just 3.23 secs are required for a complete segmentation on a workstation (Octane/SI R10000 175 MHz/1Mb cache).

8 Concluding Remarks

This chapter has presented a framework useful to segment very general images, i.e. containing both smooth and textured regions. In particular, while as regards classical edge detection we achieve results comparable to the best variational approaches, we noticeably improve texture segmentation. Though we use competing windows, our approach, achieving edge detection, does not present the *seeds*'s drawback [54] showing a very low computing time. The proposed technique is invariant to textures rotations and translations but its main drawback consists of performing at only one scale level. In fact the threshold w_1 is tuned once and for all. Since *textures live at many scale levels* [30], it follows that different textures may be considered equal using such a constraint. Moreover, this presentation has focused on many open problems of image segmentation. In particular texture segmentation under noise seems to be a further challenge for image segmentation experts.

Acknowldgements

The authors would like to thank Riccardo March for his helpful suggestions and Mr. Marmite for making his test image available (Fig. 8 in the text) (http://Animalpicturesarchive.com (mm_Leopard-captured_by-_Mr_Marmite.jpg). This work has been partially supported by the Italian National Research Council: Progetto "Multimediality" 5% Murst.

References

[1] A. Ait-Kheddache and S. Rajala, Texture classification based on higher-order fractals, *Proc. IEEE* (1988) pp. 1112-1119.

[2] N. Ahuja, and A Rosenfeld, Mosaic models for textures, *IEEE Trans. Pattern Anal. Mach. Intell. PAMI-3* No. 1 (1981) pp. 1-11.

[3] J. R. Bergen, and E. H. Adelson, Early vision and texture perception, *Nature* Vol. 333 (1988) pp. 363-364.

[4] J. R. Bergen, Theories of visual texture perception, In D. Regan (ed.), *Vision and Visual Dysfunction* (New York, Macmillian, 10B, 1991) pp. 114-134.

[5] J. R. Bergen, and B. Julesz, Rapid discrimination of visual patterns, *IEEE Trans. on System Man and Cybernetics* Vol. 13 (1993) pp. 857-863.

[6] A. Blake and A. Zissermann, *Visual Reconstruction*, (MIT Press, 1987).

[7] P. Brodatz, *Textures: A Photographic Album for Artists and Designers*, (New York: Dover, 1966).

[8] F. W. Campbell, and J. G. Robson, Application of fourier analysis to the visibility of gratings, *J. Physiol.* No. 197 (1968) pp. 551-566.

[9] A. Chambolle and P. L. Lions, Image Recovery via Total Variation Minimization and Related Problems, *Research note 95-09* (1995) CEREMADE, Univ. Paris-Dauphine, Paris, France.

[10] M. Chipot, R. March and D. Vitulano, Numerical analysis of oscillations in a non convex problem related to image selective smoothing, *Journal of Computational and Applied Mathematics* No. 136 (2001) pp. 123-133.

[11] M. Clark, A. C. Bovik, W. Geisler, Texture segmentation using gabor modulation/demodulation, *Pattern Recognition* No. 6 (1987) 261-267.

[12] J. M. H. du Buf, M. Kardan and M. Spann, Texture feature performance for image segmentation, *Pattern Recognition* No. 23 (1990) pp. 291-309.

[13] S. Geman, D. E. McClure, Statistical methods for tomographic image reconstruction, (Proc. of the 46th Session International Statistical Institute Bulletin ISI, Vol. 52, 1987).

[14] D. Geman, G. Reynolds, Constrained restoration and the recovery of discontinuities, *IEEE PAMI*, 14 (1992) pp. 367-383.

[15] J. Hadamard, *Lectures on the Cauchy Problem in Linear Partial Differential Equations*, (Yale University Press, 1923).

[16] R. M. Haralick, Statistical and structural approaches to texture, (*Proc. IEEE* 67,1979) pp. 786-804.

[17] S. L. Horowitz, and T. Pavlidis, Picture segmentation by a directed split-and-merge procedure, (*Proc. 2th ICPR*,1974) pp. 424-433.

[18] H. A Kaizer, Quantification of Textures on Aerial Photographs, *Laboratories Technical Note 221* (1955) Boston University Research, Boston, Massachussets.

[19] B. Kartikeyan, and A. Sarkar, A unified approach for image segmentation using exact statistics, *Computer Vision Graphics Image Processing* 48 (1989) pp. 217-229.

[20] G. Koepfler, C. Lopez,and J. M. Morel, A multi scale algorithm for image segmentation by variational method, *SIAM J. Numer. Anal.*, Vol. 31 No. 1 (1994) pp. 282-299.

[21] A. K. Jain, *Fundamentals of Digital Image Processing*, (Prentice Hall, Englewood Cliffs, 1986).

[22] A. K. Jain, and F. Farrokhnia, Unsupervised texture segmentation using Gabor filters, *Pattern Recognition* 24 (1991) pp. 1167-1186.

[23] L. Lazaroaía, A Study in a BV Space of a Denoising-Deblurring Variational Problem, *prepubl. no. 439*, (1996) Laboratory J-A Dieudonné, URA 168 du CNRS, University of Nice-Sophia Antipolis, France.

[24] T. S. Lee, D. Mumford, and A. Yuille, Texture segmentation by minimizing vector-valued energy functionals: The coupled-membrane model, (*Proc. of Computer Vision ECCV '92*, Vol. 588, LNCS Springer May 1992).

[25] X. Liu, D. L. Wang, and R. Ramirez, Boundary detection by contextual non-linear smoothing, *Pattern Recognition* 33 (2000) pp. 263-280.

[26] R. Lovell, W. R. Uttal, T. Shepherd, and S. Dayanand, A model of visual texture discrimination using multiple weak operators and spatial averaging, *Pattern Recognition* Vol. 25 No. 10 (1992) pp. 1157-1170.

[27] S. Y. Lu, K. S. Fu, A syntactic approach to texture analysis, *Comput. Graphics Image Process.* 7 (1978) pp. 303-330.

[28] S. Mallat, W. L. Hwang, Singularity Detection and Processing with Wavelets, *IEEE Tarns. on Information Theory* Vol. 38 No. 2 (1992) pp. 617-643.

[29] S. Mallat, *A Wavelet Tour of Signal Processing*, (Academic Press, 1998).

[30] D. Marr, *Vision, Freeman & Co.*, (San Francisco, CA, 1982).

[31] B. H. McCormick, and S. N Jayaramamurthy, A decision theory method for the analysis of texture, *Int. J. Comput. Inf. Sci* 4 (1974) pp. 1-38.

[32] MeasTex Image Texture Database, http://www.cssip.elec.uq.edu.au/guy/meastex/meastex.html,1998.

[33] O. Mitchell, C. Myers, and W. Boyne, A max-min measure for image texture analysis, *IEEE Trans. Comput.* 25 (1977) pp. 408-414.

[34] MIT Vision and Modeling Group, http://www.media.mit.edu/vismod/, 1998.

[35] D. Mumford and J. Shah, Optimal approximation by piecewise smooth functions and associated variational problems, *Commun. Pure Appl. Math* Vol. 42 (1989) pp. 577-685.

[36] P. Perona and J. Malik, Scale-space and edge detection using anisotropic diffusion, *IEEE Trans. on PAMI* Vol. 12 No. 7 (1990) pp. 16-27.

[37] T. Randen, and J. H. Husøy, Filtering for texture classification: A comparative study, *IEEE Trans. on PAMI* Vol. 21 No. 4 (1999) pp. 291-310.

[38] J. S. Read, and S. N. Jayaramamurthy, Automatic generation of texture feature detectors, *IEEE Trans. Comput.* (1972) 21 pp. 803-812.

[39] T. R. Reed, and J. H. M. du Buf, A review of recent texture segmentation and feature extraction techniques, *CVGIP: Image Understanding* Vol. 57 (1993) pp. 359-372.

[40] R. Romano, D. Vitulano, A multichannel model for robust image segmentation under noise, in V. Skala (Ed.) *Proc. of WSCG*, (Plzen 2001) pp. 79-86.

[41] R. Romano, D. Vitulano, A Variational Representation for Efficient Noisy Segmentation, in V. Skala (Ed.) *Proc. of WSCG* (Plzen 2002) pp. 41-44.

[42] A. Rosenfeld, E. Thurston, Edge and curve detection for visual scene analysis, *IEEE Trans. Comput.* (1971) 20 pp. 562-569.

[43] A. Rosenfeld, *Visual Texture Analysis: An Overview, Tech. Rep. TR-406* (1975), Computer Science Center, University of Maryland.

[44] S. M. Smith and J. M. Brady, SUSAN: a new approach to low level image processing, *Int. J. Comp. Vision*, 23 (1997) pp. 45-78.

[45] H. Tamura, S. Mori, and T. Yamawaki, Textural features corresponding to visual perception, *IEEE Trans. Systems Man Cybern.* SMC-8 (1978) pp. 460-473.

[46] S. Teboul, L. Blanc-Féraud, G. Aubert and M. Barlaud, Variational approach for edge-preserving regularization using coupled PDE's, *IEEE Trans. on Image Processing* Vol. 7 No. 3 (1998) pp. 387-397.

[47] A. Tikhonov and V. Arsénine, *Méthodes de Résolution de Problèmes Mal Posés*, (MIR, Moscow, USSR, 1976).

[48] M. Unser, Local linear transforms for texture measurements, *Signal Processing* 11 (1986) pp. 61-79.

[49] L. Van Gool, and P. Dewaele, A. Oosterlinck, Survey: Texture analysis, *Computer Vision Image Processing* 29 (1985) pp. 336-357.

[50] J. Weickert, Theoretical foundations of anisotropic diffusion in image processing, in W. Kropatsch, R. Klette, F. Solina (Eds.) Theoretical Foundations of Computer Vision, Vol. 11 *Springer, Wien* (1996) pp. 1539-1553.

[51] M. Wertheimer, Laws of organization in perceptual forms, in W. B. Ellis (ed.) *A sourcebook of Gestalt Psychology* (Harcourt Brace and Company, 1938).

[52] R. Wilson, and G. H. Granlung, The uncertainty principle in image processing, *IEEE Trans. Pattern Analysis and Machine Intelligence* Vol. 6 No. 6 (1984) pp. 758-767.

[53] Y. L. You, W. Xu, A. Tannenbaum, and M. Kaveh, Behavioral analysis of anisptropic diffusion in image processing, *IEEE Trans. Image Processing* 5 (1996) pp. 1539-1553.

[54] S. C. Zhu, and A. Yuille, Region competition: unifying snakes, region growing, and bayes/MDL for multiband image segmentation, *IEEE Trans. on PAMI* Vol. 18 No. 9 (1996) pp. 884-900.

Set Analysis of Coincident Errors and Its Applications for Combining Classifiers

Dymitr Ruta and Bogdan Gabrys
Applied Computational Intelligence Research Unit
Division of Computing and Information Systems
University of Paisley High Street, Paisley PA1-2EP, UK
E-mail: {ruta-ci0, gabr-ci0}@paisley.ac.uk

Contents

1 Introduction

There is a common agreement in many recent publications related to pattern recognition that dependency among classifier outputs plays a key role

D. Chen and X. Cheng (eds.),
Pattern Recognition and String Matching, 647–671.

in combining classifiers [1]-[7]. Diversity, independence, disagreement and most recently negative dependency are the terms often used to express a desirable relation among classifiers to ensure the maximum improvement of the fusion system [4]-[7]. In this variety of concepts the idea is the same: how to measure relationship among classifiers from their outputs so that it is possible to say something about the combined classifier performance? Recent investigations indicate that error coincidences seem to be the most valuable information in this pursuit [7]-[9].

Coincident events are traditionally described by probabilistic models. For the case of independent events, their coincidence can be easily calculated from the product of the individual probabilities of events. However in the realistic pattern recognition situations it is very naive to assume independence among classifier outputs. Realistically when many classifiers are applied for the same pattern recognition task they turn up to be strongly dependent and so are their errors showing similar patterns of misclassification [8]. To model dependence or diversity among classifier outputs a number of various measures have been proposed including correlation, disagreement and diversity [5], [17], [7], [6]. The common belief is that the more diverse classifiers, the better the performance of the combining system. Such diversity should result in different patterns of misclassification allowing classifiers to compensate each others failures and result in improved performance of the whole system.

Surprisingly, vast majority of the diversity measures have been shown completely uncorrelated with the performance of the combiner and only those based on error coincidences showed slight progress in this domain [7]. Based on these findings we decided to look more extensively at the problem of error coincidences among classifiers. In the past, probabilistic framework dominated this types of problems [10]. However, excluding the independent case, probabilistic analysis of the dependent events of error occurrences would require exponentially complex calculations of conditional probabilities of all combinations of events and becomes unmanageable even for small number of classifiers [10]. What is more, probability estimations of higher order dependencies can be unreliable due to sparsity of examples unless vast amount of data is provided. To avoid these problems some studies consider simplifications in a form of using only lower order dependencies, possibly applying them to approximate higher order dependencies and completely ignoring or assuming independence for the highest order dependencies [10].

Alternatively, coincident errors can be represented by means of sets [11]. In this approach the errors from a single classifier are mapped into corresponding set of indices of misclassified samples. If more than one classifier misclassifies a particular sample then the index of this sample becomes an element in the intersection of sets corresponding to misclassifying classifiers. Using such representation, all available information related to error coincidences can be visualized as a complex architecture of overlapping sets resembling Venn Diagrams comprehensively discussed in [12]. Such collection of classifier sets represents a simple and coherent source of complete information about error coincidences. It encapsulates all conditional probability measures but at the same time preserves information about individual indices of misclassified samples, which would have been lost in the probabilistic representation of error coincidences.

Moreover, as we show in this work, applying a simple algebra on the cardinalities of different subsets of the collection of classifier sets leads to the derivation of various measures of dependency presented independently in the past. This relates for instance to the long-lasting debate of which information to choose: mean classifier error or pairwise diversity measure (e.g. *double fault* measure discussed in [5], [7]) to optimally select classifiers for a combining method [13]. We show that both types of information stem from common relation apparent in the set representation of coincident errors and for that reason instead of being competitive, they can be supportive as they both bring some new descriptive information. We further consider potential applications of the set representation of error coincidences for majority voting operating on binary outputs (correct/incorrect), which proved as simple yet quite powerful combining method [14]-[17]. We show a novel definition of majority voting error expressed by the predefined coincidence levels and investigate its relation with cardinalities of lower and higher order error coincidences and their combinations.

The remainder of the paper is organized as follows. Section 2 gives a theoretical basis of a set-collection representation of error coincidences including definitions of two types of coincidence levels presented in the subsection 2.1, and an optimized method of collection generation and operations on error coincidences briefly described in section 2.2. In section 3 we show how the set analysis of error coincidences relates to the majority voting combiner and its error. Section 4 provides the results from the experiments with a number of real datasets and classifiers showing some numerical properties of the collection representation and correlations between majority voting error and different combinations of coincidence levels. Finally, summary, conclusions and further potential applications of the presented ideas are given in section 5.

2 Set Representation of Coincident Errors

Given a system of M trained classifiers $D = \{D_1, ..., D_M\}$ applied for the classification of N data samples let $y_i = (y_{i1}, ..., y_{iM})$ denote a joint output of a system for the ith input sample x_i, where y_{ik} refers to individual output of the hth classifier for that sample and $i = 1, ..., N$; $k = 1, ..., M$. In this work we consider only binary outputs assuming $y_{ik} = 1$ as an error and $y_{ik} = 0$ as a correct classification. The outputs form a binary matrix of outputs $Y^{N\times M}$, which is a starting point of our considerations. For the purpose of combining, there is no need to keep the information about all the outputs but only about errors and their distribution among all the classifiers [7]. This fact makes sets analysis particularly attractive for the description of errors and the way the errors are shared by different classifiers. From the set analysis standpoint each classifier D_k can be associated with a set S_k, containing the indices of misclassified data samples. We call such sets the *classifier sets.* In a very common situation of more than one error for the same sample, its index has to be shared by corresponding classifier sets. We denote a complex system of overlapping classifier sets obtained in this way by $S = \{S_1, ..., S_M\}$ and call it shortly a *collection.* The set representation of coincident errors can be now formally expressed as a mapping from a binary matrix of outputs into a collection:

$$Y \longrightarrow S \Longleftrightarrow Y^{N\times M} \rightarrow \{S_1, ..., S_M\} \tag{1}$$

The collection can be visualized by means of Venn Diagrams [12]. Figure 1 shows an example of such diagram for given outputs from 3 classifiers. For a larger number of classifier sets, visualization of coincident errors is more complicated as it is difficult to visually represent all combinations of classifiers given in the form of exclusive subsets of classifier sets. Figure 2 shows some examples of Venn Diagrams for more than 3 classifier sets. Venn Diagrams and their construction represent a complex mathematical problem on its own and some further related details can be found in [12]. For the purpose of this work only some properties of the Venn Diagrams and graphs associated with them will be used.

2.1 General and Exclusive Coincidences

Given a collection of classifier sets, the coincidences of errors from a specific combination of classifiers $\{D_{i_1}, ..., D_{i_k}\}$ can be viewed as subsets representing

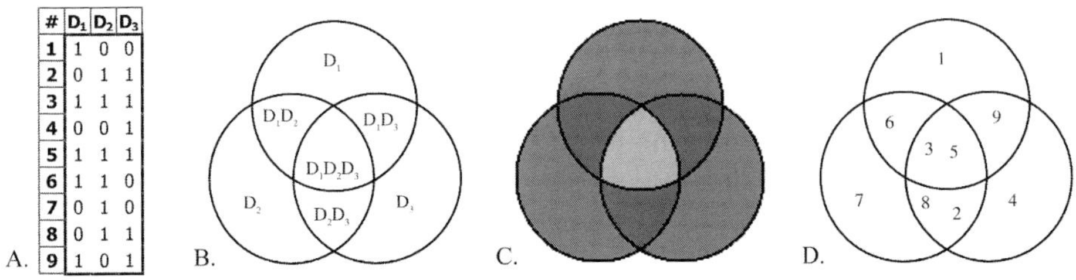

Figure 1: Visualization of a set representation of coincident errors. A: Binary outputs from 3 classifiers (0-correct, 1-error). B,C: Venn Diagrams showing all mutually exclusive subsets. D: Venn Diagram with the indices of samples put in the appropriate subsets positions.

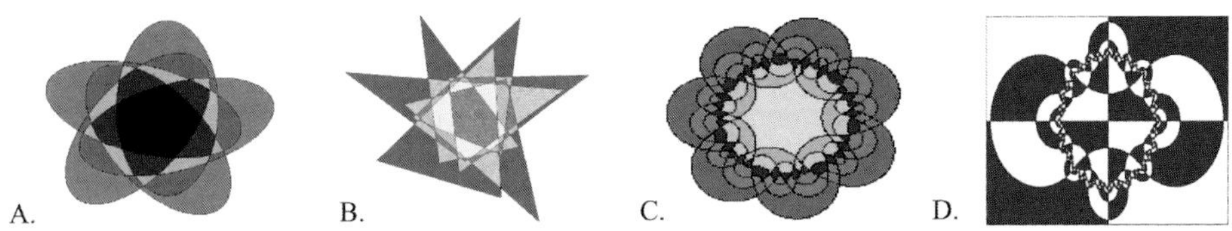

Figure 2: Venn Diagrams for more than 3 classifiers. A: 5 congruent ellipses. B: 6 triangles. C: 7 symmetrical sets - Grunbaum construction. D: bipartite plot of 8 sets - Edward's construction. See [12] for further details.

intersections of corresponding classifier sets:

$$C^G(\{D_{i_1}, ..., D_{i_k}\}) = C^G_{i_1,...,i_k} = \bigcap_{j=1}^{k} S_{i_j} \tag{2}$$

where $i_i, ..., i_k$ represent indices of classifiers and $|\{D_{i_1}, ..., D_{i_k}\}| = k$ [1] We call such type of coincidence the kth-order general coincidence of the errors from classifier subset $\{D_{i_1}, ..., D_{i_k}\}$. We call it general as it does not depend on the other classifier sets, which may also overlap the considered intersection subset. On the other hand much different but not less important information related to the same subset of classifiers is given if considered intersection subset is not overlapped by any other classifier set from the collection S. We refer to such intersection as the kth-order exclusive coincidence

[1]The $|S|$ stands for cardinality of a set S and means the number of elements in this set.

and define it formally by:

$$C^E_{i_1,\ldots,i_k} = \bigcap_{j=1}^{k} S_{i_j} - \bigcup_{j=k+1}^{M} S_{i_j} \tag{3}$$

Both coincidences are graphically shown in Figure 3-A for a simple example of 3 classifier sets. To be able to consider relationships between these two types of coincidences effectively, we define a kth-level of coincidence as a sum of all different coincidences of the same order and type as shown in the following formula:

$$L_k = |C_{i_1\ldots i_k}| + |C_{i_1\ldots i_{k-1}i_{k+1}}| + \ldots + |C_{i_{M-k}\ldots i_M}| \tag{4}$$

Note that the number of coincidences to be summed is:

$$\{C_{i_1\ldots i_k}, \ldots, C_{i_{M-k}\ldots i_M}\}| = \binom{M}{k} \tag{5}$$

An important fact about the introduced coincidences is that they both represent complete information about error relations among classifiers and in a sense complement each other. In different applications however one type of coincidence may be more useful than another for different reasons.

We consider now the relations between general and exclusive coincidence levels. Each kth-level of one type of coincidence can be decomposed into a sum of equal and higher levels of the other type of coincidences as defined by a pair of formulas:

$$L^G_k = \sum_{i=k}^{M} \binom{i}{k} L^E_i \qquad L^E_k = \sum_{i=k}^{M} (-1)^{i-k} \binom{i}{k} L^G_i \tag{6}$$

An obvious consequence of the definition of exclusive coincidence, is that all their levels sum up to the cardinality of the union of classifier sets:

$$\sum_{i=k}^{M} L^E_i = \left| \bigcup_{i=k}^{M} S_i \right| \tag{7}$$

From (2) and (3) it is also clear that any exclusive coincidence is always smaller or equal to the corresponding general coincidence of the same order. Consequently given (4), the same relation applies to coincidence levels:

$$\left| C^E_{i_1\ldots i_k} \right| \leq \left| C^G_{i_1\ldots i_k} \right| \quad \overset{(4)}{\Longrightarrow} \quad L^E_k \leq L^G_k \tag{8}$$

Note that for the highest level where $k = M$, both levels are equal: $L_M^E = L_M^G$. For the application purposes we consider a sum of higher levels of exclusive coincidences. It can be shown that adding L_i^E from certain $i = k$ to the highest level ($i = M$) gives the following simple expression:

$$\sum_{i=k}^{M} L_i^E = \sum_{i=k}^{M}\sum_{j=i}^{M}(-1)^{j-i}\binom{i}{j}L_i^G = \sum_{i=k}^{M}(-1)^{i-k}\binom{i-1}{k-1}L_i^G \quad (9)$$

Effectively, sequential summing and subtraction of consecutive general coincidence levels can easily represent the sum of exclusive coincidence levels.

2.1.1 Example: 3 classifier sets

Given 3 classifier sets $\{S_1, S_2, S_3\}$ we consider two scenarios, when different types of coincidence information is given and the other is to be retrieved.

1. *General coincidences are given.* In this case only cardinalities of different combinations of set intersections are necessary. This allows to calculate general coincidence levels according to (4) as shown below:

$$\begin{aligned} L_1^G &= |C_1^G| + |C_2^G| + |C_3^G| = |S_1| + |S_2| + |S_3| \\ L_2^G &= |C_{12}^G| + |C_{13}^G| + |C_{23}^G| = |S_1 \cap S_2| + |S_1 \cap S_3| + |S_2 \cap S_3| \\ L_3^G &= |C_{123}^G| = |S_1 \cap S_2 \cap S_3| \end{aligned} \quad (10)$$

   For the example shown in Figure 1, general coincidence levels would give the following values: $L_1^G = 5 + 6 + 6 = 17, L_2^G = 3 + 3 + 4 = 10, L_3^G = 2$. From (9) and given general coincidence levels (10), the sums of exclusive coincidence levels can be retrieved as follows:

$$\begin{aligned} L_1^E &= L_1^G - 2L_2^G + 3L_3^G \\ L_2^E &= L_2^G - 3L_3^G \\ L_3^E &= L_3^G \end{aligned} \quad (11)$$

   Again for the example in Figure 1, the values for exclusive coincidence levels would be: $L_1^E = 17 - 2 \cdot 10 + 3 \cdot 2 = 3, L_2^E = 10 - 3 \cdot 2 = 4, L_3^E = 2$.

2. *Exclusive coincidences are given.* Optimally they can be extracted directly from the binary matrix of outputs Y, introduced in section 2. As in previous case the respective coincidence levels can be calculated

according to (4):

$$\begin{aligned} L_1^E &= |C_1^E| + |C_2^E| + |C_3^E| \\ L_2^E &= |C_{12}^E| + |C_{13}^E| + |C_{23}^E| \\ L_3^E &= |C_{123}^E| \end{aligned} \tag{12}$$

For the example shown in Figure 1, exclusive coincidence levels would give the following values: $L_1^E = 1 + 1 + 1 = 3, L_2^E = 1 + 1 + 2, L_3^E = 2$, and further transformed into general coincidence levels according to (6):

$$\begin{aligned} L_1^G &= L_1^E + 2L_2^E + 3L_3^E \\ L_2^G &= L_2^E + 3L_3^E \\ L_3^G &= L_3^E \end{aligned} \tag{13}$$

with values: $L_1^G = 3 + 2 \cdot 4 + 3 \cdot 2 = 17, L_2^G = 4 + 3 \cdot 2 = 10, L_3^G = 2$ for the example in Figure 1.

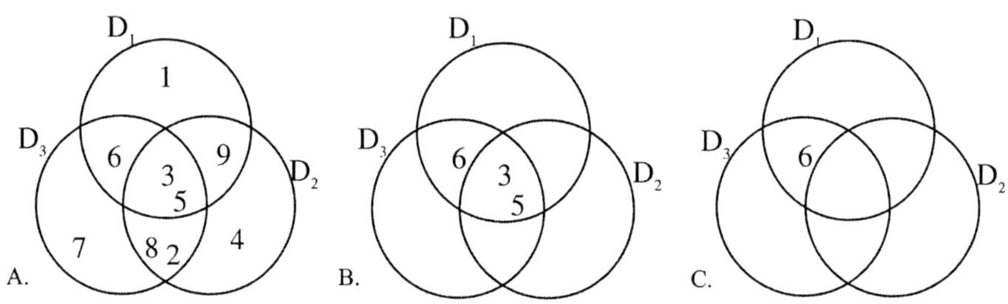

Figure 3: Two types of error coincidences for the classifier D_1 and D_3 of the ensemble $\{D_1, D_2, D_3\}$. A: Example of error indices distribution. B: General coincidences $C_G(\{D_1, D_3\}) = \{3, 5, 6\}$. C: Exclusive coincidences $C_E(\{D_1, D_3\}) = \{6\}$.

2.2 Collection Generation

The question might be asked: is the collection representation of error coincidences any better than binary matrices? We show that the answer to this question is a very decisive *yes*. First immediate advantage of the collection is that it stores only valuable information related to errors. For the purpose of combined error modeling, the superiority of information related to errors rather than correct outputs was recently confirmed in [7] and formulated as asymmetry of the diversity measure with respect to the output change. Furthermore, whatever the purpose of using error coincidences, the information regarding any specific coincidence has to be accessible as fast as possible and the collection of classifier sets offers further advantages in comparison to the binary matrices. Rather than extracting all different combinations of classifier set intersections as it happens for the binary matrix of outputs, the collection offers much simpler access to any coincidence information, exploiting inclusion relation among different coincidences. Inclusion properties can be very effectively implemented for both quick generation of a collection data structure and extraction of any information in a graph propagation manner.

Generation of the collection of classifier sets is not a straightforward process and it requires a specific data structure. As shown in the previous section general coincidences of higher orders are non-increasing in value. In practice they are decreasing very quickly as long as the classifier errors are not strongly correlated. Exploiting this fact we show a simple mechanism for rapid generation of all general coincidences. One observation is here important. Namely, a general coincidence of the kth-order can be most effectively obtained from the intersection of two general coincidences of the order $k-1$ as they have the most narrowed set of elements that we have to deal with. Starting from the lowest coincidence levels it is possible to devise a recurrent algorithm rapidly generating all general coincidence levels. Both the algorithm and an associated graph are shown in Figure 4.

Assuming that we have easier access to exclusive coincidences, we can immediately retrieve the general coincidences according to (6). Graphs associated with Venn Diagrams can be effectively used for an illustration of this procedure as shown in Figure 5. General coincidence of the kth-order is found as a sum of all exclusive coincidences (nodes) found on all the paths of ordered graph propagation starting from the highest order exclusive coincidence down towards the respective exclusive coincidence of the kth-order included in the considered general coincidence.

3 Majority Voting

Majority voting (MV) is an example of a simple fusion operator that can be applied for combining multiple classifiers with both soft and binary outputs [14], [15]. Moreover, it can be applied practically for any classifiers as their outputs can always be mapped, if necessary, to the binary representation.

A.

```
function gc=GenSetCon(c)
if size(coll)==2 then   %if 2 sets
   ints = c(1)∩c(2)     %intersection
   gc= |ints|           %cardinality
else
   M=size(coll);        %number of sets
   gc=[];
   for i=1:M-1
      ints=[];
      for j=i+1:M
         ints=[ints, c(i)∩c(j)];
      end;
      gc=[gc; {|ints|}];
      gc=[gc; GenSetCon(ints)];
   end;
end;
```

B.

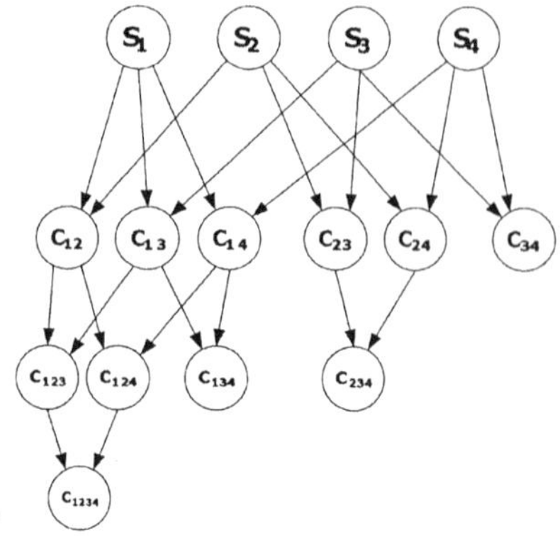

Figure 4: Collection generation. A: Algorithm. B: Visualization of the collection generation process.

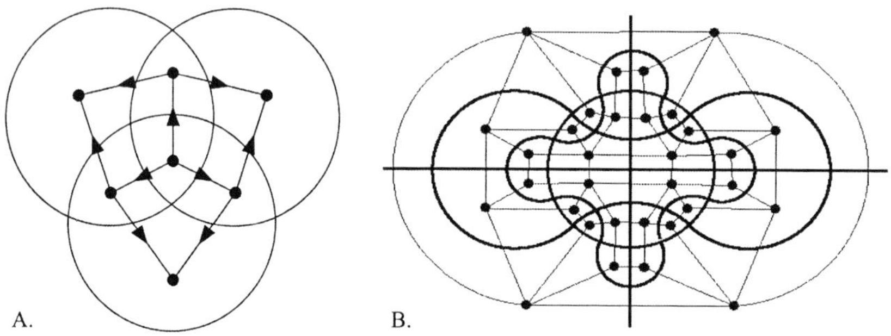

Figure 5: Graphs associated with Venn Diagrams. A: An ordered graph of exclusive coincidences for 3 classifier sets. B: Unordered graph for Edward's construction of 5 sets. To order the graph, all vertices have to be directed towards lower order coincidence.

Given a system of M classifiers: $D = \{D_1, ..., D_M\}$ applied for N input data, x_i, $i = 1, ..., N$ we can obtain the binary matrix of outputs $Y^{N\times M}$ (0-correct, 1-error) introduced in Section 2. The decision of majority voting combiner for a single i^{th} data sample can be obtained according to the following formula:

$$y_i^{MV} = \begin{cases} 0, & \text{if } \sum_{j=1}^{M} y_{ij} \geq \lceil M/2 \rceil \\ 1, & \text{otherwise} \end{cases} \tag{14}$$

For the whole binary matrix the estimate of the combined error would be then:

$$e_{MV} = \frac{1}{N}\sum_{i=1}^{N} y_i^{MV} \tag{15}$$

A more detailed definition of MV including the rejection rule observed for $\sum_{j=1}^{M} y_{ij} = M/2$ when M is even can be found in [14]. However, this work is not concerned with a detailed study of MV itself and in further analysis, without any loss of generality, we assume odd M.

3.1 Set Definition of Majority Voting Error

What is interesting about majority voting with respect to the presented set representation of error coincidences is that the definition of the MV error can be explicitly expressed by means of both types of coincidence levels as shown in the following relations:

$$e_{MV} = \frac{1}{N}\sum_{i=\lceil M/2\rceil}^{M} L_i^E = \frac{1}{N}\sum_{i=\lceil M/2\rceil}^{M} (-1)^{i-k}\binom{i-1}{k-1} L_i^G \tag{16}$$

What appears from the above definition is that not surprisingly, only higher levels of exclusive coincidences decide about the level of majority voting. Considering general coincidences this relation is no longer that explicit as although only higher levels of general coincidences are taken to the definition (16), the lower levels also include this information and may play an important part in the potential modelling of the majority voting error. The question is thus, whether a single coincidence levels or their combinations of any type may convey a meaningful information about the performance of majority voting. This problem is extensively addressed in the experimental section.

3.2 Relation with Diversity Measures

In many recent publications the relation between various diversity measures and the performance of majority voting combiner has been studied [6], [7]. In our work [7] we showed that two diversity measures: 'double fault' and 'fault majority' had especially high correlation to the majority voting error. We recall now briefly the definitions of these two measures.

- *Double Fault.* The measure can be described simply as an estimate of the probability of coincident errors for a pair of classifiers, which can be formalized by:

$$DF = \frac{N^{00}}{N} \tag{17}$$

where N^{00} denotes number of samples misclassified at the same time by both classifiers and N is a total number of samples.

- *Fault Majority.* The measure was developed on the basis of error distribution Z introduced in [15] and decomposed error distribution presented in [7]. The error distribution is defined as $Z = [z_0, ..., z_M]$ where z_k denotes the ratio of the number of samples misclassified by exactly k classifiers to the total number of samples. In the decomposed version error distribution can be separated into M components: $Z_i = [z_{i0}, ..., z_{iM}]$ where z_{ij} stands for the degree to which i^{th} classifier participate separately in j^{th} level of Z. For further details regarding error distributions refer to [15] and [7]. Having these definitions the measure of fault majority is defined by:

$$FM = \sum_{j=\lceil M/2 \rceil}^{M} \sum_{i=1}^{\lceil M/2 \rceil} Z_{ij} \tag{18}$$

The strength of both measures has been identified as resulting from the asymmetry of their definitions with respect to the change of classifier outputs and greater emphasis put on measuring coincidences of errors. The set analysis of error coincidences presented in this paper can be used to generalize the problem of error focused diversity measures and provide further tools to construct other effective diversity measures. For instance it can be noticed that the mean classifier error is an equivalent to the 1st-order general coincidence level: L_1^G and the *double fault* measure is nothing else but a normalized 2nd-order general coincidence level: L_2^G in our set representation presented above. Moreover, the most successful *fault majority* measure has an explicit relation with exclusive coincidence levels and the whole concept of discrete error distributions shown in [7], [15] on which it is based, represent normalized exclusive coincidence levels introduced in section 2.1. The usefulness of higher order coincidences for modelling of combiner performance is rather unclear and rarely discussed in the past mainly due to high complexity of such analysis. Nevertheless in the experimental section we examine in detail the relationship between different variations of coincidence levels and majority voting performance. The results of these experiments indicate significant potential applications of the set-based analysis of coincident errors for variety of purposes related to majority voting combiner.

4 Experiments

The experiments have been carried out in two groups. In the first simple experiment we intend to evaluate the performance of the collection data structure in terms of the time of accessing information about error coincidences and compare it against traditional but optimized method of coincidences retrieval from the binary matrix of outputs. The purpose of a second experiment is to comprehensively examine the relationship between majority voting error and different levels of error coincidences. In particular we want to reevaluate the significance of mean classifier error and 'double fault' diversity measure (here 1st and 2nd-order coincidence levels) in relation to the majority voting performance and learn if the higher order coincidence levels may bring some new valuable information improving this relation. To maintain the generality of our findings we chose 11 out of 19 commonly used classifiers and applied them for a classification of 8 real datasets taken mostly from the UCI Repository of Machine Learning Databases. A short description of classifiers and datasets is provided in Table 1 and Table 2 respectively.

4.1 Experiment 1

To prove the efficiency of set representation of error coincidences we compared the time of extracting the cardinalities of all possible general coincidences from the collection data structure and from the binary matrix of outputs. For that purpose we generated a series of artificial binary matrices of outputs $Y^{10000 \times k}$ where $k = \{3, 5, 7, 9, 11\}$ equivalent to the outputs from $k = 3, 5, ..., 11$ classifiers with mean error of 40%. For each of these matrices we generated the respective collection data structure and for both matrix and collection we measured the time needed to calculate cardinalities of all possible combinations of general error coincidences extracted from the binary matrix of outputs and from collection according to the algorithm shown in Figure 4. The results are shown in Table 3.

Although complexity of the process remains the same, the time of extracting information about all general coincidences is shorter for the collection representation of error coincidences. The time savings become more significant for the larger number of classifiers. The results confirm an advantage of the set representation of error coincidences in terms of accessibility and managebility of vital information about error coincidences.

Classifier name	Description
klclc	Linear classifier by KL expansion of cov. matrix
kljlc	Linear classifier by KL expansion on the joint data
loglc	Logistic linear classifier
fisherc	Minimum least square linear classifier
ldc	Normal densities based linear classifier
nmc	Nearest mean linear classifier
nmsc	Scaled nearest mean linear classifier
perlc	Linear classifier by linear perceptron
persc	Linear classifier by non-linear perceptron
pfsvc	Pseudo-Fisher support vector classifier
qdc	Normal densities based quadratic classifier
udc	Uncorrelated normal dens. based quadratic classifier
knnc	k-nearest neighbor classifier
parzenz	Parzen density based classifier
treec	Binary decision tree classifier
lmnc	Feed forward NN by Levenberg-Marquardt
rbnc	Radial basis NN classifier
rnnc	Random NN classifier
bpxnc	Feed forward NN classifier by backpropagation

Table 1: Description of classifiers used in the experiments

Dataset	samples	features	classes
Wine	178	13	3
Iris	150	4	3
Thyroid	215	5	3
Texture	5500	40	11
Biomed	194	5	2
Liver	345	6	2
Satimage	6435	36	6
Chromo	1143	8	24

Table 2: Description of datasets used in the experiments

4.2 Experiment 2

In this experiment we examined relation between majority voting error and individual/combined coincidence levels of both types. First step towards

Classifiers	$Time_{BinMtx}$ [s]	$Time_{Coll}$ [s]
3	0.02	0.01
5	0.16	0.12
7	0.84	0.58
9	4.49	2.96
11	22.77	14.58

Table 3: Comparison of the time needed to extract cardinalities of all general coincidences from a binary matrix of outputs and a collection for different number of classifiers.

this goal was to prepare binary matrices of classification outputs. Initially we applied 19 different classifiers for 8 real datasets split 100 times randomly into equally populated training and testing set. Descriptions of both classifiers and datasets used are provided in Table 1 and Table 2 respectively. The classifier outputs obtained for only testing set were hardened and stored in large binary matrices of outputs. It very quickly became apparent that error coincidence analysis of 19 classifiers is computationally very expensive and virtually impossible to perform using available resources. For this reason only 11 classifiers have been chosen guided by the minimum average *double fault* (see section 3.2 for details) measure and the binary matrices reduced accordingly.

In the next step, out of these 11 classifiers we considered all combinations of 3, 5, 7 and 9 classifiers for which all exclusive and general coincidence levels have been calculated and stored together with the majority voting error associated with each combination. We measured dependency between majority voting error and different coincidence levels separately for different sizes of the classifier team. Effectively for each series of all k-element combinations of classifiers we obtained k correlation coefficients between majority vote errors and k series of 1st to k^{th}-order coincidence levels measured separately. Figure 6 shows the evolution of the correlation coefficients along increasing levels of general error coincidences for all examined datasets. Clearly lower levels of general coincidence are better related to the majority voting error than higher levels. There is however no consistent rule which of the lower levels is the most informative in terms of correlation to the majority voting error. For the extreme example of the *Wine* dataset the 1st-level of general coincidence is surprisingly completely uncorrelated with majority voting but already the 2nd-level reaches a very high value of the correlation coefficient (above 0.9). On the other hand in the case of the *Liver* dataset the 1st-level of general coincidence is the most correlated with combined error and the correlations for the following levels are falling dramatically to values close to zero starting from the 3rd-level onwards. Taking into account all the datasets from Figure 6, 1st and 2nd general coincidence levels share the position of the most correlated information to the majority voting error. However, in the case of the *Texture* dataset, for the combinations of 9 classifiers, the maximum correlation seems to be at the 3rd-level and the shapes of the correlation curves suggest that for a larger number of classifiers this maximum could possibly be shifted towards higher coincidence levels. The tendency of increased correlation coefficient for the fixed level of general coincidence but rising number of classifiers is also evident for all datasets. The exception to this rule is observed only for the 1st level where this tendency is opposite. Effectively, the emerging conclusion could be that for a small number of classifiers the 1st coincidence level (equivalent to the mean classifier error) has a greater chance to be better correlated with majority voting than for large number of classifiers where the 2nd and possibly higher levels are more likely to have the maximum correlation with the majority voting error.

The equivalent relations between majority voting error and exclusive coincidence levels, are shown for just 2 datasets in Figure 7. The plots represent two patterns of correlation curves that we persistently observed for all datasets. For some datasets the correlation curves tend to peak around the middle level and fall on both sides as shown for the *Texture* dataset, whereas for other datasets the correlation curves were completely shapeless oscillating chaotically around zero or slightly above zero correlation level. The first pattern is easy to explain, as the exclusive coincidence levels are the ones taken directly to the sum forming majority voting error definition. The middle levels are positioned at the decision boundary of majority voting and are the first taken to the error definition. The second pattern, very commonly observed among the datasets, proves however that exclusive coincidence levels on their own represent rather insufficient information for effective modelling of majority voting error.

Furthermore, apart from the relation to individual coincidence levels we also examined the MV error relation to the three types of sums of coincidence levels:

1. k^{th} sum of type 1, where the sum of coincidence levels from 1 to k out of M levels has been used.

2. k^{th} sum of type 2, where the sum of coincidence levels from k to M out of M levels has been used.

3. k^{th} sum of type 3, where the sum of coincidence levels from $\lceil M/2 \rceil$ to M out of M classifiers according to the formula (9) has been used.

These sums have been designed as simple examples of the combinations of coincidence levels and are intended to indicate whether the joint coincidence information could improve the correlation with majority voting error. Figure 8 shows the results for type 1 sums for general coincidence levels illustrated for 4 representative datasets. For comparison the correlation curves for individual general coincidence levels (depicted in Figure 6) are also shown here using thin lines. What is striking for the *Liver* dataset is the fact that the sum of the first two general coincidence levels is better than the 1st level individually despite drastic fall of correlation for the 2nd level individually. In other words the pairwise coincidence information brings such vital additional information to the mean classifier errors that jointly they exhibit higher correlation to the majority voting error. If correlation curves are falling slowly with increasing coincidence level, the sum of more than two first levels may be optimal. For the *Satimage* dataset the sum of the first 3 and for the *Thyroid* the sum of the first 5 general coincidence levels gave the maximum correlation with the majority voting error. In the extreme cases like for the *Wine* dataset the sum of coincidence levels is not optimal due to a disastrous contribution of a completely uncorrelated 1st-level. Nevertheless addition of further general coincidence levels substantially improves correlation coefficients. Summarizing, the sum of type 1 of the general coincidence levels may substantially improve correlation with majority voting error comparing to the correlation with individual coincidence levels. Furthermore, it prevents to a certain degree, situations of completely uncorrelated measure that can happen for individual general coincidences as shown for the *Liver* and *Wine* datasets. The gain in the correlation to the majority voting error is likely to continue by adding further general coincidence levels when correlation curves for individual coincidence levels are falling slowly.

Individual examples of the sums of type 2 and type 3 are shown in Figure 9 but they did not bring any significant results. This is somewhat surprising as in the special cases they become exactly the definition of majority voting error represented by exclusive and general coincidence levels respectively. The unit correlations between type 2 sums of exclusive coincidence levels and majority voting errors are observed for summing the higher levels starting from ($\lceil M/2 \rceil$ for M classifiers) as shown for the *Satimage* dataset in Figure 9. For the case of general coincidences the unit correlations between type 3 sums of general coincidence levels and MV errors are observed for summing the higher levels starting from the middle according to (9) as shown for the *Biomed* dataset in the same figure. However only small variations from these precisely defined sums result in a total decorrelation of their values with the majority voting error.

5 Summary, Conclusions and Future Work

In this paper we attempted to show that relations among errors of multiple classifiers could be very effectively modelled by a collection of overlapping sets holding indices of misclassified samples for each classifier. Comparing this set representation of error coincidences with complete binary matrix of classifier outputs we proved experimentally that the collection of classifier sets offers faster access to the information about error coincidences. This important advantage has been achieved by excluding redundant information related to the correct classification outputs and efficiently exploiting inclusion relation among coincident errors for the algorithm of fast coincidence retrieval. Comparing with probabilistic models of error coincidences our set representation covers all possible probabilistic measures with the additional advantage of keeping indices of individual samples misclassified by particular classifiers. Additionally the error coincidences represented by a collection of sets can be, to a certain degree, visualized by Venn Diagrams, examples of which have been shown in this work, contributing to a greater understanding of the complexity of the error relations.

Enormous number of subsets arising from the collection of overlapping sets inspired us to define the two complementary types of error coincidences: exclusive and general. Exclusive error coincidences measure simultaneous error occurrences of selected classifiers having correct outputs from all other classifiers. General coincidences measure joint error occurrences for selected classifiers only, and ignore the background information from other classifiers. To handle such distinct information arising from a vast number of subsets and also for the application purposes we defined respective coincidence levels. The levels of coincidences numerically describe the sum of cardinalities of all coincidence subsets of the same number of classifier sets. By performing a simple algebra on the coincidence levels we derived definitions of a number of diversity measures operating on error coincidences. Not only mean classifier error (1st general coincidence) and *double fault* measure (2nd general coincidence level) but also complex *fault majority* measure and even majority voting error itself is shown as a quantity derivable from simple operations on coincidence levels.

Developing the findings from our previous work [7] related to the suc-

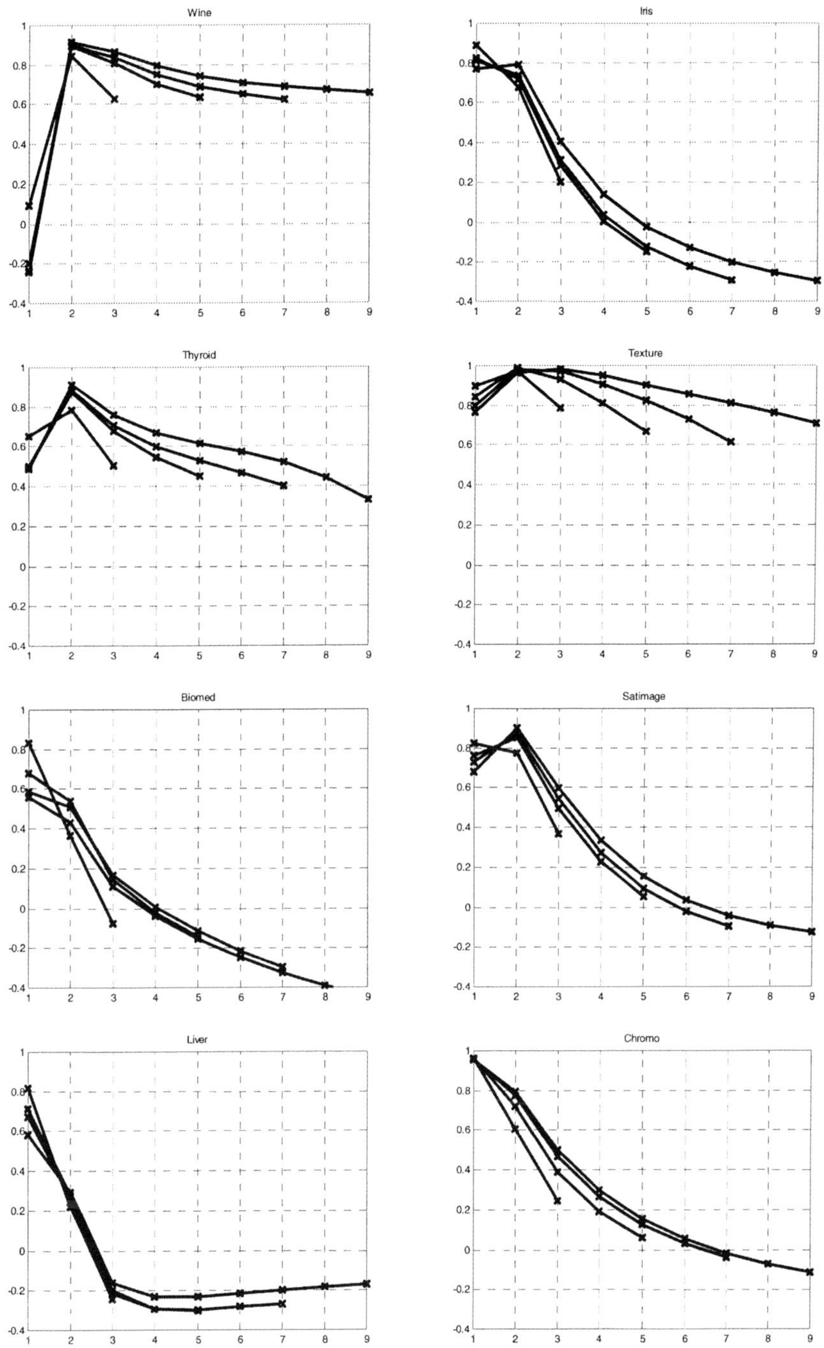

Figure 6: Correlation coefficients between MVE and general coincidences grouped in series of 3,5,7,9 out of 11 classifiers for 8 considered datasets.

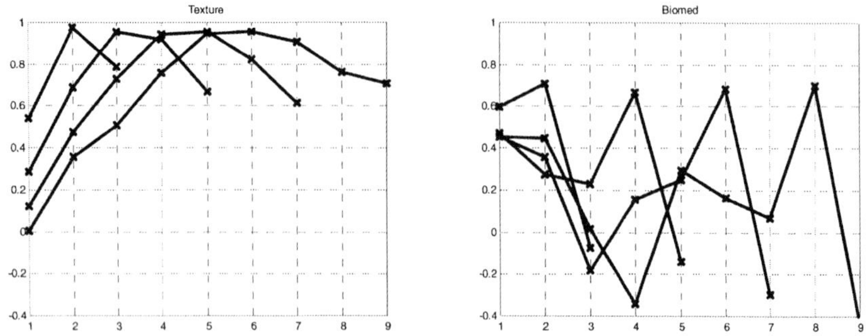

Figure 7: Correlation coefficients between majority voting error and individual exclusive coincidences grouped in series of 3, 5, 7, 9 out of 11 classifiers for 2 representative datasets showing two patterns of the relationship observed for all 8 examined datasets.

cessful asymmetric diversity measures, we investigated the relations between majority voting errors and individual and combined error coincidence levels examined by the correlation coefficients. For a number of real datasets we showed experimentally that general coincidences represent the information much better related to the majority voting errors than exclusive coincidences. What is even more precious the lowest levels of general coincidences, which represent the least complex measures, appear to be the most informative in terms of the relation to the majority voting error. Unfortunately, the linearly complex 1st-level of general coincidence representing simply mean classifier error is not always the best choice. It appears that the 2nd-level of general coincidence (*double fault*) would show the best average correlation with the majority voting error, which confirms the findings from [7]. Further analysis revealed that for some datasets the 3rd general coincidence level exhibited the higher correlation with the majority voting error and for a larger number of classifiers the optimal level is quite likely to be shifted towards higher coincidence levels. Another novel finding is that combining general coincidence levels in a form of simple sums very often shows further improvement of the correlation with majority voting error which are rising even if added level is on the falling individual tendency. This proves that different coincidence levels bring some portions of a unique valuable information related to the majority voting, which jointly can be used to achieve better correlated measures. Only for the cases when the first level is much less correlated than the second level, the sum of the first two general coinci-

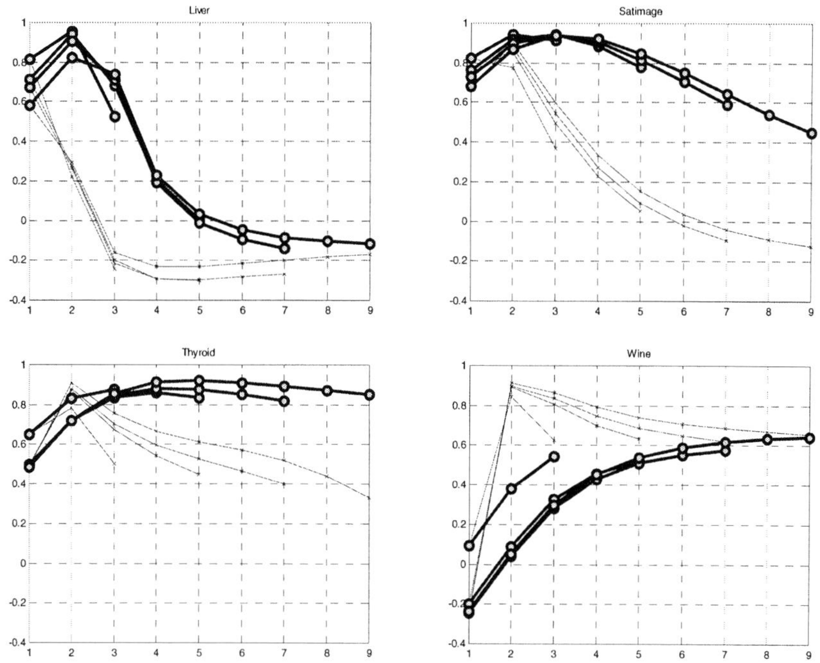

Figure 8: Correlation coefficients between majority voting error and type 1 sum (from 1st to k^{th} level) of general coincidence levels here showed in bold lines. For comparison, correlation curves of the individual general coincidences are also shown in thin lines. 4 out of 8 examined datasets with most representative patterns of the relationship are presented.

dence levels is not the optimal choice. For a smoother individual correlation curve, the sum of even more levels shows an optimal behaviour. The a priori prediction of the evolution of the correlation curves is however difficult and possibly needs additional information related to the dataset and classifiers.

Summarizing, lower levels of general coincidences both individually and in simple combinations show high correlations commonly exceeding 0.9 with the majority voting errors. This information is vital for a variety of applications related to the majority voting and possibly other combining methods. The above set analysis of error coincidence provides a good starting point for construction of simple yet well correlated with majority voting error diversity measure. The high correlation with combiner error means that it is possible to use the measure for selection of the best combination of classi-

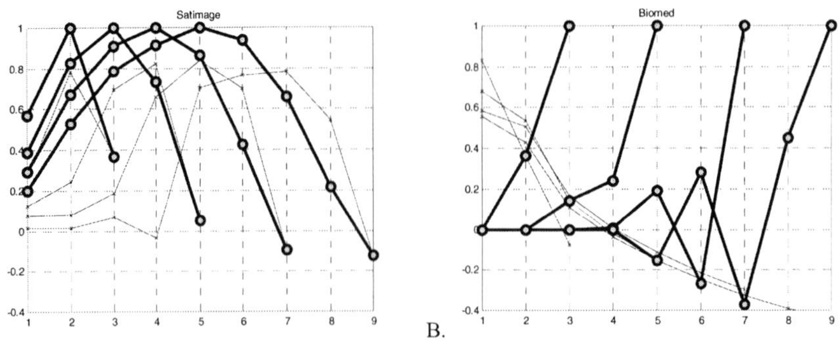

A. B.

Figure 9: Correlation coefficients between majority voting error and type2 and 3 sums of error coincidences. A: type 2 sum (from k^{th} to M^{th} level) of exclusive coincidence levels here showed in bold lines. For comparison, correlation curves of the individual exclusive coincidences are also shown in thin lines. B: type 3 sum of general coincidence levels with correlation curves of the individual general coincidences shown for comparison in thin lines.

fiers or more securely a number of best combinations as shown in Figure 10. Furthermore we intend to apply the above set analysis of error coincidences for the low cost approximation of the value of the majority voting error according to the novel definition (15), which requires estimation of any type of coincidence levels. The information about error coincidences can be also applied for the analysis of the limits of majority voting errors. Specifically, it would be interesting to analyze the level to which the combined error limits shrink with increasing knowledge of error coincidences.

References

[1] A.J.C. Sharkey, *Combining Artificial Neural Nets: Ensemble and Modular Multi-net Systems*, (Springer-Verlag, Berlin Heidelberg New York, 1999).

[2] A.J.C. Sharkey, N.E. Sharkey, Combining Diverse Neural Nets. *The Knowledge Engineering Review* Vol.12 No.3 (1997) pp. 231-247.

[3] G. Rogova, Combining the results of several neural network classifiers.*Neural Networks* Vol.5 No.5 (1994) pp. 777-781.

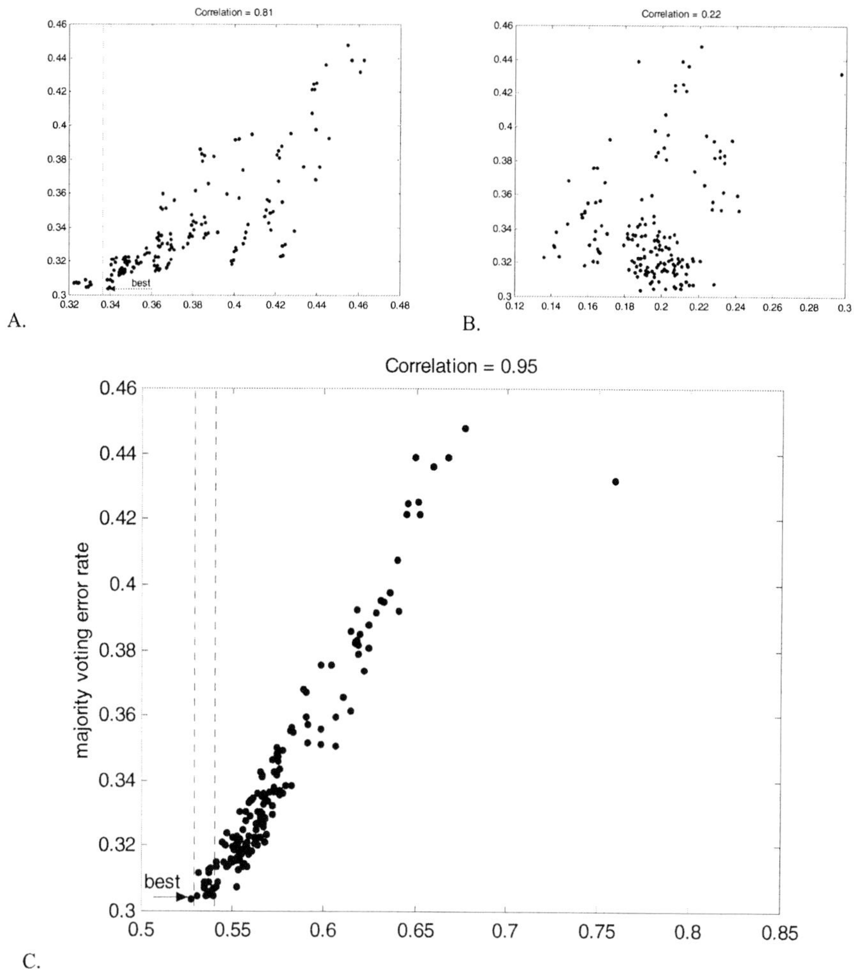

Figure 10: Illustration of the importance of correlation coefficients for classifier selection in the example of the relation between majority voting error and general coincidence levels of 3 out of 11 classifiers applied for the *Liver* dataset. A: Relation of the first general coincidence levels. B: Relation of the second general coincidence levels. C: Relation of the sum of the first and second general coincidence level with majority voting error.

[4] D. Partridge, N. Griffith. Strategies for improving neural net generalization. *Neural Computing & Applications* Vol.3 (1995) pp. 27-37.

[5] L.I. Kuncheva, C.J. Whitaker, Ten Measures of Diversity in Classifier Ensembles: Limits for Two Classifiers. *IEE Workshop on Intelligent Sensor Processing*, Birmingham (2001), 10/1-10/6.

[6] C.A. Shipp, L.I. Kuncheva, Relationships between combination methods and measures of diversity in combining classifiers. *Information Fusion*, accepted.

[7] D. Ruta, B. Gabrys, Analysis of the Correlation Between Majority Voting Errors and the Diversity Measures in Multiple Classifier Systems. *In Proceedings of the SOCO/ISFI2001 Conference*, ISBN: 3-906454-27-4, Abstract page 50, Paper No. #1824-025, Paisley, UK, 2001.

[8] D. Petridge, W.J. Krzanowski, Software diversity: practical statistics for its measurements and exploitation. *Information & Software Technology* Vol.39 (1997) pp. 707-717.

[9] B. Littlewood, D.R. Miller, Conceptual modeling of coincident failures in multiversion software. *IEEE Transactions on Software Engineering* Vol.15 No.12 (1989) pp. 1596-1614.

[10] H.-J. Kang, K. Kim, J.H. Kim, Optimal approximation of discrete probability distribution with kth-order dependency and its application to combining multiple classifiers. *Pattern Recognition Letters* Vol.18 (1997) pp. 515-523.

[11] K. Devlin, *Joy of sets: fundamentals of contemporary set theory*, 2nd eds., (Springer-Verlag New York 1993).

[12] F. Ruskey, A Survey of Venn Diagrams. *The Electronic Journal of Combinatorics* eds. March 2001, (the electronic version available at: http://www.combinatorics.org/Surveys/ds5/VennEJC.html).

[13] G. Giacinto, F. Roli, An approach to the automatic design of multiple classifier systems. *Pattern Recognition Letters* Vol.22 (2001) pp. 25-33.

[14] L. Lam, C.Y. Suen, Application of majority voting to pattern recognition: an analysis of its behavior and performance. *IEEE Transactions on Systems, Man, and Cybernetics* Vo.27 No.5 (1997) pp. 553-568.

[15] D. Ruta, B. Gabrys. A theoretical analysis of the limits of majority voting errors for multiple classifier systems. To appear in *Pattern Analysis and Applications.*

[16] R. Battiti, A.M. Colla, Democracy in neural nets: voting schemes for classification. *Neural Networks* Vol.7 No.4 (1994) pp. 691-707.

[17] L.I. Kuncheva, C.J. Whitaker, C.A. Shipp, R.P.W Duin, Limits on the majority vote accuracy in classifier fusion. *IEEE Transactions on Pattern Analysis and Machine Intelligence*, accepted.

Enhanced Neighbourhood Specifications for Pattern Classification

J. S. Sánchez
Department of Computer Languages and Information Systems
University Jaume I, Av. Vicent Sos Baynat S/N, 12006 Castellón, Spain
E-mail: `sanchez@uji.es`

A. I. Marqués
Department of Business Administration and Marketing
University Jaume I, Av. Vicent Sos Baynat S/N, 12006 Castellón, Spain
E-mail: `imarques@uji.es`

Contents

D. Chen and X. Cheng (eds.),
Pattern Recognition and String Matching, 673–702.

1 Introduction

Automatic classification is an active research area in artificial intelligence, pattern recognition, and machine learning. Its goal is to make a computer automatically label attribute vectors as examples of two or more different categories or classes. One of the most widely studied approaches corresponds to the well known Nearest Neighbour (NN) decision rule, proposed by Fix and Hodges [19], formally analysed by Cover and Hart [9], and then investigated both theoretically and experimentally by many researchers. On the other hand, a lot of successful applications have been reported in a multitude of real-world domains [10].

In its classical manifestation, given a set of n previously labelled prototypes (attribute vectors) or training set (TS), the NN classifier assigns any new pattern to the class indicated by the label of the most *similar* prototype in the TS. The degree of similarity between two samples is established by means of appropriate distance metrics defined on the feature space. Thus, the smaller the distance between the two attribute vectors, the higher their mutual similarity.

A slight improved version of the NN rule consists of considering k neighbours of the pattern to classify. Then, a new sample is assigned to the class most frequently represented among the k closest prototypes in the TS, according to a certain similarity measure. Apart from their conceptual and implementational simplicity, the NN decision rules are optimal under the theoretical assumptions of infinite number of training prototypes (that is, when $n \to \infty$). Cover and Hart [9] demonstrated that the asymptotic classification error rate of the k-NN rule tends to the optimal Bayes error as $k \to \infty$ and $k/n \to 0$. Moreover, if $k = 1$, the NN error is bounded by approximately twice the Bayes error [14].

Nevertheless, in many practical settings, such a theoretical behaviour can hardly be achieved because of certain inherent weaknesses that significantly reduce the applicability of k-NN classifiers in real-world tasks. For example, memory requirements can be prohibitive in domains with a lot of prototypes (e.g., in data mining applications), and also classification costs (in terms of response time) are too high if the decision rule has to compare a given unclassified pattern with millions of training examples.

Several works have focused on alleviating the high computational and storage loads of these classifiers by means of fast NN search algorithms [3, 15, 20, 33, 41, 43, 47, 59, 60]. Other techniques, namely *condensing*, try to minimize these practical problems by selecting a sufficiently small subset of prototypes that leads to approximately the same performance than the NN rule using the whole TS [6, 11, 16, 22, 25, 48, 55, 62].

Another drawback for NN classifiers refers to the fact that the TS may contain noisy, mislabelled and atypical prototypes, which generally lead to a significant decrease in performance. Wilson [61] proposed the first algorithm to tackle this problem: it utilizes the k-NN rule to estimate the class label of all prototypes in the TS and discards those whose label does not agree with the class associated with the largest number of the k neighbours. Subsequently, important research has been carried out to deal with this critical issue [2, 5, 14, 17, 26, 36, 49] both in pattern recognition and machine learning domains. In general, *editing* algorithms try to eliminate erroneously labelled prototypes from the original TS and "clean" possible overlapping among regions from different classes, usually yielding significant improvements in classification performance.

Joint use of editing and condensing tools, generally referred as to *prototype selection* techniques [14], leads to a composite approach in which both computational efficiency and classification accuracy are improved [12]. In fact, editing and condensing are two closely related and complementary procedures: condensing makes sense only when the classes are clustered and well-separated, which constitutes the aim of the editing algorithms.

On the other hand, NN classification accuracy significantly drops in environments where many attributes describing the prototypes are irrelevant; such attributes or features inappropriately affect the values returned by most similarity metrics. Langley and Iba [38] show that the number of prototypes needed to reach given accuracy grows exponentially with the number of irrelevant attributes. A considerable amount of research has been addressed to reduce sensitivity of NN rules to irrelevant features by selecting a subset of attributes [4, 21, 29, 34, 37, 46, 54] and also by assigning different weights to each feature [1, 8, 31, 39].

Recently, significant research in NN classification has focused on using distinct specifications of the neighbourhood concept inherent in all similarity-based classifiers. The ultimate goal of such alternative neighbourhoods is to overcome some of the aforementioned problems related with NN classification. In this context, several empirical studies [49, 50, 56, 64] have shown that in many cases, the degree of similarity among the prototypes in a TS cannot be appropriately estimated by taking into account only some distance measure, but also considering other geometrical relationships among them.

This chapter concentrates on an alternative neighbourhood concept, namely *surrounding neighbourhood*, proposed in the last years and applied to very different domains. We focus primarily on contributions to pattern classification and prototype selection. Accordingly, Section 2 introduces the rationale of surrounding neighbourhood and describes some particular implementations. The following two sections report practical applications of such a neighbourhood realization in the framework of pattern recognition. Finally, Section 5 includes the main points of this chapter and some comments on topics for further research.

2 Surrounding Neighbourhood

In this section, we explore the general concept of neighbourhood in the context of pattern classification and compare two alternative definitions: the plain or nearest neighbourhood (that is, neighbourhood defined from the minimum distance) and the novel surrounding neighbourhood. Furthermore, some particular realizations of the surrounding neighbourhood are here reported.

The concept of neighbourhood has widely been used in a large number of robotics, computer vision, pattern recognition, computational geometry and image processing problems [13, 24, 28, 63]. The neighbours of an input point p in a d-dimensional feature space can be loosely viewed as the set of points located in the proximity of p. In fact, the neighbours of a certain point p are usually assumed as those points that are the closest to p in terms of Euclidean or other norm-based distance metric.

However, neighbourhood can intuitively be understood as a property subject to two complementary constraints. Firstly, the neighbours of a point p should be as *close* to it as possible. And secondly, the neighbours should also be located as *symmetrically* around p as possible. As the nearest neighbourhood takes only the first condition into account, the neighbours selected may not be well distributed around the test point in the feature space. In general, when the neighbours of a point are homogeneously distributed around it, it seems that more local information is provided to appropriately classify the given point.

A number of different approaches have already been proposed in order to undertake this problem [7, 44, 57, 64]. All these methods have in common the fact of considering both proximity and symmetry so as to define the general concept of neighbourhood. Thus, they try to search for neighbours close enough (in the basic distance sense), but also homogeneously or symmetrically distributed around a given point. For this reason, some researchers have recently referred to this heuristic concept as *surrounding neighbourhood* [50, 64]. In particular, this chapter introduces three different approaches to such a neighbourhood representation: the graph neighbourhood [49], the nearest centroid neighbourhood [7], and the Zhang's neighbourhood [64].

2.1 Graph Neighbourhood

Let $X = \{x_1, \ldots, x_n\}$ be a finite set of n points in R^d, where d denotes the dimensionality of the feature space. From a general point of view, a proximity graph, say $G = (V, E)$, is defined as an undirected graph with the set of vertices $V = X$, and the set of edges, E, such that $(x_i, x_j) \in E$ if and only if x_i and x_j satisfy some mutual neighbourhood property. In such a case, it is said that x_i and x_j are *graph neighbours*. Thus the set of graph neighbours of a given point constitutes its *graph neighbourhood*. The graph neighbourhood of a subset, $S \subseteq V$, consists of the union of all the graph neighbours of every node in S.

The Gabriel Graph (GG) and the Relative Neighbourhood Graph (RNG), along with the Delaunay Triangulation (DT) and the Minimum Spanning Tree (MST) are probably the most prominent types of proximity or closest-point graphs [45]. As an important property, the DT is a supergraph of the GG, this constitutes a supergraph of the RNG, and this is a supergraph of the MST. Thus, they form the following hierarchical relationship:

$$MST \subseteq RNG \subseteq GG \subseteq DT$$

It is also important to remark that, with only minor modifications to the algorithms for computing the DT, any one of the other graphs can be easily and efficiently constructed [58]. Jaromczyk and Toussaint [30] reported a complete survey of the most relevant properties of these and other proximity graphs.

The DT of a set of points X is defined as the dual graph of the *Voronoi diagram* associated with the set X, which constitutes a decomposition of R^d into a number of convex regions or *Voronoi cells*, each of which defines the "region of influence" of a given point in its interior [45]. A Voronoi cell is a d-dimensional polytope that can be defined as the intersection of a finite number of closed half-spaces and is therefore, supported by a finite number of hyperplanes. Consider all triangles formed by the points such that the circumcircle of each triangle is empty of other points. The set of edges of these triangles gives the DT of the points. Therefore, two points in a DT are connected with an edge if and only if the boundaries of their Voronoi cells are adjacent.

In this chapter, we concentrate our examination only on the GG and the RNG because it has recently been done considerable work on the application of these particular proximity graphs to several classification problems. In fact, it has been demonstrated that they generally obtain a higher classification accuracy than other conventional NN-based approaches. On the other hand, there exist more efficient algorithms for the computation of the GG and the RNG in high dimensional spaces than for the construction of the DT [30, 45].

The complexity of a brute-force algorithm [56] for computing the GG (or the RNG) over an input data set with n points in d-space is $O(dn^3)$, which is absolutely prohibitive for all practical purposes. There also exists a heuristic method [56] with an expected computational cost close to $O(dn^2)$, although it has to be admitted that this is still of little value for several real-time applications.

2.1.1 Gabriel Graph

Let $d(\cdot, \cdot)$ be the Euclidean distance between two points in R^d, then the set of edges in a GG is defined as follows:

$$(x_i, x_j) \in E \Leftrightarrow d^2(x_i, x_j) \leq d^2(x_i, x_k) + d^2(x_j, x_k) \quad \forall x_k \in X, k \neq i, j$$

In this case, x_i and x_j are said to be *Gabriel neighbours* (GN). In other words, two points are GNs if and only if there is no other point from X laying in a hypersphere Γ_{x_i,x_j} (namely, diameter hypersphere) centered at their middle point and whose diameter is equal to the distance between them.

Let $B(p, r)$ denote an open hypersphere centered at p with radius r, i.e., $B(p, r) = \{q : d(p, q) < r\}$, then the diameter hypersphere Γ_{x_i,x_j} can be defined as:

$$\Gamma_{x_i,x_j} = B(\frac{x_i + x_j}{2}, \frac{d(x_i, x_j)}{2})$$

Now, the set of edges in a GG over a set of points X can be expressed using an equivalent definition as follows:

$$(x_i, x_j) \in E \Leftrightarrow \Gamma_{x_i,x_j} \cap X = \emptyset$$

From the set of GNs of a given point p, it is possible to define a surrounding region, here called the *Gabriel surrounding neighbourhood*, Γ_p, as the union of the diameter hyperspheres between p and all its GNs, x_i:

$$\Gamma_p = \cup \Gamma_{p,x_i} \quad \forall x_i \in X \ / \ \Gamma_{p,x_i} \cap X = \emptyset$$

2.1.2 Relative Neighbourhood Graph

In a similar fashion, the set of edges in a RNG can be obtained in the following way:

$$(x_i, x_j) \in E \Leftrightarrow d(x_i, x_j) \leq max\,[d(x_i, x_k), d(x_j, x_k)] \quad \forall x_k \in X, k \neq i, j$$

Now its corresponding geometric interpretation is based on the concept of lune Λ_{x_i,x_j}, which is defined as the disjoint intersection between two hyperspheres centered at x_i and x_j and whose radii are equal to the distance between them.

$$\Lambda_{x_i,x_j} = B(x_i, d(x_i, x_j)) \cap B(x_j, d(x_i, x_j))$$

Thus two points are *relative neighbours* (RN) if and only if their lune does not contain other points of the set X in its interior.

$$(x_i, x_j) \in E \Leftrightarrow \Lambda_{x_i,x_j} \cap X = \emptyset$$

In this case, the *relative surrounding neighbourhood* for a given point p, say Λ_p, can be defined as the union of the lunes between p and all its RNs, x_i:

$$\Lambda_p = \cup \Lambda_{p,x_i} \quad \forall x_i \in X \ / \ \Lambda_{p,x_i} \cap X = \emptyset$$

2.2 Nearest Centroid Neighbourhood

A second approach to surrounding neighbourhood comes from the concept of *nearest centroid neighbourhood* [7], which was initially applied to detect the border points and the interior points of a dot pattern, as well as to select a representative subset from a set of data.

In plain words, the nearest centroid neighbourhood concept is based on the direct application of those constraints mentioned at the beginning of this section. That is, by the *proximity criterion*, the k neighbours of a sample p must be as near as possible. At the same time, by the *symmetry criterion*, their centroid must be also as close to p as possible.

Let p be a given point whose k nearest centroid neighbours (NCN) should be found in a set of points $X = \{x_1, \ldots, x_n\}$, then the NCNs can be obtained as follows [7]:

1. The first NCN of p is its NN, say q_1.
2. The i-th neighbour, q_i, $i \geq 2$, is such that the centroid of this and all previously selected neighbours, $q_1, \ldots, q_{i-1}$ is the closest to p.

This iterative procedure clearly does not minimize the distance to the centroid because it gives precedence to the individual distances instead. However, because of the centroid criterion, the spatial distribution of neighbours is taken into account. On the other hand, proximity of the k NCNs to the input point is guaranteed due to the incremental nature of the way in which they are obtained from the first NN.

Let q_1 be the NN of a given point p, and let $C(S, y)$ denote the centroid of the set S and the point y. Then, the *centroid surrounding neighbourhood* for a given k, say $\Upsilon_k(p)$, can be defined as the set of prototypes that satisfy the following conditions:

1. $\Upsilon_1(p) = \{q_1\}$.
2. $\Upsilon_k(p) = \Upsilon_{k-1}(p) \cup \{x_i\}$, such that $d(p, C(\Upsilon_{k-1}(p), x_i)) \leq d(p, C(\Upsilon_{k-1}(p), x_j))$, $\forall x_i, x_j \in X - \{\Upsilon_{k-1}(p)\}$, $i \neq j$.

This kind of surrounding neighbourhood can be efficiently computed in $O(kn)$ time in any dimension. More specifically, the computation of one NCN requires at most n centroid and distance computations as well as n comparisons to find the minimum of the distances [7].

2.2.1 Nearest Centroid Neighbourhood with Local k

The greedy nature of obtaining successive NCNs broadens the search space, still with the centroid criterion in mind, even for the price of increasing computational requirements. This however seems difficult without an adverse effect on the proximity criterion.

Grabowski [23] has recently proposed a slight modification of the original concept of nearest centroid neighbourhood as a solution to such a relative drawback. For a given pattern p, all its k NCNs are computed and then, the first $k(p) \leq k$ neighbours, whose distance from the centroid $c_{k(p)}$ to p is minimal over all distances from centroids c_j, $(j = 1, \ldots, k)$ to p, are selected. In other words, this modification aims at minimizing the centroid criterion in a local manner.

2.3 Zhang's Neighbourhood

The surrounding neighbourhood implementations described in previous sections come from the pattern recognition domain. Nevertheless, researchers from the machine learning area have also utilised a similar concept in various applications. For example, Zhang et al. [64] proposed an algorithm to compute the so-called k surrounding neighbours of a given point. Here, this particular realization is called *Zhang's neighbourhood* so that it can be distinguished from the other surrounding representations.

The idea of the Zhang's approach is to select k prototypes that are not only close to a given sample, but also well distributed around it in the feature space. In other words, all selected neighbours should be close to the given prototype, but they should not be too close to each other. As can be seen, this rationale matches the concepts of graph neighbourhood and nearest centroid neighbourhood, in the sense of trying to select a set of prototypes widely distributed around the given point.

Let p be a new point whose k Zhang's neighbours (ZN) should be selected from a set of points $X = \{x_1, \ldots, x_n\}$, then these ZNs can be obtained as follows [64]:

1. The first ZN of p is its NN, say z_1.
2. The i-th neighbour, z_i, $i \geq 2$, is such that satisfies the following two conditions:
   - $d(z_i, z_{i-1}) > d(p, z_i)$
   - $d(x_j, p) \geq d(p, z_i) \quad \forall x_j \in X - \{z_k\},\ k = 1, \ldots, i$

The first statement implies that at each iteration, the two selected prototypes are not too close to each other, and the second condition means that they are "sufficiently" close to the new point p. As can be seen from the above procedure, like in the case of the NCN algorithm, proximity of the k neighbours to the input point is guaranteed because the first ZN is also selected from the first NN.

3 Classification Methods

Like the traditional nearest neighbourhood concept, the surrounding approaches can be applied to a number of pattern classification problems. In fact, the ultimate goal of the surrounding neighbourhood is to overcome some of the practical drawbacks outlined for the NN-based techniques and more specifically, to outperform the classification accuracy of such decision rules. Thus, this section summarises some classification schemes derived from the different realizations of surrounding neigbourhood introduced in Section 2.

The general idea behind these classifiers is to make a decision about the class membership of a given unclassified pattern after knowing the geometrical distribution of the training prototypes around it on the feature space. It must be pointed out that, even though there is no theoretical justification to use neighbours around (and also close) to a sample instead of NNs, this can effectively help in many practical situations (e.g., finite sample size case), in which prototypes do not fully represent the underlying statistics and/or the particular distance employed (irrelevant in the asymptotic case) exhibits some undesirable properties.

3.1 Classifiers Based on Proximity Graphs

Let $X = \{x_1, \ldots, x_n\}$ be a set of n training prototypes in a d-dimensional feature space, belonging to L different classes, and let q a new pattern to classify, then a classifier based on some kind of proximity graphs can be expressed as follows [50]:

1. Search for all the graph neighbours of q, say $X^q = \{x_1^q, \ldots, x_m^q\}$, where $m \leq n$.
2. Assign q to the class with a majority of votes from its m graph neighbours in the set X^q (resolve ties randomly).

Note that, considering the GG and the RNG, we have two different but analogous approaches, namely GN and RN decision rules, respectively. On the other hand, it is worth mentioning that the corresponding classifiers with a reject option [27] and other extensions [14] could also be defined exactly in the same way as in the k-NN rule.

It has empirically been demonstrated that the GN and RN classifiers generally outperform the k-NN decision rule, specially in domains where the number of training prototypes is not very large [50]. The reasons of such a fact would be two-fold: the classifiers based on this kind of surrounding neighbourhood seem to be more reliable because the estimate criterion employed takes into account the distribution of neighbours around the given unclassified sample and secondly, the computational cost to construct any of these graphs is still too high to consider the corresponding GN or RN rules in problems with a very large TS.

3.2 k-NCN Classification Rules

Several classification schemes derived from the nearest centroid neighbourhood have been proposed. The aim of using this kind of surrounding neighbourhood matches the general goals of the classifiers based on proximity graphs. In fact, the k-NCN decision rule differs from the GN and RN classifiers only in the neighbourhood criterion adopted.

This general strategy has been performed in a wide variety of implementations. Nevertheless, all of them primarily focus on the geometrical distribution of k neighbours around the test sample, instead of their distance to it.

The first and most trivial application of the nearest centroid neighbourhood to classification refers to the so-called k-NCN rule [50]. Let q be a certain pattern to classify, then the plain k-NCN approach can be defined in the following way:

1. Find the k NCNs of q.
2. Assign q to the class with a majority of votes among its k NCNs (resolve ties randomly).

Several examples of its practical employment in different real-world problems [17, 42] have demonstrated a certain increase in performance compared with the classification accuracy achieved by the conventional k-NN decision rule.

Just like in the case of the k-NN rule, also for the k-NCN classifier the number of neighbours k must be estimated with respect to each particular TS, preferably by means of the leaving-one-out method.

3.2.1 Averaged k-NCN Rule

A first extension to the k-NCN classification rule is meant to resolve the k-NCN ties [51]. The "tie-breaking" modification can be defined as follows: a given unclassified pattern is assigned to the class that has received at least a half of the k votes (say, $l = k/2$). If none of the possible classes receives at least l votes, it is computed the average distance between the test sample and its NCNs from a same class. In such a case, the given pattern will be finally assigned to the class with the least average distance among all classes voted.

This alternative is similar to that defined for the k-NN rule with a reject option [27]. However, it mainly differs from such a classifier in that the reject option is here substituted for an additional distance test.

1. Find the k NCNs of q.
2. If there exists a class with at least $l = k/2$ votes, then assign q to such a class.
3. If not, compute the average distance between q and each one of the classes voted. Assign q to that class with the least average distance.

3.2.2 Edited k-NCN Rule

A new proposal [51] corresponds to a matter of editing the training prototypes because it considers only the subset of NCNs that are correctly labelled according to the NN classifier. After computing the k NCNs of a given test sample, we select those whose NN among all prototypes in the TS has their same class label assigned. The advantage of using an editing strategy comes from the fact that it tends to remove outliers and retain only prototypes grouped into clustered regions with no overlapping among regions from different classes.

This classifier combines the surrounding neighbourhood provided by the plain k-NCN rule with the conventional nearest neighbourhood, here used in the editing stage.

1. Find the k NCNs of q.
2. Select the subset of NCNs, $X^q = \{x_1^q, \ldots, x_j^q\}$, $j \leq k$, correctly labelled according to the NN rule.
3. Assign q to the class with a majority of votes among those j NCNs (resolve ties randomly).

3.2.3 Pairwise k-NCN Rule

Grabowski [23] proposed a modification of the k-NCN classifier, which is analogous to the scheme introduced by Józwik and Vernazza [32] for the conventional k-NN rule. Thus, instead of dealing with a problem of L different classes, it is decomposed into $L \cdot (L-1)$ two-class problems. Then, the given unclassified point is submitted to all component classifiers and the final decision is obtained by simple voting. Such an ensemble of classifiers is called *pairwise k-NCN* or *parallel net of binary k-NCN classifiers*.

Nevertheless, it is to be noted that preliminary experiments conducted by Grabowski [23] suggest that the classification accuracy obtained by the pairwise approach are insignificantly better than those achieved by the plain k-NCN decision rule.

4 Prototype Selection Techniques

As already mentioned in Section 1, prototype selection aims at obtaining a TS in which both computational efficiency and classification accuracy are improved. While editing approaches try to eliminate erroneously labelled prototypes from the original TS and clean possible overlapping among different class regions, condensing schemes reduce the TS set size without an important accuracy degradation. The heuristic nature of condensing algorithms contrasts with the strong statistical foundation of the most popular editing techniques.

The first work on editing corresponds to that of Wilson [61], and then many others has followed [2, 14, 17, 35, 36, 49, 55]. Differences among most approaches refer to the classification rule employed for editing purposes along with the error estimate and the stopping criterion considered [18]. The error estimate is only used in this context to decide which prototypes must be removed from the original TS. The stopping criterion provides the possibility of applying editing in a iterative way.

On the other hand, Hart [25] proposed to minimize the number of training prototypes by by selectively discarding the redundant part of the TS. The basic idea is that prototypes in the TS may be very similar and some do not add extra information and thus may be eliminated. In brief, Hart's scheme consists of discarding those prototypes from the TS that are not necessary to correctly classify the rest of training patterns.

This section provides several editing and condensing algorithms based on some kind of surrounding neighbourhood. More specifically, proximity graphs have been applied both to editing and condensing [49, 56], and the nearest centroid neighbourhood has been used in many different editing procedures [17, 52]. Furthermore, an algorithm involving a combined application of editing and condensing both based on proximity graphs is also presented [49]. This combined technique includes a fast recomputation of the corresponding graph structure in a very simple way.

4.1 Editing Through Proximity Graphs

Most of classical editing algorithms consist of discarding prototypes with a sufficiently heterogeneous neighbourhood in terms of class labels. Thus it seems obvious that proximity graphs can be used as well to obtain an appropriate edited set of prototypes.

The graph-based approach [49] consists of applying the general idea of Wilson's editing algorithm [61] but using the GN or RN of each prototype, instead of the classical nearest neighborhood, in order to estimate mislabelled patterns. In a few words, the simplest graph editing can be summarised as follows: after computing the graph neighborhood of each training prototype, discard those patterns that are misclassified by their graph neighbors (instead of their k-NN). In practice, a first approach to this general editing scheme can be written as follows:

1. Construct the proximity graph corresponding to the original TS, say X.
2. For each $x_i \in X$ do:
   - Discard x_i if there is a majority of graph neighbors from a different class.

A further refinement of this general idea consists of taking not only the graph neighbours of a point, but also the neighbours of the graph neighbours from its same class (i.e., a subset of some of the second level graph neighbours). Actually, we are trying to add reliability to the detection of outliers close to the boundaries among classes. Therefore, the previous algorithm can be modified as follows:

1. Construct the proximity graph corresponding to the original TS, say X.
2. For each $x_i \in X$ misclassified by its graph neighbours do:
   - Consider the subgraph, S, given by x_i and all its graph neighbours from its same class.
   - Discard x_i if the graph neighbourhood of S has a majority of neighbours from some different class than x_i

4.2 Toussaint's Condensing

Toussaint et al. [56] proposed a condensing algorithm based on the employment of the GG or RNG of the TS. It is inspired from the *Voronoi condensing algorithm*, which finds a reduced TS by using the Voronoi diagram of the TS. The Voronoi condensing scheme is the only procedure that reduces the TS in such a way that the NN boundaries, defined by the reduced set and the original TS are exactly the same, i.e., the reduced set is decision boundary consistent and therefore, also TS consistent.

Nevertheless, the Voronoi condensing algorithm requires the construction of the Voronoi diagram, which is still a time consuming process [45]. Any algorithm in the worst case will take at least $O(n^{d/2})$ time. This practical drawback leads to consider alternative approximate methods by using the GG and RNG. Algorithmically, the GG or RNG condensing scheme can be written as follows:

1. Construct the proximity graph corresponding to the original TS, say X.
2. Visit each node of the proximity graph and mark the visited node if all its graph neighbours are not from the same class as the node visited.
3. Discard all prototypes in X corresponding to nodes in the proximity graph that are not marked.

4.3 A Combined Editing-Condensing Algorithm

The algorithms just introduced can be extended in a number of ways. One of the most obvious consists of obtaining an edited-condensed set using the same graph structure. In fact, editing and condensing are two closely related and complementary techniques [14]: condensing makes sense only when the classes are clustered and well-separated, which constitutes the main objective of any editing algorithm.

There exist some practical advantages if we are to apply both editing and condensing using graph neighbourhood-based approaches. In particular, some computation can be saved if part of the proximity information utilised for editing can be reused for condensing. Accordingly, Sánchez et al. [49] presented a simple way to apply graph editing-condensing while keeping the computational burden very close to the cost of computing the proximity graph only.

To employ Toussaint's condensing after the graph editing algorithms using the same graph structure, the edges must be recomputed when the discarded prototypes along with all their edges are removed from the graph. After this trivial step, it may be necessary to add new edges between pairs of non-neighbour nodes. From the definition of graph neighbours, we can conclude that only pairs of nodes whose zone of influence (that is, diameter hypersphere or lune) held an eliminated node can have a new edge. Otherwise, the reason that made a pair of points non-neighbours still holds after editing.

Bearing this in mind, it is possible to store for each pair of non-neighbour nodes the first detected node inside its zone of influence when the graph was computed for the first time. This makes selection of new edges much faster because only the (few) pairs of nodes whose marked prototype has been discarded during editing need to perform a search through the whole set. Note that the nodes are arranged in random order and therefore, the "first detected node" can be anynode. Thus, the algorithm can be written as follows:

1. Construct the proximity graph, $G = (V, E)$, corresponding to the original TS X and, for each pair of points $(x_i, x_j) \in E$, mark the first node that lies inside its zone of influence.
2. Graph-based editing.
3. Construct the subgraph, $G' = (V', E')$, corresponding to non-discarded nodes.
4. For each pair x_i, x_j from V', $(x_i, x_j) \notin E'$ and whose stored node is not in V', put an edge if no other node from V' lies inside its zone of influence.
5. Graph-based condensing with the recomputed graph.

It is worth pointing out that modification in Step 1 requires no extra time, while the Step 4 is expected to be applied to a very reduced number of edges, as confirmed by experimental tests [49].

4.4 k-NCN Editing Strategies

An editing algorithm was proposed in [17] as a way of improving the behaviour of classical editing in the finite sample size case. This technique corresponds to a slight modification of the original work of Wilson and basically consists of using the leaving-one-out error estimate with the k-NCN classification rule [50].

From a practical point of view, the k-NCN classifier is thought to obtain a more accurate information about prototypes in the TS and more specially, for those being close to decision boundaries. In general, when applied to editing, this results in a practical improvement of the corresponding procedure [17]. The simplest k-NCN editing algorithm can be expressed as follows:

1. Let $S = X$.
2. For each x_i in X do:
    - Discard x_i from S if it is misclassified using the k-NCN rule with prototypes in $X - \{x_i\}$.

4.4.1 Fusion of NN and k-NCN classifiers for editing

An extension to this editing procedure corresponds to that proposed by Sánchez et al. [52]. It combines information about proximity with that about the spatial distribution of prototypes in the TS. This kind of classifier fusion tries to guarantee that the prototypes used for editing are correctly labelled according to the NN rule. Although, in general, it is not possible to remove only prototypes lying in wrong class regions, this technique can be seen as a way of reducing the number of "correct" prototypes discarded during the editing process.

In a few words, after searching for the k NCNs of a training prototype, only those whose NN belongs to their same class will be considered for editing. Therefore, this can be understood as a matter of combining two editings: first, editing the k NCNs by using the NN rule and second, editing a prototype by means of the $j \leq k$ NCNs not discarded in the previous stage. This idea can be algorithmically summarised as follows:

1. Let $S = X$.
2. For each x_i in X do:
    - Identify the k NCNs of x_i in $X - \{x_i\}$.
    - Select the $j \leq k$ NCNs correctly labelled according to the NN rule.
    - Discard x_i from S if it is misclassified using the j-NCN rule with prototypes in $X - \{x_i\}$.

Alternatively, one might use the general k-NN decision rule (instead of the particular 1-NN) for editing the set of the NCNs (Step 2.b of the algorithm). On the other hand, although this scheme is based on the leaving-one-out estimate, other error estimates as holdout and B-fold cross-validation are also directly applicable.

4.4.2 Iterative k-NCN Editing

This second alternative basically consists of editing repeatedly the already edited TS until no more prototypes are removed. The idea of this moderate extension to the k-NCN algorithm is that, if a single application of editing can generally improve the classification accuracy, it is expected that successive deletions of outliers can mean an additional increase in performance [55]. Thus the iterative editing procedure can be written in the following way:

1. Let $S = \emptyset$.
2. While $S \neq X$
   - Let $S = X$.
   - Assign to X the result of applying the plain k-NCN editing to S.

An exhaustive empirical analysis carried out by Sánchez et al. [53] shows that this iterative procedure outperforms the simple k-NCN editing for all values of the parameter k. In their work, they also conclude that the highest differences are found with small values of k. On the other hand, from those experiments, there is enough evidence to state that the iterative procedure "cleans" better the overlap between different class regions than the plain k-NCN editing and as a by-product, the NN decision boundaries approximated by the iterative scheme are much simpler than those of the plain algorithm. Consequently, in practice, the computational burden to classify new input patterns will be lower when using the iterative edited set than in the case of employing the NCN edited set.

5 Experimental Results

In this section, a comparative study between some surrounding approaches and a set of well-known NN-based schemes is provided. All the experiments have been carried through over a synthetic database [40] with high overlap among regions from different classes.

Such a database consists of a collection of seven data sets that correspond to the same problem, but with dimensionality ranging from 2 to 8. All the features are designed to be equally effective in terms of their discrimination potential, thus making the analysis a function of only the size of the subset used and not of its specific attributes. The patterns belong to two different classes, which represent multivariate normal distributions with zero mean and standard deviation 1 and 2 in all dimensions, respectively. Each class contains a total number of 2,500 samples.

All the results here reported have been averaged over five different random partitions (i.e., 2,500 training prototypes and 2,500 test patterns) of each of the seven initial data sets, so as to obtain the overall classification accuracy estimates.

5.1 Comparison of Classification Approaches

First, the behaviour of different classification schemes is here studied: proximity graph and k-NCN classifiers in contrast with the k-NN rules. In the case of the k-NN and k-NCN approaches, typical settings for the parameter k (ranging from 1 to 11) have been tried and the ones leading to the best performance have finally been included in the figures.

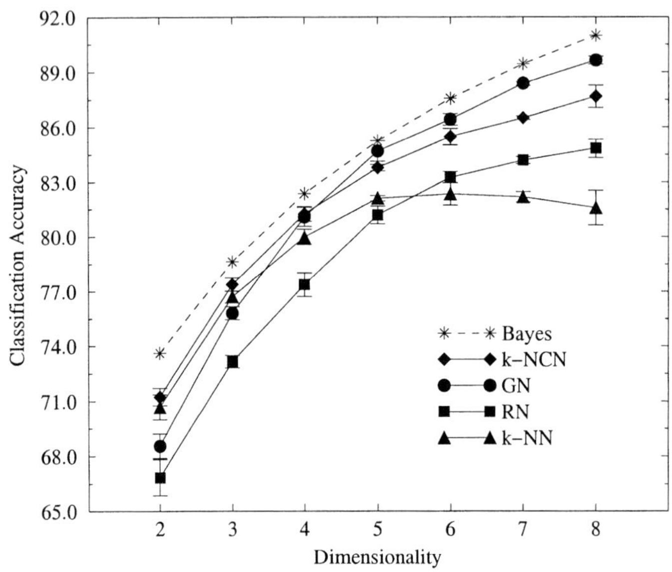

Figure 1: Accuracy of different classifiers with varying dimensionalities.

From Figure 1, it is possible to draw a number of interesting conclusions. The GG-based approach and the k-NCN rule are clearly the best in most dimensionalities, although differences between them are not statistically significant and therefore, almost nothing can be said about the relative merits of each one of them. Nevertheless, perhaps the most important is that they are not too far from the optimal *Bayes error*.

On the other hand, while the *Bayes* classifier as well as the k-NCN and both proximity graph-based decision rules yield better results as the dimensionality increases, the performance of the k-NN approach clearly degrades

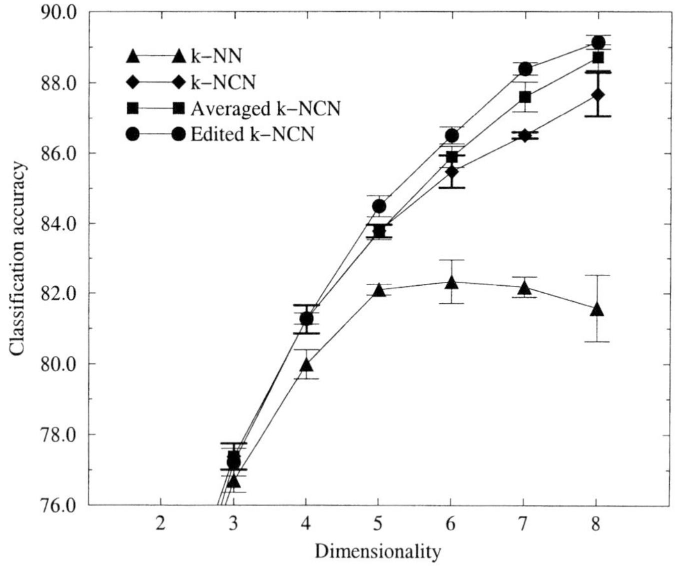

Figure 2: Classification accuracy of k-NCN and k-NN rules.

from dimensionality 5. Finally, it is apparent that the RN classification scheme has a worse behaviour than the other surrounding approaches, although it shows an improvement in performance similar to those of the GN and the k-NCN rules when dimensionality increases.

In Figure 2, the averaged k-NCN and the edited k-NCN rules have been compared to the plain k-NCN scheme. Results corresponding to the k-NN classifiers have also been included for comparison purposes. As can be seen, all k-NCN rules achieve a classification accuracy significantly higher than that of the classical k-NN rule. It appears that the k-NCN classifier and its extensions are less sensitive to the size to intrinsic dimensionality ratio.

5.2 Analysis of Prototype Selection Algorithms

The second group of experiments consists of applying the NN rule to each of the seven test sets, where the training portion has been preprocessed using several prototype selection techniques. In addition to proximity graph-based

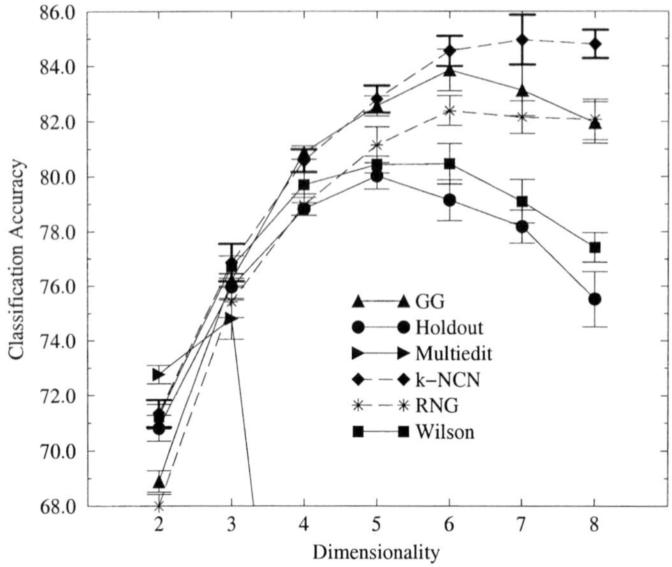

Figure 3: Classification accuracy of the different edited NN rules.

and NCN-based editing, Wilson's editing, the Holdout-based method and the Multiedit algorithm have also been used in these experiments.

From each trial, the edited set size and the classification accuracy have been calculated. Thus, the reduction of training samples (i.e., edited set size / original training set size) gives a direct measure of the amount of computational savings due to the number of prototypes removed by each technique. On the other hand, accuracy provides a check on the ability of the different algorithms to select the most "efficient" prototypes.

As can be seen in Figure 3, all editing procedures provided quite similar performance up to dimension 4. Nevertheless, there is a clear separation in their behavior as the dimensionality increases above 5. It is worth mentioning that the Multiedit algorithm gave the best result for this experiment only in the case of the lowest dimensionality (close to 73%), as is to be expected from its theoretical analysis [14]. The most significant point to be noted in the present study is the fact that the best overall classification re-

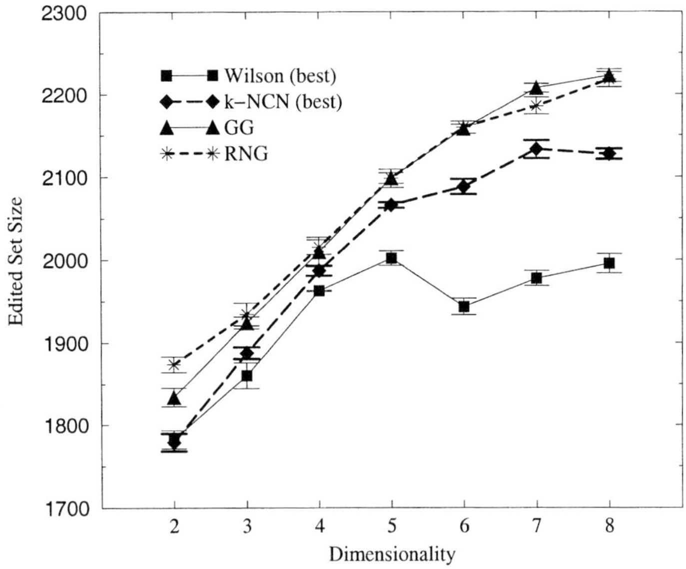

Figure 4: Number of prototypes retained by the different editing algorithms with varying dimensionalities.

sult for higher dimensionalities correspond to that of the NCN-based editing approach.

Figure 4 shows the averaged number of prototypes (and standard deviations) retained by the best editing procedures. In this case, Wilson's editing removes a larger number of prototypes than the surrounding approaches. Nevertheless, taking into account both accuracy and reduction rate simultaneously, it is worth noting that the NCN approach results in the best balanced trade-off between classification performance and final edited set size.

While Wilson's editing obtains the highest set size reduction and the lowest classification accuracy, the GG-based editing achieves the best recognition results but also the minimum set size reduction rate. Finally, the RNG-based approach also retains a very important number of prototypes, very close to that of the GG-based editing.

6 Summary

Neighbourhood or similarity based classification techniques have been applied to many real-world environments. To classify a given pattern, these schemes firstly analyse the degree of similarity between it and a subset of prototypes from the TS and then, assign it to the class represented by the most similar examples. The k-NN decision rule constitutes one of the most widely used and simplest approaches to such non-parametric classifiers. Nevertheless, these methods suffer from some important drawbacks that make difficult their application in practice: large memory and time requirements, sensitivity to noisy and mislabelled prototypes, and also to irrelevant features.

Traditionally, neighbourhood classification research has been devoted to the development and analysis of algorithms for overcoming the different NN drawbacks. In recent years, however, several researchers [23, 49, 56, 64] have focused on finding alternative neighbourhood specifications. Surrounding neighbourhood is an interesting example of such similarity definitions, in which not only proximity but also symmetry of prototypes must be analysed.

Surrounding neighbourhood can be obtained from a number of realizations. In this chapter, the nearest centroid neighbourhood, the graph neighbourhood, and the Zhang's neighbourhood have been introduced. All these have been employed in very distinct classification and prototype selection algorithms. Empirical studies have shown that in general, this kind of neighbourhood outperforms the conventional representation given by the nearest neighbourhood.

There are still many issues that could be investigated in the framework of alternative neighbourhood representations for pattern classification. In particular, one of the most important drawbacks related with the approaches introduced in the present chapter refers to the fact that the computations needed to calculate the surrounding neighbourhood of a pattern from the prototypes stored in a large TS can be prohibitively expensive, specially in applications where classification time is important. Thus it seems necessary to propose efficient algorithms for implementing such neighbourhood realizations, or even to define new specifications.

As an immediate challenge, the application of surrounding neighbourhood algorithms to other pattern recognition and machine learning domains should be done. More specifically, our present work is primarily addressed to investigate the potential of these methods when applied to problems where one class is much more represented than the others in the TS. It has been observed that this situation, which arises in several practical situations, may produce an important deterioration of the classification accuracy, specially with patterns belonging to the less represented class.

Acknowledgements

This work was partially supported by the grant TIC2000-C03-03 from the Spanish CICYT (Department of Science and Technology).

References

[1] A. Akkus and H.A. Güvenir, k Nearest neighbor classification on feature projections, in L. Saitta (ed.) *Proc. of the 13th International Conference on Machine Learning*, (Morgan Kaufmann, Bari, Italy, 1996) pp. 12-19.

[2] R. Barandela and E. Gasca, Decontamination of training samples for supervised pattern recognition methods. in F.J. Ferri et al. (eds.) *Advances in Pattern Recognition*, (Springer Verlag, Alicante, Spain, 2000) pp. 621-630.

[3] S.O. Belkasim, M. Shridhar, and M. Ahmadi, Pattern classification using an efficient KNNR, *Pattern Recognition* Vol.25, (1992) pp. 1269-1274.

[4] A.L. Blum and P. Langley, Selection of relevant features and examples in machine learning, *Artificial Intelligence* Vol.97, (1997) pp. 245-271.

[5] C.E. Brodley and M.A. Friedl, Identifying mislabeled training data, *Journal of Artificial Intelligence Research* Vol.11, (1999) pp. 131-167.

[6] C.L. Chang, Finding prototypes for nearest neighbor classifiers. *IEEE Trans. on Computers* Vol. 23 (1974) pp. 1179-1184.

[7] B.B. Chaudhuri, A new definition of neighborhood of a point in multidimensional space, *Pattern Recognition Letters* Vol.17, (1996) pp. 11-17.

[8] S. Cost and S. Salzberg, A weighted nearest neighbor algorithm for learning with symbolic features, *Machine Learning* Vol.10, (1993) pp. 57-78.

[9] T.M. Cover and P.E. Hart, Nearest neighbor pattern classification, *IEEE Trans. on Information Theory* Vol.13, (1967) pp. 21-27.

[10] B.V. Dasarathy, *Nearest Neighbor Norms: NN Pattern Classification techniques*, (IEEE Computer Society Press, Los Alamos, CA, 1991).

[11] B.V. Dasarathy, Minimal consistent subset (MCS) identification for optimal nearest neighbor decision systems design, *IEEE Trans. on Systems, Man, and Cybernetics* Vol.24, (1994), pp. 511-517.

[12] B.V. Dasarathy, J.S. Sánchez, and S. Townsend, Nearest neighbor editing and condensing tools - synergy exploitation, *Pattern Analysis and Applications* Vol.3, (2000) pp. 19-30.

[13] M. de Berg, M. van Kreveld, M. Overmars, and O. Schwarzkopf, *Computational Geometry: Algorithms and Applications*, (Springer-Verlag, Berlin, 1997).

[14] P.A. Devijver and J. Kittler, *Pattern Recognition: A Statistical Approach*, (Prentice Hall, Englewood Cliffs, NJ, 1982).

[15] A. Djouadi and E. Bouktache, A fast algorithm for the nearest-neighbor classifier, *IEEE Trans. on Pattern Analysis and Machine Intelligence* Vol.19, (1997) pp. 277-282.

[16] A. Faragó, T. Linder and G. Lugosi, Fast nearest-neighbor search in dissimilarity spaces, *IEEE Trans. on Pattern Analysis and Machine Intelligence* Vol.15, (1993) pp. 957-962.

[17] F.J. Ferri, J.S. Sánchez, and F. Pla, Editing prototypes in the finite sample size case using alternative neighbourhoods, in A. Amin et al. (eds.) *Advances in Pattern Recognition*, (Springer-Verlag, Sydney, Australia, 1998) pp 620-629.

[18] F.J. Ferri, J.V. Albert, and E. Vidal, Considerations about sample-size sensitivity of a family of edited nearest-neighbor rules, *IEEE Trans. on Systems, Man, and Cybernetics-Part B: Cybernetics* Vol.29, (1999) pp. 667-672.

[19] E. Fix and J.L. Hodges, Discriminatory analysis, non-parametric discrimination, consistency properties, *Project 21-49-004, Report No. 4, Contract AF41(128)-3*, (USAF School of Aviation Medicine, Randolph Field, TX, 1951).

[20] K. Fukunaga and P.M. Narendra, A branch and bound algorithm for computing k-nearest neighbors, *IEEE Trans. on Computers* Vol. 24 (1975) pp. 750-753.

[21] K. Fukunaga and P.M. Narendra, A branch and bound algorithm for feature subset selection, *IEEE Trans. on Computers* Vol. 26 (1977) pp. 917-922.

[22] G.W. Gates, The reduced nearest neighbor rule, *IEEE Trans. on Information Theory* Vol.18, (1972) pp. 431-433.

[23] S. Grabowski, Experiments with the k-NCN decision rule, in *Proc. of the IX Konferencja Sieci i Systemy Informatyczne*, (Lodz, Poland, 2001) *(in press)*.

[24] D.J. Hand, J.N. Kok, and M.R. Berthold, *Advances in Intelligent Data Analysis*, (Springer Verlag, Berlin, 1999).

[25] P.E. Hart, The condensed nearest neighbor rule, *IEEE Trans. on Information Theory* Vol.14, (1968) pp. 515-516.

[26] K. Hattori and M. Takahashi, A new edited k-nearest neighbor rule in the pattern classification problem, *Pattern Recognition* Vol.33, (2000) pp. 521-528.

[27] M.E. Hellman, The nearest neighbor classification rule with a reject option, *IEEE Trans. on Systems, Science, and Cybernetics* Vol.6, (1970) pp. 179-185.

[28] A.K. Jain *Fundamentals of Digital Image Processing*, (Prentice Hall, Englewood Cliffs, NJ, 1989).

[29] A. Jain and D. Zongker, Feature selection: evaluation, application and small sample performance, *IEEE Trans. on Pattern Analysis and Machine Intelligence* Vol.19, (1997) pp. 153-158.

[30] J.W. Jaromczyk and G.T. Toussaint, Relative neighborhood graphs and their relatives, *Proc. IEEE* Vol.80, (1992) pp. 1502-1517.

[31] G.H. John, R. Kohavi, and K. Pfleger, Irrelevant features and the subset selection problem, in W.W. Cohen and H. Hirsh (eds.) *Proc. of the 11th International Conference on Machine Learning*, (Morgan Kaufmann, New Brunswick, NJ, 1994) pp. 121-129.

[32] A. Józwik and G. Vernazza, Recognition of leucocytes by a parallel k-NN classifier, in *Lecture Notes of ICB Seminar*, (Warsaw, Poland, 1988) pp. 138-153.

[33] B.S. Kim and S.B. Park, A fast nearest neighbor finding algorithm based on the ordered partition, *IEEE Trans. on Pattern Analysis and Machine Intelligence* Vol. 8 (1986) pp. 761-766.

[34] R. Kohavi and G.H. John, Wrappers for feature subset selection, *Artificial Intelligence* Vol.97, (1997) pp.273-324.

[35] J. Koplowitz, and T.A. Brown, On the relation of performance to editing in nearest neighbor rules., *Pattern Recognition* Vol.13, (1981) pp. 251-255.

[36] L.I. Kuncheva, Editing for the k-nearest neighbors rule by a genetic algorithm, *Pattern Recognition Letters* Vol.16, (1995) pp. 809-814.

[37] L.I. Kuncheva and L.C. Jair, Nearest neighbor classifier: simultaneous editing and feature selection, *Pattern Recognition Letters* Vol.20, (1999) pp. 1149-1156.

[38] P. Langley and W. Iba, Average-case analysis of a nearest neighbor algorithm, in R. Bajcsy (ed.) *Proc. of the 13th International Joint Conference on Artificial Intelligence*, (Morgan Kaufmann, Chambéry, France, 1993) pp. 889-894.

[39] C.X. Ling and H. Wang, Computing optimal attribute weight settings for nearest neighbor algorithms, *Artificial Intelligence Review* Vol.11, (1997) pp. 255-272.

[40] C.J. Merz and P.M. Murphy, *UCI Repository of Machine Learning Databases.* (Univ. of California, Irvine, http://www.ics.uci.edu/mlearn, 1998).

[41] L. Micó, J. Oncina and R.C. Carrasco, A fast branch and bound nearest neighbour classifier in metric spaces, *Pattern Recognition Letters* Vol. 17 (1996) pp. 731-739.

[42] F. Moreno-Seco, J.M. Iñesta, L. Micó, and J. Oncina, Fast k-neighbour classification of human vertebrae levels, in J.S. Sánchez and F. Pla (ed.), *Pattern Recognition and Image Analysis II*, (Publicacions de la Universitat Jaume I, Castellón, Spain, 2001) pp. 343-348.

[43] S.A. Nene and S.K. Nayar, A simple algorithm for nearest neighbor search in high dimensions, *IEEE Trans. on Pattern Analysis and Machine Intelligence* Vol. 19 (1997) pp. 989-1003.

[44] J.F. O'Callaghan, An alternative definition for neighborhood of a point, *IEEE Trans. on Computers* Vol.24, (1975) pp. 1121-1125.

[45] F.P. Preparata and M.I. Shamos, *Computational Geometry. An introduction*, (Springer, New York, 1985).

[46] P. Pudil, J. Novovicova and J. Kittler, Floating search methods in feature selection, *Pattern Recognition Letters* Vol. 15 (1994) pp. 1119-1125.

[47] V. Ramasubramanian and K.K. Paliwal, Fast nearest-neighbor search algorithms based on approximation-elimination search, *Pattern Recognition* Vol. 33 (2000) pp. 1497-1510.

[48] G.L. Ritter, H.B. Woodritz, S.R. Lowry and T.L. Isenhour, An algorithm for selective nearest neighbor rule. *IEEE Trans. on Information Theory* Vol. 21 (1975) pp. 665-669.

[49] J.S. Sánchez, F. Pla, and F.J. Ferri, Prototype selection for the nearest neighbour rule through proximity graphs, *Pattern Recognition Letters* Vol.18, (1997) pp. 507-513.

[50] J.S. Sánchez, F. Pla, and F.J. Ferri, On the use of neighbourhood-based non-parametric classifiers, *Pattern Recognition Letters* Vol.18, (1997) pp. 1179-1186.

[51] J.S. Sánchez, F. Pla, and F.J. Ferri, Improving the k-NCN classification rule through heuristic modifications, *Pattern Recognition Letters* Vol. 19, (1998) pp. 1165-1170.

[52] J.S. Sánchez, R. Barandela, R. Alejo, and A.I. Marqués, Performance evaluation of prototype selection algorithms for nearest neighbor classification, in D.L. Borges and S.-T. Wu (eds.) *Proc. of the 14th. Brazilian Symposium on Computer Graphics and Image Processing*, (IEEE Computer Society Press, Florianópolis, Brazil, 2001) pp. 44-50.

[53] J.S. Sánchez, R. Barandela, A.I. Marqués, R. Alejo, and J. Badenas, Analysis of new techniques to obtain quality training sets, *Pattern Recognition Letters*, (2001) *(in press)*.

[54] W. Siedlecki and J. Sklansky, On automatic feature selection, *International Journal of Pattern Recognition and Artificial Intelligence* Vol.2, (1988) pp. 197-220.

[55] I. Tomek, An experiment with the edited nearest neighbor rule, *IEEE Trans. on Systems, Man, and Cybernetics* Vol.6, (1976) pp. 448-452.

[56] G.T. Toussaint, B.K. Bhattacharya, and R.S. Poulsen, The application of Voronoi diagrams to nonparametric decision rules, in L. Billard (ed.) *Proc. of the 16th Symposium on the Interface: Computer Science and Statistics*, (Atlanta, GA, 1984) pp. 97-108.

[57] J.I. Toriwaki and S. Yokoi, Voronoi and related neighbors on digitized two dimensional space with applications to texture analysis, in G.T. Toussaint (ed.) *Computational Morphology*, (North-Holland, 1988) pp. 207-228.

[58] M. Tüceryan and T. Chorzempa, Relative sensitivity of a family of closest-point graphs in computer vision applications, *Pattern Recognition* Vol.24, (1991) pp. 361-373.

[59] E. Vidal, An algorithm for finding nearest neighbours in (approximately) constant average time, *Pattern Recognition Letters* Vol.4, (1986) pp. 145-147.

[60] J.M. Vilar, Reducing the overhead of the AESA metric-space nearest neighbour searching algorithm, *Information Processing Letters* Vol. 56 (1995) pp. 265-271.

[61] D.L. Wilson, Asymptotic properties of nearest neighbor rules using edited data, *IEEE Trans. on Systems, Man, and Cybernetics* Vol.2, (1972) pp. 408-421.

[62] D.R. Wilson and T.R. Martinez, Reduction techniques for instance-based learning algorithms. *Machine Learning* Vol. 38 (2000) pp. 257-286.

[63] T. Young and K.-S. Fu, *Handbook of Pattern Recognition and Image Processing*, (Academic Press, Orlando, FL, 1986).

[64] J. Zhang, Y. Yim, and J. Yang, Intelligent selection of instances for prediction functions in lazy learning algorithms, *Artificial Intelligence Review* Vol.11, (1997) pp. 175-191.

Algorithmic Synthesis in Neural Network Training for Pattern Recognition

Konstantinos Sirlantzis
Department of Electronics
University of Kent, Canterbury, Kent, CT2 7NT
E-mail: `K.Sirlantzis@ukc.ac.uk`

Contents

D. Chen and X. Cheng (eds.),
Pattern Recognition and String Matching, 703–739.

1 Introduction

1.1 Neural Networks for Pattern Recognition

A large number of theoretical results establish the potential of Artificial Neural Networks (ANNs) as universal function approximators. ANNs importance in this context is that they offer a very powerful and flexible apparatus for representing non-linear mappings which can be applied in a wide range of application areas. If for example this mapping is defined on the conditional expectations of the output given the available data (input variables) — in other words, in terms of an average over a random quantity — the problem belongs in the class of "regression" [2, p. 5]. In this case the outputs represent values of continuous variables. On the other hand, when the task is to assign new inputs to one of a number of discrete classes or categories on the basis of known attributes, the problem belongs to the class of "classification" [22, p. 2]. The function to be approximated in the latter case is the conditional probability function of the output to be assigned a particular class label given its attributes vector.

Classification typically includes two–class (binary) problems as well as multi–class problems with a diverse variety of applications [28]. These extend from character recognition [9], speech recognition and understanding [19], radar/sonar signal classification [12], to medical diagnosis [4] and texture analysis [36], to name but a few. Furthermore, in [20], the variety of the examples used extends from credit risk identification in finance to satellite image analysis and understanding. This text is also a rich source of comparative results about the effectiveness of neural network classifiers with alternative machine intelligence and conventional statistical discriminators.

As pointed out by Bishop [2, p. 6] "many of the key issues which need to be addressed in tackling *pattern recognition problems* are common both to classification and regression". Although the concept of pattern recognition often means different things to different people, if we define pattern recognition as the "science that concerns the description or classification (recognition) of measurements" [28, p. 2], it comes as a natural realisation the fact that artificial neural networks were from the beginning, and still are, considered mainly as efficient 'pattern recognisers'.

1.1.1 Memorisation vs Generalisation

When dealing with the issue of learning we have to address the issues of *memorisation* and *generalisation* [26, p. 42]. This distinction becomes crucial when the models of our interest are characterised by broad generality, in terms of the classes of relationships that can represent, like the neural networks models do. Memorisation, can be considered as the ability to fit the features of the given data optimally with respect to a certain error criterion. Generalisation, on the other hand is, in essence, the ability of performing well for items not in the training set, or in other words, to perform equally well with respect to both the 'seen' and the 'unseen' cases.

The relative weight between generalisation and memorisation is problem dependent, though. For classification tasks with fully determined classes, such as the **XOR** or the **parity** problem, the issue of generalisation does not arise at all. Gemman *et al.* [8, pp. 18–33] characterise this case as "degenerate", and further illustrate and discuss the total generalisation error and its bias and variance components as functions of the number of hidden nodes—i.e. the degree of modelling flexibility—of a neural network. They also demonstrate that there exists a characteristic difference with the case of an 'ambiguous', as they call it, classification task due to noise inherent in the available data. In general, for any task which is subject to 'uncertainty', generalisation ability is the most desirable feature, dictating the choice of which method to use. As Ripley [26, p. 53] insightfully quotes from [23, p. 47] "it remains to solve the problem of 'designing fast training algorithms which minimize the true error instead of minimizing the apparent error'. Here the 'true' error refers to the rate on future examples.". Consequently, the performance of a learning procedure, with respect to this goal, passes through the search of suitable evaluation techniques. In summary, the ability of neural networks to generalise depends on the network architecture and size, the quality and quantity of the data, and the learning algorithm.

2 A Framework for Algorithmic Synthesis

2.1 Supervisor and Searcher Co-operation (SSC)

A logical conclusion indicated by our previous discussion for the design of algorithms able to successfully address the diverse range of issues arising in neural network training as *an optimisation problem in the presence of noise*, is that a framework based on a **synthesis** mechanism is needed. This should facilitate the integration of desirable characteristics of existing algorithms into a novel apparatus exhibiting qualitatively emerging behaviour different that the mere addition of its constituent parts. It should also be generic with respect to the types of algorithms that can be applied on. Finally, it should allow the behaviour of the resulting algorithms to adapt according to the characteristics of the problem at hand. The **Supervisor and Searcher Co-operation (SSC) Framework**, which we use in this work, promises such an algorithm development tool, and it is the subject of this chapter to describe and explore its properties in practice, in the context of neural network training for pattern recognition.

In real world applications, various types of noises are present in most of the available data. Here the noises are conceived in a broad sense. In some cases, these noises can be ignored, but more often than not, they play an important role in the mathematical modelling of the application. We are interested in the noisy unconstrained minimisation problem formally defined as follows:

$$\min_{x \in R^n} F(x),$$

where it is difficult or impossible to evaluate $F(x)$ exactly. Then, let $f(x)$ be an estimator of $F(x)$ in the sense

$$f(x) = F(x) + \epsilon,$$

and F is the underlying exact mathematical model. F is a continuous function on R^n and is bounded below. We are interested in finding (local) minimizers of f (that is, of F).
For given $x_0, x_1, ..., x_l$, assume that we have an iterative algorithm, called search engine (SE):

$$x_{k+1} = x_k - se_k(x_k, x_{k-1}, ..., x_{k-l}, k, f), \ k = l, l+1, ...$$

The above notation for se_k emphasise its dependence on k and the values of $\{f(x_{k-i})\}_{i=0}^m$, $\{\nabla f(x_{k-i})\}_{i=0}^m$, $\{H(x_{k-i})\}_{i=0}^m$, etc., where $H(x)$ is the Hessian matrix of f at the point x. It is important to bear in mind that x_k, se_k may all depend on the starting data x_0, se_0.

The essential idea adopted in [17, 18, 32] is to employ an alternative globally convergent but robust iterative algorithm to supervise and therefore to safeguard the convergence of the SE algorithm. This supervising algorithm will be referred to as the supervisor (SR). Then, one may obtain a globally convergent **and** robust algorithm. Assume that this supervisor algorithm (SR) reads: Given $x_0, x_1, ..., x_l$

$$x_{k+1} = x_k - sr_k(x_k, x_{k-1}, ..., x_{k-l}, k, f), \ k = l, l+1, ...$$

Again here x_k, sr_k depend on the starting data.

In general, a SR algorithm is slower but robust, and a SE algorithm is faster but only locally convergent. Therefore, they have to co-operate in order to work efficiently. In [17, 18] a *supervision principle* is proposed, based on the co-operation of the supervisor and the searcher (SSC). According to this principle the supervisor intervenes only when there is an indication that the performance of the search engine is not satisfactory, while the search engine undertakes most of the (solution) searching work. For the resulting algorithm, to a large extend, global convergence may be ensured by the supervisor but the speed is decided by the search engine.

There are various ways to implement the co-operation principle. The following may be seen as one of the simplest:
Assume $f \geq 0$. Given $x_0, x_1, ..., x_l$, define $(k = l, l+1, l+2, ...)$ the following (SSC) algorithm:

$$x_{k+1} = x_k - sr_k \text{ if } T_k f(x_k - sr_k) \leq f(x_k - se_k),$$

otherwise

$$x_{k+1} = x_k - se_k,$$

where $\{T_k\}$ is a given sequence of nonnegative real numbers.

This new algorithm may exhibit a behaviour significantly different from those of its "parent" algorithms . The algorithm actually switches between its two component algorithms at a not pre-specified frequency. The sequence $\{T_k\}$ decides the strength of the co-operation. It is clear that the behaviours of a SSC algorithm depend not only on those of the corresponding SR and SE algorithms, but also on the degree of supervision, which is determined by the sequence $\{T_k\}$. For instance, if we take $T_k = 0$ or $T_k = \infty$, there will be no action of the search engine or the supervisor respectively (assuming $f(x_k - sr_k) > 0$). Also by setting $se_k = 0$, the algorithm becomes the SR algorithm. Although such degenerate instances present little interest, they do indicate the wide range of algorithms covered by the SSC type algorithms. It seems clear that taking larger T_k will force a SSC algorithm to use more SE iterations, and therefore may increase its overall speed. However, if T_k is too large, the supervision may become too weak, and therefore the resulting SSC algorithm may not be robust enough or even convergent globally. In applications, normally one uses $T_k \equiv T > 0$. If T is smaller, then the supervisor may do most of the work, if T is larger, the search engine may be used all the way. From our experience, $T = 1$ is always the safest value. Main theoretical results about the convergence and speed properties of the algorithms developed using the Supervisor and Searcher Co-operation Framework can be found in [17, 18, 31, 32]

2.2 Learning Algorithms for ANNs

The neural networks training problem is often paralleled to that of non-parametric non-linear regression, while in other cases to that of the construction of discriminators in classification (see for example [2] and [26]). In *supervised* learning, which is of our particular interest here, it is assumed that we have a set of data pairs $D = \{(x_i, y_i)\}, \quad i = 1, ..., N\}$, also called training set, where $\{y_i\}$ are the *desired* outputs of the network. If we consider the neural network as a directed graph then the input and output vectors are represented as input and output respectively of *nodes* organised hierarchically in the so called input and output *layers*. Between them there may exist one or more layers of nodes called *hidden* layers and hidden nodes respectively. The nodes have assigned some *activation function* used to calculate the *activation level* or output of the node with respect to its input. Thus, the nodes are the main processing units of the neural network. In this hierarchical structure a particular combination of the outputs of the nodes of each layer is propagated to become the input of each node of the next layer. Then, a vector of adjustable parameters $w = [w_1, ..., w_M]^T$, called *weights*, represent the strength of the connections of this propagation procedure. These are the main parameters of the model, which are adjusted during the *training* (or *learning*) phase. The aim is to estimate the parameter values of a particular function defined by the neural network *architecture*, which will produce optimal performance with respect to a predefined measure. This is, in fact, an unconstraint problem of optimization of the performance function. However, it could be a rather complex one mainly due to the uncertainty associated with the data on which the definition of the performance function itself is based.

The most well known and widely used method for the training of the multilayer feedforward neural networks described above, is the *Error Backpropagation* (EBP) or *Generalized Delta Rule.* This was introduced by Werbos [37] but only found widespread use after it was independently re-invented by Rumelhardt et al. [27]. Our special interest in this learning algorithm is because it is one of the more suitable and general rules that can be applied to Multilayer Perceptrons (MLPs), which are the most widely used feedforward neural networks for regression and classification purposes. Its power resides in its ability to efficiently handle the learning procedure in *ordered-systems*, such as the multilayer feedforward networks, where the arrangement of the nodes admits a hierarchical structure. In these the flow of information is allowed only in one direction, from input to output layers. This is usually called a *forward pass.* The error associated with each one of the nodes is calculated by applying the chain rule of differentiation and *backpropagated* in the opposite direction, that is from the output to the input layers. This is called a *backward pass.* Finally, the weight values are updated according to the calculated errors. A full cycle of training, often called *epoch*, comprises a forward and a backward pass through all the available training data pairs and the corresponding weight updates. As we shall discuss in more detail bellow there are two ways to perform the weight updates. The first, called *online* training, corresponds to one update after a single pairs is propagated through the network, while the second, called *batch* training, corresponds to a single update after all the available data has been processed.

According to [24] the term 'Backpropagation' refers usually to two different things. First, it describes, as we have already mentioned, a particular method to calculate the derivatives of the training error with respect to current weight values. Second, it defines an optimisation (training) algorithm—sometimes equivalent to gradient descent, as we shall see below—which is using the calculated derivatives to minimise the error function.

In our previous set up, we may define:
y_k^t = the desired output of pattern pair t corresponding to the kth output unit,
s_k^t = the network output of the kth output unit,
w_{ij} = the weight of the link from the ith unit (being either input or hidden) to the jth unit(being either hidden or output). Notice here that we do not refer to any particular kind of connection weights, to allow for general multilayer networks.
Then for a network with one hidden layer we can define:
$\sigma_i, i = 1, \dots, p.$ is the ith input unit activation state,
$\bar{s}_j, j = 1, \dots, q.$ is the jth hidden unit activation state,
And:
$\sigma_i = \tilde{f}(x_i)$,
$\bar{s}_j = \bar{f}(\bar{h}_j)$ where $\bar{h}_j = \sum_i \bar{w}_{ij}\sigma_i$, and $\bar{w}_{ij}$ is the connection weight between the ith input and the jth hidden units,
$s_k = f(h_k)$ where $h_k = \sum_j w_{kj}\bar{s}_j$, and w_{jk} is the connection weight between the jth hidden unit and the kth output unit.
A commonly used set of activation functions are the following:

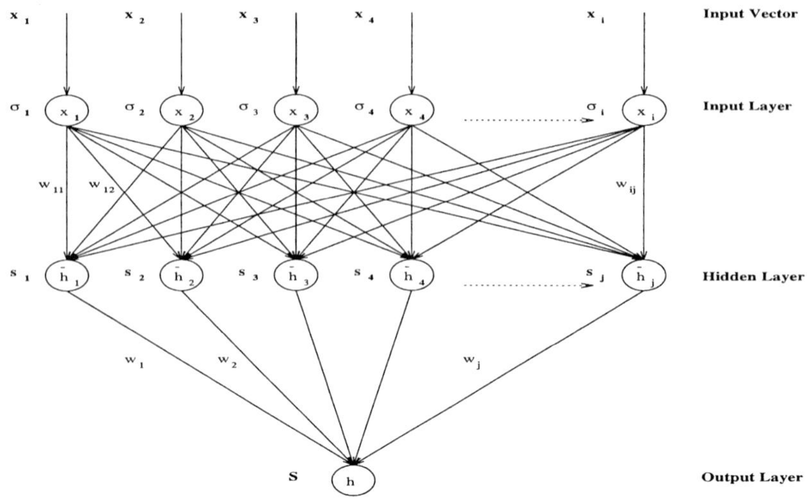

Figure 1: Schematic representation of a Feedforward Neural Network with one hidden layer.

$\tilde{f}(z) = z$ (identity input unit activation function),
$\bar{f}(z) = 1/(1+e^{-z})$ or $tanh(z)$ (sigmoid hidden unit activation function),
$f(z) = z$ (identity output unit activation function).
See Figure 1 for a graphical representation of the structure of an MLP with one node S in the output layer.

First, to consider Backpropagation as a method for calculating the error derivatives, we define the per-pattern t error function as follows:

$$E^t = \sum E^t_k$$

where as above k indexes the output nodes. Then the total or epoch error is:

$$E = \sum_t E^t, \quad \text{or} \quad E = \frac{1}{N} \sum_t E^t \tag{1}$$

where for $t = 1, \ldots, N$, the whole set of examples forms the *epoch*. Then for any weight w the error derivative is:

$$\frac{\partial E}{\partial w} = \sum_t \frac{\partial E^t}{\partial w}$$

And:

$$\frac{\partial E^t}{\partial w} = \sum_k \frac{\partial E^t}{\partial a_k} \frac{\partial a_k}{\partial w}$$

where a_k denotes the input to node k. Then, in our previous setting of a single hidden layer network, for $\bar{w}_{ij}$ and w_{jk} being the input-to-hidden and hidden-to-output weights respectively the error derivatives can be calculated as follows:

$$\frac{\partial E}{\partial w_{jk}} = \frac{\partial E}{\partial s_k} \frac{\partial s_k}{\partial \bar{s}_j} \frac{\partial \bar{s}_j}{\partial w_{jk}}$$

and:

$$\frac{\partial E}{\partial \bar{w}_{ij}} = \frac{\partial E}{\partial s_k} \frac{\partial s_k}{\partial \bar{s}_j} \frac{\partial \bar{s}_j}{\partial \sigma_i} \frac{\partial \sigma_i}{\partial \bar{w}_{ij}}$$

In these formulas the index t has been dropped for simplicity. After defining a suitable error function simple manipulations of the above formulas give the exact expressions used to update the weights in every case. The sigmoid functions mentioned above are usually chosen due to the simplicity of their first derivatives which will finally be used in these exact expressions.

The advantage of backpropagation applied in an *ordered system*, such as feedforward networks, is that it allows us to determine the change δw of every weight w independently and hence possibly in parallel, by simply applying the differentiation chain rule to the appropriate error function. Its disadvantage is that it requires the latter function to be continuously differentiable.

Now, considering the Backpropagation as a training algorithm (denoted henceforth BPR) which aims to minimize E by applying adjustments to the weights, the corresponding learning rule can be defined as follows:

$$w^{new} = w^{old} + \delta w^{old}$$

Note that 'old' and 'new' here denotes an update in general either in the on-line or the batch mode of the algorithm. In order to find a minimum (ideally global) on the surface defined by the corresponding error function, a gradient descent path is followed by defining:

$$\delta w = -\eta \frac{\partial E}{\partial w}$$

where the change *stepsize* η is a suitably chosen small positive value (usually less than 1) named the *learning rate* in the neural networks jargon. The method works as the gradient always points in the direction of strongest change of the function, hence, the negative gradient follows a descent direction on the error surface.

A large number of modifications and extensions to the basic Backpropagation algorithm have been proposed from the early years of its appearance to present date. They mainly attempt to address the issues of low speed, oscillations of the weights, and convergence to local minima, although in some cases there are also concerns about theoretical guarantees for global convergence. The first and most commonly used extension to the basic BPR is the addition of a *momentum* term to the weight update formula which aims to restrict oscillations and improve speed. Subsequent modifications proposed techniques to adapt either or both of the learning rate and momentum parameters dynamically during the evolution of training, often taking into account the topology of the error function surface. Among them first to be mentioned is the Delta–bar–Delta [14] rule which proposes individual changes to the stepsizes for each weight according to sign changes between the current partial derivative and an exponential average of the partial derivatives of past iterations. The Resilient Backpropagation (known as RPROP) [25] monitors the changes in the sign of the gradients only to propose an increase in the learning rates when no change is observed and a decrease according to an exponential schedule in the opposite case. Quickprop (QPROP) [7] introduces independent stepsizes for each direction of the gradient (every weight) as a result of an approximate secant solution to a local quadratic approximation of the error surface.

We decided to choose the Quickprop algorithm as the one to form, along with the basic BPR, the group against which the SSC training algorithm proposed here will be tested. The main properties of QPROP supporting this choice are:
(*i*) its simplicity [7],
(*ii*) it uses second–order information,
(*iii*) it uses different learning rates for every weight in the network so it belongs to the more sophisticated end of the spectrum. In contrast, the SSC algorithm we propose here uses a global learning rate for all the weights (as we shall see in the next section), and finally,
(*iv*) whenever QPROP was included in comparative studies, either with the Backpropagation algorithm and its variants, or a variety of deterministic type algorithms (e.g. those based on the Conjugate Gradient numerical

optimisation procedure), it was reported either to outperform them or at least to exhibit equal speed and generalisation ability [7, 12].

2.3 Definition of the Algorithms

We are now in the position to formally define the three algorithms selected to be used in the experiments presented here. Namely we introduce the standard Backpropagation, the Quickpropagation, and the SSC-based SABB. The formalism we adopt helps to identify the similarities with the framework introduced in Section 2.1. We also use the standard notation for the neural network parameters with w_k denoting the weight vector at iteration k. Note that all the algorithms we used update the weights in *batch* mode, so the values of weights at iteration k correspond to those resulting after an update was performed at the end of the $k$the *epoch.*

BPR The first algorithm is the classical standard Backpropagation training algorithm (henceforth denoted BPR to identify it from the Error Backpropagation (EBP) method for derivative calculation discussed previously) defined as follows:
Given an initial weight vector w_0,

$$w_{k+1} = w_k - Zg_k, \; k = 0, 1, 2, ...,$$

where $Z = \eta I$ is a diagonal matrix and I denotes the unit matrix. Also, g_k is, as usual, an estimator of the gradient $\nabla E(D, w_k)$ at w_k, with $E(D, w_k)$ the network error function as defined by Eq. 1 in page 710. This notation is used to emphasise the dependence of the error function to the training set D. The same notation for the error function is used for the definitions of the remaining algorithms as well. As already mentioned previously, η is typically a small positive constant, usually less than 1.

QPROP The Quickpropagation can be defined as the following gradient-based algorithm:
Given an initial weight vector w_0,

$$w_{k+1} = w_k - B_k g_k, \; k = 0, 1, 2, \dots ,$$

where

$$B_k = \begin{cases} B_0 & \text{if } k = 0, \text{or} \Delta w_{i,k} = 0 \\ B'_k & \text{otherwise} \end{cases} ,$$

where $B_0 = lI$, I is the unit matrix, l is a small positive constant, usually less than 1, indicated as "alternative stepsize" in the Tables presented in the next section, and $w_{i,k}$ and $g_{i,k}$ are the ith components of w_k and $\nabla E(D, w_k)$ (as defined previously) respectively. For $k \geq 1$ $B'_k = (b'_{ij,k})$, with

$$b'_{ij,k} = \begin{cases} 0 & i \neq j \\ b'_{i,k} & i = j \end{cases},$$

where

$$b'_{i,k} = \text{sign}(b_{i,k}) \min(b|\frac{\Delta w_{i,k}}{g_{i,k}}|, |b_{i,k}|),$$

and

$$b_{i,k} = (w_{i,k} - w_{i,k-1})/(g_{i,k} - g_{i,k-1}),$$

with $\Delta w_{i,k} = w_{i,k} - w_{i,k-1}$, and b a constant with most commonly suggested values $1.5 \leq b \leq 3.5$. We followed the advise given in [7] and used a value of $b \approx 1.75$, which was reported to give the best results for a class of benchmark problems within the neural networks training context.

SSC-SABB The SSC framework, described previously, can be adapted to the neural network training formalism in the following way. First, we have to redefine the Supervisor algorithm (SR), which, in this case, is the classical Stochastic Approximation (SA) algorithm:
Given an initial weight vector w_0,

$$w_{k+1} = w_k - t_k g_k, \; k = 0, 1, 2, ...$$

where $g_k = \nabla E(D, w_k)$, an estimate of the gradient of the network error function $E(D, w_k)$. In the following we assume, as usual, the following assumptions to hold for the gain sequences $\{t_k\}$:
i) $t_k > 0$;
ii) $\sum_{k=0}^{\infty} t_k = +\infty$;

Then the search engine (SE) can be formulated based on the following gradient algorithm:
Given w_0

$$w_{k+1} = w_k - r_k g_k, \; k = 0, 1, 2, ...,$$

where $r_0 = 1$ and for $k \geq 1$

$$r_k = |w_k - w_{k-1}|^2/((w_k - w_{k-1})^T(\nabla E(D, w_k) - \nabla E(D, w_{k-1})).$$

where the step length r_k is called the BB stepsize proposed by Barzilai and Borwein in [1]. We are now in the position to define the SSC-based algorithm in the neural network training context as follows:
Let $T_k > 0$ be given for $k = 0, 1, 2, \ldots$. Then define the following algorithm (SSC-SABB):

$$w_{k+1} = w_k - t_k g_k \text{ if } T_k E(D, w_k - t_k g_k) \leq E(D, w_k - r_k g_k), \quad (k = 1, 2, \ldots),$$

otherwise

$$w_{k+1} = w_k - r_k g_k.$$

In our computations, $\{|r_k|\}$ is normally forced to be bounded to avoid instabilities.

2.4 Error Functions

The role of the objective or error function used in the learning process is, as already noted, to define the difference between good and bad performances with respect to the targets and thus to guide the search for a solution. In this sense it has a fundamental effect on the outcome, so it is important to choose the function accurately reflecting the experiments' design goals. Furthermore, it was already pointed out that whether the error is evaluated over the training or the test set, or on both, has implications for the type of performance we are interested to assess, be it memorisation or generalisation.

The most commonly used error function is the so-called 'sum–of–squares' which, given the notation introduced previously, can be written for the batch training mode as:

$$E = \sum_t \sum_k (y_k^t - s_k^t)^2$$

where, as usual, t indexes the patterns and k the output nodes. Assuming linear activation functions in the output nodes $g(a_k) = a_k$, where a_k is the input to the kth output node, the error derivative with respect to a_k is:

$$\frac{\partial E}{\partial a_k} = \sum_t (y_k^t - s_k^t)$$

This function is obtained by the maximum likelihood principle under the assumption that the targets are generated from a smooth deterministic function with added Gaussian noise [2]. Although this is a reasonable starting point for regression problems, it is clearly inappropriate for the case of classification where the targets are binary variables and the activation of the output nodes is often interpreted as probabilities.

For the two–class problems we usually have a single target assuming the values 0 or 1 and a single corresponding output node. The probability model used for the target in such cases is a particular case of the binomial family called the Bernouli distribution. The error function corresponding to this model is the so–called "cross entropy" function [13, 33]:

$$E = \sum_t y^t \ln s^t (1 - y^t) \ln(1 - s^t)$$

In [2, pp. 82–84] it is shown that for the activation of an output node to be interpreted as a probability it is appropriate to use the logistic activation function, $g(a) = [1 + \exp(-a)]^{-1}$, which is often associated also with the *logistic discrimination* in statistics. Then the derivative of the error function can be written, after some algebraic manipulations, as:

$$\frac{\partial E}{\partial a} = \sum_t (y^t - s^t)$$

Following the same line of reasoning in the multiclass case, where the targets can belong to 1-of-k classes and the network has k output nodes, an error function analogue to the "cross entropy", based on the assumption of binomially distributed target values, is the following:

$$E = \sum_t \sum_k y_k^t \ln \left(\frac{s_k^t}{y_k^t} \right)$$

In this case, the appropriate activation function to use, so that the output values sum up to unity, is a generalisation of the logistic, often called the "softmax" function [3] defined as follows:

$$g(a_k) = \frac{\exp(a_k)}{\sum_{k'} \exp(a_{k'})}$$

with $k' \neq k$, which can be rewritten as:

$$g(a_k) = [1 + \exp(-A_k)]^{-1}$$

with

$$A_k = a_k - \ln\{\sum_{k' \neq k} \exp(a_{k'})\}$$

Again after some algebraic manipulations the error derivative turns up to be :

$$\frac{\partial E}{\partial a_k} = \sum_t (y_k^t - t_k^t)$$

It is easy to see that considering all three cases we examined there is a natural pairing between the error functions and the corresponding activation functions. These functions have a number of desirable properties that justify their popularity. First, they are smooth and easily differentiable. Second, they are well–understood and they allow valuable theoretical study by simplifying analysis considerably. Although, as pointed out in [20, p. 86], "it might seem more natural to use a percentage misclassification error measure in classification problems" it is advantageous to use the above functions during the learning process and utilise the recognition rate as an additional measure of evaluation. This is exactly the strategy we adopted in the experiments, and we shall present in the next section.

3 Empirical Investigations

In this section we present a series of empirical investigations aiming to assess the SSC algorithm we propose for training feedforward neural networks in the context of *pattern recognition*. The properties we are interested in are efficiency and robustness in training under various types of noises, but mainly and most importantly the *generalisation* ability of the resulting trained network, since the latter constitutes the ultimate goal of learning. Furthermore, we compare our proposal with the two algorithms chosen as benchmarks, namely the standard batch Backpropagation (BPR) and the Quickprop (QPROP).

3.1 Experimental Design

For every problem we examine and for all the algorithms we repeat the training process ten times, initialising the weights to random values uniformly distributed in the interval $[-0.01, +0.01]$. This is a common practice in the

Table 1: Parameters of algorithms tested; η: learning rate of BPR; l: alternative stepsize for QPROP; t_k: supervisor's stepsize; T: switching parameter for SABB.

Algorithms	*Ver.*	*Parameters*	
BPR	1	$\eta = 0.001$	
	2	$\eta = 0.1$	
QPROP	1	$l = 0.001$	
	2	$l = 0.1$	
SSCSABB	1	$t_k = 0.001/k$	$T = 1$
	2	$t_k = 0.1/k$	$T = 1$
	3	$t_k = 10.0/k$	$T = 1$
	4	$t_k = 0.1$	$T = 1$

neural network community. The results obtained from these repetitions (also called 'runs') provide the basis to perform valid statistical analyses of the performance characteristics of the training algorithms we are interested in. The quantities included in the following Tables correspond to mean values calculated over the whole set of runs and, where appropriate, are annotated by the corresponding standard deviations. It should be noted here that all versions of the algorithms start from the same initial weight vector for the same run. For example, for the first of the ten runs all versions of BPR, QPROP, and SSC-SABB start training using the same vector of randomly assigned weights. However, this vector differs from the one used in the second training run. This is done in order to reduce the variance of the resulting performance measure estimator and hence the variance of the estimators of their differences. As a consequence of the same argument, we also decided to use the same partition of the available data to disjoint training and test sets for all the algorithms during the same run. Doing so we restrict the parameters influencing the differences observed in performance only to the effects due to the use of different training methods.

The experiments we conduct aim to evaluate the performance, first, with respect to a range of values for the adjustable parameters of the SSC-SABB algorithm, namely the 'preference switch' T and the supervisor algorithm's stepsize t_k. The objective is to explore the level of sensitivity to these parameters under various conditions (different tasks), and provide possible guidelines for effective values. At the same time, we try different values for the control parameters of the benchmark algorithms, in order to facilitate a comparative view of their sensitivity as well. For BPR we use two different learning rates η, while for the QPROP two alternative stepsizes l. The full set of different versions we have used for each algorithm along with the corresponding parameter values are included in Table 1.

In order to observe the effect these parameters have in the *training speed* and *accuracy* of the algorithms, we report the error levels obtained at the *middle* and at the *end* of the simulation runs. Although in some of our examples we use the cross–entropy as error function (see section 2.4) during training, due to the nature of the problems (classification), as we shall discuss below, in the relevant Tables, for uniformity of exposition we present the corresponding Mean Squared Error (MSE) in all cases.

The first two of the four classification tasks we employed represent the well-known problem of Parity, a two class problem in which all possible cases can be enumerated (details will be presented below). They were included here to illustrate the difference between *memorisation* and *generalisation* even when no noise is present in the data. In the first example, the sample available for training includes all possible cases so the issue of generalisation does not arise. However, in the second task, half of the cases are kept as a test set so generalisation is relevant since the networks are required to deduce the input-output mapping from the incomplete information of the training set. Although noise is not incorporated in the input-target pairs, the uncertainty induced by restricting the information available gives rise to the issue of generalisation which can be measured by the performance in the test set.

The last two classification tasks are well-known real world problems which involve different types of noises in the measurements. In the first, the objective is to identify mines from rocks based on sonar signals corrupted by measurement noise. The second task is a digit pattern recognition problem in which the input patterns consist of Fourier coefficients extracted as features of the raw binary digit images. The uncertainty, and hence the noise component in the data, in this case, is introduced by the feature extraction process and the corresponding loss of information.

In the following, we first discuss in detail each one of the above tasks, the procedures used to generate the corresponding data sets, some of their characteristics, as well as the architectures and the training parameters of the neural networks employed to tackle them. Additionally, we comment, based on exploratory inspection of the corresponding *learning curves* (i.e. plots of the MSE vs Epochs) , on the differences observed between the

Table 2: Parameters used in the simulation runs for the **Classification** problems; i-h-o: number of input,hidden, output nodes

Problems	*Architecture* *i-h-o*	*Max Nr.* of Epochs	*Convergence* *Criterion (MSE)*
Parity 8	8-16-1	90	0.0001
Parity 8 Generalisation	8-16-1	90	0.001
Sonar	60-24-2	2500	0.12
Digits	76-40-10	2500	0.25

behaviour of the algorithm proposed here and that of the two algorithms used as benchmarks. Finally, we proceed in presenting the statistical results obtained according to the experimental schedule described previously and further discuss a series of statistical analyses based on them. As a final note, it is worth reminding here that the main interest of our investigations is focussed on the generalisation ability of the learning process as opposed to the memorisation of the training sample.

3.2 The Benchmarking Problems

Again, as a general remark, it is useful to note here, that all the networks trained to address the classification tasks we examine in this study had identity activation functions for their input nodes and symmetrical sigmoid nonlinearities (hyperbolic tangent) for their hidden nodes. However their output nodes were assigned logistic sigmoid activation functions, so that their outputs can be interpreted as class membership probabilities. The error function used for the two-class problems was the cross-entropy, while for the multi-class case its multivariate analogue (see Section 2.4). The network architectures corresponding to each of the problems along with the values for other parameters used during the performance assessment experiments are summarised in Table 2.

3.2.1 Parity

The parity problem is one of the best established benchmarks for neural network learning methods. It can be expressed as the sum (modulo 2) of n binary inputs. In other words, the target value is 'true' (1) if and only if an

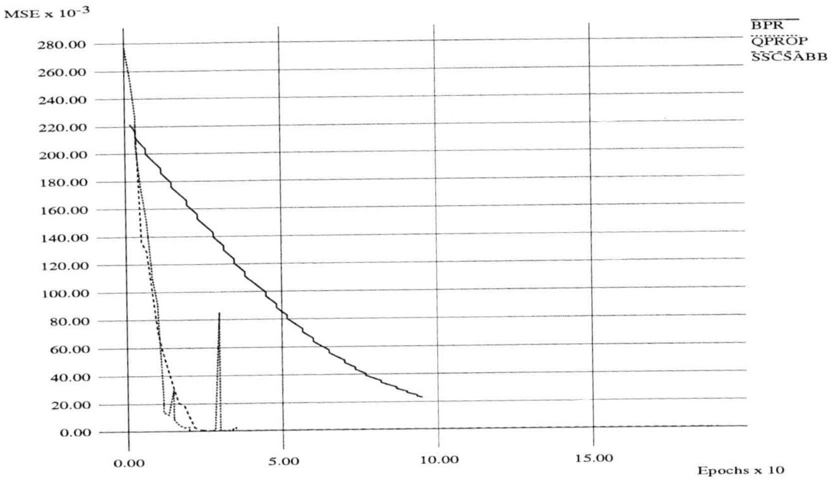

(a)

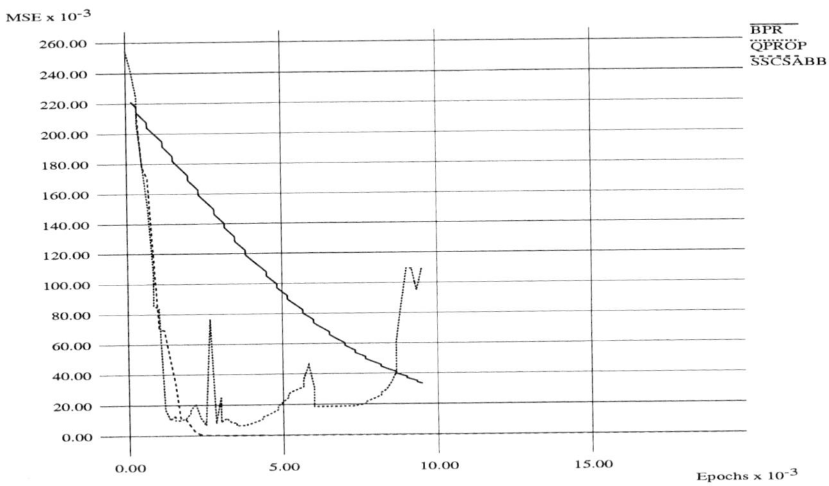

(b)

Figure 2: Parity 8 problem: Typical examples of learning curves of the QPROP and SSCSABB , in the training (a) and test(b) sets, compared to the corresponding curve for BPR in the test set.

odd number of inputs are 'true', otherwise is 'false' (0). A special case of this task where there are only two binary input variables is the well-known 'exclusive or' (XOR) problem. Despite the simple underlying rule the parity class of problems is surprisingly hard to be addressed by neural networks trained by the backpropagation of error [34]. The particular significance of the parity task as an impossible problem for linear first-order perceptrons was pointed out by Minsky and Papert [21]. It belongs in the category of the so-called "statistically neutral" problems, for which the probability (i.e. observed frequency) to assign any particular input to a specific output class is always the chance value. For the two-class problem like the one we use here this value is 0.5. It has been argued that exactly this fact prevents learning methods based on extracting relationships from examples, such as the neural networks, from performing well in the parity problems [10]. In this study we use one of the harder versions which involves 8 input variables (characterised in the following as 'Parity 8'). The exhaustive set of cases includes $2^8 = 256$ patterns (input-target pairs). In [7] it is reported that QPROP performs favourably, in terms of speed, in the two-input equivalent 'XOR' problem even in comparison to a neural network trained by a "BFGS" based algorithm. We follow the advice given in the CMU Benchmark Learning Archive [5] that for n input parity problems the networks should have at least n hidden nodes. So, we train networks with 8 input, 16 hidden, and 1 output nodes, to a maximum of 90 epochs.

When training involves all possible cases the issue of generalisation is not relevant. Therefore, in order to test our algorithm in a much harder problem we constructed a second task from the same data set which involves performance in unseen data. This problems is denoted in the following as 'Parity 8 Generalisation'. In this case the original data from the parity problem for 8 inputs was divided to two disjoint sets of equal size, one for training and one for testing. It has been reported recently in [10] that networks trained with the standard Backpropagation algorithm exhibit poor generalisation ability even when minimal numbers of patterns are excluded from the training set. This can be verified from Fig. 2(b) where the corresponding learning curve (solid line) decreases much slower than those of the QPROP (dotted line) and the SSC-SABB (dashed line). Furthermore, BPR fails to achieve error levels similar to those achieved by the other two algorithms both in the training (Fig. 2(a)) and the test (Fig. 2(b)) sets. Another interesting observation from these graphs concerns the behaviour of QPROP. In the training set it achieves the same level of MSE as SSC-SABB and with the same speed, although it seems to be somewhat unstable (observe the spikes in the corresponding line). However, in the test set (Fig. 2(b)) it shows a completely different behaviour. After it reaches a minimum level the error increases significantly up to the end of the training period and the learning curve assumes a 'U' shape, characteristic of the case of overtraining. In contrast, the generalisation error of the SSC-SABB decreases steadily as training continues and saturates at a very low level. Conclusions and possible reasons for this difference in behaviour between the two algorithms will be explored in the following section where statistical analyses of the results are discussed.

3.2.2 Sonar, Mines vs Rocks

This is a classical benchmarking data set for evaluating learning algorithms in the field of artificial neural networks. Since its first use in [11], it has been used extensively by researchers in the field [26]. However, due to the fact that there numerous ways to construct evaluating experiments with this data (see relevant comments in the CMU Learning Benchmark Archive [5]) comparisons are not straightforward. The task is to discriminate between sonar signals bounced off a metal cylinder (mine) from those obtained from a rock with roughly the same shape. The original data set consists of 208 cases each of which comprises a 60-dimensional vector of continuous real values serving as inputs and one enumerated variable, acting as target.

The data used in the present study was obtained from the CMU Learning Benchmark Archive [5] and is the one characterised as the 'aspect angle dependent' set. Since the original sonar measurements did not include aspect angle information the contributors used clustering to create two disjoint sets in each of which a representative number of samples from all aspect angles would be included. Both sets contain 104 cases. We employ one of the sets to train networks with 60 input, 24 hidden, and two output nodes. We allowed training to continue for 2500 epochs. Fig. 3 illustrates the training and generalisation behaviour of QPROP and SSC-SABB as compared to the generalisation obtained from the networks trained by BPR (solid line in both subfigures). It is not difficult to observe that in this task also, QPROP tends to overtrain the networks and hence its predictive performance degrades. Once more SSC-SABB shows resilience to this phenomenon and its generalisation ability saturates at a level equal to that achieved by BPR, but at a much faster rate. In [11], a recognition rate of 89.2% in the test set is reported to have been achieved by a network with similar architecture to ours, and trained by a modified version of the BPR. As we shall see below,

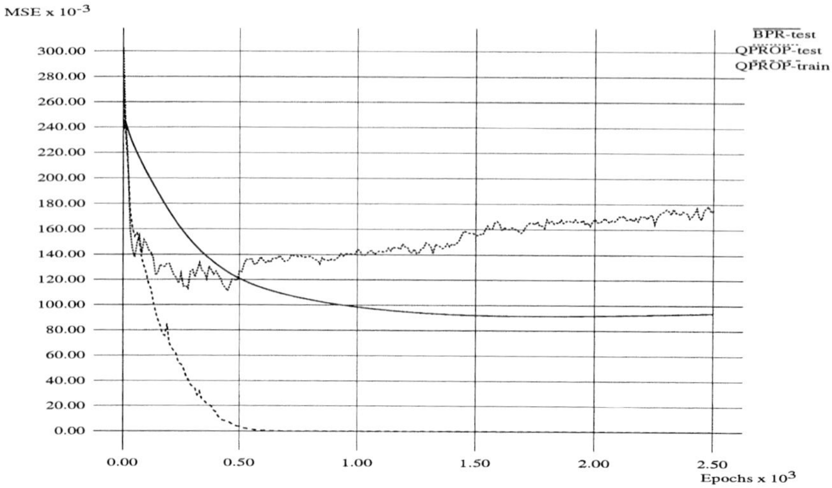

(a)

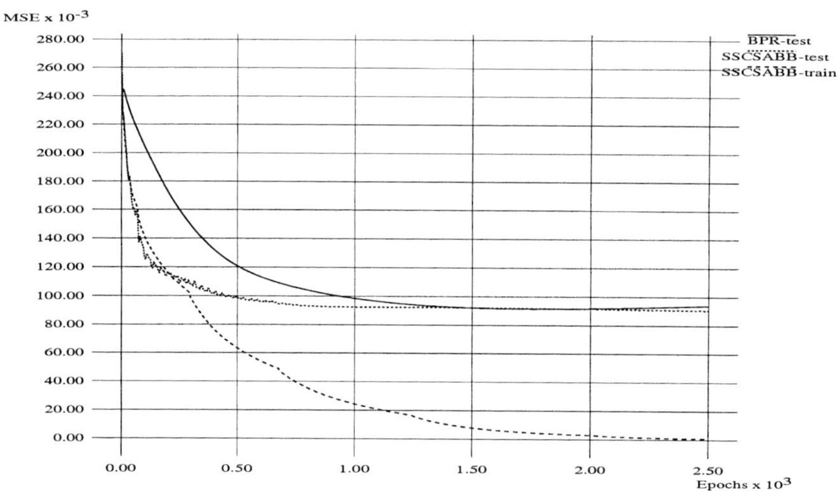

(b)

Figure 3: Sonar problem: Typical examples of learning curves of the QPROP and SSCSABB, in the training (a) and test (b) sets, compared to the corresponding curve for BPR in the test set.

where we discuss similar type of results, our algorithm achieved comparable performance.

3.2.3 Handwritten Digit Recognition

In this example we employ a more complex and realistic task from the character recognition domain. The dataset consists of 76 Fourier coefficients extracted from each of 2000 binarised images of handwritten numerals (0–9), with an original resolution of 30 x 48 pixels, obtained from Dutch utility maps. More details about this dataset can be found in [35] under the name 'mfeat'. In the same form the set was used in [6]. Samples are distributed equally over the classes (200 samples per class) and form a non-trivial 10-class pattern recognition problem. A training set was generated by randomly selecting 125 samples per class while the remaining 750 samples (75 per class) were used as a testing set.

All the networks trained for this task had 76 input, 40 hidden, and 10 output nodes. Training was continued for 2500 epochs. Fig. 4 shows characteristic learning curves of the QPROP and the SSCSABB obtained on the training and test sets. Additionally, for comparison, it includes, in both subfigures, the curve corresponding to BPR for the test set. It is not difficult to observe in Fig. 4(a) that the generalisation error of QPROP (dotted line) after reaching a minimum value starts to increase as the training error (dashed line) decreases further. This illustrates clearly the results of memorisation of the training patterns (overtraining). Fig. 4(b) demonstrates that the same does not hold for the SSC-SABB, which exhibits a behaviour qualitatively more similar to that of BPR, and succeeds to avoid overtraining.

3.3 Statistical Analysis and Discussion

Additionally to our previous discussion we also performed a series of statistical analyses and hypotheses tests in order to explore the significance of the performance differences we observed. Here, the picture formed by the corresponding results presented, appears to be clearly in favour of the SSC-SABB algorithm. We start by reporting again the average MSEs over the ten training runs for each of the four benchmarking problems.

Results in Tables 3 and 4 refer to the training set while these in Tables 5 and 6 to the test set. Note that for the conventional Parity problem (denoted as 'Parity 8' in the Tables) where all possible cases are used for training, there are no results corresponding to the test set. In this task,

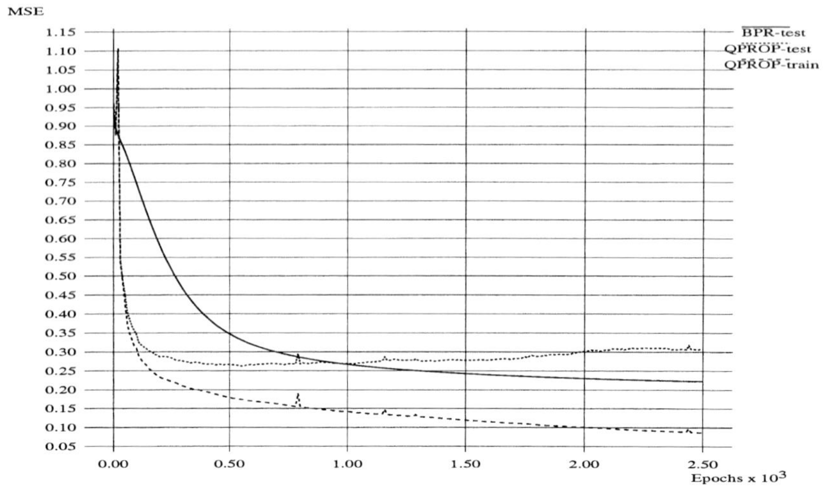

(a)

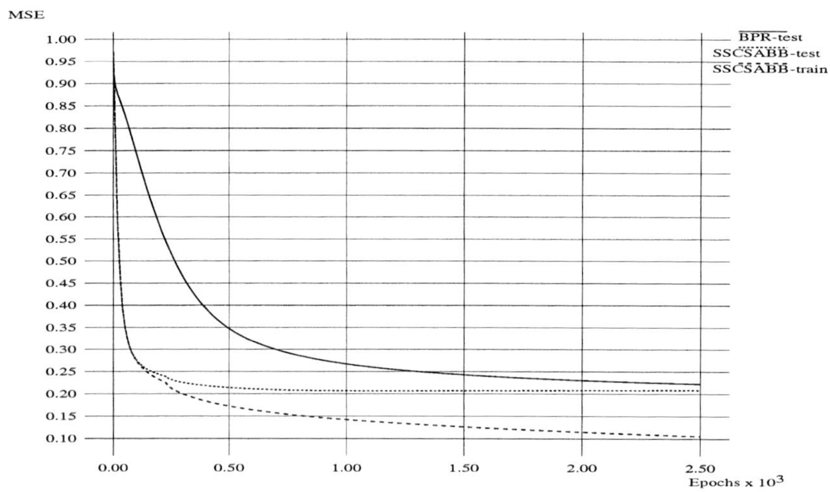

(b)

Figure 4: Handwritten Digit recognition: Typical examples of learning curves of the QPROP (a) and SSCSABB (b), in the training and test sets, compared to the corresponding curve for BPR in the test set.

Table 3: Mean Squared Error (MSE), in the **training set**, and its Standard Deviation (Std) in parentheses **at the middle** of the simulation runs for the algorithms tested.

Mean Squared Error (Standard Deviation)					
Algorithms	*Ver.*	*Parity 8*	*Parity 8 Generalis.*	*Sonar*	*Digits*
BPR	1	0.27365 (0.03482)	0.26945 (0.04023)	0.24063 (0.01465)	0.89120 (0.00832)
	2	0.11417 (0.01849)	0.10756 (0.03091)	0.05263 (0.00456)	0.25099 (0.00347)
QPROP	1	0.01264 (0.03873)	0.09267 (0.29303)	0.02494 (0.07879)	0.12445 (0.00581)
	2	0.00741 (0.01869)	0.00000 (0.00000)	0.00103 (0.00308)	0.12614 (0.00587)
SSCSABB	1	0.00000 (0.00000)	0.00000 (0.00000)	0.04645 (0.08325)	0.13415 (0.00501)
	2	0.00000 (0.00000)	0.00000 (0.00000)	0.02586 (0.05028)	0.13655 (0.00822)
	3	0.00004 (0.00013)	0.00000 (0.00000)	0.04738 (0.06194)	0.15305 (0.00989)
	4	0.00000 (0.00000)	0.00000 (0.00000)	0.00101 (0.00093)	0.12428 (0.00203)

as already mentioned, measurement of generalisation is not relevant. The Tables provide the means to identify which of the versions of each algorithm are favourable as well as to monitor the evolution of the training error and the generalisation error (often called predictive error in statistics). It is not difficult to realise from these Tables that, for the BPR, version 2 which has a learning rate of 0.1 (see Table 1) is the one exhibiting the best performance. In all problems tested the training error of BPR decreases considerably from the middle to the end of the training period. The same observation is, however, true only for the 'Parity 8–Generalisation' problem when the MSE in the test set is considered. It is important to note here that the latter (generalisation error) does not increase as training continues from the middle to the end of the run. The same, however, cannot be argued for QPROP as well. In all the problems included in Tables 5 and 6 the generalisation error of QPROP is, noticeably, higher at the end of the training period than

Table 4: Mean Squared Error (MSE), in the **training set**, and its Standard Deviation (Std) in parentheses **at the end** of the simulation runs for the algorithms tested.

Mean Squared Error (Standard Deviation)					
Algorithms	*Ver.*	*Parity 8*	*Parity 8 Generalis.*	*Sonar*	*Digits*
BPR	1	0.27054 (0.03421)	0.26563 (0.03860)	0.23105 (0.01319)	0.87256 (0.00957)
	2	0.03362 (0.00852)	0.02977 (0.01358)	0.01018 (0.00172)	0.21153 (0.00253)
QPROP	1	0.00003 (0.00009)	0.03961 (0.12527)	0.02520 (0.07969)	0.09165 (0.00935)
	2	0.00000 (0.00000)	0.00000 (0.00000)	0.00000 (0.00000)	0.09323 (0.00821)
SSCSABB	1	0.00000 (0.00000)	0.00000 (0.00000)	0.04099 (0.08609)	0.10679 (0.00421)
	2	0.00000 (0.00000)	0.00000 (0.00000)	0.01733 (0.05311)	0.11122 (0.00996)
	3	0.00000 (0.00000)	0.00000 (0.00000)	0.03888 (0.06222)	0.12658 (0.01038)
	4	0.00000 (0.00000)	0.00000 (0.00000)	0.00001 (0.00001)	0.09319 (0.00388)

in the middle. It can be seen from the corresponding entries in Tables 3 and 4 that for all the problems, except for the Digits recognition, QPROP's performance appears to be sensitive to the choice of the alternative learning rate l, with the version having the larger value being favourable. The same observation holds only for 'Parity 8–Generalisation' task with respect to predictive error (Tables 5 and 6). We do not have any apparent explanation for this change in the behaviour of QPROP in this class of problems.

Let us, now, consider SSC-SABB. We employ only versions with switching parameter values $T = 1$. This is because it has been argued in the literature that more conservative (even slower) algorithms, such as the BPR, can perform better than faster algorithms in classification tasks [24, pp. 152,156]. The argument is based on the observation that in such tasks high accuracy in the approximation of the minimum in error is not required as long as the network outputs identify unambiguously the correct class. Fur-

Table 5: Mean Squared Error (MSE), in the **test set**, and its Standard Deviation (Std) in parentheses **at the middle** of the simulation runs for the algorithms tested.

Mean Squared Error (Standard Deviation)				
Algorithms	*Ver.*	*Parity 8 Generalis.*	*Sonar*	*Digits*
BPR	1	0.26891 (0.04299)	0.24775 (0.01923)	0.89036 (0.00934)
	2	0.11608 (0.02705)	0.09976 (0.00543)	0.25707 (0.00485)
QPROP	1	0.10067 (0.28455)	0.19131 (0.03971)	0.26142 (0.00746)
	2	0.02562 (0.06739)	0.19132 (0.04215)	0.26158 (0.01740)
SSCSABB	1	0.00000 (0.00000)	0.11915 (0.05052)	0.20792 (0.00374)
	2	0.00000 (0.00000)	0.10240 (0.02062)	0.20807 (0.00382)
	3	0.00002 (0.00004)	0.11889 (0.03316)	0.20985 (0.00514)
	4	0.00000 (0.00000)	0.10129 (0.00772)	0.20784 (0.00379)

Table 6: Mean Squared Error (MSE), in the **test set**, and its Standard Deviation (Std) in parentheses **at the end** of the simulation runs for the algorithms tested.

Mean Squared Error (Standard Deviation)				
Algorithms	*Ver.*	*Parity 8 Generalis.*	*Sonar*	*Digits*
BPR	1	0.26566 (0.04160)	0.23771 (0.01783)	0.87165 (0.01112)
	2	0.04350 (0.01195)	0.09499 (0.00458)	0.23049 (0.00443)
QPROP	1	0.04119 (0.13026)	0.19931 (0.03702)	0.28847 (0.01847)
	2	0.02570 (0.05097)	0.20605 (0.04203)	0.28235 (0.01734)
SSCSABB	1	0.00000 (0.00000)	0.12208 (0.04916)	0.20891 (0.00413)
	2	0.00000 (0.00000)	0.10544 (0.01977)	0.20850 (0.00330)
	3	0.00000 (0.00000)	0.11924 (0.03086)	0.20861 (0.00525)
	4	0.00000 (0.00000)	0.10337 (0.00959)	0.21080 (0.00483)

thermore, avoiding overtraining appears to be much more important in classification tasks so that algorithms incorporating some kind of mechanism to saturate before memorising the particular training patterns are preferable. The Stochastic Approximation algorithm (SA), which forms the supervising

Table 7: Best performing version of the algorithms tested, **at the middle** of the simulation runs, in the **training and test sets** respectively.

Best performances								
Algorithms Compared	*Parity 8*		*Parity 8 Generalis.*		*Sonar*		*Digits*	
	Train	Test	Train	Test	Train	Test	Train	Test
BPR	0.114 (2)	-	0.107 (2)	0.116 (2)	0.052 (2)	0.099 (2)	0.250 (2)	0.257 (2)
QPROP	0.007 (2)	-	0.000 (2)	0.025 (2)	0.001 (2)	0.191 (1)	0.124 (1)	0.261 (1)
SSCSABB	0.000 (1-4)	-	0.000 (1-4)	0.000 (1-4)	0.001 (4)	0.101 (4)	0.124 (4)	0.207 (4)

part of the SSC algorithm, can be considered to have such a mechanism.

Unlike the other two algorithms, all versions of SSC-SABB seem to tackle the Parity task very easily since they produce virtually zero MSE from the middle of the training period both on the training and the test sets. Diversification in the behaviour of the SSC-SABB versions is observed, however, in the other two tasks, the Sonar problem and the Digits recognition. The version with constant stepsize (version 4) seems to be favourable with respect to training error for all the problems. However, the same picture is not retained when the error in the test set is considered (Tables 5 and 6), where all versions exhibit similar performance. From the same Tables, no considerable increase in the generalisation error can be observed, unlike the case of QPROP we discussed above. This provides support to our previous hypothesis that the supervisor algorithm along with the switching mechanism form a successful device to tackle the problem of overtraining.

The results corresponding to the most successful versions for each algorithm from the previous Tables are summarised in Tables 7 and 8 (rounded to three decimal points). Version number indicators are included in parentheses. An examination of the entries in these Tables verifies the superiority of the SSC-SABB for all the tasks with respect both to the training efficiency and generalisation. They also provide clear empirical evidence in support of the points raised in our previous discussion about the issue of overtraining.

We proceed by discussing a series of hypotheses testing analyses in order to define the statistical significance of the pairwise differences in perfor-

Table 8: Best performing version of the algorithms tested, **at the end** of the simulation runs, in the **training and test sets** respectively.

Best performances									
Algorithms Compared	*Parity 8*		*Parity 8 Generalis.*		*Sonar*		*Digits*	
	Train	Test	Train	Test	Train	Test	Train	Test
BPR	0.033 (2)	-	0.029 (2)	0.043 (2)	0.010 (2)	0.094 (2)	0.211 (2)	0.230 (2)
QPROP	0.000 (2)	-	0.000 (2)	0.025 (2)	0.000 (2)	0.199 (1)	0.091 (1)	0.282 (2)
SSCSABB	0.000 (1-4)	-	0.000 (1-4)	0.000 (1-4)	0.000 (4)	0.103 (4)	0.093 (4)	0.208 (2)

Table 9: T–statistic values for the differences between the mean MSEs of the best performing version of the algorithms tested, **at the middle** of the simulation runs, in the **training and test sets** respectively (critical value $t_{\{1-0.05/2,18\}} = 2.109$).

T–Statistic								
Algorithms Compared	*Parity 8*		*Parity 8 Generalis.*		*Sonar*		*Digits*	
	Train	Test	Train	Test	Train	Test	Train	Test
BPR–QPROP	12.84	-	11.00	3.94	29.65	-7.22	59.15	-1.55
BPR–SSCSABB	19.53	-	11.00	13.57	35.06	-0.51	99.67	25.28
SSCSABB–QPROP	1.25	-	1.37	1.20	0.03	7.04	0.09	20.25

mance observed among the versions of the algorithms selected to be included in Tables 7 and 8. To this end, we calculate and present in Tables 9 and 11 the t-statistic values corresponding to the differences of the means of MSEs over the 10 runs for the pairs of algorithms shown in the first column of the Tables. Since the number of the runs is 10 the critical region, corresponding to the t-statistic with 18 degrees of freedom (10+10-2) and at a significance level of 0.05, is [-2.109, 2.109]. Values out of this interval suggest rejection of the null hypothesis (H_0) of equality for the means. This is

Table 8: Best performing version of the algorithms tested, **at the end** of the simulation runs, in the **training and test sets** respectively.

Best performances								
Algorithms Compared	*Parity 8*		*Parity 8 Generalis.*		*Sonar*		*Digits*	
	Train	Test	Train	Test	Train	Test	Train	Test
BPR	0.033 (2)	-	0.029 (2)	0.043 (2)	0.010 (2)	0.094 (2)	0.211 (2)	0.230 (2)
QPROP	0.000 (2)	-	0.000 (2)	0.025 (2)	0.000 (2)	0.199 (1)	0.091 (1)	0.282 (2)
SSCSABB	0.000 (1-4)	-	0.000 (1-4)	0.000 (1-4)	0.000 (4)	0.103 (4)	0.093 (4)	0.208 (2)

Table 9: T–statistic values for the differences between the mean MSEs of the best performing version of the algorithms tested, **at the middle** of the simulation runs, in the **training and test sets** respectively (critical value $t_{\{1-0.05/2,18\}} = 2.109$).

T–Statistic								
Algorithms Compared	*Parity 8*		*Parity 8 Generalis.*		*Sonar*		*Digits*	
	Train	Test	Train	Test	Train	Test	Train	Test
BPR–QPROP	12.84	-	11.00	3.94	29.65	-7.22	59.15	-1.55
BPR–SSCSABB	19.53	-	11.00	13.57	35.06	-0.51	99.67	25.28
SSCSABB–QPROP	1.25	-	1.37	1.20	0.03	7.04	0.09	20.25

mance observed among the versions of the algorithms selected to be included in Tables 7 and 8. To this end, we calculate and present in Tables 9 and 11 the t-statistic values corresponding to the differences of the means of MSEs over the 10 runs for the pairs of algorithms shown in the first column of the Tables. Since the number of the runs is 10 the critical region, corresponding to the t-statistic with 18 degrees of freedom (10+10-2) and at a significance level of 0.05, is [-2.109, 2.109]. Values out of this interval suggest rejection of the null hypothesis (H_0) of equality for the means. This is

Table 12: Hypothesis testing for differences between the average Mean Squared Errors of the best performing version of the algorithms tested, **at the end** of the simulation runs, in the **training and test sets** respectively (0: accept null hypothesis H_0, 1: reject H_0, where the alternative hypothesis is that the first algorithm performs better than the second, and (*) denotes equality of performances).

Algorithms Compared	Hypothesis Testing							
	Parity 8		*Parity 8 Generalis.*		*Sonar*		*Digits*	
	Train	Test	Train	Test	Train	Test	Train	Test
QPROP–BPR	1	-	1	0*	1	0	1	0
SSCSABB–BPR	1	-	1	1	1	0	1	1
SSCSABB–QPROP	0*	-	0*	0*	0	1	0*	1

column has achieved statistically significant lower error than the second algorithm. Hence, it can be easily verified that the previous comparison both for the Sonar and Digits tasks turns in favour of SSC-SABB. In the rest of the cases, it is indicated that the SSC-SABB performs equally to QPROP. Additionally, SSC-SABB clearly outperforms BPR.

The next comparisons are performed with respect to robustness as expressed by the number of runs the learning algorithm achieved a prespecified level of generalisation error (MSE in the test set). The MSE values used as convergence criterion for every task are shown in Table 2. From Table 13 we can safely conclude that our training algorithm (SSC-SABB) outperforms the other two both in robustness and in speed.

However, the ultimate test for a classification system is the rate of misclassification it achieves when presented with unknown patterns. Consequently, Table 14 presents the average percentage of classification error rates for all the versions of the algorithms measured over the test sets at the end of the training period. It is worth pointing out that even the worst performing version of SCC-SABB outperforms the best version of either of the other algorithms. Finally, because in the Parity problems QPROP also presented satisfactory behaviour at the end of the training phase, we present in Table 15 the same type of results for the Parity problems but after only 10 epochs of training. It is easily observed that SSC-SABB clearly excels

Table 13: Number of successful runs (out of 10) in the **test sets**, with the corresponding mean number of epochs and their standard deviations in parentheses (rounded to the nearest integer).

Convergence statistics					
Algorithms	*Ver.*	*Parity 8 Generalis.*	*Parity 8*	*Sonar*	*Digits*
BPR	1	0	0	0	0
	2	10 (529/26)	1 (496/-)	10 (675/107)	10 (1510/151)
QPROP	1	10 (40/16)	10 (38/7)	3 (185/42)	7 (457/157)
	2	10 (33/14)	10 (69/82)	2 (230/14)	7 (310/108)
SSCSABB	1	10 (15/2)	10 (13/5)	8 (75/44)	10 (151/32)
	2	10 (17/2)	10 (12/3)	9 (136/59)	10 (141/22)
	3	10 (26/9)	10 (18/8)	7 (187/142)	10 (216/75)
	4	10 (17/2)	10 (14/5)	10 (55/22)	10 (114/17)

Table 14: Average classification error rates and the corresponding standard deviations, in the **test set**, **at the end** of simulation runs.

Classification Error Rates (%)									
Algorithms	*Ver.*	*Parity 8*		*Parity 8 Generalis.*		*Sonar*		*Digits*	
		Mean	Std	Mean	Std	Mean	Std	Mean	Std
BPR	1	51.80	3.03	46.64	4.38	40.87	4.79	76.34	1.88
	2	0.04	0.04	0.16	0.11	11.63	3.15	16.06	1.65
QPROP	1	0.00	0.00	4.45	0.44	22.11	4.01	18.88	1.75
	2	0.00	0.00	2.66	0.70	21.15	3.99	18.50	1.74
SSCSABB	1	0.00	0.00	0.00	0.00	17.21	3.49	14.92	1.59
	2	0.00	0.00	0.00	0.00	12.98	3.27	14.82	1.59
	3	0.00	0.00	0.00	0.00	15.77	3.50	14.80	1.59
	4	0.00	0.00	0.00	0.00	11.44	3.13	14.58	1.58

in both tasks with considerably significant magnitude for the performance differences.

Table 15: Average classification error rates and the corresponding standard deviations, in the **test set**, after 10 epochs.

Classification Error Rates (%)					
Algorithms	*Ver.*	*Parity 8*		*Parity 8 Generalisation*	
		Mean	Std	Mean	Std
BPR	1	52.11	3.04	48.13	4.40
	2	33.13	2.92	30.00	3.74
QPROP	1	48.67	3.08	48.59	4.33
	2	20.78	2.14	23.05	3.23
SSCSABB	1	0.00	0.00	1.41	0.52
	2	0.00	0.00	0.08	0.08
	3	20.82	1.99	9.06	1.24
	4	0.00	0.00	0.00	0.00

4 Concluding Remarks

In summary, the algorithm proposed in this study (SSC-SABB) for application to Neural Network training produced in all cases and tasks tested performance superior to that of BPR at least by an order of magnitude. In the pattern recognition problems we used, it excels in all cases both in training speed and generalisation due, we believe, its resistance to over-training as a result of the supervision process imposed by the switching mechanism. In general, we can safely conclude that the neural networks trained with the SSC-SABB are not only excellent classifiers but also at least as competent function approximators as those trained by algorithms (e.g. QPROP) reported to be among the most successful [29]. This was supported in this study by a series of elaborate statistical analyses ranging from initial exploratory comparisons to statistical hypothesis testing for the observed differences in performances.

This study makes an original contribution to the novel research area emerging in a number of science and engineering fields, that of combination, or **synthesis**, of methods. Examples of the new concepts developed and investigated in this area, which seems to flourish in the last few years, can be found for instance in the field of pattern recognition, image analysis, or biometrics, in the form of the design of Multiple Classifier Systems. In these, effective methods are sought to combine the outputs of a variety of individual classifiers (either at the decision or raw output (posterior probabilities of classes)) in order to obtain systems with improved performance. An extensive collection of current work in this area can be found in [15]. A second example of these ideas can be drawn from the combination of various types of neural networks into *ensembles* or *committees* (see, for instance, [2, pp. 365–369]) and [30] or alternatively, in the same field, the fusion of neural networks and fuzzy logic methodologies [16]. In most of the above cases a common characteristic is that the participant component algorithms (be it a statistical classifier or a trained network) are combined after they have produced individually some kind of output. The additional innovative contribution of the material reported in the present work, about algorithms developed within the Supervisor Searcher Co-operation framework, is that it investigates a method to perform a **synthesis** at the algorithmic (*as opposed to output*) level. Therefore, it goes a step further in providing an understanding of possible ways to successfully integrate in a single scheme diverse desirable properties of existing methods.

References

[1] J. Barzilai and M. Borwein, Two point step size gradient methods, *IMA J. Numerical Analysis* Vol.8 (1988) pp. 141–148.

[2] C. M. Bishop, *Neural Networks for Pattern Recognition*, (Oxford, Clarendon Press, 1995).

[3] J.S. Bridle, Probabilistic interpretation of feedforward classification network outputs, with relationships to statistical pattern recognition in F. Fogelman Soulié and J. Hérault, (eds) *Neurocomputing: Algorithms, Architectures and Applications*, (New York, Springer–Verlag, 1990) pp. 227–236. .

[4] S.-Y. Cho and T.W.S. Chow, A fast heuristic global learning algorithm for multilayer neural networks, *Neural Processing Letters*, Vol.9 (1999) pp. 177–187.

[5] CMU, Learning Benchmark Archive, http://www.cs.cmu.edu/.

[6] R. Duin and D. Tax, Experiments with classifier combining rules, in J. Kittler and F. Roli (eds) *First International Workshop on Multiple Classifier Systems*, (Springer, 2000) pp. 16–29.

[7] S.E. Fahlman, An empirical study of learning speed in Back-Propagation neural networks, Technical Report CMU-CS-88-162, (CMU, USA, 1988).

[8] S. Geman, E. Bienenstock, and E. Doursat, Neural networks and the bias/variance dilemma, *Neural Computation* Vol.4 (1992) pp. 1–58.

[9] A. Ghorbani and L. Bayat, Accelerated backpropagation learning: Extended dynamic parallel tangent optimization algorithm, in H. Hamilton, (ed.) *Lecture Notes in Artificial Intelligence*, (Berlin, Springer-Verlag, 2000) pp. 524–532.

[10] A. Ghorbani and K. Owrangh, Stacked generalization in neural networks: Generalization on statistically neural problems, in *Proc. IEEE/INNS-IJCNN*, (Washington DC, USA, 2001) pp. 1715–1720.

[11] R.P. Gorman and T.J. Sejnowski, Analysis of hidden units in a layered network trained to classify sonar targets, *Neural Networks* Vol.1 (1988) pp. 75–89.

[12] M. Hasenjager and H. Ritter, Perceptron learning revisited: The sonar targets problem, *Neural Processing Letters* Vol.10 (1999) pp. 17–24.

[13] J.J. Hopfield, Learning algorithms and probability distributions in feed–forward and feed–back networks, *Proceedings of the National Academy of Sciences* Vol.84 (1987) pp. 8429–8433.

[14] R.A. Jacobs, Increased rates of convergence through learning rate adaptation, *Neural Networks* Vol.1 (1988) pp. 295–307.

[15] J. Kittler and F. Roli (eds), *Multiple Classifier Systems*, volume 2096 of *LNCS* (Springer, 2001).

[16] C.-T. Lin and G. Lee, *Neural Fuzzy Systems: A Neuro-Fuzzy Synergism to Intelligent Systems*, (New York, Prentice Hall, 1996).

[17] WB. Liu and Y.H. Dai, Minimization algorithms based on Supervisor and Searcher Co-operation: I-faster and robust gradient algorithms for minimization problems with strong noise. *JOTA* Vol.111 No.2 (2001) pp. 359–379.

[18] WB. Liu and K. Sirlantzis, SSC Minimization Algorithms, In C. Floudas and P. Pardalos (eds) *Encyclopedia of Optimization*, (Boston, Kluwer Academic Publishers, 2001) pp. 253–257.

[19] G. Magoulas, M. Vrahatis, and G. Androulakis, Improving the convergence of the backpropagation algorithm using learning rate adaptation methods, *Neural Computation* Vol.11 (1999) pp. 1769–1796.

[20] D. Michie, D. J. Spiegelhalter, and C. C. Taylor (eds), *Machine Learning, Neural and Statistical Classification*, (New York, Ellis Horwood, 1994).

[21] M. L. Minsky and S. A. Papert, *Perceptrons. An Introduction to Computational Geometry*, (Cambridge, MA, The MIT Press, expanded edition, 1988).

[22] R. M. Neal, *Bayesian Learning for Neural Networks*, volume 118 in Lecture Notes in Statistics (New York, Springer, 1996).

[23] S. Raudys and A.K. Jain, Small sample size problems in designing artificial neural networks, in I.K. Sethi and A.K. Jain (eds) *Artificial Neural Networks and Statistical Pattern Recognition*, (Amsterdam, North–Holland, 1991) pp. 33–50.

[24] R. Reed and R. Marks, *Neural Smithing: Supervised Learning in Feedforward Artificial Neural Networks*, (Cambridge, Massachusetts, MIT Press, 1999).

[25] M. Riedmiller, Advanced supervised learning in multi-layer perceptrons - from backpropagation to adaptive learning algorithms, *Computer Standards and Interfaces* Vol.16 (1994) pp. 265–278.

[26] B. Ripley, Statistical aspects of neural networks, chapter 2 in O.E. Brandorff-Nielsen, J.L. Jensen, and W.S. Kendall (eds) *Networks and Chaos – Statistical and Probabilistic Aspects*, volume 50 of *Monographs on Statistics and Applied Probability*, (London, Chapman and Hall, 1993) pp. 40–123.

[27] D.E. Rumelhardt, G.E. Hinton, and R.T. Williams, Learning representations by back–propagating errors, *Nature* Vol.323 (1986) pp. 533–536.

[28] R. J. Schalkoff, *Pattern Recognition: Statistical, Structural and Neural Approaches*, (New York, Wiley, 1992).

[29] W. Schiffman, M. Joost, and R. Werner, Optimization of the backpropagation algorithm for training multilayer perceptrons, technical report, Institute of Physics, University of Koblenz (Koblenz, 1994).

[30] A.J.C. Sharkey (ed.), *Multi-Net Systems*, (Berlin, Springer-Verlag, 1999).

[31] K. Sirlantzis, J.F. Feng, and WB. Liu, Minimization Algorithms based on Supervisor and Searcher Co-operation: III - novel algorithms for stochastic optimization, (submitted to JOTA, 2001).

[32] K. Sirlantzis and WB. Liu, The Supervisor-Searcher Co-operation framework: A class of powerful algorithmns for stochastic optimisation, in *Euro 2001-European Operational Research Conference*, (Rotterdam, The Netherlands, Association of European Operational Research Societies, 2001) p. 218.

[33] S.A. Solla, E. Levin, and M. Fleisher, Accelerated learning in layered neural networks, *Complex Systems* Vol.2 (1988) pp. 625–640.

[34] C. Thornton, Parity: The problem that won't go away, in G. McCalla (ed.) *Proceedings of AI-96, Toronto, Canada*, (Berlin, Springer-Verlag, 1996) pp. 362–374.

[35] UCI, Machine Learning Repository, http://www.ics.uci.edu/ mlearn/.

[36] M. Vrahatis, G. Magoulas, and V. Plagianakos, Globally convergent modification of the quickprop method, *Neural Processing Letters* Vol.12 (2000) pp. 159–169.

[37] P. Werbos, *Beyond Regression: New Tools for Prediction and Analysis in Behavioral Sciences*, (PhD thesis, Harvard University, Cambridge, MA, 1974).

Binary Strings and multiclass learning problems

Terry Windeatt, Reza Ghaderi
School of Electronics, Computing and Maths
University of Surrey, Guildford, Surrey, United Kingdom GU2 7XH
E-mail: `t.windeatt@surrey.ac.uk`

Contents

D. Chen and X. Cheng (eds.),
Pattern Recognition and String Matching, 741–763.

References

1 Abstract

The Output Coding technique for solving multi-class learning problems was originally proposed with rows of an error-corecting code matrix acting as code words to represent the classes. We summarise the requirements on design of the binary strings in the code matrix and consider alternate combining strategies. For shorter codes, it is shown that both code design and combining strategy can affect generalisation. Examples are presented on synthetic data, on natural benchmark data and on an application in face verification.

2 Introduction

Traditionally, the approach that has been used in the design of pattern classification systems is to experimentally assess the performance of several classifiers with the idea that the best one will be chosen. However, the theory of ensemble classifiers represents a departure from the traditional strategy and has been developed in the past decade to address the problem of designing a system with improved accuracy. Recognising that each classifier may make different and perhaps complementary errors, the aim is to pool together the results from all classifiers to find a composite system that outperforms any individual classifier. In this way a single complex classifier may be replaced by a set of relatively simple classifiers. In the context of a system of multiple classifiers these constituent classifiers are sometimes referred to as base classifiers.

It is known that certain conditions need to be satisfied to realise the improvement from combining multiple classifiers, in particular that the base classifiers be not too highly correlated [23]. If each classifier solves a slightly different problem, then composite performance may improve. Various learning techniques have been devised with the aim of creating diverse sub-problems, thereby reducing correlation between classifiers before combining. These correlation reduction techniques are of various types and include: (i) reducing dimension of training set to give different feature sets, (ii) incorporating different types of base classifier, (iii) designing base classifiers with different parameters for same type of classifier, (iv) re-sampling training set so each classifier is specialised on different subset, and (v) coding multi-class outputs to create complementary two-class problems. The well-known techniques of Bagging [3] and Boosting [7] are in category (iv), while Output Coding, described in Section 3, is in category (v).

Bagging and the original form of Boosting are both voting algorithms and achieve impressive results for some problems by combining hard-level classifications. By hard-level we mean that a single-hypothesis decision is taken for each base classifier, in contrast with soft-level which implies a measure of confidence associated with the decision. Bagging (from Bootstrap Aggregating) forms replicate training sets by sampling with replacement, and combines the resultant classifications with a majority vote. Boosting, which combines with a weighted vote is more complex than Bagging in that the distribution of the training set is adaptively changed based upon the performance of sequentially constructed classifiers. Each new classifier is used to adaptively filter and re-weight the training set, so that the next classifier in the sequence has increased probability of selecting patterns that have been previously mis-classified.

The Output Coding method does not re-sample the training set as with Bagging and Boosting, but rather creates diverse two-class sub-problems by relabelling the training patterns according to a set of binary strings. In this way, each classifier is solving a different sub-problem of the original problem. A useful way of trying to understand correlation reduction by class re-labelling is to interpret the implied partitioning of the input space, as shown in the example in Section 6 using synthetic data. There are several motivations for decomposing a multi-class problem into separate and complementary two-class problems. Firstly some accurate and efficient two-class classifiers do not naturally scale up to multi-class. Attention can then be focused on developing an effective technique for the two-class case, without having to consider explicitly the design and automation of the multi-class classifier. Secondly, it is hoped that the parameters of a simple parallel machine run several times are easier to set than a complex machine run once and may facilitate more efficient solutions. Finally, solving different 2-class sub-problems, perhaps repeatedly with random perturbation, may help to reduce error evident in the original problem.

3 Output coding

First consider the need for a suitable Output Coding with reference to the output representation used with a Multi-layer Perceptron (MLP) network, which is a popular base classifier. A single multiple output MLP can handle a multiclass problem directly. The standard technique is to use a k-dimensional binary target string that represents each one of k classes using a single binary value at the corresponding position, for example $[0,...0,1,0,...0]$ which is sometimes referred to as one-per-class (OPC) encoding. The reason that a single multiclass MLP is not a suitable candidate for use with Output Coding is that all nodes share in the same training, so errors are far from independent and there is not much benefit to be gained from combining [16]. However if a 2-class MLP is used as base classifier, independence among classifiers can be achieved by the problem decomposition defined by the string, as well as by injection of randomness through the starting weights. Of course, no guarantee can be given that a single MLP with superior performance will never be found, but the assumption is that even if one exists its parameters would be more difficult to determine.

An alternative to OPC is distributed output coding [21], in which k binary strings are assigned to k classes on the basis of meaningful features corresponding to each bit position. For this to provide a suitable decomposition some domain knowledge is required so that each classifier output can be interpreted as a binary feature which indicates the presence or otherwise of a useful feature of the problem at hand. The strings are treated as code words and a test pattern is assigned to the class that is closest to the corresponding code word. It is this method of string matching, which is analogous to the assignment stage of error-correcting coding, that provides the motivation for employing error-correcting codes in classification.

3.1 Error-Correcting Output Coding (ECOC)

The Error-Correcting Output Coding (ECOC) method [6], described in Section 3.2, is a two stage method. The second stage is the combining step, based on error-correcting principles under the assumption that the learning task can be modelled as a communication problem, in which class information is transmitted over a channel [5]. In this model, errors introduced into the process arise from various sources including the learning algorithm, features and finite training sample. For a two-class problem, classification errors can be one of two types, either predicted class Ω_1 for target class Ω_2 or predicted class Ω_2 for target class Ω_1. From the transmission channel viewpoint, we would expect that the one-per-class (OPC) and distributed output coding matrices would not perform as well as the ECOC matrix, because of inferior error-correcting capability.

The first stage of the ECOC method provides the decomposition strategy, which is the re-labelling of patterns defined by the Output Coding matrix. Each re-labelled class contains patterns from a number of the classes from the original problem. There exists another category of methods for decomposing a complex multiclass problem into simpler two-class sub-problems. In this other category, each base classifier is capable of distinguishing between a pair of classes, which means that the final decision boundary is made up of decision boundaries that each originate from just two classes. Methods like *binary decision clustering* [24] and *pairwise coupling* [9] are included in this category. The main advantage is that sub-problems are normally simpler, which means that the decision boundary of each sub-problem is less complex. However, each sub-problem is based on only a subset of patterns and this may cause some difficulties in training. Furthermore all boundaries between pairs of classes are needed to reconstruct the original boundary, which can lead to a large number of base classifiers ($b = \frac{k!}{2!.(k-2)!}$ for a k class problem). In contrast the ECOC approach requires fewer but more powerful classifiers with the ability to repeat parts of the decision boundary, which can reduce variability as shown in [13].

3.2 ECOC algorithm

In the ECOC method, a $k \times b$ code word matrix Z has one string (code word) for each of k classes, with each column defining one of b sub-problems that use a different labelling. Specifically, for the jth sub-problem, a training pattern with target class w_i ($i = 1...k$) is re-labelled either as class Ω_1 or as class Ω_2 depending on the value of Z_{ij} (typically 0 or 1). One way of looking at the re-labelling is to consider that for each column the k classes are arranged into two super-classes Ω_1 and Ω_2. The jth ($j = 1...b$) classifier is trained using re-labelled patterns from the jth sub-problem.

For example, in a five-class problem if the *jth* column of Z is given by $[10110]^T$ the relabelling associated with the *jth* classifier means that training patterns with target classes represented by the first, third and fourth rows of Z comprise superclass Ω_1 and the remaining training patterns comprise superclass Ω_2.

A test pattern is applied to the b trained classifiers giving

$$\mathbf{y} = [y_1, y_2, \cdots y_b]^T \tag{1}$$

in which y_j is the output (usually real-valued) of jth base classifier. The distance between $\mathbf{y}$ and code word for each class is given by

$$L_i^1 = \sum_{j=1}^{b} |Z_{ij} - y_j| \tag{2}$$

and the pattern is assigned to the class w_i corresponding to closest code word
$ArgMin_i(L_i^1)$.

Originally the matching was based on Hamming Distance, but when it could be shown that ECOC produced good probability estimates [14] the combining scheme was changed to soft-level, equation (2) which represents the L^1 norm.

4 String Design

When the Output Coding technique was first developed it was believed that the code matrix should be designed using error-correcting principles to enable it to generalise well [5, 6]. Various codes have since been proposed, but most code matrices that have been investigated previously are binary and problem-independent, that is pre-designed. Random codes have received much attention, and were first mentioned in [5] as performing well in comparison with error-correcting codes. In [6] random, exhaustive, hill-climbing search and BCH coding methods were used to produce code matrices for different column lengths. Random codes were investigated in [20] for combining Boosting with Output Coding, and it was shown that a random code with a near equal column split of labels was theoretically better. Random codes were also shown in [11] to give Bayesian performance if pairs of code words were equidistant, and it was claimed that a long enough random code would not be outperformed by a pre-defined code. Although various heuristics have been employed to produce better binary problem-independent codes there appears to be little evidence to suggest that performance significantly improves by a clever choice of code, except that OPC is usually inferior [1, 5, 6]. Recently a three-valued code [1] was suggested which allows specified classes to be omitted from consideration (don't care for third value), thereby permitting integrated representation of methods such as *pairwise coupling* [9]. Another recent development is described in [4] in which problem-dependent codes are investigated and it is claimed that designed continuous codes show more promise than designed discrete codes.

In the Output Coding framework the choice of code matrix may affect performance. Before looking at properties of the code matrix that reduce generalisation error, we first note that there are three kinds of errors that can be distinguished in this framework. First there is the *Bayesian error*, which is due to overlapped classes and cannot be removed by combining, error-correcting or any other method. Secondly there is *base classifier added error*, which is related to the training of base classifiers, and which we hope to remove by the error-correcting capability of code words. Finally there is the *combining error*, which is introduced by the way in which we choose to combine the outputs of base classifiers, and is present even if all classifiers are perfect (behave like Bayesian classifier).

4.1 Maximum Distance Code

It seems reasonable that the greater the Hamming Distance between code words the more likely that error will be reduced, since the code matrix maps the problem into a different space in which classes are more easily separated. In addition, sub-problems are more independent and likely to benefit from combining if Hamming Distance between columns is maximised, remembering that a column and its complement represent identical classification problems [6]. Therefore the first two desirable properties of the code matrix are maximum Hamming distance between rows and between columns.

4.2 Equidistant code and Bayes rule

Consider that the decomposition of a multiclass classification problem into binary sub-problems in Output Coding can be interpreted as a transformation between spaces from the original output $\mathbf{q}$ to $\mathbf{y}$, given in matrix form by

$$\mathbf{y} = Z^T\mathbf{q} \tag{3}$$

where $\mathbf{q}$ are individual class probabilities.

From the distance-based decision rule in (2) and using (3)

$$L_i^1 = \sum_{j=1}^{b} |(\sum_{l=1}^{k} q_l Z_{lj}) - Z_{ij}| \tag{4}$$

and knowing that $\sum_{l=1}^{k} q_l = 1$ we have

$$L_i^1 = \sum_{j=1}^{b} |\sum_{l=1}^{i-1} q_l Z_{lj} + q_i Z_{ij} + \sum_{l=i+1}^{k} q_l Z_{lj} - Z_{ij}| \quad l = 1, 2...k \tag{5}$$

and therefore

$$L_i^1 = (1 - q_i) \sum_{j=1}^{b} \sum_{l=1}^{k} |Z_{ij} - Z_{lj}| \tag{6}$$

Consider the situation that the Hamming Distance between ith and jth code word is equal for any i and j, $i \neq j$. Then from (6), we see that when all pairs of code words are equidistant, minimising L_i^1 implies maximising posterior probability which is equivalent to Bayes rule (as also derived in [10, 27])

$$ArgMax_i(q_i) = ArgMin_i(L_i^1) \tag{7}$$

Therefore, any variation in Hamming Distance between pairs of code words will reduce the effectiveness of the combining strategy. The third desirable property of the code matrix is equal Hamming Distance between all pairs of rows.

4.3 Equidistant code and unbiased distance

Consider the case that there is a single output for each classifier (y_j) so that we can assume that a threshold exists which separates the members of the two *super groups* of classes. If the threshold for all classifiers is half way between the binary limit values (0.5 for 0-1 binary), there will be a fair comparison of distance to each code word. However if there exists some base classifier bias that causes the threshold to move away from the mid-position, then the comparison will not be fair unless there are an equal number of 1's in each code word. Therefore the third desired property of the code matrix is equal number of 1's in all code words.

Now we show that an equidistant code automatically satisfies this property. Let Z be an equidistant code with d bits error-correcting capability, $\sum_{l=1}^{b} |Z_{il} - Z_{jl}| = 2d+1$ for any pair i, j. Since Hamming Distance between pair i, j is the sum of the number of 1s in row i and row j minus number of common 1s between i, j we may write

$$\sum_{l=1}^{b} Z_{il} + \sum_{l=1}^{b} Z_{jl} - 2\sum_{l=1}^{b} Z_{il}Z_{jl} = 2d + 1 \tag{8}$$

similar equations can be written for pair i, k and pair j, k

$$\sum_{l=1}^{b} Z_{il} + \sum_{l=1}^{b} Z_{kl} - 2\sum_{l=1}^{b} Z_{il}Z_{kl} = 2d + 1 \tag{9}$$

$$\sum_{l=1}^{b} Z_{jl} + \sum_{l=1}^{b} Z_{kl} - 2\sum_{l=1}^{b} Z_{jl}Z_{kl} = 2d + 1 \tag{10}$$

From (8), (9),(10) after re-arranging

$$\sum_{l=1}^{b} Z_{il}Z_{jl} = \sum_{l=1}^{b} Z_{kl}Z_{jl} = \sum_{l=1}^{b} Z_{kl}Z_{il} = m \tag{11}$$

where m is number of common 1s in code word, and

$$\sum_{l=1}^{b} Z_{il} = \sum_{l=1}^{b} Z_{kl} = \sum_{l=1}^{b} Z_{jl} = n \tag{12}$$

where n is the number of 1s in each row.

Therefore if Z is an equidistant matrix, there is the same number of 1s in any row, and the same number of common 1s between any pair of rows, and Z provides an unbiased distance measurement. For example, if ZZ^T can be written in the form

$$ZZ^T = \begin{bmatrix} n & m & \cdots & m \\ m & n & \cdots & m \\ \cdots & \cdots & \cdots & \cdots \\ m & m & \cdots & n \end{bmatrix} \tag{13}$$

then properties (11) and (12) are automatically satisfied since

$$ZZ^T = \sum_{l=1}^{b} Z_{il}Z_{lj}^T = \sum_{l=1}^{b} Z_{il}Z_{jl} = \begin{cases} n & \text{if i=j} \\ m & \text{otherwise} \end{cases}$$

4.4 Summary of Constraints

The main constraints for designing set of binary strings that make up the code matrix are therefore as follows

- minimum Hamming Distance between rows (error-correcting capability)
- variation of Hamming Distance between rows (effectiveness of combining)
- number of columns (repetition of different parts of sub-problems)

- Hamming Distance between columns and complement of columns (independence of base classifiers)

From the theory of error-correcting codes [19] we know that finding a matrix with long code words, and having maximum and equal distance between all pairs of rows is complex. In the example in Section 6, random, equidistant and non-equidistant code matrices are compared as number of columns is varied. It shows that as code length increases the difference in performance between equidistant and random codes becomes less significant. Equidistant codes can be produced by using the BCH method [19], which employs algebraic techniques from Galois field theory. The number of rows is over-produced (BCH requires number to be power of 2), before using properties (11) and (12) to select a subset of strings. Of course these properties provide only necessary conditions for equidistant strings and so cannot be used to generate them in isolation.

4.5 Code Sensitivity

There is another issue, in addition to those described in Section 4.4, that may affect performance. Since in practice the overlap between classes can be different, we might suspect that certain code words are better suited to some super-classes more than others. Although it would seem a good idea to use a labelling in which distance between labels is based on distance between classes (e.g. Mahalanobis distance), the additional constraint is likely to be difficult to implement. Taking a different approach [27] the following modification to ECOC is suggested, in which all code words are used for each class to reduce sensitivity to code word selection.

In training, circularly shift one row of ECOC matrix k times, and for each single shift train b base classifiers. In testing, compute the distance $L^1_{n_i}$ for class w_i between b-dimensional base classifier outputs and string corresponding to class w_i. After averaging the final distance measurement, assign pattern to class w_i given by $Argmin_i \sum_{n=1}^{k} L^1_{n_i}$.

5 Alternate Combining Strategies

A variety of combining strategies have been proposed for Output Coding besides L^1 norm, referred to here as L1-OC. Other combining strategies discussed in this section include three approaches for recovering individual class probabilities, which are Least Squares (LS-OC), Inverse Hamming Distance (IHD-OC) and Linear estimation (Lin-OC). For these three methods the classification decision is based on maximising the probability estimate. One further method is also described, Dempster-Shafer (DS-OC) based on maximising probability mass function [22]. The original Hamming Distance combining strategy (HD-OC) was shown in [13] to be analogous to majority vote over the classes. Besides L1-OC, LS-OC has been the most extensively investigated. The justification for LS-OC in terms of probability estimation was reported in [14], and in [10] LS-OC was extended by incorporating ridged regression when b is small.

5.1 Least squares (LS-OC)

Recovering individual class probabilities from super-class probabilities is easily accomplished if the individual probability estimates are exact and if the k independent equations that result from equation (3) can be solved. In practice, base classifiers will not produce correct probabilities but we can model the error assuming that the expected value of the error is zero for unbiased probability estimates [8]. In general Z^T is not a square matrix, and so does not have an inverse, but $\mathbf{q}^*$, an optimal value of $\mathbf{q}$ in equation (3), can be found using the method of least squares

$$\mathbf{q}^* = (ZZ^T)^{-1}Z\mathbf{y} \tag{14}$$

using a cost function such as $(\mathbf{y} - Z^T\mathbf{q})^T(\mathbf{y} - Z^T\mathbf{q})$.

5.2 Inverse Hamming Distance (IHD-OC)

From equation (2) we may write the matrix equation

$$\mathbf{L^1} = \Delta\mathbf{q} \tag{15}$$

where $\mathbf{L^1} = [L_1^1, L_2^1, ...L_k^1]$ and Δ is a matrix whose element Δ_{ij} represents the Hamming Distance between row i and row j. Since the elements of Δ are non-negative, Δ can be inverted to find an estimate for $\mathbf{q}$.

5.3 Linear estimation (Lin-OC)

The assumption here is that there is a linear relationship between outputs of the base classifiers and individual class probabilities.

$$q_i = \sum_{j=1}^{b} W_{ij} y_j \qquad i = 1...k \tag{16}$$

where

$$\sum_{j=1}^{b} W_{ij} = 1 \qquad i = 1...k \tag{17}$$

Although there are several ways to estimate the weights, a simple approach is to count, for each column, the number of times that a training pattern belongs both to class w_i and to superclass Ω_1. The count is repeated for all classes ($i = 1, ...k$) and normalised as in (17).

5.4 Dempster-Shafer (DS-OC)

The Dempster-Shafer approach to combining provides a mathematical framework which may be regarded as a generalised form of Bayesian statistics [22]. It is based upon the concept of probability mass function (basic probability assignment) which assigns a value to a subset of propositions. We assume that there are k elemental propositions, one each for the k classes to which a pattern can be assigned. In the context of Output Coding each column of the code matrix Z defines a partitioning into two superclasses as described in Section 3.2. A superclass is a subset of the k classes and represents one of 2^k possible subsets in Dempster-Shafer theory. Therefore, each column can be regarded as providing evidence for class membership, which can be combined with evidence from previous columns.

The mass function of the jth base classifier is set to $m_j(A_j) = [y_j \quad 1 - y_j]$ for the two superclasses Ω_1 and Ω_2, where y_j is the jth base classifier output (defined in (1)), and A_j represents the decomposition defined by the jth column of Z. The evidence for the $(j+1)th$ base classifier can be combined recursively with evidence from the previous j classifiers by invoking Dempster's rule of combination to calculate the orthogonal sum $m_j \oplus m_{j+1}$

$$m_{j+1} = [m_j \oplus m_{j+1}] = 1/K \sum_{A_j \cap A_{j+1}} m_j(A_j) m_{j+1}(A_{j+1}) \tag{18}$$

where normalisation factor K is given by $1 - \sum_{A_j \cap A_{j+1} = \{\}} m_j(A_j) m_{j+1}(A_{j+1})$. Initially with $j = 1$ equation (18) gives $m_2 = m_1 \oplus m_2$, and equation (18) is applied recursively for $j = 1, ...b - 1$ to find the mass function for the combination of b classifiers. The decision strategy is to assign the class corresponding to the maximum probability mass function.

6 Examples

6.1 Synthetic Data

An artificial five-class overlapping Gaussian problem is defined by five groups of two-dimensional random vectors having normal distribution, shown in Table 1. The classifiers are not trained, but using the parameters from Table 1, the posterior probability of super-class membership is computed. The Bayes rate for this five class problem is 74.28%, and the Bayes (optimum) decision boundary is shown in figure 1.

class	w_1	w_2	w_3	w_4	w_5
μ_i(mean)	[0,0]	[3,0]	[0,5]	[7,0]	[0,9]
σ_i^2(variance)	1	4	9	25	64

Table 1: Distribution parameters of data used in artificial benchmark

The following code matrices are used in these experiments:

C1: $k \times k$ one-per-class (OPC: 1s on main diagonal 0 elsewhere))
C2: $k \times 7$ matrix with randomly chosen binary elements
C3: $k \times 7$ BCH code (minimum distance of 3, non-equal)
C4: $k \times 7$ BCH code with equal distance of 4
C5: $k \times 15$ matrix with randomly chosen binary elements
C6: $k \times 15$ BCH code with equal distance of 8

To determine robustness of the Output Coding framework, Gaussian random noise with different mean (μ_n) and variance (σ_n^2), is added to the soft-level (real-valued) output of base classifiers to simulate noisy imperfect classifiers. Decision boundaries in figure 2(a) (code C1) and figure 2(b) (code C4) demonstrate the difference between one-per-class (OPC) and the ECOC concept. By comparing with the Bayes boundary shown in figure 1, we can see that each classifier concentrates on different part of the input space. So each classifier has a less complex problem to handle compared with the complete boundary that would, for example, be handled by a single multiclass classifier. Figure 2(b) also shows how some parts of the boundary appear in more than one classifier, and so are learned repeatedly, as investigated in [13, 25]. For added noise ($\mu_n = 0$, $\sigma_n^2 = 0.25$) figure 3(a) and figure 3(b) show respectively the C1 and C4 decision boundaries of the ensemble as number of individual classifiers is increased. A comparison shows that

Code	L1	LS	IHD	Lin	DS
C1*	0.91	0.91	0.91	0.91	0.91
C2	0.89	0.83	0.61	0.58	0.90
C3	0.84	0.84	0.83	0.82	0.84
C4*	0.92	0.92	0.92	0.92	0.92
C5	0.97	0.97	0.97	0.94	0.97
C6*	0.97	0.97	0.97	0.97	0.99

Table 2: Mean value over five databases of ratio of classification rate of specified coding/combining (* equidistant) strategy to maximum classification rate over all strategies for a particular database

code C4, by virtue of repeated learning of parts of decision boundaries, more effectively reduces variance.

Figure 4 shows a plot of the percentage match with Bayes rate as level of noise is increased and demonstrates the robustness of ECOC when it has been modified to reduce sensitivity to code word selection (Section 4.5). The error-correcting capability for codes C2, C4, C6, C5 is 1,3,7,2 bits respectively. From figure 4 the following points can be observed:

- if no noise is added, the equidistant matrices C4, C6 match Bayes rate perfectly
- if the number of classifiers is the same, the equidistant code performs slightly better than random code, shown by comparing C5 with C6 ($b = 15$) and C4 with C2 ($b = 7$)
- if longer random code is used, repetition of sub-problems gives improved performance even if error-correction capability is reduced, shown by comparing C5 with C4

Figure 5 compares original and modified decision boundaries when noise ($\mu_n = 0$, $\sigma_n^2 = 0.25$) is added to each individual classifier, and confirms that error variance has been reduced.

6.2 Popular Benchmark Data

Simple two-dimensional data shown in the example in Section 6.1 is not necessarily representative of real data, so it is interesting to see performance on some popular benchmarks (See [2] for a description of the datasets).

The base classifier is a conventional single hidden layer MLP trained by Levenberg-Marquardt algorithm, with default training parameters.

There are fifty epochs of training and the number of hidden nodes for the datasets is (zoo/1, car/1, vehicle/5, glass/2, satellite/2). Table 2 gives a comparison of codes and combining strategies over the five datasets for codes C1 to C6, where each entry denotes the mean ratio of classification rate at the specified code/combining strategy to the best classification rate for that problem over all strategies. From table 2, we can observe that longer codes generally perform better but combining strategy appears to have little impact on performance. However, for code C2 the results for IHD-OC and Lin-OC demonstrate that these two combining strategies are more sensitive to shorter codes than the others. For 7-bit code, equidistant performs best (C2 and C3 compared with C4). For the 15-bit code, the individual results show that random is better for zoo (error rate 5.2% compared with 6.5%) and for car (error rate 25.5% compared with 27.2%). while equidistant is better for the other three datasets, but there is no difference in mean ratio.

On seven data-sets (zoo, vehicle, segment, iris, glass, cmc, car) with ten independent runs the improvement in test error rate of modified (Section 4.5) over original ECOC is 5% mean and 30% standard deviation averaged over the seven data-sets. Further details may be found in [26].

6.3 Face Verification

Automation of a system that performs personal identity verification may use a variety of biometric modalities including facial features, voice characteristics, iris scan, fingerprints. Facial images are a popular source of biometric information since they are relatively easy to acquire, and provide discriminatory features routinely used by humans for recognition. However automated face verification systems often have poor levels of performance and improving them is known to be a difficult task.

The extended M2VTS (XM2VTS) database contains 295 subjects, who were recorded in four separate sessions uniformly distributed over a period of 5 months, and within each session a number of shots were taken including both frontal-view and rotation sequences. Some sample profiles are shown in figure 6, and further details of this database can be found in [18]. [1] The experimental protocol (known as Lausanne evaluation protocol) provides a framework within which the performance of vision-based person authentication systems running on XM2VTS database can be measured. The protocol

[1] *http://www.ee.surrey.ac.uk/Research/VSSP/xm2fdb.html*

assigns 200 clients and 95 impostors, with two shots of each session for each subject's frontal or near frontal images [17], giving 3 training, 3 evaluation and 2 test images for each client. The impostor set is partitioned into 25 evaluation and 70 test impostors. Details of the this protocol can be found in [15]. [2]

Each image is first projected to a lower dimensional feature space using Principal Components and Linear Discriminant Analysis (PCA + LDA) so that each pattern is represented by a vector with 199 elements. The base classifier is a MLP with one hidden layer containing 199 input nodes, 35 hidden nodes, two output nodes and with fixed learning rate, momentum and number of epochs. The dual output is mapped to a value between 0 and 1 to give an estimation of probability of super-class membership. Further details of the system can be found in [12].

There are 200 clients, so from the identification viewpoint it is a 200 class problem. An equidistant 200x511 code matrix is generated by the BCH method, as described in Section 4.4, with Hamming Distance between any pair of rows being 256 bits. In the ECOC method, the classifier outputs represent estimates of super-class probabilities and these are the estimates used to represent clients and impostors for face verification. Specifically, each client i is represented by a set Y_i of N ECOC classifier output vectors, i.e.

$$Y_i = \{\underline{y}_i^l | l = 1, 2, ..., N\} \tag{19}$$

where N is the number of ith client patterns available for training.

The verification task reduces to ascertaining whether classifier outputs representing an image are jointly consistent with a claimed identity. In order to test the hypothesis that the client claim is authentic it is necessary to adopt a test statistic such as the average distance between $\underline{y}$ and the elements of set Y_i. The metric in (2) is preferred to Euclidean distance since it is believed to be more robust to outliers, that is

$$d_i(\underline{y}) = \frac{1}{N} \sum_{l=1}^{N} \sum_{j=1}^{b} |y_j^l - y_j| \tag{20}$$

where y_j is the jth binary classifier output for the test pattern, and y_j^l is the jth classifier output for the lth member of class i. The distance in (20) is checked against a decision threshold, which can be set using the evaluation set, to determine if the client's claim is accepted, Using this ECOC framework the best performance to date has been achieved using this

[2] *http://www.idiap.ch/~m2vts/Experiments/xm2vtsdb_protocol_october.ps*

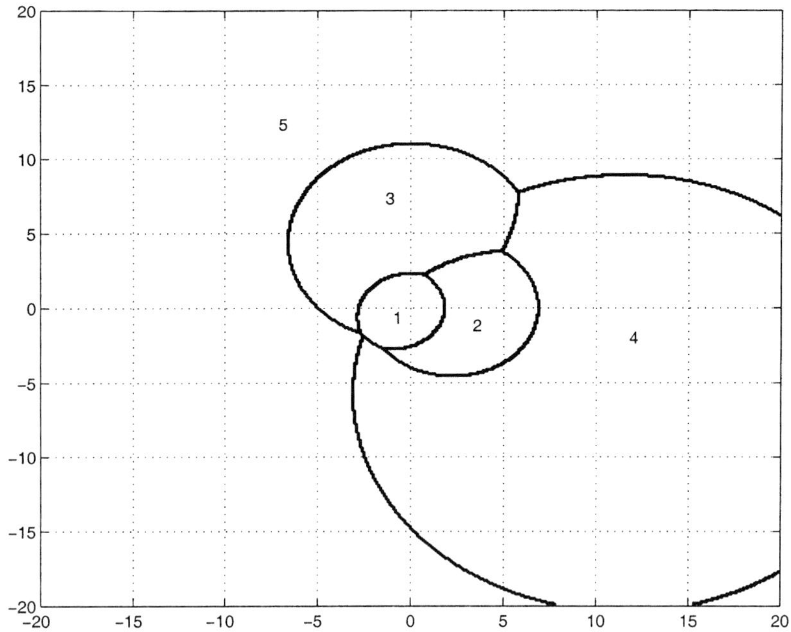

Figure 1: Optimum decision boundary for five class artificial data

publicly available benchmark (False Acceptance 0.75 % and False Rejection 1.25%).

7 Conclusion

Output Coding is a method for solving multiclass learning problems that uses a set of binary strings to decompose the multiclass into complementary two-class problems. A two-class classifier is then sufficient to solve the multiclass problem. Although error-correcting codes were originally proposed for designing binary strings, random codes achieve similar performance to optimal codes provided they are long enough. It is shown here that while a variety of combining strategies are possible in the Output Coding framework performance is not very dependent on the choice.

Acknowledgments The support received for the face database from Omniperception Ltd, EPSRC grant GR/M61320 and EU framework Project

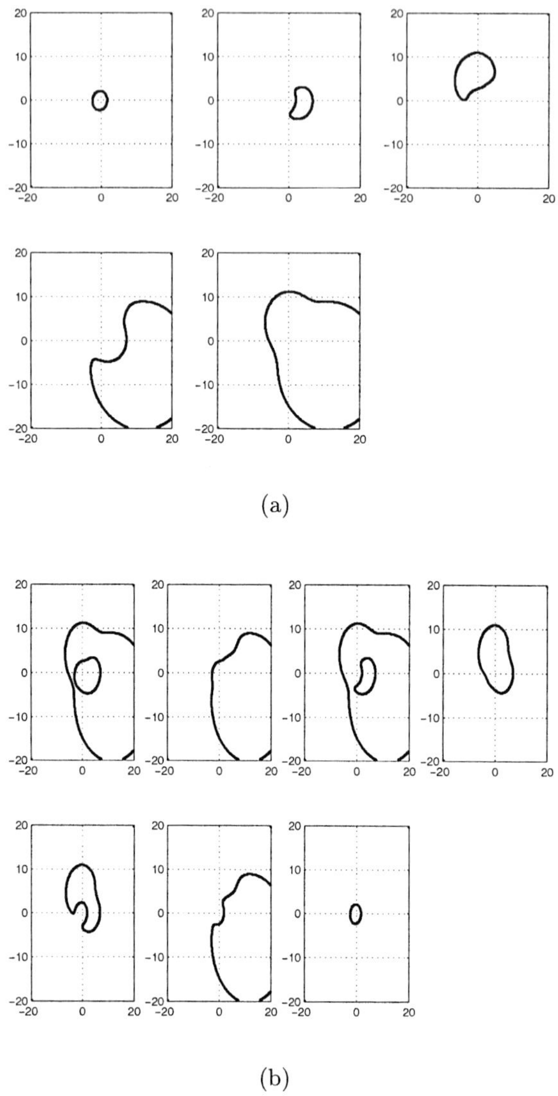

Figure 2: Individual decision boundaries made by several base classifiers (a) C1 Code 5 classifiers (b) C4 code 7 classifiers

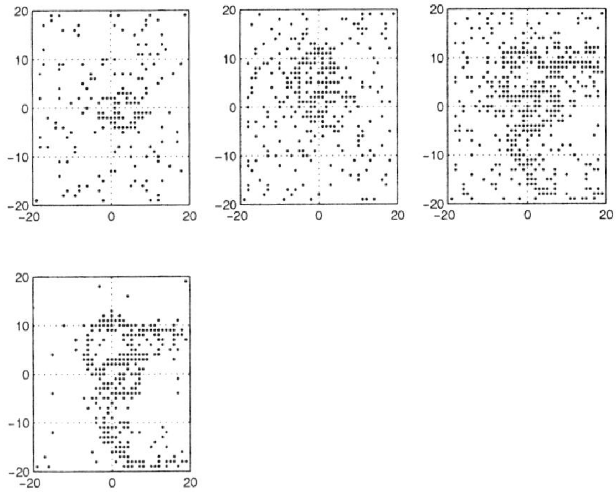

(a)

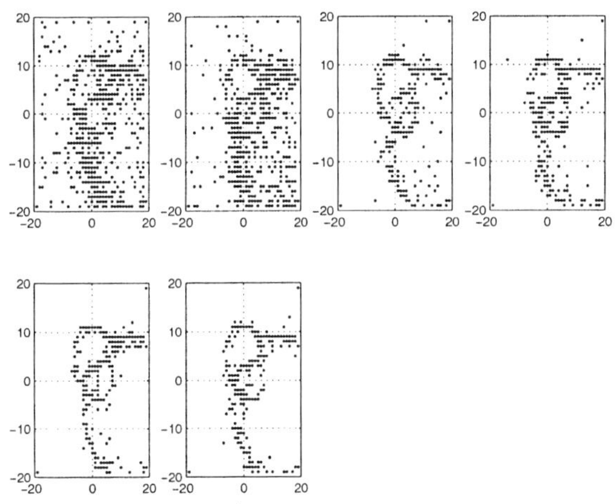

(b)

Figure 3: Decision boundaries of ensemble as number of base classifiers is increased in presence of noise ($\mu_n = 0$, $\sigma_n^2 = 0.25$). The noise-free Bayes boundary is shown in figure 1. (a) C1 code 2,3,4,5 classifiers (b) C4 code 2,3,4,5,6,7 classifiers (numbered left to right)

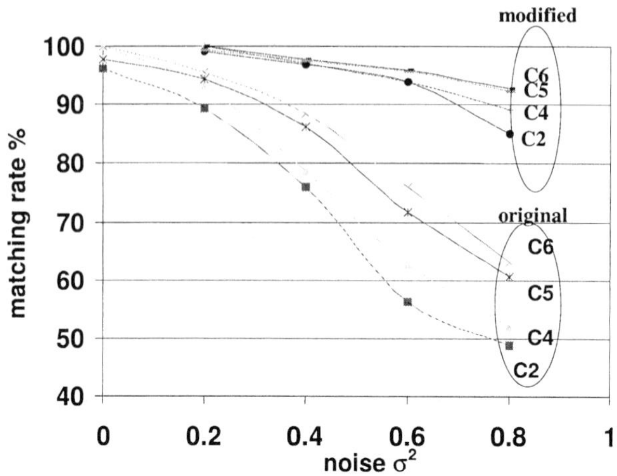

Figure 4: Percentage match with Bayes rate for increasing levels of Gaussian noise modified ECOC (top 4 curves) and original method

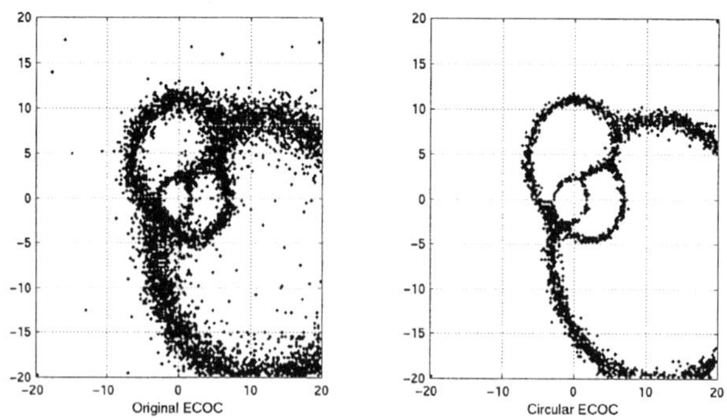

Figure 5: Comparison of decision boundaries for added noise when ECOC has been modified (right) to make it less sensitive to code word selection

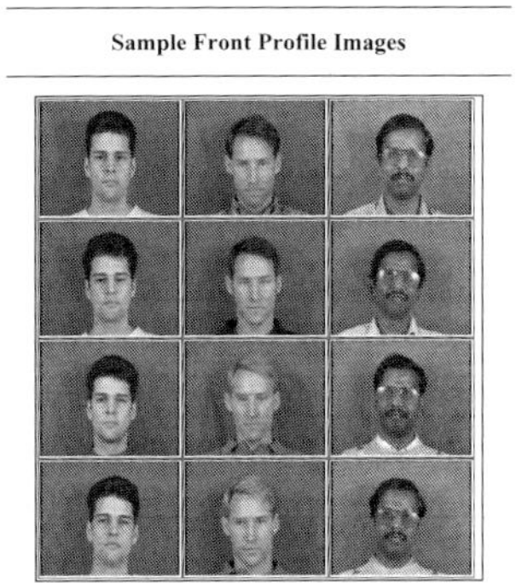

Figure 6: Images from Extended M2VTS Database

Banca is gratefully acknowledged.

References

[1] E.L. Allwein, R.E. Schapire, and Y. Singer. Reducing multi-class to binary: A unifying approach for margin classifiers. *Machine learning research*, 1:113–141, 2000.

[2] C.L. Blake and Merz C.J. Uci repository of machine learning databases. Technical report, Irvine, Univ. of Calif., Inf. and Comp. Scienc, 1998.

[3] L. Breiman. Bagging predictors. *Machine Learning*, 24(2):123–140, 1997.

[4] K. Crammer and Y. Singer. On the learnability and design of output codes for multiclass problems. *Machine Learning*, to appear.

[5] T.G Dietterich and G. Bakiri. Error-correcting output codes: A general method for improving multiclass inductive learning programs. In *Proceedings of the Ninth National Conference on Artificial Intelligence (AAAI-91)*, pages 572–577. AAAI Press, 1991.

[6] T.G. Dietterich and G Bakiri. Solving multi-class learning problems via error-correcting output codes. *Journal of Artificial Intelligence Research*, 2:263–286, 1995.

[7] Y. Freund and R.E. Schapire. A decision-theoretic generalisation of on-line learning and application to boosting. *Journal of computer and system science*, 55:119–139, 1997.

[8] R. Ghaderi and T. Windeatt. Least squares and estimation measures via error correcting output code. In *2nd Int. Workshop Multiple Classifier Systems,Lecture notes in computer science, Springer-V erlag*, pages 148–157, Jul 2001.

[9] T. Hastie and R Tibshirani. Classification by pairwise coupling. *The annals of statistics*, 2:451–471, 1998.

[10] G. M. James. *Majority Vote Classifiers: Theory and Applications.* PhD thesis, Dept. of Statistics, Univ. of Stanford, Calif., May 1998.

[11] G. M. James and T. Hastie. The error coding method and PICT's. *Computational and Graphical Statistics*, 7:377–387, 1998.

[12] J. Kittler, R. Ghaderi, T. Windeatt, and G. Matas. Face verification using error correcting output codes. In *Computer Vision and Pattern Recognition CVPR01*, volume 1, pages 755–760, Hawaii, December 2001. IEEE Press.

[13] E.B. Kong and T.G. Diettrich. Error-correcting output coding corrects bias and variance. In *12th Int. Conf. of Machine Learning*, pages 313–321, San Fransisco, 1995. Morgan Kaufmann.

[14] E.B. Kong and T.G. Diettrich. Probability estimation via error-correcting output coding. In *Int. Conf. of Artificial Inteligence and soft computing*, Banff,Canada, 1997.

[15] J Luettin and G. Maitre. *Evaluation Protocol For The Extended M2VTS Database (XM2VTS).* Dalle Molle Institute for Perceptual Artificial Intelligence, P.O. Box 592 Martigny, Valais, Switzerland, July 1998. IDIAP-Com 98-05.

[16] F. Masulli and G. Valentini. Effectiveness of error correcting output codes in multiclass learning problems. In J.Kittler and F.Roli, editors, *Multiple Classifier Systems, MCS2000*, pages 107–116, Cagliari, Italy, 2000. Springer Lecture Notes in Computer Science.

[17] J. Matas, M. Hamouz, M. Jonsson, J. Kittler, Y. Li, C. Kotroupolous, A. Tefas, I. Pitas, T. Tan, H. Yan, F. Smeraldi, J. Bigun, N. Capdevielle, w. Gerstner, S. Ben-Yacoub, Y. Abduljaoued, and Y. Majoraz.

Comparison of face verification results on the xm2vts database. In A. A Sanfeliu, J.J Villanueva, M. Vanrell, R. Alqueraz, J. Crowley, and Y. Shirai, editors, *Proceedings of the 15th ICPR*, volume 4, pages 858–863, Los Alamitos, USA, September 2000. IEEE Computer Soc Press.

[18] K. Messer, J. Matas, J. Kittler, J. Luettin, and G. Maitre. XM2VTSDB: The extended M2VTS database. In *Proc. of AVBPA'99*, pages 72–77, 1999.

[19] W.W. Peterson and J.R. Weldon. *Error-Correcting Codes.* MIT press, Cambridge,MA, 1972.

[20] R.E. Schapire. Using output codes to boost multiclass learning problems. In *14th International Conf. on Machine Learning*, pages 313–321. Morgan Kaufman, 1997.

[21] T.J. Sejnowski and C.R. Rosenberg. Parallel networks that learn to pronounce english text. *Complex systems*, 1:145–168, 1987.

[22] G. Shafer. *A mathematical theory of evidence.* Princeton Univ Press, Princeton, New Jersey, 1976.

[23] K. Tumer and J. Ghosh. Error correlation and error reduction in ensemble classifiers. *Connection Science, special issue on combining artificial neural ne tworks: ensemble approaches*, 8(3):385–404, 1996.

[24] C.L. Wilson, P.J. Grother, and C.S. Barnes. Binary decision clustering for neural-network-based optical character recognition. *Pattern Recognition*, 29(3):425–437, 1996.

[25] T. Windeatt and R. Ghaderi. Binary codes for multi-class decision combining. In *14th Annual International Conference of Society of Photo-Optical Instrumentation Engineers (SPIE)*, volume 4051, pages 23–34, Florida,USA, April 2000.

[26] T. Windeatt and R. Ghaderi. Multi-class learning and error-correcting code sensitivity. *Electronics Letters*, 36(19):1630–1632, Sep 2000.

[27] T. Windeatt and R. Ghaderi. Binary labelling and decision level fusion. *Information Fusion*, 2(2):103–112, 2001.

15/11/08

19 No to 6.0

Box No 7298 - 3
7299 - 16
7300 - 12
7301 - 10

41